# Bacteria *versus* Antibacterial Agents

## AN INTEGRATED APPROACH

# Bacteria *versus* Antibacterial Agents

## AN INTEGRATED APPROACH

*Oreste A. Mascaretti*
Department of Organic Chemistry
Faculty of Biochemistry and Pharmacy
Universidad Nacional de Rosario
Rosario, Argentina

ASM PRESS
WASHINGTON, D.C.

Address editorial correspondence to ASM Press, 1752 N St. NW, Washington, DC 20036-2904, USA

Send orders to ASM Press, P.O. Box 605, Herndon, VA 20172, USA
Phone: (800) 546-2416 or (703) 661-1593
Fax: (703) 661-1501
E-mail: books@asmusa.org
Online: www.asmpress.org



**Library of Congress Cataloging-in-Publication Data**

Mascaretti, Oreste A.
Bacteria versus antibacterial agents : an integrated approach / Oreste A. Mascaretti.
p. ; cm.
Includes bibliographical references and index.
ISBN 1-55581-258-9 (hardcover)
1. Antibacterial agents. 2. Antibiosis.
[DNLM: 1. Anti-Infective Agents. 2. Drug Resistance, Bacterial. 3. Bacteria—drug effects. QV 250 M395b 2003] I. Title.

RM409.M375 2003
616.9'2061—dc21

2003002553

10 9 8 7 6 5 4 3 2 1


*Printed in the United States of America*

**Cover illustration:** *Salmonella enterica* serovar Typhi, the bacterium that causes typhoid fever. The salmonellae are gram-negative, non-spore-forming, facultatively anaerobic bacilli in the family *Enterobacteriaceae*. Adults with typhoid fever are treated with ciprofloxacin or ceftriaxone. The patient may be switched to oral trimethoprim-sulfamethoxazole or amoxicillin as his or her condition improves and antibiotic susceptibility data become available. Source: Public Health Image Library, Centers for Disease Control and Prevention.

*This book is dedicated to the memory of Dr. Guillermo O. Cobeñas, a rural medical doctor in Salliquelo, Buenos Aires Province, Argentina, where I was born and raised and where I received my elementary and secondary education. Dr. Cobeñas introduced me to what it means to be a university graduate. He embodied professional excellence and was a humanitarian, always committed to give his best to his patients.*

# Contents

# Foreword

Since they were first introduced in the 1940s, antibiotics have transformed medicine; most infectious diseases can be effectively controlled by the appropriate use of the correct drugs. Bacterial diseases which frequently decimated the world's population throughout history, such as plague, typhus, cholera, and tuberculosis, are little more than history to most people living in industrialized countries. The situation is such that antibiotics are taken for granted by the physicians who prescribe them and the public that consumes them. For most of the latter half of the past century, the production and sale of antibiotics have been highly competitive, with billions of dollars a year in sales. More recently, the competition has intensified in another sense; bacteria are responding to antibiotics by a variety of avoidance tactics, and the next phase of antibiotic development will take on a different focus: restoring their efficacy. This will require considerable research effort.

Thus, the importance of antibiotics to public health and well-being worldwide cannot be underestimated, and the availability of a text that describes all aspects of these magical small molecules is welcomed. *Bacteria versus Antibacterial Agents: an Integrated Approach* is such a text, and it will be welcomed as a source for university and medical school use.

In addition to the discovery and production of antibiotics, studies of the modes of action by which these compounds act on target cells have revealed fascinating biochemical stories. This information has been important in elucidating the mechanisms of replication, transcription, translation, and cell wall synthesis in bacteria and other living organisms. It also provided the experimental and conceptual basis for what is now known as chemical biology, wherein complex biological processes are dissected by analyzing inhibition by small molecules and using mutants refractory to inhibition as a means to identify key components of reactions.

Subsequent studies by three-dimensional structure analyses provided a detailed understanding of inhibitor-receptor interactions and the underlying structure-activity relationships. Such has been the case for many of the inhibitors of ribosome function and DNA synthesis.

Along with its many notable successes in therapy and in basic science, the use of antibiotics has had its downside; the target organisms can employ a variety of genetic and biochemical strategies to evade inhibition by small molecules, and there exists a real threat that many antibiotics will become useless as a result. New approaches to the discovery of novel inhibitors are desperately needed, and a great deal of effort is going into attempts to find new active molecules or to redesign old ones in order to stem the threat of emerging resistant pathogens.

These goals can be realized only by an understanding of the biology of the target organism, the biochemical action of the antibiotic, and the potential resistance mechanisms, not to mention the participation of the host response. Oreste Mascaretti touches on all of these matters in this book and has done a real service in providing a readable and well-illustrated account of the world of antibiotics.

Finally, while this book focuses on antibacterials, the same principles of chemistry, microbiology, genetics, and biochemistry apply in the development of antivirals and anticancer agents. The work done on antibacterials that is so clearly described here has meaningful lessons beyond infectious diseases.

*Julian Davies*
Department of Microbiology and Immunology
University of British Columbia
Vancouver, British Columbia, Canada

# Preface

This book presents an integrated approach to both the basic concepts and the most recent developments in the field in a form which is suitable for the needs of those who are studying clinical and molecular microbiology, biotechnology, biochemistry, chemical biology, bioorganic chemistry, pharmacy, or any of the related biomedical sciences.

The book was designed as an introductory text on antibacterial agents with broad coverage of the subject. I have tried to provide a comprehensive and concise overview of important subject areas. Supplanting a more detailed discussion of each topic, a list of references is included at the end of each chapter. The reference lists are limited mostly to some leading books, recent reviews, and a few classic articles where appropriate.

The understanding of the mechanisms of bacterial resistance has rapidly advanced in recent years, opening the way for the design of new antibacterial agents tailored to overcome the mechanisms that bacteria use to resist the action of traditional antibacterial agents.

In compiling diverse information into a single and comprehensive book, I have attempted to give a clear presentation of cutting-edge knowledge. The key features I present are as follows:

- The basics of bacterial cell structure and function for a better understanding of cellular metabolism and mechanisms of antibiotic action, as well as the mechanisms developed by prokaryotic cells to overcome the action of antibiotics
- The characteristic features of bacterial pathogenicity
- How antibacterial drugs reach their targets in gram-positive and gram-negative bacteria
- How the human immune system is involved in a wide range of immune responses in the battle against bacterial infections
- The genetic basis of resistance to antibacterial drugs

- The biochemical mechanisms of action of antibacterial drugs
- Improved penicillins, cephalosporins, tetracyclines, quinolones, macrolides, and glycopeptides developed by pharmaceutical companies to restock the antibacterial arsenal
- Recent advances in the research on and development of new classes of antibacterial drugs to combat the rising tide of drug-resistant bacterial infections

In preparing this textbook, I have striven for a coherent overall presentation. I have had the chapters read by experts (see the acknowledgments) who helped me to eliminate errors. However, any shortcomings of the book are solely my responsibility. There is no doubt that I have omitted some topics and given too much room to others, but the field of antibiotics covers such a broad area of interest that I had to make a personal judgment on what had to be included and what could be left out.

Where appropriate, mechanistic aspects of the action of antibacterial agents and β-lactamase inhibitors are discussed to help the reader understand the underlying principles of action on a particular target in bacteria. Additionally, each chapter on a group or subgroup of antibiotics or other antibacterial class also surveys the topic of structure-activity relationships.

The most recent and most significant developments in new antibacterial agents with novel modes of action are presented by using selected examples. Key terms are printed in bold type, and their definitions are given, when they are first encountered in the text.

I hope that that medicinal and bioorganic chemists, molecular biologists, microbiologists, geneticists, and immunologists each will find relevant topics from disciplines that they are not too familiar with. I expect that some of the information provided in this book will be useful in the development of new antibiotics and synthetic antibacterial agents, and I also hope to see more interdisciplinary cooperation between scientists in these different disciplines.

Any comments or suggestions for improvements in future editions would be gratefully received.

*Oreste A. Mascaretti*
Department of Organic Chemistry
Facultad de Ciencias Bioquímicas y Farmacéuticas
Universidad Nacional de Rosario
Suipacha 531
S 2000 LRK Rosario
Argentina
Fax: 54-(0)341-4370477
E-mail: masca@citynet.net.ar

# Acknowledgments

I express my gratitude for the constructive advice that I received from the following scientists: Eduardo L. Setti (Celera, San Francisco, Calif.), Lina Quatrocchio (Roche Bioscience, Palo Alto, Calif.), Nancy D. Hanson (Center for Reseach in Anti-infectives and Biotechnology, Creighton School of Medicine, Creighton University, Omaha, Nebr.), Heinz G. Floss (Department of Chemistry, University of Washington, Seattle, Wash.), Ronald J. Dworkin (Infectious Disease Reference Laboratory, Providence Portland Medical Center, Portland, Oreg.), Julian Davies (Department of Microbiology and Immunology, University of British Columbia, Vancouver, British Columbia, Canada), Stephen H. Zinner (Department of Medicine, Harvard Medical School, Cambridge, Mass.), Moreno Galleni (Centre for Protein Engineering, Institut de Chimie, University of Liège, Liège, Belgium), André Bryskier (Anti-infective Diseases Group, Aventis Pharma, Romainville, France), Nafsika H. Georgopapadakou (Newbiotics Inc., San Diego, Calif.), Joyce Sutcliffe (Rib-X Pharmaceuticals, Inc., New Haven, Conn.), Michael W. Russell (Department of Microbiology/Oral Biology, University of Buffalo, The State University of New York, Buffalo, N.Y.), Christopher N. C. Body (Department of Chemical Engineering, Stanford University, Stanford, Calif.), James P. Nataro (Department of Pediatrics, University of Maryland School of Medicine, Baltimore, Md.), Alexander S. Mankin (Center for Pharmaceutical Biotechnology, University of Illinois, Chicago, Ill.), George M. Eliopoulos (Department of Medicine, Massachusetts General Hospital, Boston, Mass.), Carlos A. Fossati (Departamento de Inmunología, Facultad de Ciencias Exactas, Universidad Nacional de La Plata, La Plata, Argentina), and Eleonora García Véscovi and Fernando C. Soncini (Departamento de Microbiología, Facultad de Ciencias Bioquímicas, Universidad Nacional de Rosario, Rosario, Argentina). I also thank Ada Yonath (Weizmann Institute, Rehovot, Israel) for her comments on the discussion of the structural basis for the antibiotic activity of the ketolide ABT-773 in chapter 18 and Christopher Schofield (Dyson Perrins Laboratory, Oxford University, Oxford, United Kingdom) for providing the photo of the monument in the Botanic Garden of Oxford University, which appears in the epilogue.

Thanks are also due to the following ASM Press staff members: Jeff Holtmeier (director) for the confidence he put in me, Kenneth April (senior production editor) for coordinating the publication process, and Jennifer Adelman for her assistance while I was writing the book. I also thank Yvonne Strong for copyediting the manuscript, Paula Ellison for editing and sizing the figures, Russell Burnett for proofreading the book, Susan Schmidler for designing the cover and interior, and Mary Boss, Sara Gryske, and others on the staff of Impressions Book and Journal Services for project management.

chapter 1



# Structure and Function of Prokaryotic and Eukaryotic Cells

This chapter is not intended to be a comprehensive discussion of molecular biology of the cell. Rather, it focuses on an overview of the basic concepts of eukaryotic and prokaryotic cell structure and function. The practical importance of each topic is described in the context of understanding the invasion of eukaryotic hosts by pathogenic bacterial (prokaryotic) cells. The great diversity of bacteria that inhabit the human body can be found as members of the normal flora, as opportunistic pathogens, or as true pathogens. Each bacterium has its own unique physiology and metabolic pathways that allow it to survive and reproduce in its particular habitat. Proper identification and characterization (intracellular or extracellular) of pathogenic bacterial species that cause infection in humans is critical to the decision about treatment with antibacterial agents.

The ability of bacteria to change rapidly, acquire new genes, and undergo mutations presents continual challenges to diagnostic bacteriologists as they isolate and characterize bacteria that cause disease in humans. Bacterial taxonomy refers to the classification and grouping of bacteria. It is based on phenotypic (observable) and genotypic (genetic) similarities and differences. The formal levels of bacterial classification, in successively smaller subsets, are as follows: kingdom, division, class, order, family, tribe, genus, and species. For example, *Staphylococcus* (genus) *aureus* (species) belongs to the family *Micrococcaceae*.

## Cells as the Basic Units of Life

Microscopic examination of any organism reveals that it is composed of membrane-enclosed structures called cells. The **cell** is the fundamental structural and functional unit of all living organisms. All cells have either a nucleus or a nucleoid, in which the **genome** (the complete set of genes, composed of deoxyribonucleic acid [DNA]) is stored and replicated.

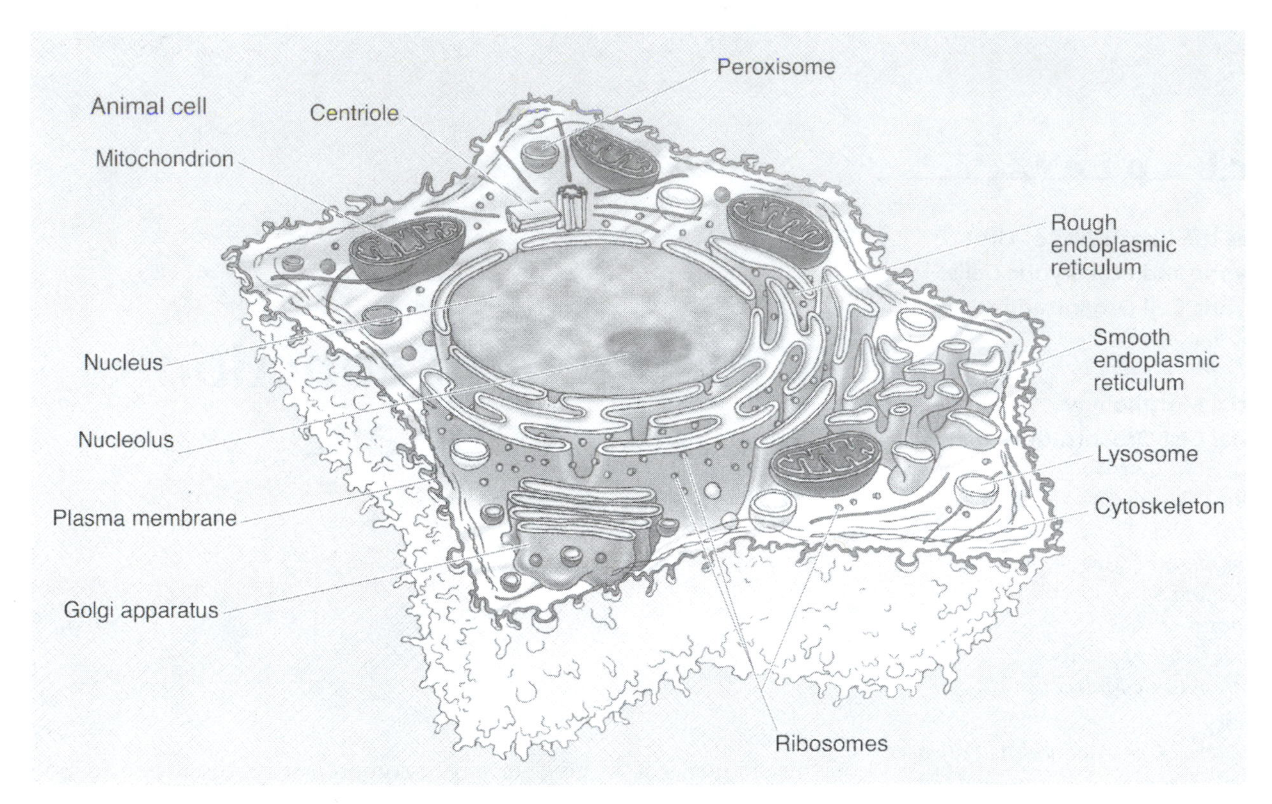


**Figure 1.1** Structure of a representative animal cell. Reprinted from G. M. Cooper, *The Cell: a Molecular Approach* (ASM Press, Washington, D.C., and Sinauer Associates, Inc., Sunderland, Mass., 2000), with permission from the American Society for Microbiology.

The contents enclosed by the plasma membrane (Fig. 1.1) constitute the **cytoplasm.** The liquid portion of the cytoplasm is called the **cytosol.** Cells are composed of small molecules, macromolecules, and organelles. Of the many different types of molecules in the various organelles and the cytosol that constitute the living cells, water is by far the most abundant. Therefore, most cellular components are essentially in an aqueous environment. Except for water, most of the molecules found in the cell are macromolecules, which can be classified into four different categories: lipids, carbohydrates, proteins, and nucleic acids.

## Eukaryotic and Prokaryotic Cells

Cells are divided into two main classes, initially defined by whether they possess a nucleus. **Eukaryotic cells** have a **nucleus** in which the genetic material is separated from the cytoplasm. Eukaryotes (from Greek *eu,* meaning "true," and *karyon,* meaning "nucleus") can be found in algae, fungi (e.g., yeasts and molds), protozoa, plants, and animals. **Prokaryotic cells,** which include the various types of bacteria (eubacteria) and archaea (archaebacteria), lack a nuclear envelope. Prokaryotes (from Greek *pro,* meaning "before") have a **nucleoid** (the region of the cell containing DNA) rather than an enveloped nucleus.

The basis for this nomenclature lies in the organizational complexity of eukaryotic cells on the one hand and the relative simplicity of prokaryotic cells on the other. Prokaryotes have a relatively simple structure and are invariably unicellular (although they may form colonies of independent cells). Eukaryotes, which may be multicellular as well as unicellular, are more complex than prokaryotes. Table 1.1 summarizes the principal differences between prokaryotic and eukaryotic cells.

As mentioned above, algae, protozoa, and fungi are microbial eukaryotes that belong to the group called **protists.** Their main characteristics are summarized below.

**Algae.** The term "algae" has long been used to denote all organisms that produce $O_2$ as a product of photosynthesis. Many algal species are unicellular microorganisms. Other algae may form extremely large multicellular structures.

**Protozoa.** Protozoa are unicellular nonphotosynthetic protists.

**Table 1.1** Comparison of prokaryotic and eukaryotic cell organization[a]

| Characteristic | Prokaryotic cell | Eukaryotic cell |
|---|---|---|
| Size | Generally small (1–10 μm) | Generally large (5–100 μm) |
| Genome | A single circular piece of DNA with nonhistone protein; genome in nucleoid, not surrounded by membrane | Multiple chromosomes; DNA complexed with histone and nonhistone proteins in chromosomes; chromosomes in nucleus with membrane envelope |
| Extrachromosomal DNA | Plasmids, small circular pieces of DNA containing accessory information, may be present in the cytoplasm | In mitochondria |
| Cell division | By binary fission (no mitosis) | By mitosis and meiosis including mitotic spindle; centrioles in many species |
| Membrane-bounded organelles | Absent | Mitochondria, chloroplasts (in plants, some algae), endoplasmic reticulum, Golgi complexes, lysosomes (in animals) |
| Site of protein production | Ribosomes free in the cytoplasm or attached to the cell membrane; 70S in size, consisting of a 50S and a 30S subunit. | Ribosomes, 80S in size, consisting of a 60S and 40S subunit; the ribosomes are covered by a membrane (rough endoplasmic reticulum), smooth endoplasmic reticulum, or Golgi apparatus, where secreted proteins are packaged and transported to the cell surface |
| Cell wall | Present, imparts rigidity | Usually absent except for fungi, which contain chitin in the cell wall; absorption, |
| Nutrition | Absorption | ingestion; photosynthesis in some species |
| Energy metabolism | No mitochondria; oxidative enzymes bound to plasma membrane, great variation in metabolic pattern | Oxidative enzymes packaged in mitochondria, more unified pattern of oxidative metabolism |
| Cytoskeleton | None | Complex, with microtubules, intermediate filaments, actin filaments |
| Intracellular movement | None | Cytoplasmic streaming, endocytosis, phagocytosis, mitosis, vesicle transport |

[a] Modified from D. L. Nelson and M. M. Cox, *Lehninger Principles of Biochemistry,* 3rd ed. (Worth Publishers, New York, N.Y., 2000), with permission from the publisher.

**Fungi.** The fungi are nonphotosynthetic protists growing as a mass of branching filaments (hyphae) known as a **mycelium.** The mycelial form are called **molds.** The **yeasts** do not form a mycelium but are easily recognized as fungi by their sexual reproductive processes and by the presence of transitional forms.

It is worthwhile to note that viruses are supramolecular complexes that can replicate themselves in appropriate host cells. A **virus** consists of a nucleic acid molecule (DNA or ribonucleic acid [RNA]) surrounded by a protein coat (capsid) and sometimes a membranous envelope. Viruses lack cellular structure and cannot carry out independent metabolic reactions, commonly regarded as hallmarks of life. Viruses exist in two states. Outside the host cells they exist as nonliving particles called **virions.** Once a virus or its nucleic acid component enters a specific host cell, it becomes an intracellular parasite. Viruses propagate by hijacking the metabolic and genetic machinery of living cells and diverting it to the formation of new viruses. The host cell may be an animal, plant, or bacterial cell. Viruses specific for bacteria are known as **bacteriophages,** or **phages.** Viruses differ in shape and complexity of structure. The human immunodeficiency virus is relatively simple in structure, but it causes acquired immunodeficiency syndrome (AIDS) by destroying cells central to the human immune response.

## Eukaryotic Cell Organization

Eukaryotes include most multicellular organisms and many unicellular species. Eukaryotic cells are generally 10 to 100 μm in diameter and thus have $10^3$ to $10^6$ times the volume of typical prokaryotic cells. Microscopic examination of a eukaryotic cell reveals that it is composed of membrane-enclosed structures. The enclosing membrane is called the cell membrane or the plasma membrane.

## The Plasma Membrane

All cells—both prokaryotic and eukaryotic—are surrounded by a **plasma membrane,** which defines the boundaries of the cell and separates its internal contents from the environment. The plasma membrane is similar to a semipermeable sheet that covers the entire cell. This membrane consists of both lipids and proteins. Its fundamental structure is the phospholipid bilayer, which is impermeable to most water-soluble molecules. The passage of ions and most biological molecules across the plasma membrane is therefore mediated by proteins, which are responsible for the selective traffic of molecules into and out of the cell. The external surface of a cell is in contact with other cells, the external fluid, solutes, and nutrient molecules.

The plasma membrane of all eukaryotic cells contains many **transporters,** proteins that cross the membrane and carry nutrients into the cell and various products out. Eukaryotic cells also have surface membrane proteins (**signal receptors**) with highly specific binding sites for extracellular signaling molecules (receptor ligands). Integral membrane proteins are inserted into the lipid bilayer, whereas peripheral proteins are bound to the membrane indirectly by protein-protein interactions. Most integral membrane proteins are transmembrane proteins, with portions exposed on both sides of the lipid bilayer. The extracellular portions of these proteins are usually glycosylated, as are the peripheral membrane proteins bound to the external face of the membrane (Fig. 1.2).

**Endocytosis** is a mechanism for transporting components of the surrounding medium into the cytoplasm. **Exocytosis** is the inverse of endocytosis, in which a vesi-

**Figure 1.2** Fluid mosaic model of the cytoplasmic membrane structure. Reprinted from G. M. Cooper, *The Cell: a Molecular Approach* (ASM Press, Washington, D.C., and Sinauer Associates, Inc., Sunderland, Mass., 2000), with permission from the American Society for Microbiology.

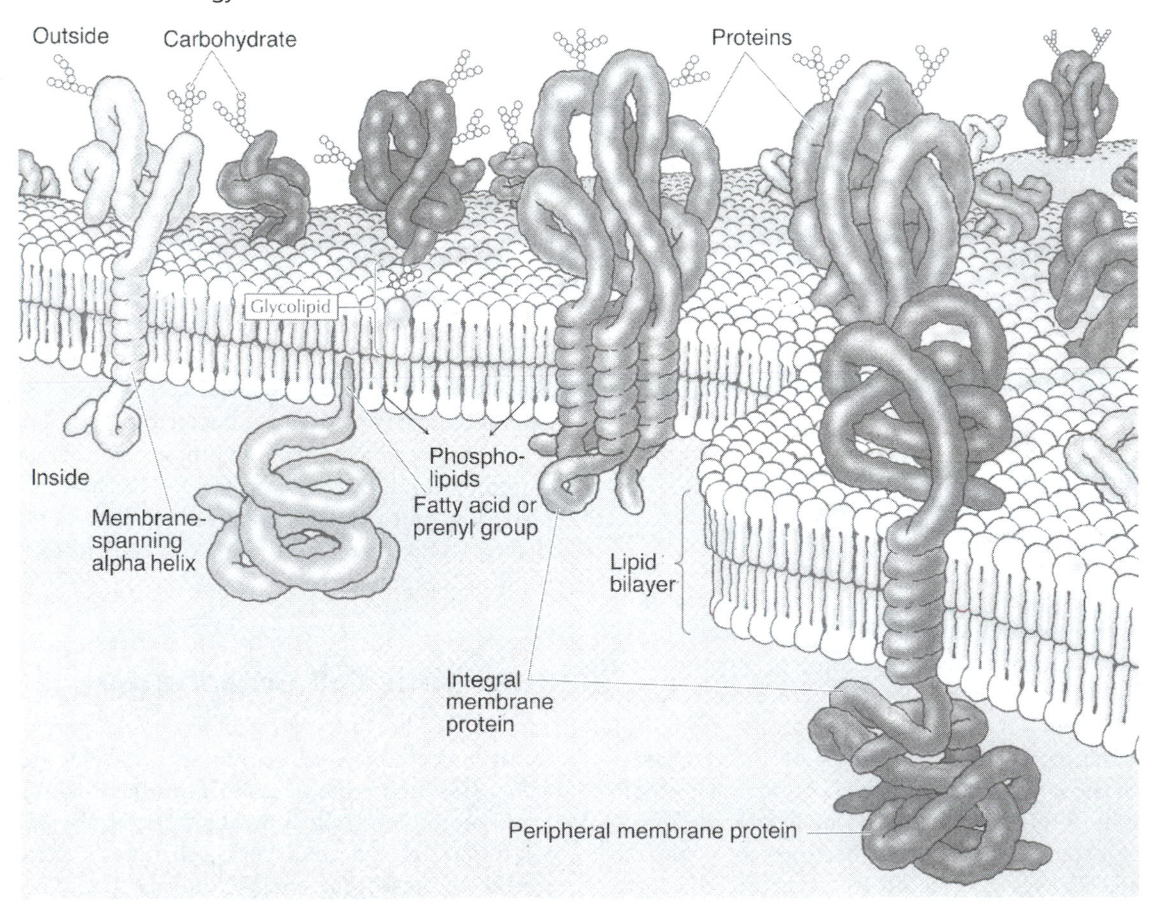

cle in the cytoplasm moves to the inside surface of the plasma membrane. **Phagocytosis** is a cell-initiated form of endocytosis in which the material carried into the cell (within a phagosome) is particulate, such as a cell fragment or even a small cell.

## Cytoplasmic Structures

Within the cytoplasm, there are a number of macromolecules and larger structures. Some of the structures are membranous and are called **organelles.** Organelles commonly found in animal cells include the nucleus, the endoplasmic reticulum, the Golgi apparatus, lysosomes, the mitochondria, peroxisomes, and ribosomes (Fig. 1.1; Table 1.2).

Besides organelles, animal cells contain a collection of filamentous structures termed the **cytoskeleton,** which is important in maintaining the three-dimensional integrity of the cell.

### THE NUCLEUS

The eukaryotic nucleus is quite complex in structure and biological activity compared to the relatively simple prokaryotic nucleoid. The nucleus is structurally defined by the **nuclear envelope,** a double membrane that is topologically a single sheet with two layers. The nuclear envelope separates the contents of the nucleus from the cytoplasm and provides the structural framework of the nucleus (Fig. 1.1). The nucleus serves both as the repository of genetic information and as the control center of the cell. The genomes of most eukaryotes are larger and more complex than those of prokaryotes.

An understanding of gene structure and function is fundamental to an appreciation of eukaryotic nuclear organization. In molecular terms, a **gene** is the entire nucleic acid sequence that is necessary for the synthesis of a polypeptide or an RNA (e.g., ribosomal and transfer RNAs [rRNA and tRNA]). Within the nucleus, the genetic information (encoded in the base sequences of DNA molecules) is **transcribed** into molecules of RNA, which are transported to the ribosomes in the cytoplasm, where the **translation** of RNA into proteins takes place.

The inner membrane of the nuclear envelope is lined with filamentous proteins that constitute the nuclear lamina. The chromosomes of eukaryotic cells contain linear DNA macromolecules arranged as a double helix. Eukaryotic DNA macromolecules are associated with basic proteins called histones that bind to the DNA by ionic interactions and coiled into a dense mass called **chromatin.** The DNA molecules are visible under a light microscope only when the cell is undergoing division. The **nuclear pore complexes** are the only channels through which small polar molecules, ions, and macromolecules (proteins and RNAs) can travel between the nucleus and the cytoplasm. The **nucleolus** is the most important substructure within the nucleus (Fig. 1.1). This is the site of rRNA transcription and processing and of ribosome assembly. The nucleolus, which is not surrounded by a membrane, is organized around the chromosomal regions that contain the genes for the 5.8S, 18S, and 28S rRNAs.

A unique feature of the nucleus is that it disassembles and re-forms each time most cells divide. At the beginning of mitosis, the chromosomes condense, the nucleolus disappears, and the nuclear envelope breaks down, resulting in the release of most of the contents of the nucleus into the cytoplasm. At the end of mitosis, the chromosomes decondense and the nuclear envelopes re-form around the separated sets of daughter chromosomes.

**Table 1.2** Functions of animal eukaryotic organelles[a]

| Organelle(s) | Function(s) |
|---|---|
| Plasma membrane | Mechanical cell boundary, selectively permeable barrier with transport systems; mediates cell-cell interactions and adhesion to surfaces |
| Cytoplasmic matrix | Environment for other organelles; location of many metabolic processes |
| Microfilaments, intermediate filaments, and microtubules | Cell structure and movement; forms the cytoskeleton |
| Endoplasmic reticulum | Transport of materials; protein and lipid synthesis |
| Ribosomes | Protein synthesis |
| Golgi apparatus | Packaging and secretion of materials for various purposes; lysosome formation |
| Lysosomes | Intracellular digestion |
| Mitochondria | Energy production through use of the tricarboxylic acid cycle, electron transport, oxidative phosphorylation, and other pathways |
| Nucleus | Repository for genetic information; control center for cell |
| Nucleolus | rRNA synthesis; ribosome construction |

[a] Reprinted from L. M. Prescott, J. P. Harley, and D. A. Klein, *Microbiology,* 4th ed. (McGraw-Hill, Boston, Mass., 2000), with permission from the publisher.

## THE ENDOPLASMIC RETICULUM

The **endoplasmic reticulum** (ER) (Fig. 1.1) is a highly convoluted, three-dimensional network of membrane-enclosed spaces extending throughout the cytoplasm and enclosing a subcellular compartment separate from the cytoplasm. Ribosomes attach to the outer (cytoplasmic) surface of the ER. The attachment of thousands of ribosomes gives the rough ER its granular appearance and hence its name. In other regions of the cell, the ER is free from ribosomes. This smooth ER, which is physically continuous with the rough ER, is the site of lipid biosynthesis and a variety of other important processes.

## THE GOLGI APPARATUS

Nearly all eukaryotic cells have **Golgi apparatuses** (also called Golgi complexes), which are systems of membranous sacs or cisternae, arranged as flattened stacks. The Golgi apparatus is structurally and functionally asymmetric. One side (called the *cis*-Golgi apparatus) faces the rough ER (and the nucleus), and the other side (called the *trans*-Golgi apparatus) faces the plasma membrane. Newly synthesized proteins move from the ER to the Golgi apparatus, fusing with the *cis* side. As the proteins pass through the Golgi apparatus to the *trans* side, enzymes in the complex modify the protein molecules by adding sulfate, carbohydrate, or lipid moieties to the side chains of certain amino acids. One of the functions of this modification is to ensure that the protein reaches its proper destination as it leaves the Golgi apparatus in a transport vesicle budding from the *trans* side.

## LYSOSOMES

**Lysosomes,** found only in animal cells, are spherical vesicles bounded by a single membrane bilayer (Fig. 1.1). They contain enzymes capable of digesting proteins, polysaccharides, nucleic acids, and lipids. They function as cellular recycling centers, breaking down molecules brought into the cell by endocytosis and fragments of foreign cells brought in by phagocytosis. These materials selectively enter the lysosome by fusion of the lysosomal membrane with endosomes or phagosomes and are then degraded to their simpler components (small peptides, amino acids, monosaccharides, etc.).

## MITOCHONDRIA

**Mitochondria** (singular, mitochondrion) are visible in most eukaryotic cells under electron microscopy. The mitochondria are surrounded by two membranes, an outer membrane and an inner membrane, which are separated by an intermembrane space (Fig. 1.1). The inner membrane forms numerous folds called **cristae,** which extend into the interior (or matrix) of the organelle.

Mitochondria are responsible for generating most of the useful energy derived from the breakdown of lipids and carbohydrates. The matrix contains the mitochondrial enzymes responsible for the reactions of oxidative metabolism, tricarboxylic acid cycle activity, and the generation of adenosine triphosphate (ATP) by electron transport and oxidative phosphorylation. In addition, the mitochondrial matrix contains the specific mitochondrial DNA, which encodes tRNAs, rRNAs and some mitochondrial proteins. The assembly of mitochondria thus involves proteins encoded by their own genomes and translated within the organelle as well as proteins encoded by the nuclear genome and imported from the cytosol.

## PEROXISOMES

Peroxisomes are small, membrane-enclosed organelles that contain enzymes involved in a variety of metabolic reactions, including several aspects of energy metabolism. Some of the oxidative reactions, such as the oxidation of a fatty acid, are accompanied by the production of hydrogen peroxide ($H_2O_2$). Because hydrogen peroxide is harmful to the cell, peroxisomes contain the enzyme **catalase,** which catalyzes the decomposition of hydrogen peroxide to water ($H_2O$).

## RIBOSOMES

Ribosomes are the sites of protein synthesis in both prokaryotic and eukaryotic cells. Eukaryotic ribosomes can either be associated with the ER or be free in the cytoplasm. Ribosomes are usually designated according to their apparent sedimentation coefficients, which are characterized by the **Svedberg unit** (**S**) and reflect the rate at which a molecule sediments in a solvent. Both prokaryotic and eukaryotic ribosomes are composed of two distinct subunits, each containing characteristic proteins and rRNAs.

The general structures of prokaryotic and eukaryotic ribosomes are similar, although they differ in some details (Fig. 1.3). The small subunit (40S) of eukaryotic ribosomes is composed of 18S rRNA and approximately 30 proteins; the large subunit (60S) contains the 28S, 5.8S, and 5S rRNAs and about 45 proteins. The large subunit (70S) of prokaryotic ribosome contains 5S and 23S rRNA and about 33 proteins, while the small subunit of prokaryotic ribosome contains 16S rRNA and 21 proteins.

## THE CYTOSKELETON AND CELL MOVEMENT

The membrane-enclosed organelles discussed above constitute one level of the organizational substructure of eukaryotic cells. A further level of organization is provided by the **cytoskeleton,** which consists of a network

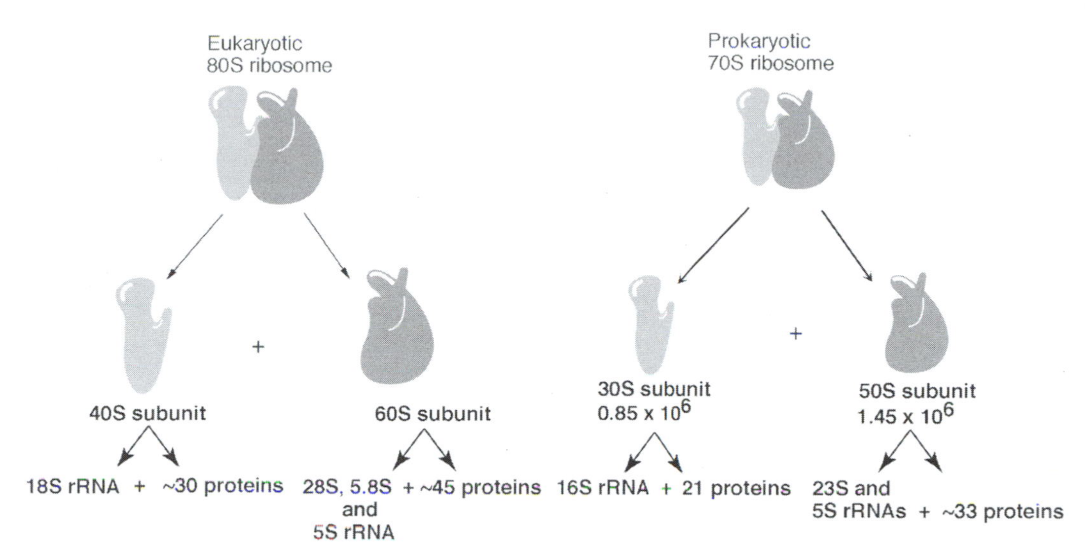


**Figure 1.3** Schematic three-dimensional model and comparison of eukaryotic and prokaryotic ribosomes.

of protein filaments extending throughout the cytoplasm of all eukaryotic cells.

There are three general types of cytoplasmic filaments: actin filaments, intermediate filaments, and microtubules. They differ in width (from about 6 to 25 nm), composition, and specific function, but they all provide a structural framework for the cell. Aside from playing this structural role, the cytoskeleton is responsible for the motion of organelles through the cytoplasm or the movement of whole cell.

The protein actin assembles in the presence of ATP into long helical, noncovalent polymers, called **actin filaments** or **microfilaments.**

**Intermediate filaments** are a family of structures with dimensions (diameter, 8 to 10 nm) that are between those of actin filaments (about 7 nm) and microtubules (about 25 nm). One function of intermediate filaments is to provide internal mechanical support for the cell and to position its organelles. Whereas actin filaments and microtubules are polymers of single types of proteins (actin and tubulin), intermediate filaments are composed of a variety of proteins that are expressed in different types of cells.

**Microtubules** are rigid, hollow rods approximately 25 nm in diameter. They assemble spontaneously from their monomeric subunits, α- and β-tubulin (Fig. 1.4). In cells, microtubules undergo continuous polymerization and depolymerization by addition of tubulin subunits primarily at one end and dissociation at the other end.

Microtubules are present throughout the cytoplasm but are concentrated in specific regions at certain times. For example, after sister chromatids separate and move to the opposite poles of a cell during mitosis, a highly organized array of microtubules (the mitotic spindle) provides the framework and probably the driving force for the separation of these daughter chromosomes.

A full account of the eukaryotic cell structure and function as well as the cellular processes in terms of chemical reactions (chemistry of eukaryotic cells) can be found in Cooper (2000) (see the references listed at the end of this chapter).

## Bacterial Morphology

Most commonly encountered bacteria have one of two shapes. **Cocci** (singular, **coccus**) are roughly spherical

**Figure 1.4** Structure of microtubules. Dimers of α- and β-tubulin polymerize to form microtubules, which are composed of 13 protofilaments assembled around a hollow core. Reprinted from G. M. Cooper, *The Cell: a Molecular Approach* (ASM Press, Washington, D.C., and Sinauer Associates, Inc., Sunderland, Mass., 2000), with permission from the American Society for Microbiology.

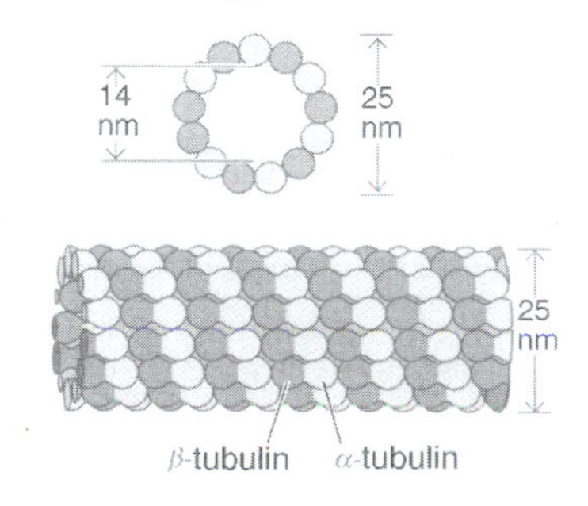

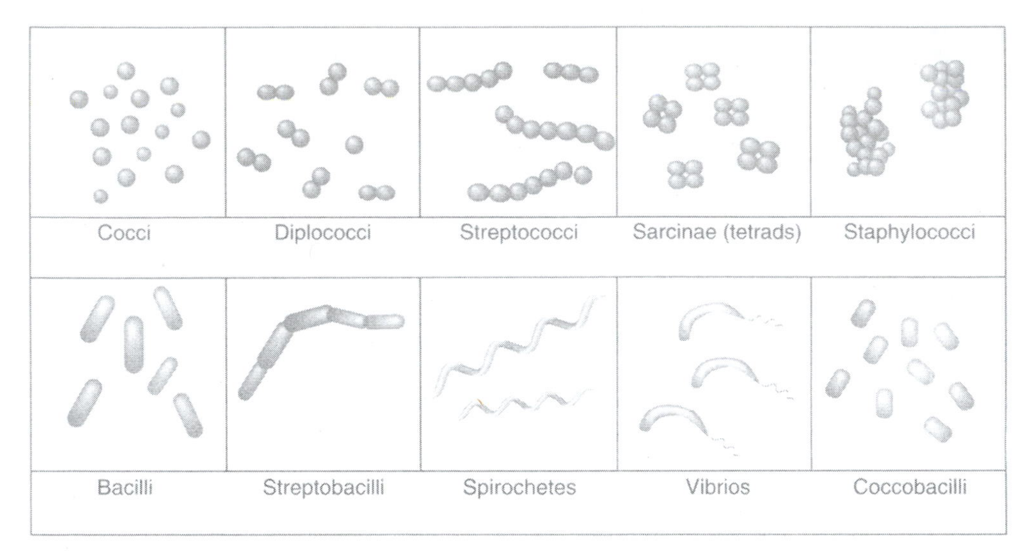


**Figure 1.5** Morphologies of some commonly encountered bacteria. Reprinted from T. S. Walker, *Microbiology* (The W. B. Saunders Co., Philadelphia, Pa., 1998), with permission from the publisher.

cells (Fig. 1.5). They can exist as individual cells but also can be associated in characteristic arrangements. **Diplococci** (singular, **diplococcus**) arise when cocci divide and remain together to form pairs. Long chains of cocci result when cells adhere after repeated divisions in one plane; this pattern is seen in the genera *Streptococcus, Enterococcus,* and *Lactococcus. Staphylococcus* cells divide in random planes to generate irregular grapelike clumps. Divisions in two or three planes can produce symmetrical clusters of cocci. *Micrococcus* cells often divide in two planes to form a square group of four cells called a tetrad. In the genus *Sarcina,* cocci divide in either two or three planes to produce tetrads or cubic packets of eight cells.

The other common bacterial shape is the rod-shaped cell, often called a **bacillus** (plural, **bacilli**). *Bacillus megaterium* is a typical example of a bacterium with a rod shape. Although many rods do occur singly, they may remain together after division to form pairs or chains (e.g., *Bacillus megaterium* is found in long chains). A few rod-shaped bacteria, the **vibrios,** are curved to form distinctive comma shapes or incomplete spirals.

## Bacterial Cell Organization

The prokaryotic cell is smaller and simpler than the eukaryotic cell at every level, with one exception: the layers that surround the prokaryotic cell are more complex. A variety of structures are found in prokaryotic cells, and their major functions are summarized in Table 1.3. Figure 1.6 provides illustrations of some of these structures.

### The Bacterial Cell Wall

The bacterial cell wall is a rigid structure that maintains the shape of the cell. The intact cell wall is necessary for survival of the bacteria because its mechanical strength prevents damage or bursting (osmotic lysis) of the relatively fragile underlying cytoplasmic membrane of the cell as a result of the high osmotic pressure inside it.

There are two basic types of organization of the outer bacterial layer, which can be distinguished by the Gram stain procedure developed by Hans Christian Joachim Gram in 1884 at the University of Berlin. This staining method consists of treating a dried and fixed film of bacteria on a glass microscopic slide with the basic dye gentian violet followed by a 3% potassium iodide-iodine ($KI-I_2$). This solution acts as a mordant, to form a water-insoluble dark purple complex in the cell. The cells are then treated (washed) with alcohol or acetone (or a mixture of the two solvents), which removes the color from (destains) some bacteria (the **gram-negative** ones) but not the **gram-positive** ones, which remain dark purple.

A few years after the development of the initial Gram stain, the pathologist Carl Weigert added another step to the Gram procedure that involves a counterstain with the dye safranin, which turns the gram-negative cells pink or red and leaves gram-positive bacteria dark purple.

The true structural difference between these two groups of bacteria became clear with the advent of the transmission electron microscope. The layers that surround the prokaryotic cell are referred to collectively as the cell envelope. The structure and organization of the cell envelope differ in gram-positive and gram-negative

**Table 1.3** Functions of prokaryotic structures[a]

| Structure(s) | Function(s) |
|---|---|
| Plasma membrane | Selectively permeable barrier; mechanical boundary of cell; nutrient and waste transport; location of many metabolic processes |
| Ribosomes | Protein synthesis |
| Inclusion bodies | Storage of carbon, phosphate, and other substances |
| Nucleoid | Localization of genetic material (DNA) |
| Cell wall | Gives bacteria shape and protection from lysis in dilute solutions |
| Capsules and slime layers | Resistance to phagocytosis, adherence to surfacers |
| Fimbriae and pili | Attachment to surfaces, bacterial mating |
| Flagella | Movement |
| Endospore | Survival under harsh environment conditions |
| Gas vacuole | Buoyancy for floating in aquatic environments |
| Periplasmic space (in gram-negative bacteria) | Contains hydrolytic enzymes and binding proteins for nutrient processing and uptake |

[a] Reprinted from L. M. Prescott, J. P. Harley, and D. A. Klein, *Microbiology*, 4th ed. (McGraw-Hill, Boston, Mass., 2000), with permission from the publisher.

bacteria. It is this difference that defines these two major assemblages of bacterial species. Simplified diagrams of these two types of cell envelopes are presented in Fig. 1.7 and 1.8.

Aside from bacterial classification based on differential Gram staining, some mycobacteria have an acid-fast cell wall and prokaryotes of the *Mycoplasma* and *Ureaplasma* genera are unique in lacking a cell wall and containing sterols in their cell membranes.

**Acid-fast staining** is another differential staining procedure. *Mycobacterium* species do not bind to simple stains readily and must be stained by a harsher treatment such as by heating with a mixture of basic fuchsin and phenol (the Ziehl-Neelsen method). Once basic fuchsin has penetrated with the aid of heat and phenol, acid-fast cells are not easily decolorized by an acid-alcohol wash and hence remain red. This is due to the high lipid content (30 to 60% by weight) of the cell walls. Substances

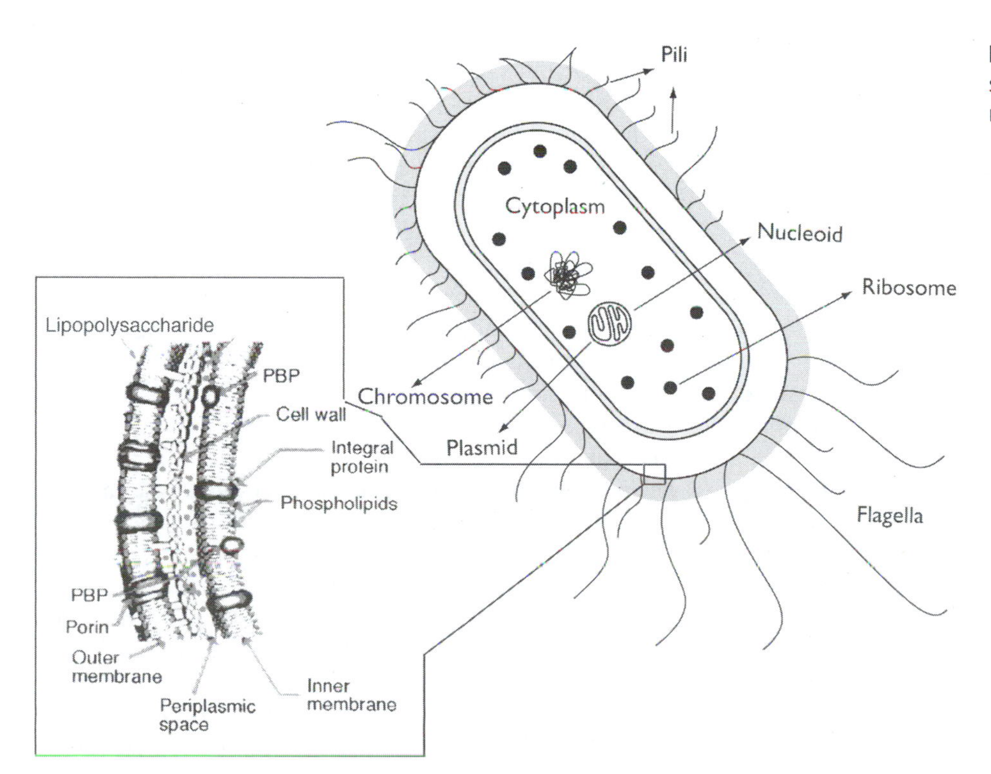


**Figure 1.6** Diagrammatic representation of a gram-negative bacterial cell.

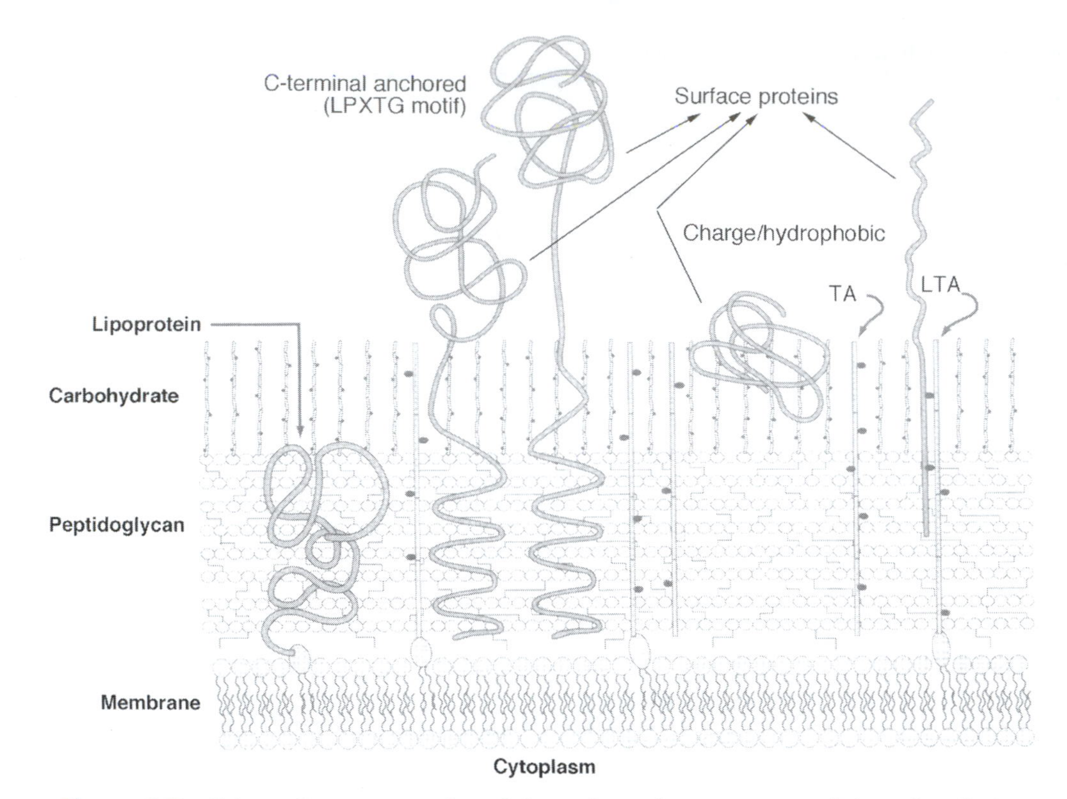


**Figure 1.7** Schematic representation of the major surface structures of the cell wall of gram-positive bacteria. Abbreviations: TA, teichoic acid; LTA, lipoteichoic acid. Reprinted from V. A. Fischetti, R. P. Novick, J. F. Ferretti, D. A. Portnoy, and J. L. Rood (ed.), *Gram-Positive Pathogens* (ASM Press, Washington, D.C., 2000), with permission from the American Society for Microbiology.

such as glycolipids, glyceropeptidolipids, phospholipids, lipooligosaccharides, and even mycolic acid—a group of branched-chain hydroxy lipids—appear responsible for acid fastness. Non-acid-fast bacteria are decolorized by acid-alcohol and thus are stained blue by a methylene blue counterstain. This method identifies *Mycobacterium tuberculosis* and *M. leprae,* the pathogens responsible for tuberculosis and leprosy, respectively.

## POLYMERS IN BACTERIAL CELL WALLS

The gram-positive cell wall possesses an extensive multilayered structure of homogeneous **peptidoglycan** or **murein** that is typically between 15 and 30 nm thick, lying outside the plasma membrane. Biophysical studies have indicated that glycan chains are not randomly oriented in the peptidoglycan layer but, rather, are arranged in parallel. Since one layer of peptidoglycan is about 1 nm thick, this implies that the cell wall of a bacterium such as *Bacillus subtilis,* which is 25 nm thick, consists of 25 layers of peptidoglycan. A certain proportion of all peptidoglycan polymers are cross-linked to adjacent strands so that a large macromolecular murein sacculus completely surrounds the cell.

In addition, gram-positive cell walls contain **teichoic acids,** polymers of glycerol or ribitol joined by phosphate groups unless grown under phosphate limitation. Amino acids such as D-alanine or sugars such as glucose are attached to the glycerol and ribitol groups. The teichoic acids are connected either to the peptidoglycan itself by a covalent bond with the hydroxyl groups of *N*-acetylmuramic acid or to plasma membrane lipids; in the latter case, they are called **lipoteichoic acids.** In some gram-positive bacteria teichoic acid is absent and the wall contains teichuronic acid, a phosphate-free polysaccharide containing carboxyl groups. This is particularly the case for bacilli grown in continuous culture in a medium where phosphate is the growth-limiting factor. The addition of phosphate to the medium causes a rapid return of teichoic acid and loss of teichuronic acid.

The structure of peptidoglycan and teichoic acids and the biosynthesis of peptidoglycan are discussed below. The bacterial cell wall contains polymers which are chemically very different from anything found in the host cells; therefore, one might expect that the biosynthesis of bacterial cell wall components could be inhibited selectively through interference with steps which

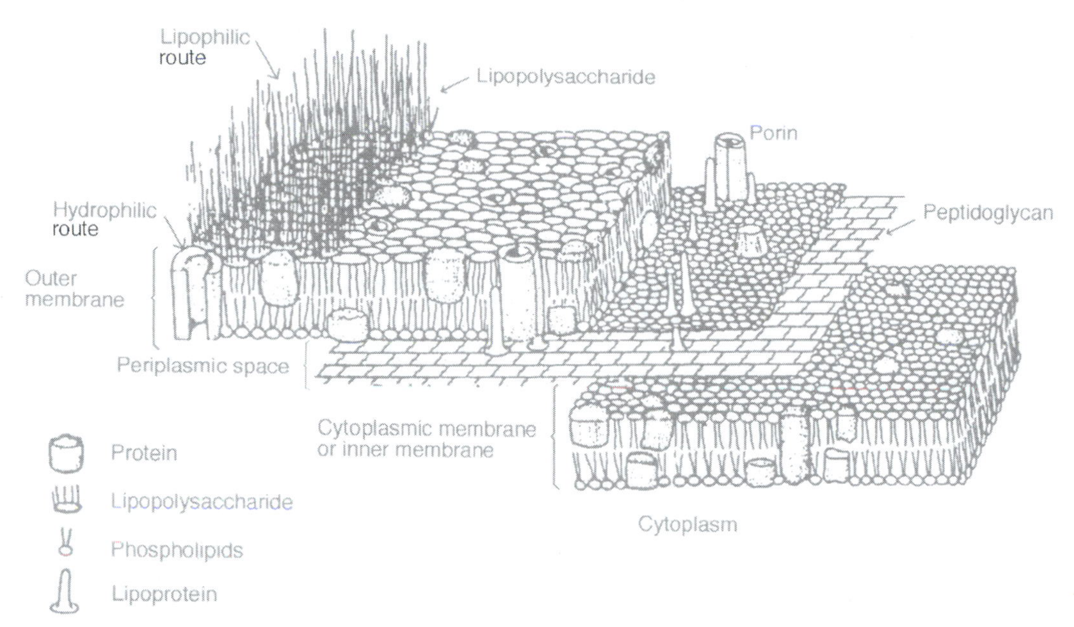


**Figure 1.8** Schematic representation of the major surface structures of the cell wall of gram-negative bacteria. Reprinted from L. E. Bryan (ed.), *Antimicrobial Drug Resistance* (Academic Press, Inc., San Diego, Calif., 1984), with permission from the publisher.

have no counterpart in the host metabolism. The β-lactams and glycopeptides (vancomycin and teicoplanin) are good examples of selective inhibitors of this kind. As discussed below, these inhibitors interfere with later steps in the biosynthesis of the peptidoglycan polymer.

From electron microscopic analysis of the gram-positive cell envelope and some chemical and immunological analysis, a schematic representation of the gram-positive cell wall has emerged (Fig. 1.7). The structure differs significantly from the gram-negative cell wall in two ways: (i) the presence of a thicker and more cross-linked peptidoglycan and (ii) the lack of an outer membrane. Because of these differences, surface molecules in gram-positive organisms differ from those in gram-negative organisms, in which specialized systems are required to transport and anchor molecules through the outer membrane.

Generally, surface proteins in gram-positive bacteria can be separated into three categories: (i) those that are anchored by their C-terminal ends, (ii) those that bind by way of charge or hydrophobic interaction (some proteins are bound ionically to the lipoteichoic acid), and (iii) those that bind via their N-terminal region (lipoproteins).

## CELL WALLS OF GRAM-NEGATIVE BACTERIA

*Escherichia coli* is the best studied gram-negative prokaryotic cell. The *E. coli* cell is about 2 μm long and not quite 1 μm in diameter. The cell envelope in *E. coli*, as in all gram-negative bacteria, is composed of three layers: the peptidoglycan or murein layer, the outer membrane, and the cytoplasmic membrane. Between the outer membrane and the inner (or cytoplasmic) membrane and encompassing the thin peptidoglycan layer, there is an area called the **periplasmic space** (Fig. 1.8). This division of layers is artificial. For instance, cytoplasmic membrane proteins with large periplasmic domains belong, at least operationally, to two layers. Furthermore, during growth, vectorial transport of proteins and peptidoglycan precursors occurs continuously.

In gram-negative bacteria, the peptidoglycan network is present as one or two layers and represents less than 10% of the cell dry weight (*E. coli*). The peptidoglycan layer is the essential shape-maintaining layer, and it can also be considered the assembly scaffold of the outer membrane components. The outer membrane contains proteins, phospholipids, and lipopolysaccharides.

## PEPTIDOGLYCAN CHEMICAL STRUCTURE

Both gram-positive and gram-negative bacteria are surrounded by a cell wall that is composed mainly of peptidoglycan, a strong net-like polymer responsible for maintaining the shape and size of the bacterial cell and for resisting the high intracellular osmotic pressure. The basic structure of peptidoglycan has now been determined for many species, and although there is considerable variation in the details of composition, the general structure remains the same. As the name implies, peptidoglycan consists of glycan chains with peptide substituents that are cross-linked.

Peptidoglycan consists of repeating (β1→4)-linked *N*-acetylglucosamyl-*N*-acetylmuramyl units cross-linked through short peptide chains. The fundamental disaccharide unit is shown in Fig. 1.9. The *N*-acetylmuramic acid residues are linked to short peptides whose composition varies from one bacterial species to another. In some species, the L-lysine residues are replaced by *meso*-diaminopimelic acid, an amino acid that is found in nature only in prokaryotic cell walls. The polymer also contains D-amino acids, which are also characteristic constituents of prokaryotic cell walls. The peptide chains of the peptidoglycan are cross-linked between parallel polysaccharide backbones.

Most gram-negative bacteria have a peptidoglycan structure in which *meso*-diaminopimelic acid at position 3 is directly linked through its free amino group with the free carboxyl group of the terminal D-alanine (at position 4) of an adjacent peptide chain structure (Fig. 1.10a). In gram-positive bacteria the *meso*-diaminopimelic acid at position 3 is replaced by L-lysine and the peptide subunits of the glycan chain are cross-linked by interpeptide bridges. In *Staphylococcus aureus*, the peptide interbridge is formed by a pentaglycine unit (Fig. 1.10b). There is high diversity among gram-positive peptidoglycan chemotypes (more than 100 chemotypes are known) that involves subtle different amino acid substituents in the peptide stems and/or different linkage units between the stems.

**Figure 1.9** Structure of one of the repeating units of the peptidoglycan cell wall structure, the glycan tetrapeptide. Each monosaccharide derivative (G) is attached to other monosaccharide derivatives (M) by β1→4 glycosidic bonds. The structure illustrated is that found in *E. coli* and most other gram-negative bacteria. In some bacteria, other amino acids are found.

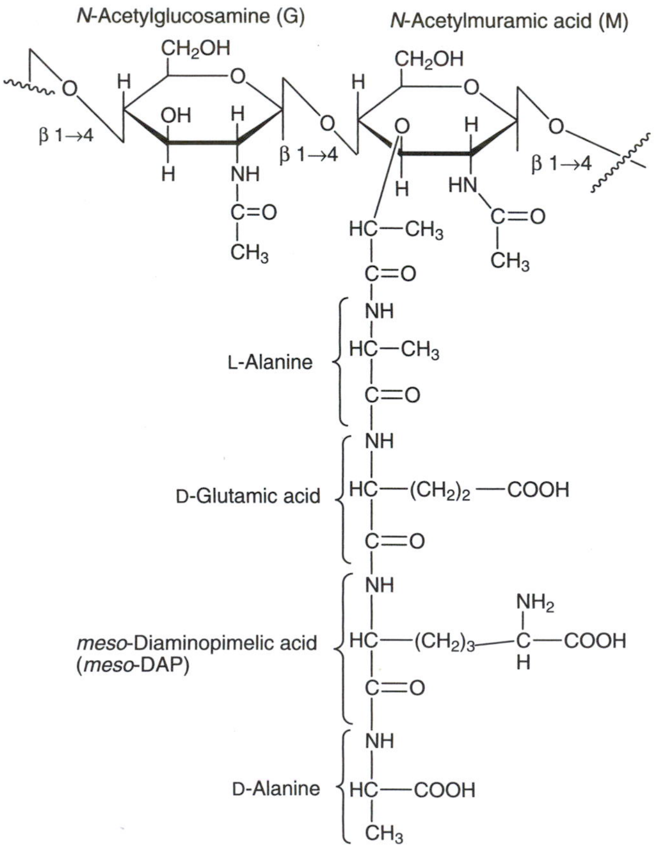

## Peptidoglycan Biosynthesis

The biosynthesis of a peptidoglycan unit proceeds by a series of cytoplasmic and membrane reactions. The various steps of the pathway have been identified in a variety of bacteria, and a general scheme which seems to be valid for both gram-positive and gram-negative bacteria has been established (Fig. 1.11).

For convenience, three groups of successive reactions of the biosynthetic pathway can be considered, each of which occurs at a different cellular site: (i) formation of the uridine diphosphate–*N*-acetylmuramic acid (UDP-MurNAc) peptides, (ii) formation of the lipid intermediates, and (iii) the transpeptidation reaction leading to the cross-linked peptidoglycan polymer. Figures 1.12 to 1.15 diagram how the cell wall of *S. aureus* is biosynthesized.

*The Cytoplasmic Stage.* The first stage is catalyzed by soluble enzymes in the cytoplasm. Synthesis of the pepti-

**Figure 1.10** Peptidoglycan cross-links. (a) *E. coli* peptidoglycan; (b) *S. aureus* peptidoglycan. The abbreviations and structures of amino acid in the figure are given in appendix A.

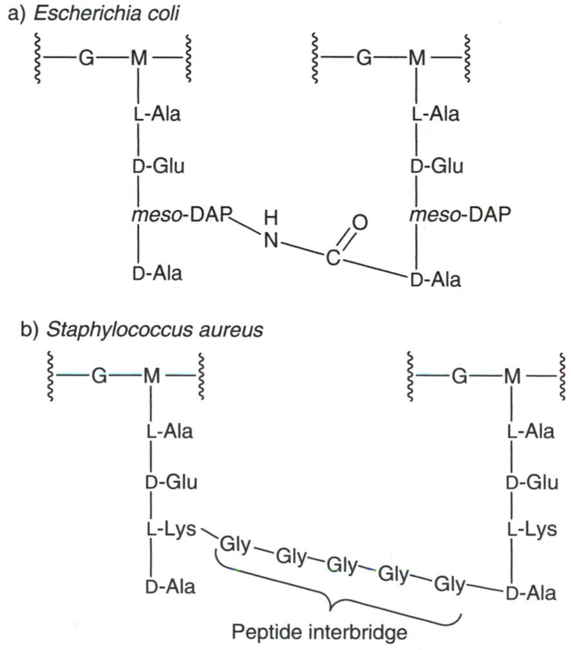

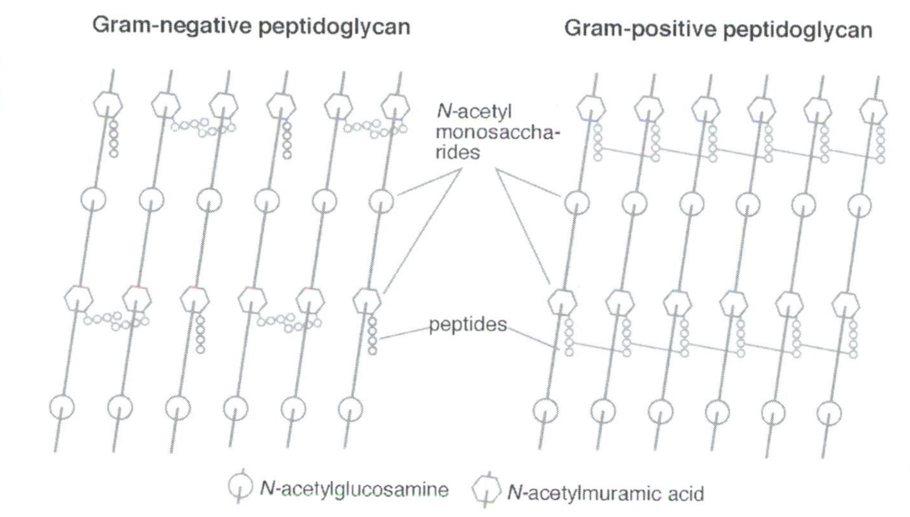


**Figure 1.11** Diagrammatic representation of peptidoglycan structures with adjacent glycan strains cross-linked directly from the carboxyl-terminal D-alanine to the ε-amino group of an adjacent tetrapeptide or through a peptide interbridge.

doglycan begins with the formation of UDP-*N*-acetylmuramic acid from UDP-*N*-acetylglucosamine (UDP-GNAc), catalyzed by the enzymes MurA and MurB. Next, phosphoenolpyruvate is added to position 3 of the *N*-acetylglucosamine residue, which is catalyzed by a transferase to yield UDP-GNAc-3'-enolpyruvyl ether, and the subsequent reaction involves reduction of the enolpyruvyl ether moiety to D-lactate, which is catalyzed by a reductase to yield UDP-MurNAc (Fig. 1.12).

The conversion of UDP-MurNAc acid to UDP-MurNAc pentapeptide occurs by sequential addition to the UDP-MurNAc lactyl group of L-alanine, D-glutamic acid, L-lysine, and the preformed dipeptide D-alanyl–D-alanine. In each case, the addition is catalyzed by a specific ligase and requires a divalent cation ($Mn^{2+}$ or $Mg^{2+}$) and the hydrolysis of ATP for activity.

The synthesis of the dipeptide D-alanyl–D-alanine begins with the conversion of L- to D-alanine, catalyzed by the enzyme alanine racemase, followed by conversion to D-alanyl-D-alanine, catalyzed by D-alanyl–D-alanine synthetase. The UDP-MurNAc pentapeptide always ends in the residue D-alanyl–D-alanine (Fig. 1.13).

*The Membrane Stage.* The second stage of peptidoglycan biosynthesis involves the transfer of *N*-acetylglucosamine and *N*-acetylmuramyl pentapeptide from the intracellular sites to the membrane, where they can be incorporated into the growing peptidoglycan. The membrane steps involve first a **translocase,** which catalyzes the transfer of the phospho-MurNAc pentapeptide moiety of UDP-MurNAc pentapeptide to the membrane acceptor undecaprenylpyrophosphate, yielding MurNAc-pentapeptide–pyrophosphoryl undecaprenol (Fig. 1.14). Construction of this pentapeptide is catalyzed by the enzymes MurC, MurE, and MurF in both gram-negative and gram-positive bacteria.

The 55-carbon-atom isoprenoid lipid carrier molecule called **undecaprenol** or **bactoprenol** (composed of 11 isoprene units) is anchored to the cytoplasmic membrane. Undecaprenyl phosphate, acting as a lipophilic carrier, plays a central role in this second stage. *N*-Acetylglucosamine is then added in a **transglycosylation** reaction to form a β1→4 glycosidic bond, resulting in undecaprenylpyrophosphoryl-*N*-acetylmuramoylpentapeptide-*N*-acetylglucosamine (GNAc-MurNAc disaccharide unit). Five glycine molecules are added as five individual glycyl-tRNAs. In *S. aureus,* the next stage of peptidoglycan biosynthesis involves building a pentaglycine strand on the ε-amino group of the lysine side chain. Similar interpeptide bridges are found in other gram-positive organisms including *S. carnosus,* which builds an interpeptide bridge using glycine, *S. epidermidis* and *S. haemolyticus,* which build interpeptide bridges using glycine and serine, *Streptococcus pneumoniae,* which builds an interpeptide bridge using serine and alanine, and *Enterococcus faecium,* which builds interstrand bridges of aspartic acid and asparagines. The gram-positive bacterium *B. subtilis* does not contain an interpeptide bridge but, like the gram-negative *E. coli,* contains a *meso*-diaminopimelic acid moiety in the third position of the pentapeptide that is responsible for the cross-linking between strands.

The FemXAB family of enzymes is responsible for the sequential building of this interpeptide strand in *S. aureus.* FemX is responsible for addition of the first glycine to the ε-amino group of the lysine side chain. FemA is responsible for addition of the next two glycine residues, and FemB is responsible for addition of the final two glycine residues. Recently, the first protein structure from this family, *S. aureus* FemA, was solved at 2.1-Å resolution by X-ray crystallography (Benson et al., 2002). This last reaction completes the formation of the

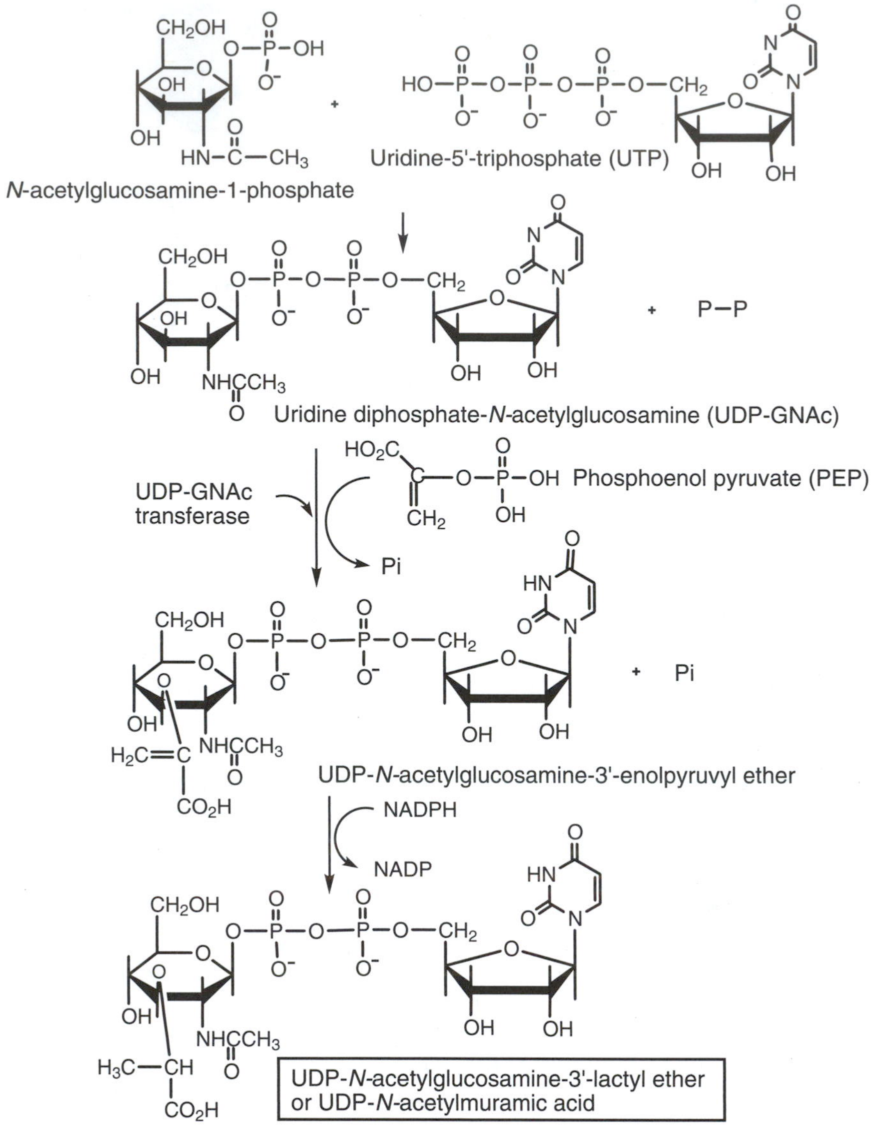

**Figure 1.12** Reactions involved in the biosynthesis of UDP-MurNAc acid in *S. aureus.* Pi, inorganic phosphate.

fundamental disaccharide unit that is added to a peptidoglycan acceptor (Fig. 1.14).

*The Extracellular Stage.* The third stage of the biosynthetic process is also catalyzed by membrane-bound enzymes. Each bacterium has one or more **transpeptidases** that catalyze the cross-linking of lineal chains of peptidoglycan units. Transpeptidases are also known as **penicillin-binding proteins** (PBPs) because they are the primary targets of β-lactam antibiotics.

The nascent peptidoglycan is attached to the preexisting peptidoglycan by the formation of cross-links between peptide chains of both polymers. Only when cross-links are formed does peptidoglycan become an insoluble matrix capable of maintaining the structural integrity of the wall.

Detailed studies of transpeptidases from a range of bacteria have established that the transpeptidation reaction begins with the cleavage of the bond between D-alanyl–D-alanine to form an acyl-D-alanyl enzyme, releasing the terminal D-alanine. This reaction is followed by a second reaction in which this intermediate reacts either with the amino group of the terminal glycine from the pentaglycine (nucleophilic attack) to form a new peptide bond, joining two linear chains of peptidoglycan polymer (transpeptidation reaction) (Fig. 1.15), or with water in a reaction catalyzed by a carboxypeptidase through the formation of an acyl-D-alanine carboxypeptidase, releasing D-alanine. In bacilli, the preexisting wall is the donor of the free amino group for the formation of the peptide bridge and the carboxyl group comes from the nascent lipopolysaccharide unit.

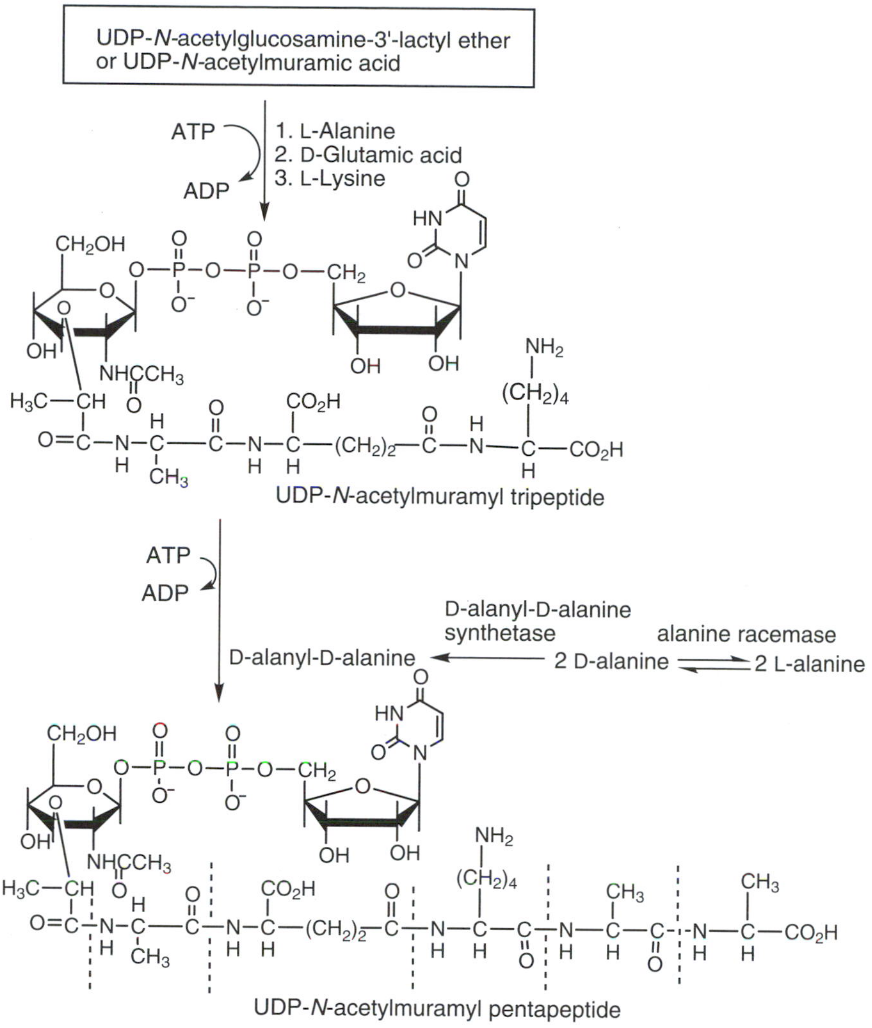

**Figure 1.13** Later stage of the reactions shown in Fig. 1.12. The reactions involved in the biosynthesis of UDP-MurNAc pentapeptide (first stage) in *S. aureus* are shown.

The complex series of enzyme-catalyzed reactions involved in the biosynthesis of the cross-linked peptidoglycan polymer affords many potential sites of inhibition for specific antibiotics. In chapters 7 to 10, 13 and 14 various types of inhibitors (β-lactam antibiotics, glycopeptide antibiotics, bacitracin, D-cycloserine, and phosphonomycin) are discussed in detail.

## Secondary Cell Wall Polymers

Attached to the peptidoglycan polymer, especially to the *N*-muramyl residues, are a variety of secondary polymers. In the simplest case, as exemplified by *B. subtilis* and *B. licheniformis*, these are teichoic (a highly phosphorylated polymer) and teichuronic (phosphate is replaced by uronic acid) acids. Some cell walls, such as in *Streptococcus pneumoniae*, contain teichoic acid (linked to peptidoglycan muramyl moieties) and lipoteichoic acids linked to the cytoplasma membrane by the lipid substituent.

*Teichoic Acids.* Teichoic acids are the most abundant of the secondary wall polymers isolated from gram-positive bacteria. The walls of gram-positive bacteria contain glycerol or ribitol phosphates that are linked by phosphodiester bonds and extend from the cell envelope into the environment. These polyol chains are called teichoic acids. The repeating unit may be glycerol joined by 1,3 or 1,2 linkages or ribitol joined by 1,5 linkages. Most teichoic acids contain substituent groups on the polyol chain, including D-alanine (ester linked), GNAc, etc., usually attached to position 2 or 3 of glycerol or position 3 or 4 of ribitol. These substituents can act as specific antigenic determinants. Their structures are illustrated in Fig. 1.16. The teichoic acids are covalently linked to peptidoglycan. So far, they have not been found in gram-negative bacteria.

By virtue of their negatively charged groups, these divalent anionic polymers can bind to divalent cations, particularly $Mg^{2+}$. Although teichoic acids have long been known, the function is still a matter of speculation.

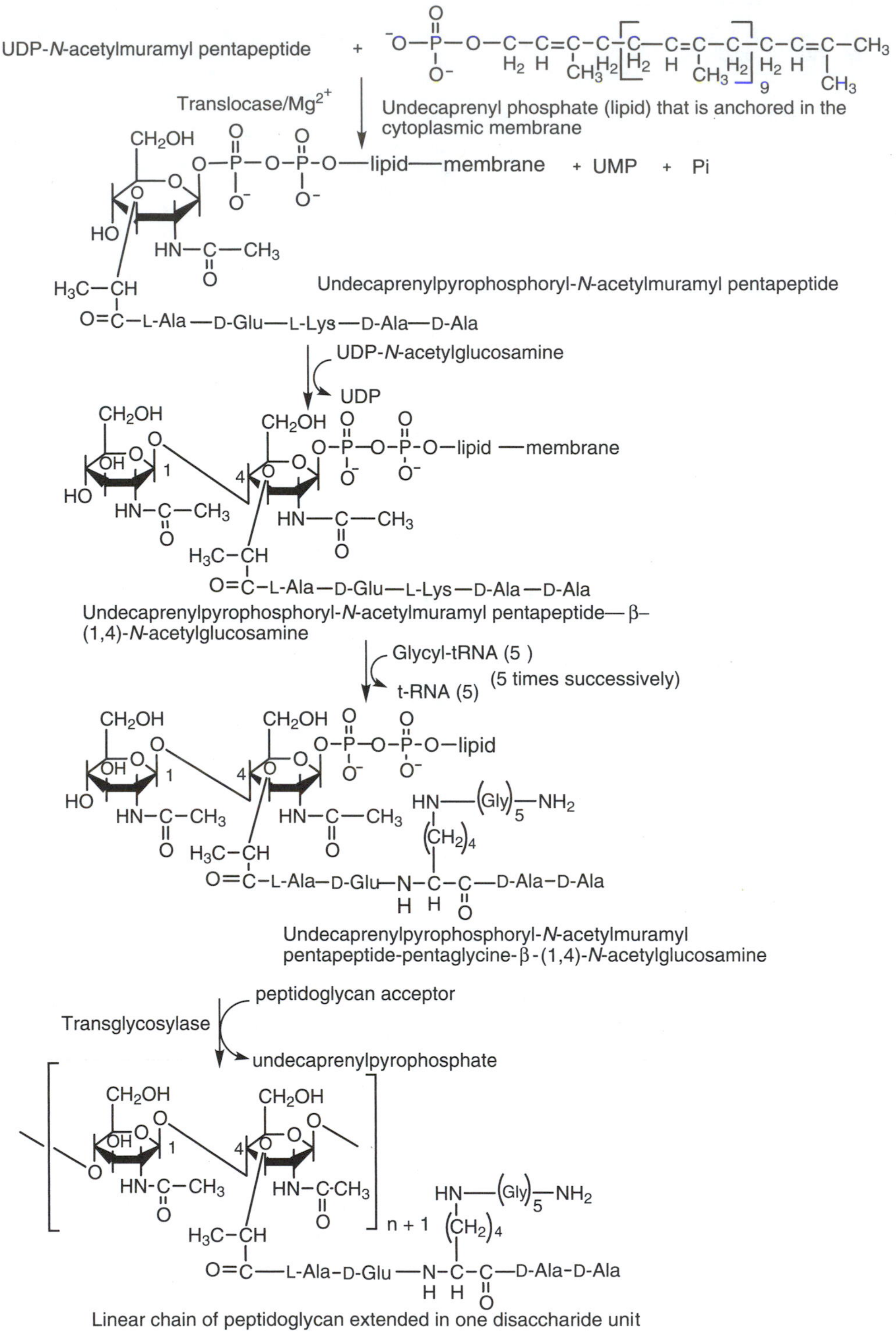

UDP-*N*-acetylmuramyl pentapeptide + Undecaprenyl phosphate (lipid) that is anchored in the cytoplasmic membrane

Translocase/$Mg^{2+}$

Undecaprenylpyrophosphoryl-*N*-acetylmuramyl pentapeptide + UMP + Pi

UDP-*N*-acetylglucosamine → UDP

Undecaprenylpyrophosphoryl-*N*-acetylmuramyl pentapeptide—β-(1,4)-*N*-acetylglucosamine

Glycyl-tRNA (5 ) → t-RNA (5) (5 times successively)

Undecaprenylpyrophosphoryl-*N*-acetylmuramyl pentapeptide-pentaglycine-β-(1,4)-*N*-acetylglucosamine

Transglycosylase: peptidoglycan acceptor → undecaprenylpyrophosphate

Linear chain of peptidoglycan extended in one disaccharide unit

**Figure 1.14** Later stage of the reaction shown in Fig. 1.13. The reactions involved in the biosynthesis of linear chain of peptidoglycan units in *S. aureus* are shown.

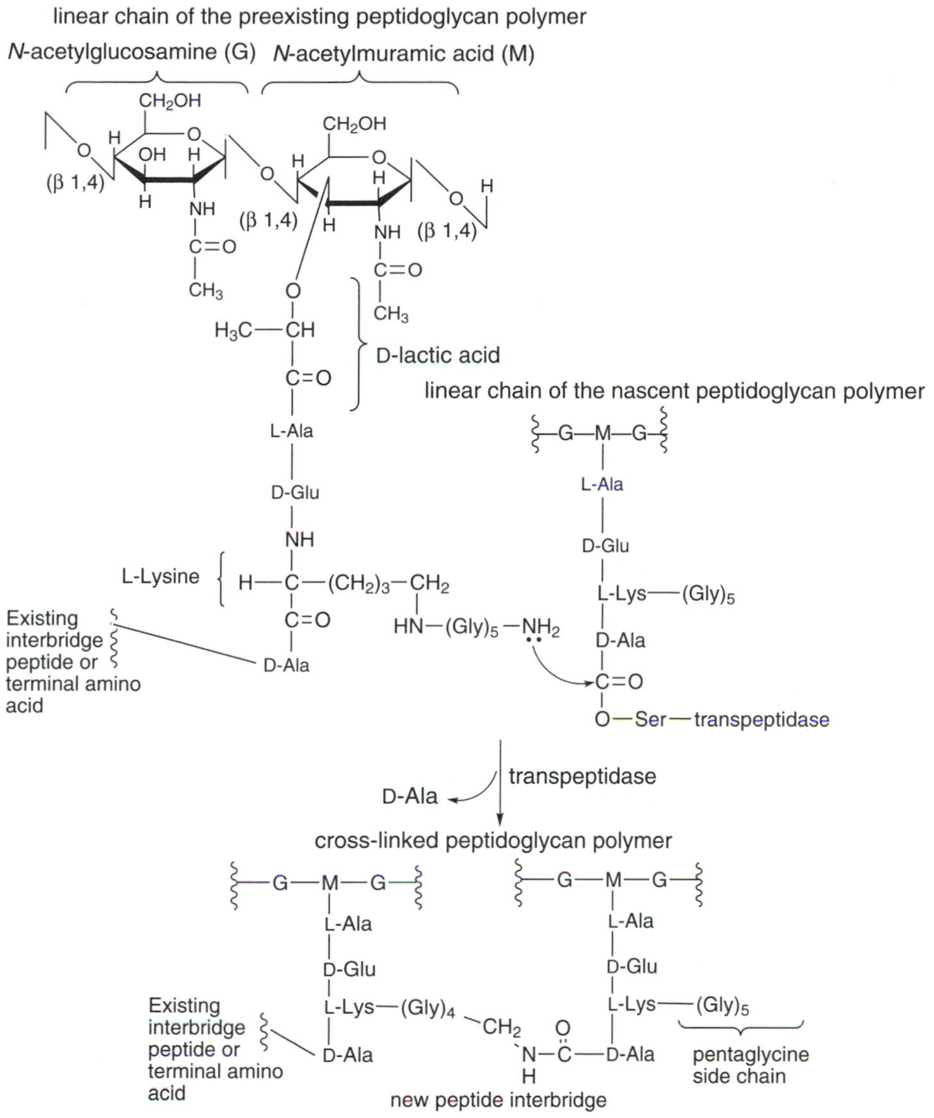

**Figure 1.15** Final stage of the biosynthesis of peptidoglycan in *S. aureus.*

*Lipoteichoic Acids.* Most gram-positive bacteria also contain a second form of teichoic acid, which is not covalently linked to the peptidoglycan. Instead, it is attached to lipids in the cytoplasmic membrane, traverses the cell wall, and extends into the environment. These acids are known as lipoteichoic acids and are usually composed of highly substituted chains of glycerol phosphate. Lipoteichoic acids are major antigenic determinants in some bacteria. For example, the D-polysaccharide of *Enterococcus faecalis* is a lipoteichoic acid.

*Teichuronic Acids.* The second group of anionic polymers present in walls of gram-positive bacteria is the teichuronic acids. This group of acid polysaccharides was recognized first as a minor component of the cell wall of *B. licheniformis* and subsequently in *Micrococcus luteus.* It has since been shown that when a number of species of gram-positive bacteria are grown in media with inorganic phosphate as the limiting nutrient, teichoic acids cease to be formed and their place is taken by teichuronic acids. One component is always a uronic acid, which may be either glucuronic acid or the uronic acid derived from an amino sugar such as aminomannuronic or aminoglucuronic acid. Like the teichoic acids, teichuronic acids are linked covalently to the glycan chains in the peptidoglycan by a single bond (Fig. 1.17).

## The Outer Membrane

### GENERAL CHARACTERISTICS

The outer membrane of gram-negative bacteria plays a significant role in a variety of functions; it serves as a diffusion barrier to extracellular solutes and interacts with

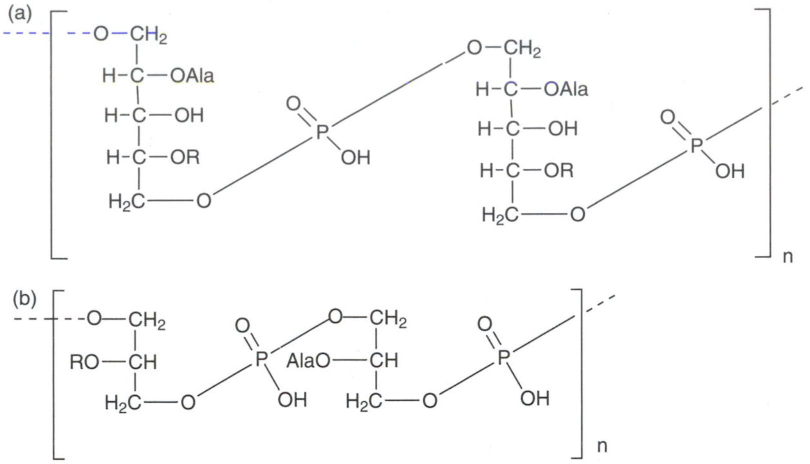

**Figure 1.16** General structures of some teichoic acids. (a) Ribitol teichoic acid. R = β-glucosyl residues in *B. subtilis* and α- or β-linked *N*-acetylglucosamine in various strains of *S. aureus*. (b) Glycerol teichoic acid. R = α-glucosyl residues in *B. subtilis.*

the bacterial environment. This membrane is composed of a bilayer containing phospholipids, lipopolysaccharide, and outer membrane proteins (Fig. 1.8).

The bilayer peptidoglycan polymer is surrounded by the **outer membrane,** which is a bilayer structure (Fig. 1.8 and 1.18). The composition of its inner leaflet resembles that of the cytoplasmic membrane, while the phospholipids of the outer leaflet are replaced by lipopolysaccharide (LPS) molecules (Fig. 1.8 and 1.18). The outer membrane has special channels consisting of protein molecules called **porins.** The porins are present in large numbers in the outer membrane and form water-filled channels that permit the diffusion of small hydrophilic solutes across the outer membrane.

The outer membrane plays a major role as a barrier to a wide variety of molecules. This makes gram-negative bacteria less permeable than gram-positive ones. Thus, the permeability of the outer membrane plays a major role in the resistance of many gram-negative bacteria to antibacterial agents by preventing access of these molecules to their site of action in the cytoplasmic membrane or the cytoplasm. It also plays a major role in determining which molecules are exported from the cell and how the cells interact with extracellular molecules and other cells in the environment.

**Figure 1.17** Generalized structure of the teichuronic acid present in the wall of *B. licheniformis.*

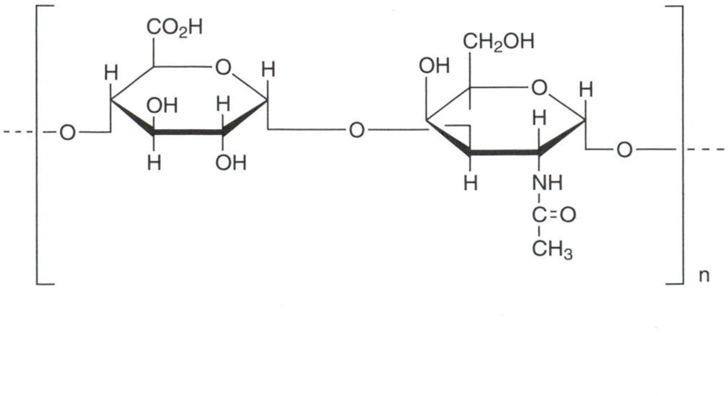

The outer membrane of gram-negative bacteria provides structures and receptors that affect adhesion to host cells, resistance to phagocytosis, and susceptibility to bacteriophages.

## OUTER MEMBRANE CONSTITUENTS

### Structures of Lipopolysaccharides

LPSs are macromolecules that are present exclusively in the outer leaflet of the outer membrane of gram-negative bacteria. LPSs are powerful immunomodulators in infected hosts, and may cause endotoxic shock. Most of them share a common architecture but vary considerably in structural motifs from one genus, species, and strain to another.

LPS molecules (Fig. 1.19) consist of a bisphosphorylated lipid (**lipid A**) forming the matrix of the outermost membrane leaflet, which is stabilized by divalent cations, and a hydrophilic polysaccharide, extending outward from the bacterium (Fig. 1.18). The polysaccharide consists generally of two distinct regions, a **core oligosaccharide** containing 10 to 12 sugars and a polysaccharide chain of repeating units, the **O-specific chain.**

LPS is of considerable importance, not only because of its role in the pathogenicity of many gram-negative bacteria but also from both the immunological and taxonomic points of view.

LPS is extremely toxic to mammalian cells, and has been called the **endotoxin** of gram-negative bacteria because it is firmly bound to the cell surface. Although the entire LPS molecule is called endotoxin, it is actually the lipid A portion of LPS that is the endotoxin. In animals, injection of LPS results in dramatic systemic toxic effects such as fever, hypotension, and rapid death.

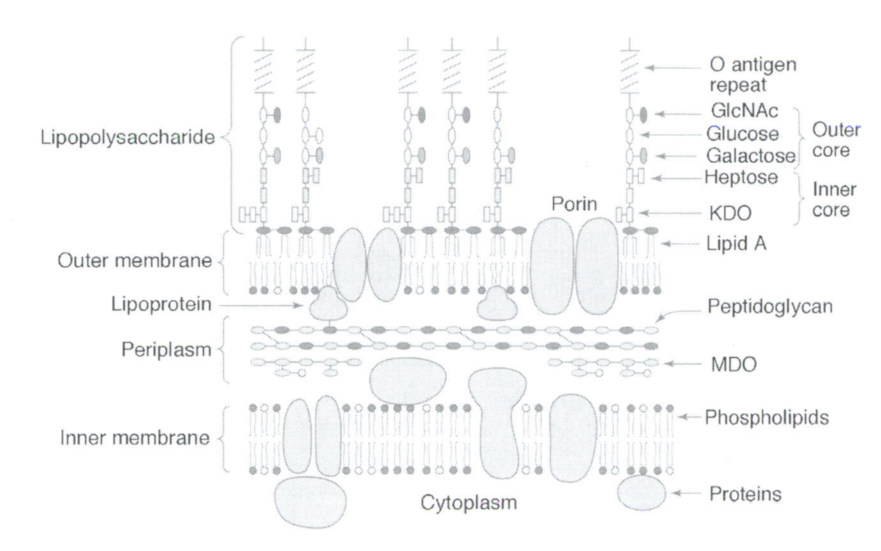


**Figure 1.18** Molecular representation of the envelope of a gram-negative bacterium. Ovals and rectangles represent sugar residues, whereas circles depict polar head groups of the glycerophospholipids (phosphatidylethanolamine and phosphatidylglycerol). KDO and MDO represent membrane-derived oligosaccharides. The core region shown is that of *E. coli* K-12, a strain that does not normally contain an O-antigen repeat unless transformed with an appropiate plasmid. Reprinted from C. R. H. Raetz, *J. Bacteriol.* **175:**5745–5753, 1993, with permission from the American Society for Microbiology.

Humans infected with a large number of circulating gram-negative bacteria develop sepsis and are in danger of dying of septic shock.

*The O Antigen.* The **O antigen** is also known as the bacterial somatic antigen because it is so immunologically prominent. It is composed of a series of repeating units of linear trisaccharides or branched tetra- or pentasaccharides, and it extends from the cell surface to the environment (Fig. 1.19). The O antigen may be up to 50 units in length, but the lengths of individual O antigens on a single bacterium are highly variable. The O-chain structure, when it is present, confers serotype specificity on a species or strain of bacteria.

O antigen serves as a receptor that allows some bacteria to adhere to host surface structures. It moderately affects the ability of bacteria to resist phagocytosis by polymorphonuclear leukocytes. Some bacteria that do not have O antigen are typically more susceptible to phagocytosis than are bacteria that have O antigen.

*The LPS Core.* The O antigen is attached to lipid A via the LPS core. The core can be divided into two portions, known as the **inner core** and **outer core.** The outer-core region contains heptose residues (Hep) which are often substituted by phosphate, pyrophosphate, or diphosphoethanolamine. This region usually consists of neutral or amino hexoses such as D-glucose, D-galactose, D-glucosamine, D-galactosamine, or *N*-acetyl derivatives. The inner core is composed of a sugar known as 2-keto-3-deoxyoctulosonic acid (KDO). These two monosaccharides are apparently unique to LPS. In the inner core these monosaccharides are also attached to phosphorylethanolamine and pyrophosphorylethanolamine (Fig. 1.20).

Wild-type enterobacterial species with O chains are termed "smooth" because of the morphology of their colonies. Enterobacterial mutants producing rough-appearing colonies and lacking LPS O chains are accordingly termed "rough." Strains that express only lipid A and the inner core are called "severe rough" strains. Other bacteria never express O antigen. The LPS of these naturally rough bacteria is called lipooligosaccharide (LOS) to denote the natural absence of O antigen.

The only function known for the core so far is to anchor the O antigen to lipid A.

*Lipid A.* Lipid A structures from the *Enterobacteriaceae* and many other bacteria have a common backbone consisting of a ($\beta$1→6)-linked D-glucosamine disaccharide carrying ester- and amide-linked fatty acids, and phosphate groups at positions C-1 and C-4 (Fig. 1.19). When the phosphate groups have cationic substituents such as L-arabinosamine or phosphoethanolamine, the bacterium has enhanced resistance to several cationic peptides. The polysaccharide is linked to the lipid A through KDO at C-6 of the nonreducing glucosamine residue. The lipid A disaccharide of certain bacteria is composed of D-2,3-dideoxyglucose. Lipid A, the component of LPS responsible for endotoxic activity, is always present.

## Major Proteins of the Outer Membrane

The proteins of the outer membrane fall essentially into two major categories: (i) lipoproteins and (ii) porins. Porins are a family of transmembrane proteins providing channels in the outer membrane through which low-molecular-weight hydrophilic substances can diffuse.

*Lipoproteins.* In all gram-negative bacteria except *Pseudomonas* species, the outer membrane is anchored

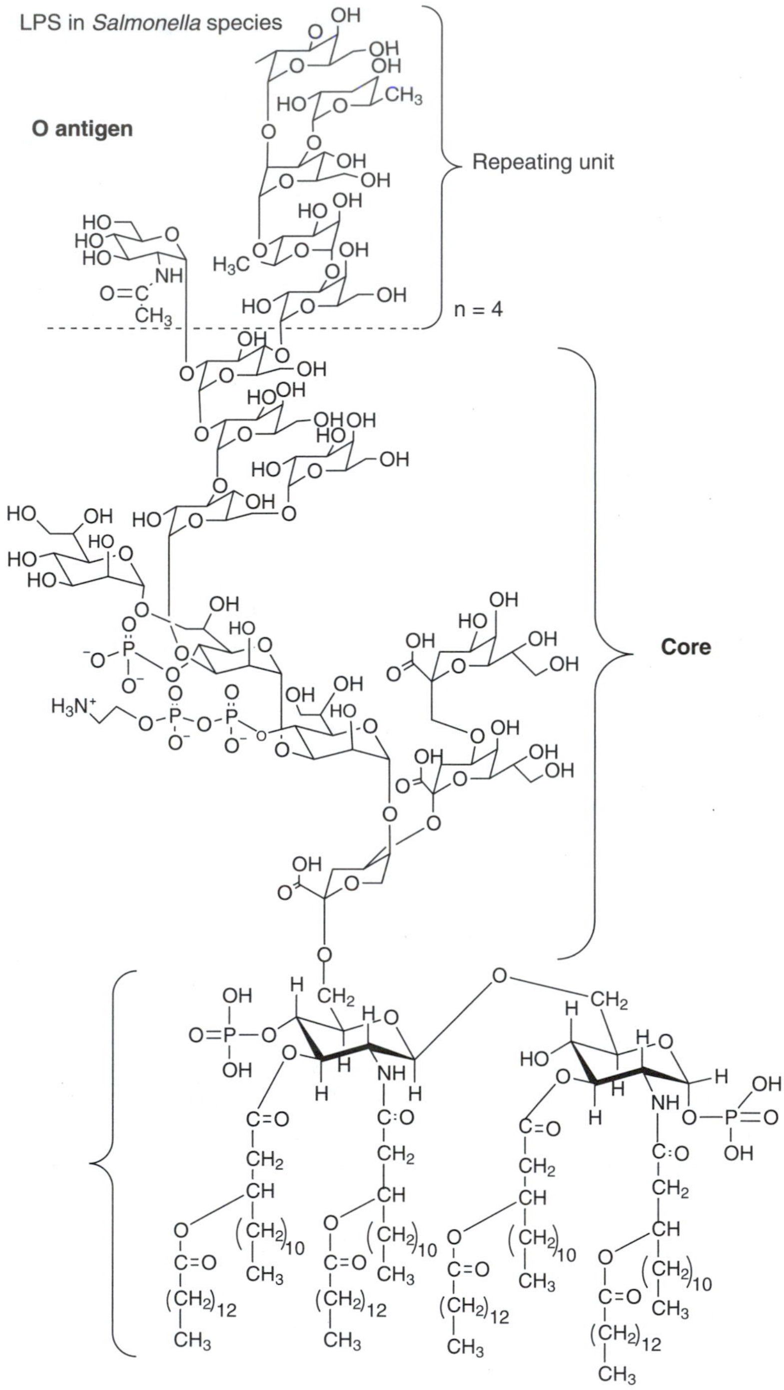

**Figure 1.19** Diagram showing the molecular structure of LPS in *Salmonella* species. LPS is composed of three major regions: the type-specific O antigen, which extends into the external environment from the bacterial surface; the core, which consists of inner and outer cores and which anchors the O antigen to the membrane; and lipid A, which is a phospholipid-like molecule that is embedded in the outer leaflet of the outer membrane.

to peptidoglycan layers via a lipoprotein called Braun's lipoprotein.

Braun has shown that a specific lipoprotein, which is 12 to 14 nm long and composed of 57 amino acids, is covalently linked to the peptidoglycan layer of several enteric bacteria in such a way that it extends outward toward the outer membrane (Fig. 1.8). The lipid component, consisting of a diglyceride thioether linked to a terminal cysteine, is noncovalently inserted in the outer membrane.

Lipoprotein is the most abundant protein in the cell envelope of gram-negative cells (ca. $7.0 \times 10^5$ copies per cell). The function of the lipoprotein (inferred from the behavior of lipoprotein-lacking mutants) is to stabilize the outer membrane and anchor the peptidoglycan layer.

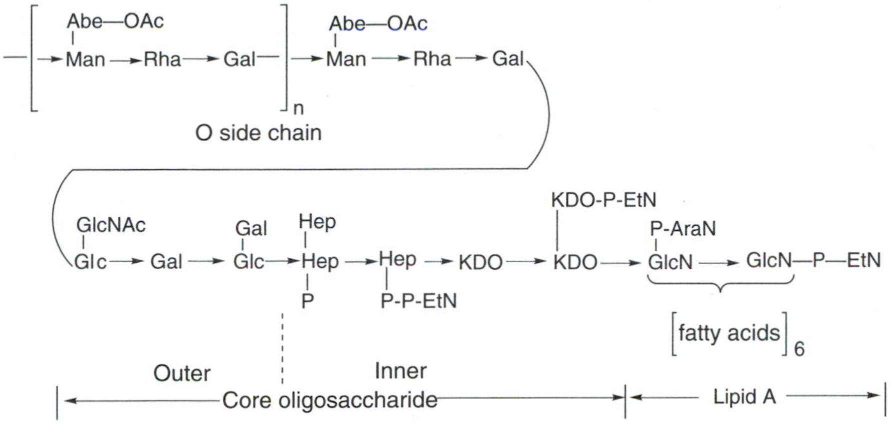

**Figure 1.20** Structure of *S. enterica* serovar Typhimurium LPS. The molecule is made up of the side chain (Abe, abequose; Man, D-mannose; Rha, L-rhamnose; OAc, O-acetyl) linked to the core oligosaccharide (Gal, D-galactose; GlcNAc, *N*-acetylglucosamine; Glc, D-glucose; Hep, L-glycero-D-mannoheptose; EtN, ethanolamine) and lipid A (GlcN, D-glucosamine; AraN, 4-aminoarabinose). The fatty acid substituents are present in both amide and ester linkages.

*Porins.* Porins are homotrimeric proteins, with subunit sizes ranging from 30 to 50 kDa, that span the outer membrane to form narrow water-filled channels. These channels allow the passage of low-molecular-weight hydrophilic (polar) solutes with an exclusion limit of about 600 to 700 Da but exclude nonpolar molecules of comparable size. Larger molecular mass hydrophilic antibacterial agents, such as vancomycin, bacitracin, polymyxins, and streptogramins, are excluded or pass through the porins relatively slowly, which accounts for their lack of antibacterial activity against gram-negative bacteria.

Porins are generally divided into two classes: nonspecific porins, e.g., OmpC (outer membrane protein C) and OmpF, which permit the general diffusion of small polar molecules (<600 Da), and specific porins (e.g., LamB), which facilitate the diffusion of specific molecules. In *E. coli,* three porins, PhoE, OmpF, and OmpC, have been studied. Porins of *E. coli* and *S. enterica* serovar Typhimurium are trimeric proteins that cluster and penetrate both faces of the outer membrane.

The permeability of the outer membrane varies widely among gram-negative species. For example, the outer membrane of *Pseudomonas aeruginosa* is 100 times less permeable than that of *E. coli.* Only about 100 to 300 porins per cell form functional channels, resulting in a very low uptake of antibacterial agents by this route. Moreover, the presence of charged amino acids in the protein/channel interface further restricts the movement of even small hydrophilic compounds.

The role of porins in the uptake of β-lactam antibiotics has been investigated in detail by Nikaido and coworkers. Using a series of *E. coli* mutants engineered to contain only one type of porin, they determined the effect of charge and hydrophobicity on the permeability of a series of penicillin and cephalosporins. The presence of a net positive charge was shown to enhance the rate of diffusion across outer membranes, whereas an increase in lipophilicity resulted in a decreased rate of diffusion. Thus, the outer membrane acts as a barrier to the uptake of β-lactam antibiotics.

The first X-ray structure of a porin was determined by the group of Georg Schulz and Wolfram Welte at Freiburg University in Freiburg, Germany, who succeeded in growing crystals of a porin from *Rhodobacter capsulatus* that diffracted to 1.8 Å resolution. Each subunit of the trimeric porin molecule from *R. capsulatus* folds into a 15-strand up-and-down antiparallel β barrel in which all β strands form hydrogen bonds with their neighbors.

The structures of a porin from *R. capsulatus* and OmpF and PhoE from *E. coli* provide a wealth of information about the structure and function of porins. Figure 1.21 shows a ribbon diagram of OmpF from *E. coli.*

The general diffusion pore matrix porin (OmpF) and osmoporin (OmpC), encoded by *ompF* and *ompC,* respectively, are regulated by osmotic pressure and temperature. Porins which exhibit some specificity aside from their general diffusion properties include phosphorin (PhoE), which is derepressed under phosphate starvation, and maltoporin (LamB), which is induced by maltose. The OmpF, OmpC, and PhoE porins are highly homologous. Although they have similar solute exclusion sizes ($M_r$, ~600), they differ in other aspects of their channel characteristics. The OmpF and OmpC porins are weakly cation selective, whereas PhoE is weakly anion selective.

The crystal structures of matrix porin and phosphoporin both reveal trimers of identical subunits, each consisting of a 16-strand antiparallel β-barrel containing a pore. A long loop inside the barrel contributes to a constriction of the channel where the charge distribution affects ion selectivity. The structures explain functional characteristics at the molecular level and their alterations by known mutations.

Trimeric maltoporin (LamB protein) facilitates the diffusion of maltodextrins across the outer membrane of

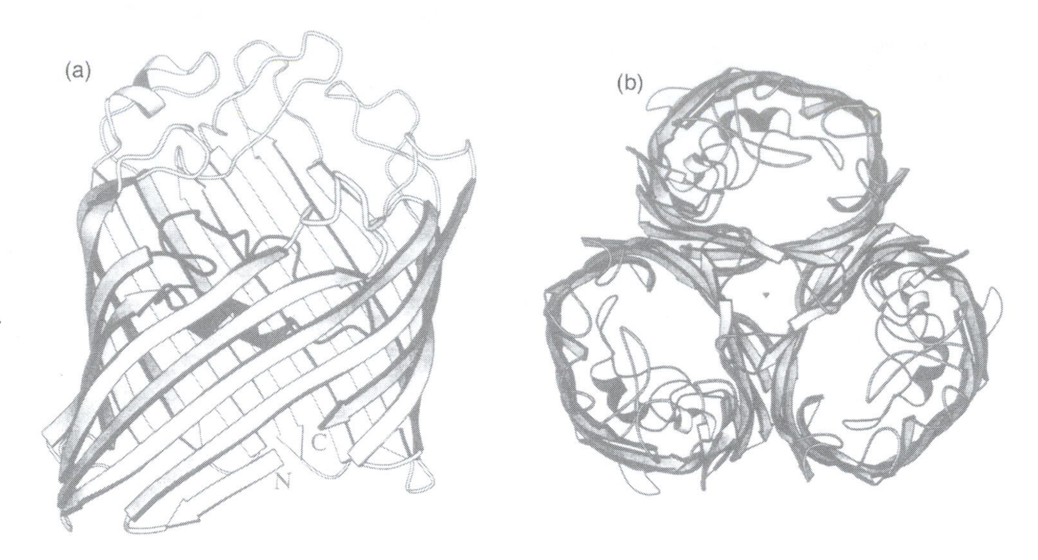


**Figure 1.21** Ribbon diagram of the OmpF porin from *E. coli.* (a) View from within the plane of the membrane onto the face of the barrel closest to the threefold axis. (b) View of the trimer looking along the threefold axis (triangle) from outside the cell. Reprinted from S. W. Cowan, *Curr. Opin. Struct. Biol.* **3:**501–507, 1993, with permission from the publisher.

gram-negative bacteria. The crystal structure of maltoporin from *E. coli,* determined at a resolution of 3.1 Å, reveals an 18-strand, antiparallel β-barrel that forms the framework of the channel. Three inwardly folded loops contribute to a constriction about halfway through the channel.

In 1999, Doménech-Sánchez et al. reported the identification and characterization of a new porin gene of *Klebsiella pneumoniae* and its role in β-lactam antibiotic resistance. They observed that besides the *ompK36* and *ompK35* porin genes, a new porin gene designated *ompK37* is also present in all clinical isolates tested. The *ompK37* porin gene was cloned, sequenced, and overexpressed in *E. coli.* Functional characterization suggested a narrower pore for the OmpK37 porin than for the *K. pneumoniae* porins OmpK36 and OmpK35. This was found to correlate with resistance to certain β-lactam antibiotics, since a *K. pneumoniae* strain expressing OmpK37, but not OmpK36 or OmpK35, was less susceptible to β-lactam antibiotics than was the same strain expressing either porin OmpK36 or OmpK35.

For further information on the three-dimensional structure of porins, the reader should check the excellent reviews by Cowan and by Schulz and the original publications by Cowan et al. (1995) and Doménech-Sánchez et al. (1999).

## The Periplasmic Space

In gram-negative bacteria, the space between the outer membrane and the inner membrane is called the **periplasmic space.** It contains the peptidoglycan layers and a gel-like solution of proteins. The periplasmic proteins include binding proteins for specific substrates (e.g., amino acids, polysaccharides, vitamins, and ions), hydrolytic enzymes (e.g., alkaline phosphatase and 5'-nucleotidase) that break down nontransportable substrates into transportable ones, and certain enzymes, such as β-lactamases that hydrolyze β-lactam antibiotics.

## The Cytoplasmic Membrane

The bacterial cytoplasmic membrane, also called the inner membrane in gram-negative bacteria, is a thin structure (5 to 10 nm thick) that surrounds the cell, and can be seen only under the electron microscope. This structure separates the interior of the cell (cytoplasm) from its environment.

The membranes of prokaryotes are distinguished from those of eukaryotes by the absence of sterols; the

**Figure 1.22** Chemical structures of cholesterol (a steroid) and diploptene (a hopanoid).

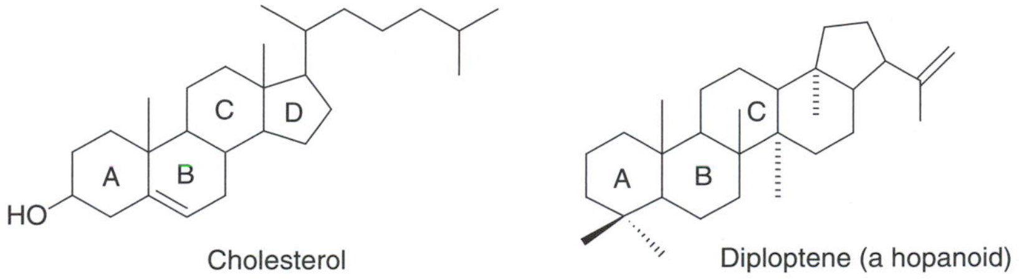

Cholesterol

Diploptene (a hopanoid)

only exception is that of mycoplasmas, which incorporate host sterols, such as cholesterol, into their membrane. Some bacterial membranes do contain pentacyclic sterol-like molecules called hopanoids (Fig. 1.22). Like steroids in eukaryotes, the hopanoids probably stabilize the bacterial membrane.

The cytoplasmic membrane is composed of proteins (60 to 70%), lipids and phospholipids (20 to 30%), and small amounts of carbohydrates. Although the composition of the inner membrane of gram-positive bacteria is similar to that of the inner membrane of gram-negative bacteria, the cytoplasmic membranes of gram-positive organisms possess a class of macromolecules that is not present in gram-negative membranes. Many membranes of gram-positive bacteria contain the membrane-bound lipoteichoic acid.

Membrane-associated **phospholipids** are structurally asymmetric with polar and nonpolar ends (Fig. 1.23) that are called **amphipathic.** The polar ends (e.g., phosphoethanolamine) interact with water and are **hydrophilic;** the nonpolar **hydrophobic** ends (fatty acids) are insoluble in water, pointing inward toward each other, and they tend to associate with one another in a hydrophobic environment. This property of phospholipids enables them to form a bilayer membrane. Membrane proteins float within the lipid bilayer.

The most widely accepted current model for membrane structure is the fluid mosaic model of Singer and Nicholson (Fig. 1.2). This model distinguishes between two types of membrane proteins. **Peripheral proteins** are loosely connected to the membrane and make up about 20 to 30% of total membrane proteins. About 70 to 80% of membrane proteins are **integral proteins,** which, like membrane lipids, are amphipathic: their hydrophobic regions are buried in the lipid while the hydrophilic portions project from the membrane surface. Some of these proteins even extend all the way through the lipid layer.

## FUNCTIONS OF THE CYTOPLASMIC MEMBRANE

The major functions of the cytoplasmic membrane are as follows:

- to serve as the site of selective permeability and carrier-mediated transport of solutes
- to mediate electron transport and oxidative phosphorylation in aerobic species
- to excrete hydrolytic exoenzymes
- to bear the enzymes and carrier molecules that function in the biosynthesis of the cell wall polymers, membrane lipids, and the DNA
- to bear receptors and other proteins of the chemotactic and other sensory transduction systems

Apart from the biosynthesis of peptidoglycan (already presented), it is beyond the scope of this book to discuss further details of the major functions of bacterial cytoplasmic membranes.

## ANTIBACTERIAL AGENTS THAT AFFECT THE CYTOPLASMIC MEMBRANE

The **polymyxins** are polycationic decapeptides produced by various species of *Bacillus*. They consist of detergent-like cyclic peptides that selectively cause disorganization of membrane structures containing phosphatidylethanolamine.

## MESOSOMES

Mesosomes are invaginations of the cytoplasmic membrane in the shape of vesicles, or tubular vesicles (Fig. 1.24). They are seen in both gram-positive and gram-

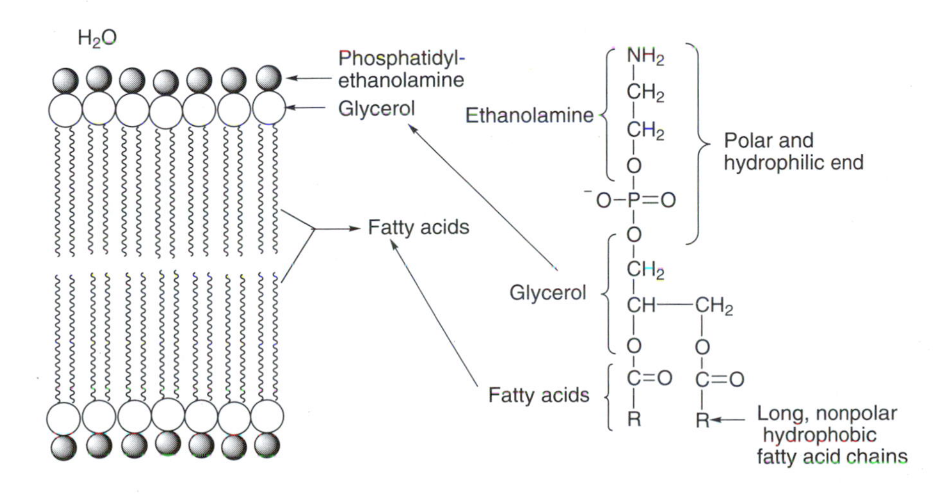


**Figure 1.23** (Left) Fundamental structure of a phospholipid bilayer. (Right) Structures of a polar membrane phospholipid, phosphatidylethanolamine, which forms the hydrophilic region, and fatty acids (long chains), which are esterified to glycerol in the hydrophobic region.

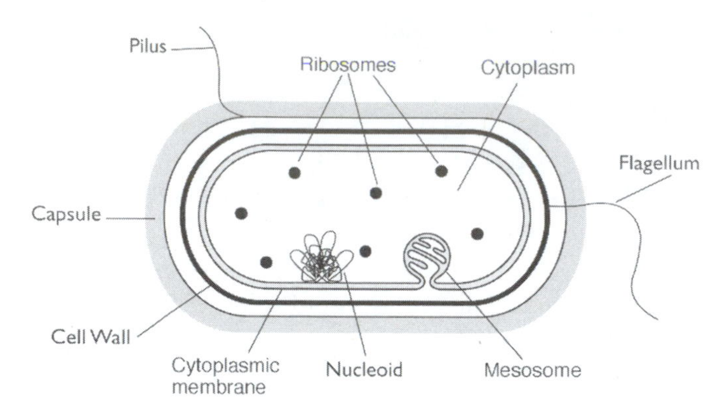


**Figure 1.24** Morphology of a gram-positive bacterium showing the mesosome, as well as other structures.

negative bacteria, although they are generally more prominent in the former. Despite years of research on mesosomes, their exact function is still unknown.

## Capsules and Slime Layers

Many gram-positive and gram-negative bacteria are surrounded by a mucopolysaccharide or peptide layer of protective material called a **capsule** (Fig. 1.24). When the capsular material is so loosely associated with the bacterium that it can be easily washed away, it is called a **slime layer**.

Capsules and slime layers usually are composed of polysaccharides such as glucose, amino sugars, rhamnose, 2-keto-3-deoxygalactonic acid, and uronic acids of various sugars. A few capsules, such as that of *Bacillus anthracis* (the anthrax bacillus), are composed of polypeptides, mainly poly-D-glutamic acid. Capsules are clearly visible under the light microscope when negative stains or special capsule stains are employed. They also can be studied by electron microscopy.

The capsule contributes to the invasiveness of pathogenic bacteria. Encapsulated cells are protected from phagocytosis by host polymorphonuclear leukocytes. *S. pneumoniae* provides a classic example. When it lacks a capsule, it is destroyed easily and does not cause disease, whereas the encapsulated variant quickly kill mice. Most encapsulated bacteria cause acute pyogenic (pus-forming) infections because of the preponderance of polymorphonuclear leukocytes at the infection site. Capsules are the most important of the bacterial antiphagocytic structures. Sometimes capsules prevent opsonization by interfering with complement activation or deposition of complement split products.

Capsules or slime layers allow some bacteria to adhere to host surfaces. Pili are the key adhesin for most mucosal bacteria, but some oral streptococci persist on dental surfaces by expressing a slime layer. Capsules also allow some bacteria to persist in their environment by keeping them from drying out when they are exposed to surfaces such as catheter surfaces.

## Endospores

A small number of gram-positive bacteria can form a special resistant dormant structure called **endospores.** Endospores (so called because the spore is formed within the cell) are readily seen under the light microscope as strongly refractible bodies.

Endospore-forming bacteria are found most commonly in the soil, and virtually any samples of soil have some endospores present. These forms of resting bacteria resist the bactericidal effects of heat, drying, radiation, freezing, and toxic chemicals (such as disinfectants). The sporulation process begins when nutritional conditions become unfavorable, depletion of nitrogen or carbon (or both) being the most significant factor. Sporulation involves the production of many new structures, enzymes, and metabolites along with the disappearance of many vegetative-cell components.

Endospores play a major role in the epidemiology of some diseases. Some types of infections occur only when endospores are introduced into a site where they can germinate, such as a wound site. There are two clinically important types of endospore-forming bacteria. The first consists of members of the aerobic genus *Bacillus* and includes *B. anthracis, B. cereus,* and *B. subtilis.* The second consists of members of the anaerobic genus *Clostridium* and includes *C. botulinum, C. difficile, C. perfringens,* and *C. tetani.* Among the diseases caused by these two types of organisms are anthrax, food poisoning, botulism, tetanus, gas gangrene, and pseudomembranous colitis.

For further information on bacterial sporulation, properties of endospores, and the germination process, the reader should consult a textbook of general microbiology.

## Bacterial Cell Appendages

Extending from the surfaces of many bacteria are appendages that mediate locomotion or adherence to host surface. There are two major protein cellular appendages, **flagella** and **pili.** Flagella are locomotive appendages. The various types of pili (fimbriae) act as cellular adhesins, are involved in the transfer of DNA during conjugation, or mediate resistance to phagocytosis.

### FLAGELLA

Bacterial flagella (singular, flagellum) are long (15 to 20 μm long), thin (12 to 20 nm in diameter) appendages. Most motile bacterial species move actively by using flagella. Four types of arrangements are known. Monotrichous (*trichous* means "pertaining to hair") bacteria have a single flagellum at one end of the pole, called a polar flagellum; amphitrichous bacteria have a single flagellum at each pole (*amphi* means "on both sides"); lophotrichous bacteria have a cluster of flagella at one or both ends (multiple polar flagella); and peritrichous bacteria have flagella distributed over the entire cell.

### Flagellar Ultrastructure

Transmission electron microscope studies have shown that the bacterial flagellum is composed of three parts: (i) the basal body, which is embedded in the cell; (ii) the hook, a short, curved segment which links the filament to its basal body as a flexible coupling; and (iii) the filament itself (Fig. 1.25), which extends from the surface to the tip. The filament is a hollow, rigid cylinder constructed of a single protein called flagellin, which ranges in molecular mass from 30 to 60 kDa. The filament is a rigid helix, and the bacterium moves when this helix rotates. The rotatory motion of the flagellum is imparted from the basal body, which functions as a motor.

### Chemotaxis

The flagellum is a highly effective swimming device. Swimming is quite a task for the bacterium because the surrounding water seems thick and viscous. Bacteria do not always swim aimlessly but, rather, are attracted by nutrients such as sugars and amino acids and are repelled by many harmful substances and bacterial waste products. Movement toward chemical attractants and away from repellents is known as **chemotaxis.**

This chapter gives only a brief description of the ultrastructure and function of the flagellum. For a more comprehensive treatment, the reader should consult a textbook of general microbiology or biology of microorganisms.

### PILI (FIMBRIAE)

Many gram-negative bacteria possess rigid surface appendages called **pili** (singular, pilus) or **fimbriae** (singular, fimbria). They are shorter and thinner than flagella. They are composed of protein subunits termed pilin. Most piliated bacteria are gram negative; only a few piliated gram-positive bacteria have been identified. Two classes can be distinguished: common pili and sex pili.

**Common pili** are thinner than sex pili and are usually present in large numbers on a cell. The most important function of common pili is to serve as **adhesin,** and their presence is responsible for the ability of many bacteria to adhere to and colonize mucosal surfaces of the host cell. **Sex pili** are thicker than common pili, and each "male" bacterium (a bacterium with sex pili) typically exhibits only a few such pili. Sex pili are encoded on conjugative plasmids, and they facilitate the transfer of these plasmids from donor bacteria to recipient bacteria.

Figure 1.26 shows two cells of *E. coli.* The one on the left has many common pili and a single sex pilus attaching it to the nonpiliated one on the right.

## The Cytoplasmic Matrix

### THE NUCLEOID

The prokaryotic nucleoid, the equivalent of the eukaryotic nucleus, can be seen by light microscopy in material stained with the Feulgen stain. It is Feulgen positive, indicating the presence of DNA. Chromosomal DNA is located in an irregularly shaped region of the cytoplasm, and it tends to aggregate as a distinct structure.

Electron microscopy reveals the absence of a nuclear membrane and a mitotic apparatus. The prokaryotic chromosome almost always consists of a single continuous circular molecule of double-stranded DNA, although some bacteria are known to have linear chromosomal DNA. Careful electron microscopy studies often have shown the nucleoid in contact with either the mesosome or the cytoplasmic membrane. Nucleoids have been isolated intact and free from membranes. Chemical analysis reveals that they are composed of DNA (about 60%), some RNA, and a small amount of protein. No histones are present.

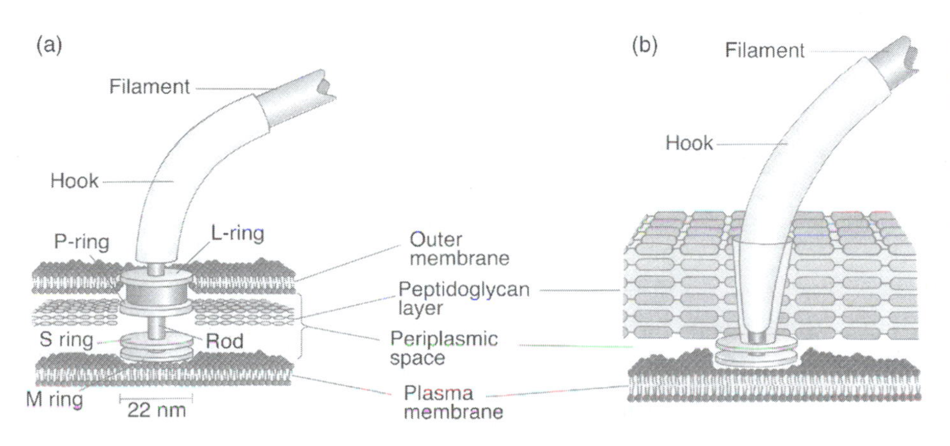


**Figure 1.25** Ultrastructure of bacterial flagella. Flagellar basal bodies, hook, and filament in gram-negative (a) and gram-positive (b) bacteria are shown. Reprinted from L. M. Prescott, J. P. Harley, and D. A. Klein, *Microbiology,* 4th ed. (McGraw-Hill, Boston, Mass., 1999), with permission from the publisher.

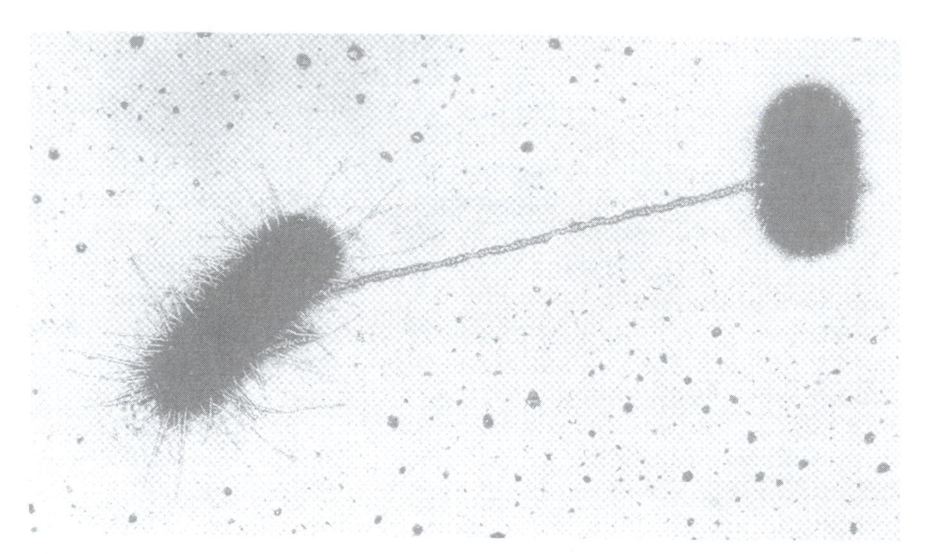

**Figure 1.26** Photomicrograph of two *E. coli* cells shows a bacterial mating (conjugation) process. The cell at the left has several common pili and a sex pilus connected to the right cell. The cell at the right lacks common pili. Reprinted from T. S. Walker, *Microbiology* (The W. B. Saunders Co., Philadelphia, Pa., 1998), with permission from the publisher.

The total amount of DNA in the chromosome of a bacterium such as *E. coli* is about $4.6 \times 10^9$ bp. Linearized it is about 1 mm long, yet the *E. coli* cell is only about 2 to 3 μm long. To package this much DNA into the cell requires that the DNA be extensively folded and twisted. This is called **supercoiling DNA** (Fig. 1.27).

Bacteria reproduce asexually and are typically **haploid** in genetic complement. This means that all the cell's genetic information is present on this single chromosome. As is well known, bacteria multiply by binary fission. After the appropriate increases in cell dimensions, septa appear, starting at the circumference and proceeding inward, until the two daughter cells separate.

In many bacteria, extrachromosomal DNA has been found as small DNA molecules called **plasmids.** These are small circular, double-stranded DNA molecules that can exist and replicate independently of the chromosome or may be integrated with it; in either case, they are inherited or passed on to the progeny. Plasmid genes can render bacteria resistant to antibacterial agents. Some bacterial plasmids may be present in as many as 30 copies in one bacterial cell.

**Figure 1.27** The bacterial chromosome and supercoiling. (a) The DNA double helix in the shape of a closed circle. (b) Supercoiled form of circular DNA.

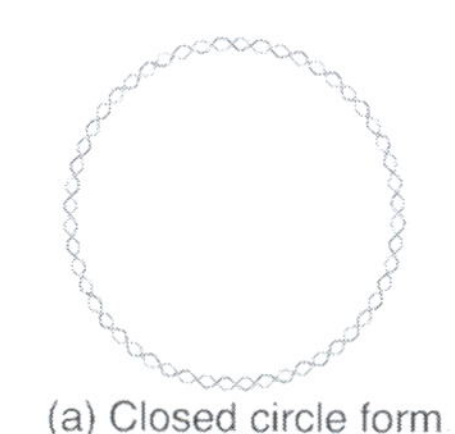


(b) Supercoiled form

## Structure of DNA

Genetic information is stored as a base sequence in the DNA. The length of a DNA molecule is usually expressed in thousands of base pairs, or kilobase pairs (kb).

*The Watson-Crick Structure: B-DNA.* In 1953 the scientists James Watson and Francis Crick proposed the double-helix structure of B-DNA. According to the Watson-Crick model, bacterial DNA is a double-stranded molecule, with **complementary bases,** adenine plus thymine (A-T) and cytosine plus guanine (C-G), paired by hydrogen bonding. The base pairs are stacked in the interior of the DNA double helix (Fig. 1.28). Each of the four bases is covalently linked to phospho-2'-deoxyribose to form a **nucleotide.** The negatively charged phosphodiester backbone of DNA faces the outside of the double helix.

Each turn of the helix contains about 10 bp. The sequence of bases in the two strands is therefore complementary, while the polarity of the strands is in opposite directions, i.e., $5' \rightarrow 3'$ and $3' \rightarrow 5'$.

## Replication of DNA

Because DNA contains the genetic information of the cell, its replication and division must lead to the formation of two complementary identical DNA chromosomes.

*Semiconservative Mechanism of DNA Replication.* The semiconservative mechanism of DNA replication assumes that during replication the two strands of the

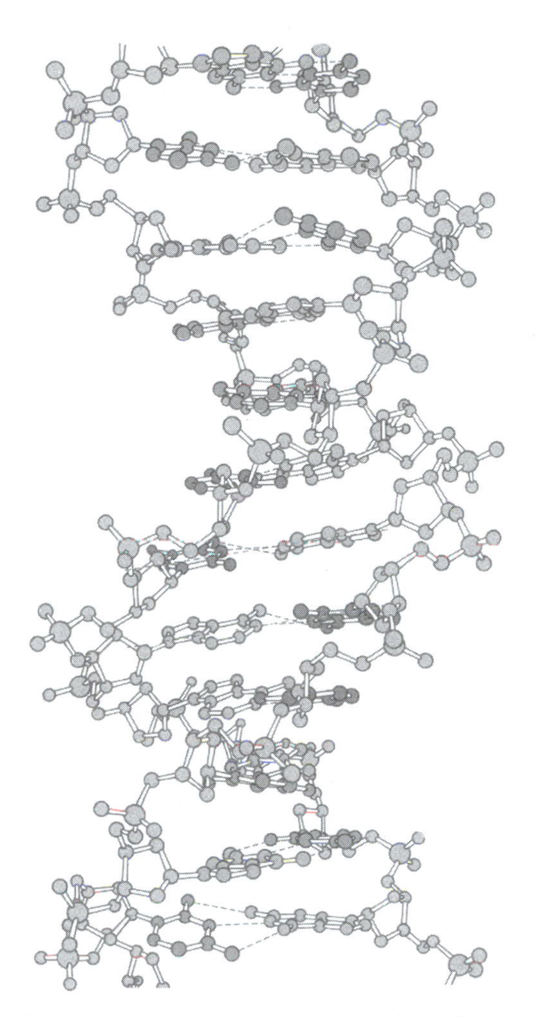

**Figure 1.28** Diagrammatic representation of the strucure of the B form of DNA. The model shows the complementarity of the bases A-T and C-G and the antiparallel orientation of the two strands of polynucleotides.

double helix unwind and separate. Each single strand then serves as a template for the synthesis of a new complementary strand (Fig. 1.29). **DNA polymerases** are involved in this synthesis by joining the correct sequence of new nucleotides to the preexisting template strand of DNA by base pairing, thus producing a new polynucleotide strand.

## THE BACTERIAL GENOME

As mentioned above, most bacterial genes are carried on the bacterial chromosome, a single circular DNA molecule containing about 4,000 kbp. Many bacteria also contain additional genes on plasmids that range in size from several kilobase pairs to 100 kbp. DNA circles (chromosome and plasmids) which contain genetic information necessary for their own replication are called **replicons.**

The study of bacterial infection is being revolutionized by the development of a technology that allows the sequencing and analysis of the entire bacterial chromosome.

In 1995, *Haemophilus influenzae* became the first free-living organism for which the entire genome sequence was published. In 2000 the DNA sequences of both chromosomes of the cholera pathogen *Vibrio cholerae* (a bacterium that is the etiological agent of cholera) were published in *Nature.* In the same journal in 2000, the complete genome sequence of *Pseudomonas aeruginosa* was reported. This sequence is of interest because it provides insights into the role of this bacterium as a pathogen and because it offers new information about the relationships between genome size, genetic complexity and ecological versatility in bacteria. At $6.3 \times 10^6$ Mbp, the *P. aeruginosa* genome is markedly larger than most of the 25 sequenced bacterial genomes determined over a 5-year period (1995 to 2000).

Currently, about 70 microbial genomes have been sequenced since the initial sequencing of the *H. influenzae* genome. These prokaryotic genomes include pathogenic bacteria such as *Streptococcus pyogenes, S. pneumoniae, Salmonella enterica* serovar Typhi CT18, *Yersinia pestis, Chlamydia trachomatis, C. pneumoniae, Borrelia burgdorferi, Treponema pallidum, Mycobacterium tuberculosis,* and enterohemorrhagic *E. coli;* these are among many pathogen genome sequences in the public domain.

*S. aureus* is one of the major causes of community-acquired and hospital-acquired infections. Keiichi Hiramatsu and colleagues, from Juntendo University, Tokyo, Japan, have sequenced the whole genome of N315, a methicillin-resistant *S. aureus* (MRSA) strain, and

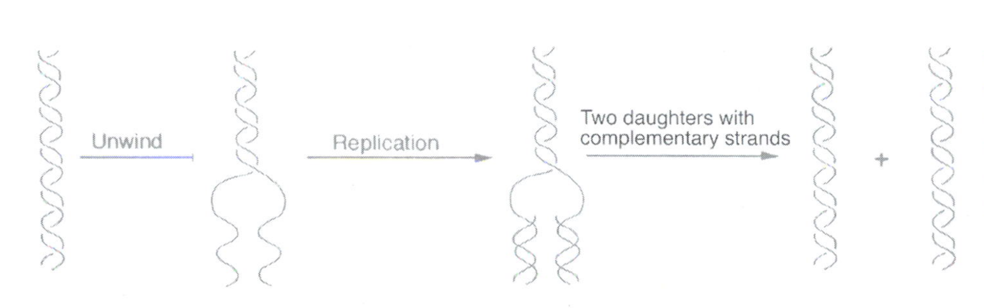


**Figure 1.29** Diagrammatic representation of semiconservative DNA replication. The replication fork of DNA and the synthesis of two progeny strands are shown. Each copy contains one new and one old strand.

Mu50, an MRSA strain with vancomycin resistance, according to a report published in *The Lancet* (Kuroda et al., 2001). The researchers found that the genomes contained a complex mixture of genes, many of which appeared to be the result of lateral gene transfer. Antibiotic resistance genes were carried largely by plasmids or by mobile genetic elements, which include a unique resistance island. Three classes of new pathogenicity islands were identified, including a toxic shock syndrome toxin island family, exotoxin islands, and enterotoxin islands. Gene clusters from the last two islands were closely linked to other gene clusters that appeared pathogenic. The analysis also revealed 70 new virulence candidates.

The collection of complete DNA sequence information provided new insights into the molecular architecture of bacterial cells, revealing the basic genetic and metabolic structures that support the viability of the organisms. Genomic information has also revealed new avenues for inhibition of bacterial growth and viability, expanding the number of possible drug targets for new antibacterial agents.

### RNA

RNA in bacteria is present in single-stranded form. RNA molecules can fold back upon themselves in regions where complementary base pairing can occur to form a variety of highly folded structures. In DNA the sugar is D-2-deoxyribose; in RNA the sugar is D-ribose. In DNA the purine bases are adenine (A) and guanine (G) and the pyrimidine bases are thymine (T) and cytosine (C). In RNA, uracil (U) replaces thymine, so that the complementary bases that determine the structure of RNA are A-U and C-G.

The most general function of RNA is to provide communication of DNA gene sequences as **messenger RNA** (mRNA) to ribosomes. mRNA contains the genetic information of DNA in a single-stranded molecule complementary in base sequence to a portion of the base sequence of DNA within ribosomes, which contain rRNA. rRNA molecules, of which several distinct types are known, are important structural components of the ribosome. mRNAs are translated into the amino acid structure of proteins via tRNA. The tRNA molecules translate the genetic information from nucleotides into amino acids, the building blocks of proteins.

Figure 1.30 shows a diagrammatic representation of the cloverleaf secondary structure of tRNA. The pattern of folding observed in RNA is called its secondary structure. This cloverleaf structure contains several arms, which are composed of a loop or a loop with hydrogen-bonded stem. The 5' end and the region near the 3' end of the RNA molecule are base-paired, forming the amino acid stem. The carboxyl group of the amino acid is covalently linked to either the 2'- or 3'-hydroxyl group of the ribose of the adenylate at the 3' end. If the amino acid is initially attached to the 2'-hydroxyl group, it is shifted to the 3'-hydroxyl group in an additional step. The amino acid must be attached to the 3' position in order to serve as a substrate for protein synthesis.

### RIBOSOMES

The ribosome is the universal organelle for the synthesis of proteins in cells. Ribosomes are small particles composed of protein and rRNA, which are visible under the transmission electron microscope. A single prokaryotic cell such as *E. coli* contains about 30,000 ribosomes. In prokaryotes the small subunit (30S particle) comprises the 16S rRNA and 21 proteins and the large subunit (50S particle) consists of two rRNAs, the 5S rRNA and the 23S rRNA, together with about 35 proteins. Functionally, the small subunit is devoted to the recognition of mRNA and tRNA while the large subunit is the center of protein synthesis. A prokaryotic ribosome measures about 200 Å in diameter.

Recently, crystal structures of each prokaryotic subunit and of the whole prokaryotic ribosome, the 70S particle, have been published. The resolution of the structures is still rather low (4.5 to 7.8 Å). The resolution, while not high enough to resolve and identify individual RNA bases or protein amino acids, is sufficient to distinguish protein $\alpha$-helices from RNA; thus, the images reveal some of the overall arrangements of the proteins with respect to the RNAs.

The ribosome has been the subject of intense modeling activity. Several antibiotics inhibit protein synthesis—either at the 50S level, like the macrolides, or at the 30S level, like the aminoglycosides. However, bacteria develop resistance to virtually any available antibiotic. The knowledge at the atomic level of the antibiotic binding sites on the bacterial ribosome is still scanty, and in the future we hope to obtain X-ray crystal structures of 70S ribosome complexed with antibiotics. This should provide crucial new structural data to guide efforts in antibacterial drug design necessary to overcome the crisis in antibacterial therapy resulting from the spread of resistance.

## Summary

This chapter has essentially dealt with the structure and function of the major components of the bacterial cells. Bacteria occur as single cells or as cell associations.

The bacterial cell structures described are listed below.

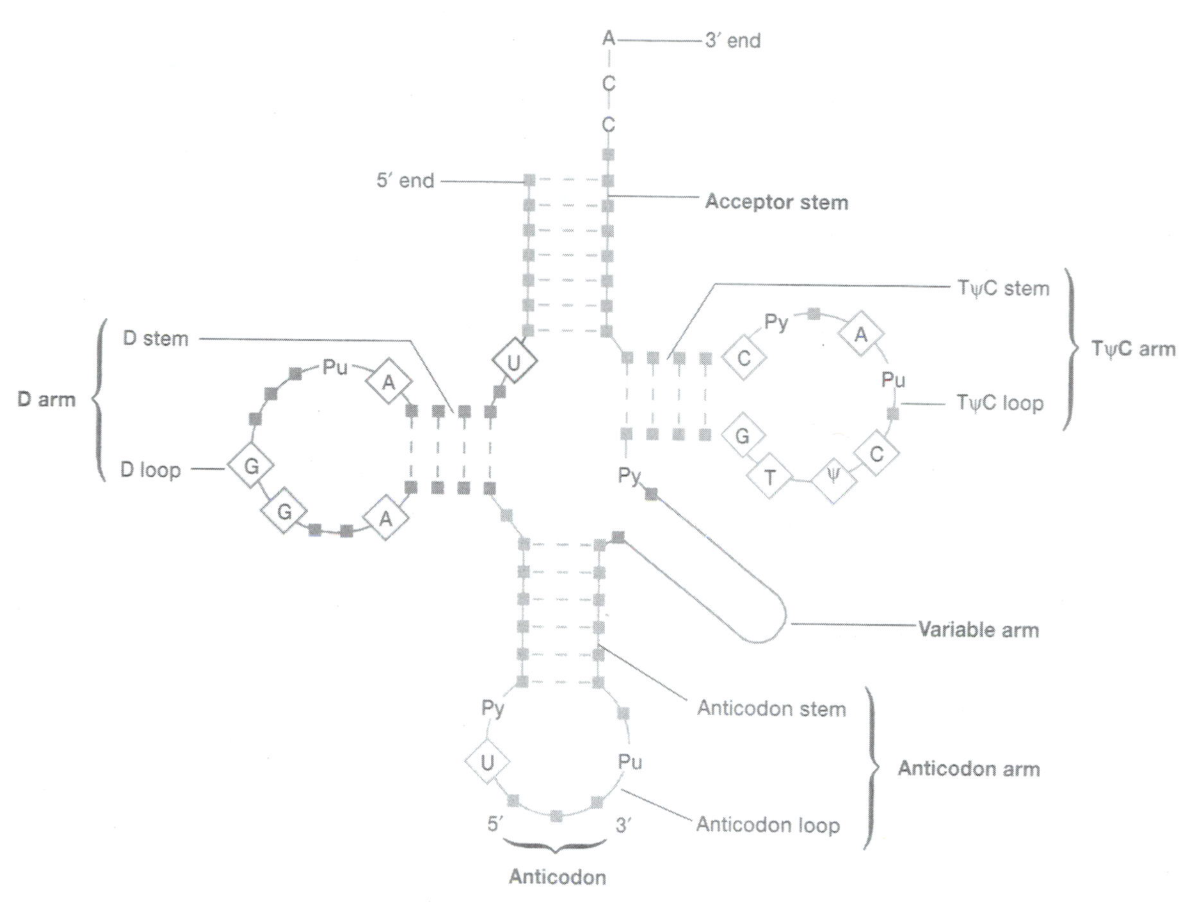


**Figure 1.30** Cloverleaf secondary structure of tRNA. Watson-Crick base pairing is indicated by dashed lines between nucleotide residues. The molecule is divided into an acceptor stem and four arms. Reprinted from L. M. Prescott, J. P. Harley, and D. A. Klein, *Microbiology,* 4th ed. (McGraw-Hill, Boston, Mass., 1999), with permission from the publisher.

## The Bacterial Cell Wall and Its Polymers

The bacterial cell wall is a unique structure which surrounds the cytoplasmic membrane. Bacterial cell walls are constructed from a variety of macromolecules and polymers. Structurally, the wall is necessary for maintaining the cell's characteristic shape and countering the effect of osmotic pressure. Peptidoglycan (or murein) is a cross-linked biopolymer. It is the major macromolecule that surrounds and protects bacterial cells.

## Bacterial Membranes, Porins, and the Periplasmic Space

Bacterial cell membranes consist primarily of phospholipids and proteins in a fluid mosaic structure in which the phospholipids form a bilayer. The outer membrane, which is composed of LPS, phospholipids, and proteins, forms a layer of the cell wall of gram-negative bacteria which acts as a penetration barrier for antibiotics and various substances. The porins are the major proteins present in the outer membrane and form water-filled channels that permit the diffusion of small hydrophilic solutes across the outer membrane. The periplasm is a specific cellular compartment of the cell envelope of gram-negative bacteria confined by the cytoplasmic membrane and the outer membrane.

## The Cytoplasm

The cytoplasm is defined as everything enclosed by the bacterial cytoplasmic membrane. The cytosol is the liquid portion of the cytoplasm.

## The Nucleoid and Genetic Elements

The nucleoid is the site of DNA and RNA synthesis. The chromosome is the main genomic element of bacteria as a single larger circular DNA molecule. It contains the genes for all "essential" functions and structures of the bacterial cell. Bacterial plasmids are extrachromosomal DNA genomes that replicate autonomously and in a controlled manner. Ribosomes are the sites of protein synthesis.

## Appendages

Bacterial flagella are membrane-embedded, filamentous organelles that are utilized by bacteria to move through liquid or semisolid media. Bacterial fimbriae, also known as pili, are hairlike proteinaceous structures extending from the cell surface that are involved in adherence. Sex pili are used by some bacteria for mating.

## REFERENCES AND FURTHER READING


### General (Books)

**Brooks, G. F., J. S. Butel, and S. A. Morse.** 1998. *Jawetz, Melnick, & Adelberg's Medical Microbiology,* 21st ed. Appleton & Lange, Stamford, Conn.

**Cooper, G. M.** 2000. *The Cell: A Molecular Approach,* 2nd ed. ASM Press, Washington, D.C., and Sinauer Associates, Inc., Sunderland, Mass.

**Fischetti, V. A., R. P. Novick, J. J. Ferretti, D. A. Portnoy, and J. I. Rood** (ed.). 2000. *Gram-Positive Pathogens.* ASM Press, Washington, D.C.

**Lengeler, J. W., G. Drews, and H. G. Schlegel.** 1999. *Biology of the Prokaryotes.* Georg Thieme Verlag, Stuttgart, Germany.

**Madigan, M. T., J. M. Martinko, and J. Parker.** 2000. *Brock Biology of Microorganisms,* 9th ed. Prentice-Hall, Upper Saddle River, N.J.

**Nelson, D. L., and M. M. Cox.** 2000. *Lehninger Principles of Biochemistry,* 3rd ed. Worth Publishers, New York, N.Y.

**Prescott, L. M., J. P. Harley, and D. A. Klein.** 1999. *Microbiology,* 4th ed. McGraw-Hill, Boston, Mass.

**Salyers, A. A., and D. D. Whitt.** 2001. *Microbiology.* Fitzgerald Science Press, Bethesda, Md.

**Walker, T. S.** 1998. *Microbiology.* The W. B. Saunders Co., Philadelphia, Pa.

**White, D.** 2000. *The Physiology and Biochemistry of Prokaryotes.* Oxford University Press, New York, N.Y.

### Original Articles and Reviews

**The cell wall**

**Beveridge, T. J.** Bacterial cell wall. Accepted for publication in *Encyclopedia of Life Sciences.* Nature Publishing, London, United Kingdom.

**Ghuysen, J. M., and R. Hakenbeck.** 1994. *Bacterial Cell Wall.* Elsevier, Amsterdam, The Netherlands.

**Volker-Höltje, J.** 2000. Cell walls, bacterial, p. 759–771. *In* J. Lederberg (ed.), *Encyclopedia of Microbiology,* 2nd ed., vol. 1. Academic Press, Inc., San Diego, Calif.

**Ward, J. B.** 1990. Cell wall structure and function, p. 553–607. *In* C. Hansch, P. G. Sammes, and J. B. Taylor (ed.), *Comprehensive Medicinal Chemistry,* vol. 2. Pergamon Press, Oxford, United Kingdom.

**Peptidoglycan**

**Benson, T. E., D. B. Prince, V. T. Mutchler, K. A. Curry, A. M. Ho, R. W. Sarver, J. C. Hagadorn, G. H. Choi, and R. L. Garlick.** 2002. X-ray crystal structure of *Staphylococcus aureus* FemA. *Structure* **10:**1107–1115.

**van Heijenoort, J.** 2001. Recent advances in the formation of the bacterial peptidoglycan monomer unit. *Nat. Prod. Rep.* **18:**503–519.

**Fimbriae and pili**

**Li, X., and H. L. T. Mobley.** Bacterial pili and fimbriae. Accepted for publication in *Encyclopedia of Life Sciences.* Nature Publishing, London, United Kingdom.

**Mulvey, M. A., K. W. Dodson, G. E. Soto, and S. J. Hultgren.** 2000. Fimbriae, pili, p. 361–379. *In* J. Lederberg (ed.), *Encyclopedia of Microbiology,* 2nd ed., vol. 2. Academic Press, Inc., San Diego, Calif.

**Flagella**

**Aizawa, S. I.** 2000. Flagella, p. 380–389. *In* J. Lederberg (ed.), *Encyclopedia of Microbiology,* 2nd ed., vol. 2, Academic Press, Inc., San Diego, Calif.

**Manson, M. D., J. P. Armitage, J. A. Hoch, and R. M. Macnab.** 1998. Bacterial locomotion and signal transduction. *J. Bacteriol.* **180:**1009–1022.

**Morgan, D. G., and S. Khan.** Bacterial flagella. Accepted for publication in *Encyclopedia of Life Sciences.* Nature Publishing, London, United Kingdom.

**Genome**

**Fleischmann, R. D., M. D. Adams, O. White, R. A. Clayton, E. F. Kirkness, A. R. Kerlavage, C. J. Bult, J. F. Tomb, B. A. Dougherty, J. M. Merrick, K. McKenney, G. Sutton, W. FitzHugh, C. Fields, J. D. Gocayne, J. Scott, R. Shirley, L.-I. Liu, A. Glodek, J. M. Kelley, J. F. Weidman, C. A. Phillips, T. Spriggs, E. Hedblom, M. D. Cotton, T. Utterback, M. C. Hanna, D. T. Nguyen, D. M. Saudek, R. C. Brandon, L. D. Fine, J. L. Fritchman, J.L. Fuhrmann, N. S. M. Geoghagen, C. L. Gnehm, L. A. McDonald, K. V. Small, C. M. Fraser, H. O. Smith, and J. C. Venter.** 1995. Whole-genome random sequencing and assembly of *Haemophilus influenzae. Science* **269:**496–512.

**Guild, B. C.** 1999. Genomics, target selection, validation, and assay considerations in the development of antibacterial screens. *Annu. Rep. Med. Chem.* **34:**227–239.

**Heidelberg, J. F., J. A. Eisen, W. C. Nelson, R. A. Clayton, M. L. Gwinn, R. J. Dodson, D. H. Haft, E. K. Hickey, J. D. Peterson, L. Umayan, S. R. Gill, K. E. Nelson, T. D. Read, H. Tettelin, D. Richardson, M. D. Ermolaeva, J. Vamathevan, S. Bass, H. Qin, I. Dragoi, P. Sellers, L. McDonald, T. Utterback, R. D. Fleishmann, W. C. Nierman, O. White, S. L. Salzberg, H. O. Smith, R. R. Colwell, J. J. Mekalanos, J. C. Venter, and

**C. M. Fraser.** 2000. DNA sequence of both chromosomes of the cholera pathogen *Vibrio cholerae*. *Nature* **406**:477–483.

**Jenks, P. J.** 1998. Sequencing microbial genomes—what will it do for microbiology? *J. Med. Microbiol.* **47**:375–382.

**Kuroda, M., T. Ohta, I. Uchiyama, T. Baha, H. Yuzawa, I. Kobayashi, L. Cui, A. Oguchi, K. Acki, Y. Nagai, J. Lian, T. Ito, M. Kanamori, H. Matsumaru, A. Maruyama, H. Murakami, A. Hosoyama, Y. Mizutani-Ui, N. K. Takahashi, T. Sawano, R. Inoue, C. Kaito, K. Sekimizu, H. Hirakawa, S. Kuhara, S. Goto, J. Yabuzaki, M. Kanehisa, A. Yamashita, K. Oshima, K. Furuya, C. Yoshino, T. Shiba, M. Hattori, N. Ogasawara, H. Hayashi, and K. Hiramatsu.** 2001. Whole genome sequencing of methicillin-resistant *Staphylococcus aureus*. *Lancet* **357**:1225–1240.

**Moir, D. T., K. J. Shaw, R. S. Hare, and G. F. Vovis.** 1999. Genomics and antimicrobial drug discovery. *Antimicrob. Agents Chemother.* **43**:439–446.

**Stover, C. K., X. Q. Pham, A. L. Erwin, S. D. Mizoguchi, P. Warrener, M. J. Hickey, F. S. L. Brinkman, W. O. Hufnagle, D. J. Kowalik, M. Lagrou, R. L. Garber, L. Goltry, E. Tolentino, S. Westbrock-Wadman, Y. Yuan, L. L. Brody, S. N. Coulter, K. R. Folger, A. Kas, K. Larbig, R. Lim, K. Smith, D. Spencer, G. K. S. Wong, Z. Wu, I. T. Paulsen, J. Reizer, M. H. Saier, R. E. W. Hancock, S. Lory, and M. V. Olson.** 2000. Complete genome sequence of *Pseudomonas aeruginosa* PA01, an opportunistic pathogen. *Nature* **406**:959–964.

**Strauss, E. J., and S. Falkow.** 1997. Microbial pathogenesis: genomics and beyond. *Science* **276**:707–712.

**The nucleoid**

**Robinow, C., and E. Kellenberger.** 1994. The bacterial nucleoid revisited. *Microbiol. Rev.* **58**:211–232.

**The outer membrane**

**Braun, V.** 1975. Covalent lipoprotein from the outer membrane of *Escherichia coli*. *Biochim. Biophys. Acta* **415**:335–377.

**Caroff, M., D. Karibian, J. M. Cavaillon, and N. Haefner-Cavaillon.** 2002. Structural and functional analyses of bacterial lipopolysaccharides. *Microbes Infect.* **4**:915–926.

**Hancock, R. E. W.** 1991. Bacterial outer membranes: evolving concepts. *ASM News* **57**:175–182.

**Koebnik, K. R., K. P. Locher, and P. Van Gelden.** 2000. Structure and function of bacterial outer membrane proteins: barrels in a nutshell. *Mol. Microbiol.* **37**:239–253.

**Nikaido, H.** 1989. Outer membrane barrier as a mechanism of antimicrobial resistance. *Antimicrob. Agents Chemother.* **33**:1831–1836.

**Nikaido, H., and M. Vaara.** 1985. Molecular basis of bacterial outer membrane permeability. *Microbiol. Rev.* **49**:1–32.

**Noland, B. W., J. M. Newman, J. Hendle, J. Badger, J. A. Christopher, J. Tresser, M. D. Buchanan, T. A. Wright, H. J. Müller-Dieckmann, K. S. Gajiwala, and S. G. Buchanan.** 2002. Structural studies of *Salmonella typhimurium* ArnB (PmrH) aminotransferase: a 4-amino-4-deoxy-L-arabinose lipopolysaccharide-modifying enzyme. *Structure* **10**:1569–1580.

**Porins**

**Calamita, G.** 2000. The *Escherichia coli* aquaporin-Z water channel. *Mol. Microbiol.* **37**:254–262.

**Cowan, S. W.** 1993. Bacterial porins: lessons from three high-resolution structures. *Curr. Opin. Struc. Biol.* **3**:501–507.

**Cowan, S. W., T. Schirmer, G. Rummel, M. Steiert, R. Ghosh, R. A. Pauptit, J. N. Jansonius, and J. P. Rosenbusch.** 1992. Crystal structures explain functional properties of two *E. coli* porins. *Nature* **358**:727–733.

**Doménech-Sánchez, A., S. Hernández-Allés, L. Martínez-Martínez, V. J. Benedí, and S. Alberti.** 1999. Identification and characterization of a new porin gene of *Klebsiella pneumoniae*: its role in β-lactam antibiotic resistance. *J. Bacteriol.* **181**:2726–2732.

**Hancock, R. E. W.** 1987. Role of porins in outer membrane permeability. *J. Bacteriol.* **169**:929–933.

**Misuno, T., M. Y. Chou, and M. Inouye.** 1983. A comparative study on the genes for three porins of the *Escherichia coli* outer membrane. *J. Biol. Chem.* **258**:6932–6940.

**Nikaido, H.** 1992. Porins and specific channels of bacterial outer membranes. *Mol. Microbiol.* **6**:435–442.

**Schirmer, T., T. A. Keller, Y.-F. Wang, and J. P. Rosenbusch.** 1995. Structural basis for sugar translocation through maltoporin channels at 3.1 Å resolution. *Science* **267**:512–514.

**Schirmer, T., and J. P. Rosenbusch.** 1991. Proterozoic and eukaryotic porins. *Curr. Biol.* **1**:539–545.

**Schulz, G. E.** 1994. Structure-function relationships in porins as derived from a 1.8 Å resolution crystal structure, p. 343–352. *In* J. M. Ghuysen and R. Hakenbeck (ed.), *Bacterial Cell Wall.* Elsevier, Amsterdam, The Netherlands.

**Schulz, G. E.** 1993. Bacterial porins: structure and function. *Curr. Biol.* **3**:701–707.

**Weis, M. S., and G. E. Schulz.** 1992. Structure of porin refined at 1.8 Å resolution. *J. Mol. Biol.* **227**:493–509.

**Ribosomes**

**Ban, B., P. Nissen, J. Hansen, P. B. Moore, and T. A. Steitz.** 2000. The complete atomic structure of the large ribosomal subunit at 2.4 Å resolution. *Science* **289**:905–920.

**Cate, J. H., M. M. Yusupov, G. Z. Yusupova, T. N. Earnest, and H. F. Noller.** 1999. X-ray crystal structures of 70S ribosome functional complexes. *Science* **285**:2095–2104.

**Clemons, W. M., J. L. C. May, B. T. Wimberly, J. P. McCutcheon, M. S. Capel, and V. Ramakrishnan.** 1999. Structure of a bacterial 30S ribosomal subunit at 5.5 Å resolution. *Nature* **400**:833–847.

**Garrett, R.** 1999. Mechanics of the ribosome. *Nature* **400**:811–812.

**Garrett, R. A., S. R. Douthwaite, A. Liljas, A. T. Matheson, P. B. Moore, and H. F. Nolle (ed.).** 2000. *The Ribosome: Structure, Function, Antibiotics, and Cellular Interactions.* ASM Press, Washington, D.C.

**Squires, C. L.** 2000. Ribosome synthesis and regulation, p. 127–139. *In* J. Lederberg (ed.), *Encyclopedia of Microbiology,* 2nd ed., vol. 4. Academic Press, Inc., San Diego, Calif.

**Tocilj, A., F. Schlunzen, D. Janell, M. Gluhmann, H. A. S. Hansen, J. Harms, A. Bashan, H. Bartels, I. Agmon, F. Franceschi, and A. Yonath.** 1999. The small ribosomal subunit from *Thermus thermophilus* at 4.5 Å resolution: Patter fittings and identification of a functional site. *Proc. Natl. Acad. Sci. USA* **96:**14252–14257.

**Westhof, E., and N. Leontis.** 2000. Atomic glimpses on a billion-year old molecular machine. *Angew. Chem. Int. Ed. English* **39:**1587–1591.

**Yusupov, M. M., G. Z. Yusupova, A. Baucom, K. Lieberman, T. N. Earnest, J. H. D. Cate, and H. F. Noller.** 2001. Crystal structure of the ribosome at 5.5 Å resolution. *Science* **292:**883–902.

**tRNA**

**Soll, D., and U. L. RajBhandary (ed.).** 1995. *tRNA: Structure, Biosynthesis, and Function.* ASM Press, Washington, D.C.

**RNA polymerase**

**DeHaseth, P. L., M. L. Zupancic, and M. T. Record.** 1998. RNA polymerase-promoter interactions: the coming and goings of RNA polymerase. *J. Bacteriol.* **180:**3019–3025.

chapter 2



# Bacterial Pathogenesis

The topic of bacterial pathogenesis is beyond the scope of this book and is not discussed in detail. Rather, the discussion focuses on presenting an integrated aspect of the key events during a bacterial infection. Bacterial pathogens (prokaryotic cells) and their human hosts (eukaryotic cells) are engaged in a constant struggle for survival, and the balance between bacterial strategies for resisting the host immune response and the action of antibacterial agents determines the outcome of bacterial infections. As discussed in chapters 3 and 5, bacteria have developed a variety of mechanisms for surviving the actions of antibacterials as well as immunologic defenses.

The purpose of this chapter therefore is to serve as an introduction of some of the most relevant points and to highlight features that are particularly essential for an understanding of bacterial invasion of eukaryotic cells, immunity to extracellular and intracellular bacteria, evasion of immune mechanisms by extracellular and intracellular bacteria, antibacterial mechanisms of action, and bacterial resistance to antibacterial agents.

## Definitions

In this section some important terms in bacterial pathogenesis are defined.

The term **infectious disease** applies when signs and symptoms result from an infection process which leads to damage to the host or altered physiology.

A **bacterial pathogen** is usually defined as any bacterium that has the capacity to cause disease. Its ability to cause disease is called **pathogenicity.**

**Virulence** provides a quantitative measure of the pathogenicity or the likelihood of causing disease. For example, encapsulated pneumococci are more virulent than nonencapsulated pneumococci.

**Virulence factors** refer to the properties (i.e., gene products) that enable a microorganism to establish itself on or within a host of a particular species and enhance its potential to cause disease. Virulence factors include bacterial toxins, cell surface proteins that mediate bacterial attachment, cell surface carbohydrates and proteins that protect a bacterium, and hydrolytic enzymes that may contribute to the pathogenicity of the bacterium. Identification of virulence factors often requires some knowledge about physiology and the molecular mechanism of pathogenesis for a given bacterium.

The definition of a **chaperone** according to the *Oxford Dictionary of Biochemistry and Molecular Biology* is "any of a functional class of unrelated families of proteins that assist the correct non-covalent assembly of other polypeptide-containing structures *in vivo*, but are not components of these assembled structures when they are performing their normal biological functions."

**Lectins** are proteins that bind mono- and oligosaccharides specifically and reversibly but are devoid of catalytic activity (i.e., are not enzymes) and, in contrast to antibodies (immunoglobulins), are not products of an immune response.

**Enterotoxins** are exotoxins that act on the small intestine, generally causing massive secretion of fluid into the intestinal lumen, leading to diarrhea.

It is useful to distinguish pathogens that regularly cause disease in some proportion of susceptible individuals with an apparent intact defense system from other microorganisms which rarely cause diseases in people with intact host defense systems but can cause disease in immunocompromised persons. The term "**opportunistic**" refers to this last category of pathogens.

## Bacterial Invasion of Human Cells

Our body surfaces are defended by surface epithelia that provide mechanical, chemical, biochemical and microbiological barriers to infection, as described in chapter 3. These epithelia comprise the skin and the linings of the body's tubular structures, such as the gastrointestinal, respiratory, and genitourinary tracts. Infections occur only when the bacterial pathogen can colonize or cross these barriers.

Bacterial pathogens most commonly enter the body via epithelial barriers presented by the mucosa of the respiratory, digestive, and urogenital tracts or through damaged skin, such as in burn patients, where infection is a major cause of mortality and morbidity. Less often, insects or hypodermic needles introduce bacteria directly into the blood.

The survival strategy of bacterial pathogens requires their multiplication. Some bacteria multiply at and remain on the surface of the eukaryotic cell, whereas others multiply within the eukaryotic cells of human and animal hosts (Table 2.1 and Fig. 2.1). **Extracellular bac-**

**Table 2.1** Selected examples of bacterial pathogenesis and their location with respect to eukaryotic cells[a]

| Bacterial pathogen | Major diseases induced | Interaction with host cells |
|---|---|---|
| Extracellular bacteria | | |
| *Staphylococcus* species | Skin and tissue infections | Adherence to extracellular matrix |
| *Streptococcus* species | Otitis media, pharyngitis, scarlet fever, meningitis | Adherence to extracellular matrix |
| *Bordetella pertussis* | Whooping cough | Adherence to cells |
| *Neisseria gonorrhoeae* | Gonorrhea | Adherence to cells |
| *Neisseria meningitidis* | Meningitis | Adherence to cells |
| *Helicobacter pylori* | Ulcers, gastritis | Adherence to cells |
| *Escherichia coli* | Diarrheas, meningitis, urinary tract infections | Adherence to cells |
| *Yersinia* species | Plague, mesenteric lymphadenitis, diarrhea | Adherence to cells and matrix |
| *Vibrio cholerae* | Cholera | Adherence to cells |
| Intracellular bacteria | | |
| Macrophages | | |
| *Legionella pneumoniae* | Legionnaires' disease | Within a vacuole |
| *Mycobacterium tuberculosis* | Tuberculosis | Within a vacuole |
| *Mycobacterium leprae* | Leprosy | Within a vacuole |
| Macrophage and epithelial cells | | |
| *Salmonella* species | Typhoid fever, gastroenteritis | Within a vacuole |
| *Shigella* species | Dysentery, gastroenteritis | Intracytoplasmic |
| *Listeria monocytogenes* | Listeriosis, meningitis | Intracytoplasmic |
| *Chlamydia* species | Trachoma, sexually transmited diseases, pneumonia | Within a vacuole |

[a] Reprinted from B. B. Finlay and P. Cossart, *Science*, **276**:718–725, 1997, with permission from the publisher.

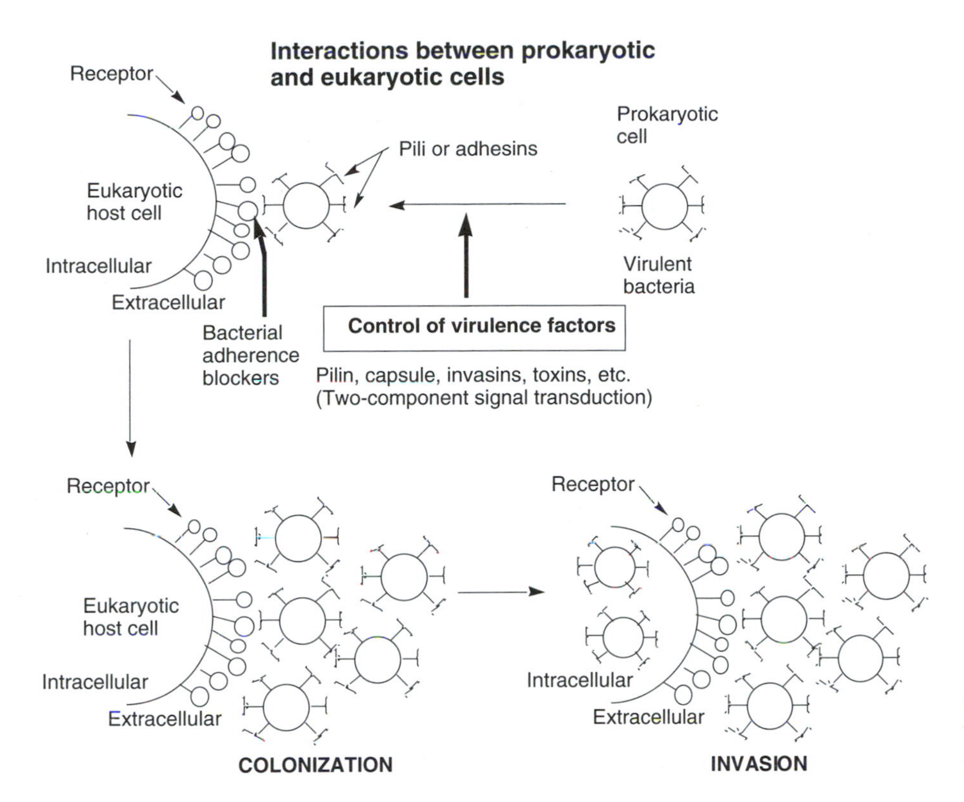


**Figure 2.1** Diagrammatic representation of the steps in bacterial invasion of a human cell.

teria are capable of replicating outside the host cells, e.g., in the circulation, in extracellular connective tissues, and in various tissue spaces such as airways, genitourinary tract, and intestinal lumens. **Intracellular bacteria** can resist antibiotics and other antibacterial agents with poor ability to penetrate eukaryotic cell membranes. The intracellular and extracellular concentrations of macrolides are generally very high. The accumulation of macrolides within phagocytes and other cells involved in immunity is thought to explain in part their excellent clinical effect. In contrast, β-lactams and aminoglycosides generally show poor or slow cellular penetration.

The key pharmacokinetic features required for an antibiotic to possess activity against infection of endothelial or epithelial tissues by obligate intracellular bacteria are the localization and duration of useful concentrations of the antibiotic in these tissues.

## Steps in Bacterial Invasion

The evolution of an infectious disease caused by bacteria in an individual involves complex interactions between the prokaryotic cell and the eukaryotic host cell target. The infectious cycle comprises entry of the pathogen, establishment and multiplication, avoidance of host defenses, damage, and exit.

The key events during a bacterial infection cycle include different stages, executed by pathogenic bacteria using conserved macromolecular systems. The specificity of the bacterial effector molecules used and the resulting effects on the host vary from pathogenic bacteria to bacteria, but the underlying steps are as follows:

- adherence (attachment) to the eukaryotic cell
- entry of the bacteria into the body
- avoidance of host immune mechanisms of defense (the ability of pathogenic bacteria to colonize and invade mammalian hosts depends on their ability to evade host defenses; bacteria and host defenses interact continuously during infection)
- tissue damage or functional impairment and colonization of the host tissues
- mechanisms to resist the actions of antibacterial agents that are administered to the host (these mechanisms develop as a result of natural selection)

Two types of systems that contribute to the infectious cycle are described here: adhesins, which allow the bacteria to become established, and toxins, which damage the host and permit the bacteria to evade host immunity mechanisms.

## Bacterial Adhesion

The first interaction between a pathogenic bacterium and the host entails **bacterial adhesion** to the surface of the eukaryotic target cells. A variety of molecules and macromolecular structures, collectively known as **adhesins,** recognize and interact with receptors on the surface of the eukaryotic host cells. A number of studies have examined the development of inhibitors of bacterial adherence.

## PILUS AND NONPILUS ADHESINS

Frequently, adhesins are assembled into pili or fimbriae that extend from the bacterial surface. Alternatively, the adhesins are directly associated with the microbial cell surface. These adhesions can be performed by the afimbrial adhesins (also called nonfimbrial adhesins).

### Pili or Fimbriae

The previous chapter described the composition of pili from many subunits of a major structural protein (pilin) packed tightly into a helical array.

Pili are fibrous organelles that are expressed on the surface of gram-negative bacteria and mediate attachment to host tissues. The structural and functional characterization of these organelles, particularly their role in triggering signals in both the bacterium and the host on attachment, has begun to reveal the molecular mechanism of bacterial diseases.

Many gram-negative pathogens assemble structurally and functionally diverse adhesive pili on their surface by a major mechanism: the chaperone-usher pathway (Fig. 2.2). The assembly of adhesive pili requires periplasmic chaperones (bacterial proteins). The role of the outer membrane proteins, known as ushers, in uncapping chaperones from their target proteins allows the ordered assembly of the subunits into surface structures.

### Host Cell Adhesion Mechanisms

Uropathogenic *Escherichia coli* adheres to erythrocytes and uroepithelial cells of the kidneys and urinary tract.

**Figure 2.2** Model of P-pilus biogenesis and PapG recognition of three Gal-(α1→4)-Gal-containing receptors. Reprinted from S. J. Hultgren, S. Abraham, M. Caparon, P. Falk, J. W. St Geme, and S. Normark, *Cell* **73:**887–901, 1993, with permission from the publisher.

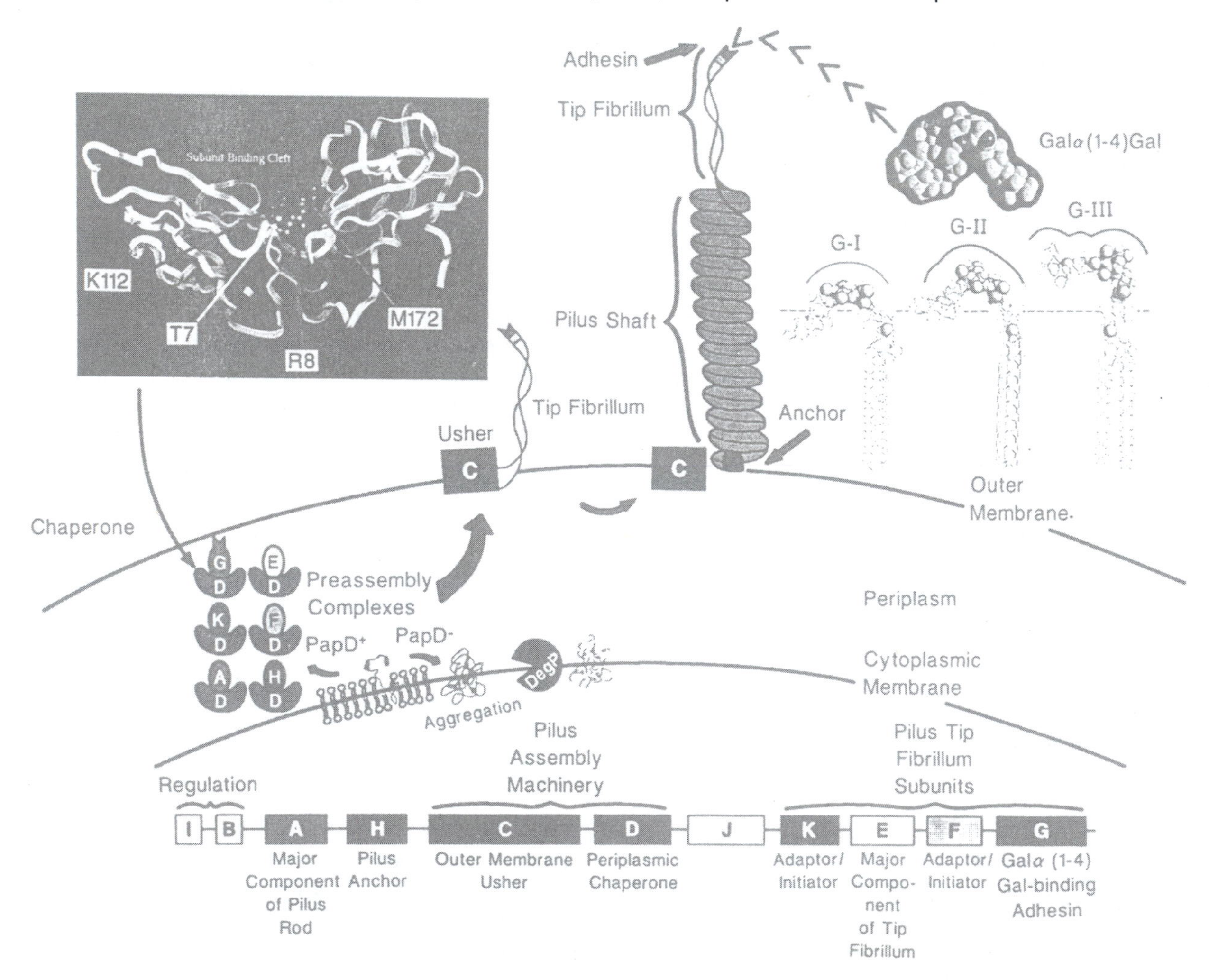

Studies have revealed that this specific binding event occurs between PapG adhesin proteins located at the tip of P (pyelonephritis) pili and Gal-(α1→4)-Gal saccharide epitopes in the globoseries of glycolipids of host cells. This saccharide structure is linked by a β-glycosidic bond to a ceramide group anchoring the receptor in the host cell membrane (Fig. 2.3). The receptor-binding domain of PapG has been determined to lie in the N terminus of the protein. All animals cells contain glycosphingolipids, which have the lipid (ceramide) moieties embedded in the cell membrane and the hydrophilic carbohydrate moieties protruding into the surrounding medium.

Bacterial adhesion leading to infection can be divided into three distinct categories: specific adhesion to host cell surface molecules, specific adhesion to extracellular matrix (proteins, glycoproteins, and proteoglycans), and adhesion to biomaterial surfaces of medical devices.

### Nonpilus Adhesins

The crystal structures of the extracellular domains of two related bacterial outer membrane proteins, invasin from *Yersinia pseudotuberculosis* and intimin from the enteropathogenic *E. coli,* have recently been solved. Both adhesins are rigid rods composed of a series of immunoglobulin-like domains capable of extending from the surface of the bacterium. At the end of the rod is an incomplete lectin-like domain that contains the receptor-binding site, but this domain lacks a full carbohydrate-binding fold. Instead, the receptor is bound at one side of this domain. For a comprehensive account of the molecular basis of bacterial adhesion and the molecular mechanisms of pathogenesis, the reader is encouraged to check the excellent reviews cited at the end of the chapter.

Note that these topics are included here because it is thought that therapies which can prevent or disrupt these molecular mechanisms have great potential for development as alternatives to antibiotics for the prevention and treatment of infections. The increase in the incidence of antibiotic-resistant bacterial pathogens has pointed the need to understand the strategies of pathogenic bacteria at the molecular level in order to reveal new targets for the design of new and better drugs.

Recently, two publications have appeared, one of which, by Entenäs et al. (2000), describes the stereoselective synthesis of optically active β-lactams, designed to be suitable compounds for inhibiting pilus formation in uropathogenic *E. coli,* and the other, by Kihlberg and Magnusson (1996), describes the use of synthetic oligosaccharides and their deoxy and deoxyfluoro derivatives as inhibitors of the bacterial adhesins of *E. coli* and *Streptococcus suis.* In addition, synthetic peptides were identified that inhibit the chaperone PapD, which is required for the assembly of protein subunits into the *E. coli* P pilus. Both adhesin-receptor interactions and subunit-chaperone interactions constitute targets for the design of novel synthetic antibacterial agents as an alternative to the use of antibiotics to treat bacterial infections.

## Toxins

The term "toxins" is derived from the Greek *toxikon* (bow poison) and refers to the poisonous material applied to arrows used by ancient Greek warriors.

### Bacterial Toxins

Bacterial toxins are produced by microorganisms and affect human and animal cells at distance from the bacteria themselves.

The term **endotoxin** describes the toxic activities associated with bacterial envelope components. It usually refers to the activities of the lipid A moiety of lipopolysaccharide (LPS) of gram-negative bacteria (Fig. 1.19 and 1.20). The term **exotoxin** is used to describe pro-

**Figure 2.3** Saccharides of the globoseries glycolipids.

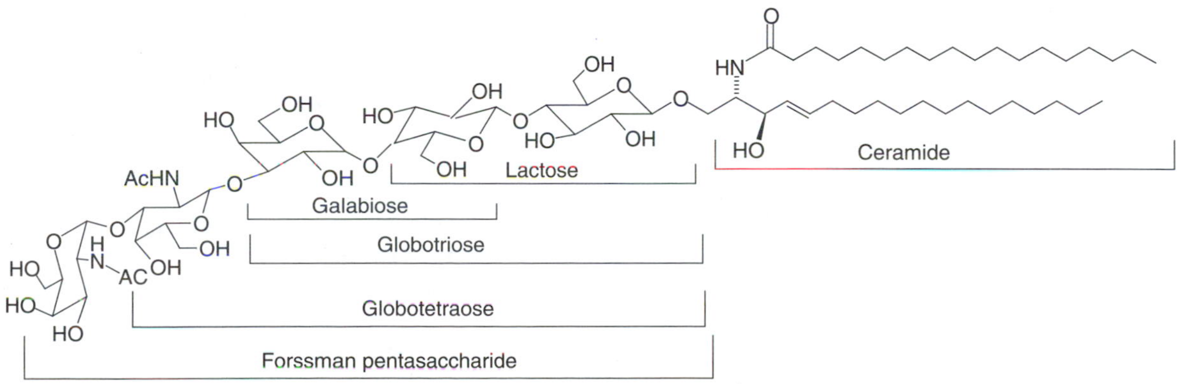

teinaceous bacterial products that are released extracellularly from the bacterium during exponential growth and are toxic for target cells. As described above, bacteria cause diseases primarily by colonizing surfaces or by invading host cells. This section focuses on bacteria that causes disease primarily by producing exotoxins.

The observation that culture supernatant free from bacteria fully reproduced the symptoms of diseases such as diphtheria, tetanus, cholera, and botulism led to the conclusion that in these instances, bacterial toxins were the only factors needed by bacteria to cause a disease.

Exotoxins exert their effects in a variety of ways, for instance, by

- inhibiting protein synthesis
- inhibiting nerve synapse function
- disrupting membrane transport
- damaging plasma membranes, or
- acting as superantigens and overstimulating the immune system.

Some of the pathogenic bacteria that produce exotoxins are presented in Table 2.2. To have an effect, an exotoxin must interact with a specific molecule that serves as a receptor on the surface of the target cell. In many exotoxins, the receptor-binding domain is on one polypeptide chain whereas the toxic function is on a second chain. Antibodies that bind to the receptor-binding site on the toxin molecule can prevent the toxin from binding to the cell and thus protect the cell from toxic attack. This protective effect of antibodies is called **neutralization.**

Most toxins are active at nanomolar concentrations. To neutralize toxins, antibodies must be able to diffuse into the tissues and bind to the toxin rapidly and with high affinity.

Figure 2.4 is a schematic representation of four groups of bacterial toxins. Group 1 toxins act by binding receptors on the cell membrane and sending a signal to the cell. Group 2 toxins act by forming pores in the cell membrane, thus perturbing the cell permeability barrier. Group 3 toxins are A/B toxins, composed of a binding domain (B subunit) and an enzymatically active effector domain (A subunit). Following receptor binding, the toxins are internalized and located in endosomes, from which the A subunit can be transferred directly to the cytoplasm by using a pH-dependent conformational change (mechanism 3.1 in Fig. 2.4) or can be transported to the Golgi apparatus and to the endoplasmic reticulum, from which the A subunit is finally transferred to the cytoplasm (mechanism 3.2 in Fig. 2.4). Group 4 toxins are injected directly from the bacterium into the cell by a contact-dependent secretion system (type III or type IV secretion system).

**Table 2.2** Some examples of common diseases caused by bacterial toxins[a]

| Disease | Organism | Toxin | Effects in vivo |
|---|---|---|---|
| Tetanus | *Clostridium tetani* | Tetanus toxin | Blocks inhibitory neuron action, leading to clonic muscle contraction |
| Diphtheria | *Corynebacterium diphtheriae* | Diphtheria toxin | Inhibits protein synthesis, leading to epithelial cell damage and myocarditis |
| Gas gangrene | *Clostridium perfringens* | Clostridial toxin | Phospholipase activation, leading to cell death |
| Cholera | *Vibrio cholerae* | Cholera toxin | Activates adenylate cyclase and elevates cyclic AMP levels in cells, leading to changes in intestinal epithelial cells, resulting in loss of water and electrolytes |
| Anthrax | *Bacillus anthracis* | Anthrax toxic complex | Increases vascular permeability leading to edema, hemorrhage, and circulatory collapse |
| Botulism | *Clostridium botulinum* | Botulinum toxin | Blocks release of acetylcholine, leading to paralysis |
| Whooping cough | *Bordetella pertussis* | Pertussis toxin | ADP-ribosylation of G proteins, leading to lymphoproliferation |
| | | Tracheal cytotoxin | Inhibits cilia and causes epithelial cell loss |
| Scarlet fever | *Streptococcus pyogenes* | Erythrogenic toxin<br>Leucocidin, streptolysins | Vasodilation leading to scarlet fever rash<br>Kill phagocytes, allowing bacterial survival |
| Food poisoning | *Stapyhlococcus aureus* | Staphylococcal enterotoxin | Acts on intestinal neurons to induce vomiting; also a potent T-cell mitogen (SE superantigen) |
| Toxic shock syndrome | *Staphylococcus aureus* | Toxic shock syndrome toxin | Causes hypotension and skin loss; also a potent T-cell mitogen (TSST-1 superantigen) |

[a]Reprinted from C. A. Janeway, P. Travers, M. Walport, and M. Shlomchik, *Immunobiology*, 5th ed. (Garland Publishing, Taylor & Francis Group, New York, N.Y., 2001), with permission from the publisher.

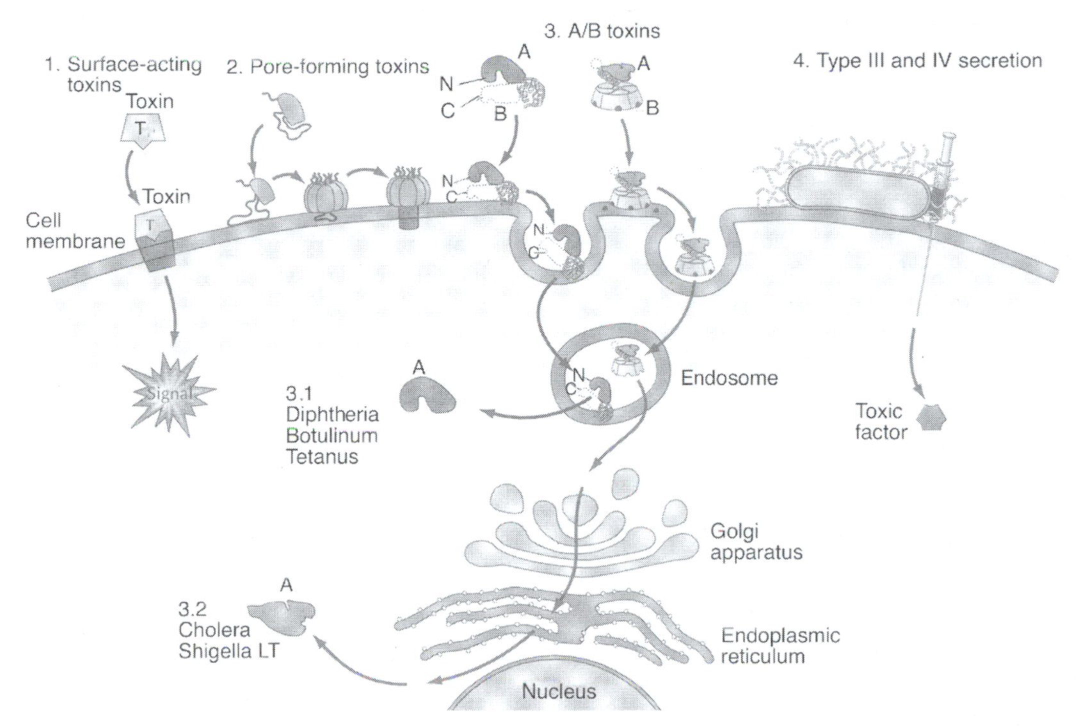


**Figure 2.4** Schematic representation of the four groups of bacterial toxins. Reprinted from P. Cossart, P. Boquet, S. Normark, and R. Rappuoli (ed.), *Cellular Microbiology* (ASM Press, Washington, D.C., 2000), with permission from the American Society for Microbiology.

## Vaccines

The importance of using virulence factors as antigens in the development of vaccines has been recognized and exploited for many years. Most vaccines inhibit colonization of the host by the pathogen or prevent pathological effects generally by inhibiting a toxin.

Vaccines have been developed for some of the exotoxin-mediated diseases and continue to be important in prevention of disease. The vaccines, called **toxoids,** are made from exotoxins, which are modified so that they are no longer toxic. Thus, vaccines do not kill bacterial pathogens per se but, rather, prevent colonization or inhibit a toxin.

## Superantigens

Superantigens are another type of protein toxin. They are produced mostly by *Staphylococcus aureus* and *Streptococcus pyogenes*. They are bivalent molecules that bind to distinct molecules, the major histocompatibility complex class II molecule and the variable part of the T-cell receptor. The cross-linking between the major histocompatibility complex and the T-cell receptor can activate T cells even without a specific peptide. Therefore, superantigens are potent polyclonal activators of T cells. In vivo, the potent polyclonal activation results in a massive release of cytokines such as interleukin-1 and tumor necrosis factor (see chapter 3), which are thought to play an important role in diseases such as the toxic syndrome induced by toxic shock syndrome toxin 1 (TSST-1), vomiting and diarrhea caused by staphylococcal enterotoxins, and the exanthemas caused by the pyrogenic streptococcal exotoxins.

## Endotoxins

As discussed in chapter 1, gram-negative bacteria have LPS in the outer membrane of their cell wall. The LPS (Fig. 1.19) is called an endotoxin because it is bound to the bacterium and is released when the microorganism undergoes lysis. The toxic component of the LPS is the lipid portion, called lipid A.

When the LPS is injected into an animal, it causes a variety of physiological effects. Fever is an almost universal symptom because endotoxin stimulates the host cell to release proteins called endogenous pyrogens, which affect the temperature-controlling center of the brain. Additionally, the animal may develop diarrhea, experience a rapid decrease in lymphocyte, leukocyte, and platelet numbers, and enter a generalized inflammatory state. Large doses of endotoxins can cause death, primarily through hemorrhagic shock and tissue necrosis. However, the toxicity of endotoxins is much lower than that of exotoxins.

## Pathogenicity Islands

The virulence properties of many pathogenic bacteria are due to proteins encoded by large gene clusters called pathogenicity islands. Pathogenicity islands (PAIs) are large elements, frequently longer than 30 kb, that reintegrated into the chromosome of bacterial pathogens and carry one or more of the following virulence determinants:

- PAIs are encountered in a number of strains of pathogenic gram-negative bacteria, where they play an important role in virulence, but are absent in nonpathogenic bacteria.
- PAIs carry mobile genes and are associated with tRNA genes and/or insertion sequence (IS) elements. Genomic islands have been acquired by horizontal gene transfer.
- PAIs have a different G+C content from the host genome.

PAIs were first studied in detail in uropathogenic and enteropathogenic *E. coli* strains. Remember that most *E. coli* strains are nonpathogenic and constitute part of the normal gastrointestinal flora. However, other *E. coli* strains are pathogenic and can cause bladder infections, meningitis, or diarrhea. Pai I (this abbreviation is used when one is referring to a specific, named pathogenicity island) (70 kb) of uropathogenic strain 536 encodes a hemolysin. Pai II (190 kb) encodes P fimbriae as well as hemolysin. Pai I and II have a G+C content of 41%, compared with the host genome of 51%. Pai III or LEE (locus of enterocyte effacement) (35 kb) is a type II secretion system that mediates the injection of proteins (known as effector proteins). These are specialized systems for delivering specific virulence factors directly into the cytoplasm of the target host cell. Pai IV and Pai V are the sites of integration of conjugative transposons. Pai IV encodes hemolysin adhesion-associated Pap pili, whereas Pai V encodes hemolysin and cytotoxic necrotizing factor.

PAIs are also found in a number of other gram-negative pathogens. For example, five large PAIs have been found in strains of *Salmonella enterica* serovar Typhimurium.

## Bacterial Biofilms

In this chapter, various aspects of bacterial pathogenicity have been presented from the point of view that these microorganisms exist as isolated single cells suspended in an aqueous environment (i.e., the planktonic mode); however, most in vivo populations of bacteria grow as adherent biofilms. The most important features of a biofilm are that it forms at interfaces (solid-liquid, liquid-air, and solid-air) and that the bacteria within it display a marked decrease in susceptibility to antibacterial agents and host defense systems compared with their planktonic counterparts.

Examples of organisms that commonly form biofilms include *Pseudomonas aeruginosa, Staphylococcus epidermidis, S. aureus, E. coli, Lactobacillus* spp., and *Streptococcus* spp. Life-threatening infection caused by *P. aeruginosa* biofilms in cystic fibrosis patients is a well-known example. Bacterial biofilm formation is especially common in areas where indwelling medical devices are used, for example, catheters and prosthetic joints.

Biofilm cells typically show very slow growth and are under some sort of nutrient limitation. One obvious difference between planktonic cells and biofilms is the presence of a polysaccharide matrix enveloping the community, which retards diffusion of antibiotics into the biofilm. However, retarded diffusion alone would not explain resistance to antibacterial agents. At present the resistance is not well understood, but it has been suggested that (i) it can be due to inactivation of the antibiotic by enzymes trapped in the biofilm matrix, for example, inactivation of piperacillin by *P. aeruginosa* biofilms; (ii) that the reduced growth rate of bacteria in biofilms renders them less susceptible to the antibacterial agent; and (iii) that altered gene expression within the biofilm can result in a phenotype with reduced susceptibility to the agent. With regard to host defenses, bacteria within biofilms are less easily phagocytosed and less susceptible to complement (see chapter 3) than are their planktonic counterparts.

## Concluding Remarks

To summarize this brief description of virulence factors, inhibition of these factors is one potential therapeutic strategy in the search for novel targets for new antivirulence drugs such as vaccines and inhibitors of bacterial adhesion and of LPS synthesis.

### REFERENCES AND FURTHER READING

#### General (Books)

**Cossart, P., P. Boquet, S. Normark, and R. Rappuoli (ed.).** 2000. *Cellular Microbiology.* ASM Press, Washington, D.C.

**Kaper, J. B., and J. Hacker (ed.).** 1999. *Pathogenicity Islands and Other Mobile Virulence Elements.* ASM Press, Washington, D.C.

**Mims, C. A., A. Nash, and J. Stephen.** 2000. *Mims' Pathogenesis of Infectious Disease.* Academic Press, Inc., San Diego, Calif.

**Oelschlaeger, T. A., and J. Hacker.** 2000. *Bacterial Invasion into Eukaryotic Cells.* Kluwer Academic/Plenum Publishers, New York, N.Y.

**Raoult, D.** 1993. *Antimicrobial Agents and Intracellular Pathogens.* CRC Press, Inc., Boca Raton, Fla.

**Salyers, A. A., and D. D. Whitt.** 2002. *Bacterial Pathogenesis: a Molecular Approach,* 2nd ed. ASM Press, Washington, D.C.

## Original Articles and Reviews

### General

**Donnenberg, M. S.** 2000. Pathogenic strategies of enteric bacteria. *Nature* **406:**768–774.

**Falkow, S.** 1991. Bacterial entry into eukaryotic cells. *Cell* **65:**1099–1102.

**Finlay, B. B., and P. Cossart.** 1997. Exploitation of mammalian host cell functions by bacterial pathogens. *Science* **276:**718–725.

**Finlay, B. B., and S. Falkow.** 1997. Common themes in microbial pathogenicity revisited. *Microbiol. Mol. Biol. Rev.* **61:**136–169.

**Finlay, B. B., and S. Falkow.** 1989. Common themes in microbial pathogenicity. *Microbiol. Rev.* **53:**210–230.

**Janeway, C. A., P. Travers, M. Walport, and M. Shlomchik.** 2001. *Immunobiology: the Immune System in Health and Disease,* 5th ed. Garland Science Publishing, New York, N.Y.

### Virulence factors

**Goldschmidt, R. M., M. J. Macielag, D. J. Hlasta, and J. F. Barrett.** 1997. Inhibition of virulence factors in bacteria. *Curr. Pharm. Design* **3:**125–142.

### Intracellular organisms

**Maurin, M., and D. Raoult.** 1997. Intracellular organisms. *Int. J. Antimicrob. Agents* **9:**61.

**McOrist, S.** 2000. Obligate intracellular bacteria and antibiotic resistance. *Trends Microbiol.* **8:**483–486.

### Bacterial pili

**Abraham, S. N., A. B. Jonsson, and S. Normark.** 1998. Fimbriae-mediated host pathogen cross-talk. *Curr. Opin. Microbiol.* **1:**75–81.

**Choudhury, D., A. Thompson, V. Stojanoff, S. Langermann, J. Pinkner, S. J. Hultgren, and S. D. Knight.** 1999. X-ray structure of the FimC-FimH chaperone-adhesin complex from uropathogenic *Escherichia coli. Science* **285:**1061–1066.

**Forest, K. T., S. A. Dunham, M. Koomey, and J. A. Tainer.** 1999. Crystallographic structure reveals phosphorylated pilin from *Neisseria:* phosphoserine sites modify type IV pilus surface chemistry and fibre morphology. *Mol. Microbiol.* **31:**743–752.

**Hultgren, S. J., and S. Normark.** 1991. Chaperone-assisted assembly and molecular architecture of adhesive pili. *Annu. Rev. Microbiol.* **45:**383–415.

**Sauer, F. G., A. M. A. Mulvey, J. D. Schilling, J. J. Martinez, and S. J. Hultgren.** 2000. Bacterial pili: molecular mechanisms of pathogenesis. *Curr. Opin. Microbiol.* **3:**65–72.

**Sauer, F. G., K. Fütterer, J. S. Pinkner, K. W. Dodson, S. J. Hultgren, and G. Waksman.** 1999. Structural basis of chaperone function and pilus biogenesis. *Science* **285:**1058–1061.

### Pilus and nonpilus bacterial adhesins

**Hamburger, Z. A., M. S. Brown, R. R. Isberg, and P. J. Bjorkman.** 1999. Crystal structure of invasin: a bacterial integrin-binding protein. *Science* **286:**291–295.

**Hultgren, S. J., S. Abraham, M. Caparon, P. Falk, J. W. St Geme, and S. Normark.** 1993. Pilus and nonpilus bacterial adhesins: assembly and function in cell recognition. *Cell* **73:**887–901.

**Hultgren, S. J., C. H. Jones, and S. Normark.** 1996. Bacterial adhesins and their assembly, p. 2730–2756. *In* F. C. Neidhardt, R. Curtiss III, J. L. Ingraham, E. C. C. Lin, K. B. Low, B. Magasanik, W. S. Reznikoff, M. Riley, M. Schaechter, and H. E. Umbarger (ed.), Escherichia coli *and* Salmonella: *Cellular and Molecular Biology,* 2nd ed. ASM Press, Washington, D.C.

**Jacques, M.** 1996. Role of lipo-oligosaccharides and lipopolysaccharides in bacterial adherence. *Trends Microbiol.* **4:**408–410.

**Luo, Y., E. A. Frey, R. A. Pfuetzner, A. L. Creagh, D. G. Knoeche, C. A. Haynes, B. B. Finlay, and N. C. Strynadka.** 2000. Crystal structure of enteropathogenic *Escherichia coli* intimin-receptor complex. *Nature* **405:**1073–1077.

**Soto, G. E., and S. J. Hultgren.** 1999. Bacterial adhesins: common themes and variations in architecture and assembly. *J. Bacteriol.* **181:**1059–1071.

**Striker, R., U. Nilsson, A. Stonecipher, G. Magnusson, and S. J. Hultgren.** 1995. Structural requirements for the glycolipid receptor of human uropathogenic *Escherichia coli. Mol. Microbiol.* **16:**1021–1029.

### Lectins

**Lee, Y. C., and R. T. Lee.** 1995. Carbohydrate-protein interactions: basis of glycobiology. *Acc. Chem. Res.* **28:**321–327.

**Sharon, N., and H. Lis.** Lectins. Accepted for publication in *Encyclopedia of Life Sciences.* Nature Publishing, London, United Kingdom.

**Sharon, N., and H. Lis.** 1989. Lectins as cell recognition molecules. *Science* **246:**227–234.

### Adhesion and pilus assembly inhibitors

**Entenäs, H., G. Soto, S. J. Hultgren, G. R. Marshall, and F. Almqvist.** 2000. Stereoselective synthesis of optically active β-lactams, potential inhibitors of pilus assembly in pathogenic bacteria. *Organic Lett.* **2:**2065–2067.

**Kihlberg, J., and G. Magnusson.** 1996. Use of carbohydrates and peptides in studies of adhesion of pathogenic bacteria and in efforts to generate carbohydrate-specific T cells. *Pure Appl. Chem.* **68:**2119–2128.

### Bacterial exotoxins

**Alouf, J. E., and J. H. Freer.** 1999. *Bacterial Protein Toxins,* 2nd ed. Academic Press, Ltd., London, United Kingdom.

**Boquet, P., P. Munro, C. Fiorentini, and I. Just.** 2000. Toxins from anaerobic bacteria: specificity and molecular mechanism of action. *Curr. Opin. Microbiol.* **1:**66–74.

**Bacterial endotoxins**

**Kastowsky, M., T. Gutberlet, and H. Bradaczek.** 1992. Molecular modeling of the three-dimensional structure and conformational flexibility of bacterial lipopolysaccharide. *J. Bacteriol.* **174:**4798–4806.

**Raetz, C. R. H.** 1993. Bacterial endotoxins: extraordinary lipids that activate eucaryotic signal transduction. *J. Bacteriol.* **175:**5745–5753.

**Raetz, C. R. H.** 1990. Biochemistry of endotoxins. *Annu. Rev. Biochem.* **59:**129–170.

**Rietschel, E. T., and H. Brade.** 1992. Bacterial endotoxins. *Sci. Am.* **1992**(Aug.):26–33.

**The chaperone/usher pathway**

**Knight, S. A., J. Berglund, and D. Choudhury.** 2000. Bacterial adhesins: structural studies reveal chaperone function and pilus biogenesis. *Curr. Opin. Chem. Biol.* **4:**653–660.

**Sauer, F. G., M. Barnhart, D. Choudhury, S. D. Knight, G. Waksman, and S. J. Hultgren.** 2000. Chaperone-assisted pilus assembly and bacterial attachment. *Curr. Opin. Struct. Biol.* **10:**548–556.

**Thanassi, D. G., E. T. Saulino, and S. J. Hultgren.** 1998. The chaperone/usher pathway: a major terminal branch of the general secretory pathway. *Curr. Opin. Microbiol.* **1:**223–231.

**Pathogenicity islands**

**Hensel, M.** 2000. *Salmonella* pathogenicity island 2. *Mol. Microbiol.* **36:**1015–1023.

**Hentschel, U., and J. Hacker.** 2001. Pathogenicity islands: the tip of the iceberg. *Microbes Infect.* **3:**545–548.

**Winstanley, C., and C. A. Hart.** 2001. Type III secretion systems and pathogenicity islands. *J. Med. Microbiol.* **50:**116–127.

**Bacterial biofilms**

**Brooun, A., S. Liu, and K. Lewis.** 2000. A dose-response study of antibiotic resistance in *Pseudomonas aeruginosa* biofilms. *Antimicrob. Agents Chemother.* **44:**640–646.

**Costerton, J. W., Z. Lewandowski, D. E. Caldwell, D. R. Korber, and H. M. Lappin-Scott.** 1995. Microbial biofilms. *Annu. Rev. Microbiol.* **49:**711–745.

**Costerton, J. W., P. S. Stewart, and E. P. Greenberg.** 1999. Bacterial biofilms: a common cause of persistent infections. *Science* **284:**1318–1322.

**Watnick, P., and R. Kolter.** 2000. Biofilm, city of microbes. *J. Bacteriol.* **182:**2675–2679.

chapter 3



# Essentials of the Immune System

This chapter presents the basic principles of the immune response of the host through the innate (or nonspecific) and adaptive (or specific) immune systems, particularly as they relate to bacterial infection. It also introduces the concepts for an integral understanding of host-bacterium relationships, focused on intracellular and extracellular bacteria, as well as antibacterial drug-bacterium relationships, with emphasis on the intracellular location of some bacteria that allow them to resist antibiotics and synthetic antibacterial agents with poor ability to penetrate eukaryotic cell membranes, such as the β-lactam compounds. For more detailed discussions of fundamental immunology, readers are referred to texts on immunology or chapters on immune responses in texts on microbiology.

Host-bacterium relationships may be altered by antibacterial drugs in three major ways: alteration of the tissue response, alteration of the immune response, and alteration of the microbial flora. As an example, the inflammatory response of the tissue to an infection may be altered if the antibacterial agent suppresses the multiplication of the bacteria but does not eliminate them from the body, and in this way an acute process may be transformed into a chronic one. Conversely, the suppression of inflammatory reactions in tissues by impairment of cell-mediated immunity in recipients of tissue transplants and in patients receiving antineoplastic therapy or experiencing an immunocompromising disease (e.g., AIDS) causes enhanced susceptibility to infections and impaired responsiveness to antibacterial drugs.

Historically, immunity meant protection from diseases, more specifically, infectious disease caused by microorganisms such as viruses, bacteria, pathogenic fungi and protozoans.

In the struggle against microbial invaders, human hosts protect themselves by the action of specific cells and molecules collectively called the **immune system.** Thus, the physiological function of the immune system is to defend the host against infectious microbes. However, even noninfectious foreign substances can elicit immune responses. Therefore, a more inclusive definition of immunity is "a reaction to foreign substances," including microbes, as well as macromolecules such as proteins and polysaccharides.

**Immunology** is the study of immunity in a broader sense and of the cellular and molecular events that occur after an organism encounters microbes and other foreign macromolecules. **Immunity** can be natural (often called innate or nonspecific) or acquired (adaptive or specific). Natural immunity is not acquired through contact with an antigen. It is nonspecific and includes barriers to infectious agents (for example, skin and mucous membranes), natural killer (NK) cells, phagocytosis, inflammation, and a variety of other nonspecific factors.

## The Immune System: an Overview

The immune system consists of innate immunity and adaptive (specific) immunity. Adaptive immunity, which occurs after exposure to a foreign antigen (e.g., a foreign microorganism) is specific and is mediated by either antibodies (molecules) or lymphoid cells (cellular) (Fig. 3.1). Unlike innate immune responses, adaptive immune responses are reactions to specific antigenic challenges and display four characteristic attributes:

- antigenic specificity
- diversity
- immunologic memory
- self/nonself recognition

## Immunodeficiencies

Defects in any components of the immune system can result in its failure to recognize and respond properly to antigens. Such **immunodeficiencies** can make a person unusually prone to infections. Most genetic errors associated with these immunodeficiencies are located on the X chromosome and produce primary or congenital immunodeficiencies. Other immunodeficiencies can be acquired as a consequence of infections by immunosuppressive microorganisms such as human immunodeficiency virus (HIV).

## Immunocompromised Hosts

The term **immunocompromised hosts** describes individuals with nonspecific and/or specific immunity defects. These individuals are at increased risk of infection with a variety of microorganisms, including microorganisms that are not pathogenic for healthy individuals (i.e., opportunistic agents). Immunocompromised hosts include persons with leukemia, lymphoma, multiple myeloma, cancer metastases, or bone marrow transplants; patients receiving immunosuppresive or cytotoxic drug therapy; those with advanced diabetes mellitus; and those with HIV infection, which causes acquired immunodeficiency syndrome (AIDS) by destroying one type of immune cells, the CD4 T lymphocytes, involved in the immune response. Because of this preexisting host damage from HIV infection, AIDS patients do not have effective resistance to an infection; they generally die after contracting some infectious agent.

## Innate Immunity

Innate immunity is the defense mechanism that exists before an infection occurs. It reacts in essentially the same way in response to repeated infections.

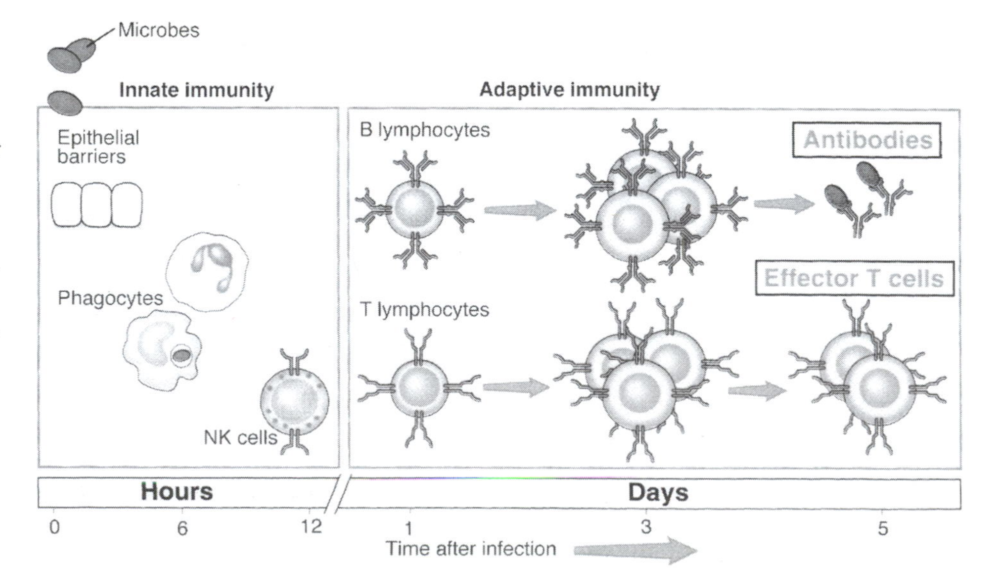


**Figure 3.1** Innate and adaptive immunity. The mechanisms of innate immunity provide the initial defense against infections. Only selected mechanisms are shown; for example, the complement system, an important component of the innate immunity, is not included. The kinetics of the innate and adaptive immunity are approximations and may vary in different infections. Reprinted from A. K. Abbas, A. H. Lichtman, and J. S. Pober, *Cellular and Molecular Immunobiology,* 4th ed. (The W. B. Saunders Co., Philadelphia, Pa., 2000), with permission from the publisher.

Innate immunity refers to natural defenses against pathogenic organisms that cause infection. Innate immunity provides the first line of defense during the critical period just after exposure of the host to a pathogenic microorganism and is immediately triggered when the protective barriers are breached. Generally, most of the microorganisms encountered by a healthy individual are readily cleared within few days by nonspecific defense mechanisms before they activate the adaptive immune system. If an invading microorganism eludes the innate mechanism or is not cleared by it, the specific immune response of adaptive immunity is triggered. In contrast to innate immunity, adaptive immunity takes days to become effective while the antigenic message is acting and specific weapons are being developed.

Innate immunity has four principal components:

- natural anatomical barriers also known as physical barriers (skin and mucous membranes)
- physiological barriers (temperature, low pH, etc.)
- phagocytosis
- inflammation

## Anatomical Barriers

Before a bacterium invades the host and causes infection, it must attach to and penetrate the surface of epithelial layers of the body. Bacteria enter the body by active or passive means. Whatever the point of entry, they have to cross physical barriers.

### SKIN

The intact skin constitutes an effective mechanical barrier against invading microorganisms. The skin consists of two different layers: a thin outer layer, the epidermis, and a thicker layer, the dermis. The epidermis contains several layers of tightly packed epithelial cells. The dermis, which is composed of connective tissue, contains blood vessels, hair follicles, sebaceous glands, and sweat glands. The sebaceous glands produce an oily secretion called sebum, which consists of lactic and fatty acids, which maintain the pH of the skin between 3 and 5; this pH inhibits the growth of most microorganisms. A few bacteria that metabolize sebum live as commensals on the skin and sometimes cause a severe form of acne.

Because very few microorganisms have the innate ability to penetrate the skin, most of them must gain access by some physical means such as via an arthropod vector, trauma, surgical incision, or intravenous catheter.

### MUCOUS MEMBRANES

The mucous membranes of the respiratory, digestive, and urogenital systems withstand microorganisms because the intact stratified squamous epithelium and mucous secretions form a protective covering that resists penetration and traps many microorganisms.

Bacteria adhere to mucous membranes via interactions between fimbriae or pili in the bacteria and certain glycoproteins or glycolipids that are expressed only by epithelial cells of the mucous membranes of particular tissues. For this reason, some tissues are susceptible to bacterial invasion whereas others are not.

In the respiratory tract, a film of mucus covers the surface and is constantly being driven upward toward the natural orifices by ciliated cells. Bacteria tend to stick to this film. In addition, mucus and tears contain lysozyme, which has antibacterial properties. Lysozyme is an enzyme that catalyzes the hydrolysis of the $\beta 1 \rightarrow 4$ glycosidic bond connecting *N*-acetylmuramic acid and *N*-acetylglucosamine in the peptidoglycan, especially in gram-positive bacteria.

### THE GASTROINTESTINAL TRACT

Once bacteria as well as other microorganisms reach the stomach, many are killed by its gastric juice (a mixture of hydrochloric acid, enzymes, and mucus). The very high acidity of gastric juice (pH 2 to 3) is usually sufficient to destroy most organisms and their toxins. However, many organisms are protected by food particles and reach the small intestine.

Once in the small intestine, pathogens often are damaged by various pancreatic enzymes, bile, enzymes in intestinal secretions, and secretory immunoglobulin A (IgA) antibody. In addition, the normal microbiota of the large intestine is very important in preventing the establishment of pathogenic organisms.

### THE GENITOURINARY TRACT

Under normal circumstances the kidneys, urethra, and urinary bladder of mammals are sterile, and the urine in the bladder is also sterile. However, in both males and females, a few bacteria are usually present in the distal portion of the urethra. The factors responsible for the sterility of the genitourinary tract are complex.

The vagina has another unique defense. Under the influence of estrogens, the vaginal epithelium produces increased amounts of glycogen, which is degraded to lactic acid by the acid-tolerant bacterium *Lactobacillus acidophilus*. Thus, an acidic environment (pH 3 to 5) unfavorable to most organisms is established.

### THE EYES

Constant bathing of the eyes by tears is an effective means of protection. Foreign substances are continuously diluted and washed away via tear ducts into the nasal cavity. Tears also contain a large amount of lysozyme.

THE NORMAL INDIGENOUS MICROBIOTA

It should be remembered that mucous membranes of the body carry a constant normal microbial flora (nonpathogenic bacteria) that hinders the establishment of pathogenic microorganisms. Some means by which normal microbiota may inhibit pathogenic colonization include

- producing bacteriocins that are toxic to other bacteria
- competing with potential pathogens for space and nutrients
- inhibiting infections by preventing pathogens from attaching to host surfaces
- influencing specific clearing mechanisms that help the body or a particular area to get rid of pathogens

### Phagocytosis

Bacteria and other microorganisms that enter the lymphatic systems, lungs, or bloodstream are engulfed and killed by a variety of **phagocytic cells** (monocytes, macrophages, tissue macrophages, and neutrophils). They do this via surface receptors that allow them to attach nonspecifically to a variety of microorganisms.

**Phagocytosis** is the ingestion of particulate material, which may include the whole bacterium. The steps involved in the phagocytosis process of bacteria include

1. Attachment of a bacterium to long membrane evaginations
2. Ingestion of bacterium and formation of a phagosome, which moves toward a lysosome
3. Fusion of the phagosome and lysosome to form a phagolysosome, which releases lysosomal enzymes (a variety of hydrolases, phospholipase A, ribonuclease [RNase], deoxyribonuclease [DNase], and proteases) into the phagosome; collectively, these enzymes participate in the destruction of the entrapped bacteria
4. Digestion of ingested material
5. Release of digestion products from the cell

The organelle lysosome is synthesized in the Golgi apparatus and endoplasmic reticulum. Besides the oxygen-independent lysosomal hydrolases, macrophages and neutrophil lysosomes contain oxygen-dependent enzymes that can produce reactive oxygen intermediates, such as superoxide radical ($O_2^-$), hydrogen peroxide ($H_2O_2$), singlet oxygen ($^1O_2$), and hydroxyl radical ($OH^{\cdot}$).

The phagocytic process can be greatly enhanced by **opsonization,** the process by which bacteria and viruses are coated by complement proteins and/or an antibody. This process increases the efficiency of phagocytosis of these microbes. The function of the complement system (C3b) is discussed later in this chapter.

### Inflammation

Tissue damage caused by a wound or by an invading pathogenic microorganism induces a complex sequence of events collectively known as the inflammatory response.

Inflammation is a general nonspecific reaction to tissue injuries, such as those caused by pathogens, wounds, foreign particles, and other noxious stimuli such as toxins. The characteristic **inflammatory response** results in redness, swelling, pain, and heat at the site of infection. The most important outcome of the inflammatory response is the immediate localization of the pathogen, often by the production of a fibrin clot at the site of the inflammation.

In summary, the release of inflammatory mediators (cytokines, interleukin-1 [IL-1], and tumor necrosis factor alpha [TNF-α]), as well as other mediators, elicits changes in local blood vessels. This begins with dilation of local arterioles and capillaries. A second effect is to induce changes in expression of various adhesion molecules on endothelial cells and on leukocytes. The reader is referred to immunology texts for additional information on inflammation process.

**Fever** is an abnormal increase in body temperature. It is the most common systemic manifestation of the inflammatory response. Although it can be caused by noninfectious disease, most fever is caused by infection. One of the reasons why fever occurs during many infections is that certain products of pathogenic organisms are pyrogenic (fever inducing). The best-studied pyrogenic agents are the endotoxins of gram-negative bacteria. The fever may be beneficial to the host since a slight temperature increase induces phagocytic and antibody responses. However, high temperatures (40°C, 104°F) usually benefit the pathogen because host tissues are further damaged.

## Adaptive Immunity

Adaptive immunity is capable of recognizing and selectively eliminating specific foreign microorganisms and molecules (i.e., foreign antigens). Unlike nonspecific immune responses, specific immune responses are reactions to specific antigens. The body recognizes these substances as not belonging to itself and develops a specific immune response, leading to destruction or neutralization of the substances.

The immune response can consist of antibody-mediated (humoral or molecular) and cell-mediated (cellular) responses that work in a concerted fashion. An encounter with a bacterium usually elicits a complex variety of responses. To illustrate the mechanisms of specific host defense, an overview of these is given in Color Plate 3.1 (see color insert), and details are presented later in this chapter.

Although innate immunity and adaptive immunity are often considered separately for convenience and for the sake of understanding, it is important to recognize that they frequently work together. For example, macrophages are phagocytic but also produce important cytokines that help to induce the adaptive immune response. Complement components of the innate immune system are activated by antibodies that are molecules of the adaptive system.

Adaptive immunity can be passive or active.

### Passive Immunity

**Passive immunity** is transmitted by antibody or lymphocytes preformed in another host.

### Active Immunity

**Active immunity** is induced after contact with antigens (e.g., bacteria or their products). This contact may occur as a consequence of clinical or subclinical infection, immunization with live or killed infectious agent or their antigens, exposure to microbial products (e.g., toxins and toxoids), or transplantation of foreign cells. In all these instances, the host actively produces antibodies, and lymphoid cells acquire the ability to respond to the antigens.

### Mechanisms of the Adaptive Immune Response

After entering the host and interacting with the nonspecific defense system described above, a bacterial cell or its major antigens are the target of the specific immune response.

An **antigen** (antibody generator) is a nonself foreign substance (from a bacterium), such as protein, nucleoprotein, lipid, and carbohydrate. Specialized **antigen-presenting cells,** such as **macrophages,** roam the body, ingesting the antigens they find and fragmenting them into antigenic peptides.

Pieces of these peptides are joined to the **major histocompatibility complex** (MHC) molecules and are displayed on the surface of the cell. Other white blood cells, called **T lymphocytes,** have receptor molecules that enable them to recognize a different peptide-MHC combination. T cells activated by that recognition divide and secrete **lymphokines,** which are chemical signals that mobilize other components of the immune system. One set of cells that respond to those signals comprises the **B lymphocytes,** which also have receptor molecules of single specificity on their surface. Unlike the receptors of T cells, however, those of B cells can recognize parts of antigens free in solution, without MHC molecules. When activated, the B cells divide and differentiate into plasma cells that secrete an **antibody** (also known as immunoglobulin). By binding to antigens, they find the antibodies scavenging cells. Some T and B cells become memory cells that persist in the circulation and boost the readiness of the immune system to eliminate the same antigen if it eventually come in contact again.

Two independent X-ray crystal structures offer powerful glimpses of one of the most important interactions in the immune system. In 1996, within a month of one another, Garcia et al. and Garboczi et al. published structures of the complex formed by the receptor of cytotoxic T lymphocytes (T cells) and an MHC molecule. Both crystal structures are of complexes containing class I MHC molecules.

In 2000, the Nobel Prize in medicine was awarded to Rolf M. Zinkernagel and Peter C. Doherty for the work they did in the early 1970s when they first identified that, in order for the immune system to be activated, an element from the T cell must recognize not only the foreign antigen but also a protein produced by the organism being infected (MHC). Subsequent work revealed that it is not the whole antigen but, rather, a small peptide fragment of it that the T cell recognizes.

### Immunogens and Antigens

Immunogens are substances that when administered to an individual, induce an immune response. Antigens are substances that react with either antibodies or antigen-specific receptors.

Some substances recognized by immune systems are not true immunogens. For example, **haptens** are low-molecular-weight substances that combine with specific antibody molecules but do not themselves induce antibody formation. Haptens include molecules such as sugars, amino acids, and other low-molecular-weight organic compounds.

The antibodies or antigen-specific receptors do not react with the antigen macromolecules as a whole but only with parts (or sectors) of the molecule called its **antigenic determinants** or **epitopes.**

### The Major Histocompatibility Complex

The MHC was originally recognized for its role in triggering T-cell responses that caused the rejection of transplanted tissue. It is now known that MHC class I and class II molecules bind peptide antigens and present them to T cells. The T-cell receptors recognize only the antigen presented by MHC molecules on another cell, the antigen-presenting cell. Th cells generally recognize antigens combined with class II molecules, whereas Tc cells generally recognize an antigen combined with class I molecules.

Two classes (classes I and II) of polymorphic MHC genes code for human leukocyte antigens (HLA), which act as peptide receptors and are critical to antigen presentation. Each MHC molecule consists of an extracellular peptide-binding cleft or groove and is anchored to the cell by the transmembrane and cytoplasmic domains (Fig. 3.2 and 3.3).

The class I glycoproteins are encoded by HLA-A, HLA-B, and HLA-C genes. These glycoproteins consist of two noncovalently linked polypeptide chains: an MHC-encoded α chain (or heavy chain) with molecular weight of 44,000 to 47,000 and noncovalently bound with a non-MHC-encoded polypeptide subunit with molecular weight of 12,000 called **$\beta_2$-microglobulin** (Fig. 3.2). Class I molecules are found in virtually all nucleated cells in the body. The heavy chain has a "binding groove" for peptides to be recognized by T cells.

The class II MHC molecules are heterodimer glycoproteins, consisting of α and β chains (Fig. 3.3). There are three kinds of class II genetic loci in humans (*DR*, *DP*, and *DQ*). Class II glycoproteins have a restricted tissue distribution and are found chiefly in macrophages, B cells, and other antigen-presenting cells.

The peptide-binding cleft of class I MHC molecules is formed by the $\alpha_1$ and $\alpha_2$ segments of the heavy chain, and that of class II MHC molecules is formed by the $\alpha_1$ and $\beta_1$ segments of the two chains. The Ig-like domains of class I and class II molecules contain the binding sites for the cell coreceptors CD8 and CD4, respectively.

## Cells and Tissues of the Immune System

The cells of the specific immune system are normally present as circulating cells in the blood and lymph, in lymphoid organs, and in virtually all tissues.

### BLOOD AND LYMPH

Blood consists of cellular and extracellular components. It contains many cells and many molecules involved in the immune response. The most numerous cells in human blood are **erythrocytes** (red blood cells), which are nonnucleated cells that carry oxygen from the lungs to the tissues. **Leukocytes** (white blood cells) include a variety of phagocytic cells such as monocytes and lymphocytes, which are involved in antibody production and cell-mediated immunity.

**Lymph** is a fluid similar to blood but lacks red blood cells.

### BONE MARROW

The bone marrow is the site of generation of all circulating blood cells in adults, including immature lymphocytes, and is the site of B-cell maturation. During fetal development, the generation of all blood cells (red and

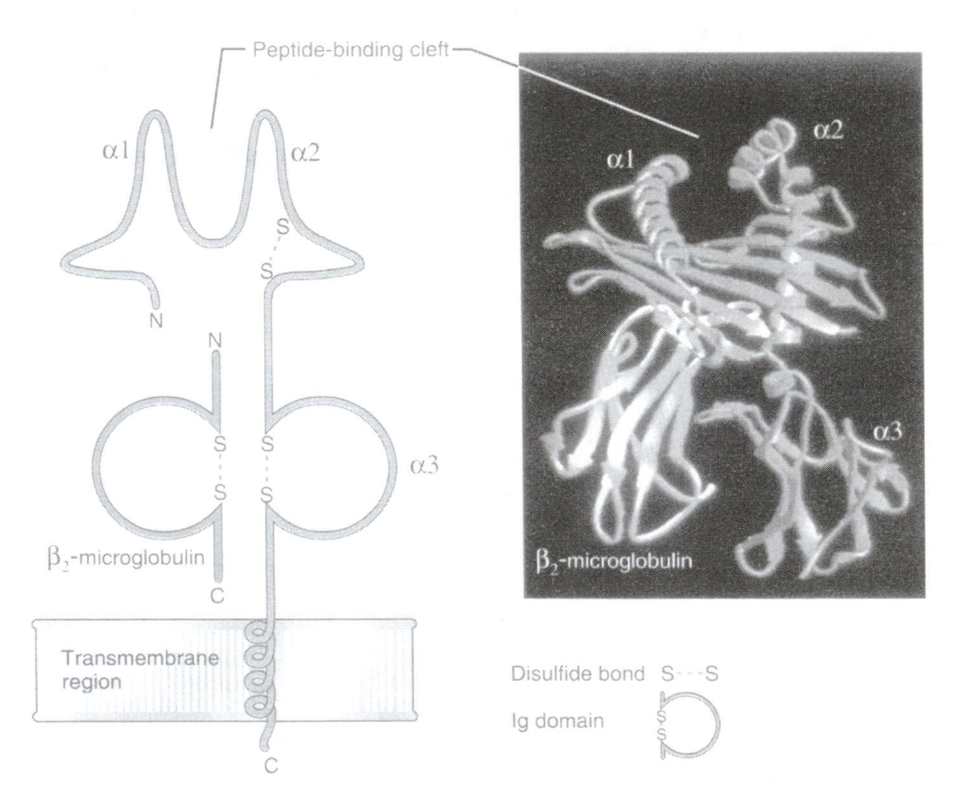


**Figure 3.2** Structure of a class I MHC molecule. The schematic diagram (left) illustrates the different regions of the MHC molecule (not drawn to scale). Class I molecules are composed of a polymorphic α chain noncovalently attached to nonpolymorphic $\beta_2$-microglobulin. The α chain contains glycosylated carbohydrate residues (not shown). The ribbon diagram (right) shows the structure of the extracellular portion of the HLA-B27 molecule resolved by X-ray crystallography. Reprinted from A. K. Abbas, A. H. Lichtman, and J. S. Pober, *Cellular and Molecular Immunobiology,* 4th ed. (The W. B. Saunders Co., Philadelphia, Pa., 2000), with permission from the publisher.

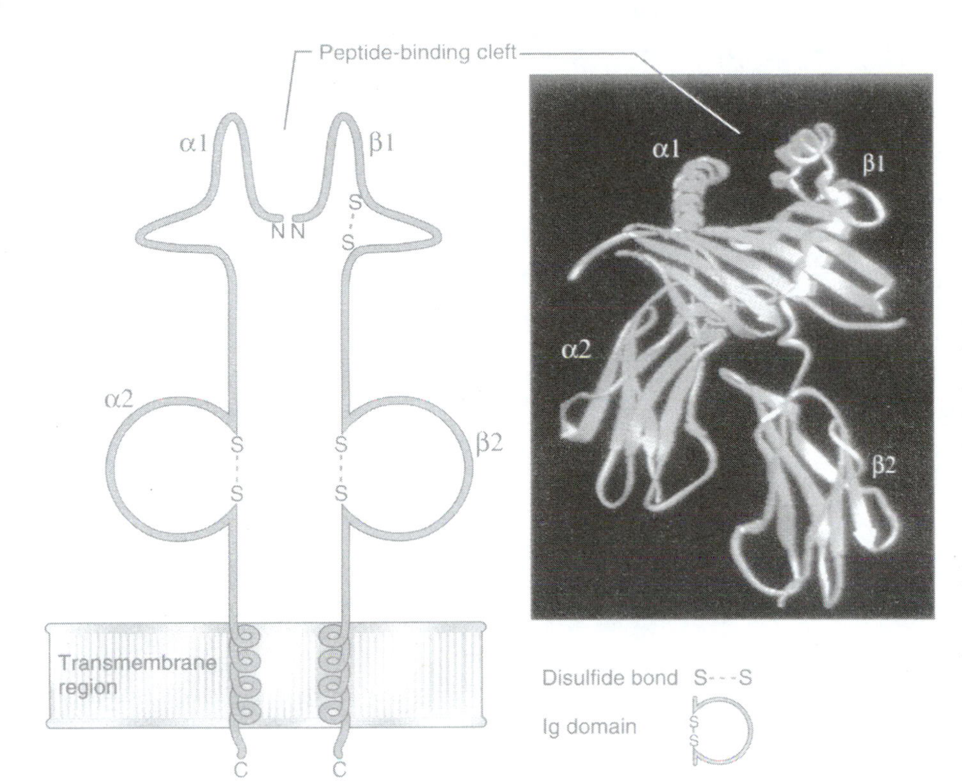


**Figure 3.3** Structure of a class II MHC molecule. The schematic diagram (left) illustrates the different regions of the MHC molecule (not drawn to scale). Class II molecules are composed of a polymorphic α chain noncovalently attached to a polymorphic β chain. Both chains contain glycosylated carbohydrate residues (not shown). The ribbon diagram (right) shows the structure of the extracellular portion of the HLA-DR1 molecule resolved by X-ray crystallography. Reprinted from A. K. Abbas, A. H. Lichtman, and J. S. Pober, *Cellular and Molecular Immunobiology,* 4th ed. (The W. B. Saunders Co., Philadelphia, Pa., 2000), with permission from the publisher.

white cells), called **hematopoiesis,** occurs initially in blood islands of the yolk sac and the para-aortic mesenchyme and later in the liver and spleen. This function is taken over gradually by the bone marrow and by the marrow of the flat bones. Blood cells originate from a type of cell called the **hematopoietic stem cell,** which becomes committed to differentiate along particular lineages.

## THE THYMUS

The thymus is the site of T-cell maturation.

## CELLULAR COMPONENTS

The principal cells of the adaptive immune system belong to two groups of cells: the lymphocytes and the antigen-presenting cells (APCs), also known as accessory cells.

### Lymphoid Cells

**Lymphocytes** are one of the many types of white blood cells produced in the bone marrow during hematopoiesis (Fig. 3.4). They are the central cells of the adaptive immune system and are responsible for the attributes of diversity, specificity, memory, and self/nonself recognition.

There are three types of lymphocytes—B cells, T cells, and NK cells—although only B and T cells have the two defining characteristic of the specific immune response, **antigen specificity** and **memory.**

Some lymphocytes migrate through the circulatory or lymphatic systems to the secondary lymphoid tissues (the spleen, mucosal lymphoid tissue, and lymph nodes), where they produce lymphocyte colonies. These lymphoid tissues are the sites where lymphocytes interact with antigens (Fig. 3.5).

*Classes of Lymphocytes.* *B lymphocytes.* The letter B in the term "B lymphocyte" was originally derived from the first letter of "bursa of Fabricius," a specialized appendage of the cloaca of birds and chickens where these lymphocytes differentiate. B cells are distributed by the blood and account for up 20 to 30% of the circulating lymphocytes. B cells are primarily responsible for the development of humoral (antibody-mediated) immunity. **B cells** or **B lymphocytes** mature in the liver before birth and after birth in the bone marrow.

**Naive lymphocytes** are lymphocytes which have never encountered their specific antigens and thus have never responded to them. All lymphocytes that leave the central lymphoid organs are naive lymphocytes; those from the thymus are naive T cells, and those from the bone marrow are naive B cells (Fig. 3.6).

*T lymphocytes.* The undifferentiated lymphocytes that migrate to the thymus undergo special processing and become **T cells** or **T lymphocytes** (the letter T refers

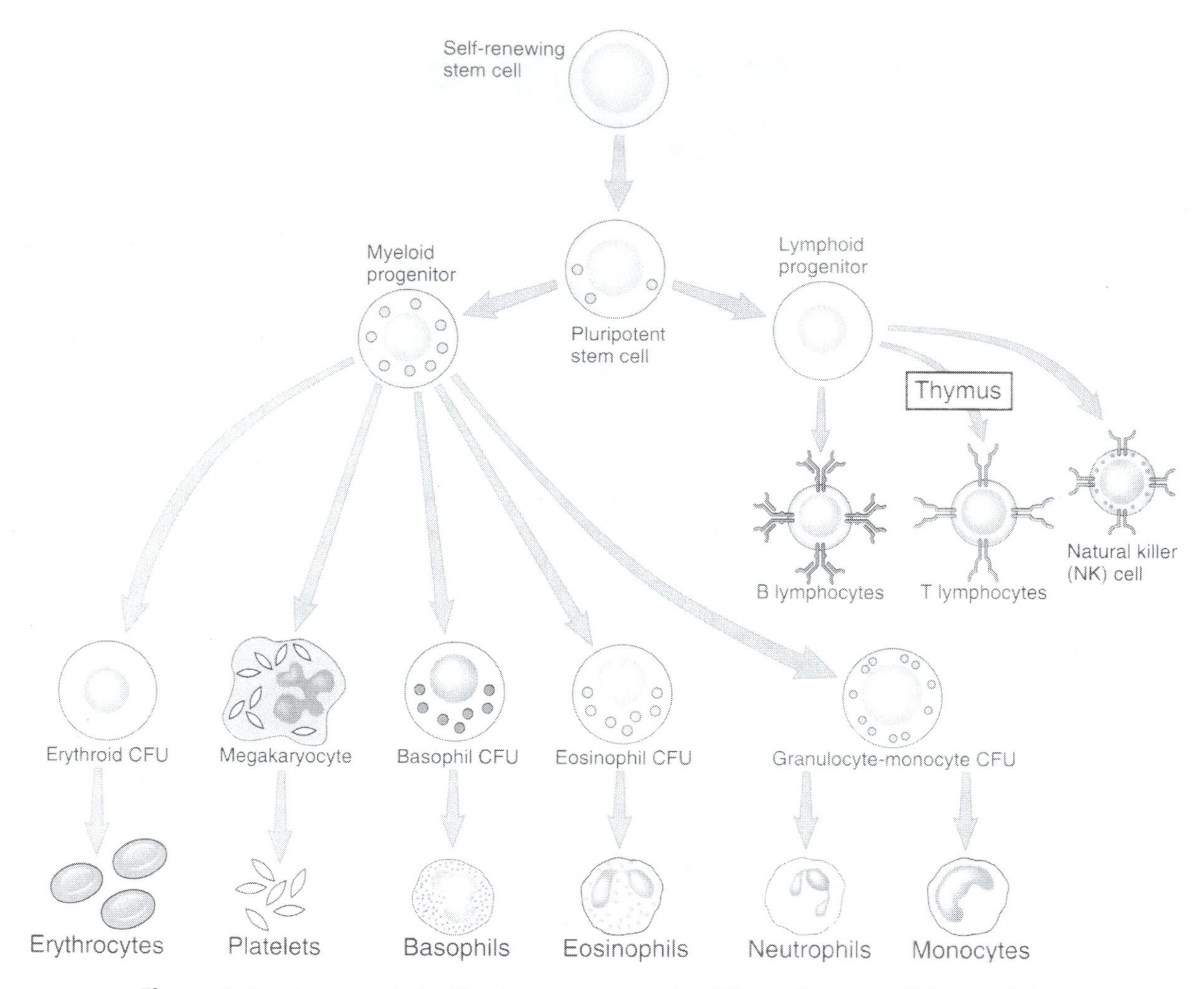


**Figure 3.4** Hematopoiesis. The development of the different lineages of blood cells is depicted in this "hematopoietic tree." Reprinted from A. K. Abbas, A. H. Lichtman, and J. S. Pober, *Cellular and Molecular Immunobiology,* 4th ed. (The W. B. Saunders Co., Philadelphia, Pa., 2000), with permission from the publisher.

to the thymus). Some T cells are transported away from the thymus and enter the bloodstream, where they comprise 70 to 80% of the circulating lymphocytes.

T lymphocytes, the effector cells of cell-mediated immunity, are further subdivided into functionally distinct populations, the **helper (Th) T cells** and **cytolytic (or cytotoxic) (Tc) T cells.** T cells do not produce antibody molecules (Fig. 3.6).

*Natural killer (NK) cells.* There is a third class of lymphoid cell called the NK cells, which do not have characteristics of either T or B cells. NK cells are involved in specific immunity against viruses and other intracellular microbes (Fig. 3.6).

*Activation and Mechanism of Action of Lymphocytes.*
*Activation of B lymphocytes.* A mature B cell can be activated by an encounter with an antigen that expresses epitopes recognized by its cell surface Ig or indirectly by an intimate interaction with a helper T cell. Once mature, B cells are stimulated by antigens or helper-T-cell-derived signals, including CD40 ligand (CD40L) and cytokines. On activation, the B cells are called activated B lymphocytes.

*B-cell activation by antigens.* The response of B cells to antigens was understood once the cell surface Igs were identified as the B cell antigen receptors.

The activation of antigen-specific B cells is initiated by binding of antigen to membrane immunoglobulin

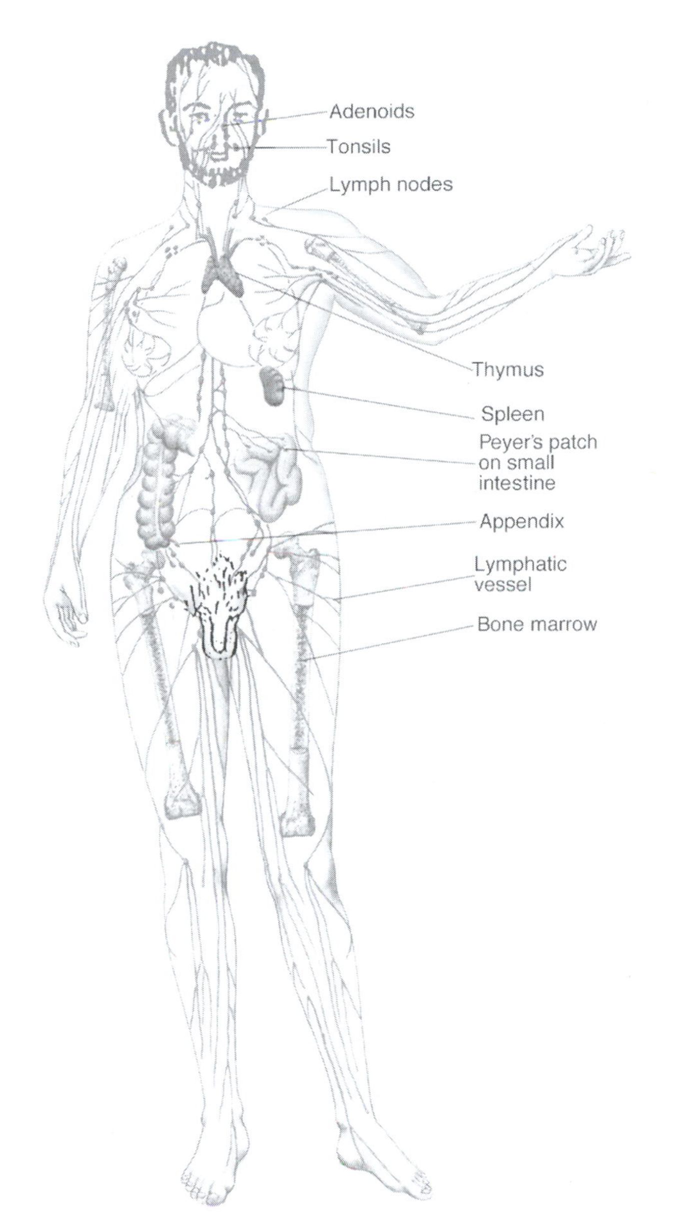


**Figure 3.5** Overall view of the lymphatic system, showing the locations of major organs.

molecules, which are the antigen receptors of mature B cells. Antigen contact with the specific B cell triggers the transmembrane signaling function of the **B-cell antigen receptor (BCR).** This, in turn, induces early events in B-cell activation. After antigen binding, the BCR molecules are internalized, leading to antigen uptake and degradation in endosomes or lysosomes. B-cell activation prepares the cell to divide and to differentiate either into **antibody-secreting cells** or into **memory cells.**

*Helper T-cell contact-mediated activation of B cells.* When antigen-derived peptides bind the cleft of class II MHC molecules, the antigen is taken up by macrophages and/or dendritic cells and processed into small peptide fragments, which then display the resulting peptide-class II MHC complexes on the cell surface. Then naive helper T cells become activated. These activated helper T cells search for and eventually find B cells. This pathway requires the expression of CD40 ligand by the T cell and receipt of this signal by CD40 on the B cell. The cytokines are secreted directly into the narrow space between the interacting lymphocytes (Fig. 3.7). This is a very simplified form of presenting B-cell activation; for more details about this topic, the reader should refer to immunology texts.

*Memory B cells.* Primary responses result from the activation of naive B cells, whereas secondary responses are due to stimulation of expanded clones of memory B cells. In the secondary response, memory cells give rise to greater immediate antibody production.

*B cells and antibody production.* Newly generated B cells initially express membrane IgM and thereafter express IgD. Class switching during an immune response leads to the appearance of memory B cells that express other classes of Ig (IgG, IgA, or IgE).

*The B1 subset of B cells.* B1 cells express IgM antibodies encoded by germ line antibody genes, mature independently of the bone marrow, generally recognize multimeric sugar and lipid antigens of microbes, and are T-cell independent.

*T cells.* T cells constitute the second major class of lymphocytes. T-cell precursors migrate from the bone marrow to the thymus, where some mature into antigen-specific T cells. These T cells then migrate to the secondary lymphoid organs and tissues of the body. As mentioned above, T cells may be subdivided into two distinct classes, the helper (Th) T cells and the cytolytic (or cytotoxic) (Tc) T cells. Th cells provide help for B cells by cell surface signaling and by producing cytokines which are critical for B-cell growth and differentiation.

Th cells that express the membrane glycoprotein CD4 ($CD4^+$ cells) are restricted to recognize antigen bound to class II MHC molecules, whereas Tc cells express CD8 ($CD8^+$ cells), a dimeric membrane glycoprotein, are restricted to recognition of antigen bound to class I MHC molecules, and recognize viral antigens presented on the surface of infected cells.

*T-cell receptors.* T-cell receptors (TCR) are found only on the T-cell membrane. TCR do not recognize

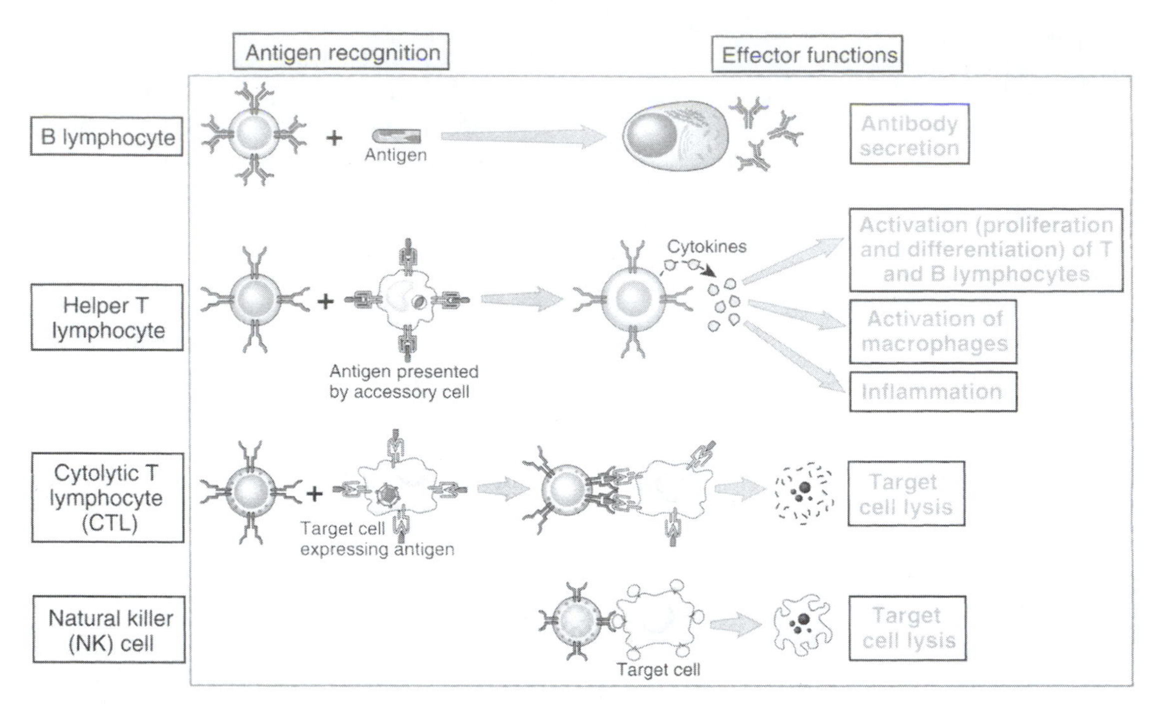


**Figure 3.6** Classes of lymphocytes. B lymphocytes recognize soluble antigens and develop into antibody-secreting cells. Helper T lymphocytes recognize antigens on the surface of host accessory cells and secrete cytokines, which stimulate different mechanisms of immunity and inflammation. Cytolytic T lymphocytes recognize antigens on target cells and lyse these targets. NK cells use receptors that are not fully identified to recognize and lyse targets. Reprinted from A. K. Abbas, A. H. Lichtman, and J. S. Pober, *Cellular and Molecular Immunobiology,* 4th ed. (The W. B. Saunders Co., Philadelphia, Pa., 2000), with permission from the publisher.

**Figure 3.7** Bidirectional molecular interactions between B and T lymphocytes. In the model for the role of multiple ligand-receptor pairs in T-cell-dependent B-cell activation, helper T cells recognize the antigen in the form of peptide-MHC complexes, CD40 ligand then binds to CD40 receptors in the B cells, and together with the liberation of cytokines mediated by T cells, B cells initiate proliferation and differentiation. B7-1 and B7-2 are now designated CD80 and CD86, respectively. Reprinted from A. K. Abbas, A. H. Lichtman, and J. S. Pober, *Cellular and Molecular Immunobiology,* 4th ed. (The W. B. Saunders Co., Philadelphia, Pa., 2000), with permission from the publisher.

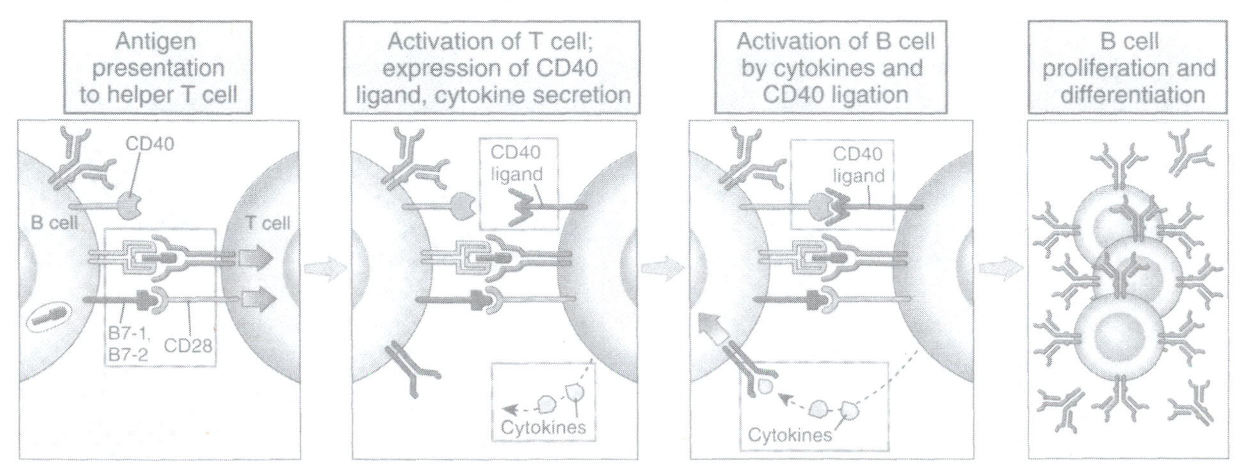

native antigens; they recognize only short peptides derived from the antigen (e.g., a bacterium) and bound to class II MHC molecules on the surface of APCs. There are two ways in which foreign proteins can get into APCs: (i) by entry of the whole bacterium or (ii) by being endocytosed (engulfed) by phagocytes. The proteins of the bacterium are degraded by cellular proteases, and the resulting peptides are then "loaded" onto class II MHC molecules. These complexes, after being transported to the surface of the cell, are the substances recognized by antigen-specific Th cells. Naive T lymphocytes migrate to encounter the processed antigen.

*Th-cell activation.* T cells express several integral membrane proteins, other than the members of the TCR complex, which play crucial roles in the responses of these cells to antigen recognition. These proteins are collectively called accessory molecules. Activation of T lymphocytes is a consequence of multiple ligand-receptor interactions that occur at the interface of the T cell and an APC (Fig. 3.8). The primary responses of the T cells are the recognition of peptide-MHC complexes on APCs in the lymphoid nodes. The principal consequences of the activation of naive cells are proliferation of the antigen-specific clone and differentiation of the progeny into effector and memory cells. Effector T cells enter the circulation, locate antigens processed in peripheral tissues (not limited to lymphoid organs), and become reactivated. On antigen recognition, effector cells of the $CD4^+$ subset secrete different sets of cytokines and perform distinct functions (Fig. 3.9). Th cells are of two types, type 1 (Th1) and type 2 (Th2). Th1 cells produce IFN-$\gamma$, IL-2, and tumor necrosis factor beta (TNF-$\beta$), which mediate macrophage activation, whereas Th2 cells produce IL-4, IL-5, IL-6, and IL-10, which act as growth and/or differentiation factors for B cells. $CD8^+$ cytolytic T lymphocytes kill cells that display class I MHC-associated foreign antigen. A similar heterogeneity in the cytokine profile among $CD8^+$ cytotoxic T cells (Tc1 and Tc2) has been reported.

## Antigen-Presenting Cells

APCs (**accessory cells**) are cells that do not express clonally distributed receptors for antigens but do participate in initiating the lymphocyte response to antigens. The major APC populations in the immune system are composed of mononuclear phagocytes and dendritic cells.

*Mononuclear Phagocytes.* The mononuclear phagocyte system consists of cells that have a common lineage whose primary function is phagocytosis. The cells of the mononuclear phagocyte system originate in the bone marrow, circulate in the blood, mature, and become activated in various tissues. The first cells that enter the peripheral blood after leaving the bone marrow are incompletely differentiated and are called **monocytes.** Once they settle in tissues, these cells mature and become **macrophages.**

Mononuclear phagocytes function as APCs in the recognition and activation phases of adaptive immune responses. Their main functions as APCs are to display antigen in a form that can be recognized by T lymphocytes and to produce membrane and secret proteins that serve as second signals for T-cell activation. Their effector functions in innate immunity are to phagocytose microbes and to produce cytokines that recruit and activate other inflammatory cells.

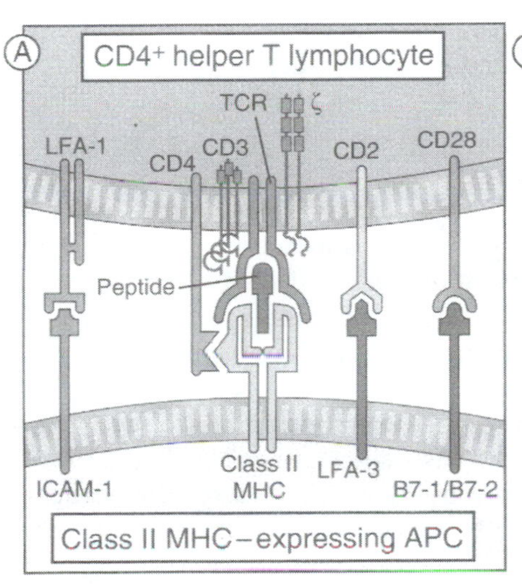


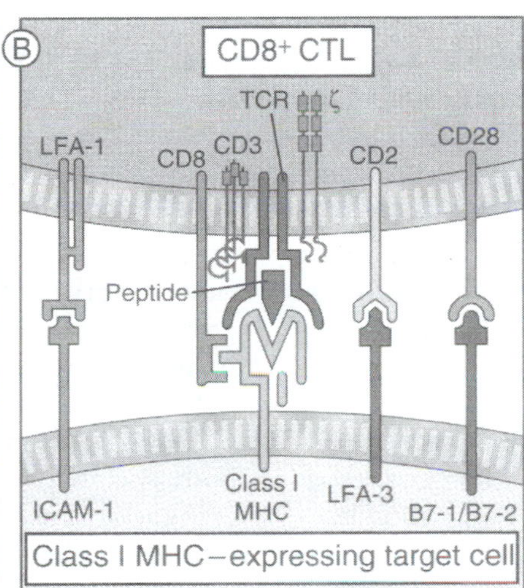


**Figure 3.8** Accessory molecules of T lymphocytes. The interaction of a $CD4^+$ helper T cell with an APC or of a $CD8^+$ cytolytic T lymphocyte (CTL) with a target cell involves multiple T-cell membrane proteins that recognize different ligands on the APC or target cell. Abbreviations: ICAM-1, intercellular cell adhesion molecule 1; LFA-1, lymphocyte function antigen 1. Reprinted from A. K. Abbas, A. H. Lichtman, and J. S. Pober, *Cellular and Molecular Immunobiology,* 4th ed. (The W. B. Saunders Co., Philadelphia, Pa., 2000), with permission from the publisher.

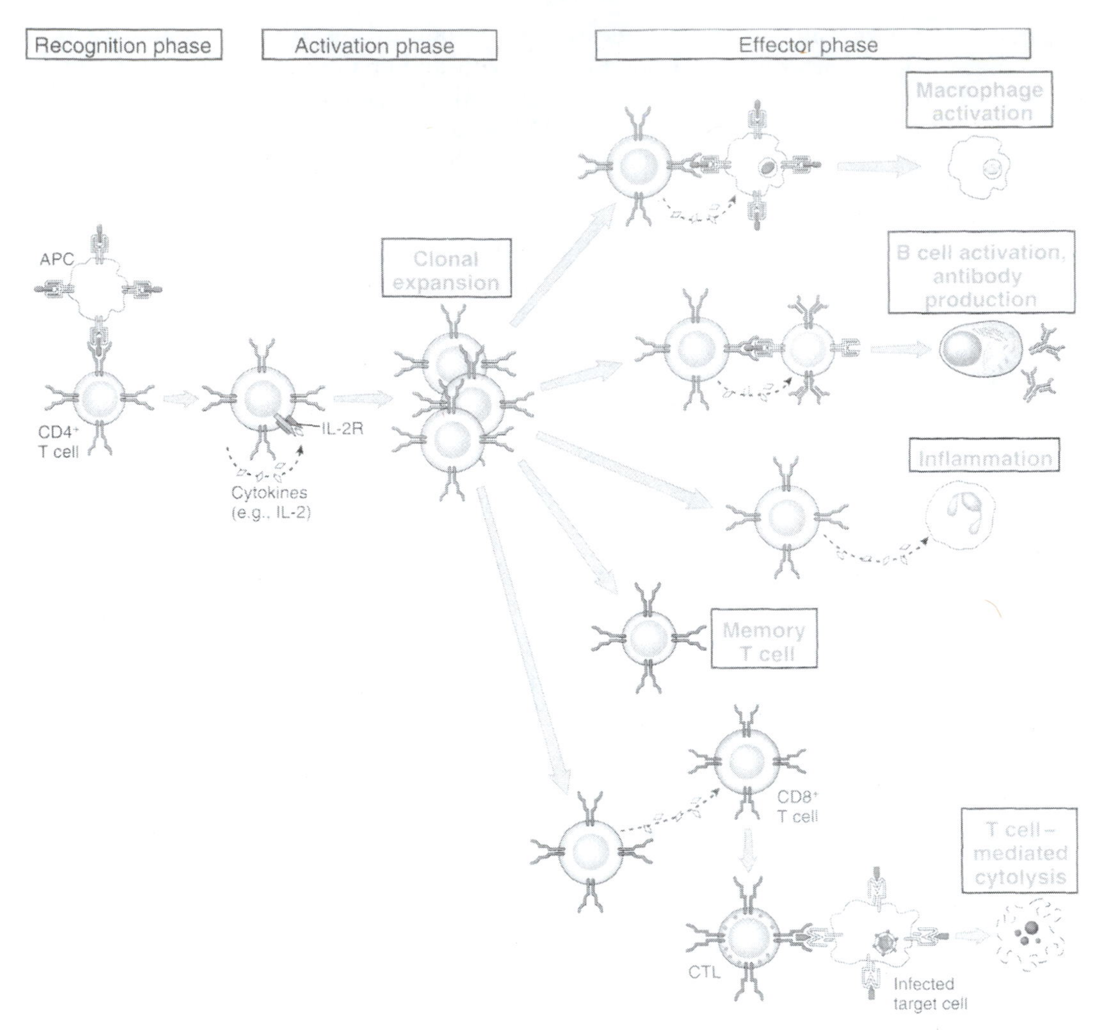


**Figure 3.9** Phases of T-cell responses. In the recognition and activation phases of T-cell responses, antigen recognition by a T cell, in this case a naive CD4+ cell, induces cytokine (e.g., IL-2) secretion, clonal expansion as a result of IL-2-induced autocrine cell proliferation, and differentiation of the T cells into effector cells or memory cells. In the effector phase of the response, the effector cell responds to antigen by producing cytokines that have several actions, such as activation of macrophages, B lymphocytes, and CD8+ T cells and induction of inflammation. Reprinted from A. K. Abbas, A. H. Lichtman, and J. S. Pober, *Cellular and Molecular Immunobiology,* 4th ed. (The W. B. Saunders Co., Philadelphia, Pa., 2000), with permission from the publisher.

*Dendritic Cells.* Dendritic cells are APCs that play important roles in the induction of the T-lymphocyte response to protein antigens. All dendritic cells are thought to arise from bone marrow precursors.

## Cytokines

The term **cytokine** is a generic term for the soluble proteins or glycoproteins released by one cell population that acts as an intercellular chemical mediator or signaling molecule.

When released from mononuclear phagocytes, these proteins are called **monokines;** when released from T lymphocytes, they are called **lymphokines;** when produced by a leukocyte and on another leukocyte, they are called **interleukins.** When their effect is to stimulate the growth and differentiation of immature leukocytes in the bone marrow, they are called **colony-stimulating factors** (CSFs).

Table 3.1 lists selected cytokines that are involved in both specific and nonspecific immunity. The topic of cytokines is not discussed in detail here; the reader is referred to chapter 12 of *Kuby Immunology* or any other immunology text.

**Table 3.1** Selected important cytokines

| Name[a] | Major cellular source(s) | Selected biological effect(s) |
|---|---|---|
| IFN-α, IFN-β | Phagocytes, fibroblasts | Antiviral, pyrogenic |
| IFN-γ | T cells, NK cells | Activation of mononuclear phagocytes |
| TNF-α | Phagocytes, NK cells | Cell activation, fever, cachexia, antitumor activity |
| TNF-β, LT | T cells, B cells | Activation of leukocytes, antitumor activity |
| TGF-β | T cells, macrophages | Leukocyte growth regulation, angiogenesis |
| IL-1 | Phagocytes | Cell activation, fever, cachexia |
| IL-2 | T cells | T-cell growth and activation |
| IL-3 | T cells | Hematopoiesis |
| IL-4 | T cells, mast cells | Isotype (class) switching to IgE |
| IL-6 | T cells, macrophages | Lymphocyte growth, differentiation |
| IL-8 | Macrophages | Chemotactic for neutrophils |
| IL-10 | Helper T cells | Inhibition of macrophages |
| IL-12 | B cells, macrophages | Differentiation of Th1 cells, activation of NK cells |
| GM-CSF | T cells, phagocytes, etc. | Hematopoiesis of the granulocyte and monocyte lineage |
| M-CSF | Macrophages, etc. | Differentiation to monocytes |
| G-CSF | T cells, phagocytes, etc. | Differentiation to granulocytes |

[a] LT, lymphotoxin; TGF-β, transforming growth factor β; GM-CSF, granulocyte-macrophage colony-stimulating factor.

## Antibodies

**Antibodies** (immunoglobulins) are a group of glycoproteins present in the blood serum and tissue fluids of mammals, which can combine with antigenic determinants. As mentioned above, *B lymphocytes are the only cells that synthesize antibody molecules.*

### GENERAL FEATURES OF ANTIBODY STRUCTURE

All antibody molecules have a common structure composed by four chains connected to each other by disulfide bonds (Fig. 3.10). This structure consists of two identical **light (L) chains,** which usually consist of about 220 amino acids and have a mass of approximately 25 kDa, and two identical **heavy (H) chains,** which consist of about 440 amino acids and have a mass of about 50 to 70 kDa. Both light and heavy chains contain two different regions. The **constant regions** ($C_L$ and $C_H$) have amino acid sequences that do not vary significantly between antibodies of the same subclass. The **variable regions** ($V_L$ and $V_H$) from different antibodies do have different sequences. This variability accounts for the capacity of different antibodies to bind to a tremendous number of structurally diverse antigens.

Detailed comparison of the amino acid sequences of $V_L$ and $V_H$ domains revealed that the sequence variability is concentrated in several **hypervariable (HV) regions.** The hypervariable regions form the antigen-binding site of the antibody molecule. Because the antigen-binding site is complementary to the structure of the epitope, the hypervariable regions are also called **complementarity-determining regions.**

Figure 3.11 is a computer-generated model of the antibody IgG structure showing the arrangement of the four polypeptide chains and the carbohydrate chain.

### ANTIBODY FUNCTIONS

All antibody molecules are functional. The Fab region is concerned with binding to an antigen, whereas the Fc region mediates binding to host tissues, various cells of the immune system, some phagocytic cells, or the first component of the complement system (Fig. 3.10).

### ANTIBODY CLASSES

Antibodies can be separated into five major classes on the basis of their physical, chemical, and immunological properties: IgG, IgA, IgM, IgD, and IgE (Fig. 3.12). **IgG** is the major antibody in human serum, accounting for 80% of the antibody pool. It is present in blood plasma and tissue fluids. The IgG class acts against bacteria and viruses by opsonizing the invaders and neutralizing toxins. It is also one of the two antibody classes that activate complement by the classical pathway (IgM is the other). IgG is the only antibody molecule that is able to cross the placenta and provide acquired passive immunity in utero and to the neonate at birth.

IgG antibodies are further divided into four immunologically distinct subclasses called IgG1, IgG2, IgG3, and IgG4. The structural characteristics that distinguish these subclasses from each other are the amino acid con-

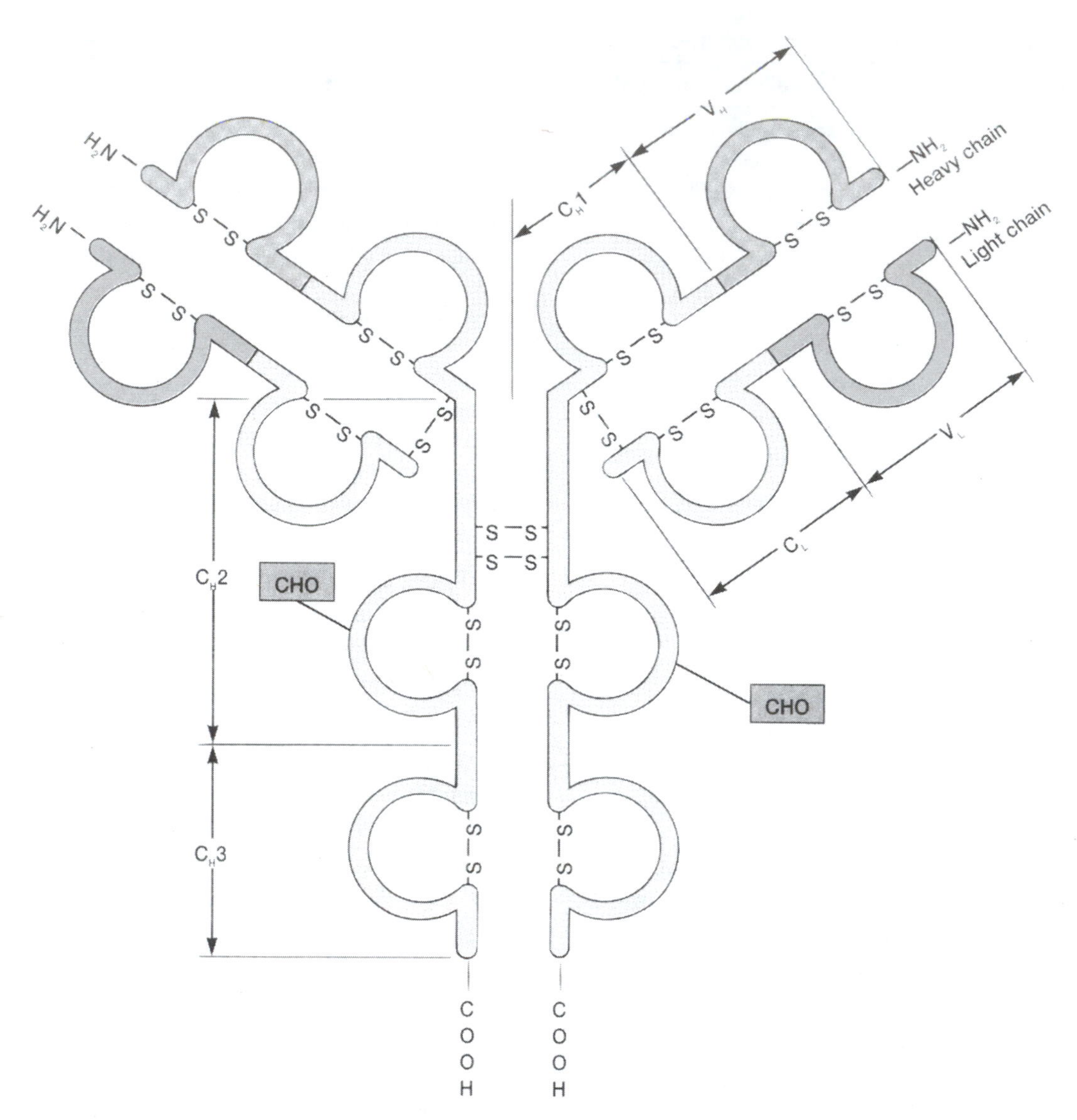


**Figure 3.10** Schematic diagram of an antibody (Ig) molecule. Within the antibody unit structure, intrachain disulfide bonds create loops that form domains. All light chains contain a single variable domain ($V_L$) and a single constant domain ($C_L$). Heavy chains contain a variable domain ($V_H$) and either three or four constant domains ($C_H1$, $C_H2$, $C_H3$, and $C_H4$). Reprinted from L. M. Prescott, J. P. Harley, and D. A. Klein, *Microbiology,* 4th ed. (McGraw-Hill, Boston, Mass., 1999), with permission from the publisher.

tent in their chain composition and the number and arrangement of interchain disulfide bonds (Fig. 3.12). Differences in biological function have been noted in these subclasses. For example, IgG2 antibodies are opsonic and develop in response to polysaccharide antigens. IgG1 and IgG3 bind to monocytes and macrophages and activate complement most effectively. The IgG4 antibodies function as skin-sensitizing antibodies.

**IgM** accounts for about 10% of the antibody pool. It is usually a polymer (pentamer) of five monomeric units, each composed of two heavy chains and two light chains (Fig. 3.13). IgM is the first antibody made during B-cell maturation and the first secreted into serum during a primary antibody response. Since IgM is so large, it does not leave the bloodstream or cross the placenta. It agglutinates bacteria, activates complement by the classical pathway, and enhances the ingestion of pathogens by phagocytic cells.

**IgA** accounts for about 15% of the antibody pool. Most (90 to 95%) of IgA is present in the serum as a monomer of two heavy and two light chains. However, 5

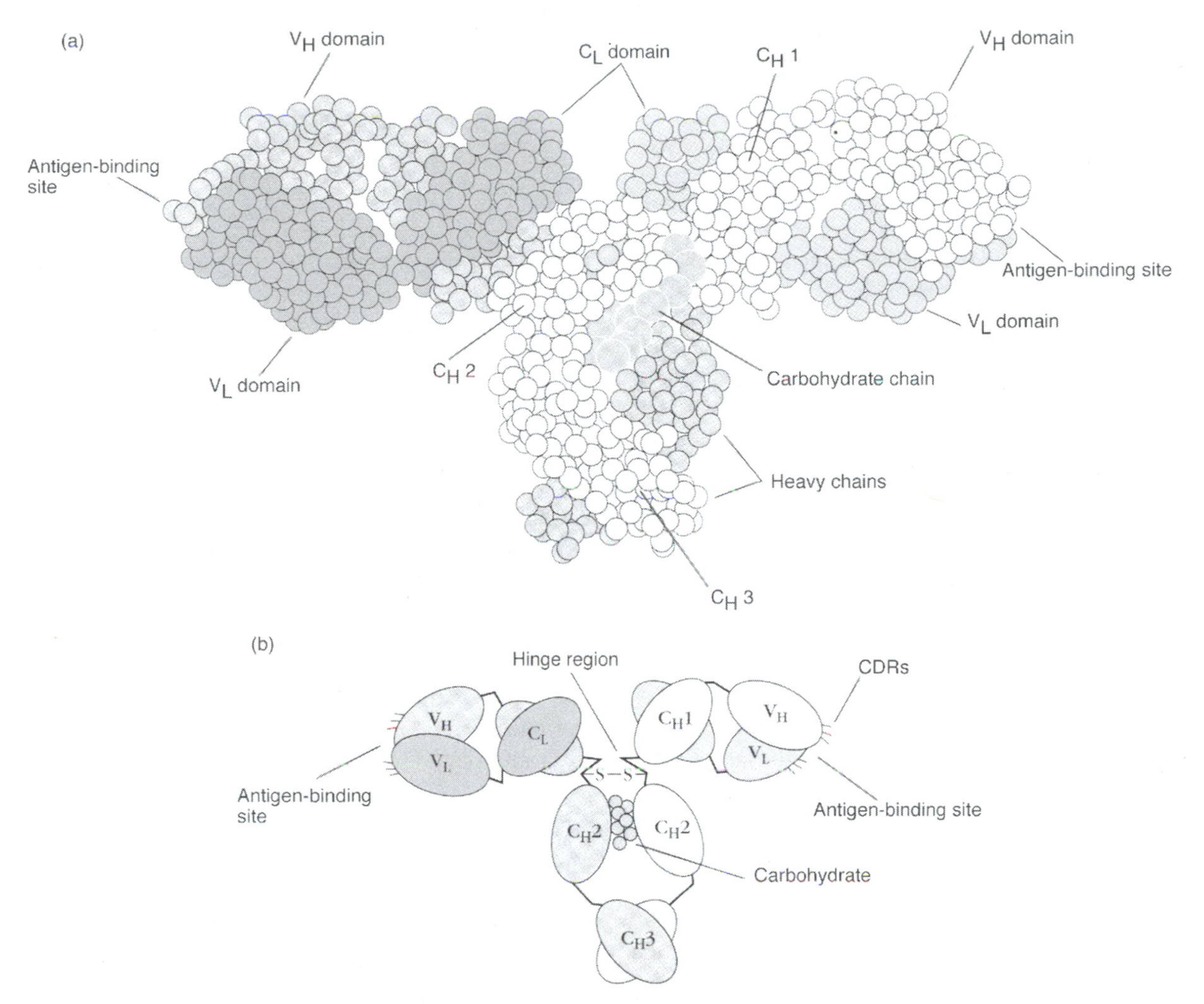


**Figure 3.11** Interactions between domains in the separate chains of an Ig molecule are critical to its quaternary structure. (a) Model of an IgG molecule based on X-ray crystallographic analysis, showing the association between domains. Each solid ball represents an amino acid; the larger balls represent carbohydrate. (b) A schematic diagram showing the interacting heavy- and light-chain domains. Note that the $C_H2/C_H2$ domains protrude because of the presence of carbohydrate in the interior. CDRs, complementarity-determining regions. Reprinted from R. A. Goldsby, T. J. Kindt, and B. A. Osborne, *Kuby Immunology,* 4th ed. (W. H. Freeman & Co., New York, N.Y., 2000), with permission from the publisher.

to 10% of IgA occurs in the serum as a polymerized dimer held together by a J chain (Fig. 3.13). When transported from the mucosa-associated lymphoid tissue to mucosal surfaces, IgA acquires a protein termed the secretory component. **Secretory IgA** is the primary antibody of the secretory immune system. This system is found in the gastrointestinal tract, upper and lower respiratory tracts, and genitourinary system. Secretory IgA is also found in saliva, tears, and breast milk. In these fluids and related body areas, secretory IgA plays a major role in protecting surface tissues against infectious microorganisms by the formation of an immune barrier.

**IgD** is an antibody found in trace amounts in the blood serum. It has a monomeric structure (Fig. 3.13). IgD antibodies do not fix complement and cannot cross the placenta, but they are abundant on the surface of B cells and bind antigens, thus signaling B cells to start antibody production.

**IgE** makes up only a small percentage of the total antibody pool (Fig. 3.13). The classic skin-sensitizing and anaphylactic antibodies belong to this class. IgE molecules have four constant-region domains and two light chains. The Fc portion of IgE can bind to special Fc receptors on mast cells and basophils. When two IgE

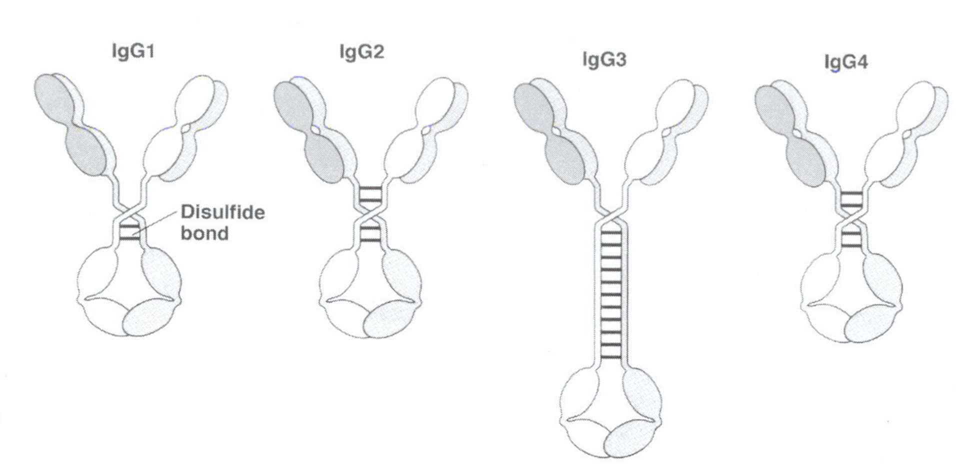


**Figure 3.12** General structures of the four subclasses of human IgG, which differ in the number and arrangement of interchain disulfide bonds (thick black lines) linking the heavy chains. A notable feature of human IgG3 is its 11 interchain disulfide bonds. Reprinted from R. A. Goldsby, T. J. Kindt, and B. A. Osborne, *Kuby Immunology,* 4th ed. (W. H. Freeman & Co., New York, N.Y., 2000), with permission from the publisher.

molecules on the surface of these cells are cross-linked by binding to the same antigen, the cells degranulate. This degranulation releases histamine and other pharmacological mediators of anaphylaxis. Thus, although IgE is present in small amounts, this class of antibody has very potent biological capabilities.

## The Complement System

The complement system was discovered many years ago as a heat-labile component of human blood plasma that augmented the opsonization of bacteria by antibodies and helped other antibodies to kill bacteria. This activity was said to "complement" the antibacterial activity of antibody, hence the name "complement." It is now known that the **complement system** is composed of many serum proteins that play major roles in the human defensive immune response.

The complement cascade is made up of at least 20 complement proteins designated C1 to C9, factor B, factor D, factor H, factor I, C4b-binding proteins, C1 INH complex, S protein, and properdin. The complement system acts in a cascade fashion, such that the activation of one component results in the activation of the next. Within plasma and other body fluids, complement proteins are in an inactive state.

#### OVERVIEW OF COMPLEMENT ACTIVATION

There are two main pathways of complement activation: the classical and the alternative pathways (Fig. 3.14).

The alternative pathway is activated by C3b binding to various activating surfaces, such as microbial cell walls, to lipopolysaccharide (LPS) of the bacterial cell outer membrane, or to some endotoxins. The classical pathway is activated by C1 binding to the antigen-antibody complexes. Cell-associated C3b becomes a component of the enzyme that cleaves C5 (C5 convertase) and initiates the late steps of complement activation. The late steps of both pathways are the same (not shown here), and the complement activated by the two pathways serves the same functions. For a detailed discussion of the activation of the complement system, the reader is referred to the references at the end of this chapter.

#### FUNCTIONS OF THE COMPLEMENT SYSTEM

The principal effector functions of the complement system in innate and specific humoral immunity are to promote phagocytosis of microbes, stimulate inflammation, and induce lysis of the microbes.

### Opsonization and Phagocytosis

As noted above, phagocytes have an intrinsic ability to bind directly to microorganisms via nonspecific cell surface receptors, form phagosomes, and digest the microorganisms. This phagocytic process can be greatly enhanced by opsonization by the complement component C3b or by both antibody and C3b (Fig. 3.15).

## Extracellular and Intracellular Bacteria

Chapter 2 mentioned that the development of a bacterial infectious disease in an individual involves complex interactions between the bacteria and the host cells. The goal of the invading bacterium is to propagate itself. The function of the immune system is to protect the host against microbial infections.

It was also mentioned that the key events during bacterial infection include

- entry of the bacteria
- invasion and colonization of host tissues
- evasion of host immunity
- tissue injury or functional impairment

It was also mentioned that some bacteria produce disease by liberating toxins, even without extensive colonization of the host tissues.

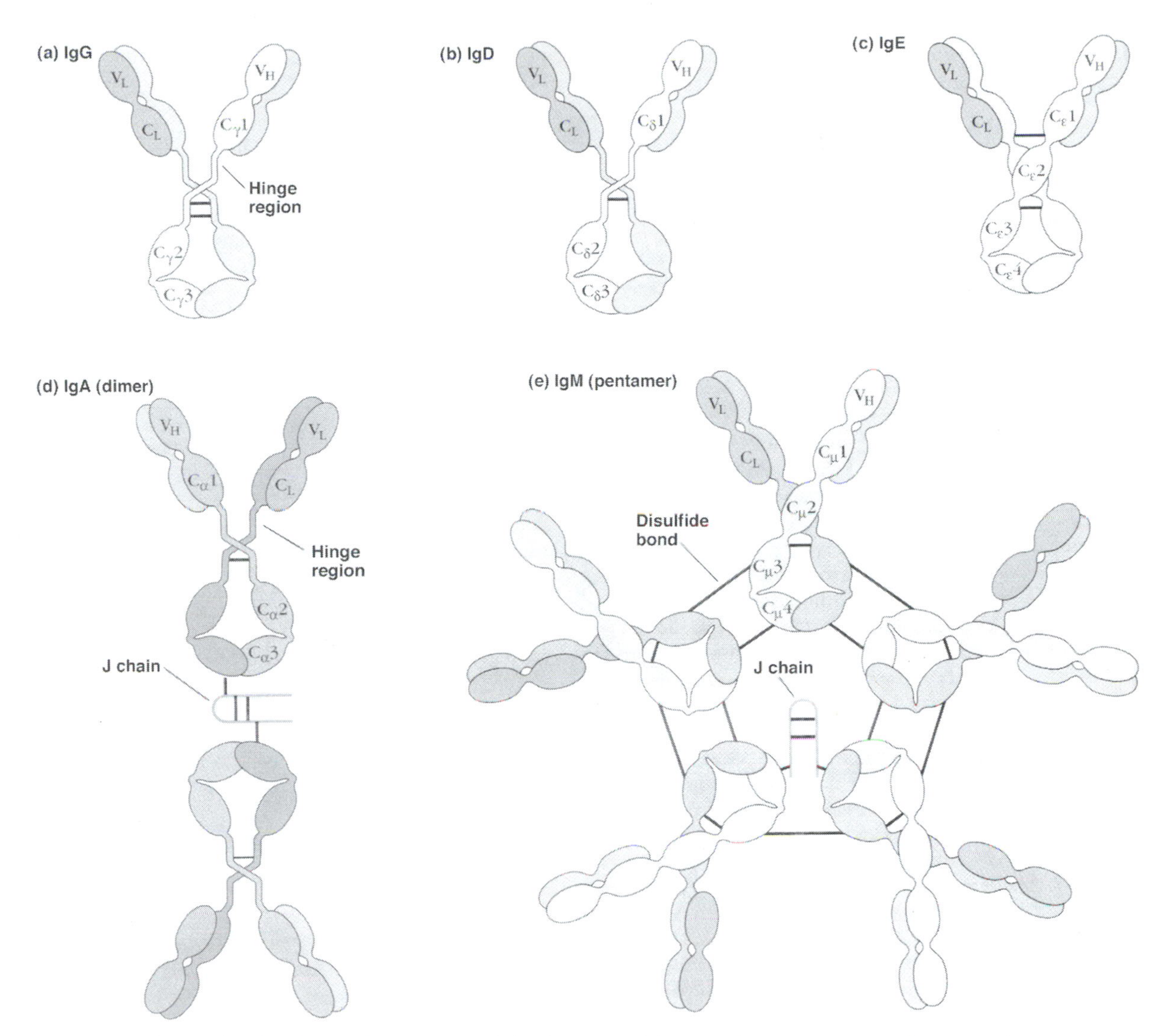


**Figure 3.13** General structures of the five major classes of secreted antibody. The polymeric forms of IgM and IgA contain a polypeptide called the J chain that is linked by two disulfide bonds to the Fc region in two different monomers. Serum IgM is always a pentamer; most serum IgA exists as a monomer, although some dimers, trimers, and even tetramers sometimes are present. Intrachain disulfide bonds and disulfide bonds linking light and heavy chains are not shown. Reprinted from R. A. Goldsby, T. J. Kindt, and B. A. Osborne, *Kuby Immunology,* 4th ed. (W. H. Freeman & Co., New York, N.Y., 2000), with permission from the publisher.

Based on the pathogenesis of infection and the resulting immune response the bacteria can be categorized into two general types: (i) those causing intracellular infections and (ii) those causing extracellular infections. Here the discussion focuses on these two types of pathogenic bacteria.

## Extracellular Bacteria

Extracellular bacteria are capable of replicating outside of host cells, for example in the circulation, in connective tissues, and in tissue spaces such as the airways and intestinal lumens. Pathogenic extracellular bacteria are generally able to colonize mucosal tracts. Extracellular bacteria include numerous pathogens. Table 3.2 lists many bacteria that cause extracellular infection in humans, along with the diseases they cause and some of their major virulence factors.

Extracellular bacteria multiply in the epithelial surface at the site of entry into the body. This occurs, for instance, in diphtheria and streptococcal infections of

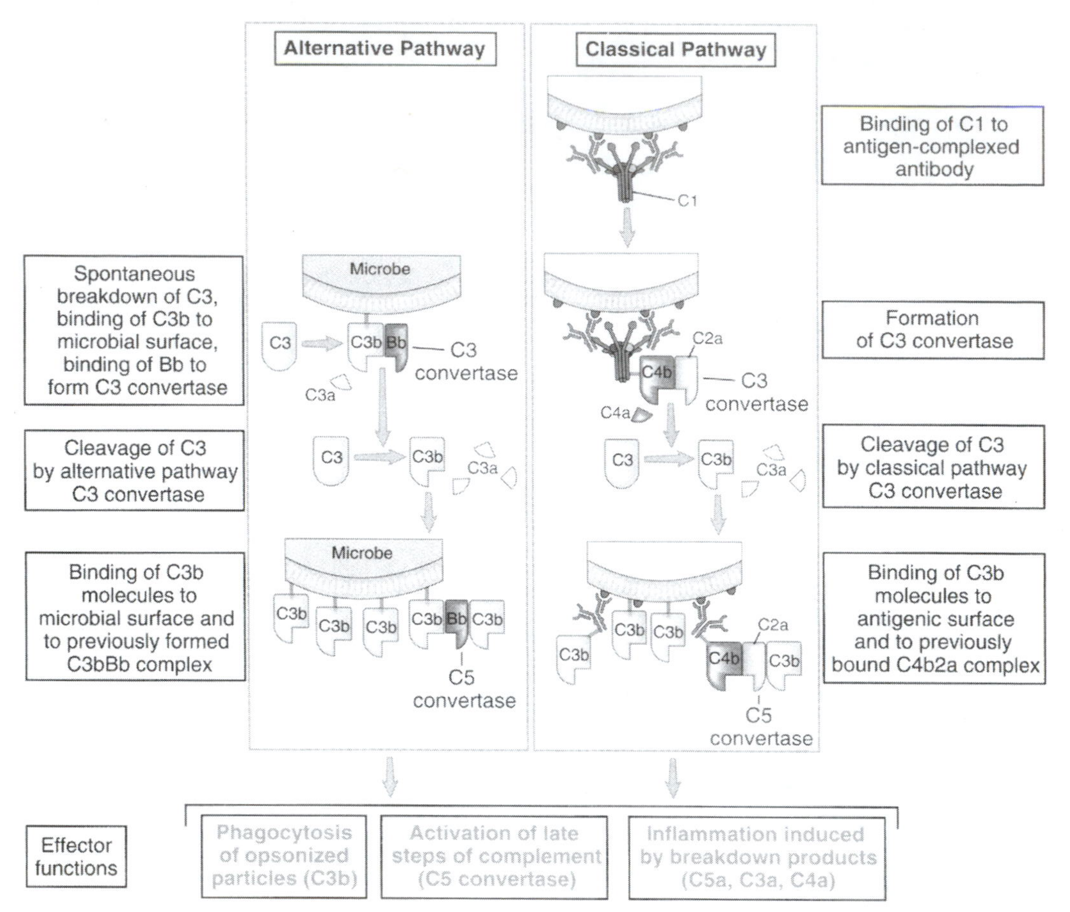


**Figure 3.14** Overview of complement activation (see the text for an explanation). Reprinted from A. K. Abbas, A. H. Lichtman, and J. S. Pober, *Cellular and Molecular Immunobiology*, 4th ed. (The W. B. Saunders Co., Philadelphia, Pa., 2000), with permission from the publisher.

the throat, gonococcal infections of the conjunctiva or urethra, and most *Salmonella* infections of the intestine.

Many different species of extracellular bacteria can be pathogenic because they induce a localized inflammation response which results in tissue destruction at the site of infection or because they produce toxins which have diverse pathological effects. The toxins may be endotoxins, such as the lipopolysaccharide of gram-negative bacteria, or exotoxins, such as diphtheria toxin, which is secreted by the bacteria. In this case, the toxin exerts a toxic effect on the host cell by blocking protein synthesis.

## INNATE IMMUNITY TO EXTRACELLULAR BACTERIA

The principal mechanisms of innate immunity in response to extracellular bacteria are complement activation, phagocytosis, and the inflammatory response.

## ADAPTIVE IMMUNE RESPONSE TO EXTRACELLULAR BACTERIA

The humoral immune response is the main protective response against extracellular bacteria. We have already seen that the adaptive immunity is triggered when an infection eludes the innate defense mechanisms and generates a threshold dose of antigens. This antigen then initiates a specific immune response, which becomes effective only after several days (5 to 7 days), the time required for the T-cell-dependent B-cell response to proliferate and differentiate. The antibodies act in several ways against extracellular bacteria, including inactivating bacterial toxins. The effector mechanisms used by antibodies to combat infections caused by these bacteria include neutralization, opsonization, phagocytosis, and activation of complement by the classical pathway (Fig. 3.16).

**Antibodies** perform several functions that serve to eliminate bacteria.

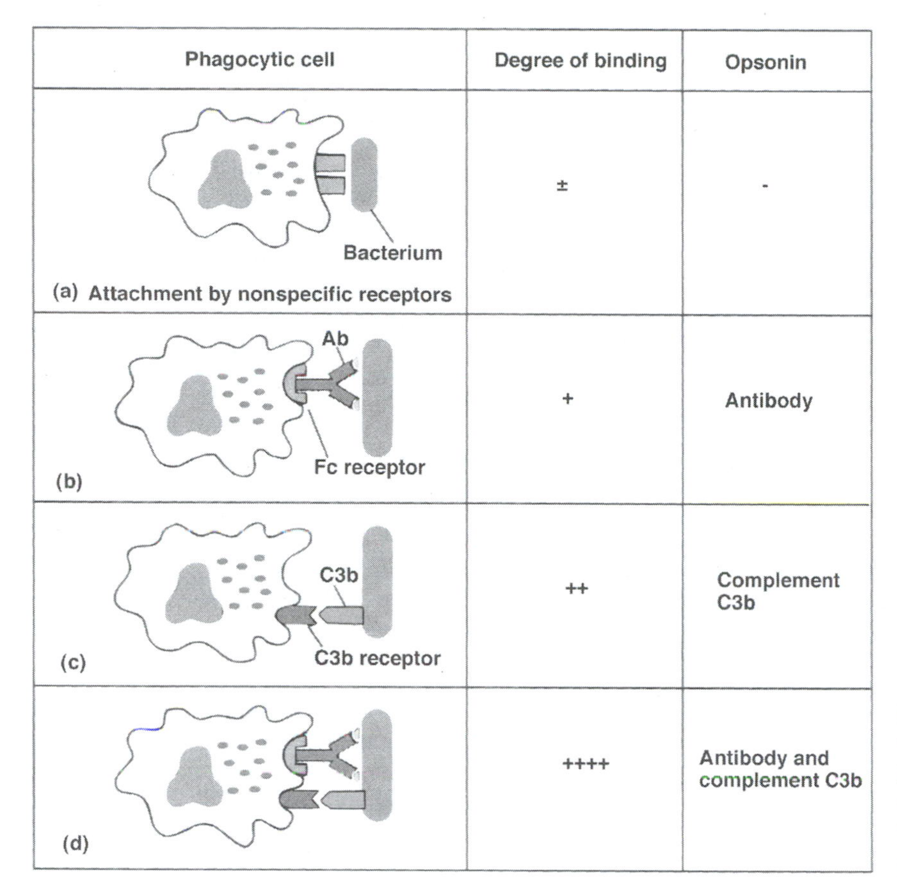

Phagocytic cell
Degree of binding
Opsonin
Bacterium
(a) Attachment by nonspecific receptors
±
-
Ab
Fc receptor
(b)
+
Antibody
C3b
C3b receptor
(c)
++
Complement C3b
(d)
++++
Antibody and complement C3b


**Figure 3.15** Schematic diagram of the opsonization of a bacterium. Reprinted from L. M. Prescott, J. P. Harley, and D. A. Klein, *Microbiology,* 4th ed. (McGraw-Hill, Boston, Mass., 1999), with permission from the publisher.

**Neutralization** is mediated by high-affinity IgG and IgA isotypes. These antibodies bind and neutralize bacterial toxins and prevent their binding to target cells.

**Opsonization** is carried out by some subclasses of IgG. IgG antibodies opsonize bacteria and enhance phagocytosis by binding receptors on monocytes, macrophages, and neutrophils that further promote phagocytosis.

**Complement activation** by IgM and subclasses of IgG occurs when both IgM and IgG antibodies activate the complement system, leading to the production of the bactericidal membrane attack complex and the liberation of by-products that are mediators of acute inflammation.

The protein antigens of extracellular bacteria also activate $CD4^+$ helper T cells, which produce cytokines that stimulate antibody production, induce local inflammation, and enhance the phagocytic and bactericidal activities of macrophages. IFN-$\gamma$ is the principal cytokine responsible for macrophage activation, while TNF and lymphotoxin trigger inflammation.

## EVASION OF IMMUNE MECHANISMS BY EXTRACELLULAR BACTERIA

The virulence of extracellular bacteria has been linked to a number of mechanisms that favor tissue invasion and colonization and resist innate immunity.

Pathogenic bacteria have developed various means of evading or subverting normal host defense mechanisms. These include

- genetic variation in the structure of surface antigens
- genetic alteration in the structure of LPS
- resistance to phagocytosis
- inhibition of complement activation

One mechanism used by bacteria to evade humoral immunity is antigenic variation. Some surface antigens of many bacteria such as gonococci and *Escherichia coli* are in their pili. The major antigen of the pili is a protein of approximately 35 kDa called pilin. The pilin genes of gonococci undergo extensive gene conversions to produce antigenically distinct pilin molecules. The ability to alter antigens helps the bacteria to evade attack by pilin-specific antibodies. In other bacteria, such as *Haemophilus influenzae,* changes to produce glycosyl synthetases lead to chemical alterations in the structure of surface lipopolysaccharide and other polysaccharides, enabling the bacteria to evade humoral immune responses to these antigens. Bacteria with polysaccharide capsules also resist phagocytosis and therefore are more virulent than homologous strains lacking a

**Table 3.2** Examples of bacteria commonly associated with extracellular disease[a]

| Species | Diseases | Important virulence structures and molecules[b] | Special adaptations critical to host infection |
|---|---|---|---|
| *Neisseria gonorrhoeae* | Urethritis, cervicitis, salpingitis, endometritis, prostatitis, arthritis, proctitis, pharyngitis | LPS, fimbriae, peptidoglycan, unidentified adhesins and invasins | Phase and antigenic variation, molecular mimicry of human antigens |
| *Neisseria meningitidis* | Meningitis, meningococcemia, arthritis, pneumonia, asymptomatic carriage | Capsular polysaccharide, LPS, fimbriae, membrane proteins | Phase and antigenic variation, molecular mimicry of human antigens |
| *Haemophilus influenzae* type b | Meningitis, sepsis, arthritis, epiglottitis, asymptomatic carriage | Capsular polysaccharide, LPS, fimbriae, peptidoglycan | Asymtomatic colonization, phase and antigen variation, molecular mimicry of human antigens |
| *Streptococcus pneumoniae* | Pneumonia, otitis media, meningitis, sinusitis | Capsular polysaccharide, neuramidase, hyaluronidase | Symptomatic colonization, genetic transformation permitting continual generation of new genotypes |
| *Staphylococcus aureus* | Impetigo, folliculitis, boils, cellulitis, wound infections, toxic shock, osteomyelitis, food poisoning, bacteremia | Tissue-degrading enzymes, enterotoxins, toxins, capsule | Resistant to dehydration, asymptomatic colonization, regulation of virulence factor expression |
| *Vibrio cholerae* | Diarrhea | Cholera toxin, fimbriae | Bacterial dispersal via cholera toxin, which induces copious watery diarrhea |
| *Clostridium tetani* | Tetanus | Tetanus toxins | Opportunistic infection by a spore-forming soil anaerobe |

[a] Reprinted from W. E. Paul, *Fundamental Immunology*, 4th ed. (Lippincott-Raven, Philadelphia, Pa., 1999), with permission from the publisher.
[b] LPS, lipopolysaccharide.

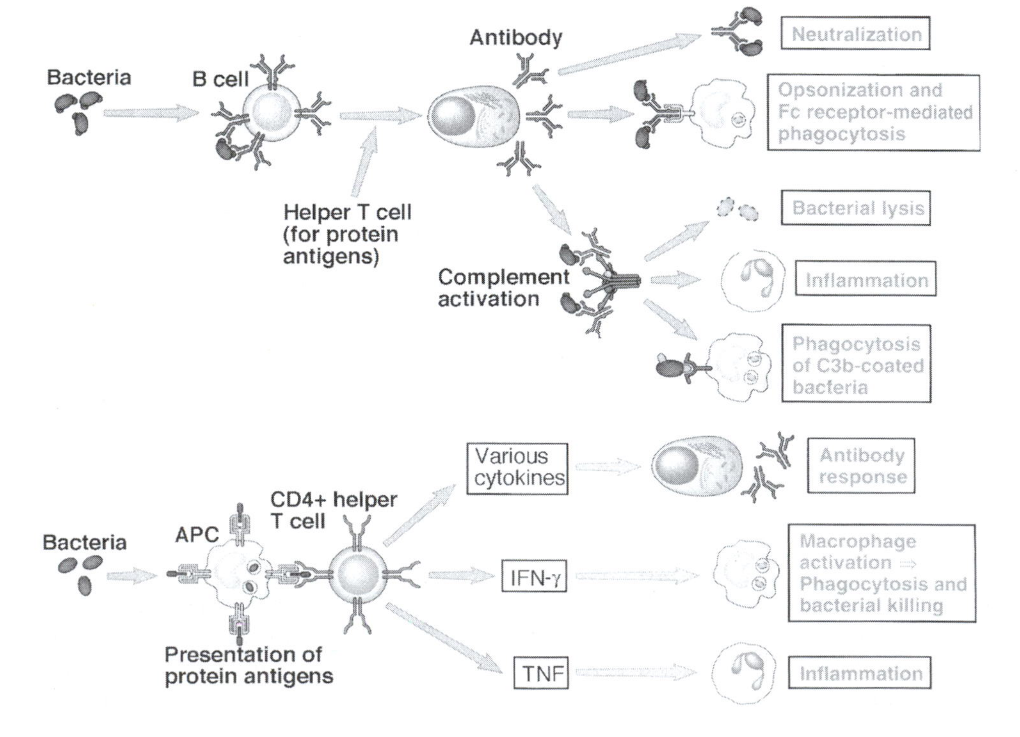


**Figure 3.16** Specific immune response to extracellular bacteria. Reprinted from A. K. Abbas, A. H. Lichtman, and J. S. Pober, *Cellular and Molecular Immunobiology*, 4th ed. (The W. B. Saunders Co., Philadelphia, Pa., 2000), with permission from the publisher.

capsule. The capsules of many gram-positive and gram-negative bacteria contain sialic acid residues that inhibit complement activation by the alternative pathway.

## Intracellular Bacteria

As their name implies intracellular bacteria reside within host cells at some stage of infection. Intracellular bacteria can be divided into two groups depending on whether growth inside cell is required for the replication and survival: (i) **obligate intracellular bacteria,** which multiply only within eukaryotic cells; and (ii) **facultative intracellular bacteria,** which do not depend exclusively on the intracellular habitat but survive both extracellularly and intracellularly.

Table 3.3 lists several facultative and obligate intracellular bacteria of medical importance, the disease they cause, their preferential target cell, their preferred location in host cells, and their preferred port of entry. Some diseases caused by obligate intracellular bacteria, such as epidemic typhus or trachoma, have been known since ancient times. The medical significance of other such bacteria is only now being appreciated; hence, some of these bacteria are considered responsible for emerging infectious diseases. Much of the difficulty associated with recognition of these bacteria as agents of human disease is due to their obligate intracellular nature. Without a bacteriological medium to support their growth, isolation requires tissue culture or animal inoculation. Even this has been problematic in certain cases since some intracellular bacteria are restricted to cell types that support their growth.

Intracellular bacteria are internalized by both phagocytic and nonphagocytic cells, in which they multiply. The ability of these bacteria to enter nonphagocytic host cells, such as epithelial cells, fibroblasts, hepatocytes, and endothelial cells, requires specific uptake mechanisms. Although the molecular details of the uptake mechanism differ among intracellular bacteria, the first event following the specific interaction between the bacterial cell and the host cell is always the formation of a primary phagosome.

**Table 3.3** Major infections of humans caused by facultative and obligate intracellular bacteria[a]

| Pathogen | Disease | Preferred target cell | Preferred location in host cell | Preferred port of entry |
|---|---|---|---|---|
| Facultative intracellular bacteria | | | | |
| *Mycobacterium tuberculosis, M. bovis* | Tuberculosis | Macrophages | Phagosome | Lung |
| *Mycobacterium leprae* | Leprosy | Macrophages, Schwann cells | Phagolysosome (?) | Nasopharyngeal mucosa |
| *Salmonella enterica* | Typhoid fever | Macrophages | Phagosome | Gut |
| *Brucella* spp. | Brucellosis | Macrophages | Phagolysosome | Mucosa |
| *Legionella pneumophila* | Legionnaires' disease | Macrophages | Phagosome | Lung |
| *Listeria monocytogenes* | Listeriosis | Macrophages, hepatocytes | Cytosol | Gut |
| *Francisella tularensis* | Tularemia | Macrophages | Phagosome | Skin, mucosa, lung |
| Obligate intracellular bacteria | | | | |
| *Rickettsia rickettsii* | Rocky Mountain spotted fever | Endothelial cells, smooth muscle cells | Cytosol | Blood vessel |
| *Rickettsia prowazekii* | Endemic typhus | Endothelial cells | Cytosol | Broken skin, mucosa |
| *Rickettsia typhi* | Typhus | Endothelial cells | Cytosol | Blood vessel |
| *Rickettsia tsutsugamushi* | Scrub typhus | Endothelial cells | Cytosol | Blood vessel |
| *Coxiella burnetii* | Q fever | Macrophages, lung parenchymal cells | Late phagosome | Lung |
| *Chlamydia trachomatis* | Urogenital infections, conjunctivitis,trachoma, lymphogranuloma venereum | Epithelial cells | Phagosome | Eye, urogenital mucosa |
| *Chlamydia psittaci* | Psittacosis | Macrophages, lung parenchymal cells | Phagosome | Lung |
| *Chlamydia pneumoniae* | Pneumonia | Lung parenchymal cells | Phagosome | Lung |

[a] Reprinted from W. E. Paul, *Fundamental Immunology,* 4th ed. (Lippincott-Raven, Philadelphia, Pa., 1999), with permission from the publisher.

### IMMUNITY TO INTRACELLULAR BACTERIA

Intracellular bacteria are endowed with the capacity to survive and replicate inside mononuclear phagocytes. Since these bacteria can find a niche that is inaccessible to circulating antibodies, their elimination requires immune mechanisms that are very different from the mechanisms of defense against extracellular bacteria.

#### Innate Immunity to Intracellular Bacteria

The innate response to intracellular bacteria consists mainly of phagocytes and NK cells. Phagocytes, initially neutrophils and later macrophages, ingest and attempt to destroy bacteria as part of the innate immune response. However, pathogenic intracellular bacteria are resistant to degradation within phagocytes. Intracellular bacteria activate NK cells either directly or by stimulating macrophage production of IL-12, a powerful NK-cell-activating cytokine. The NK cells produce IFN-γ, which in turn activates macrophages and promotes killing of phagocytosed bacteria. Thus, NK cells provide an early defense against intracellular bacteria, before the development of adaptive immunity.

#### Adaptive Immune Response to Intracellular Bacteria

The major protective immune response against intracellular bacteria is cell-mediated immunity. Cell-mediated immunity consists of two types of reactions: (i) macrophage activation by the T-cell-derived signal CD40 ligand, and (ii) production of IFN-γ, which results in killing of phagocytosed bacteria and lysis of infected cells by cytolytic T lymphocytes.

#### Evasion of Immune Mechanisms by Intracellular Bacteria

Different intracellular bacteria have developed various strategies to resist elimination by phagocytes. The principal ones are inhibition of phagolysosome formation, scavenging of reactive oxygen intermediates, and disruption of the phagosome membrane and escape into the cytoplasm.

Intracellular pathogens employ diverse strategies to survive. For example, *Mycobacterium tuberculosis, M. leprae,* and *Legionella pneumophila* inhibit phagosome fusion with lysosomes, thereby preventing exposure to toxic lysosomal contents. In contrast, *Listeria monocytogenes* and *Shigella flexneri* lyse the phagosomal membrane and escape into the cytoplasm. A third group of bacteria (*M. leprae* and *Salmonella enterica* serovar Typhimurium) are found within the macrophage phagolysosomal compartment, where they apparently resist inactivation by lysosomal factors. Resistance to phagocytosis is one of the reasons why intracellular bacteria tend to cause chronic infections that may last for years and are difficult to eradicate.

For a more detailed discussion of the topics of immunity and its evasion by extra- and intracellular bacteria, see the references at the end of the chapter.

## Interactions between Bacteria, Immune Defenses, and Antibacterial Drugs

In order to be cured of a bacterial infection, a patient must have an operative immunological defense mechanism (i.e., must be immunocompetent). If antibacterial therapy is given, it will help fight invading bacteria; however, antibacterials alone, without the contribution of the immune response, are seldom able to overcome an infection.

Patients with impaired immunity, such as those who have AIDS or are neutropenic because of cancer chemotherapy, are at particular risk of developing infection caused by microorganisms and protozoal infections, which then may become systemic. The introduction of more aggressive antineoplastic chemotherapy regimes in individuals with solid-organ malignancies has resulted in increasing numbers of these patients experiencing profound neutropenia (i.e., granulocyte counts of $<0.1 \times 10^9$/liter), which puts them at risk of infection.

In persons with defective cell-mediated immunity, the treatment of infections caused by intracellular bacteria such as *Salmonella* spp., *L. pneumophila, L. monocytogenes, Mycobacterium* spp., and *Brucella* spp. is often very difficult.

In order to be effective, antibacterial agents have to reach their target in the bacteria. In a given individual, the route of administration, the pharmacological characteristics of the drug, pharmacokinetics, and localization of the infection play important roles in determining whether the antibiotic and other antibacterial agents reach the bacteria. As discussed in subsequent chapters, hydrophilic antibiotics, such as ampicillin and ceftazidime, use porins to pass through the outer membrane of gram-negative bacteria. Once the antibacterial agent has penetrated the membrane, it binds to its target. For most antibacterial agents, the target is well known. Less well known, usually, is exactly how binding to the target results in death of the bacteria.

## Antibacterial Penetration into Eukaryotic Cells

Since phagocytic cells (such as monocytes and macrophages) are the primary cells encountered by a bacterium in an infected host, it is not surprising that most intracellular bacteria have acquired the ability to multiply within these cells. The ability of some bacteria to multiply within phagocytic cells is an effective way of escaping from the antibacterial activity of antibiotics and other antibacterial agents. Thus, bacteria most often survive within phagocytic cells or within nonphagocytic cells such as endothelial cells. Intracellular bacteria that infect phagocytic cells have developed mechanisms which allow them to resist phagolysosomal activity.

Cellular penetration and subcellular distribution of antibiotics govern their activity against intracellular bacteria. Survival within eukaryotic cells plays a significant role in bacterial pathogenicity and explains difficulties in controlling primary infections and preventing relapses. Thus, antibiotics are screened for their intracellular pharmacokinetic and pharmacodynamic properties. Data on the activity of antibiotics on intracellular bacteria should describe the in vivo situation in human cells.

*L. pneumophila* accounts for 90% of cases of Legionnaires' disease. The principal clinical illness is pneumonia. Recommendations for the treatment of Legionnaires' disease are based largely on clinical experience of the first recognized outbreak in Philadelphia in 1976; in a retrospective review, patients treated with erythromycin or tetracycline had a 50% lower mortality rate than did patients treated with β-lactams. Subsequently, erythromycin became the treatment of choice. More recently, a number of new antibacterial agents have appeared. However, a formal assessmeant of their comparative efficacy in treatment has proved difficult. Assessment of the efficacy and potential utility of antibacterial agents against *Legionella* spp. is based on four methods. First, in vitro susceptibility testing can be performed to screen for active agents. *Legionella* spp. may be grown in either buffered yeast extract broth or buffered charcoal-yeast extract agar, both supplemented with α-ketoglutarate. The MICs of an antibiotic can vary with these growth media. A number of antibacterial agents, including β-lactams, are highly active in vitro. However, β-lactam antibacterials do not penetrate the intracellular compartment; since legionellae are intracellular pathogens, in vitro extracellular susceptibility does not always correspond to in vivo activity.

*L. pneumophila* has been cultivated in vitro in a number of cell lines to help determine the intracellular action of antibacterials. Results obtained by this method generally give a good correlation with animal studies. One exception is gentamicin, which is active against bacteria grown in cell lines but inactive in animal models and human disease. For an important article on the treatment of Legionnaires' disease, see Dedicoat and Venkatesan (1999).

It is essential to remember these aspects of intracellular bacteria when drawing conclusions about the treatment of infectious diseases from in vitro studies of the antibacterial agent, since it is only one aspect of a complicated process.

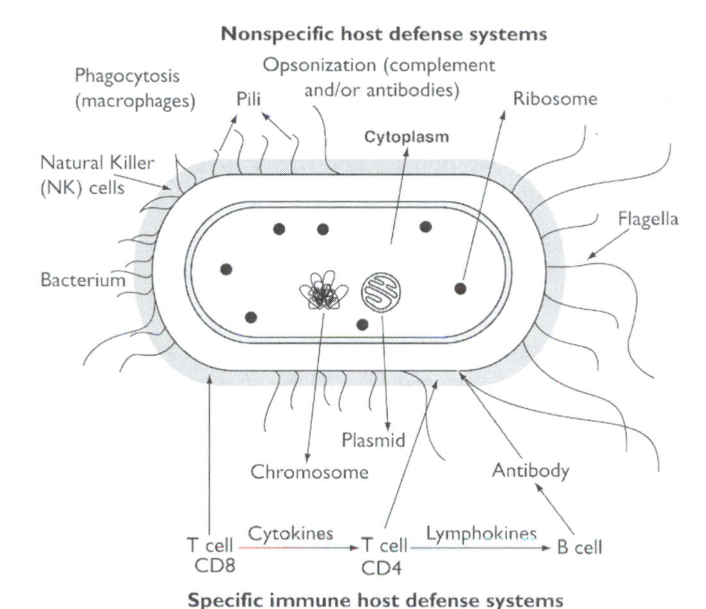


**Figure 3.17** Schematic presentation of the major host defense mechanisms against bacteria.

## Schematic Integrated Overview of Mechanisms of Host Defense against Bacteria

Figure 3.17 shows the main mechanisms of nonspecific and specific immune responses of human host cells to a pathogenic bacterium.

### REFERENCES AND FURTHER READING

#### General (Books)

**Abbas, A. K., A. H. Lichtman, and J. S. Pober.** 2000. *Cellular and Molecular Immunobiology,* 4th ed. The W. B. Saunders Co., Philadelphia, Pa.

**Goldsby, R. A., T. J. Kindt, and B. A. Osborne.** 2000. *Kuby Immunology,* 4th ed. W. H. Freeman & Co, New York, N.Y.

**Janeway, C. A., P. Travers, M. Walport, and M. Shlomchik.** 2001. *Immunobiology,* 5th ed. Garland Publishing, New York, N.Y.

**Paul, W. E.** 1999. *Fundamental Immunology,* 4th ed. Lippincott-Raven, Philadelphia, Pa.

**Prescott, L. M., J. P. Harley, and D. A. Klein.** 1999. *Microbiology,* 4th ed. McGraw-Hill, Boston, Mass.

#### Cellular Immunity

**Rolph, M. S., and S. H. E. Kaufmann.** 2000. Cellular immunity, p. 729–743. *In* J. Lederberg (ed.), *Encyclopedia of Microbiology,* 2nd ed., vol. 1. Academic Press, Inc., San Diego, Calif.

### Immunity to Bacteria

**Kaufman, S. H. E.** 1998. The immune response to infectious agents. *Methods Microbiol.* **25**:1–19.

**Kennedy, R. C., and A.M. Watts.** Immunity to bacteria. Accepted for publication in *Encyclopedia of Life Sciences.* Nature Publishing, London, United Kingdom.

### Indigenous Microflora

**Berg, R. D.** 1996. The indigenous gastrointestinal microflora. *Trends Microbiol.* **4**:430–435.

### Major Histocompatibility Complex

**Brown, J. H., T. S. Jardetzky, J. C. Gorga, L. J. Stern, R. G. Urban, J. L. Strominger, and D. C. Wiley.** 1993. Three-dimensional structure of the human class II histocompatibility antigen HLA-DR1. *Nature* **364**:33–39.

**Engelhard, V. H.** 1994. Structure of peptide associated with class I and class II MHC molecules. *Annu. Rev. Immunol.* **12**:181–207.

**Gao, G. F., and B. K. Jakobsen.** 2000. Molecular interactions of coreceptor CD8 and MHC class I: the molecular basis for functional coordination with the T-cell receptor. *Immunol. Today* **21**:630–636.

**Gao, G. F., J. Tormo, U. C. Gerth, J. R. Wyer, A. J. McMichael, D. I. Stuart, J. I. Bell, E. Y. Jones, and B. K. Jakobsen.** 1997. Crystal structure of the complex between human CD8αα and HLA-A2. *Nature* **387**:630–634.

**Garboczi, D. N., P. Ghosh, U. Utz, Q. R. Fan, W. E. Biddison, and D. C. Wiley.** 1996. Structure of the complex between human T-cell receptor, viral peptide and HLA-A2. *Nature* **384**:134–141.

**Garcia, K. C., M. Degano, R. L. Stanfield, A. Brunmark, M. R. Jackson, P. A. Peterson, L. Teyton, and I. A. Wilson.** 1996. An αβ T cell receptor structure at 2.5 Å and its orientation in the TCR-MHC complex. *Science* **174**:209–221.

**Madden, D. R.** 1995. The three-dimensional structure of peptide-MHC complexes. *Annu. Rev. Immunol.* **12**:587–622.

**Penn, D. J.** Major histocompatibility complex (MHC). Accepted for publication in *Encyclopedia of Life Sciences.* Nature Publishing, London, United Kingdom.

**Rock, K. L., and A. L. Goldberg.** 1999. Degradation of cell proteins and the generation of MHC class I-presented peptides. *Annu. Rev. Immunol.* **17**:739–779.

**Scott, C. A., P. A. Peterson, L. Teyton, and I. A. Wilson.** 1998. Crystal structures of two I-A-peptide complexes reveal that high affinity can be achieved without larger anchor residues. *Immunity* **8**:319–329.

**York, I. A., and K. L. Rock.** 1996. Antigen processing and presentation by the class I major histocompatibility complex. *Annu. Rev. Immunol.* **14**:369–396.

### B Cells

**Bondada, S., and R. L. Chelvarajan.** B lymphocytes. Accepted for publication in *Encyclopedia of Life Sciences.* Nature Publishing, London, United Kingdom.

### T Cells

**Fields, B. A., and R. A. Mariuzza.** 1996. Structure and function of the T-cell receptor: insights from X-ray crystallography. *Immunol. Today* **17**:331–336.

**Garcia, K. C., L. Teyton, and I. A. Wilson.** 1999. Structural basis of T cell recognition. *Annu. Rev. Immunol.* **17**:369–397.

**Goldrath, A. W., and M. J. Bevan.** 1999. Selecting and maintaining a diverse T-cell repertoire. *Nature* **402**:255–262.

**Janeway, C. A.** 2000. T lymphocytes, p. 583–591. *In* J. Lederberg (ed.), *Encyclopedia of Microbiology,* 2nd ed., vol. 4. Academic Press, Inc., San Diego, Calif.

**Li, H., A. Llera, E. L. Malchiodi, and R. A. Mariuzza.** 1999. The structural basis of T cell activation by superantigens. *Annu. Rev. Immunol.* **17**:435–466.

**Mosmann, T. R., and S. Sad.** 1996. The expanding universe pf T-cell subsets: Th1, Th2 and more. *Immunol. Today* **17**:138–146.

**Romagnani, S.** 1996. Understanding the role of Th1/Th2 cells in infection. *Trends Microbiol.* **4**:470–473.

### Macrophages

**Bogdan, C.** Macrophages. Accepted for publication in *Encyclopedia of Life Sciences.* Nature Publishing, London, United Kingdom.

### Phagocytes

**Labro, M. T.** 1998. Antibacterial agents—phagocytes: new concepts for old in immunomodulation. *Int. J. Antimicrob. Agents* **10**:11–21.

**Labro, M. T.** 2000. Interference of antibacterial agents with phagocyte functions: immunomodulation or "immuno-fairy tales." *Clin. Microbiol. Rev.* **13**:615–650.

### Neutrophils

**Kuijpers, T. W., and D. Roos.** Neutrophils. Accepted for publication in *Encyclopedia of Life Sciences.* Nature Publishing, London, United Kingdom.

### Bacterial Capsules and Evasion of Immune Responses

**Merino, S., and J. M. Tomás.** Bacterial capsules and evasion of immune responses. Accepted for publication in *Encyclopedia of Life Sciences.* Nature Publishing, London, United Kingdom.

### Immunoglobulins (Antibodies)

**Garman, S. C., J. P. Kinet, and T. S. Jardetzky.** 1998. Crystal structure of the human high-affinity IgE receptor. *Cell* **95**:951–961.

### Natural Killer Cells

**Colonna, M., A. Moretta, F. Vély, and E. Vivier.** 2000. A high resolution view of NK-cell receptors: structure and function. *Immunol. Today* **21**:428–431.

**Valiante, N. M.** Natural killer (NK) cells. Accepted for publication in *Encyclopedia of Life Sciences.* Nature Publishing, London, United Kingdom.

### Antigens

**Matsuda, K., N. Yamamoto, and M. Saito.** Antigens: lipids. Accepted for publication in *Encyclopedia of Life Sciences.* Nature Publishing, London, United Kingdom.

**Ravindranath, M. H., and D. L. Morton.** Antigens: carbohydrates. Accepted for publication in *Encyclopedia of Life Sciences.* Nature Publishing, London, United Kingdom.

**Wang, J. O., and T. Watanabe.** Antigen presentation to lymphocytes. Accepted for publication in *Encyclopedia of Life Sciences.* Nature Publishing, London, United Kingdom.

### Epitopes

**Muller, C. P.** Epitopes. Accepted for publication in *Encyclopedia of Life Sciences.* Nature Publishing, London, United Kingdom.

### Complement

**Arlaud, G. J., and M. G. Colomb.** Complement: classical pathway. Accepted for publication in *Encyclopedia of Life Sciences.* Nature Publishing, London, United Kingdom.

**Sunyer, J. O., and J. D. Lambris.** Complement. Accepted for publication in *Encyclopedia of Life Sciences.* Nature Publishing, London, United Kingdom.

**Zipfe, P. F.** Complement: alternative pathway. Accepted for publication in *Encyclopedia of Life Sciences.* Nature Publishing, London, United Kingdom.

### Intracellular Bacteria

**Dedicoat, M., and P. Venkatesan.** 1999. The treatment of Legionnaires' disease. *J. Antimicrob. Chemother.* **43:**747–752.

**Hackstadt, T.** 1998. The diverse habitats of obligate intracellular parasites. *Curr. Opin. Microbiol.* **1:**82–87.

**Kaufmann, S. H. E.** 1993. Immunity to intracellular bacteria. *Annu. Rev. Immunol.* **11:**129–163.

**Kaufmann, S. H. E.** 1995. Immunity to intracellular microbial pathogens. *Immunol. Today* **16:**338–342.

**Kaufmann, S. H. E.** 1999. Immunity to intracellular bacteria, p. 1335–1371. *In* W. E. Paul (ed.), *Fundamental Immunology,* 4th ed. Lippincott-Raven, Philadelphia, Pa.

**Maurin, M., and D. Raoult.** 1997. Intracellular organisms. *Int. J. Antimicrob. Agents* **9:**61–70.

**Rakita, R. M.** 1998. Intracellular activity, potential clinical use of antibiotics. *ASM News* **64:**570–575.

**Raoult, D.** 1993. *Antimicrobial Agents and Intracellular Pathogens.* CRC Press, Inc., Boca Raton, Fla.

### Extracellular bacteria

**Nahm, M. H., M. A. Apicella, and D. E. Briles.** 1999. Immunity to extracellular bacteria, p. 1373–1386. *In* W. E. Paul (ed.), *Fundamental Immunology,* 4th ed. Lippincott-Raven, Philadelphia, Pa.

chapter 4



# Molecular Genetics of Bacteria

The main purpose of this chapter is to provide an outline of molecular bacterial genetics as an aid to understanding the origins and nature of bacterial resistance to antibacterial agents through mutation, bacterial recombination, and transfer of bacterial plasmids and transposable elements. Also, the three types of bacterial gene transfer (transformation, transduction, and conjugation) are discussed.

## Definitions

Some knowledge of basic terminology is necessary at the beginning of this survey of general principles.

**Genetics** is the science of heredity. It is concerned with the physical and chemical properties of the hereditary material, how this material is transmitted from one generation to the next, and how the information it contains is expressed during the development of an individual. The unit of heredity is the **gene,** which is a segment of DNA that carries in its nucleotide sequence information for a specific biochemical or physiologic property. The term **genome** refers to all genes present in a cell. The circular DNA molecule of a bacterium that carries most of its normal genes is commonly called a **chromosome.** Most bacteria have only one chromosome. This means that there is only one unique DNA molecule per cell, which carries most of the normal genes. Often, the term "genome" describes the chromosome of bacteria that contain a single chromosome. The designation "chromosomal DNA" distinguishes this molecule from **plasmid DNA,** an autonomous self-replicating extrachromosomal circular DNA which carries genes that are not always required for growth of the bacterium.

DNA and plasmids can form right-handed (negatively supercoiled) or left-handed (positively supercoiled) supercoils. In *Escherichia coli,* the degree of supercoiling is determined by the competing action of DNA gyrase, which produces negative superhelical turns via an ATP-dependent reaction involving a double-stranded break, which produces a two-step decrease in the linking number, and topoisomerase I, which relaxes DNA via a single-stranded break, resulting in a single-step increase in the linking number. The balance between these two enzymes' activities is illustrated in Fig. 4.1.

Circular DNA molecules (chromosome and plasmids) which contain genetic information necessary for their own replication are called **replicons. Transposons** are genetic elements that contain several kilobase pairs of DNA including the information necessary for their migration from one genetic locus to another.

The **phenotype** is the appearance or the collection of observable characteristics of an organism. In contrast, the **genotype** is the specific set of genes it possesses. Generally, strains of a bacterial species are compared with an arbitrarily chosen **wild-type** strain, which is the strain normally found in nature. A **mutation** is any change in the genomic DNA sequence with respect to the wild-type strain. A strain with a mutation is called a **mutant.** The process of generating mutations is called **mutagenesis.** In nature, mutations sometimes arise spontaneously because errors occur during DNA replication or repair. This is called **spontaneous mutagenesis.** Mutagenesis can also be induced by the addition of chemical compounds called **mutagens** or by exposure to radiation, both of which result in chemical alterations of the DNA.

An **auxotroph** is a mutant that lacks the ability to synthesize an essential nutrient and therefore must require the addition of it or a precursor to its growth medium. The requirement for a specific amino acid is a common form of auxotrophy.

Some genes are responsible for resistance to antibacterial agents (antibiotics and synthetic antibacterials).

**Figure 4.1** Genomic elements of bacteria. The main genomic element of bacteria is a single large circular DNA molecule called the chromosome. It contains the genes for all the "essential" functions and structures of the bacterial cell. Most bacteria carry additional DNA elements. These smaller circular DNA molecules are called plasmids. Topo I, topoisomerase I. The illustrations of the chromosome and the plasmid are not to scale.

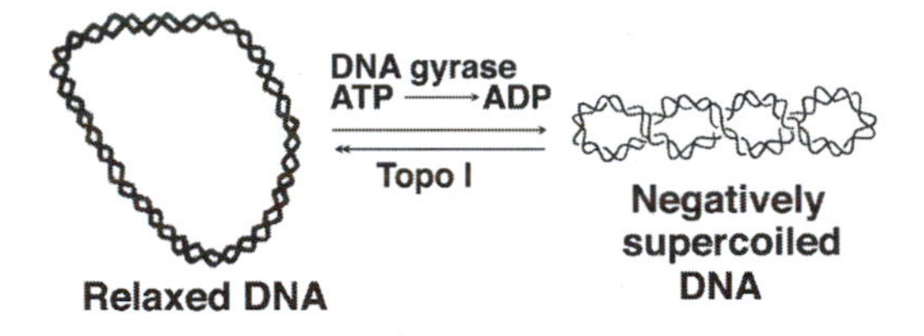


**Chromosome**

Bacterial DNA is located within the cytoplasm. It consists of a single chromosome which varies in size between different species of bacteria. It is often seen as a discrete area within the cell in electron micrographs and is called the **nucleoid**.
The *E. coli* chromosome is 4 x $10^6$ base pairs long.
The DNA is circular, tightly supercoiled.
Indispensable
Codes for:
- cell size and shape
- vital cell structures
- 'standard' metabolic activities
- DNA replication, expression
- DNA repair

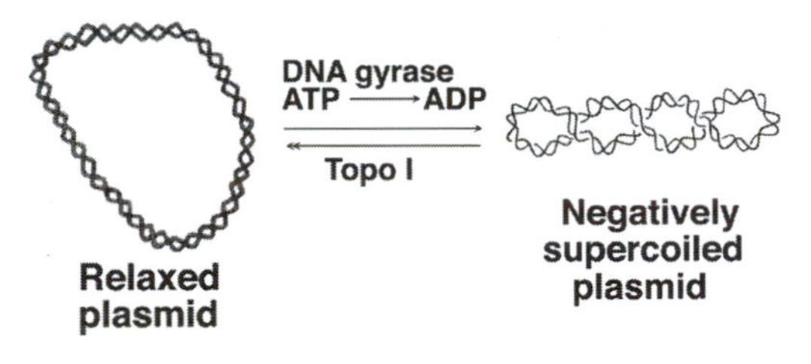


**Plasmid**

Plasmids are extrachromosomal molecules of DNA that replicate independently of the bacterial chromosome. They are normally covalently closed circular, supercoiled molecules, but there also are instances of linear plasmids in a few bacterial species. They range in size from about 2.2 to 210 kilobases (kb) (0.1 to 10% the size of the chromosome).
1 to several hundred copies/cell

Dispensable under optimal conditions
Codes for:
- fertility
- nitrogen fixation
- additional metabolic activities
- resistance to antibacterial agents, etc.
- virulence

The phenotypes of an ampicillin-resistant cell and an ampicillin-sensitive cell are written Amp$^r$ and Amp$^s$, respectively.

The **genetic code** is the correspondence between triplets of bases in RNA and amino acids in proteins.

## Replication of the Bacterial Chromosome and Cell Division

The replication of a bacterial DNA chromosome occurs during the cell division cycle (binary fission). The **cell division cycle** is the elapsed time during which a cell grows and divides into two progeny cells. **Cell division** is the process by which the larger cell splits into new cells. The original cell before cell division is called the mother cell, and the two progeny cells after division are called the daughter cells.

## Gene Expression

Gene expression is accomplished through a sequence of events in which the information contained in the base sequence of DNA is first transcribed into an RNA molecule, which is used to determine the amino acid sequence of a protein molecule (Fig. 4.2). RNA molecules are synthesized by using the base sequence in a region of one of the DNA strands as a template to make the complementary RNA molecule (transcription). This reaction is catalyzed by an enzyme called **RNA polymerase.** Protein molecules are then synthesized by using the base sequence of this RNA molecule to direct the sequential joining of amino acids in a particular order (translation). Each gene represents a particular polypeptide chain.

**Figure 4.2** Transfer of information from DNA to protein. The nucleotide sequence in DNA specifies the sequence of amino acids in a polypeptide. Bacterial DNA usually exists as a circular two-chain structure. The information contained in the nucleotide sequence of only one of the DNA chains (coding strand) is used to specify the nucleotide sequences of the mRNA molecules. This sequence information is used in polypeptide synthesis. A 3-nucleotide sequence in the mRNA molecule codes for a specific amino acid in the polypeptide chain.

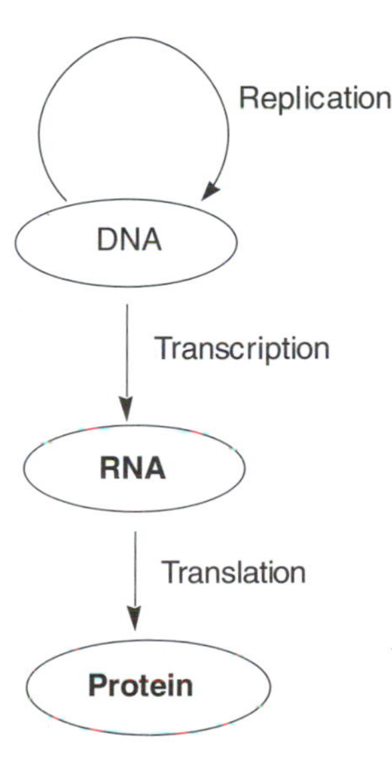


## The Genetic Code

The relationship between a sequence of DNA bases and the sequence of the corresponding protein is called the genetic code. The genetic code is read in groups of three nucleotides, each group representing an amino acid (Fig. 4.3). Each trinucleotide sequence is called a **codon.** By convention, all nucleotide sequence are written in the 5'→3' direction. A gene includes a series of codons that is read sequentially from a starting point at one end to a termination point at the other end.

The mRNA carries the information for the sequence of amino acids in a protein in the form of a genetic code (Fig. 4.3) to specify each of the 20 amino acids normally present in proteins. Because each of the three bases in a codon can be one of the four nucleotides, there are a total of $4^3 = 64$ possible combinations. This means that there is **code degeneration.** That is, there are up to six different codons for the amino acids leucine, serine, and arginine, four for proline, threonine, valine, glycine, and alanine, and three for isoleucine; others have one or two. Just two amino acids, methionine and tryptophan, have only one codon. Only 61 codons, the **sense codons,** direct amino acid incorporation into proteins. The remaining three codons (UGA, UAG, and UAA) are involved in the termination of translation and are called **stop codons** or **nonsense codons.**

The genetic code is read in nonoverlapping triplets from a fixed starting point. "Nonoverlapping" implies that each codon consists of three nucleotides and that successive codons are represented by successive trinucleotides. The start codon for most protein synthesis in bacteria is AUG. AUG corresponds to the amino acid methionine. The AUG codon is often called the **initiation codon.**

Frequently a mutant protein differs by a single amino acid from the wild-type form. The amino acid substitution can be accounted for by a single base change between the codons corresponding to the two different amino acids.

## Mutations

A **mutation** is any change in the genomic DNA sequence compared with the wild-type strain. A strain with a mutation is called a mutant (see above).

Second position

| First position (5' End) | U | | C | | A | | G | | Third position (3' End) |
|---|---|---|---|---|---|---|---|---|---|
| U | UUU | Phe | UCU | Ser | UAU | Tyr | UGU | Cys | U |
| | UUC | | UCC | | UAC | | UGC | | C |
| | UUA | Leu | UCA | | UAA | STOP | UGA | STOP | A |
| | UUG | | UCG | | UAG | | UGG | Trp | G |
| C | CUU | Leu | CCU | Pro | CAU | His | CGU | Arg | U |
| | CUC | | CCC | | CAC | | CGC | | C |
| | CUA | | CCA | | CAA | Gln | CGA | | A |
| | CUG | | CCG | | CAG | | CGG | | G |
| A | AUU | Ile | ACU | Thr | AAU | Asn | AGU | Ser | U |
| | AUC | | ACC | | AAC | | AGC | | C |
| | AUA | | ACA | | AAA | Lys | AGA | Arg | A |
| | AUG | Met | ACG | | AAG | | AGG | | G |
| G | GUU | Val | GCU | Ala | GAU | Asp | GGU | Gly | U |
| | GUC | | GCC | | GAC | | GGC | | C |
| | GUA | | GCA | | GAA | Glu | GGA | | A |
| | GUG | | GCG | | GAG | | GGG | | G |

**Figure 4.3** The genetic code. The code is presented in the RNA form. Reprinted from L. M. Prescott, J. P. Harley, and D. A. Klein, *Microbiology,* 4th ed. (McGraw-Hill, Boston, Mass., 1999), with permission from the publisher.

## Rate of Mutations in Bacteria

A typical bacterial population doubles its size every 30 min. Bacteria can reach very high population densities during unencumbered growth ($10^7$ cells/ml in blood infections and as much as $10^9$ cells/ml in tissue infections). The rate of spontaneous mutations in bacteria is relatively low ($10^{-6}$/base) per generation. This means that there is 1 chance in 1,000,000 that a mutation will arise at some location in a given gene during one cell cycle. However, considering the large sizes of the genomes (e.g., the size of the *E. coli* K-12 genome is $4.6 \times 10^6$ bp) and the high potential for bacterial population densities under favorable growth conditions, this rate of mutation is not insignificant per generation of progeny.

## Types of Mutations

Any base pair of DNA can be mutated. A **point mutation** changes only a single base pair in the sequence of DNA. When this change involves replacing one nucleotide by another, it is known as a **base pair substitution.** The consequence of such a change depends both on the nature of the change and on its location. If the change is within the coding region of a gene (i.e., the region which ultimately is translated into proteins), it may give rise to an alteration of the amino acid sequence at that point, which may affect the function of the protein. The error in the DNA is transcribed into mRNA, and this erroneous mRNA in turn is used as a template and translated into protein. The alteration may of course have little or no effect, either because the changed triplet still codes for the same amino acid or because the new amino acid is sufficiently similar to the original one that the function of the protein remains unaffected.

One kind of point mutation is, for example, if the triplet UUA codes for leucine. A single base change in the DNA can give rise to one of nine other codons, as shown in Fig. 4.4. Two of the possible changes (UUG and CUA) are completely silent, since the resulting codon still codes for leucine. Mutations that give rise to no change in the amino acids (because the genetic code is degenerate) are called **silent mutations.** Two other changes (AUA and GUA) may well have little effect on the protein since the substituted amino acids (isoleucine and valine) are similar to the original leucine (e.g., they are all hydrophobic amino acids). A second type of point mutation is the **mis-**

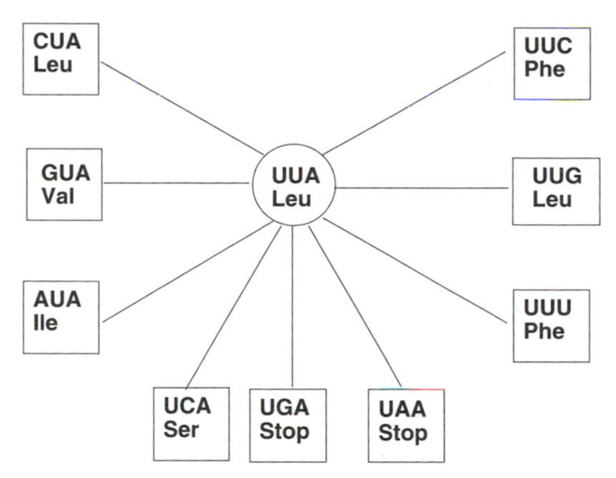


**Figure 4.4** Codons arising by single-base substitutions from UUA. Reprinted from J. W. Dale, *Molecular Genetics of Bacteria,* 3rd ed. (John Wiley & Sons, Ltd., Chichester, United Kingdom, 1998), with permission from the publisher.

**sense mutation.** The significance of the change to UCA, resulting in the substitution of serine (which is considerably different) for leucine, will depend on the role played by that amino acid (and its neighbors) in the overall function or conformation of the protein. The term "missense mutation" is used because the chemical "sense" (sequence of amino acids) in the resulting polypeptide has changed. If the change occurred at a critical point in the polypeptide chain, the protein could be inactive or have reduced activity. The outcome depends on where in the polypeptide chain the substitution has occurred and how it affects the folding or the catalytic activity of the protein (although not all proteins have catalytic activity).

A third type of point mutation causes the early termination of translation and therefore results in a shortened polypeptide. Mutations of this type are called **nonsense mutations** because they involve the conversion of a sense codon to a nonsense or stop codon. The mutation is expressed at the level of protein structure. However, at the protein level, the effect may range from complete loss of activity to no change at all.

Because the genetic code is read from one end in consecutive blocks of three bases, any deletion or insertion of a base pair results in a **reading frame shift,** and the translation of the gene is completely upset (Fig. 4.5). **Deletions** are mutations in which a region of the DNA has been eliminated. **Insertions** occur when new bases are added to the DNA. Insertions, like deletions, can involve a single base or many bases. Mutations that remove or add base pairs are called **frameshift mutations.**

## Transition and Transversion Mutations

Errors can occur during replication if a base of a template nucleotide changes from a keto form to an enol form or an amino form changes to an imino form (Fig. 4.6). These tautomeric shifts change the hydrogen-bonding characteristics of the bases.

The most common effect of these changes is a **transition mutation,** where the purine (A and G) in a base pair is replaced by the other purine base and the pyrimidine (C and T) is replaced by the other pyrimidine. Thus, an A-T pair would become G-C or a C-G pair would become T-A (Fig. 4.7).

In **transversion mutations,** a purine base is replaced by a pyrimidine base or vice versa. For example, an A-T pair would become T-A or C-G.

## Mutagens

As stated above, mutations can occur spontaneously but only at a very low rate. There are a variety of chemical agents or radiation that can increase the mutation rate and therefore are said to induce mutations. These agents are called **mutagens.**

**Figure 4.5** Shifts in the reading frame as a result of deletion or insertion mutations.

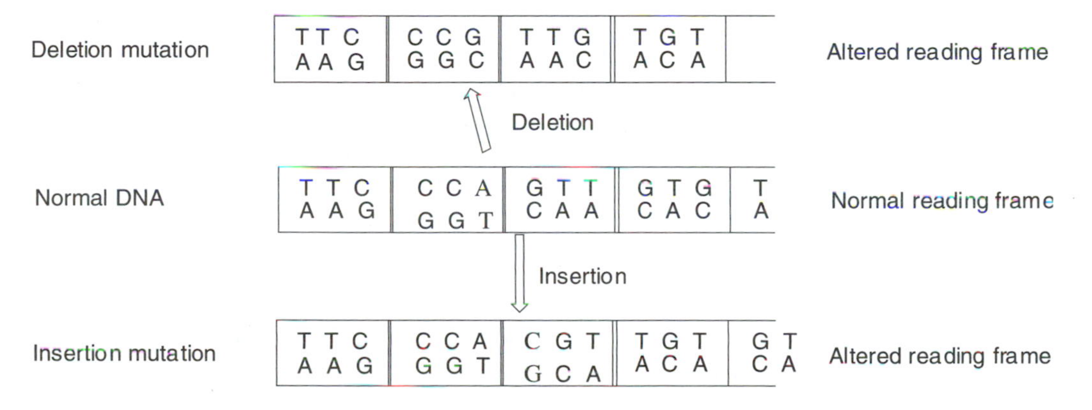

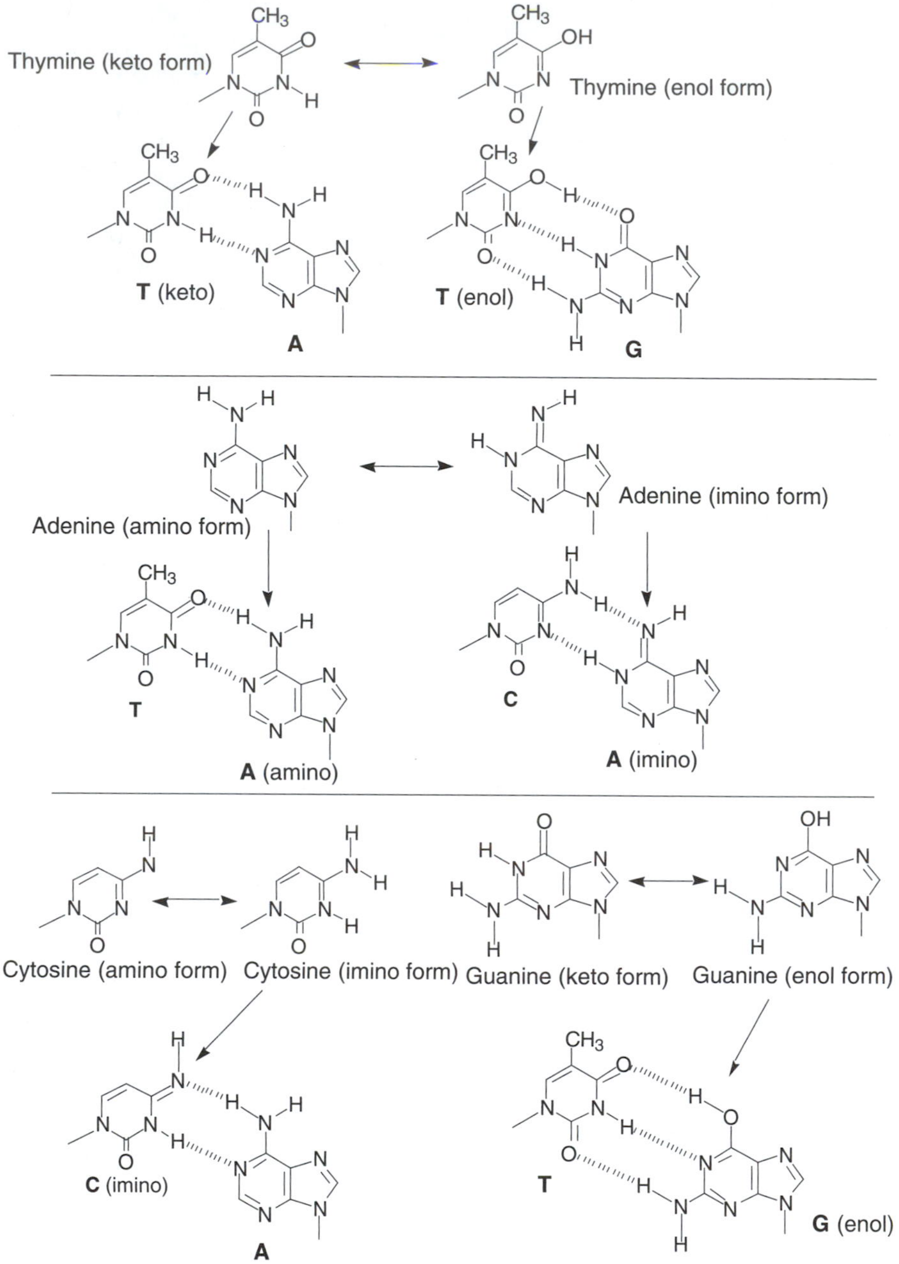

Figure 4.6 Normally, A-T and C-G pairs are formed when keto groups participate in hydrogen bonds. In contrast, enol tautomers produce A-C and G-T base pairs. A-T pairs are formed when the amino group participates in hydrogen bonding. In contrast, imino tautomers produce A-C.

## CHEMICAL AGENT-INDUCED MUTAGENESIS

Chemical mutagens can be conveniently classified according to their mechanism of action. Three common modes of mutagen actions are

- incorporation of base analogs
- reaction with DNA (alkylating agents)
- intercalation

### Mutations Caused by Incorporation of Base Analogs

Base analogs have a similar structure to normal bases and can be incorporated into the growing polynucleotide chain during replication. For example 5-bromouracil (5-BU) is an analogue of thymine in which the methyl group at the 5 position is replaced by a bromine atom, which has a similar steric size (Fig. 4.8). 5-BU undergoes a tautomeric shift from the keto form to an enol much more frequently than does a normal base. The enol form makes hydrogen bonds like cytosine and directs the incorporation of guanine rather than adenine.

### Mutations Caused by Alkylation of Bases

An example of an alkylating mutagen is 1-methyl-3-nitro-1-nitrosoguanidine, which adds a methyl group to the oxygen atom at position 6 of the guanine molecule

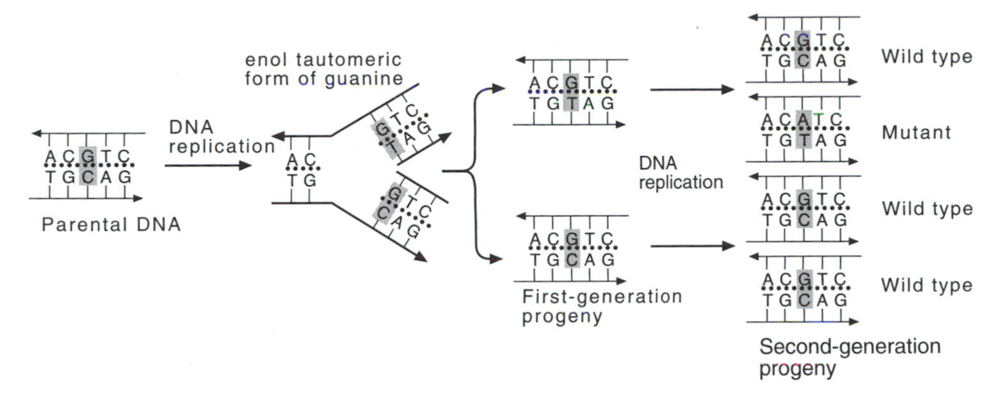


**Figure 4.7** Mutation as a consequence of tautomerization during DNA replication. The temporary enolization of guanine leads to the formation of an A-T base pair in the mutant, and a G-C to A-T transition mutation occurs. The process requires two replication cycles. The mutation occurs only if the abnormal first-generation G-T base pair is missed by repair mechanisms. Reprinted from L. M. Prescott, J. P. Harley, and D. A. Klein, *Microbiology,* 4th ed. (McGraw-Hill, Boston, Mass., 1999), with permission from the publisher.

**Figure 4.8** Mutagenesis by the base analog 5-bromouracil (5-BU). Base pairing of the normal keto form of 5-BU is shown at the top. The enol form of 5-BU (bottom) base-pairs with guanine rather than with adenine as might be expected for a thymine analog. If the keto form of 5-BU is incorporated in place of thymine, its occasional tautomerization to the enol form produces an A-T → G-C transition mutation.

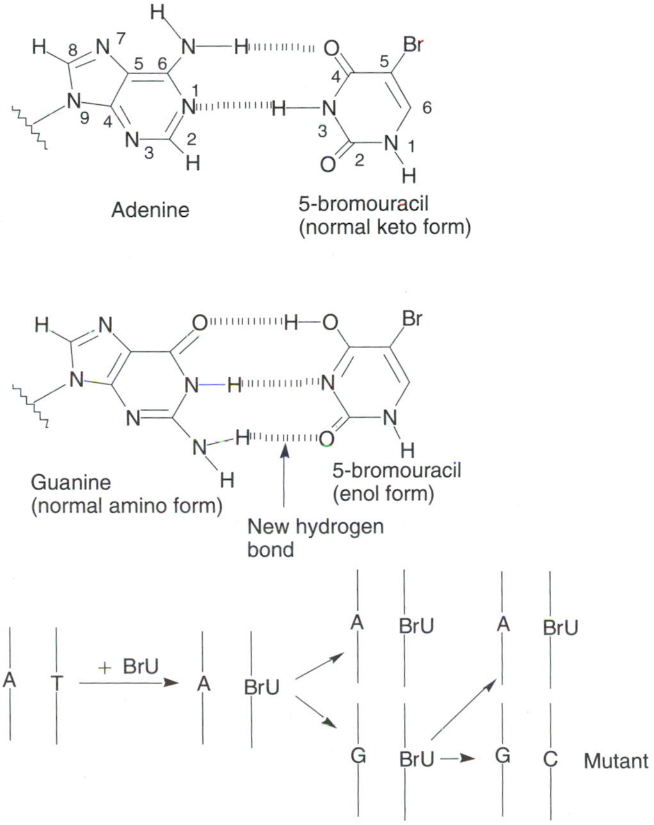

(Fig. 4.9), causing it to mispair with thymine. A subsequent round of replication could then result in a G-C→A-T transition. The alkylating agents act directly on the already formed DNA molecule, unlike base analogs, which act only after incorporation during replication.

## Action of Intercalating Agents

One group of mutagens, which are planar molecules like the acridine dyes, act as intercalating agents. These

**Figure 4.9** 1-Methyl-3-nitro-1-nitrosoguanidine mutagenesis. Mutagenesis occurs because of the methylation of guanine.

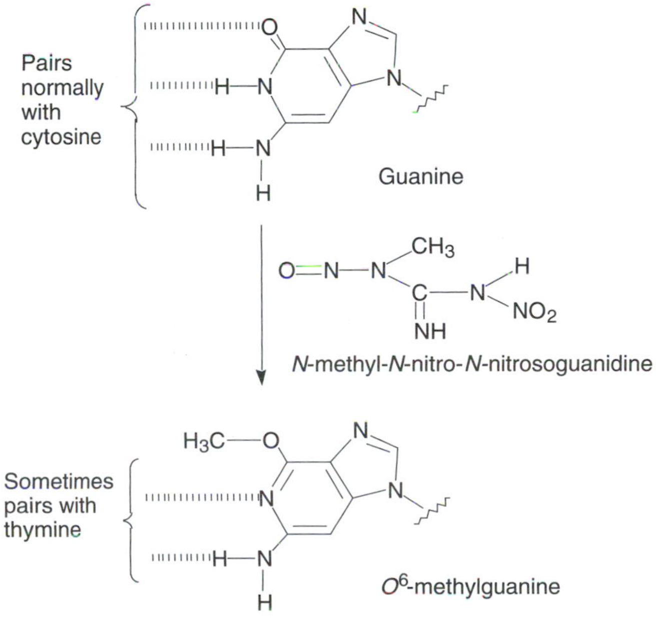

mutagens insert between two DNA base pairs, thereby pushing them apart. During replication, this abnormal conformation can lead to microinsertions or microdeletions. Ethidium bromide is also an intercalating agent that acts as a mutagen.

### RADIATION-INDUCED MUTAGENESIS

Ultraviolet (UV) radiation, gamma radiation, and X radiation are the most common types of mutagenic radiation found in nature and are used in mutagenesis experiments.

#### UV Radiation

The purine and pyrimidine bases of the nucleic acids (DNA and RNA) have a relatively weak absorption maximum at 260 nm. Therefore, the principal effect of UV radiation is the cross-linking of two adjacent pyrimidine bases on the same DNA strand, to form covalently joined dimers; these are usually **thymine dimers,** although the effect can also occur with cytosine. Figure 4.10 shows the structure of a thymine dimer and illustrates how it interrupts base-pairing between the two DNA strands, so that during DNA replication, the probability that DNA polymerase will insert an incorrect nucleotide at this position is greatly increased.

**Figure 4.10** Thymine dimers. (a) UV light cross-links the two thymidine bases on the top strand. This distorts the DNA so that these two bases no longer pair with their adenine partners. (b) The two bonds joining the two thymine residues form a cyclobutane ring.

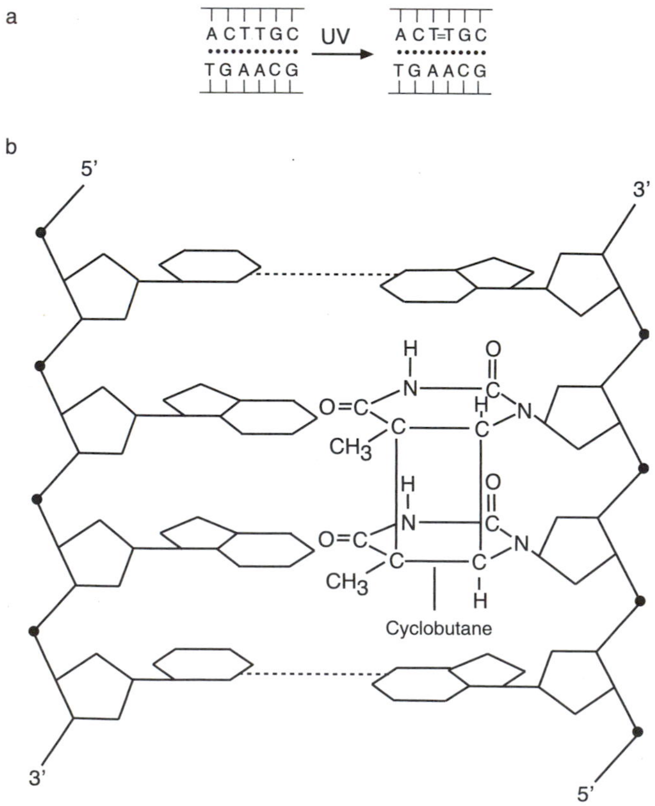

a

ACTTGC UV ACT=TGC
TGAACG TGAACG

b

5' 3'

H O
N C
O=C H N
C C
CH3
H O
N C
O=C N
C C
CH3 H
Cyclobutane

3' 5'

#### Gamma and X Radiation

The more energetic gamma rays and X rays, like the UV rays, can interact directly with the DNA molecule. However, they cause most of their damage by affecting water molecules, which ionize, bringing about mutagenic effects by generating free radicals, the most important being the hydroxyl radical, ·OH. This free radical immediately reacts with a DNA molecule, as well as other macromolecules. It can change the structure of a base, but it frequently causes a single- or double-stranded break.

## Genetic Recombination

Genetic recombination is the process in which a new recombinant chromosome, with a genotype different from that of either parent, is formed by combining genetic material from two organisms. It results in a new arrangement of genes or parts of genes and normally is accompanied by a phenotypic change. The process consisting of breakage and rejoining of DNA molecules is called a **crossover.**

By far the most common form of recombination depends on extensive sequence similarity between the two participating DNA molecules. It is called **general** or **homologous recombination.** Homologous recombination is extremely important to all organisms. However, it is also very complex. Here, the focus is on a general overview of bacterial recombination and on the introduction of both bacterial plasmids and transposable elements. The reader is referred to chapter 10 of Snyder and Champness (2003) for a more detailed discussion.

A common feature of all the forms of gene transfer between bacteria, except for the transfer of plasmids, is the requirement for the transferred piece of DNA (donor) to be inserted into the recipient chromosome (acceptor) by homologous recombination. The reason for the exception is that plasmids are independent replicons and can be replicated and inherited without inserting into the chromosome.

Homologous recombination can occur between any two DNA sequences that are the same or very similar, and it usually involves the breaking of the DNA molecules in the same region, where the sequences are similar, and the joining of one DNA to the other. Figure 4.11 shows a simplified diagram of homologous recombination. Staggered breaks are made in the two DNA molecules at the sites where their nucleotide sequences are

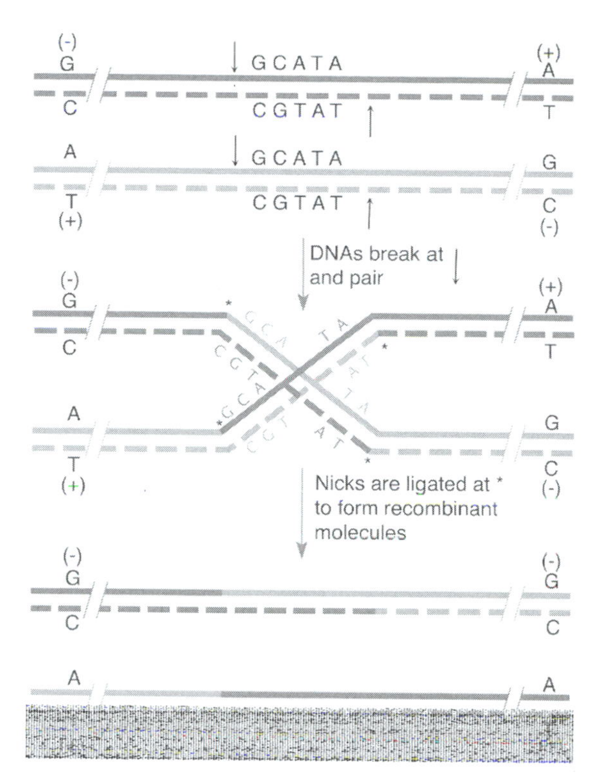


**Figure 4.11** A simplified diagram of homologous recombination. –, mutation; +, wild type. Reprinted from L. Snyder and W. Champness, *Molecular Genetics of Bacteria,* 2nd ed. (ASM Press, Washington, D.C., 2003), with permission from the American Society for Microbiology.

homologous on both molecules. The strands then cross over, and the ends are joined to form two new DNA molecules. This model is oversimplified, but it illustrates how two DNA molecules in a cell link up and exchange sequences. The detailed mechanisms by which recombination occurs are still under investigation.

## Extrachromosomal Genetic Elements

**Plasmids** are autonomous units that exist in the cell as extrachromosomal genomes. They are small, circular (with a few exceptions), double-stranded supercoiled extrachromosomal DNA molecules (Fig. 4.1), that exist independently of host chromosomes and are present in many bacteria. In many cases, they are quite small, just a few kilobases in length; however, in some organisms, notably members of the genus *Pseudomonas,* plasmids with up to several hundred kilobases are common. Most plasmids are only about 0.2 to 4% of the size of the bacterial DNA chromosome.

DNA elements, such as plasmids, transposons, insertion sequences, integrons, transducing phages, and naked DNA, are required for bacterial genetic exchange.

### TYPES OF PLASMIDS AND THEIR BIOLOGICAL SIGNIFICANCE

Plasmids may be classified in terms of their mode of existence and spread. An **episome** is a plasmid that can exist either with or without being integrated into the host chromosome. Some plasmids, **conjugative plasmids,** carry genes for pili and can transfer copies of themselves to other bacteria during conjugation. A brief summary of the types of plasmids and their properties is given in Table 4.1.

#### Fertility Factors

A plasmid called the fertility factor (**F factor**) plays a major role in conjugation in *E. coli* and was the first plasmid to be described (Fig. 4.12). The F factor is about 94.5 kb long and carries genes responsible for cell attachment and plasmid transfer between specific bacterial strains during conjugation. Most of the information required for plasmid transfer is located in the *tra* operon, which contains about 21 genes. Many of these direct the formation of sex pili, which attach the $F^+$ cell (the donor cell containing an F plasmid) to an $F^-$ cell (Fig. 1.26). Other gene products aid DNA transfer.

The F factor also has several segments, called insertion sequences, that assist plasmid integration into the host cell chromosome. Thus, the F factor is an episome that can exist outside the bacterial chromosome or be integrated into it (Fig. 4.13).

#### Resistance Factors

Plasmids often confer antibiotic resistance to the bacteria that contain them. **R factors** (or R plasmids) typically have genes that code for enzymes capable of destroying or modifying antibacterial agents. They are seldom integrated into the host chromosome. Genes coding for resistance to antibacterial agents such as ampicillin, chloramphenicol, and kanamycin have been found in plasmids. Some R factors have only a single resistance gene, whereas other can have as many as eight. Often the resistance genes are within a transposon, and thus it is possible for bacterial strains to develop multiple resistance plasmids.

Because many R factors also are conjugative plasmids, they can spread through a population, although not as rapidly as the F factors. Often, nonconjugative R factors also move between bacteria during plasmid-promoted conjugation. Thus, a whole population can become resistant to antibacterial agents. The fact that some of these plasmids are readily transferred between species further promotes the spread of resistance.

Transfer of DNA occurs most often via plasmids. Within plasmids, resistance genes are often carried by transposons, which can shuttle determinants between

**Table 4.1** Major types of plasmids[a]

| Type | Representatives | Approx size (kb) | Copy no. | Host(s) | Phenotypic features[b] |
|---|---|---|---|---|---|
| F factor[c] | F factor | 95–100 | 1–3 | *E. coli, Salmonella, Citrobacter* | Sex pilus, conjugation |
| R factors | RP4 | 54 | 1–3 | *Pseudomonas* and many other gram-negative bacteria | Sex pilus, conjugation; Ap, Km, Nm, Tc |
| | R1 | 80 | 1–3 | Gram-negative bacteria | Ap, Km, Su, Cm, Sm |
| | R6 | 98 | 1–3 | *E. coli, Proteus mirabilis* | Su, Sm, Cm, Tc, Km, Nm |
| | R100 | 90 | 1–3 | *E. coli, Shigella, Salmonella, Proteus* | Cm, Sm, Su, Tc, Hg |
| | pSH6 | 21 | | *S. aureus* | Gm, Tm, Km |
| | pSJ23a | 36 | | *S. aureus* | Pn, Asa, Hg, Gm, Km, Nm |
| | pAD2 | 25 | | *E. faecalis* | Em, Km, Sm |
| Col plasmids | ColE1 | 9 | 10–30 | *E. coli* | Colicin E1 production |
| | ColE2 | | 10–15 | *Shigella* | Colicin E2 production |
| | CloDF13 | | | *E. cloacae* | Cloacin DF13 production |
| Virulence plasmids | Ent (P307) | 83 | | *E. coli* | Enterotoxin production |
| | K88 plasmid | | | *E. coli* | Adherence antigens |
| | ColV-K30 | 2 | | *E. coli* | Siderophore for iron uptake; resistance to immune mechanism |
| | pZA10 | 56 | | *S. aureus* | Enterotoxin B |
| Metabolic plasmids | CAM | 230 | | *Pseudomonas* | Camphor degradation |
| | SAL | 56 | | *Pseudomonas* | Salicylate degradation |
| | TOL | 75 | | *P. putida* | Toluene degradation |
| | pJP4 | | | *Pseudomonas* | 2,4-Dichlorophenoxyacetic acid degradation |
| | | | | *E. coli, Klebsiella, Salmonella* | Lactose degradation |
| | | | | *Providencia* | Urease |
| | sym | | | *Rhizobium* | Nitrogen fixation and symbiosis |

[a] Reprinted from L. M. Prescott, J. P. Harley, and D. A. Klein, *Microbiology*, 4th ed. (McGraw-Hill, Boston, Mass., 1999), with permission from the publisher.

[b] Abbreviations used for resistance to antibiotics and metals: Ap, ampicillin; Asa, arsenate; Cm, chloramphenicol; Em, erythromycin; Gm, gentamicin; Hg, mercury; Km, kanamycin; Nm, neomycin; Pn, penicillin; Sm, streptomycin; Su, sulfonamides; Tc, tetracycline.

[c] Many R plasmids, metabolic plasmids, and others are also conjugative.

more and less promiscuous plasmids or into and out of the chromosome. The problem of multiresistance is increasing and becoming more complex. Nowadays, it is common to encounter members of the *Enterobacteriaceae* with five or six plasmids.

## Col Plasmids

Bacteria also harbor plasmids with genes that may give them a competitive advantage in the microbial world. **Bacteriocins** are bacterial proteins that destroy other bacteria. They usually act against closely related strains. Bacteriocins often kill cells by forming channels in the plasma membrane, thus increasing its permeability. They also may degrade DNA and RNA or attack peptidoglycan and weaken the cell wall. Col plasmids contain genes for the synthesis of bacteriocins known as **colicins.** For example, one Col plasmid produces colicins that kill *Enterobacter* species. Clearly the host is unaffected by the bacteriocin it produces. Some Col plasmids are conjugative and also can carry resistance genes.

Most bacteriocins that have been identified are peptides or proteins and are produced by gram-negative bacteria. For example, *E. coli* synthesizes colicins, which are encoded by several different plasmids (ColB, ColE1, ColE2, ColI, and ColIV) (Table 4.1). Some colicins bind to specific receptors on the cell envelope of sensitive target bacteria and cause cell lysis, attack specific intracellular sites such as ribosomes, or disrupt energy production.

Bacteriocins may give their producers an adaptive advantage in comparison with other bacteria.

## Other Types of Plasmids

Several other important types of plasmids have been discovered. Some plasmids, called **virulence plasmids,** make their hosts more pathogenic because the bacterium is

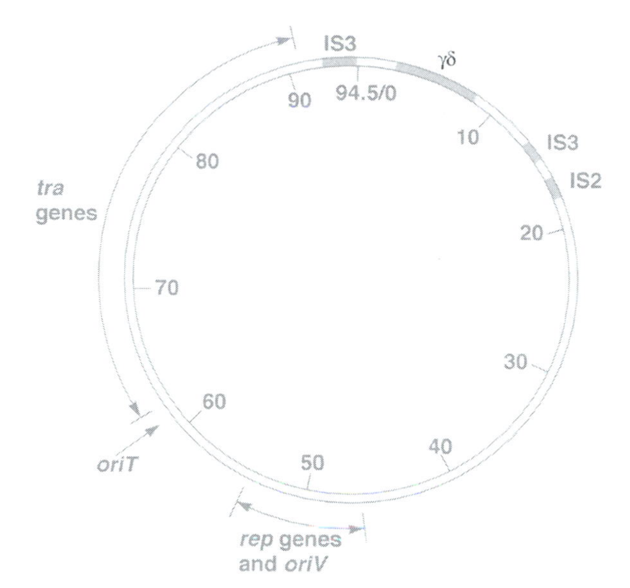


**Figure 4.12** Genetic map of the F (fertility) factor of *E. coli.* The numbers on the interior show the size of the plasmid in kilobase pairs. The approximate locations of several insertion sequences are shown. The *tra* (transfer function) genes code for proteins needed in pilus synthesis and conjugation. The *rep* (replication) genes code for proteins involved in DNA replication. *oriV,* also called *oriS* (origin of replication), is the initiation for circular DNA replication, and *oriT* (origin of transfer), is the site for initiation of rolling-circle replication and gene transfer during conjugation. Reprinted from L. M. Prescott, J. P. Harley, and D. A. Klein, *Microbiology,* 4th ed. (McGraw-Hill, Boston, Mass., 1999), with permission from the publisher.

better able to resist host defense or to produce toxins. For example, enterotoxigenic strains of *E. coli* cause traveler's diarrhea because of a plasmid that codes for an enterotoxin. **Metabolic plasmids** carry genes for enzymes that degrade substances such as aromatic compounds (e.g., toluene), pesticides (e.g., 2,4-dichlorophenoxyacetic acid), and sugars (e.g., lactose). Metabolic plasmids even carry the genes required for some strains of *Rhizobium* to induce legume nodulation and carry out nitrogen fixation.

### Replication of Plasmids

Most plasmids in gram-negative bacteria replicate in a manner similar that of the chromosome. Most plasmids in gram-positive bacteria replicate by a rolling-circle mechanism. The reader is directed to textbooks of molecular genetics of bacteria or microbiology or to specialized reviews for additional information about the replication of plasmids. Some of these references are listed at the end of this chapter.

## TRANSPOSABLE ELEMENTS

Transposable elements are segments of DNA with the ability to move from one site on a chromosome to another. Such movement is called **transposition.** DNA segments that carry the genes required for this process and consequently move around chromosomes are called **transposable elements** or **transposons.** Unlike other processes that reorganize DNA, transposition does not require extensive areas of homology between the transposon and its destination site. They were first discovered in the 1940s by Barbara McClintock during her studies on maize genetics.

The simplest transposable elements are **insertion sequences** or IS elements (Fig. 4.14). An IS element is a short sequence of DNA (around 750 to 1,600 bp long) containing only the genes for enzymes required for its transposition and bounded at both ends by identical or very similar sequences of nucleotides in reverse orientation, known as inverted repeats (Fig. 4.14c). Inverted repeats are usually about 15 to 25 bp long. Between the inverted repeats is a gene that codes for an enzyme called **transposase.** This enzyme is required for transposition and accurately recognizes the ends of the IS. The transposon itself usually encodes its own transposases, so that it carries with it the ability to hop each time it moves. For this reason, transposons have been called "jumping genes."

Transposable elements also can contain genes other than those required for transposition (for example, antibacterial resistance or toxic genes). These elements are often called **composite transposons** or elements. Composite transposons often consist of a central region containing the extra genes, flanked on both sides by IS elements that are identical or very similar in sequence (Fig. 4.14b). Transposition in prokaryotes involves a series of events, including self-replication and recombinational processes. Typically in bacteria, the original transposon remains at the parental site on the chromosome while a replicated copy inserts in the target DNA (Fig. 4.14c). This is called **replicative transposition.** Transposition of the Tn*3* transposon is a well-studied example of replicative transposition. This process requires the Tn*3* transposase enzyme, encoded by the *tnpA* gene (Fig. 4.15).

Transposable elements produce a variety of important effects. They can insert within a gene to cause a mutation or stimulate DNA rearrangement, leading to deletions of genetic material. Because some transposons carry stop codons or termination sequences, they may block translation or transcription. Transposons are also located within plasmids and participate in such processes as plasmid fusion and the insertion of F plasmids into the *E. coli* chromosome.

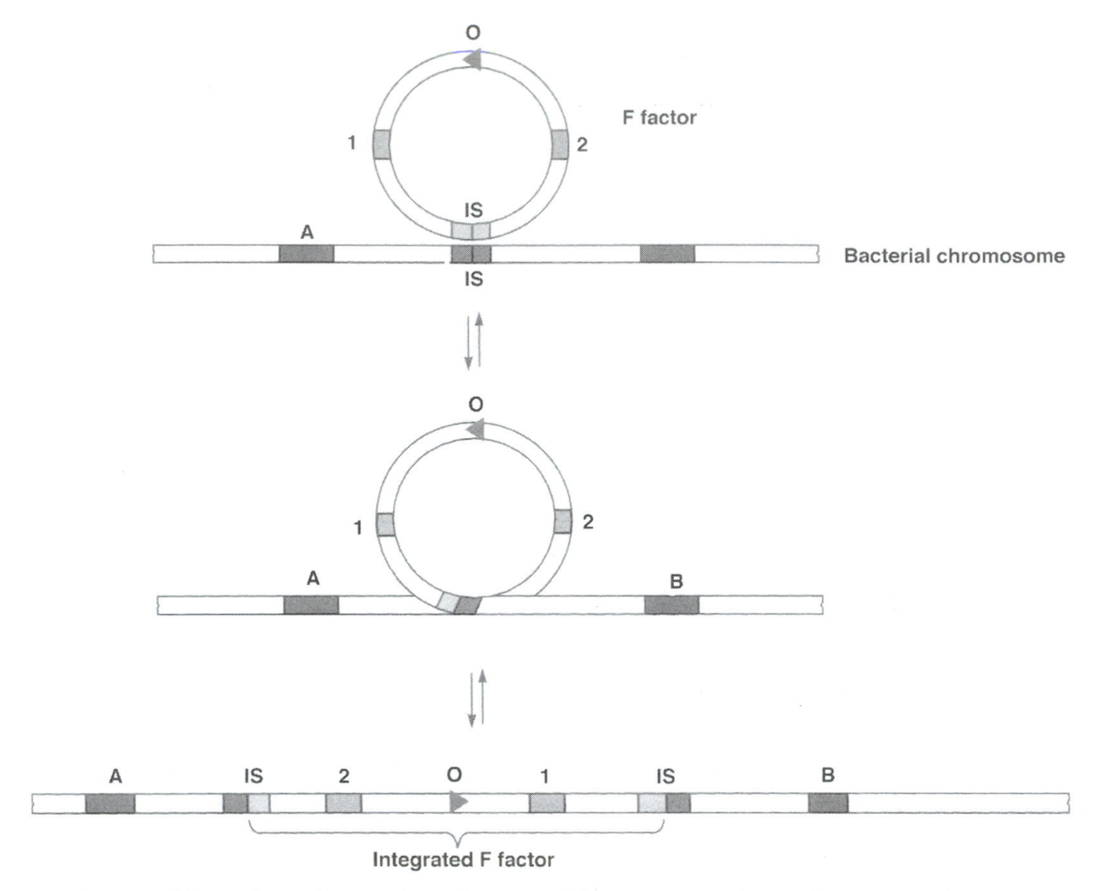


**Figure 4.13** F factor integration. The reversible integration of an F factor into a host bacterial chromosome is shown. The process begins with the association between plasmid and bacterial insertion sequences. The arrowhead labeled O indicates the site at which the oriented transfer of chromosome to the recipient cell begins. A, B, 1, and 2 represent genetic markers. Reprinted from L. M. Prescott, J. P. Harley, and D. A. Klein, *Microbiology,* 4th ed. (McGraw-Hill, Boston, Mass., 1999), with permission from the publisher.

In the above discussion of plasmids, it was noted that an R plasmid can carry genes for resistance to several drugs. Transposons have antibacterial agents resistance genes and play a major role in generating these plasmids. Consequently, these elements cause serious problems in the treatment of infectious diseases. Multiple-antibacterial-agent resistance plasmids probably often arise from transposon accumulation on a single plasmid. Because transposons also move between plasmids and primary chromosomes, drug resistance genes can exchange between plasmids and chromosomes, resulting in the further spread of antibacterial agents resistance. Clearly, transposable elements play an extremely important role in the generation and transfer of new gene combinations.

## Mobile Gene Cassettes and Integrons

An **integron** is a genetic unit that includes the determinants of the components of a site-specific recombination system capable of capturing and mobilizing genes that are contained in mobile elements called **gene cassettes.**

The cassettes are mobile elements that include a gene (most commonly an antibiotic resistance gene) and an integrase-specific recombination site that is a member of a family of sites known as 59-base elements. Cassettes can exist either free in a circularized form or integrated at the *att*1 site, and only when integrated formally is a cassette part of an integron.

It is now well established that many antibiotic resistance genes found in clinical isolates of gram-negative bacteria (particularly members of the *Enterobacteriaceae* and pseudomonads) determine resistance to a range of antibiotics (aminoglycosides, trimethoprim, chloramphenicol, penicillins, and cephalosporins). Although free circular cassettes cannot replicate, they are likely to be important participants in the process which leads initially to the association of a cassette with an integron

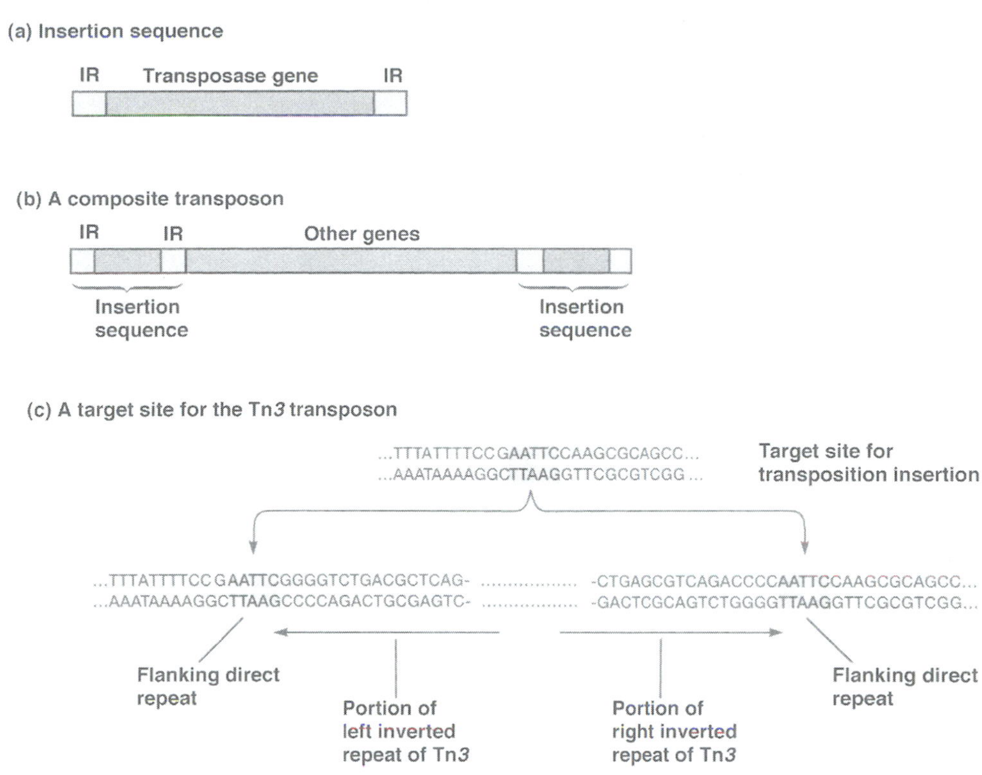


**Figure 4.14** Insertion sequences and transposons. The structures of insertion sequences (a), composite transposons (b), and target sites (c), are shown. IR, inverted repeat. In panel c, the highlighted 5-base target site is duplicated during Tn*3* transposition to form flanking direct repeats. The remainder of Tn*3* lies between the inverted repeats. Reprinted from L. M. Prescott, J. P. Harley, and D. A. Klein, *Microbiology,* 4th ed. (McGraw-Hill, Boston, Mass., 1999), with permission from the publisher.

and subsequently to the spread of resistance genes from one integron to another.

Mobile cassette genes were first identified as the various integrated antibiotic resistance genes found in integrons. In the early 1980s, studies using restriction mapping and heteroduplex analysis revealed that different sets of antibiotic resistance genes were sometimes found in the same location in otherwise closely related plasmids such as R388 and pSa or transposons such as Tn*21*, Tn*2603*, and Tn*2424*. Cassettes do not generally include a promoter and are transcribed from a promoter in the integron.

The structure of gene cassettes as a new class of mobile element is discussed in detail in recent reviews; readers are directed to these reviews for additional information.

## Gene Transfer

The bacteria exchange genetic information in nature. There are three fundamental distinct mechanisms by which such genetic transfer can occur:

- conjugation, which involves the direct transfer of DNA from one cell to another
- transformation, in which a cell takes up isolated DNA molecules or fragments of DNA from the medium surrounding it
- transduction, in which the transfer is mediated by bacterial viruses (bacteriophages)

### Bacterial Conjugation

Bacterial conjugation is a process of direct transfer of DNA from one bacterial cell to another (**cell-to-cell contact**).

Conjugation involves a **donor cell,** which contains a conjugative plasmid, and a **recipient cell,** which does not. The genes that control conjugation are contained in the *tra* region of the plasmid. Many genes of the *tra* region are required for the synthesis of the sex pilus. Only the donor cell has these pili (Fig. 1.26). Pili allow specific pairing to take place between the donor cell and the recipient cell. All conjugation in gram-negative bacteria is thought to depend on cell pairing brought about

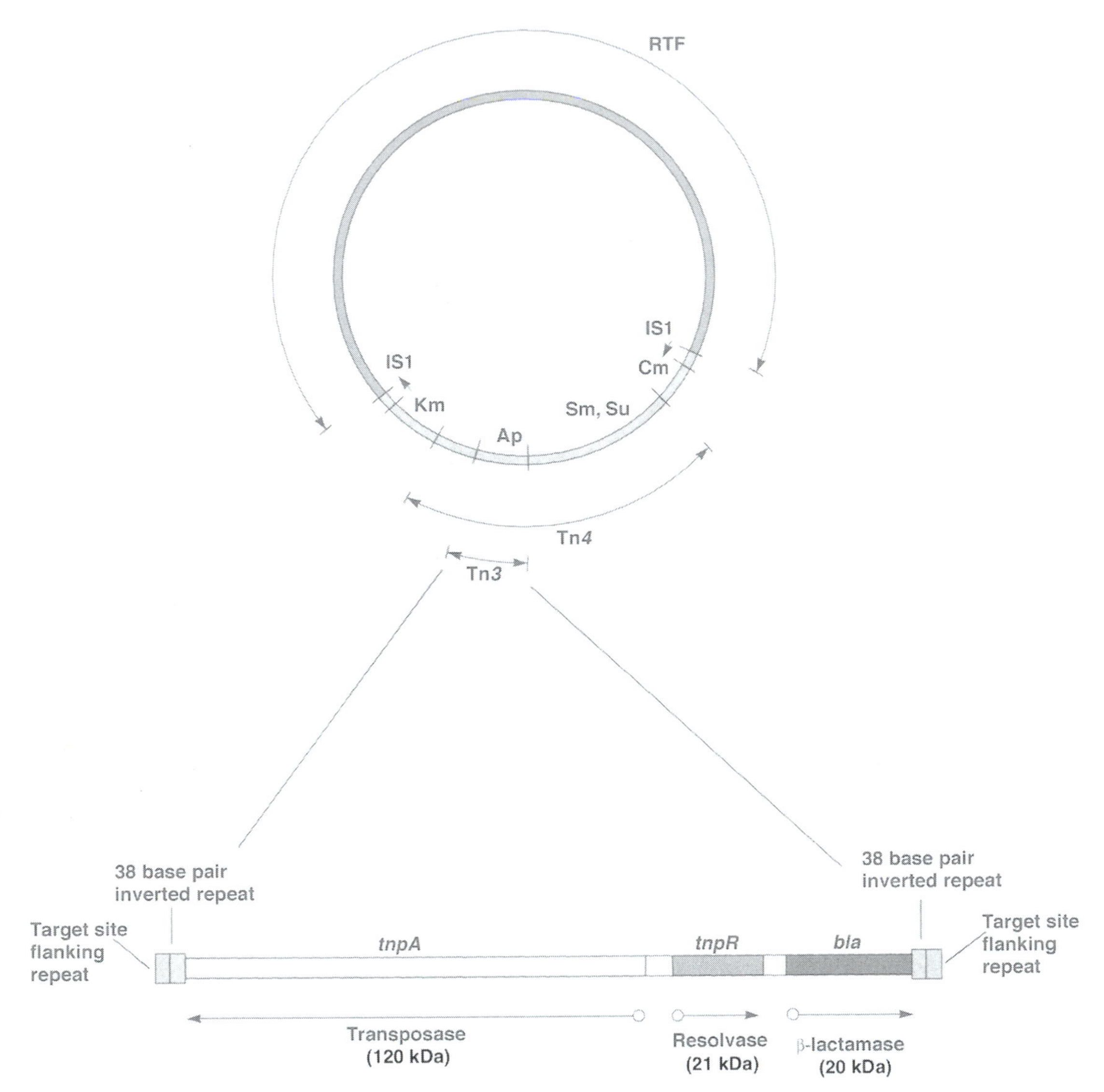


**Figure 4.15** Structures of R factors and transposons. The R plasmid carries resistance genes for five antibacterial agents: chloramphenicol (Cm), streptomycin (Sm), sulfonamide (Su), ampicillin (Ap), and kanamycin (Km). These are contained in the Tn*3* and Tn*4* transposons. The resistance transfer factor (RTF) codes for the proteins necessary for plasmid replication and transfer. The structure of Tn*3* is shown in more detail. The arrows indicate the direction of gene transcription. Reprinted from L. M. Prescott, J. P. Harley, and D. A. Klein, *Microbiology,* 4th ed. (McGraw-Hill, Boston, Mass., 1999), with permission from the publisher.

by pili. The pili make specific contact with a receptor on the recipient cell and retract, pulling the two cells together.

Transfer of plasmid DNA from the donor cell to the recipient cell is initiated by a protein that makes a single-strand break (nick) at a specific site in the DNA known as the origin of transfer (*oriT*). A plasmid-encoded helicase unwinds the plasmid DNA, and the single nicked strand is transferred to the recipient cell, starting with the 5' end generated by the nick. Concurrently, the free 3' end of the nicked strand is extended to replace the DNA transferred, by a process of replication known as a rolling circle. The nicking protein is thought to remain attached to the 5' end of the transferred DNA. DNA synthesis in the recipient cell converts the transferred single strand into a double-stranded molecule (Fig. 4.16).

Conjugation occurs most frequently in gram-negative bacteria (such as *E. coli* and related members of the

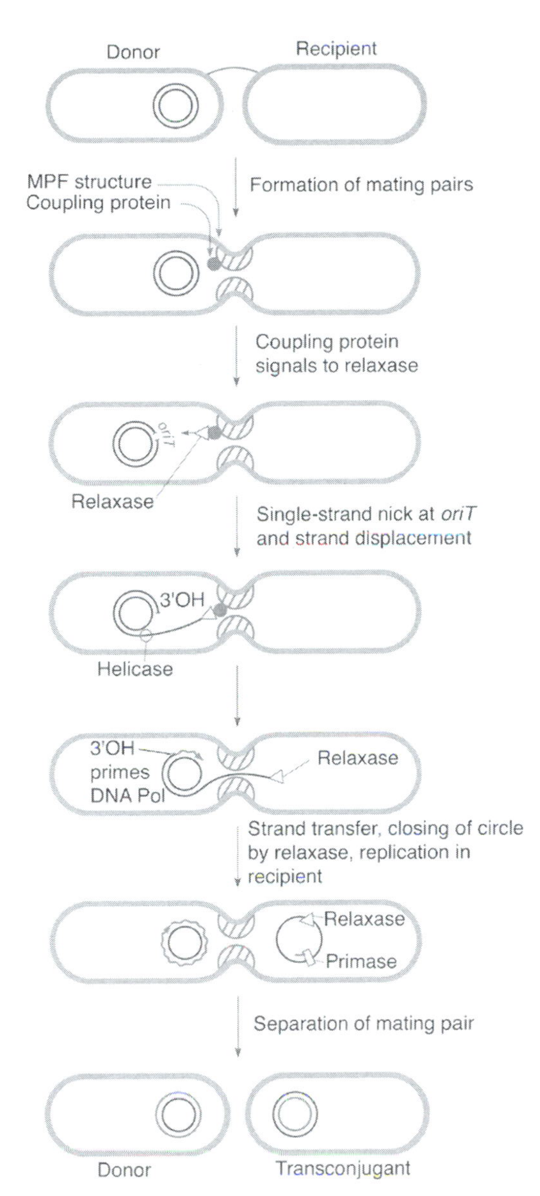


**Figure 4.16** Mechanism of DNA transfer during conjugation, showing mating-pair formation (Mpf functions). The donor cell produces a pilus, which forms on the cell surface and which may contact a potential recipient cell and bring it into close contact or may help hold the cells in close proximity after contact has been made, depending on the type of pilus. A pore then forms in the adjoining cell membrane. On receiving a signal from the coupling protein that contact with a recipient has been made, the relaxase protein makes a single-stranded cut at the *oriT* site in the plasmid. A plasmid-encoded helicase then separates the strands of the plasmid DNA. The relaxase protein, which has remained attached to the 5' end of the single-stranded DNA, is then transported out of the donor cell through the channel directly into the recipient cell, dragging the single-stranded attached DNA along with it. Once in the recipient, the relaxase protein helps recyclize the single-stranded DNA. A primase, either that of the host or plasmid encoded and injected with the DNA, then primes replication of the complementary strand to make the double-stranded circular plasmid DNA in the recipient. The 3' end at the nick made by the relaxase in the donor can also serve as a primer, making a complementary copy of the single-stranded plasmid DNA remaining in the donor. Therefore, after transfer, both the donor and the recipient bacteria end up with a double-stranded circular copy of the plasmid. Reprinted from L. Snyder and W. Champness, *Molecular Genetics of Bacteria,* 2nd ed. (ASM Press, Washington, D.C., 2003), with permission from the American Society for Microbiology.

*Enterobacteriaceae*) but is also found in some gram-positive genera such as *Streptomyces* and *Streptococcus.*

## DNA Transformation

Another way in which DNA can move between bacteria is through transformation. Transformation is the uptake by a cell of a naked DNA molecule or a fragment of DNA from the medium and the incorporation of this molecule into the recipient chromosome in a heritable form.

When bacteria lyse, they release a considerable amount of DNA into the surrounding environment. These fragments may be relatively large and contain several genes. The ability of a cell to take up a molecule of DNA is associated with the development of a **competent** state, in which a DNA acceptor on the cell and transformation-specific proteins are present. Competency is a complex phenomenon and is dependent on several conditions. Bacteria need to be in a certain stage of growth; for example, *Streptococcus pneumoniae* becomes competent during the exponential phase when the population reaches about $10^7$ to $10^8$ cells per ml. When a population becomes competent, bacteria such as pneumococci secrete a small protein called the competence factor, which stimulates the production of 8 to 10 new proteins required for transformation.

A number of gram-positive (*Bacillus subtilis, S. pneumoniae,* and *Enterococcus faecalis*) and gram-negative (*Neisseria gonorrhoeae, Neisseria meningitidis, Acinetobacter* spp., *Pseudomonas* spp., *Moraxella* spp.,

*Haemophilus influenzae, Campylobacter jejuni,* and *Helicobacter pylori*) organisms are naturally competent or naturally transformable, meaning that they readily take up extracellular DNA.

### MECHANISM OF TRANSFORMATION

There are four steps in transformation: development of competence, binding of DNA to the cell surface, processing and uptake of free DNA, and integration of the DNA into the chromosome by recombination. The mechanism of transformation (Fig. 4.17) has been extensively studied in *S. pneumoniae.* A double-stranded DNA released from lysed cells binds noncovalently to cell surface receptors. The bound double-stranded DNA is nicked and cleaved into smaller fragments by membrane-bound endonucleases, and one of the two DNA strands is exonucleotically cleaved by a membrane-bound DNase, allowing the remaining single strand to enter the cell through a membrane-spanning DNA translocation channel. The single-stranded DNA is coated with proteins to protect the DNA from further nuclease digestion. The transformed DNA integrates into the chromosome and replaces the chromosomal DNA fragments by homologous recombination; this process requires the *recA* gene. Sometimes the transformed DNA is a plasmid capable of replicating autonomously from the chromosome.

**Figure 4.17** Transformation in *Haemophilus influenzae.* Double-stranded DNA is first taken up in transformasomes. One strand is degraded, and the other strand invades the chromosome, displacing one chromosome strand. Reprinted from L. Snyder and W. Champness, *Molecular Genetics of Bacteria,* 2nd ed. (ASM Press, Washington, D.C., 2003), with permission from the American Society for Microbiology.

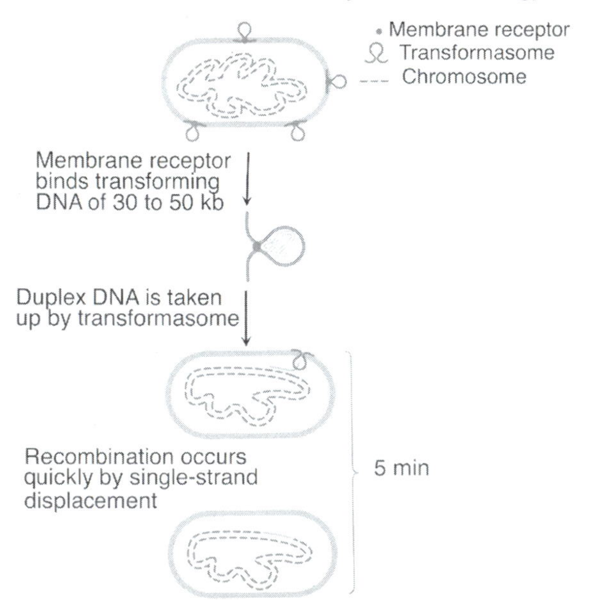


## Transduction

In the third mode of bacterial gene transfer, a bacteriophage transfers the genetic material. There are two kinds of transduction: generalized and specialized. Transduction may be the most common mechanism for gene exchange and recombination in bacteria.

### GENERALIZED TRANSDUCTION

In **generalized transduction,** virtually any genetic marker can be transferred from donor to a recipient. An example of the way in which transducing particles may be formed is given in Fig. 4.18.

When the population of sensitive bacteria is infected with a phage, the events of the lytic cycle after adsorption and penetration are initiated. Phages that reproduce

**Figure 4.18** An example of generalized transduction. A phage infects a Trp⁺ bacterium, and in the course of packaging DNA into heads, the phage mistakenly packages some bacterial DNA containing the *trp* region instead of its own DNA into a head. In the next infection, this transducing phage injects the Trp⁺ bacterial DNA instead of phage DNA into the Trp⁻ bacterium. If the incoming DNA recombines with the chromosome, a Trp⁺ recombinant transductant may arise. Only one strand of the DNA is shown. Reprinted from L. Snyder and W. Champness, *Molecular Genetics of Bacteria,* 2nd ed. (ASM Press, Washington, D.C., 2003), with permission from the American Society for Microbiology.

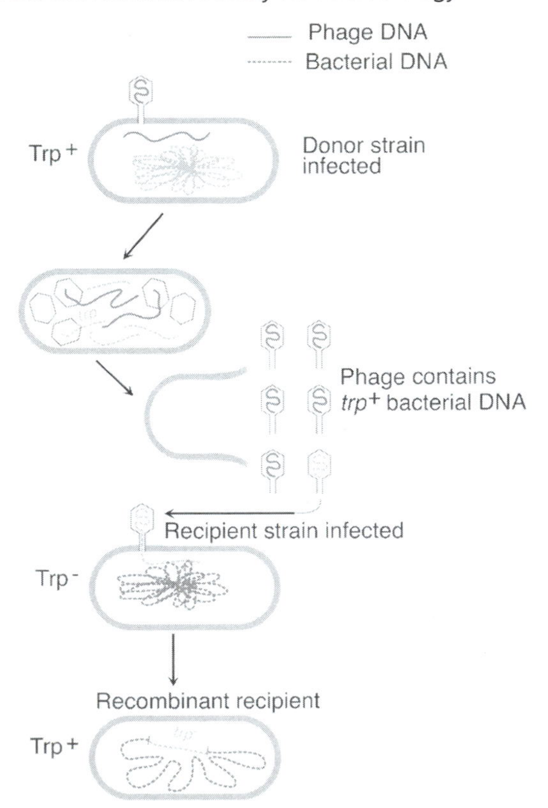

using a lytic cycle often are called **virulent bacteriophages** because they destroy the host cell. The key step involved in transduction is the packaging of DNA into the phage. This process is normally highly specific for phage DNA. However, with some phages, fragments of bacterial DNA (produced by phage-mediated degradation of the bacterial chromosome) are incorporated into a phage capsid because of an error made during the virus life cycle. The virus containing these transducing particles injects them into another bacterium (recipient cell), completing the transfer. In **specialized transduction,** the transducting particles carry only a specific portion of the bacterium genome.

## Significance of the Origin, Transmission, and Spread of Bacterial Resistance Genes

The spread of bacteria with resistance to diverse antibacterial agents is one of the most serious threats to the successful treatment of bacterial infections. Resistant mutants are spontaneously formed. Early on, these bacteria were a minority, although through the years they became a majority because they survived antibacterial action that resulted in the lysis of the sensitive population of bacteria.

Frequently, a bacterial pathogen is resistant to antibiotics because of its R plasmids. Plasmid resistance genes often code for enzymes that destroy or modify the activity of the antibacterial agents, for example, the production of β-lactamases that hydrolyze β-lactam antibiotics or enzymes that catalyze the acetylation of chloramphenicol. Once a bacterial cell possesses an R plasmid, the plasmid may be transferred to another bacterial species quite rapidly through normal gene exchange process such as conjugation, transformation, and transduction. Because a single plasmid can carry several transposable elements, a bacterial pathogen can become resistant to several antibacterial agents simultaneously. Other aspects of bacterial resistance are covered in more detail in later chapters when particular groups of antibiotics and synthetic antibacterial agents are considered.

## REFERENCES AND FURTHER READING


### General (Books)

**Dale, J. W.** 1998. *Molecular Genetics of Bacteria,* 3rd ed. John Wiley & Sons, Ltd., Chichester, United Kingdom.

**Klug, W. S., and M. R. Cummings.** 2001. *Essentials of Genetics,* 4th ed. Prentice-Hall, Upper Saddle River, N.J.

**Madigan, M. T., J. M. Martinko, and J. Parker.** 2000. *Brock Biology of Microorganisms,* 9th ed. Prentice Hall, Upper Saddle River, N.J.

**Prescott, L. M., J. P. Harley, and D. A. Klein.** 1999. *Microbiology,* 4th ed. McGraw-Hill, Boston, Mass.

**Snyder, L., and W. Champness.** 2003. *Molecular Genetics of Bacteria,* 2nd ed. ASM Press, Washington, D.C.

**Winter, P. C., G. I. Hickey, and H. L. Fletcher.** 1998. *Genetics.* Springer Verlag, New York, N.Y.

### Chromosome

**Drlica, K.** 2000. Chromosome, bacterial, p. 808–821. *In* J. Lederberg (ed.), *Encyclopedia of Microbiology,* 2nd ed., vol. 1. Academic Press, Inc., San Diego, Calif.

### Plasmids

**Khan, S. A.** 1997. Rolling-circle replication of bacterial plasmids. *Microbiol. Mol. Biol. Rev.* **61:**442–455.

**Thomas, C. M.** 2000. Plasmids, bacterial, p. 711–729. *In* J. Lederberg (ed.), *Encyclopedia of Microbiology,* 2nd ed., vol. 3. Academic Press, Inc., San Diego, Calif.

### Bacterial Genetic Exchange

**De la Cruz, F., and J. Davies.** 2000. Horizontal gene transfer and the origin of species: lessons from bacteria. *Trends Microbiol.* **8:**128–133.

**Heinemann, J. A.** 2000. Horizontal transfer of genes between microorganisms, p. 698–707. *In* J. Lederberg (ed.), *Encyclopedia of Microbiology,* 2nd ed., vol. 2. Academic Press, Inc., San Diego, Calif.

**MaGee, D. J., C. Coker, J. M. Harro, and H. L. T. Mobley.** Bacterial genetic exchange. Accepted for publication in *Encyclopedia of Life Sciences.* Nature Publishing, London, United Kingdom.

### Conjugation

**Frost, L. S.** 2000. Conjugation, bacterial, p. 847–862. *In* J. Lederberg (ed.), *Encyclopedia of Microbiology,* 2nd ed., vol. 1. Academic Press, Inc., San Diego, Calif.

### Transformation

**Wilkins, B. M., and P. A. Meacock,** 2000. Transformation, genetic, p. 651–665. *In* J. Lederberg (ed.), *Encyclopedia of Microbiology,* 2nd ed., vol. 4. Academic Press, Inc., San Diego, Calif.

### Transduction

**Masters, M.** 2000. Transduction: host DNA transfer by bacteriophages, p. 637–650. *In* J. Lederberg (ed.), *Encyclopedia of Microbiology,* 2nd ed., vol. 4. Academic Press, Inc., San Diego, Calif.

### Transposable Elements

**Bennett, P. M.** 2000. Transposable elements, p. 704–724. *In* J. Lederberg (ed.), *Encyclopedia of Microbiology,* 2nd ed., vol. 4. Academic Press, Inc., San Diego, Calif.

### Conjugative Transposition

**Rice, L. B.** 1998. Tn*916* family conjugative transposons and dissemination of antimicrobial resistance determinants. *Antimicrob. Agents Chemother.* **42:**1871–1877.

**Salyers, A. A., N. B. Shoemaker, A. M. Stevens, and L.-Y. Li.** 1995. Conjugative transposons: an unusual and diverse set of integrated gene transfer elements. *Microbiol. Rev.* **59**:579–590.

**Scott, J. R., and G. G. Churchward.** 1995. Conjugative transposition. *Annu. Rev. Microbiol.* **49**:367–397.

### Mobile Gene Cassettes and Integrons

**Bennett, P. M.** 1999. Integrons and gene cassettes: a genetic construction kit for bacteria. *J. Antimicrob. Chemother.* **43**:1–4.

**Hall, R. M., and C. M. Collis.** 1995. Mobile gene cassettes and integrons: capture and spread of genes by site-specific recombination. *Mol. Microbiol.* **15**:593–600.

**Recchia, G. D., and R. M. Hall.** 1995. Gene cassettes: a new class of mobile element. *Microbiology* **141**:3015–3027.

**Stokes, H. W., D. B. O'Gorman, G. D. Recchia, M. Parsekhian, and R. M. Hall.** 1997. Structure and function of 59-base element recombination sites associated with mobile gene cassettes. *Mol. Microbiol.* **26**:731–745.

### Mutation Frequencies

**Martinez, J. L., and F. Baquero.** 2000. Mutation frequencies and antibiotic resistance. *Antimicrob. Agents Chemother.* **44**:1771–1777.

chapter 5



# Mechanisms of Bacterial Resistance to the Action of Antibacterial Agents

## Definition of Bacterial Resistance

Currently, the published definitions of antibacterial (antibiotic) resistance vary depending on the purpose of each study.

Bacterial resistance to antibacterial agents is a quantitative measurement of the efficiency (concentration expressed in micrograms per milliliter or as inhibition zones in millimeters for the diffusion techniques) of an antibacterial agent against a specific bacterium. Available methods for the in vitro measurement of antibacterial activity are based on testing increasing concentrations of an antibacterial agent against a bacterial isolate to find out at which concentration the growth of the bacterium is inhibited. This is known as the **minimum inhibitory concentration** (MIC) of the drug. Indeed, the MIC is a relative measurement of the smallest amount of antibacterial agent required to inhibit the growth (cell division) of a bacterium.

## Background

Bacterial drug resistance is one of the most important problems in present-day antibacterial chemotherapy. Bacterial resistance to an antibacterial agent is best described and defined in comparison to its opposite, bacterial susceptibility. Bacterial resistance to antibacterial agents is a condition in which there is no susceptibility or decreased susceptibility to antibacterial agents that ordinarily cause inhibition of bacterial cell growth or cell death.

Bacterial resistance can be either natural (intrinsic) or acquired. In **acquired resistance** the bacterial population is initially susceptible to antibacterial agents but the bacteria undergo changes either by acquisition of plasmid and transposon or by chromosome mutation, and strains emerge that are less susceptible or not at all susceptible to these antibacterial drugs.

## Historical Perspective

The historic observation in 1929 by Sir Alexander Fleming of the antibacterial properties of a "mold broth filtrate" he had obtained after the growth of the fungus *Penicillium notatum* in a nutrient broth at St. Mary's Hospital in London marked the starting point of antibiosis. In 1940, a research team at Oxford University led by Florey, Chain, and Abraham isolated penicillin in a pure form for the first time. The team discovered the clinical importance of penicillin in the treatment of bacterial infection in human patients. This discovery was the starting point for the development and use of antibiotics to control infections in humans.

One of the most important landmarks in the story of antibiotics was the X-ray crystallographic determination of the structure of penicillin G (earlier named penicillin) by Hodgkin and Rogers-Low in May 1945. They established the fused bicyclic β-lactam–thiazolidine nucleus and the relative stereochemistry of the penicillin molecule.

Also, in 1940 Abraham and Chain first described a bacterial enzyme in *Escherichia coli* that destroyed penicillin. At that time, they termed it penicillinase, since the β-lactam structure of the penicillin was not then known. This observation was the first indication of bacterial resistance to the action of an antibiotic.

Over the past 63 years (1940 to 2003), the emergence and spread of bacterial resistance to the action of antibiotics and synthetic antibacterial agents are certainly the most striking examples of evolution that have arisen in bacteria. During the past decade, bacterial resistance to antibiotics has been increasing as a result of the use, overuse, and misuse of broad-spectrum antibiotics.

In 1999 at a colloquium in San Juan, P.R., the American Academy of Microbiology reviewed the fundamental relationships between the use of antimicrobial agents in human, animals, agriculture, and aquaculture and the effects on humans, animals, and the environment of this usage. Its Committee recognized the risk to human health posed by the growing problem of antibiotic resistance and concluded with recommendations for the future.

## Current Status of Bacterial Resistance

The past decade has witnessed a significant increase in the prevalence of resistance to antibacterial agents. Resistance to antibacterial agents has important implications for morbidity, mortality, and health care in hospitals, as well as in the community.

After a period when relatively few new antibiotics were introduced, we now realize, at the start of the 21st century, that the emergence of strains of antibiotic-resistant bacteria must be fought with a novel generation of antibiotics as well as synthetic antibacterials. A multidisciplinary team of chemists, microbiologists, biologists, geneticists, and representatives of other scientific disciplines are destined to play a crucial role in discovering them.

## Characteristics of Bacterial Resistance

Bacterial resistance is described in terms of either phenotypic (e.g., growth patterns) or genotypic (e.g., presence or expression of genes) characteristics of bacteria, or both, and can be categorized according to origin (intrinsic versus acquired resistance) or type (single or multiple).

Bacteria are ingenious in the manner in which they accumulate different antibacterial resistance genes. For some, these resistance traits may reside on the chromosome, on plasmids, or on integrons. Bacteria can have natural (intrinsic) resistance to antibacterials and can acquire resistance through mutations. Chromosomal genes are passed directly to daughter cells (clonal spread), whereas genes on plasmids, transposons, and bacteriophages, which might include genes originating from the chromosome, can be transmitted horizontally between bacteria of the same or other species or genera.

Once a clinically relevant level of resistance to an antibacterial agent appears in a pathogenic bacterium, the first question to be asked is that of the nature of the origin, evolution, and selection of resistance. The next question is whether the mutant bacterium will spread.

It is now widely accepted that the driving force of antibiotic resistance is the selective pressure of antibacterial agent use. Selection favors the resistant bacterium. Horizontal gene transfer occurs by the movement of antibiotic resistance genes onto plasmid and conjugative transposon vehicles, most frequently by transposition and site-specific recombination mechanisms. Clonal spread of the bacterium containing resistance or multiresistance genes has ultimately led to global dissemination and also to the possibility of transfer to new species and genera.

## Sources of Resistance Genes

Resistance genes arise by mutation and are transferred by plasmid- or transposon-mediated transfer. The major advantage of dissemination of resistance by mutations is that they are stably inherited by the progeny of the mutant bacteria (vertical transfer of resistance genes).

We have already seen that bacteria can acquire foreign DNA by three mechanisms: conjugation, transformation, and transduction. Two types of genetic elements are self-transferable by conjugation: plasmids and transposons. Conjugative plasmids can efficiently transfer among gram-positive and gram-negative bacteria belonging to different genera. Conjugative transposons of gram-positive as well as gram-negative bacteria represent another efficient mode of transfer of antibacterial resistance genes between phylogenetically distant bacteria genera.

Horizontal gene transfer among bacteria is a perpetual phenomenon that has a significant impact on bacterial evolution. Thus, the source of resistance genes is a combination of mutations that are hereditary and plasmid and transposable elements that are self-transferable by conjugation.

The ultimate level of organization for the migrant genetic information is provided by integrons. In this system, isolated gene cassettes can be integrated by site-specific recombination into a genetic unit, which, besides the recombinant system, also provides promoter expression of the incoming DNA. Integrons lead to the construction of resistance operons that are often part of transposons themselves located on self-transferable plasmids.

Table 5.1 summarizes the characteristics of different elements involved in the spread of resistance genes.

## Detection Methods

Tests are commonly used to measure the susceptibility of a bacterium to different concentrations of antibiotics (antibacterials) either in broth or agar culture media or on paper disks. As mentioned earlier in the chapter, the lowest antibacterial agent (antibiotic) concentration that inhibits bacterial growth is termed the minimum inhibitory concentration (MIC), and concentrations that inhibit 50% ($MIC_{50}$) or 90% ($MIC_{90}$) of bacterial growth can indicate shifts in the susceptibility of bacterial populations.

In this section, the meaning and consequences of sub-MIC concentrations of antibiotics are discussed. The most common method of antibiotic administration in the treatment of infectious diseases is intermittent administration; under these conditions, the pharmacokinetic curves show that antibiotic concentrations have a sinusoidal behavior. They exceed MICs only for a certain period, after which they drop to below the MIC until the next administration starts the cycle again.

Although these sub-MIC concentrations do not kill bacteria, they are still capable of modifying their physicochemical characteristics and interfering with some important bacterial cell functions such as adhesiveness, fimbriation, and host-bacterium interactions.

In 1990, the National Committee for Clinical Laboratory Standards (NCCLS) defined the sub-MIC concentration as the concentration of an antimicrobial agent that is not active against bacterial growth but is still active in altering bacterial biochemistry and shape in vitro and in vivo, and thus reducing bacterial virulence.

DNA-based techniques such as PCR are now available to detect resistance genes within bacterial populations. The application of molecular techniques offers the opportunity to determine the presence of genetic determinants within hours, in contrast to assessing the phenotypic characteristics of bacteria, which can require several days.

## General Mechanisms of Bacterial Resistance

It should be kept in mind that bacterial populations are very large and actively growing. Treating a patient with an antibiotic results in a strong selective pressure not only on the pathogen responsible for the disease but also

**Table 5.1** Characteristics of different elements involved in the spread of resistance genes

| Element | Characteristics | Role in spread of resistance genes |
|---|---|---|
| Self-transmissible plasmid | Circular, autonomously replicating element; genes needed for conjugal DNA transfer | Transfer of resistance genes; mobilization of other elements that carry resistance genes |
| Conjugative transposon | Integrated element that can excise to form a nonreplicating circular transfer intermediate; carries genes needed for conjugal DNA transfer | Same as self-transmissible plasmid |
| Mobilizable plasmid | Circular, autonomously replicating element; carries gene that allows it to use the conjugal apparatus provided by a self-transmissible plasmid | Transfer of resistance genes |
| Transposon | Can move from one DNA segment to another within the same cell | Can carry resistance genes from chromosome to plasmid or vice versa |
| Gene cassette | Circular, nonreplicating DNA segment containing only open reading frames; integrates into integrons | Carries resistance genes |
| Integron | Integrated DNA segment that contains an integrase, a promoter, and an integration site for gene cassettes | Forms clusters of resistance genes, all under the control of the integron promoter |

on the entire flora of the patient. The relative minority of naturally resistant bacteria in a population increases with continuation of the antibiotic therapy that eliminates antibiotic-sensitive cells. This process of enrichment is called **selection,** and the continuing presence of the antibiotic is said to exert a selective pressure in favor of the resistant bacteria.

Mechanisms of bacterial resistance to the action of antibacterials refer to the process that enables bacteria to overcome the harmful effects of antibiotics. Although the manner of acquisition of resistance may vary among various bacteria, only a few mechanisms of resistance have been developed by bacteria; however, these mechanisms are highly effective.

Bacteria may prevent the entry of antibiotics and synthetic antibacterial agents into the bacterial cell by changing their cell membrane composition, by decreasing the antibiotic uptake, or by increasing the efflux of the antibiotic. They may also modify antibiotics within bacterial cells by producing cellular enzymes that render the antibiotics inactive or less active. The bacterial cell also may alter the affinity of the target for the antibiotic, resulting in a weaker interaction between the target and the antibiotic. Which of these mechanisms prevails depends on the antibiotic, its target site, the bacterial species, and whether the resistance is mediated by a resistance plasmid (R plasmid) or by a chromosomal mutation.

In this chapter, a general overview of the major mechanisms of bacterial resistance is presented. However, the mechanisms of resistance determined for bacterial isolates are discussed in more detail as each group and subgroup of antibacterial agents is presented in subsequent chapters. The following schemes summarize some of the current mechanisms of bacterial resistance to the action of antibacterial agents. It seems convenient at this stage to describe some recently developed ways to counteract these mechanisms of bacterial resistance.

## Active-Efflux Mechanisms of Resistance

There has been increasing awareness of the role of efflux mechanisms as mediators of resistance, particularly of resistance to the fluoroquinolones, tetracyclines, and macrolides.

Impermeability was considered to be the main mechanism of tetracycline resistance in gram-negative bacteria, since less drug accumulated in resistant cells than in susceptible cells. Later studies showed that energy inhibitors increased the uptake of tetracycline in resistant cells. This finding suggested an **active-efflux** component to resistance. Subsequent studies showed that tetracycline-resistant cells lose accumulated drug faster than susceptible cells do and that tetracycline enters everted vesicles by an energy-dependent process. It was demonstrated that a carrier-mediated efflux of tetracycline actively transported the drug out of the cell, preventing it from inhibiting protein synthesis on the ribosome.

In 1974, studies of *E. coli* led to the discovery of a specific membrane protein called the Tet protein, which is associated with tetracycline resistance. Synthesis of this protein is induced by tetracycline and only occurs in cells bearing tetracycline resistance determinants.

While the genes for many efflux systems can occur on exogenously acquired genetic elements, there appear to be other systems intrinsic to the organisms, with genes located on the bacterial chromosome, as exemplified by those for minocycline and fluoroquinolones in *E. coli.*

In order to reach their targets (DNA gyrase and topoisomerase IV), located in the cytoplasm, fluoroquinolones must cross the cell wall and cytoplasmic membrane of gram-positive bacteria and additionally the outer membrane of gram-negative bacteria. The cell wall is thought to provide little or no barrier to the diffusion of small molecules such as fluoroquinolones, which have molecular masses of around 300 to 400 Da. For some flurorquinolone-resistant clinical isolates of many species of gram-negative bacteria, the role of energy-dependent efflux systems that *pump drug out of the cell* and act in concert with reduced diffusion due to reduced porin channels has been demonstrated. The best studied organisms have been *E. coli* and *P. aeruginosa,* but examples have also been found in several other gram-negative bacteria.

Some efflux systems deal with a narrow range of structurally related substrates; for example, the *E. coli* tetracycline exporter TetB is capable of extruding tetracycline and a narrow range of close structural analogs. However, currently it is recognized that many bacteria carry genes specifying multidrug resistance pumps that export structurally unrelated antibacterial agents. Some of them play an important role in conferring antibiotic resistance on pathogens such as *P. aeruginosa* and *Staphylococcus aureus.* There is therefore a need to understand the structures, functions, and origins of these transport systems.

### MULTIPLE ANTIBACTERIAL RESISTANCE PUMPS IN BACTERIA

Bacteria communicate with their enviroment in part via solute-specific transport systems. These systems consist of integral membrane proteins that usually span the cytoplasmic membrane of the cell multiple times as α-helices.

Antibacterial efflux has been identified in five families, primarily on the basis of amino acid sequence identity. These families are the ATP-binding cassette (ABC) superfamily, the major facilitator superfamily (MFS), the resistance-nodulation-division (RND) family, the small multidrug resistance (SMR) family, and a novel family of proteins, the multidrug and toxic compound extrusion (MATE) family. Antibacterial efflux pumps fall into the RND, MFS, and MATE groups, with the RND and MATE families so far being unique to gram-negative bacteria. Thus, MFS-type transporters dominate the efflux systems of antibacterial agents in gram-positive bacteria. These resistance efflux systems are characteristically energy dependent.

Gram-positive bacteria contain several distinct types of multidrug resistance pumps. (Table 5.2 shows some examples of bacterial multidrug efflux transporters in gram-positive and gram-negative bacteria.) These MFS transporters are usually found in the 12-transmembrane segment drug pump family (12 membrane-spanning domains).

Unlike gram-positive bacteria, gram-negative cells have an outer membrane (OM) that contains specialized proteins called porins that enable small hydrophilic molecules to enter cells but restrict the entry of many hydrophobic antibiotics (antibacterials). In gram-negative bacteria, the majority of multidrug transporters have a common three-component organization: a transporter located in the inner membrane (IM) functions with an OM channel and a periplasmic accessory protein. In this arrangement, efflux complexes traverse both the IM and OM, facilitating direct passage of the substrate from the cytoplasm or the cytoplasmic membrane to the external medium. Zgurskaya and Nikaido (2000) pointed out that direct efflux of drugs into the medium is advantageous for gram-negative bacteria since the expelled drug molecules must again cross the OM. Hence, drug efflux works synergistically with the low permeability of the OM. The synergy is apparent since some carbapenems such as meropenem select for a loss of specific basic amino acid/carbapenem channel, OprD, in the OM as well as for the overproduction of MexAB-OprM efflux complex.

Nikaido (1998) proposed a hypothetical model for gram-positive and gram-negative bacteria that is composed of transport proteins located in the cytoplasm membrane and also a hypothetical structure and mechanism of tripartite efflux pumps in gram-negative bacteria (Fig. 5.1) (for a detailed explanation, refer to Nikaido [1998]).

The importance of MDR mechanisms is very well covered in recent review articles which address various aspects of the drug efflux; the reader is therefore directed to these for comprehensive information.

## SUMMARY OF RESISTANCE DUE TO DECREASED ACCUMULATION OF ANTIBACTERIAL DRUGS INSIDE BACTERIA

Figure 5.2 summarizes the decreased accumulation due to active efflux and/or decreased uptake as well as some of the current methods, both developed and under development, to overcome resistance due to decreased accumulation.

The active efflux of antibiotics as a mechanism of resistance was discovered by Levy and coworkers in 1980 when they were studying the mechanism of tetracycline resistance in enterobacteria. Since then, it has

**Table 5.2** Some examples of bacterial multidrug efflux transporters in gram-positive and gram-negative bacteria[a]

| Transporter name | Organism | Transporter class | Thought to function with: | Antibacterial substrates |
|---|---|---|---|---|
| Gram-positive pumps | | | | |
| NorA | *S. aureus* | MFS(12) | | Fluoroquinolones, chloramphenicol |
| Bmr | *B. subtilis* | MFS(12) | | Fluoroquinolones, chloramphenicol |
| Gram-negative pumps | | | | |
| AcrB | *E. coli* | RND | AcrA, TolC | β-Lactams, erythromycin, tetracycline, chloramphenicol |
| MexB | *P. aeruginosa* | RND | MexA, OprM | β-Lactams, erythromycin, tetracycline, chloramphenicol |
| MexD | *P. aeruginosa* | RND | MexC, OprJ | Fourth-generation cephalosporins, tetracycline, fluoroquinolones, chloramphenicol |
| MexF | *P. aeruginosa* | RND | MexE, OprN | Fluoroquinolones, chloramphenicol |

[a] Reprinted from H. Nikaido, *Curr. Opin. Microbiol.* 1:516–523, 1998, with permission from the publisher.

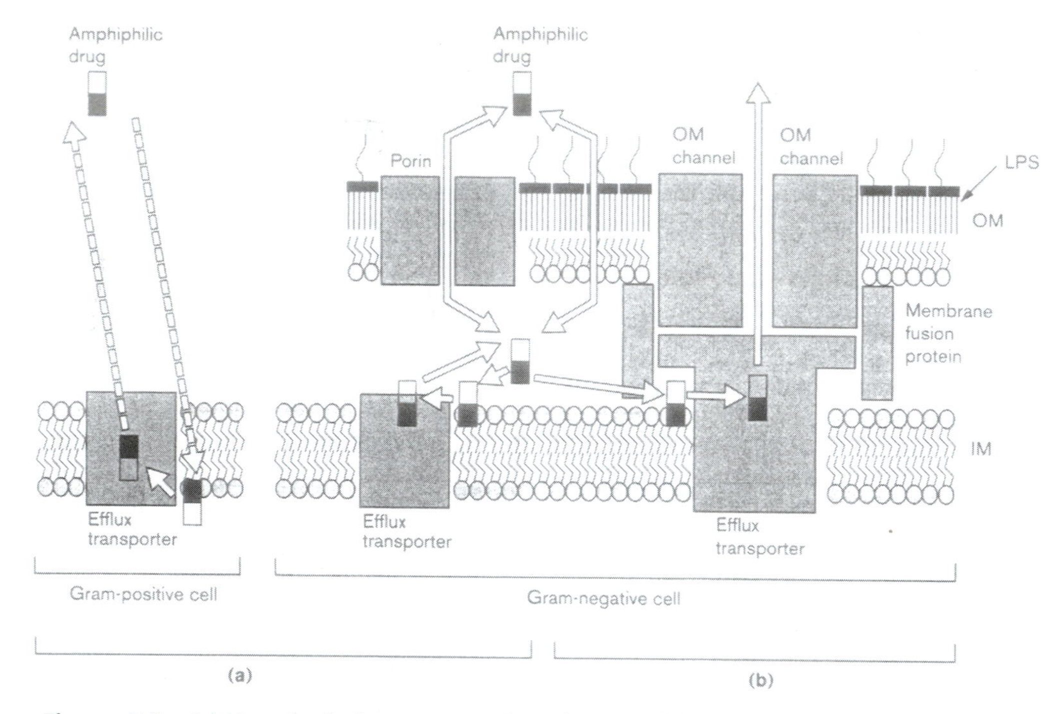


**Figure 5.1** (a) Hypothetical structure and mechanism of the transporter. In the cytoplasmic membrane in gram-positive bacteria (left), drugs enter unhindered and are pumped out into the medium. In gram-negative bacteria, drug molecules that have traversed the OM via porin channels or through its bilayer domain are extruded into the periplasmic space. (b) Complex efflux machinery that occurs only in gram-negative bacteria. The drug molecules are captured and pumped out directly into the medium by an assembly that contains, in addition to the pump, an OM channel and a membrane fusion protein. In both panels a and b, efflux of the drug molecules from the cytoplasm may occur, but it is not shown in the diagram for reasons of simplicity. LPS, lipopolysaccharide. Reprinted from H. Nikaido, *Curr. Opin. Microbiol.* **1**:516–523, 1998, with permission from the publisher.

been demonstrated that one or multiple efflux pumps participate in the efflux of antibiotics.

## Enzymatic Inactivation of Antibiotics by Modification of Their Structure

Bacterial species may become insensitive to certain antibiotics because they possess enzymes that inactivate these antibiotics. This is an important mechanism of resistance to β-lactams, aminoglycosides, and chloramphenicol.

β-Lactamases are a group of enzymes that catalyze the opening of the β-lactam nucleus of β-lactam antibiotics (penicillin, cephalosporins, cephamycins, carbapenems, and monobactams), resulting in products that can no longer inhibit their targets; these enzymes are called penicillin-binding proteins, and they catalyze the transpeptidation reaction in cell wall biosynthesis. The β-lactamase genes are located in the chromosome, in plasmids, or in transposons.

The major mechanism of inactivation of aminoglycoside antibiotics involves aminoglycoside-modifying enzymes. There are three types of aminoglycoside-modifying enzymes, each of which transfers a functional group onto the aminoglycoside structure to render the antibiotic ineffective. These enzymes are aminoglycoside acetyltransferases, which transfer the acetyl group from acetyl coenzyme A; aminoglycoside phosphoryltransferases, which transfer the phosphoryl group from ATP; and aminoglycoside nucleotidyltransferases, which transfer a nucleotide from nucleotide triphosphates. Each group of enzymes has different isozymic forms that differ in substrate specificity. At least 30 different genes for aminoglycoside-modifying enzymes are known in bacteria.

Chloramphenicol is an antibiotic that binds to bacterial ribosomes. Chloramphenicol acetyltransferase inactivates chloramphenicol by acetylating it using acetylcoenzyme A as the acetyl group donor. There are

**A** Decreased accumulation

- **Active efflux**
  - *tetracycline*
  - *Fluoroquinolones*
  - *Macrolides (14- and 15-membered macrolides)*
- **Decreased uptake**
  - *In gram-negative bacteria, particularly in P. aeruginosa; loss of essential porins (OmpF, OmpC, OmpD2)*

**B** Overcoming resistance due to decreased accumulation

- **Overcoming active efflux**
  - *Glycylcyclines (DMG-DMDOT and DMG-MINO)*
  - *Efflux-pump inhibitors (tetracycline); 13(alkyl-thio) and 13(aryl-thio) derivatives of 5-hydroxy-6-deoxytetracycline*
- **Overcoming decreased uptake**
  - *Carbapenems (D2 porin channel) and cephem-containing catechol or hydroxypyridone moieties (tonB)*
  - *Use of bacterial cell wall permeabilizers*

**Figure 5.2** Decreased accumulation of antibacterials and overcoming resistance due to it.

several *cat* genes that encode different chloramphenicol acetyltransferases.

A few other antibiotic-modifying enzymes are known. Fosfomycin is inactivated by a plasmid-determined glutathione *S*-transferase, which catalyzes the formation of a glutathione-fosfomycin adduct.

SUMMARY OF DRUG INACTIVATION-MEDIATED RESISTANCE

Figure 5.3 summarizes the mechanisms of bacterial resistance due to drug inactivation and how researchers developed some ways of overcoming these mechanisms.

## Alteration of the Target Site

Mutations in several ribosomal proteins are known to be related to antibiotic resistance. For several strains of *E. coli,* the mutated protein is known but the amino acid actually altered has not been documented. Identification of protein determinants for resistance will aid our understanding of the mechanisms of antibiotic function and resistance. Several antibiotics act by targeting protein biosynthesis, interacting with chromosomal structural proteins, rRNAs, and ribosome-associated proteins (Fig. 5.4).

Mass spectrometry can be used as a valuable tool to rapidly locate and characterize mutant proteins by using a small amount of material. Recently, Wilcom et al. (Pharmacia Corp., Kalamazoo, Mich.) used electrospray (ESP) and matrix-assisted laser desorption ionization–time-of-flight (MALDI-TOF) mass spectrometry to map out all 56 ribosomal proteins in *E. coli* on the basis of intact molecular masses. In particular, these scientists studied proteins responsible for streptomycin, erythromycin, and spectinomycin resistance in three strains of *E. coli* and characterized each mutation responsible for resistance by analyzing tryptic peptides of these proteins by MALDI-TOF and nano-ESP tandem mass spectrometry. These techniques can be used to

**Figure 5.3** Drug inactivation and overcoming resistance due to it.

**A** Drug inactivation

- β-Lactamases
- Aminoglycoside-inactivating enzymes
- Chloramphenicol acetyltransferases
- Macrolide esterases/phosphotransferases
- Renal dihydropeptidases

**B** Overcoming drug inactivation-mediated resistance

- Enzyme inhibition
  - β-Lactamase inhibitors
  - Renal dihydropeptidase inhibitor (cilastatin)
- Analogs resistant to enzyme inactivation
  - 3rd and 4th generation cephalosporins
  - 1β-Methylcarbapenems
  - New aminoglycosides (isepamicin)

- Penicillin-binding proteins (PBPs)
- rRNA point mutations
- rRNA methylation
- DNA gyrase
- Beta-lactamases
- Dihydrofolate reductase
- Dihydropteroic synthase
- Van-related proteins
- Overproduction of antibiotic target

**Figure 5.4** Target alteration.

detect small changes in the masses as well as other proteins and peptides. ESP ionization involves application of an electric field to a solution sprayed from a needle. In MALDI, gas phase ions are generated by desorption ionization of the molecule of interest from a layer of crystals formed from volatile matrix molecules.

The best studied and most frequently encountered mechanism of resistance to the macrolides is target site modification, whereby bacterial rRNA is methylated, causing reduced affinity between the antibiotic and the ribosome. This mechanism confers resistance to macrolides, lincosamides, and streptogramin B antibiotics (giving rise to the so-called $MLS_B$ phenotype), which accounts for most of the resistance in these clinical isolates.

A family of enzymes is known to catalyze the mono- or dimethylation of the N-6 amino group of adenine in a highly conserved region of 23S rRNA, which may be involved directly in the formation of peptidyltransferase centers.

## REFERENCES AND FURTHER READING

### General

**Alekshun, M. N., and S. B. Levy.** 2000. Bacterial drug resistance: response to survival threats, p. 323–366. *In* G. Storz and R. Hengge-Aronis (ed.), *Bacterial Stress Responses.* ASM Press, Washington, D.C. Also see references therein.

**American Academy of Microbiology.** 1999. *Report on Antimicrobial Resistance.* American Society for Microbiology, Washington, D.C.

**Amyes, S. G. B., and C. G. Gemmell.** 1992. Antibiotic resistance in bacteria. *J. Med. Microbiol.* **36:**4–29.

**Coleman, K., M. Athalye, A. Clancey, M. Davison, D. J. Payne, C. R. Perry, and I. Chopra.** 1994. Bacterial resistance mechanisms as therapeutic targets. *J. Antimicrob. Chemother.* **33:**1091–1116.

**Courvalin, P.** 1996. Evasion of antibiotic action by bacteria. *J. Antimicrob. Chemother.* **37:**855–869.

**Davies, J.** 1994. Inactivation of antibiotics and the dissemination of resistance genes. *Science* **264:**375–381.

**Hall, R. M., and S. R. Partridge.** 2001. Evolution of multiple antibiotic resistance by acquisition of new genes, p. 37–51. *In* D. Hughes and D. L. Andersson (ed.), *Antibiotic Development and Resistance.* Taylor & Francis, London, United Kingdom.

**Jacoby, G. A., and G. L. Archer.** 1991. New mechanisms of bacterial resistance to antimicrobial agents. *N. Engl. J. Med.* **324:**601–612.

**Levy, S. M.** 2002. *The Antibiotic Paradox,* 2nd ed. Perseus Publishing, Cambridge, Mass.

**Livermore, D. M.** 2003. Bacterial resistance, origins, epidemiology, and impact. *Clin. Infect. Dis.* **36:**811–823.

**Marchese, A., and G. C. Schito.** 2001. Role of global surveillance in combating bacterial resistance. *Drugs* **61:**167–173.

**Mobashery, S., and E. Azucena.** Mechanism of bacterial antibiotic resistance. Accepted for publication in *Encyclopedia of Life Sciences.* Nature Publishing, London, United Kingdom.

**Rosen, B. P., and S. Mobashery.** 1998. *Resolving the Antibiotic Paradox.* Kluwer Academic/Plenum Publishers, New York, N.Y.

**Rowe-Magnus, D. A., and D. Mazel.** 1999. Resistance gene capture. *Curr. Opin. Microbiol.* **2:**483–488.

**Salyers, A. A., and C. F. Amábile-Cuevas.** 1997. Why are antibiotic resistance genes so resistant to elimination? *Antimicrob. Agents Chemother.* **41:**2321–2325.

**Tenover, F. C.** 2001. Development and spread of bacterial resistance to antimicrobial agents: an overview. *Clin. Infect. Dis.* **33**(Suppl. 3):S108–S115.

**White, W.** 1998. Medical consequences of antibiotic use in agriculture. *Science* **279:**996–997.

**Williams, R. J., and D. L. Heymann.** 1998. Containment of antibiotic resistance. *Science* **279:**1153–1154.

### Mutation Frequencies and Antibiotic Resistance

**Martinez, J. L., and F. Baquero.** 2000. Mutation frequencies and antibiotic resistance. *Antimicrob. Agents Chemother.* **44:**1771–1777.

### Methods for Assessing Antimicrobial Resistance

**Bergeron, M. G.** 2000. Genetic tools for the simultaneous identification of bacterial species and their antibiotic resistance genes: impact on clinical practice. *Int. J. Antimicrob. Agents* **16:**1–3.

**Bergeron, M. G., and M. Ouellette.** 1998. Preventing antibiotic resistance through rapid genotypic identification of bacteria and of their antibiotic resistance genes in the clinical microbiology laboratory. *J. Clin. Microbiol.* **36:**2169–2172.

**Cockerill, F. R.** 1999. Genetic methods for assesing antimicrobial resistance. *Antimicrob. Agents Chemother.* **43:**199–212.

**Davison, H. C., J. C. Low, and M. E. J. Woolhouse.** 2000. What is antibiotic resistance and how can we measure it? *Trends Microbiol.* **8:**554–559.

**Greenwood, D.** 2000. Detection of antibiotic resistance in vitro. *Int. J. Antimicrob. Agents* **14:**303–309.

**Martineau, F., F. J. Picard, P. H. Roy, M. Ouellette, and M. G. Bergeron.** 1996. Species-specific and ubiquitous DNA-based assays for rapid identification of *Staphylococcus epidermidis*. *J. Clin. Microbiol.* **34:**2888–2893.

**Senda, K., Y. Arakawa, S. Ichiyama, K. Nakashima, H. Ito, S. Otsuka, K. Shimokata, N. Kato, and M. Phta.** 1996. PCR detection of metallo-β-lactamase gene (*blaIMP*) in gram-negative rods resistant to broad-spectrum β-lactams. *J. Clin. Microbiol.* **34:**2909–2913.

### Sub-MIC Concentrations

**Braga, P. C., M. Dal Sasso, and M. T. Sala.** 2000. Sub-MIC concentrations of cefodizime interfere with various factors affecting bacterial virulence. *J. Antimicrob. Chemother.* **45:**15–25.

### Efflux Mechanisms

**Alekshun, M. N., S. B. Levy, T. R. Mealy, B. A. Seaton, and J. F. Head.** 2001. The crystal structure of MarR, a regulator of multiple antibiotic resistance, at 2.3 Å resolution. *Nat. Struct. Biol.* **8:**710–714.

**Borges-Walmsley, M. I., and A. R. Walmsley.** 2001. The structure and function of drug pumps. *Trends Microbiol.* **9:**71–79.

**Köhler, T., M. Michea-Hamzehpour, S. F. Epp, and J. C. Pechere.** 1999. Carbapenem activities against *Pseudomonas aeruginosa*: respective contributions of OprD and efflux systems. *Antimicrob. Agents Chemother.* **43:**424–427.

**Koronakis, V., J. Li, E. Koronakis, and K. Stauffer.** 1997. Structure of TolC, the outer membrane component of the bacterial type I efflux system, derived from two-dimensional crystals. *Mol. Microbiol.* **23:**617–626.

**Koronakis, V., A. Sharf, E. Koronakis, B. Luisi, and C. Hughes.** 2000. Crystal structure of the bacterial membrane protein TolC central to multidrug efflux and protein export. *Nature* **405:**914–919.

**Levy, S. B.** 1992. Active efflux mechanisms for antimicrobial resistance. *Antimicrob. Agents Chemother.* **36:**695–703.

**Lewis, K.** 1994. Multidrug resistance pumps in bacteria: variations on a theme. *Trends Biochem. Sci.* **19:**119–123.

**Lewis, K., D. C. Hooper, and M. Ouellette.** 1997. Multidrug resistance pumps provide broad defense. *ASM News* **63:**605–610.

**Lomovskaya, O., M. S. Warren, and V. Lee.** 2001. Efflux mechanisms: molecular and clinical aspects, p. 65–90. *In* D. Hughes and D. L. Andersson (ed.), *Antibiotic Development and Resistance*. Taylor & Francis, London, United Kingdom.

**McKeegan, K. S., M. I. Borges-Walmsley, and A. R. Walmsley.** 2003. The structure and function of drug pumps: an update. *Trends Microbiol.* **11:**21–29.

**Murakami, S., R. Nakashima, E. Yamashita, and A. Yamaguchi.** 2002. Crystal structure of bacterial multidrug efflux transporter AcrB. *Nature* **419:**587–593.

**Nikaido, H.** 1994. Prevention of drug access to bacterial targets: permeability barriers and active efflux. *Science* **264:**382–387.

**Nikaido, H.** 1996. Multidrug efflux pumps of gram-negative bacteria. *J. Bacteriol.* **178:**5853–5859.

**Nikaido, H.** 1998. Multiple antibiotic resistance and efflux. *Curr. Opin. Microbiol.* **1:**516–523.

**Nikaido, H.** 1998. Antibiotic resistance caused by gram-negative multidrug efflux pumps. *Clin. Infect. Dis.* **27**(Suppl. 1)**:**S32–S41.

**Nikaido, H.** 2000. How do exported proteins and antibiotics bypass the periplasm in gram-negative bacterial cells? *Trends Microbiol.* **8:**481–483.

**Pao, S. S., I. T. Paulsen, and M. H. Saier.** 1998. Major facilitator superfamily. *Microbiol. Mol. Biol. Rev.* **62:**1–34.

**Paulsen, I. T., M. H. Brown, and R. A. Skurray.** 1996. Proton-dependent multidrug efflux systems *Microbiol. Rev.* **60:**575–608.

**Poole, K.** 2000. Efflux-mediated resistance to fluroquinolones in gram-positive bacteria and the mycobacteria. *Antimicrob. Agents Chemother.* **44:**2505–2509.

**Saier, M. H., I. T. Paulsen, M. K. Sliwinski, S. S. Pao, R. A. Surray, and H. Nikaido.** 1998. Evolutionary origins of multidrug and drug-specific efflux pumps in bacteria. *FASEB J.* **12:**265–274.

**Van Bambeke, F., E. Balzi, and P. M. Tulkens.** 2000. Antibiotic efflux pumps. *Biochem. Pharmacol.* **60:**457–470.

**Zgurskaya, H. I., and H. Nikaido.** 2000. Multidrug resistance mechanisms: drug efflux across two membranes. *Mol. Microbiol.* **37:**219–225.

### Target Site Alterations

**Spratt, B. G.** 1994. Resistance to antibiotics mediated by target alterations. *Science* **264:**388–393.

**Wilcox, S. K., G. S. Cavey, and J. D. Pearson.** 2001. Single ribosomal protein mutations in antibiotic-resistant bacteria analyzed by mass spectrometry. *Antimicrob. Agents Chemother.* **45:**3046–3055. See the cited references to ESP and MALDI-TOF mass spectrometry.

### Conjugative Transposition

**Scott, J. R., and G. G. Churchward.** 1995. Conjugative transposition. *Annu. Rev. Microbiol.* **49:**367–397.

chapter 6



# Antibiotics and Synthetic Antibacterial Agents

## Definition of Antibiotics

**Antibiotics** are chemical substances produced by actinomycetes, fungi, or bacteria that interfere with some functions of structure or process essential to bacterial growth (bacteriostatic) or survival (bactericidal) without harm to the eukaryotic host harboring the infecting bacteria.

## Antibiotics and Synthetic Antibacterial Agents

One of the most profound developments in the history of modern medicine has been the discovery of compounds that control bacterial infections. These compounds, known as antibiotics, are produced by living microorganisms and, at low concentrations, inhibit the growth of different species of bacteria. They are used primarily as antibacterial agents in therapy of human infectious diseases.

Antibiotic molecules are biosynthesized through chemical reactions related to primary intermediary metabolism. They are the end products (secondary metabolism) of actinomycetes, filamentous fungi, or bacteria, especially of the genus *Streptomyces*. Semisynthetic variations of these natural products are also called antibiotics. This definition distinguishes chemicals produced by the secondary metabolism of a microorganism and their partial chemical modification (semisynthesis) from antibacterial agents totally synthesized by the chemist, the so-called **synthetic antibacterial agents** (e.g., quinolones, sulfonamides, trimethoprim, oxazolidinones, and nitrofurans).

It is worth noting here that the antibiotic chloramphenicol was originally isolated from the soil actinomycete *Streptomyces venezuelae* but is now produced entirely by chemical synthesis rather than by fermentation. This is mainly due to its simple structure; however, it is still categorized as an antibiotic.

The central concept of antibiotic action is **antibiosis.** The noun "antibiotic" was first used by Waksman in 1942 to denote chemical substances that are produced by a microorganism and that inhibit the growth of or kill other microorganisms. The discovery of the antimicrobial effects of some bacteria by Pasteur in 1877 and the observation by Fleming in 1929 that an "active mold broth filtrate" (chemical substances) produced by *Penicillium notatum* inhibits *Staphylococcus* spp. were the beginning of the antibiosis concept. Antibiotics affect the structure or physiological function of bacteria. Bacteriostatic antibiotics inhibit the growth of bacteria, while bactericidal antibiotics kill them.

## Classification

In this section, the antibiotics and synthetic antibacterials are grouped on the basis of their chemical structure and by mechanism of action.

### Classification by Chemical Structure

Classification by chemical structure is most useful for establishing chemical structure-antibiotic activity relationships. Frequently, one group is subdivided into subgroups reflecting similar structural features. The following classes and chemical structures of antibiotics and synthetic antibacterials of major clinical significance are discussed in this book.

#### ANTIBIOTICS

#### β-Lactams

All β-lactam antibiotics are structurally characterized by having a four-membered cyclic amide (β-lactam) (Figure 6.1). Substructure groups are the penicillins, the cephalosporins, the carbapenems, the monobactams, and the β-lactam derivatives clavulanic acid, sulbactam and tazobactam, which are β-lactamase inhibitors.

#### Aminoglycosides

Structurally the aminoglycosides contain two or more amino monosaccharides connected by glycosidic linkages to the aminocyclitol 2-deoxystreptamine, streptamine, or streptidine (Fig. 6.2). Representative members are streptomycin, neomycin, netilmicin, kanamycin, gentamicin, tobramycin, amikacin, and dibekacin.

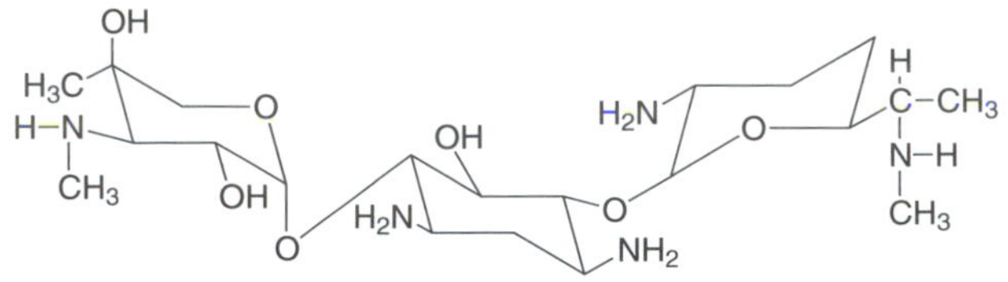

**Figure 6.1** Chemical structures of penicillins and cephalosporins.

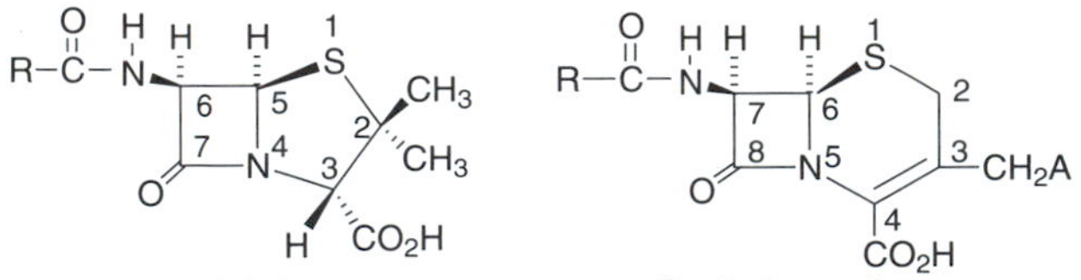

**Figure 6.2** Chemical structure of gentamicin, an aminoglycoside antibiotic.

#### Macrolides

Macrolide antibiotics are **macrocyclic lactones** to which two sugars are attached (one of these is an amino sugar) (Fig. 6.3).

Macrolide antibiotics are grouped according to the size of the macrocyclic lactone ring (the aglycone), which can be 14, 15, or 16 membered. Representative antibiotics with a 14-member ring are erythromycin A, oleandomycin, roxithromycin, clarithromycin, and fluorithromycin. Macrolides with a 16-member ring include miokamycin, rokitamycin, and midecamycin. Azithromycin is a member of the subgroup of 15-member ring macrolide antibiotics. This class is known as the **azalides.** Chemically, azalides are not macrolides, but they are considered a subgroup within the macrolides rather than a separate chemical group.

The chemical structure of erythromycin A consists of the aglycone erythronolide A, to which the dimethylamino monosaccharide D-desosamine is linked by a β-glycosidic bond at position 5 and the neutral monosaccharide L-cladinose is linked by an α-glycoside bond at position 3.

#### Ketolides

Ketolides are a new class of semisynthetic 14-member-ring macrolides, which are characterized by a keto function at

**Figure 6.3** Chemical structure of erythromycin A.

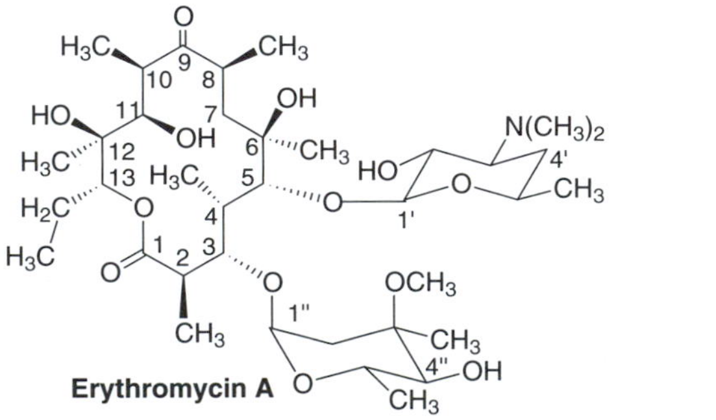

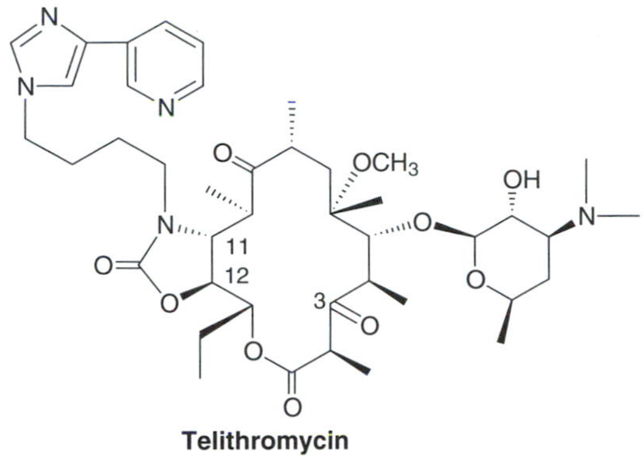

**Figure 6.4** Chemical structure of telithromycin. (Source: Hoechst Marion Roussel.)

position 3 of the erythronolide A ring, which replaces the cladinose moiety, a sugar long considered to be essential for antibacterial activity. In some ketolides, the 11 and 12 positions of the lactone ring are extended with an alkyl-aryl chain from a carbamate group (Figure 6.4). Several ketolides have been developed, and many of them are undergoing clinical trials. Telithromycin (HMR 3647) is the first ketolide introduced into clinical practice.

## Tetracyclines

Structurally the tetracyclines are a group of antibiotics that are characterized by containing a **hydronaphthacene** group, consisting of four fused rings (Fig. 6.5). Five tetracycline antibiotics are marketed in the United States: tetracycline, demeclocycline, oxytetracycline, doxycycline, and minocycline.

## Lincosamides

The lincomycin (also called lincosamide) antibiotics are characterized by an alkyl-6-amino-6,8-dideoxy-1-thio-α-D-galactooctalyranoside joined to a proline moiety by an amide linkage (Fig. 6.6). Clindamycin is the only lincosamide antibiotic marketed in the United States.

## Ansamycins

The ansamacrolides, or ansamycins, are a group of antibiotics characterized by a naphthalene structure in which nonadjacent positions are bridged by an aliphatic chain to form a macrocycle. One of the aliphatic-aromatic junctions is an amide bond (Fig. 6.7). Rifampin (known in some countries as rifampicin), a member of this group, is a semisynthetic antibiotic derived from the antibiotic rifamycin B.

**Figure 6.5** Chemical structure of tetracycline.

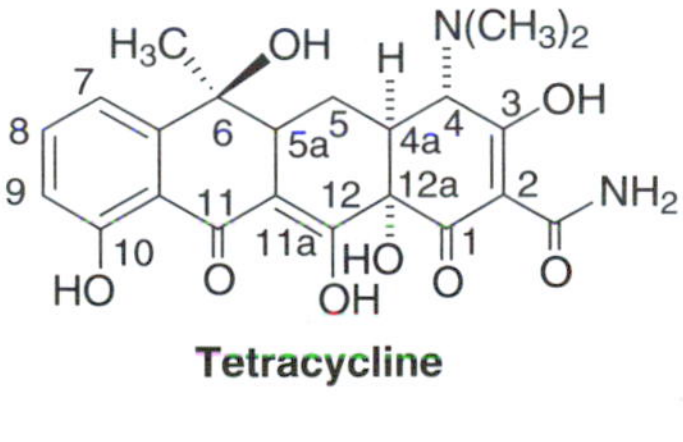

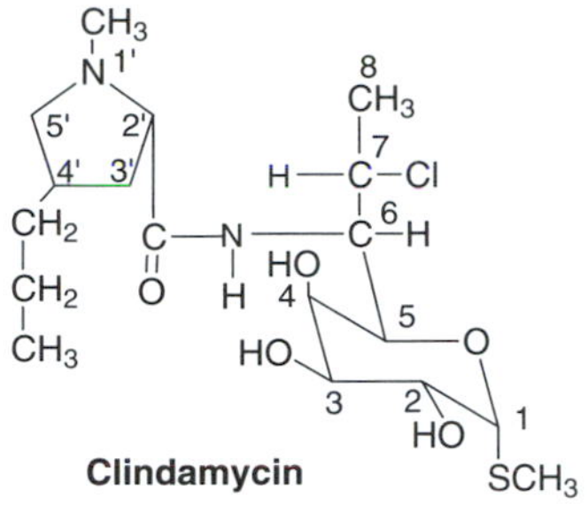

**Figure 6.6** Chemical structure of clindamycin.

## Glycopeptides

Vancomycin and teicoplanin are examples of glycopeptide antibiotics used to control bacterial infections. Structurally, these glycopeptide antibiotics consist of a cyclic peptide backbone composed of seven amino acids. When the amino acids are numbered 1 to 7 from the N to the C terminus, residues 4 and 5 are always *p*-hydroxyphenylglycines, residues 2 and 6 are derived from tyrosine, and residue 7 is a 3,5-dihydroxyphenylglycine. In vancomycin, residue 1 is *N*-methylleucine and residue 3 is asparagine (Fig. 6.8). In teicoplanin, residue 1 is *p*-hydroxyphenylglycine and residue 3 is 3,5-dihydroxyphenylglycine.

**Figure 6.7** Chemical structure of rifampin.

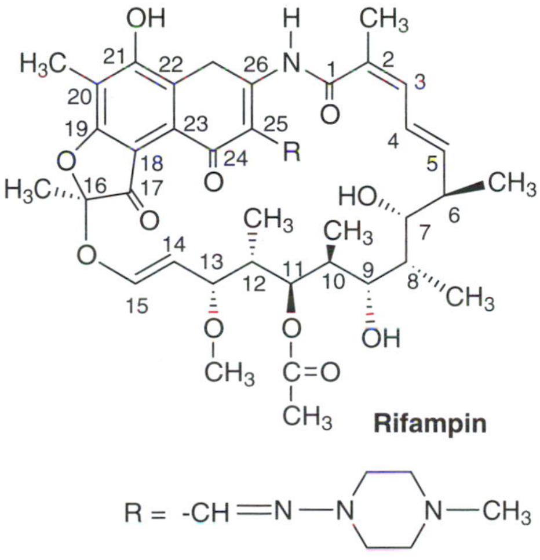

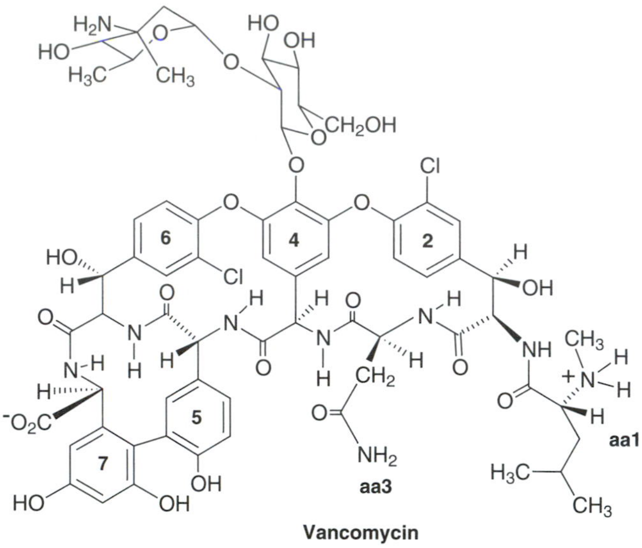

**Figure 6.8** Chemical structure of vancomycin. aa, amino acid.

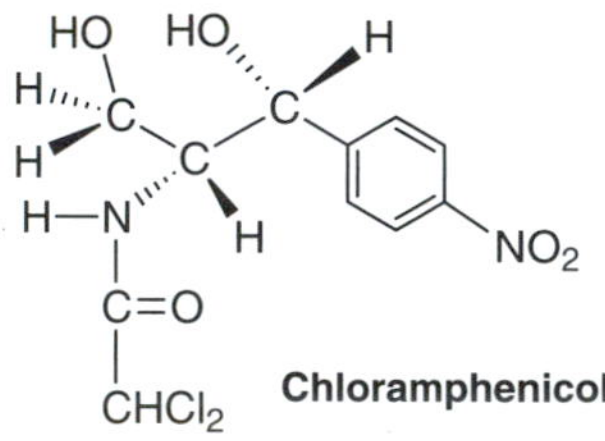

**Figure 6.9** Chemical structure of chloramphenicol.

## Chloramphenicol

Chloramphenicol is the only member of its group. Chemically, it has functional groups such as the *p*-nitrobenzene and the dichloroacetamide group on C-2 (Fig. 6.9).

## Cyclic Peptides

Of the cyclic peptide antibiotics, polymyxin B, colistin (polymyxin E), gramicidin A, and bacitracin are the only antibiotics used clinically.

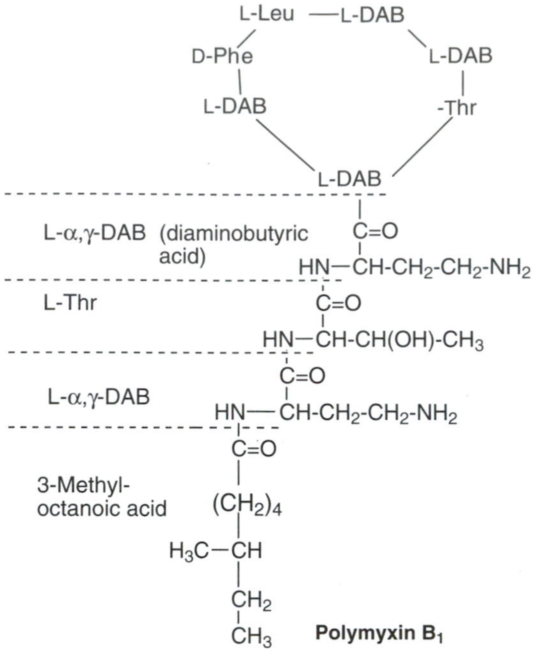

**Figure 6.10** Chemical structure of polymyxin $B_1$.

Polymyxin B contains a seven-amino-acid ring attached to a three-amino-acid tail, to which is attached a fatty acyl group (Fig. 6.10).

## Streptogramins

Streptogramins are a group of natural cyclic peptide macrolactones (Fig. 6.11). They represent a unique antibiotic class in which each member consists of two structurally unrelated molecules: group A and B streptogramins. Quinupristin (the methanesulfonate salt of 5-(*R*)-[3-(*S*)-quinuclidinyl]-thiomethyl pristinamycin IA) and dalfopristin (the methanesulfonate salt of (26-*S*-[diethylaminoethylsulfonyl] pristinamycin IIB) are two semisynthetic pristinamycin derivatives which were developed in France and which, since 1999, are also

**Figure 6.11** Chemical structures of the methanesulfonates (mesylates) of quinupristin and dalfopristin.

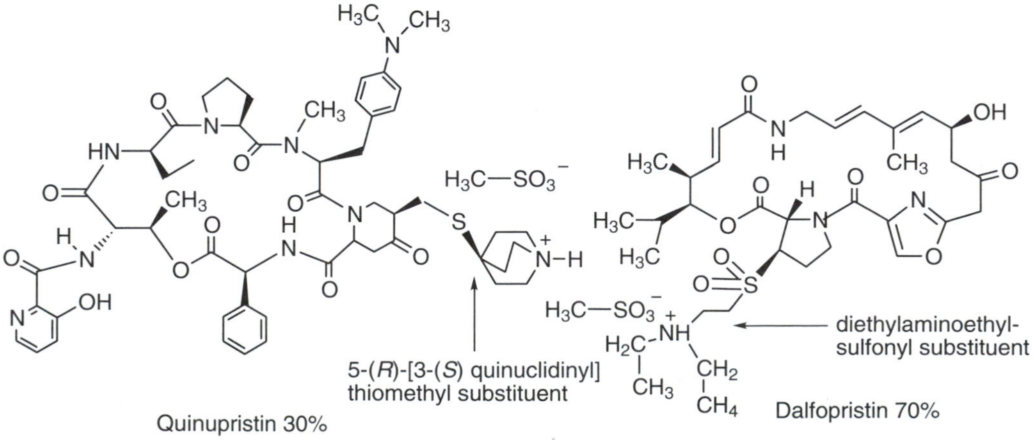

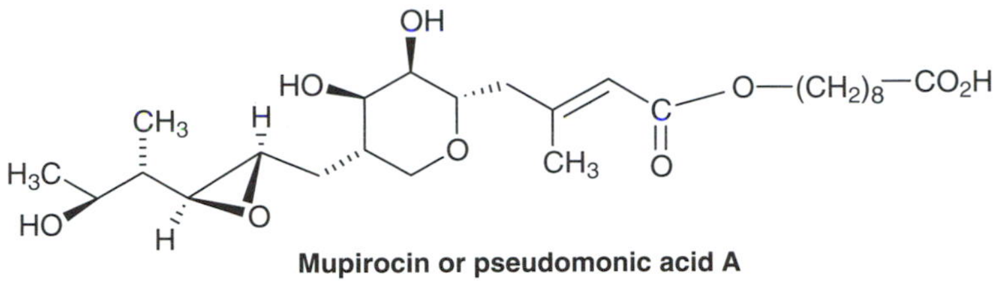

Mupirocin or pseudomonic acid A

**Figure 6.12** Chemical structure of mupirocin.

marketed in the United States as a drug containing quinupristin and dalfopristin in the ratio of 30:70 (wt/wt).

### Mupirocin

The structure of mupirocin (formerly known as pseudomonic acid) consists of monic acid with a short fatty acid side chain. The terminal portion of the molecule distal to the fatty acid resembles the amino acid isoleucine (Fig. 6.12). This antibiotic is the only member in its group.

### D-Cycloserine

D-Cycloserine is a structural analog of D-alanine. Only D-cycloserine falls into this group (Fig. 6.13).

### Fosfomycin

Fosfomycin [L-(*cis*)-1,2-epoxypropylphosphonic acid] (formerly known as phosphonomycin) is a naturally occurring antibiotic obtained from species of *Streptomyces* (Fig. 6.14). Structurally, it is the simplest of all antibiotics. Currently it is produced by chemical synthesis rather than by fermentation. It has been extensively used in many countries, such as Spain, Italy, France, and Japan. It is currently approved by the Food and Drug Administration (FDA) in the United States.

## SYNTHETIC ANTIBACTERIALS

### Oxazolidinones

The oxazolidinones represent a novel chemical class of synthetic antibacterial agents. Linezolid [3-(fluorophenyl)-2-oxazolidinone] (Fig. 6.15) is the first totally synthetic antibacterial agent in this new class. It was approved in April 2000 by the FDA.

**Figure 6.13** Chemical structure of D-cycloserine.

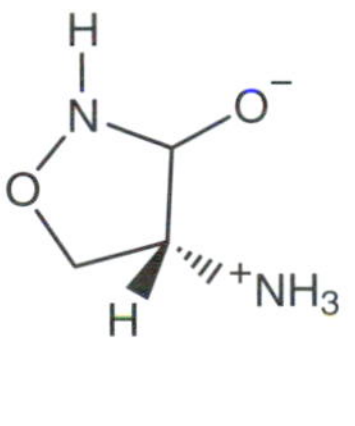

D-Cycloserine

### Quinolones

The quinolones, or fluoroquinolones, are synthetic antibacterial agents. Nalidixic acid (1,8-naphthyridonecarboxylic acid or 8-azaquinolonecarboxylic acid) was discovered in the 1960s as a by-product of the purification of the antimalarial drug chloroquine. The fluoroquinolones were introduced into therapy in the mid-1980s. Structurally, the fluoroquinolones are derivatives of 1,4-dihydro-4-oxo-3-quinolinecarboxylic acid characterized by the presence of a fluorine atom at position 6 of the quinoline ring (Fig. 6.16).

Eight fluoroquinolones are currently marketed in the United States: ciprofloxacin, enoxacin, levofloxacin, lomefloxacin, ofloxacin, sparfloxacin, gatifloxacin, and moxifloxacin.

### Sulfonamides

Sulfonamides are commonly called sulfa drugs. They are derived from sulfanilamide (*p*-aminobenzensulfonamide) and are structural analogs of *p*-aminobenzoic acid. Sulfamethoxazole (5-methyl-3-sulfonamidoisoxazole) (Fig. 6.17) is a classic example of this class of antibacterial agent.

### Diaminopyrimidines

Trimethoprim (Fig. 6.18) is the most important antibacterial agent of the diaminopyrimidine class. It is a 2,4-diaminopyrimidine derivative.

### 5-Nitroimidazoles and 5-Nitrofurans

The nitroheterocyclic compounds consist of a nitro group joined to a heterocyclic ring. Two groups of these compounds are important in therapy of infections: the 5-nitroimidazoles and the 5-nitrofurans. Metronidazole is a 5-nitroimidazole antimicrobial drug. Its chemical formula is 1-(β-hydroxyethyl)-2-methyl-5-nitroimidazole

**Figure 6.14** Chemical structure of fosfomycin.

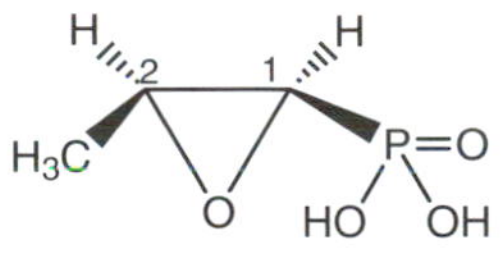

Fosfomycin

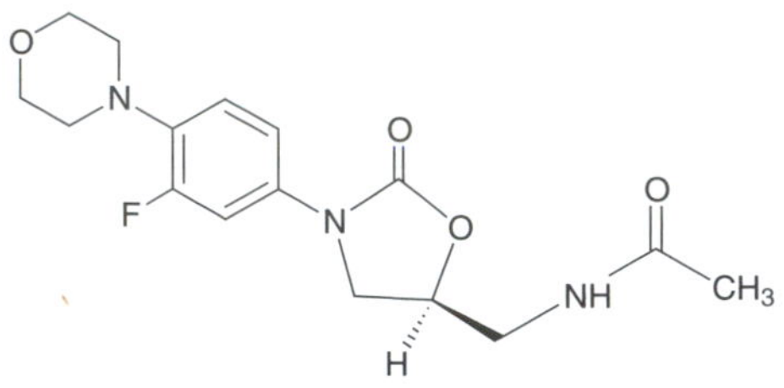

Linezolid

**Figure 6.15** Chemical structure of linezolid.

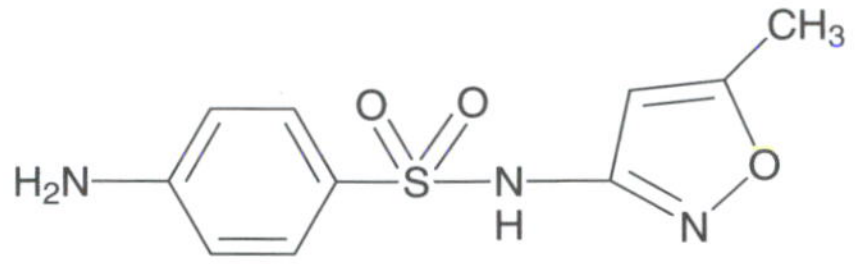

Sulfamethoxazole

**Figure 6.17** Chemical structure of sulfamethoxazole.

(Fig. 6.19). The nitrofurans are a class of synthetic antibacterial agents characterized by the presence of a 5-nitro-2-furanoyl group. Furazolidone (3-[(5-nitro-2-furanyl)methyleneamino]-2-oxazolidinone) is a representative of this group of synthetic antibacterial agents.

## Classification by Mechanism of Action

Antibacterial drugs exert their action by interfering with either the structure or metabolic pathways of bacteria. Figure 6.20 shows the sites of action of some of the antibacterial agents. The molecular mechanisms of action of specific antibiotics and synthetic antibacterials are considered in more detail when individual groups of antibacterial agents are discussed in the following chapters.

The most common mechanisms of action are

- inhibition of bacterial cell wall biosynthesis
- inhibition of protein, RNA, or DNA synthesis
- damage of membranes

### INHIBITION OF BACTERIAL CELL WALL BIOSYNTHESIS

Since bacterial cells have a high internal osmotic pressure, they require a rigid peptidoglycan polymer to protect themselves from lysis. Some of the antibiotics that interfere with the biosynthesis of peptidoglycan are β-lactams (penicillins, cephalosporins, cephamycins, monobactams, and carbapenems), glycopeptides (vancomycin and teicoplanin), D-cycloserine, fosfomycin, and bacitracin.

The most selective antibacterial agents are those that interfere with the biosynthesis of bacterial cell walls. These drugs have a high therapeutic index (the ratio of the toxic dose to the therapeutic dose) because they interfere with a biosynthetic process that is unique to prokaryotic cells (cell wall biosynthesis).

### INHIBITION OF PROTEIN SYNTHESIS

Although the mechanism of protein synthesis is similar in prokaryotes and eukaryotes, prokaryotic ribosomes differ substantially from those in eukaryotes. This explains the effectiveness of several clinically important antibiotics.

There are many clinically useful antibiotic classes that interact with rRNA or interfere with some steps in bacterial protein synthesis. Examples include aminoglycosides and tetracyclines, which bind conserved sequences within the 16S rRNA of the 30S ribosomal subunit; macrolides, ketolides, lincosamides, quinupristin-dalfopristin, and chloramphenicol, which act at the level of the 23S rRNA of the 50S ribosomal subunit; oxazolidinones, which are potent inhibitors of bacterial protein biosynthesis and prevent the formation of the *N*-formylmethionyl-tRNA–ribosome–mRNA ternary complex; and mupirocin, which inhibits the isoleucyl-tRNA synthetase.

### CELL MEMBRANE DISRUPTION

The bactericidal activity of the polymyxins and cationic antimicrobial peptides results from their interaction with the bacterial cytoplasmic membrane, causing gross disorganization of its structure.

### INHIBITION OF NUCLEIC ACID SYNTHESIS

The growth and division of the bacterial cell depend on, among other factors, DNA and RNA synthesis. Antibacterial drugs can disrupt nucleic acid synthesis in a variety

**Figure 6.16** Chemical structure of ciprofloxacin.

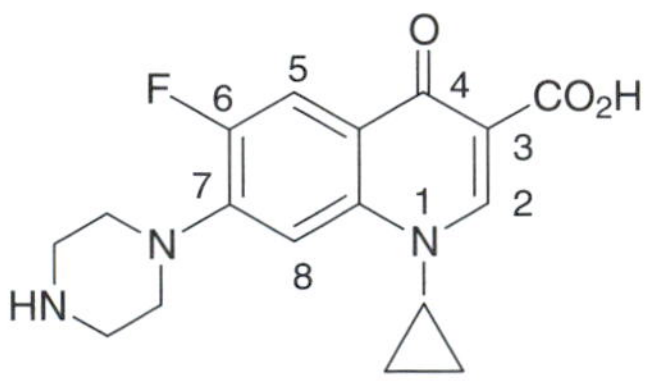

Ciprofloxacin

**Figure 6.18** Chemical structure of trimethoprim.

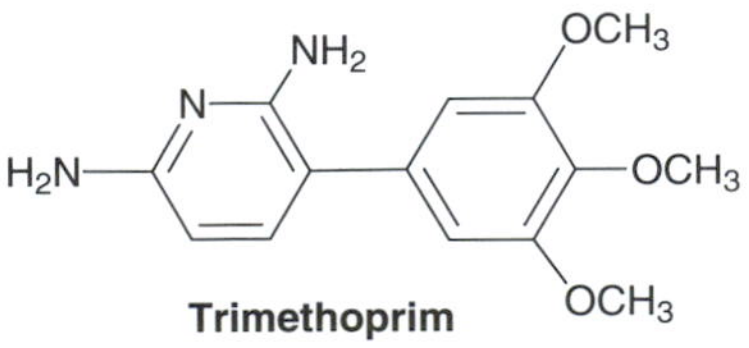

Trimethoprim

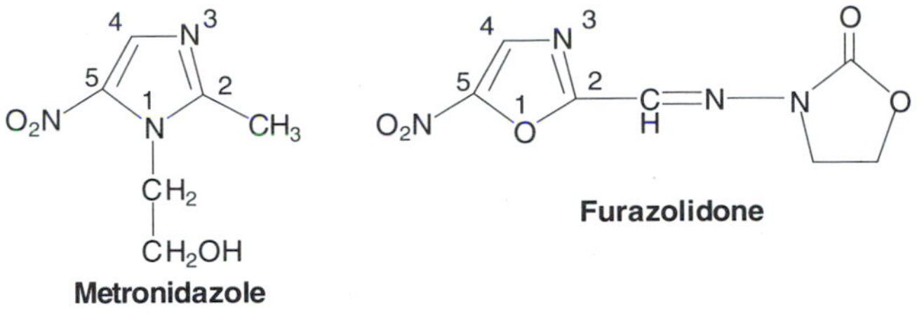

Metronidazole

Furazolidone

**Figure 6.19** Chemical structures of metronidazole and furazolidone.

of ways, both direct (e.g., the action of fluoroquinolones on DNA gyrase and topolsomerase IV) and indirect (e.g., the action of sulfonamides and trimethoprim on folic acid metabolism).

## Industrial Production of Antibiotics and Clavulanic Acid

The microorganisms used to produce antibiotics are fungi (molds), actinomycetes (filamentous bacteria), and bacteria (Table 6.1). Until 1974, antibiotics known to be produced by actinomycetes were produced almost exclusively by the genus *Streptomyces*. After 1974, other genera such as *Actinoplanes, Micromonospora,* and *Nocardia* also were found to produce antibiotics.

For growth and multiplication, microorganisms conduct a variety of biochemical and metabolic processes to obtain energy and new cell material. All industrial microbiology processes require the initial isolation of microorganisms from nature. Antibiotic-producing microorganisms are subjected to extensive genetic manipulations and modifications before they are used for antibiotic-manufacturing purposes. Most wild-type antibiotic-producing microorganisms synthesize only low concentrations of product or produce undesirable byproducts. Strain improvement is critical for the development of a successful product (Fig. 6.21). Multiple mutations were necessary to create a strain of the mold *Penicillium chrysogenum* that synthesized enough penicillin to form the basis of a commercial process. The original (wild-type) strain, NRRL 1951, produced 60 mg/liter, but today's production strains yield more than 25 g/liter.

There are two basic types of microbial metabolites: primary and secondary. A **primary metabolite** is formed during the exponential growth phase of the microorganism, whereas a **secondary metabolite** is formed near the end of the growth phase, frequently in the stationary phase of growth. Secondary metabolites are not essential for growth and reproduction. Natural-product antibiotics are secondary metabolites, formed from primary metabolites or from intermediate products. Antibiotics accumulate

**Figure 6.20** Schematic representation of the sites of action of antibacterial agents on a gram-negative bacterial cell.

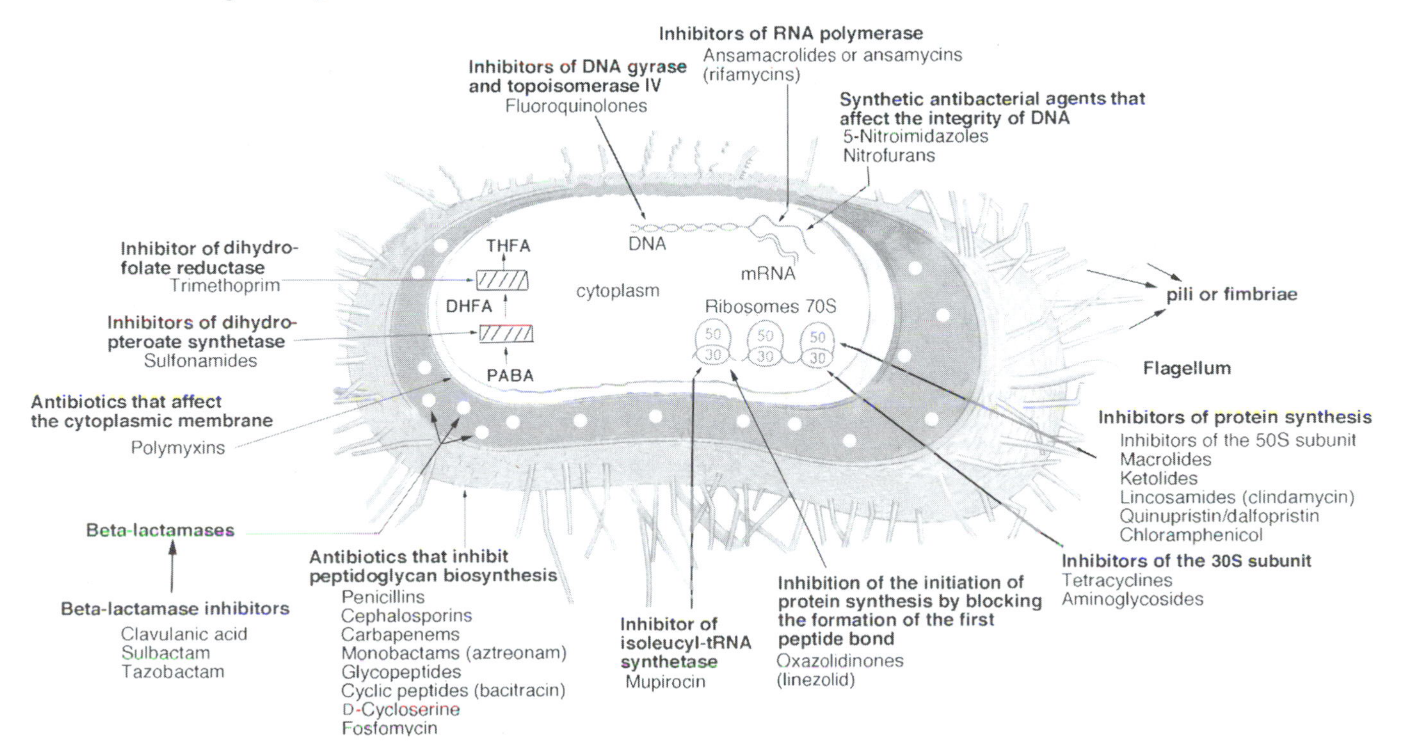

**Table 6.1** Some industrial microorganisms that produce antibiotics

| Type | Microorganism | Antibiotic produced |
|---|---|---|
| Fungi (molds) | *Penicillium chrysogenum* | Pencillins G and V |
| | *Penicillium notatum* | Penicillins G and V |
| | *Acremonium chrysogenum* (formerly *Cephalosporium acremonium)* | Cephalosporin C |
| Actinomycetes | *Streptomyces* spp. | Kanamycin, tetracycline, tobramycin, streptomycin, oxytetracycline, vancomycin, rifampin, streptogramins, and others |
| | *Streptomyces cattleya* | Imipenem |
| | *Streptomyces clavuligerus* | Clavulanic acid (a β-lactamase inhibitor) |
| | *Micromonospora purpurea* | Gentamicin |
| | *Actinoplanes teichomyceticus* | Teicoplanin |
| | *Nocardia* spp. | Cephamycins |
| | *Nocardia uniformis* | Nocardicin |
| Bacteria | *Pseudomonas fluorescens* | Mupirocin |
| | *Bacillus brevis* | Gramicidin |
| | *Bacillus subtilis* | Bacitracin |
| | *Bacillus polymyxa* | Polymyxin B |
| | *Chromobacterium violaceum* and *Gluconobacter* spp. | Monobactams |

**Figure 6.21** Strain improvement. Conventional strain improvement engineers the metabolism in a random manner. The mutated strains have to be screened or selected for the desired end point. Metabolic engineering with molecular genetic tools uses a preexisting knowledge database to produce strains that have directed and precise changes. Reprinted from A. Khetan and W.-S. Hu, p. 717–724, *in* A. L. Demain and J. E. Davies (ed.), *Manual of Industrial Microbiology and Biotechnology,* 2nd ed. (ASM Press, Washington, D.C., 1999), with permission from the publisher.

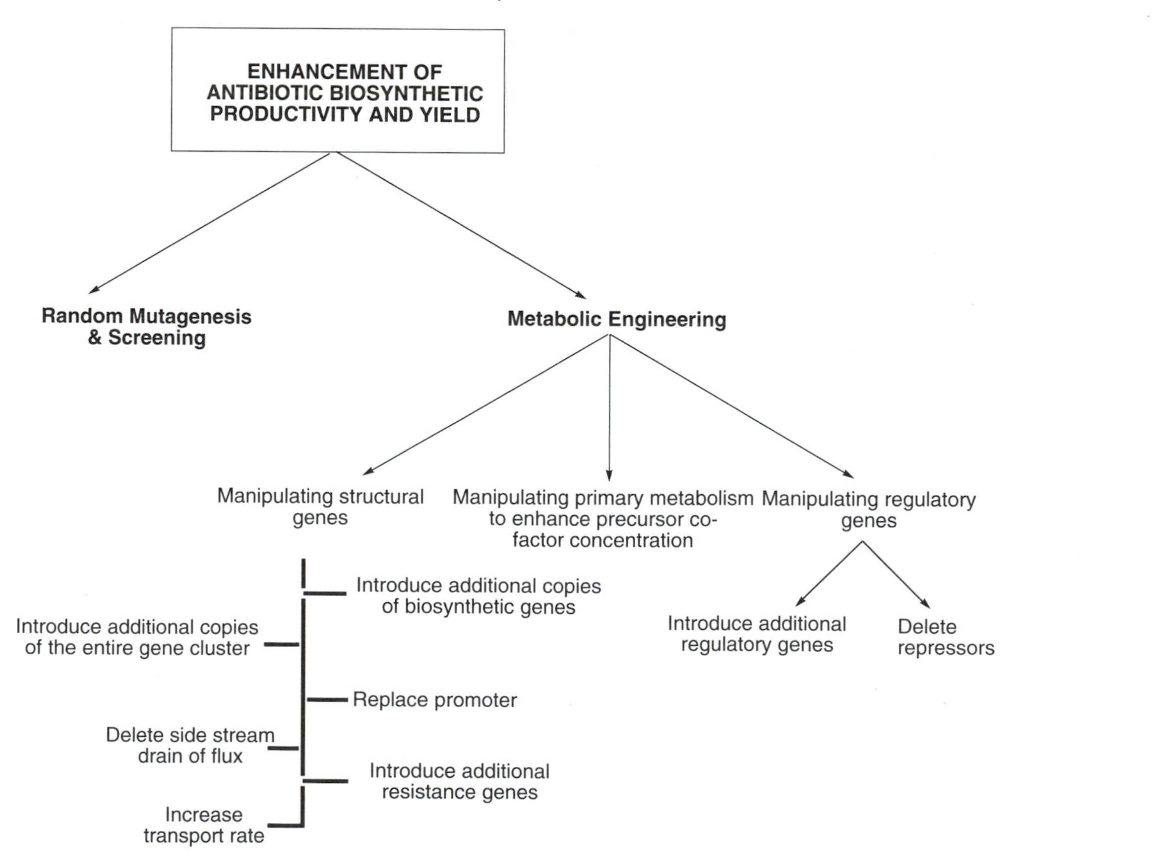

either in the culture medium or inside the cells. Some of the new commercial antibiotics have resulted from semisynthetic modifications of the naturally produced antibiotics. Secondary-metabolite yields can be improved by optimizing nutritional and physical components (temperature, aeration, etc.) and, as mentioned above, by genetic modification of the producing organism.

Development of industrial fermentation processes is usually carried out in three steps: (i) flask scale, where basic screening is carried out; (ii) pilot scale, where optimal operating conditions are established; and (iii) plant scale, where the process is brought to an economically favorable level (the vessel in which the industrial process is carried out is called a fermentor).

Fermentors can vary in size from the small 5- to 10-liter laboratory-scale fermentor to the enormous 150,000-liter industrial-scale fermentor. They are designed for operation in batch or continuous fermentation. Processes operated in batch mode require larger fermentors than do processes operated continuously or semicontinuously. The microorganism is grown at constant temperature in a sterile nutrient medium containing a carbon source, a nitrogen source, and a small amount of inorganic salts, under controlled conditions of aeration, agitation, and pH.

One of the most important and complicated aspects of industrial microbiology is the transfer of a process from small-scale laboratory equipment to large-scale commercial equipment, a procedure called **scale-up.**

## Isolation and Purification of Antibiotics Produced by Fermentation

When a maximal yield of antibiotic is obtained, the fermentation broth is processed by purification procedures tailored to the specific antibiotic being produced. Nonpolar antibiotics are usually purified by solvent extraction procedures; water-soluble compounds are commonly purified by ion-exchange methods or chemical precipitation.

## REFERENCES AND FURTHER READING

### Industrial Microbiology and Biotechnology

**Aharonowitz, Y., and G. Cohen.** 1981. The microbiological production of pharmaceuticals. *Sci. Am.*, **245**:141–152.

**Borders, D.** 1993. Antibiotics (survey), p. 107–118. *In* M. Howe-Grant (ed.), *Chemotherapeutics and Disease Control (Encyclopedia Reprint Series)*. John Wiley & Sons, Inc., New York, N.Y.

**Demain, A. L., and J. E. Davies (ed.).** 1999. *Manual of Industrial Microbiology and Biotechnology,* 2nd ed. ASM Press, Washington, D.C.

**Demain, A., and N. A. Solomon.** 1981. Industrial microbiology. *Sci. Am.* **245**:67–75.

**Gaden, E. L.** 1981. Production methods in industrial microbiology. *Sci. Am.* **245**:181–196.

**Hopwood, D. A.** 1981. The genetic programming of industrial microorganisms. *Sci. Am.* **245**:91–102.

**Lal, R., R. Khanna, H. Kaur, M. Monisha, N. Dhingra, S. Lal, K. H. Gaterman, R. Eichenlaub, and P. K. Ghosh.** 1996. Engineering antibiotic producers to overcome the limitations of classical strain improvement programs. *Crit. Rev. Microbiol.* **22**:201–255.

**Madigan, M. T., J. M. Martinko, and J. Parker.** 2000. *Brock Biology of Microorganisms,* 9th ed. Prentice-Hall, Upper Saddle River, N.J.

**Phaff, H. J.** 1981. Industrial microorganisms. *Sci. Am.* **245**:77–89.

**Strohl, W. R.** 1997. *Biotechnology of Antibiotics,* 2nd ed. Marcel Dekker, Inc., New York, N.Y.

chapter 7



# β-Lactams, Penicillin-Binding Proteins, and β-Lactamases

## β-Lactams

Chemically, a β-lactam is a four-member cyclic amide. The β-lactam compounds comprise two groups of therapeutic agents of considerable clinical importance: the β-lactam antibiotics and the β-lactamase inhibitors. The β-lactamase inhibitors are discussed in chapter 11.

### β-Lactam Antibiotics

β-Lactam antibiotics encompass a variety of natural and semisynthetic compounds which possess a single common feature: **a β-lactam ring.**

The potential utility of penicillin for controlling antibacterial infections in animals was recognized in 1940; shortly afterward its activity in humans was also discovered. The cephalosporins were first detected in 1945, when a strain of *Cephalosporium acremonium* (currently called *Acremonium chrysogenum*) produced antibiotic material that was active against both gram-negative and gram-positive organisms. The new β-lactam compound isolated was named cephalosporin C. Until 1971, penicillins and cephalosporins were the sole representatives of the β-lactam antibiotics. They are now subclassified as the classical or traditional group of β-lactam antibiotics.

Since 1971, novel natural mono- and bicyclic β-lactam structures have been discovered; these are now called nonclassical or nontraditional β-lactam antibiotics. They include the cephamycins, carbapenems, and monobactams (Fig. 7.1). The cephamycins, described almost simultaneously by Lilly and Merck Laboratories in 1971, are produced by *Streptomyces* species and have the same basic cephem ring system. The carbapenem thienamycin was isolated in 1979 from fermentation of strains of *Streptomyces cattleya.* Thienamycin, the olivanic acids, and other related derivatives are characterized by the presence of a bicyclic ring system with a carbon atom replacing the sulfur atom of the penicillins and cephalosporins (Fig. 7.1). In 1979, another entirely new class of β-lactam antibiotics was

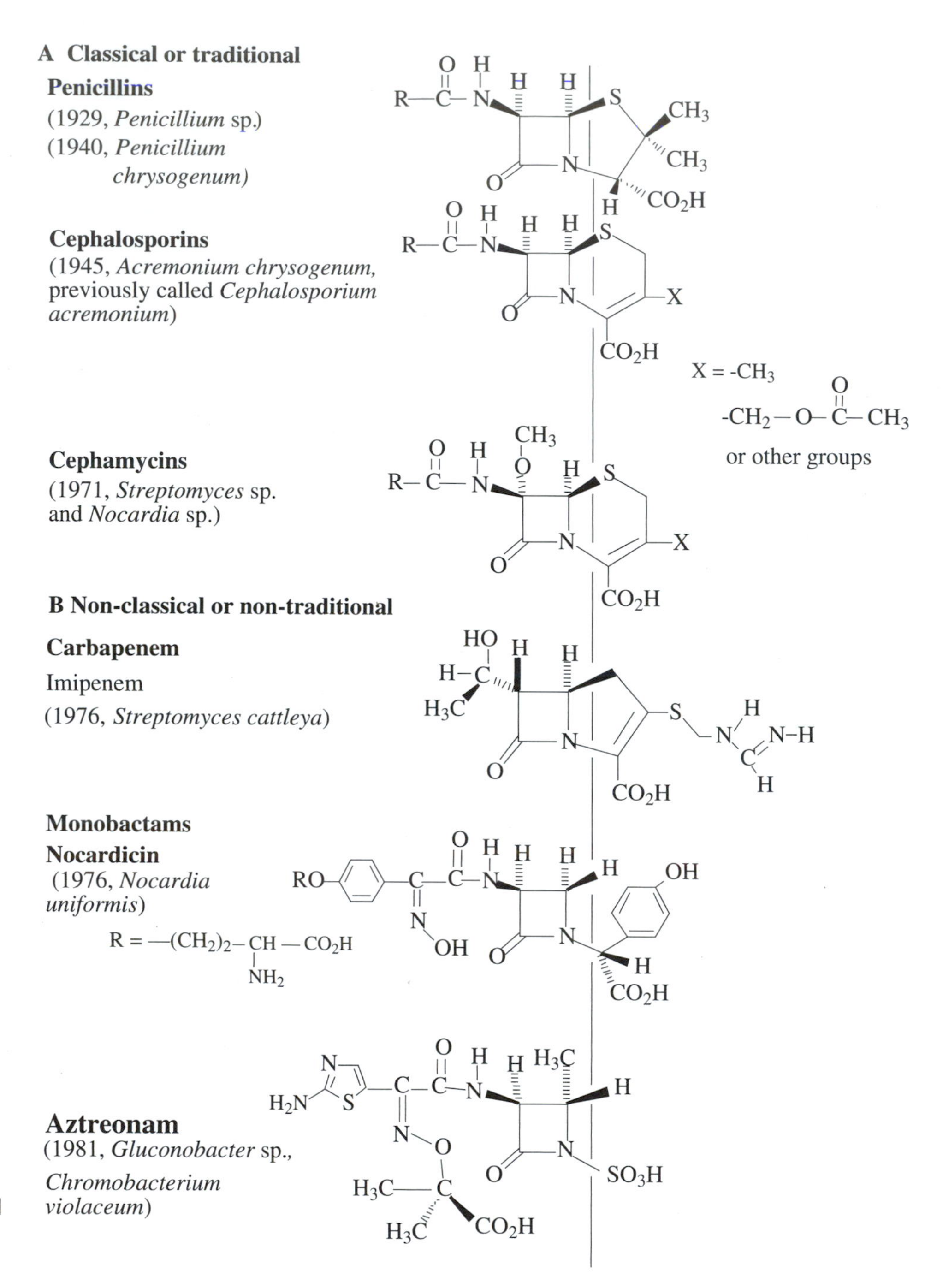


**Figure 7.1** Structures of representative classical and nonclassical β-lactam antibiotics.

discovered. The term **monobactam** (monocyclic β-lactam) was used for the first time in 1981 by Sykes et al. to describe a group of monocyclic bacterially produced β-lactam antibiotics isolated from soil bacteria such as *Chromobacterium, Agrobacterium,* and *Flexibacter.*

## NOMENCLATURE SYSTEMS AND BASIC SKELETONS

The antibiotics and synthetic antibacterial agents are identified by three different types of names:

- a chemical systematic name based on *Chemical Abstracts* rules of chemical nomenclature
- a trivial or generic name (nonproprietary, nonsystematic) that is simpler and easier to remember
- a trade name that is given by the manufacturer to distinguish them from competitive products

## PENICILLINS

The penicillin nucleus consists of the β-lactam ring fused through a nitrogen atom and the adjacent tetrahedral

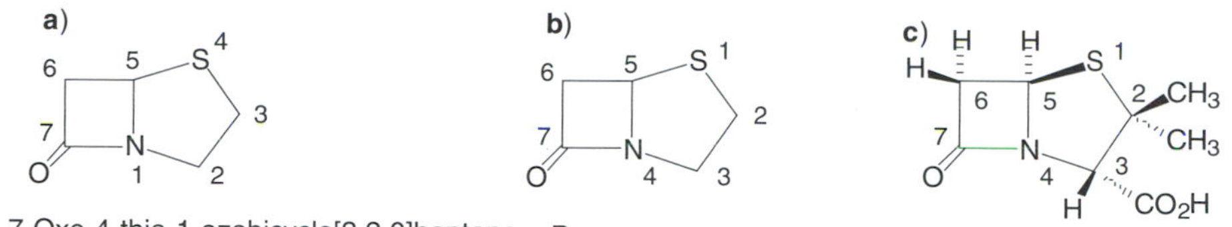

a) 7-Oxo-4-thia-1-azabicyclo[3.2.0]heptane b) Penam c) Penicillanic acid

**Figure 7.2** Basic skeleton and nomenclature systems of penicillins.

carbon atom to a 5-member thiazolidine ring. *Chemical Abstracts* has indexed the unsubstituted bicyclic nucleus as 7-oxo-4-thia-1-azabicyclo[3.2.0]heptane (Fig. 7.2a).

A simpler nomenclature system designates the unsubstituted bicyclic ring system of the penicillins as **penam** (Fig. 7.2b). Another simplification is the use of **penicillanic acid** to designate the penicillin ring system, with the substituent as indicated in Fig. 7.2c. When penicillins are named as derivatives of the 1-azabicyclo[3.2.0]heptane ring, numbering starts from the bridgehead nitrogen atom. In this book, the numbering system of the bicyclic ring is for nonsystematic names. It has been adopted to allow cross-reference correlations to the penams (e.g., penicillins), cephems (e.g., cephalosporins and cephamycins), and carbapenems (e.g., imipenem and meropenem).

Single-crystal X-ray diffraction was used to deduce the structure of penicillin when ambiguity developed in the interpretation of chemical evidence. Fig. 7.3 shows the conformation of the nucleus of the sodium salt of benzylpenicillin. The four-member ring is planar, while the five-member ring has a puckered appearance; N, C-5, S, and C-2 are nearly in the same plane, but C-3 is out of this plane and tilted away from the four-member ring. The absolute stereochemistry in all penicillins is 3(*S*):5(*R*):6(*R*) as indicated.

The stereochemistry of the substituents attached to the penam nucleus is designated by the α and β notation, similar to the notation used for steroids. Accordingly, the substituents located above a plane are designated β (concave surface) and those located below the plane are designated α (convex surface).

**Figure 7.3** Conformation of the sodium salt of benzylpenicillin in the solid state. R is $-NHCOCH_2C_6H_5$. The absolute configuration of chiral centers and the α and β positions below or above the plane are shown.

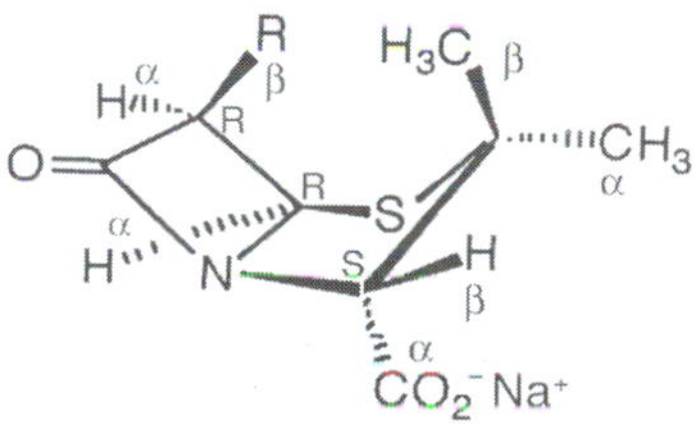

According to the nonsystematic simplification of using penicillanic acid as the basic structure and the α and β notation for the orientation of the substituents, the structure represented in Fig. 7.4 is designated 6-β-aminopenicillanic acid (commonly abbreviated to 6-APA). The structure drawn on the right is the zwitterionic representation of the left-hand structure. By analogy to the nomenclature used for the penam nucleus, the trivial names cephem, carbapenem, and clavam are used to identify the structures represented in Figure 7.5a to c. In cephalosporins, the β-lactam ring is fused to a six-member dihydrotiazine heterocyclic ring. A simplified nomenclature is the use of the names cephalosporanic acid and desacetylcephalosporanic acid for the structures in Fig. 7.5d and e.

## SPECTRAL CHARACTERISTICS

Several properties shared by some of the β-lactam antibiotics are characteristic of this group; however, in other cases the properties are very different, and it is difficult to give a clear picture of properties of the individual members and how they differ from each other.

### Infrared Spectroscopy

One of the most distinctive characteristics of the bicyclic β-lactam antibiotics is the infrared stretching frequencies for the β-lactam carbonyl group. The β-lactam carbonyl stretching occurs at 1,770 to 1,815 $cm^{-1}$ in penicillins and cephalosporins. This absorption frequency provides information about the integrity of the β-lactam ring. For monobactams, this stretching occurs at 1,750 to 1,760 $cm^{-1}$. The difference between this stretching frequency and that in penicillins and cephalosporins indicates the normal amide resonance in monobactams.

**Figure 7.4** Chemical structure of 6-β-aminopenicillanic acid in its nonionized and zwitterionic forms.

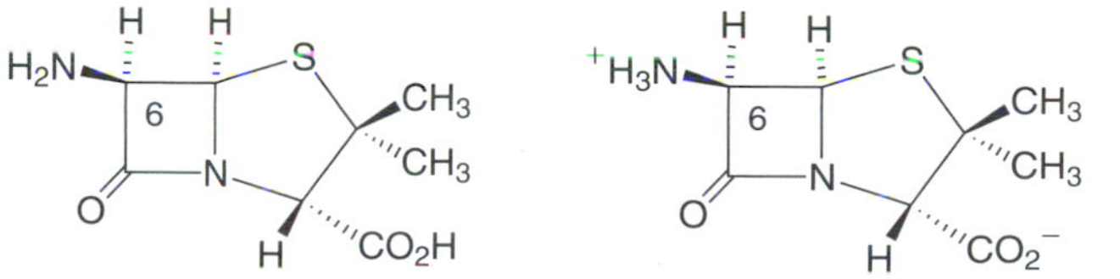

6-β-aminopenicillanic acid or 6-APA

6-β-aminopenicillanic acid or 6-APA, zwitterionic form

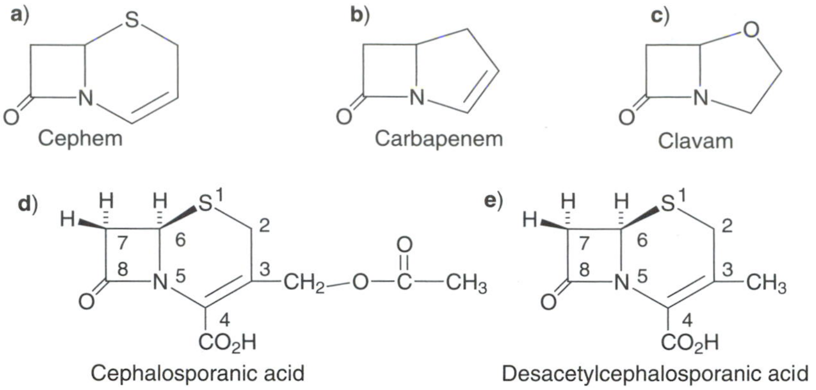

**Figure 7.5** (a to c) Basic skeleton of cephalosporins and cephamycins (a), carbapenems (b), and clavulanic acid (c). (d and e) Chemical structures of cephalosporanic acid (d) and desacetylcephalosporanic acid (e).

Fig. 7.6 shows that in a normal amide, a resonance tends to shorten the C—N bond and lengthen the C—O bond. A major requirement for maximization of this type of charge delocalization is that the three atoms connected to the nitrogen atom must be **coplanar** so that the unshared electron pair of the nitrogen atom can be involved in a π bond with the adjacent carbonyl carbon atom. In bicyclic compounds such as the penicillins, the nitrogen atom is out of this plane, as indicated by the higher frequency of the absorption band. This implies an increase in the double-bond character of the bonds between the carbon and oxygen atoms. In cephalosporins there is another factor besides the lack of planarity of the β-lactam nitrogen that may add additional contributions: the delocalization of the unshared electron pair of the nitrogen atom into the adjacent (allylic) olefinic π orbital system (enamine resonance).

## Proton Nuclear Magnetic Resonance Spectroscopy

The most characteristic features of the $^1$H nuclear magnetic resonance (NMR) spectra of penicillins and cephalosporins are the single proton doublet and quartet signals originating from the two β-lactam ring hydrogen atoms $H_A$ and $H_B$, common to both systems (Fig. 7.7). $H_A$ gives rise to a doublet [$J_{AB}(cis)$ = 4.5 Hz, or $J_{AB}(trans)$ = 1.5 to 2.0 Hz], while $H_B$ gives rise to the quartet ($J_{AB}$, same as above; $J_{AX}$ = 8 to 11 Hz). Addition of a drop of $D_2O$ to a solution of a given penicillin or cephalosporin in an organic solvent favors the deuterium exchange of the $H_X$ for $D_X$; the $H_B$ quartet then collapses to a doublet. Aside from these resonances, the $^1$H NMR spectra of penicillin derivatives are characterized by the presence of two singlets corresponding to the 2α-methyl (δ 1.23 to 1.66) and 2β-methyl (δ 1.49 to 1.84) groups and one proton singlet (δ 4.38 to 4.92), corresponding to H-3.

## The Nuclear Overhauser Effect

The nuclear Overhauser effect is a well-known method for assigning three-dimensional distributions of substituents in the pencillin molecule and in other β-lactam derivatives.

## Carbon-13 Nuclear Magnetic Resonance Spectroscopy

The $^{13}$C NMR features of all β-lactam antibiotics have been studied and are well documented.

## UV Spectroscopy

The penam moiety of penicillins is devoid of characteristic absorption spectra. The cephalosporins display absorption maxima at about 260 nm. Saturation of the double bond or isomerization (Δ-3 to Δ-2) results in loss of UV absorption. Thienamycin has a UV absorption maximun at 296 nm.

**Figure 7.6** Resonance in secondary amides, unstrained monocyclic β-lactams, and strained β-lactams.

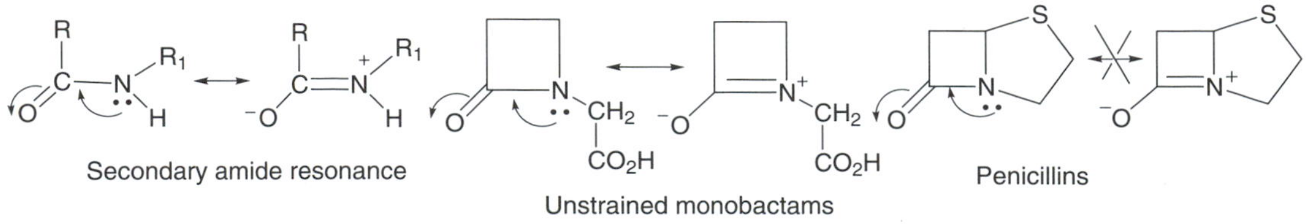

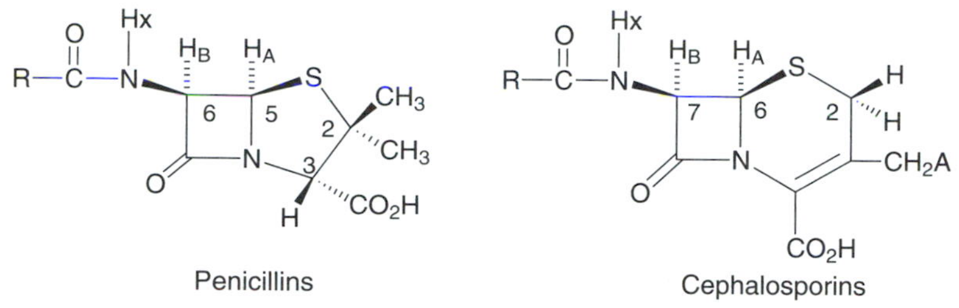

Penicillins

Cephalosporins

**Figure 7.7** Hydrogen atoms in penicillins and cephalosporins and coupling signals in the $^{1}$H NMR spectra.

CHEMICAL PROPERTIES

Looking back over the past 63 years (1940 to 2003), the development of the chemistry of penicillins and other β-lactam derivatives and their medical applications has reached levels inconceivable to the original workers.

In research starting in the 1980s and continuing today, I have been carrying out chemical modifications of the penam nucleus to synthesize potential inhibitors of serine β-lactamases and, more recently, of metallo-β-lactamases. Figure 7.8 depicts some aspects of the reactivity of the penam nucleus.

MODE OF ACTION OF β-LACTAM ANTIBIOTICS

Both gram-negative and gram-positive bacteria are surrounded by a cell wall. The major component of the cell wall is peptidoglycan. Biosynthesis of peptidoglycan is essential for bacterial survival; disruption of peptidoglycan may lead to cell lysis. All β-lactam antibiotics owe their activity to their ability to act as irreversible inhibitors of the D-alanyl-D-alanine carboxypeptidase/transpeptidases (DD-peptidases). These enzymes are collectively known as **penicillin-binding proteins (PBPs)** and are the enzymes that catalyze the cross-linking of neighboring peptidoglycan strands in the final step of bacterial cell wall biosynthesis. The β-lactam antibiotics inhibit PBPs by forming very long-lived acyl-enzyme intermediates. The reaction begins with binding of a peptidoglycan polymer, the **donor strand,** to the enzyme. It is proposed that a lysine residue acts as the catalytic base that abstracts a proton from the active serine during nucleophilic attack on the carbonyl carbon of the D-Ala–D-Ala peptide bond of the substrate, progressing through a transient tetrahedral intermediate state to a tetrahedral intermediate. The tetrahedral intermediate collapses through a further transition state to form an acyl-enzyme complex, with concomitant release of the terminal D-Ala. Many DD-peptidases are capable of carboxypeptidase and transpeptidase activity, so that the acyl-enzyme complex has two possible fates. It may simply hydrolyze, thereby releasing a peptidoglycan strand shortened by one D-Ala residue; alternatively, if a second strand of peptidoglycan is available, a transpeptidase may catalyze the formation of a peptide bond between the first peptidoglycan strand and the amino group of a glycylamine of the second, resulting in a cross-linked peptidoglycan polymer (Fig 7.9).

In 1965, Tipper and Strominger suggested that a β-lactam antibiotic, such as a penicillin or a cephalosporin, mimics the conformation of the acyl-D-Ala–D-Ala portion of the bacterial peptidoglycan (Fig. 7.10). Because β-lactams mimic the natural D-Ala–D-Ala substrate of the DD-peptidases, these enzymes are the target of β-lactam antibiotics.

## Penicillin-Binding Proteins

PBPs are a group of bacterial membrane-bound enzymes whose active sites are available in the periplasmic space.

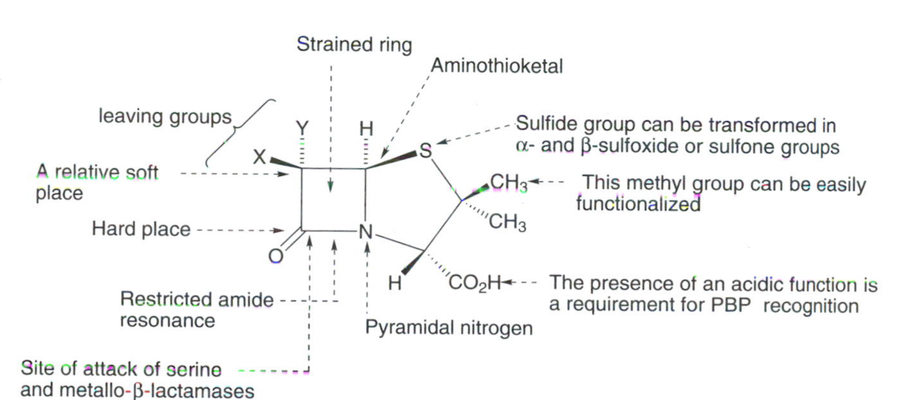


**Figure 7.8** Penicillanic acid derivative structure showing some positions where chemical transformations are made. Chemical mapping of the different positions and orientation of the molecule is very useful for structure-activity relationships.

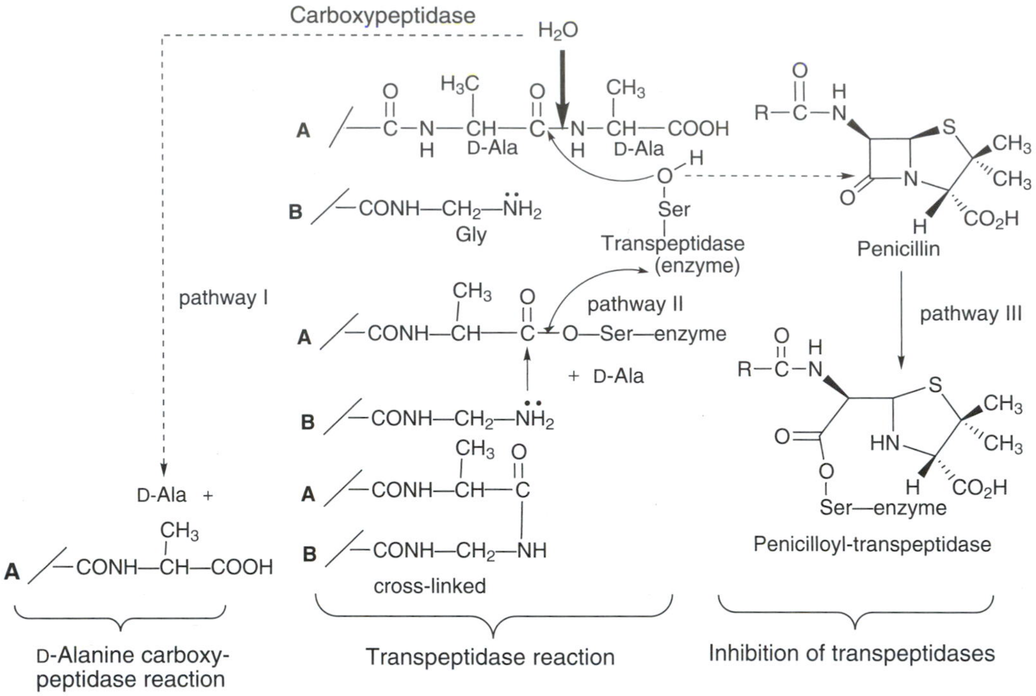

**Figure 7.9** Carboxypeptidase and transpeptidase reactions (pathways I and II). These reactions are catalyzed by bacterial PBPs. Pathway III shows the action of a penicillin. The penicilloyl-transpeptidase is more stable than the acyl-D-alanyl-transpeptidase, and consequently the transfer of the acyl group to an amino group does not take place (pathway III).

**Figure 7.10** Dreiding stereomodels of penicillin and of the acyl-D-alanyl-D-alanine end of the nascent peptidoglycan. Arrows indicate the position of the OC—N bond in the β-lactam ring of the penicillin and of the OC—N peptide bond joining the two D-alanine residues. Reprinted from D. J. Tipper and J. L. Strominger, *Proc. Natl. Acad. Sci. USA* **54:**1133, 1965, with permission from the publisher.

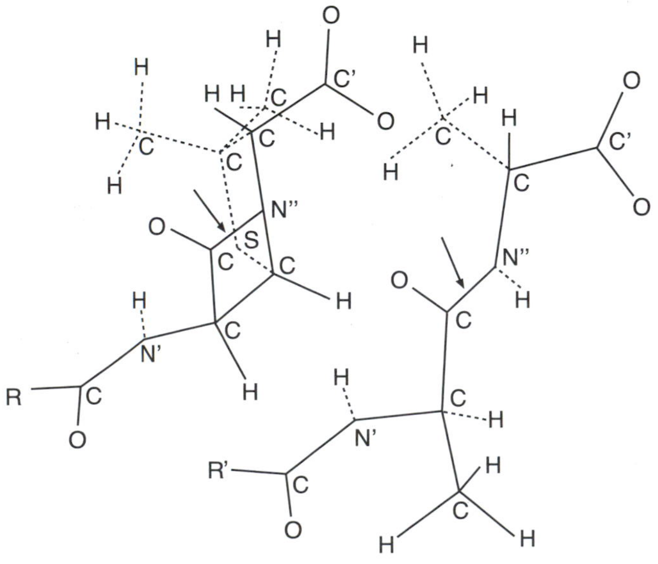

Protein bands corresponding to these enzymes can be visualized on fluorograms after sodium dodecyl sulfate-polyacrylamide gel electrophoresis with radioactive penicillin for labeling, hence the name "penicillin-binding proteins."

PBPs have the ability to catalyze the rupture of the β-lactam bond of β-lactam antibiotics and to form a serine-ester-linked penicilloyl enzyme (Fig. 7.9) This intermediate is relatively stable, since deacylation of the enzyme is very slow.

There are two groups of PBPs: low-molecular-weight and high-molecular-weight PBPs (Table 7.1). Each bacterium contains a set of PBPs which have a low degree of relatedness and all of which are targeted by β-lactam antibiotics. The affinity for a given β-lactam antibiotic is different for different PBPs, and, conversely, a given PBP has distinct affinities for different β-lactams. Therefore, point mutations leading to a reduced affinity for one β-lactam do not necessarily affect the affinity of the PBPs for another compound, allowing different β-lactams to interact with and inhibit different PBPs, as described for *Escherichia coli.*

Most bacterial species contain four to eight PBPs with molecular masses of 35 to 120 kDa. By convention, the PBPs are designated numerically in order of decreasing

**Table 7.1** Properties of the PBPs of *E. coli* K-12[a]

| PBP | Mol wt | Gene symbol | Proposed function | Examples of antibiotics with marked affinity |
|---|---|---|---|---|
| 1A<br>1Bα<br>1Bβ<br>1Bγ | 92,000<br>86,500<br>84,000<br>81,500 | *mrcA* (*ponA*)<br>*mrcB* (*ponB*) | Bifunctional enzymes with transglycosylase and transpeptidase activities. Synthesize peptidoglycan at the growing zones of the side wall. | Benzylpenicillin and most cephalosporins |
| 2 | 66,000 | *pbpA* | Bifunctional enzyme with transglycosylase and transpeptidase activity. Initiates peptidoglycan insertion at new growth sites, which are then further extended by PBP1A and PBP1B. | Mecillinam, imipenem |
| 3 | 60,000 | *fts-1* (*pbpB, sep*) | Bifunctional enzyme with transglycosylase and transpeptidase activity. Required specifically for formation of the cross-wall at cell division. | Cephalexin and many other cephalosporins, piperacillin, aztreonam |
| 4 | 49,000 | *dacB* | DD-Carboxypeptidase and/or endopeptidase. The first activity may control the extent of peptidoglycan cross-linking by transpeptidases; the second activity causes hydrolysis of cross-links during cell elongation. | Benzylpenicillin, ampicillin, imipenem |
| 5<br>6 | 42,000<br>40,000 | *dacA*<br>*dacC* | DD-Carboxypeptidases that may control the extent of peptidoglycan cross-linking by transpeptidases | Cefoxitin |
| 7 | 29,000 | *pbpG* | Unknown | Penems |

[a] Adapted from M. Howe-Grant (ed.), *Chemotherapeutics and Disease Control* (John Wiley & Sons, Inc., New York, N.Y., 1993) with permission from the publisher.

molecular mass, e.g., 1 to 7 in *E. coli* K-12 (Table 7.2). Once the number of PBPs in a species is assigned, subsequently identified enzymes are numbered as derivatives of the previously established ones. For example PBP2 of *Streptococcus pneumoniae* was resolved into three components, which were named PBP2x, PBP2a, and PBP2b.

In general, PBPs are species specific (e.g., *S. pneumoniae* PBPs are different from *E. coli* PBPs, which are different from *Staphylococcus aureus* PBPs). However, the PBPs of *S. pneumoniae* and viridans streptococcal species contain identical or closely related DNA sequences, and there is a high degree of homology between PBPs of different enterococcal species.

PBP1a and PBP1b of *E. coli* are bifunctional in the sense that they connect the disaccharide pentapeptide to a glycan chain (transglycosylase activity) and link peptide side chains from neighboring glycan chains to each other (transpeptidase activity). The bifunctional enzymatic activity of these PBPs probably results from the presence of two distinct, catalytically active centers in the same polypeptide, i.e., an amino-terminal domain catalyzing the transglycosylation reaction and a carboxy-terminal domain catalyzing the transpeptidation reaction.

PBP2 and PBP3 exhibit transpeptidase activity. Whether they can also carry out transglycosylation activity is still unclear. It is generally assumed that PBP2 is involved in cell elongation and PBP3 is involved in cell division.

Inhibition of the transpeptidase activity of the high-molecular-weight PBPs of *E. coli* by β-lactam antibiotics eventually leads to cell death. Concomitant inhibition of PBP1a and PBP1b leads to rapid cell lysis, inhibition of PBP2 leads to growth as osmotically stable spherical or ovoid forms, and inhibition of PBP3 leads to filamentation.

*E. coli* PBP4 through PBP7 are DD-carboxy/endopeptidases (Table 7.1). Serine DD-carboxypeptidases hydrolyze the D-alanyl-D-alanine bonds (Fig. 7.12). They control the extent of peptidoglycan cross-linking by limiting the number of pentapeptide units available for transpeptidation. Monofunctional serine DD-carboxypeptidases also hydrolyze peptidoglycan interpeptide bonds (endopeptidase activity). In *E. coli* PBP4, PBP5, and PBP6 do not appear to be essential for growth, and thus their interactions with β-lactam antibiotics are not considered responsible for the killing action of these antibiotics. As mentioned above, PBP1a, PBP1b,

**Table 7.2** β-Lactamase classification[a]

| Bush-Jacoby-Medeiros group | Molecular class | Preferred substrates | Inhibited by: Clavulanic acid | Inhibited by: EDTA | Representative enzymes |
|---|---|---|---|---|---|
| 1 | C | Cephalosporins | – | – | AmpC β-lactamases from gram-negative bacteria; MIR-1 |
| 2a | A | Penicillins | + | – | Penicillinases from gram-positive bacteria |
| 2b | A | Penicillins, cephalosporins | + | – | TEM-1, TEM-2, SHV-1 |
| 2be | A | Penicillin, narrow-spectrum and extended-spectrum cephalosporins and monobactams | + | – | TEM-3 to TEM-26, SHV-2 to SHV-6, *K. oxytoca* K1 |
| 2br | A | Penicillins | ± | – | TEM-30 to TEM-36, TCR-1 |
| 2c | A | Penicillins, carbenicillin | + | – | PSE-1, PSE-3, PSE-4 |
| 2d | D | Penicillins, cloxacillin | ± | – | OXA-1 to OXA-11, PSE-2 (OXA-10) |
| 2e | A | Cephalosporins | + | – | Inducible cephalosporinases from *Proteus vulgaris* |
| 2f | A | Penicillins, cephalosporins, carbapenems | + | – | NMC-A from *E. cloacae,* Sme-1 from *S. marcescens* |
| 3 | B | Most β-lactams including carbapenems | – | + | L1 from *S. maltophilia,* CcrA from *B. fragilis* |
| 4 | ND[b] | Penicillin | – | ? | Penicillinase from *P. cepacia* |

[a] Reprinted from K. Bush, G. A. Jacoby, and A. A. Medeiros, *Antimicrob. Agents Chemother.* 39:1211–1233, 1995, with permission from the publisher.
[b] ND, not determined.

PBP2, and PBP3 are the primary lethal targets of β-lactams.

## Crystallographic Structure of DD-Peptidases: Penicillin Target Enzymes

A very high resolution (1.6 Å) image of the structure of the low-molecular-weight DD-peptidase of *Streptomyces* sp. strain R61 was obtained in 1995 (Kelly and Kuzin, 1995), allowing improved mapping and refinement of its complexes with β-lactams. This DD-peptidase is an exocellular enzyme with a molecular mass of 37,500 Da (349 amino acid residues), which had served as a model for the low-molecular-weight peptidases that are normally membrane bound.

In 1996 the structure of the low-molecular-weight PBP of *S. pneumoniae,* PBP2x, was determined by X-ray crystallography at 3.5-Å resolution (Pares et al., 1996). In 2000, Gordon et al. described the structure of PBP2x at 2.4 Å. This new crystal structure provides a more accurate model on which the authors based their analysis. The structure of PBP2x revealed an active site similar to those of the class A β-lactamase. Six *S. pneumoniae* PBPs have been characterized, a single low-molecular-weight PBP and five high-molecular-weight PBPs: PBP1a, PBP1b, PBP2a, PBP2b, and PBP2x.

In 2003, studies carried out at the Institut de Physique and Institut de Chimie at the Centre d'Ingénierie des Protéines, Universite de Liège (Liège, Belgium), found that the *Streptomyces* K15 penicillin-binding DD-transpeptidase is involved in peptide cross-linking during bacterial cell wall peptidoglycan assembly. The main objective of the scientists was to gain insight into the catalytic mechanism, specifically the roles of residues Lys38, Ser96, and Cys98, belonging to the structural elements defining the active-site cleft. They investigated these aspects by site-directed mutagenesis, biochemical studies, and X-ray diffraction analysis and found that Lys38His and Ser96Ala mutations almost completely abolished penicillin binding and severely impaired the transpeptidase activities while the geometry of the active site was essentially the same as in the wild-type enzyme. It is proposed that Lys38 acts as the catalytic base that abstracts a proton from the active serine Ser35 during nucleophilic attack and that Ser96 is a key intermediate in proton transfer from the Oγ of Ser35 to the substrate leaving group nitrogen.

As mentioned in chapter 5, bacteria have developed several mechanisms of resistance, one of which is the mutation of the target enzymes to reduce their affinity for β-lactam antibiotics. A high degree of β-lactam antibiotic resistance can be achieved by modification of these five high-molecular-weight PBPs of *S. pneumoniae.* Gordon et al. (2000) also described the PBP2x-cefuroxime complex (a second-generation cephalosporin). In 2001, Dideberg et al. reported the crystal structure of PBP2x from a highly penicillin-resistant *S. pneumoniae* clinical isolate. In 2002, Lim and Strynadka reported that the crystal structure of a soluble derivative of PBP2a has been determined to 1.8 Å resolution and provides the highest-resolution structure for a high-molecular-mass PBP. Additionally, determination of the structures of the acyl-PBP complexes of PBP2a with nitrocefin, penicillin G, and methicillin allowed, for the first time, comparison of an apo and acylated resistant PBP. Analysis of the PBP2a active site in these forms revealed the structural basis of its resistance and identified features in newly developed β-lactams that are likely to be important for high affinity binding.

In 2003, Vernet et al. described the expression and biochemical characterization of the soluble extracellular region of *S. pneumoniae* PBP1b. This structural information has been assessed in an effort to understand the molecular mechanism of β-lactam antibiotic resistance.

## β-Lactamases

β-Lactamases are plasmid or chromosomally encoded bacterial enzymes that catalyze the hydrolysis of the β-lactam C—N bond of β-lactam antibiotics to give the corresponding β-amino acid devoid of antibacterial activity. β-Lactamases are produced in a constitutive or inducible manner and are secreted into the periplasmic space of gram-negative bacteria or into the outer medium of gram-positive bacteria.

Catalysis of β-lactam hydrolysis occurs by the addition of a water molecule across the β-lactam bond via one of two mechanisms, depending on the structural type of enzyme: (i) by acylating a serine at the active site of a β-lactamase or (ii) by using a water molecule coordinated to one or two zinc atoms at the active center of a metallo-β-lactamase (Fig. 7.11).

The production of β-lactamases is considered to be the most common mechanism of bacterial resistance to β-lactam antibiotics. Many β-lactamases have already been described, and every year new members of this group are discovered. β-Lactamases are often called penicillinases, cephalosporinases, or carbapenemases, depending on their specificity.

Resistance to β-lactam antibiotics was already known before penicillins (penicillins G and V) became clinically useful antibiotics. In 1940, Abraham and Chain first described a bacterial enzyme that destroyed penicillin; they termed it penicillinase, since the β-lactam structure of penicillin was not yet known. This observation was the first indication that β-lactamases could constitute a serious threat to the effectiveness of β-lactam compounds as antibacterial agents. Over the past 60 years, β-lactamases have been examined in detail by enzymologists, microbiologists, chemists, and X-ray crystallographers in an attempt to understand how they function in their various settings.

**Figure 7.11** Opening of the β-lactam ring by catalysis with a β-lactamase enzyme.

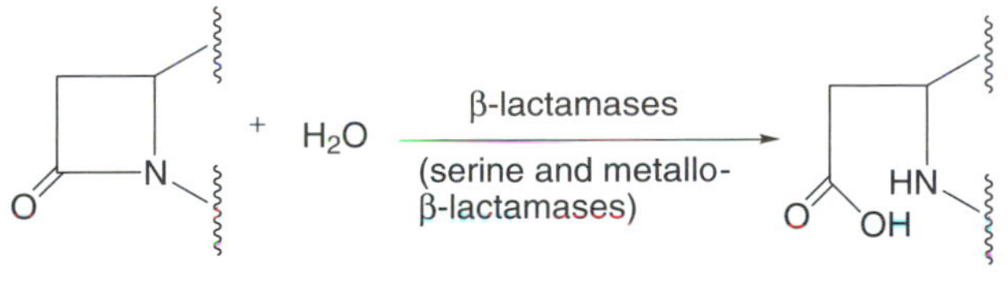

### β-Lactamase Nomenclature

β-Lactamase nomenclature has not followed rational rules. Some enzymes have three- or four-letter abbreviations. Some were named for properties related to some of the substrates they hydrolyze, such as CARB (for "activity against carbenicillin") and OXA and IMP (for "activity against oxacillins and imipenem," respectively), and some were named for their biochemical properties, such as SHV (for "sulfhydryl variable"). NMC-A stands for "nonmetallo carbapenemase of class A." The name CTX is an abbreviation of cefotaximase and refers to the potent hydrolytic activity of these enzymes against cefotaxime. The AmpC β-lactamases were named for the genes that encode them and hence for ampicillin hydrolysis. Other enzymes were named for a particular bacterium; for example, PSE refers to *Pseudomonas*-specific enzymes and Sme refers to the *Serratia marcescens* enzyme. TEM was named for the patient from whom it was isolated in 1963 in Athens, Greece, a young female called Temoniera. The names of some other enzymes refer to a hospital, for example, MIR (for "Miriam Hospital"), or a particular state in the United States (such as OHIO). The names of yet other β-lactamases refer to the pharmaceutical company and bacteria with which they are associated, such as FPM (for "Fujisawa's *Proteus mirabilis*"), and others refer to the researchers who discovered them, such as HMS (for "Harris, Matthew, Sykes"). The novice reader in this field should appreciate that the nomenclature does not always refer to the properties of these particular enzymes.

### β-Lactamase Classification

β-Lactamases have been distinguished according to function (especially substrate profile and sensitivity to inhibitors), physical properties (such as isoelectric point and molecular weight), and genetic location (plasmid or chromosome). The first classification was proposed by Jack and Richmond in 1970 and was expanded by Richmond and Sykes in 1973, in which five classes of β-lactamases from gram-negative bacteria were grouped on the basis of their substrate profiles. This classification was updated by Sykes and Matthew in 1976, who adopted the technique of isoelectric focusing to identify specific β-lactamases. More recently, in an attempt to address the increasing number of newly discovered β-lactamases, the classification by Sykes and Matthew was further modified, first by Bush in 1989 and later by Bush, Jacoby, and Medeiros in 1995, who grouped the enzymes according to their substrate and inhibitor profiles. In 1980, Ambler proposed the first classification of β-lactamases on the basis of their primary structure and amino acid sequence identity. On that basis, four classes have been identified: class A, C, and D β-lactamases are active-site serine enzymes, while class B enzymes are metalloproteins containing at least one $Zn^{2+}$ ion per active subunit.

### THE BUSH-JACOBY-MEDEIROS FUNCTIONAL CLASSIFICATION

The Bush-Jacoby-Medeiros scheme classifies the β-lactamases from naturally occurring bacterial isolates into four groups based on substrate and inhibitor profiles (Table 7.2). Group 1 includes cephalosporinases that are not well inhibited by clavulanic acid. Sequences available show that all are closely related to class C β-lactamases. Group 2 is the largest category, comprising β-lactamases that are generally well inhibited by clavulanic acid. Group 2 enzymes include many of the most common β-lactamases including the staphylococcal penicillinases and the broad-spectrum plasmid-encoded enzymes such as the TEM-1 and SHV-1 enzymes. The enzymes fall into six subgroups based on preferential hydrolysis of penicillins, cephalosporins, oximino β-lactams, cloxacillin, carbenicillins, or carbapenems. Over half are plasmid encoded. All enzymes with known sequences belong to molecular class A, except for the class D cloxacillin-hydrolyzing β-lactamases. Group 3 comprises metallo-β-lactamases (molecular class B) that hydrolyze penicillins, cephalosporins, and carbapenems. These are plasmid-encoded β-lactamases. Group 4 consists of a small group of penicillinases that are not well inhibited by clavulanic acid. Sequences are not yet known for any members of this group.

### AMBLER'S MOLECULAR CLASSIFICATION

According to Ambler (1980), β-lactamases are also grouped into four molecular classes (A to D) based on their primary sequence homology. The main division between these classes is based on the chemistry of their catalytic mechanisms and distinguishes class A, C, and D enzymes, which have serine in the active site (Fig. 7.12), from class B enzymes, which are zinc dependent.

## Serine β-Lactamases and Serine DD-Transpeptidases

The serine β-lactamases recognize the β-lactam antibiotics and β-lactam inhibitors such as clavulanic acid, sulbactam, and tazobactam among other mechanism-based inhibitors. β-Lactamases inactivate β-lactam antibiotics by hydrolyzing the lactam bond -OC—N- of the β-lactam nucleus but are themselves the targets for the inhibitory action of other β-lactam derivatives (lacking the amide side chain). The serine DD-transpeptidases involved in the biosynthesis of the peptidoglycan bind penicillins and other β-lactam antibiotics and are inactivated.

The mechanism by which both classes of enzyme cleave the cyclic amide bond involves the nucleophilic addition of the hydroxyl group of an active serine site to the carbonyl carbon and formation of a tetrahedral intermediate that collapses to an acyl-enzyme intermediate. Serine β-lactamases differ from serine DD-transpeptidases in that they catalyze the deacylation step very efficiently only with β-lactams that have an aromatic (planar) substituent joined to the secondary amide side chain. This mechanism is schematically represented by a three-step equation (Fig. 7.13).

Penicillanic acid sulfone is a typical example of a very poor inhibitor of serine DD-transpeptidases that act as an irreversible mechanism-based inhibitor of β-lactamases. In 1981, Knowles et al. proposed a chemical mechanism for the inhibition of TEM β-lactamases by penicillanic acid sulfone, which, after the initial acylation step and rearrangement reaction, ends with a residual fragment of the original molecule of the inhibitor (called the suicide

**Figure 7.12** Inactivation of penicillin by active-site serine β-lactamases.

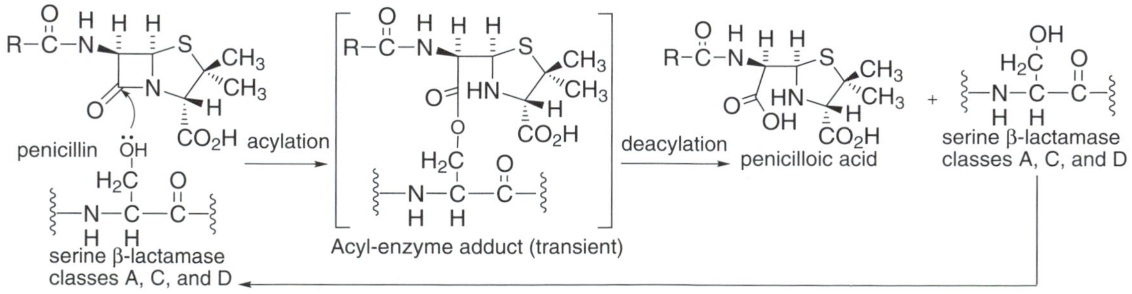

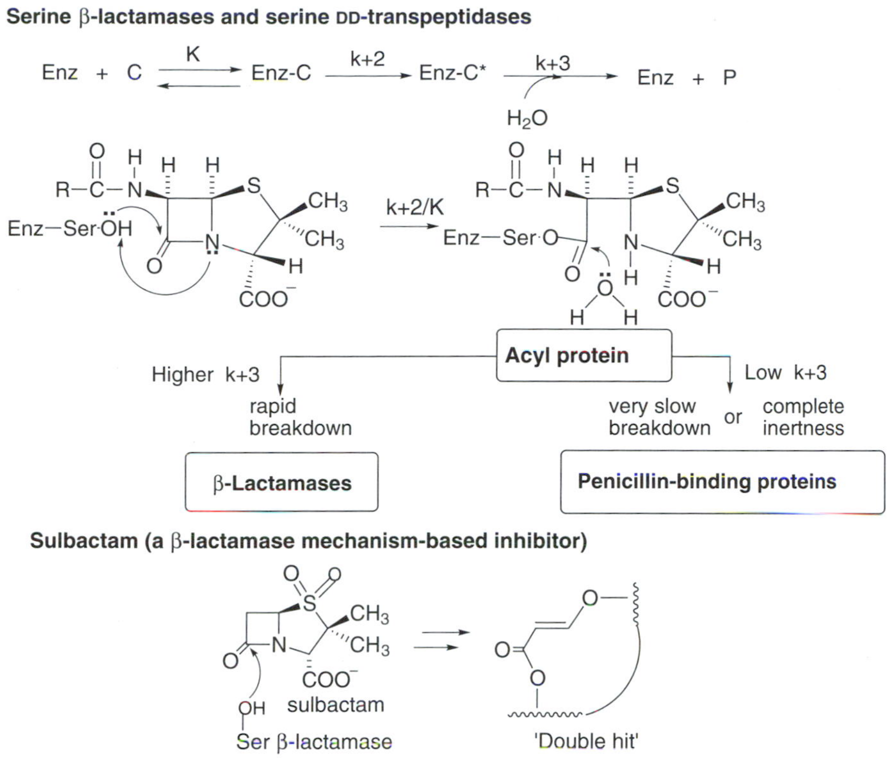

**Figure 7.13** The penicillin-interactive, serine-active-site PBPs and β-lactamases. Enz, enzyme; C, antibiotic; Enz-C, noncovalent Michaelis complex; Enz-C*, covalent acyl-enzyme; P, inactive degradation product(s) of the antibiotic.

inhibitor) being linked by covalent bonds to one oxygen atom from serine 70 and one oxygen atom from serine 130 on the β-lactamase. This is termed the "double-hit" mechanism (Fig. 7.13, bottom).

The interaction between these two families of enzymes with β-lactam antibiotics and β-lactam derivatives that act as inactivators of β-lactamases is represented in Fig. 7.13.

## EVOLUTION OF SERINE DD-TRANSPEPTIDASES AND SERINE β-LACTAMASES

The serine β-lactamases and serine DD-transpeptidases are thought to have a common evolutionary ancestor. Although the evolutionary distance may vary considerably, all these penicillin-interactive proteins and domains appear to be members of a single superfamily of active-site serine enzymes distinct from the classical serine proteases. The amino acid alignments reveal several conserved boxes that consist of strict identities or homologous amino acids. The significance of these homologies and differences is highlighted by the recent results of X-ray crystallography and site-directed mutagenesis experiments that have demonstrated the three-dimensional structural similarities between representatives of these enzymes. In short, the two families of serine enzymes share structural topologies and certain mechanistic features.

## CLASS A SERINE β-LACTAMASES

Among the four Ambler classes of β-lactamases, the class A enzymes are largely plasmid encoded and are the most numerous and widely distributed among gram-positive and gram-negative bacteria; they are the best-characterized enzymes.

### Three-Dimensional Structural Data for Class A β-Lactamases

The three-dimensional tertiary structures established at the atomic level by X-ray crystallography of eight class A β-lactamases, TEM-1 and TOHO-1 from *E. coli*, PC1 from *Staphylococcus aureus*, SHV-1 from *Klebsiella pneumoniae*, NMC-A from *Mycobacterium fortuitum*, BLIC from *Bacillus licheniformis*, SAG from *Streptomyces albus* G, and PER-1 from *Pseudomonas aeruginosa*, have been reported as of December 2000.

In 2002, M. Nukaga and colleagues reported the structure of a chromosomal extended-spectrum β-lactamase (ESBL) with the ability to hydrolyze cephalosporins including cefuroxime and ceftazidime, determined by X-ray crystallography to 1.75 Å resolution.

### The Catalytic Mechanism of Class A β-Lactamases

Several mechanisms for the catalysis by class A enzymes have been proposed on the basis of crystallographic, molecular modeling, and mutagenesis studies. A detailed understanding of the intermolecular relationships of β-lactam antibiotics and β-lactamase inhibitors that influence substituent binding and catalysis or inhibition is the cornerstone of the design of future generations of β-lactam antibiotics as well as β-lactamase inhibitors. However, despite the extensive available kinetic, structural, and mutagenesis data, the factors explaining the diversity of the specific profiles of class A β-lactamases and their detailed catalytic mechanism have not yet been thoroughly elucidated.

Ser70 (according to the numbering of Ambler) is the active-site residue in class A β-lactamases, as clearly demonstrated by using site-directed mutagenesis. Sequence alignment of class A β-lactamases has revealed highly conserved motifs consisting of three polar regions: the tetrad Ser70-Xaa-Xaa-Lys73, the triad Ser130-Asp131-Asn132, and the triad Lys234-Thr(Ser)235-Gly236. As well as these motifs, Glu166 is strictly conserved in class A β-lactamases. Its carboxylate side chain lies in the active site and interacts with several other residues and with a water molecule. Its strategic position, close to the active-site serine side chain, makes it an ideal candidate to play a critical role in the catalytic mechanism. It is generally accepted that Glu166 is specifically involved in the deacylation step.

For class A enzymes, a general base (proton abstractor) is presumed to participate in the active site by abstracting a proton from the nucleophilic hydroxyl group of Ser70 to facilitate acyl-enzyme formation, as in the serine protease mechanism.

One of the most controversial points is the role of a general base in the acylation step. One hypothesis proposes that this role is played by the conserved Glu166. A second hypothesis postulates that the Lys73 functions as the proton acceptor in the acylation reaction and that the Glu166 residue participates in the deacylation reaction.

Kinetic studies and trapping experiments with covalent substrate-enzyme or inhibitor-enzyme complexes have shown that class A β-lactamases hydrolyze substrates through a Ser70-linked acyl-enzyme intermediate. On the basis of the tertiary structure of the active-site cavity around Ser70 of the staphylococcal β-lactamase, Herzberg and Moult (1987) proposed a catalytic mechanism in which the conserved residues, Lys73, Glu166, and Lys234, participate in addition to acylated Ser70. These scientists also proposed that Lys73 catalyzes the acylation step and Glu166 catalyzes the deacylation step, which acts as a general base catalyst. They proposed that during the proton transfer from Ser70 to the β-lactam nitrogen atom, protonated Lys73 provides a potential gradient which reduces the energy barrier for the transfer and polarizes the nitrogen atom favorably to receive the proton.

Based on the the proposal of Herzberg and Moult in 1987, Adachi et al. (1991) suggested a possible catalytic mechanism for class A β-lactamases (Fig. 7.14).

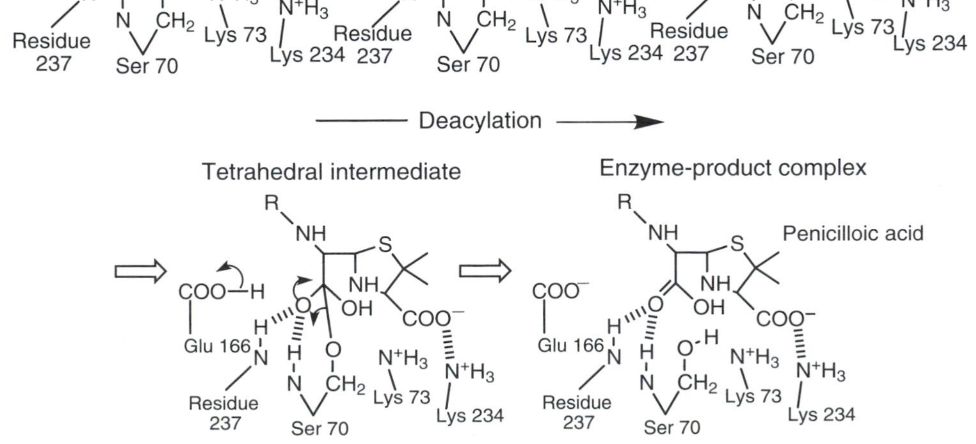

**Figure 7.14** Schematic drawing of a possible catalytic mechanism for class A β-lactamases. Through an electrostatic interaction between the ammonium group of Lys234 and the C-3 carboxylate the substrate is recognized by the enzyme in both ground-state binding and transition-state binding. Reprinted from H. Adachi, T. Ohta, and H. Matsuzawa, *J. Biol. Chem.* **266:**3186–3191, 1991, with permission from the publisher.

In March 2000, Atanasov et al. reported that protonation of the β-lactam nitrogen is the trigger event in the catalytic action of class A β-lactamases.

## CLASS C SERINE β-LACTAMASES

Of the four classes of β-lactamases, class C enzymes are the second most common. Class C β-lactamases are chromosomally encoded enzymes (Bush-Jacoby-Medeiros group 1) produced by gram-negative bacteria such as *Enterobacter* spp., *Serratia* spp., *Citrobacter* spp., and *Morganella* spp., that can hydrolyze many β-lactam antibiotics, including cephamycins and extended broad-spectrum cephalosporins. In recent years, *ampC* genes have also been found in conjugative plasmids. *ampC* genes encode a variety of enzymes, including MIR-1, CMY-1, CMY-2, BIL-1, and FOX-1, some of which are closely related to chromosomal AmpC of *Citrobacter freundii* or *Enterobacter cloacae*. Bacteria that harbor *ampC* on plasmids have antibiotic susceptibility patterns similar to those of strains overproducing chromosomally encoded β-lactamase.

MIR-1 was the first plasmid-mediated class C β-lactamase to be identified. The sequence of $bla_{\text{MIR-1}}$ has now been completely determined.

### Three-Dimensional Structural Data for Class C β-Lactamases

The structures of AmpC β-lactamases from *E. coli*, *Enterobacter cloacae* P99, and *C. freundii* have been determined by X-ray crystallography.

### The Catalytic Mechanism of Class C β-Lactamases

The overall mechanism of group C β-lactamases is well known and closely resembles that of group A β-lactamases such as TEM-1. In the first step, the catalytic Ser64 attacks the lactam carbonyl carbon, opening the β-lactam ring to form an acylated enzyme intermediate. In the second step, this acyl intermediate is attacked by an activated water molecule, releasing the hydrolyzed β-lactam (Fig. 7.15). Structural studies suggested that the conserved residue Tyr150 is the catalytic base that activates the hydrolytic water for its attack on the acyl intermediate. The details of the deacylation step remain controversial.

## CLASS D SERINE β-LACTAMASES

Class D (Bush-Jacoby-Medeiros group 2) was established in the late 1980s from the analysis of known β-lactamase sequences. Class D β-lactamases are sometimes called oxacillin-hydrolyzing enzymes because all the oxacillin-hydrolyzing β-lactamases belong to this class. Thus far, 24 Ambler class D enzymes, named OXA-1 to OXA-22, AmpS, and LCR-1, have been characterized by sequence and/or biochemical analyses. The three-dimensional structure of OXA-10 has recently been determined. The OXA-type (oxacillin-hydrolyzing) enzymes are widespread and have been found mostly in members of the *Enterobacteriaceae* and *P. aeruginosa*. They usually confer resistance to amino- and ureidopenicillin and possess high-level hydrolytic activity against oxacillin, cloxacillin, and methicillin. Their activities are weakly inhibited by clavulanic acid, but sodium chloride (NaCl) possesses a strong inhibitory activity. A 1999 review by Naas and Nordmann covers the OXA-type β-lactamases in detail.

### Three-Dimensional Structural Data for Class D β-Lactamases

The crystal structure of OXA-10 from *P. aeruginosa* was reported simultaneously by Paetzel et al. and by Maveyraud et al. and Golemi et al. in December 2000 (see references at the end of this chapter). It is the first three-dimensional structure reported for a class D enzyme. Paetzel et al. showed that there are significant differences between the folding observed in this structure and those of related class A and class C serine β-lactamases and that the structure suggests the unique, cation-mediated formation of a homodimer. Aside from

**Figure 7.15** Simplified mechanism of β-lactam hydrolysis by group C β-lactamases. Enz, enzyme.

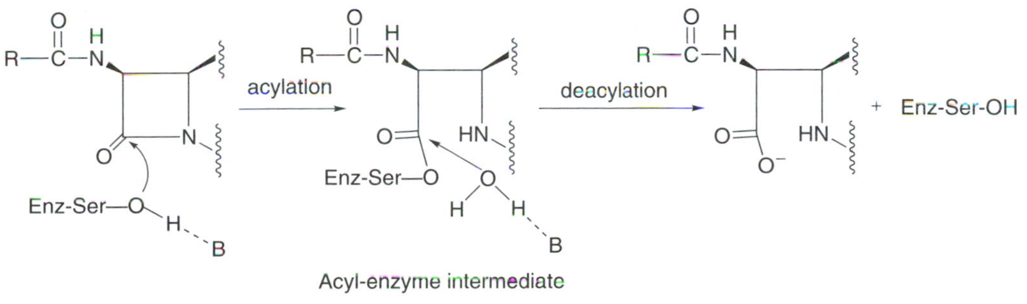

these observations, these authors hypothesized, on the basis of kinetic and hydrodynamic data, that the dimer is a relevant species in solution and is the more active form of the enzyme. The Ser67 side-chain hydroxyl group would probably act as the active-site nucleophile in the acylation step reaction. The coordinates of this structure have been stored in the Protein Data Bank (accession no. 1FOF).

In 2000, Maveyraud et al. reported the crystal structure of the OXA-10 class D β-lactamase determined at 1.66-Å resolution. This structure revealed that this β-lactamase and the class A β-lactamases, despite very low sequence similarity, have a similar overall fold. Major differences are found when comparing the molecular details of the active site of this class D enzyme to the corresponding regions in class A and C β-lactamases. The coordinates and structure factor in both the MAD phasing and the native structures can be assessed from the Protein Data Bank (accession no. 1E3U and 1E4D, respectively).

### The Catalytic Mechanism of Class D β-Lactamases

It is generally accepted that the role of Ser70 in class A, Ser64 in class C, and now Ser67 in class D β-lactamases is in nucleophilic acylation and that the role of Glu166 in the class A enzymes and of Tyr150 in the class C enzymes is general base deacylation. A comparison of the active-site regions shows that class D enzymes have no counterpart to Glu166 or Tyr150. Paetzel et al. (2000) suggested that the catalytic mechanism by which class D enzymes hydrolyze β-lactam antibiotics differs from the mechanism operating in class A and C serine β-lactamases. These authors also discussed the inhibition of class D enzymes by chloride ions.

## CLASS B METALLO-β-LACTAMASES

Metallo-β-lactamases are included in molecular class B and in group 3 of the functional (Bush-Jacoby-Medeiros) classification. They are produced by bacteria as extracellular or periplasmic enzymes. All known representatives possess two conserved metal-binding sites and require zinc ions as enzymatic cofactors. By catalyzing the hydrolysis of β-lactams, they render the corresponding strains resistant to all β-lactam antibiotics (penicillins, cephalosporins, and carbapenems). Carbapenems, e.g., imipenem, meropenem, and panipenem (this last antibiotic is available only in Japan), are the most potent agents for chemotherapy of infectious diseases caused by gram-negative rods because of their high affinity for PBP2, stability to most serine β-lactamases, including ESBLs, and excellent permeation across the bacterial outer membrane. β-Lactamase-hydrolyzing carbapenems identified as metallo-β-lactamases have been found in a number of species.

All metallo-β-lactamases isolated to date are chromosome encoded, with the exception of the IMP and VIM enzymes. Recently, a novel plasmid-mediated metallo-β-lactamase, IMP-1, which is produced in the presence of the $bla_{IMP}$ gene, was found in *P. aeruginosa, Serratia marcescens,* and other members of the family *Enterobacteriaceae,* which frequently cause nosocomial outbreaks worldwide.

During the past few years, class B β-lactamases have attracted considerable attention because of their ability to function as carbapenemases as well as their ability to utilize clavulanate and penicillanic acid sulfones (e.g., sulbactam and tazobactam), which function as mechanism-based inhibitors of serine-β-lactamases. IMP-1 and all other metallo-β-lactamases are able to hydrolyze a wide variety of β-lactam antibiotics, including imipenem, to give ring-opened β-amino acids, which are no longer effective as inhibitors of bacterial membrane-bound transpeptidase enzymes.

The increasing emergence of pathogenic bacterial strains producing metallo-β-lactamases is due mainly to rapid dissemination by horizontal gene transfer by both plasmid- and integron-borne mechanisms; this problem has resulted in growing interest in this enzyme family because these enzymes, as mentioned above, efficiently hydrolyze nearly all β-lactam antibiotics and the most widely used inhibitors of serine-β-lactamases such as clavulanic acid, sulbactam, and tazobactam.

The metallo-β-lactamases with known sequences share a small number of conserved motifs, but others show significant sequence diversity and have thus been classified into three subclasses: B1, B2, and B3. Subclass B1 exhibits a broad substrate profile, and its zinc-binding sites are composed of His116, His118, and His196 (site 1) and Asp120, Cys221, and His263 (site 2). Subclass B1 includes BcII from *Bacillus cereus,* CcrA from *Bacteroides fragilis,* IMP-1 and VIM-1 from *P. aeruginosa,* and BlaB from *Chryseobacterium meningosepticum.* In subclass B2, the zinc ligand in site 2 is conserved, whereas His116 in site 1 is replaced by Asn. Representatives of subclass B2 efficiently hydrolyze only carbapenems and are active as monozinc enzymes, whereas the binding of a second zinc ion causes noncompetitive inhibition. Enzymes of this subclass are found in *Aeromonas hydrophila* (CphA), *Aeromonas veronii,* and *Burkholderia cepacia.* In subclass B3, Cys221 is replaced by Ser, and this residue is replaced by His121 as a zinc ligand in site 2. Subclass B3 enzymes exhibit a broad spectrum of activity, with a putative preference for cephalosporins. Subclass B3 includes the L1 enzyme from *Stenotrophomonas maltophilia,* along with FEZ-1 from *Legionella gormanii* and GOB-1 from *Chryseobacterium meningosepticum.*

The question whether zinc β-lactamases are active as the mono- or dizinc enzyme is controversial, and alternative mechanisms for the two enzyme states have been proposed.

Class B β-lactamases pose a particular clinical threat because they have a broad substrate profile and are often expressed in combination with other β-lactamases.

The first metallo-β-lactamase discovered was produced by a strain of the relatively innocuous soil bacterium *Bacillus cereus*; it was identified nearly 39 years ago. Metallo-β-lactamases are now known to be expressed by at least 25 strains, including *Bacteroides fragilis, Aeromonas hydrophila, Stenotrophomonas maltophilia, Serratia marcescens, P. aeruginosa, K. pneumoniae, Aeromonas veronii, Chryseobacterium gleum, C. meningosepticum, C. indologenes, Empedobacter brevis, Myroides odoratus,* and *M. odoratimimus* (formerly designated in a single species as *Flavobacterium odoratum*). Some of these strains are human pathogens. Plasmid-mediated metallo-β-lactamases have been found in clinical isolates, raising concern about this horizontal transfer of β-lactamase activity.

## Structure and Mechanism of Metallo-β-Lactamases

During the past 4 years, X-ray crystal structures of several metallo-β-lactamases have been determined and the catalytic mechanisms for monozinc and dizinc enzymes have been elucidated. These advances have shed light on how such a diverse group of enzymes are evolving to inactivate a broad spectrum of β-lactam antibiotics so efficiently.

## Three-Dimensional Structural Data for Metallo-β-Lactamases

The first tertiary structure of metallo-β-lactamases dates back to 1987, with the low-resolution (3.5-Å) X-ray crystal structure of the Cd(II)-substituted *B. cereus* enzyme. Eight years later, a high-resolution (2.5-Å) crystal structure of the monozinc *B. cereus* enzyme was reported. This work was followed by the publication of the crystal structures of the *B. fragilis* enzyme (1.85 Å) in 1996 and its Cd(II)Cd(II) (2.15 Å) and Zn(II)Hg(II) (2.75 Å) derivatives in 1997.

In 1998, two crystal structures of the dizinc *B. cereus* enzyme (1.85 and 1.90 Å) were determined. The crystal structure of the *S. maltophilia* enzyme (1.7 Å) was revealed in 1998. Unlike the other metallo-β-lactamases, which are all monomeric, the *S. maltophilia* L1 enzyme is a tetramer. Also, the structures of the IMP-1 metallo-β-lactamase complexed with a mercaptocarboxylate inhibitor (3.1 Å) and a 2,3-(*S,S*)-benzylaryl-succinic acid inhibitor (1.8 Å) were solved in 2000 and 2001, respectively.

*S. maltophilia* (formerly *Pseudomonas maltophilia*) is an organism of increasing clinical significance through its action as an opportunistic nosocomial pathogen of immunocompromised individuals such as transplant recipients and cystic fibrosis and cancer patients. Although associated primarily with respiratory infections, it has been identified in a wide variety of bacteremias and can be a significant cause of mortality. It is intrinsically resistant to a wide range of antibiotics, and clinical strains may exhibit additional high-level multidrug resistance. Resistance of this organism to a wide range of β-lactam antibiotics is due primarily to expression of a pair of chromosomal β-lactamases, a class A active-site serine enzyme, L2, and a metallo-β-lactamase, L1. The L1 enzyme is the prototypical member of the subclass B3 of metallo-β-lactamases, which diverge markedly in sequence from enzymes such as BcII and CcrA. The crystal structure of L1 confirms that it shares the αβ/βα sandwich fold characteristic of the superfamily and has a binuclear zinc site, but the structure also reveals an active-site geometry that is considerably at variance with those of BcII, CcrA, and *P. aeruginosa* IMP-1. L1 is unique among the known metallo-β-lactamases in that it exists in solution as a compact tetramer whereas all the other enzymes appear to be monomeric. L1 hydrolyzes a wide range of penicillin, cephalosporin, and carbapenem substrates.

*Catalytic mechanism of mono- and binuclear $Zn^{2+}$ β-lactamases.* The enzyme from *B. cereus* 569/H/9 (BcII) represents a member of subclass B1 with three His ligands at one site and one Asp ligand, one Cys ligand, and one His ligand at the other site (3H1 and DCH1 sites, respectively). Various crystal structures of BcII are available, representing mononuclear and binuclear species. Although catalytic mechanisms for the enzyme with either one or two zinc ions bound have been discussed, the respective roles of the two binding sites during catalysis are still unclear. Generally, the three-histidine site is considered to be the primary catalytic site; however, the importance of the aspartic acid-cysteine-histidine site for catalysis has become obvious from studies of the Cys168Ala mutant. When only one zinc ion is bound to this mutant, it shows very low activity compared with the wild type, whereas wild-type-like activity is almost restored when a second metal ion is bound.

The *B. cereus* BcII and *A. hydrophila* CphΔ enzymes are active with one Zn(II) ion per enzyme. One high-affinity (nanomolar) and one low-affinity (micromolar) macroscopic dissociation constant for the binding of metal ions were found for the wild-type metallo-β-lactamase from *B. cereus* as well as six single-site mutants in which all ligands in the two metal-binding sites were altered. The known dissociation constants of class B β-lactamases for the formation of monozinc enzymes range from 6 pM for CphA to 0.6 nM for *B. cereus* strain 569/H/9 BcII. A second zinc ion is bound with dissociation constants in the micromolar range. In particular, the low affinity for a second zinc ion leads to the question whether physiological conditions offer sufficiently high metal ion concentrations to maintain the active metal-bound state of the enzymes.

Different catalytic mechanisms have been proposed for the mono- and binuclear enzymes. In the first case, the zinc-bound water/hydroxide is expected to be the attacking nucleophile.

## The Catalytic Mechanism of the Monozinc β-Lactamase from *B. cereus*

Sutton et al. (1987) have proposed that water molecule W1, which is so closely associated with Zn1 that it is best described as a hydroxide, is the preactivated nucleophile that attacks the β-lactam carbonyl carbon atom. In the Michaelis complex, the β-lactam carbonyl oxygen atom interacts with Zn1, which expands its coordination number to 5 and serves to polarize the C=O bond, enhancing its susceptibility to nucleophilic attack. At Zn2, one oxygen atom of the substrate's conserved carboxylate moiety replaces W2 while the other interacts with Lys171. As hydroxide W1 attacks the β-lactam carbonyl bond, the lone pair of the β-lactam nitrogen atom may interact with Zn2, maintaining its pentacoordination. As the transition state develops, with tetrahedral coordination at the β-lactam carbonyl carbon atom, Asn180 NδH and Zn1 can stabilize the oxyanion. Asp90, initially hydrogen bonded to the zinc-bound hydroxide, is ideally located to accept the proton from the hydroxide and, as β-lactam bond cleavage occurs, to protonate the nitrogen atom. Mutagenesis has demonstrated that this residue indeed performs a critical role in the *B. cereus* enzyme.

Deprotonation of the hydroxide by Asp90 implies the formation of a dianion, and mechanistic studies by Bounaga et al. (1998) indicate that a dianionic species is indeed formed, since there is an inverse second-order dependence of reactivity of the *B. cereus* enzyme on the hydrogen ion concentration at low pH.

The mechanism proposed by Bounaga et al. for the BcII monozinc enzyme was based on an insightful analysis of pH-rate profiles, inhibitor-binding studies, and solvent kinetic isotope effects that allowed these scientists to identify a kinetically significant intermediate. The $k_{cat}/K_m$ values for benzylpenicillin hydrolysis showed an inverse second-order dependence on $H^+$ concentration at low pH, suggesting that the rate may be suppressed by two protonation events. Currently, the most heavily favored mechanism that accounts for all the observations (Fig. 7.16) involves binding of the substrate β-lactam carbonyl oxygen atom to Zn(II), nucleophilic attack on the carbonyl carbon by the preformed and zinc-bound hydroxide, and subsequent stabilization of the oxyanion by Zn(II) and

**Figure 7.16** Proposed catalytic mechanism of benzylpenicillin hydrolysis by the mononuclear metallo-β-lactamase from *B. cereus.* H, histidine; Asp, aspartate. Adapted from S. Bounaga, A. P. Laws, M. Galleni, and M. I. Page, *Biochem. J.* **331:**703–711, 1998, with permission from the publisher.

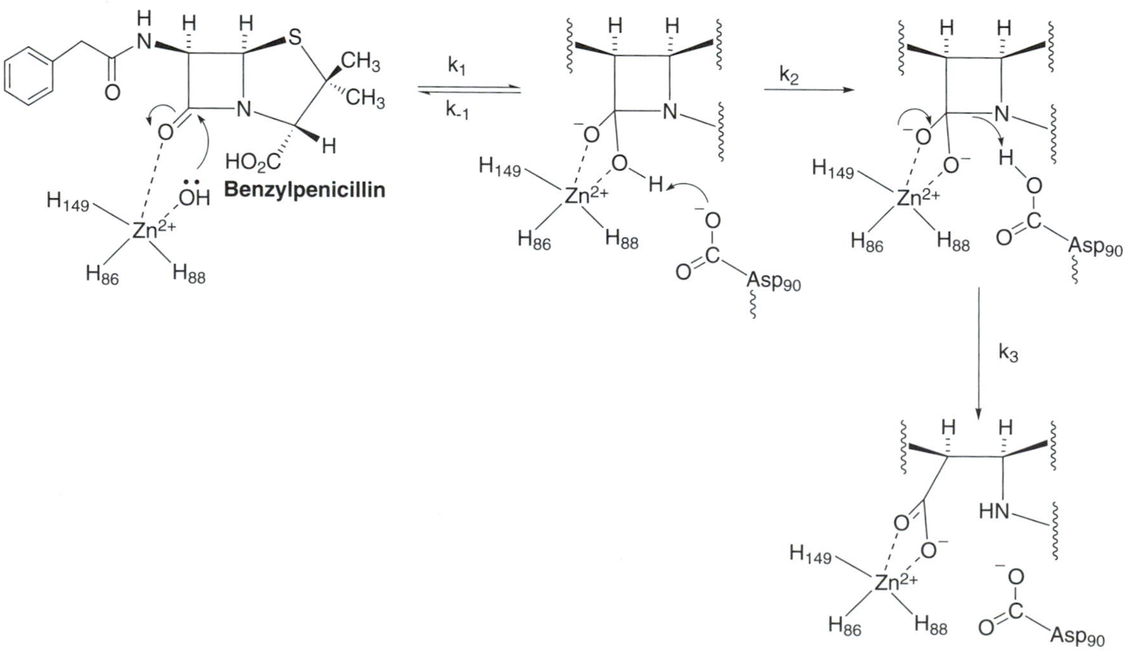

Asn180. This is followed by deprotonation by Asp90 to yield a dianionic tetrahedral intermediate. The breakdown of this species is presumably facilitated by general acid catalysis from Asp90. Formation of the dianion is thought to generate more "electron push," which can facilitate cleavage of the C—N bond. The role of Zn(II) is to preform the nucleophilic hydroxide, to act in conjunction with Asp90 and Cys168 in orienting the nucleophile for attack, and to provide Lewis acid catalysis by polarizing the carbonyl and increasing its electrophilicity.

## The Catalytic Mechanism of the Dizinc β-Lactamase from *B. fragilis*

Stopped-flow spectroscopic studies of the hydrolysis of nitrocefin (a chromogenic cephalosporin) catalyzed by the soluble *B. fragilis* enzyme CcrA (also called CfiA) revealed a mass deficit between substrate disappearance and product formation, leading to the proposal of an enzyme-bound intermediate. Using a combination of stopped-flow single-wavelength and multiple-wavelength (diode array) spectroscopy, a minimal kinetic mechanism was assigned, consisting of a linear pathway with one obligatory intermediate, in addition to the ES (enzyme-substrate) and EP (enzyme-product) complexes. Global analysis of the multiple-wavelength data resulted in an extracted spectrum of the intermediate that was considerably different from that of the tetrahedral intermediate observed during hydrolysis by the monozinc *B. cereus* enzyme. The spectrum can be essentially reproduced by deprotonating the hydrolyzed product with a strong base in organic solvents, leading to the identification of EI (the enzyme-intermediate complex) as a C—N bond-cleaved species in which the former lactam nitrogen is not protonated. This result has significant implications for the chemical mechanism, suggesting that protonation of the lactam nitrogen prior to C—N bond cleavage is not required.

In contrast to the pH dependence of the *B. cereus* metallo-β-lactamase, the $k_{cat}$ and $k_{cat}/K_m$ values for the dinuclear *B. fragilis* metallo-β-lactamase are relatively pH independent between pH 5.25 and 10.0, suggesting that no catalytically important groups have $pK_a$ values in this range. Coincubation of substrate, enzyme, and a variety of nucleophiles did not result in modification of either product or enzyme, again failing to provide evidence for an anhydride-type mechanism. Taken together, these observations led to a proposed chemical mechanism for the *B. fragilis* CcrA lactamase (Fig. 7.17). A

**Figure 7.17** Catalytic mechanism of nitrocefin hydrolysis by the dizinc metallo-β-lactamase from *B. fragilis*. H, histidine; C, cysteine; Asp, aspartate; Asn, asparaginamide. Adapted from Z. Wang, W. Fast, and S. J. Benkovic, *Biochemistry* **38:**10013–10023, 1999, with permission from the publisher.

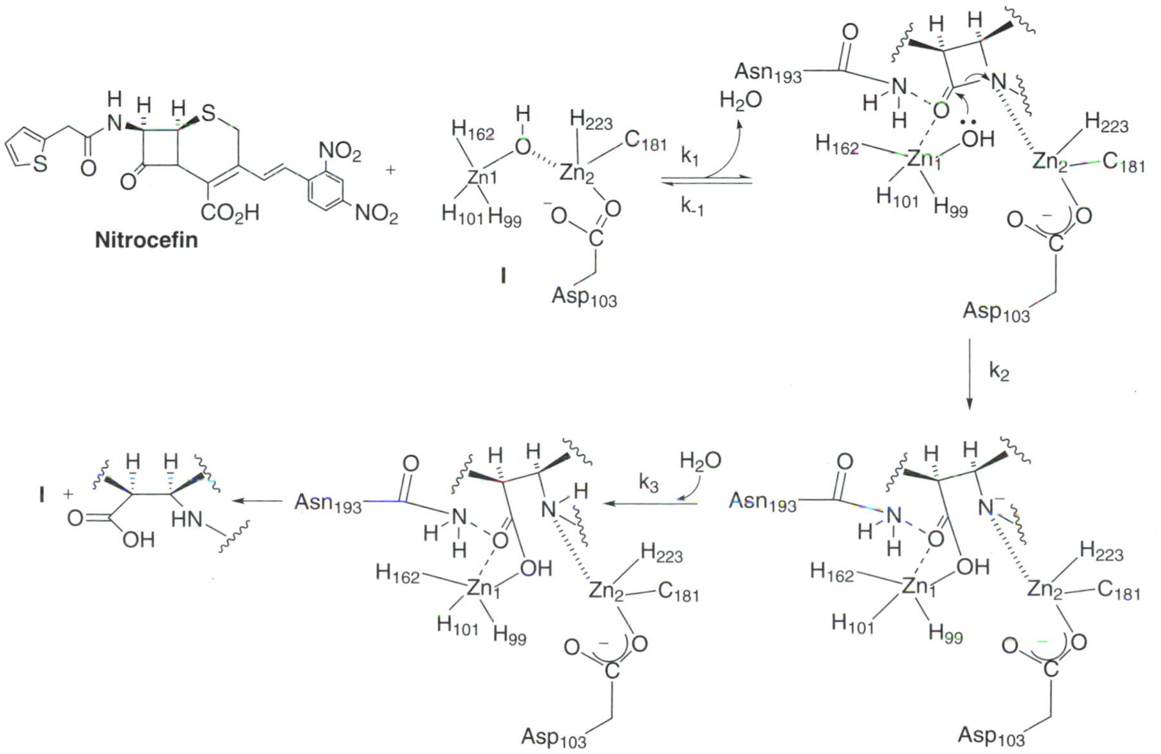

Nitrocefin
$NO_2$
$NO_2$
$CO_2H$
I
$H_{162}$ $H_{223}$ $C_{181}$ $Zn_1$ $Zn_2$ $H_{101}$ $H_{99}$
$Asp_{103}$
$Asn_{193}$
$H_2O$
$k_1$
$k_{-1}$
$k_2$
$k_3$

Zn1-bound hydroxide is preformed with a $pK_a$ of $<5.25$. Substrate binding polarizes the lactam bond by coordinating the carbonyl oxygen atom to Zn1 and the lactam nitrogen atom to Zn2. Attack of the Zn1 hydroxide leads to oxyanion stabilization by Zn1 and Asn193, followed by subsequent cleavage of the C—N bond. The lactam nitrogen is expelled as an anion, which is stabilized by Zn2 acting as a superacid and substituting for the proton seen in the mononuclear mechanism. Protonation of this intermediate is the rate-limiting step, consistent with a (H/2H) solvent isotope effect of 2.9.

It is often proposed, with support from crystal structure determinations, that a zinc-bound hydroxide acts as the nucleophile to attack bound substrates. At present, there is no definitive evidence about the exact position of the nucleophile during hydrolysis, namely, whether it resides in a terminal or bridging position at the dizinc center.

Wang et al. (1999) have suggested that the mechanistic differences between the mono- and binuclear enzymes may reflect the evolutionary advantage of incorporating a second Zn(II) ion, which enhances the C—N cleavage step. A similar intermediate was detected in nitrocefin hydrolysis by another dinuclear zinc metallo-β-lactamase (the L1 enzyme from *S. maltophilia*), although the rates of intermediate formation and decay are somewhat different. As discussed above, several models have been postulated for the cleavage of the β-lactam bond and protonation of the nitrogen atom; these vary primarily in the source of the proton required for product formation. Experimental evidence points to different conclusions for different enzymes: in the BcII enzyme from *B. cereus*, a conserved aspartate in the active site appears to shuttle a proton from the oxyanion to the β-lactam nitrogen and the major intermediate is a dianion species, whereas in CcrA from *B. fragilis*, the major intermediate in hydrolysis of the chromogenic cephalosporin nitrocefin is the deprotonated product and the rate-determining step is donation of a proton by a second zinc-bound water molecule.

## REFERENCES AND FURTHER READING

### β-Lactam Antibiotics

#### General (books)

**Flynn, E. H.** 1972. *Cephalosporins and Penicillins. Chemistry and Biology*. Academic Press, Inc., New York, N.Y.

**Howe-Grant, M.** 1993. *Chemotherapeutics and Disease Control.* John Wiley & Sons, Inc., New York, N.Y.

**Page, M. I.** 1992. *The Chemistry of β-Lactams.* Blackie Academic & Professional, London, United Kingdom.

**Salton, M., and G. D. Shockman.** 1981. *β-Lactam Antibiotics. Mode of Action, New Developments, and Future Prospects.* Academic Press, Inc., New York, N.Y.

**Sheehan, J. C.** 1982. *The Enchanted Ring. The Untold Story of Penicillin.* MIT Press, Cambridge, Mass.


#### Monobactams

**Sykes, R. B., C. M. Cimarusti, D. P. Bommer, K. Bush, D. M. Floyd, N. H. Georgopapadakou, W. H. Koster, W. C. Liu, W. L. Parker, P. A. Principe, M. L. Rathmum, W. A. Slusarchyk, W. H. Trejo, and J. S. Wells.** 1981. Monocyclic β-lactam antibiotics produced by bacteria. *Nature* **291:**489–491.


#### Spectral characteristics

**Demarco, P. V., and R. Nagerajan.** 1972. Physical-chemical properties of cephalosporins and penicillins, p. 311–369. *In* E. H. Flynn (ed.), *Cephalosporins and Penicillins. Chemistry and Biology.* Academic Press, Inc., New York, N.Y.


#### Mechanism of action

**Tipper, D. J., and J. L. Strominger.** 1965. Mechanism of action of penicillins: a proposal based on their structural similarity to acyl-D-alanyl-D-alanine. *Proc. Natl. Acad. Sci. USA* **54:**1133–1141.

**Waxman, D. J., and J. L. Strominger.** 1982. β-Lactam antibiotics: biochemical mode of action, p. 209–285. *In* R. B. Morin, and M. Gorman (ed.), *Chemistry and Biology of β-Lactam Antibiotics,* vol. 3. Academic Press, Inc., New York, N.Y.

**Waxman, D. J., and J. L. Strominger.** 1983. Penicillin-binding proteins and the mechanism of action of β-lactam antibiotics. *Annu. Rev. Biochem.* **52:**825–869.

**Waxman, D. J., R. R. Yocum, and J. L. Strominger.** 1980. Penicillins and cephalosporins are active site-directed acylating agents: evidence in support of the substrate analogue hypothesis. *Philos. Trans. R. Soc. Lond. Ser. B* **259:**257–271.


### Penicillin-Binding Proteins

**Beadle, B. M., R. A. Nicholas, and B. K. Shoichet.** 2001. Interaction energies between β-lactam antibiotics and *E. coli* penicillin-binding protein 5 by reversible thermal denaturation. *Protein Sci.* **10:**1254–1259.

**Denome, S. A., P. K. Elf, T. A. Henderson, D. E. Nelson, and K. D. Young.** 1999. *Escherichia coli* mutants lacking all possible combinations of eight penicillin binding proteins: viability, characteristics, and implications for peptidoglycan synthesis. *J. Bacteriol.* **181:**3981–3993.

**Di Giulini, A. M., A. Dessen, O. Dideberg, and T. Vernet.** 2003. Functional characterization of penicillin-binding protein 1b from *Streptococcus pneumoniae. J. Bacteriol.* **185:**1650–1658.

**Ghuysen, J. M.** 1991. Serine β-lactamases and penicillin-binding proteins. *Annu. Rev. Microbiol.* **45:**37–67.

**Ghuysen, J. M.** 1994. Molecular structures of penicillin-binding proteins and β-lactamases. *Trends Microbiol.* **2:**372–380.

**Ghuysen, J. M.** 1996. Penicillin-binding proteins. Wall peptidoglycan assembly and resistance to penicillins: facts, doubts and hopes. *Int. J. Antimicrob. Agents* **8:**45–60.

**Ghuysen, J. M., and G. Dive.** 1994. Biochemistry of the penicilloyl serine transferase, p. 103–129. *In* J. M. Ghuysen and R.

Hakenbeck (ed.), *Bacterial Cell Wall.* Elsevier, Amsterdam, The Netherlands.

**Goffin, C., and J. M. Ghuysen.** 1998. Multimodular penicillin-binding proteins: family of orthologs and paralogs. *Microbiol. Mol. Biol. Rev.* **62:**1079–1093.

**Herzberg, O., and J. Moult.** 1991. Penicillin-binding and degrading enzymes. *Curr. Opin. Struct. Biol.* **1:**946–953.

**Lee, W., M. A. McDonough, L. P. Kotra, Z. Li, N. R, Silvaggi, Y. Takeda, J. A. Kelly, and S. Mobashery.** 2001. A 1.2-Å snapshot of the final step of bacterial cell wall biosynthesis. *Proc. Natl. Acad. Sci. USA* **98:**1427–1431.

**Nanninga, N.** 1998. Morphogenesis of *Escherichia coli. Microbiol. Mol. Biol. Rev.* **62:**110–129.

**Nelson, D. E., and K. D. Young.** 2000. Penicillin binding protein 5 affects cell diameter, contour, and morphology of *Escherichia coli. J. Bacteriol.* **182:**1714–1721.

### Crystallographic structure of PBPs

**Dessen, A., N. Mouz, E. Gordon, J. Hopkins, and O. Dideberg.** 2001. Crystal structure of PBP2x from a highly penicillin-resistant *Streptococcus pneumoniae* clinical isolate. *J. Biol. Chem.* **276:**45106–45112.

**Gordon, E., N. Mouz, E. Duée, and O. Dideberg.** 2000. The crystal structure of the penicillin-binding protein 2x from *Streptococcus pneumoniae* and its acyl-enzyme form: implication in drug resistance. *J. Mol. Biol.* **299:**477–485.

**Kelly, J. A., and A. P. Kuzin.** 1995. The refined crystallographic structure of a DD-peptidase penicillin-target enzyme at 1.6 Å resolution. *J. Mol. Biol.* **254:**223–236.

**Lim, D., and N. C. J. Strydnaka.** 2002. Structural basis for the β-lactam resistance of PBP2a from methicillin-resistant *Staphylococcus aureus. Nat. Struct. Biol.* **9:**870–876.

**Pares, S., N. Mouz, Y. Pétillot, R. Hakenbeck, and O. Dideberg.** 1996. X-ray structure of *Streptococcus pneumoniae* PBP2x, a primary penicillin target enzyme. *Nat. Struct. Biol.* **3:**284–288.

**Rhazi, N., P. Charlier, D. Dehareng, D. Engher, M. Vermeire, J. M. Frère, M. Nguyen-Distèche, and E. Fonzé.** 2003. Catalytic mechanism of the *Streptomyces* K15 DD-transpeptidase/penicillin-binding protein by site-directed mutagenesis and structural analysis. *Biochemistry* **42:**2895–2906.

## β-Lactamases

### General

**Matagne, A., A. Dubus, M. Galleni, and J. M. Frère.** 1999. The β-lactamase cycle: a tale of selective pressure and bacterial ingenuity. *Nat. Prod. Rep.* **16:**1–19.

**Medeiros, A. A.** 1997. Evolution and dissemination of beta-lactamases accelerated by generations of beta-lactam antibiotics. *Clin. Infect. Dis.* **24**(Suppl. 1):S19–S45.

### Evolution of DD-peptidases and β-lactamases

**Ghuysen, J.-M.** 1988. Evolution of DD-peptidases and β-lactamases, p. 268–284. *In* P. Actor, L. Daneo-Moore, M. L. Higgins, M. R. J. Salton, and G. D. Shockman (ed.), *Antibiotic Inhibition of Bacterial Cell Surface Assembly and Function.* American Society for Microbiology, Washington, D.C.

**Massova, I., and S. Mobashery.** 1998. Kinship and diversification of bacterial penicillin-binding proteins and β-lactamases. *Antimicrob. Agents Chemother.* **42:**1–17.

**Massova, I., and S. Mobashery.** 1999. Structural and mechanistic aspects of evolution of β-lactamases and penicillin-binding proteins. *Curr. Pharm. Design* **5:**929–937.

### Functional classification of β-lactamases

**Bush, K.** 1989. Characterization of β-lactamases. *Antimicrob. Agents Chemother.* **33:**259–263.

**Bush, K.** 1989. Classification of β-lactamases: groups 1, 2a, 2b, and 2b'. *Antimicrob. Agents Chemother.* **33:**264–270.

**Bush, K.** 1989. Classification of β-lactamases: groups 2c, 2d, 2e, 3, and 4. *Antimicrob. Agents Chemother.* **33:**271–276.

**Bush, K., G. A. Jacoby, and A. A. Medeiros.** 1995. A functional classification scheme for β-lactamases and its correlation with molecular structure. *Antimicrob. Agents Chemother.* **39:** 1211–1233.

**Sykes, R. B., and M. Matthew.** 1976. The β-lactamases of gram-negative bacteria and their role in resistance to β-lactam antibiotics. *J. Antimicrob. Chemother.* **2:**115–157.

### Molecular structure classification of β-lactamases

**Ambler, R. P.** 1980. The structure of β-lactamases. *Philos. Trans. R. Soc. Lond. Ser. B* **289:**321–331.

**Huovinen, P., S. Huovinen, and G. A. Jacoby.** 1988. Sequence of PSE-2 β-lactamase. *Antimicrob. Agents Chemother.* **32:** 134–136.

**Jaurin, B., and T. Grundstrom.** 1981. *ampC* cephalosporinase of *Escherichia coli* K-12 has a different evolutionary origin from that of β-lactamases of the penicillinase type. *Proc. Natl. Acad. Sci. USA* **78:**4897–4901.

### Class A serine β-lactamases

**Bonomo, R. A., J. R. Knox, S. D. Rudin, and D. Shlaes.** 1997. Construction and characterization of an OHIO-1 β-lactamase bearing Met69Ile and Gly238Ser mutations. *Antimicrob. Agents Chemother.* **41:**1940–1943.

**Bouthors, A. T., N. Dagoneau-Blanchard., T. Naas, P. Nordmann, V. Jarlier, and W. Sougakoff.** 1998. Role of residues 104, 164, 166, 238 and 240 in the substrate profile of PER-1 β-lactamase-hydrolysing third-generation cephalosporins. *Biochem. J.* **330:**1443–1449.

**Bouthors, A. T., J. Delettré, P. Mugnier, V. Jarlier, and W. Sougakoff.** 1996. Site-directed mutagenesis of residues 164, 170, 171, 179, 220, 237 and 242 in PER-1 β-lactamase hydrolysing expanded-spectrum cephalosporins. *Protein Eng.* **12:**313–318.

**Gheorghiu, R., M. Yuan, L. M. C. Hall, and D. M. Livermore.** 1997. Bases of variation in resistance to β-lactams in *Klebsiella oxytoca* isolates hyperproducing K1 β-lactamase. *J. Antimicrob. Chemother.* **40:**533–541.

**Hata, M., Y. Fujii, M. Ishii, T. Hoshino, and M. Tsuda.** 2000. Catalytic mechanism of class A β-lactamases. The role of

Glu166 and Ser130 in the deacylation reaction. *Chem. Pharm. Bull.* **48:**447–453.

**Ma, L., Y. Ishii, M. Ishiguro, H. Matsuzawa, and K. Yamaguchi.** 1998. Cloning and sequencing of the gene encoding Toho-2, a class A β-lactamase preferentially inhibited by tazobactam. *Antimicrob. Agents Chemother.* **42:**1181–1186.

**Poirel, L., T. Naas, M. Guibert, E. B. Chaibi, R. Labia, and P. Nordmann.** 1999. Molecular and biochemical characterization of VEB-1, a novel class A extended-spectrum β-lactamase encoded by an *Escherichia coli* integron gene. *Antimicrob. Agents Chemother.* **43:**573–581.

**Tzouvelekis, L. S., E. Tzelepi, P. T. Tassios, and N. J. Legakis.** 2000. CTX-M-type β-lactamases: an emerging group of extended-spectrum enzymes. *Int. J. Antimicrob. Agents* **14:**137–142.

**Class C serine β-lactamases**

**Adediran, S. A., and R. F. Pratt.** 1999. β-Secondary and solvent deuterium kinetic isotope effects on catalysis by the *Streptomyces* R61 DD-peptidase: comparisons with a structurally similar class C β-lactamase. *Biochemistry* **38:**1469–1477.

**Bulychev, A., and S. Mobashery.** 1999. Class C β-lactamases operate at the diffusion limit for turnover of their preferred cephalosporin substrates. *Antimicrob. Agents Chemother.* **43:**1743–1746.

**Hanson, N. D., and C. C. Sanders.** 1999. Regulation of inducible AmpC beta-lactamase expression among Enterobacteriaceae. *Curr. Pharm. Design* **8:**881–894.

**Jacoby, G. A., and J. Tran.** 1999. Sequence of the MIR-1 β-lactamase gene. *Antimicrob. Agents Chemother.* **43:**1759–1760.

**Marchese, A., G. Arlet, G. C. Schito, P. H. Lagrange, and A. Philippon.** 1998. Characterization of FOX-3, an AmpC-type plasmid-mediated β-lactamase from an Italian isolate of *Klebsiella oxytoca.* *Antimicrob. Agents Chemother.* **42:**464–467.

**Patera, A., L. C. Blaszczak, and B. K. Shoichet.** 2000. Crystal structures of substrate and inhibitor complexes with AmpC β-lactamase: possible implications for substrate-assisted catalysis. *J. Am. Chem. Soc.* **122:**10504–10512.

**Pfeifle, D., E. Janas, and B. Wiedemann.** 2000. Role of penicillin-binding proteins in the initiation of the AmpC β-lactamase expression in *Enterobacter cloacae.* *Antimicrob. Agents Chemother.* **44:**169–172.

**Pratt, R. F., M. Dryjanski, E. S. Wun, and V. M. Marathias.** 1996. 8-Hydroxypenillic acid from 6-aminopenicillanic acid: a new reaction catalyzed by class C β-lactamase. *J. Am. Chem. Soc.* **118:**8207–8212.

**Tondi, D., R. A. Powers, E. Caselli, M. C. Negri, J. Blázquez, M. P. Costi, and B. K. Shoichet.** 2001. Structure-based design and in-parallel synthesis of inhibitors of AmpC β-lactamase. *Chem. Biol.* **8:**593–610.

**Trépanier, S., J. R. Knox, N. Clairoux, F. Sanschagrin, R. C. Levesque, and A. Huletsky.** 1999. Structure-function studies of Ser-289 in the class C β-lactamase from *Enterobacter cloacae* P99. *Antimicrob. Agents Chemother.* **43:**543–548.

**Class D serine β-lactamases**

**Ledent, P., X. Raquet, B. Joris, J. Van Beeumen, and J. M. Frère.** 1993. A comparative study of class D β-lactamases. *Biochem. J.* **292:**555–562.

**Naas, T., and P. Nordmann.** 1999. OXA-type beta-lactamases. *Curr. Pharm. Des.* **5:**865–879.

**Crystal structure of class A β-lactamases**

*Staphylococcus aureus PC1*

**Herzberg, O.** 1991. Refined crystal structure of β-lactamase from *Staphylococcus aureus* PC1 at 2.0 Å resolution. *J. Mol. Biol.* **217:**701–719.

**Herzberg, O., and C. C. H. Chen.** 2001. Structures of the acyl-enzyme complexes of the *Staphylococcus aureus* β-lactamase mutant Glu166Asp:Asn170Gln with benzylpenicillin and cephaloridine. *Biochemistry* **40:**2351–2358.

**Herzberg, O., and J. Moult.** 1987. Bacterial resistance to β-lactam antibiotics: crystal structure of β-lactamase from *Staphylococcus aureus* PC1 at 2.5 Å resolution. *Science* **236:**694–701.

*Bacillus licheniformis 749/C*

**Knox, J. R., and P. C. Moews.** 1991. β-Lactamase of *Bacillus licheniformis* 749/C refinement at 2 Å resolution and analysis of hydration. *J. Mol. Biol.* **220:**435–455.

**Moews, P. C., J. R. Knox, O. Dideberg, P. Charlier, and J. M. Frère.** 1990. β-Lactamase of *Bacillus licheniformis* 749/C at 2 Å resolution. *Proteins Struct. Funct. Genet.* **7:**156–171.

*Escherichia coli TEM-1*

**Jelsch, C., F. Lenfant, J. M. Masson, and J. P. Samama.** 1992. β-Lactamase TEM1 of *E. coli. FEBS Lett.* **299:**135–142.

**Jelsch, C., L. Mourey, J. M. Masson, and J. P. Samama.** 1993. Crystal structure of *Escherichia coli* TEM1 β-lactamase at 1.8 Å resolution. *Proteins Struct. Funct. Genet.* **16:**364–383.

**Strynadka, N. C. J., H. Adachi, S. E. Jensen, K. Johns, A. Sielecki, C. Betzel, K. Sutoh, and M. N. G. James.** 1992. Molecular structure of the acyl-enzyme intermediate in β-lactam hydrolysis at 1.7 Å resolution. *Nature* **359:**700–706.

**Strynadka, N. C. J., R. Martin, S. E. Jennsen, M. Gold, and J. B. Jones.** 1996. Structure-based design of a potent transition state analogue for TEM-1 β-lactamase. *Nat. Struct. Biol.* **3:**688–695.

*Enterobacter cloacae NMC-A*

**Swarén, P., L. Maveyraud, X. Raquet, S. Cabantous, C. Duez, J.-D. Pédelacq, S. Mariotte-Boyer, L. Mourey, R. Labia, M.-H. Nicolas-Chanoine, P. Nordmann, J.-M. Frère, and J.-P. Samama.** 1998. X-ray analysis of the NMC-A β-lactamase at 1.64 Å resolution, a class A carbapenemase with broad substrate specificity. *J. Biol. Chem.* **273:**26714–26721.

*Klebsiella pneumoniae SHV-1*

**Kuzin, A. P., M. Nukaga, Y. Nukaga, A. M. Nujer, R. A. Bonomo, and J. R. Knox.** 1999. Structure of the SHV-1 β-lactamase. *Biochemistry* **38:**5720–5727.

*Pseudomonas aeruginosa PER-1*
**Traniert, S., A. T. Bouthors, L. Maveyraudt, V. Guillett, W. Sougakoff, and J. P. Samama.** 2000. The high resolution crystal structure for class A β-lactamase PER-1 reveals the bases for its increase in breadth of activity. *J. Biol. Chem.* **275:**28075–28082.

*Pseudomonas aeruginosa PSE-4*
**Lim, D., F. Sanschagrin, L. Passmore, L. De Castro, R. C. Levesque, and N. C. J. Strynadka.** 2001. Insights into the molecular basis for the carbenicillinase activity of PSE-4 β-lactamase from crystallographic and kinetic studies. *Biochemistry* **40:**395–402.

*Proteus vulgaris β-lactamase*
**Nukaga, M., K. Mayama, G. V. Crichlow, and J. R. Knox.** 2002. Structure of an extended-spectrum class A β-lactamase from *Proteus vulgaris* K1. *J. Mol. Biol.* **317:**107–117.

*β-Lactamase Toho-1*
**Ibuka, A., A. Taguchi, M. Ishiguro, S. Fushinobu, Y. Ishii, S. Kamitori, K. Okuyama, K. Yamaguchi, M. Konno, and H. Matsuzawa.** 1999. Crystal structure of the E166A mutant of extended-spectrum β-lactamase Toho-1 at 1.8 Å resolution. *J. Mol. Biol.* **285:**2079–2087.

**Crystal structure of class C β-lactamases**

*AmpC β-lactamase from Escherichia coli and Enterobacter cloacae P99*
**Lobkovsky, E., P. C. Moews, H. Liu, H. Zhao, J. M. Frère, and J. R. Knox.** 1993. Evolution of an enzyme activity: crystallographic structure at 2 Å resolution of cephalosporinase from *ampC* gene of *Enterobacter cloacae* P99 and comparison with a class A penicillinase. *Proc. Natl. Acad. Sci. USA* **90:**11257–11261.

**Usher, K. C., L. C. Blaszczak, G. S. Weston, B. K. Shoichet, and S. J. Remington.** 1998. Three-dimensional structure of AmpC β-lactamase from *Escherichia coli* to a transition-state analogue: possible implications for the oxyanion hypothesis and for inhibitor design. *Biochemistry* **37:**16082–16092.

*β-Lactamase from Citrobacter freundii*
**Oefner, C., A. D'Arcy, J. J. Daly, K. Gubernator, R. L. Charnas, I. Heinze, C. Hubschwerlen, and F. K. Winkler.** 1990. Refined crystal structure of β-lactamase from *Citrobacter freundii* indicates a mechanism for β-lactamase hydrolysis. *Nature* **343:**284–288.

**Crystal structure of class D β-lactamases**

*Pseudomonas aeruginosa OXA-10*
**Golemi, D., L. Maveyroud, S. Vakulenku, S. Tranier, A. Ishiwata, L. P. Kotra, J. P. Samama, and S. Mobashery.** 2000. The first structural and mechanistic insights for class D β-lactamases: evidence for a novel catalytic process for turnover of β-lactam antibiotics. *J. Am. Chem. Soc.* **122:**6132–6133.

**Maveyraud, L., D. Golemi, L. P. Kotra, S. Tranier, S. Vakulenko, S. Mobashery, and J. P. Samama.** 2000. Insights into class D β-lactamases are revealed by the crystal structure of the OXA10 enzyme from *Pseudomonas aeruginosa. Structure* **8:**1289–1298.

**Paetzel, M., F. Danel, L. Castro, S. C. Mosimann, M. G. P. Page, and N. C. J. Strynadka.** 2000. Crystal structure of the class D β-lactamase OXA-10. *Nat. Struct. Biol.* **7:**918–925.

**Catalytic mechanism of β-lactamases: general**
**Frère, J. M., A. Dubus, M. Galleni, A. Matagne, and G. Amicosante.** 1999. Mechanistic diversity of β-lactamases. *Biochem. Soc. Trans.* **27:**58–63.

**Page, M. I.** 1999. The reactivity of β-lactams, the mechanism of catalysis and the inhibition of β-lactamases. *Curr. Pharm. Design* **5:**895–913.

**Page, M. I., and A. P. Laws.** 1998. The mechanism of catalysis and the inhibition of β-lactamases. *J. Chem. Soc. Chem. Commun.* **1998:**1609–1617.

**Catalytic mechanism of class A β-lactamases**
**Adachi, H., T. Ohta, and H. Matsuzawa.** 1991. Site-directed mutants, at position 166, of RTEM-1 β-lactamase that form a stable acyl-enzyme intermediate with penicillin. *J. Biol. Chem.* **266:**3186–3191.

**Atanasov, B. P., D. Mustafi, and M. M. Makinen.** 2000. Protonation of the β-lactam nitrogen is the trigger event in the catalytic action of class A β-lactamases. *Proc. Natl. Acad. Sci. USA* **97:**3160–3165.

**Ishiguro, M., and S. Imajo.** 1996. Modeling study on a hydrolytic mechanism of class A β-lactamases. *J. Med. Chem.* **39:**2207–2218.

**Matagne, A., and J. M. Frère.** 1995. Contribution of mutant analysis to the understanding of enzyme catalysis: the case of class A β-lactamases. *Biochim. Biophys. Acta* **1246:**109–127.

**Matagne, A., J. Lamotte-Brasseur, and J. M. Frère.** 1998. Catalytic properties of class A β-lactamases: efficiency and diversity. *Biochem. J.* **330:**581–598.

**Class B metallo-β-lactamases: general**
**Bush, K.** 1998. Metallo-β-lactamases: a class apart. *Clin. Infect. Dis.* **27**(Suppl. 1)**:**40–53.

**Cricco, J. A., and A. J. Vila.** 1999. Class B β-lactamases: the importance of being metallic. *Curr. Pharm. Design* **5:**915–927.

**Felici, A., G. Amicosante, A. Oratore, R. Strom, P. Ledent, B. Koris, L. Fanuel, and J. M. Frère.** 1993. An overview of the kinetic parameters of class B β-lactamases. *Biochem. J.* **291:**151–155.

**Galleni, M., J. Lamotte-Brasseur, G. M. Rossolini, J. Spencer, O. Dideberg, J. M. Frère, and The Metallo-β-Lactamase Working Group.** 2001. Standard numbering scheme for class B β-lactamases. *Antimicrob. Agents Chemother.* **45:**660–663.

**Gilson, H. S. R., and M. Krauss.** 1999. Structure and spectroscopy of metallo-β-lactamase active sites. *J. Am. Chem. Soc.* **121:**6984–6989.

**Payne, D. J.** 1993. Metallo-β-lactamases—a new therapeutic challenge. *J. Mol. Microbiol.* **39:**93–99.

**Payne, D. J., R. Cramp, J. H. Bateson, J. Neale, and D. Knowles.** 1994. Rapid identification of metallo- and serine β-lactamases. *Antimicrob. Agents Chemother.* **38:**991–996.

**Wang, Z., W. Fast, A. M. Valentine, and S. J. Benkovic.** 1999. Metallo-β-lactamase: structure and mechanism. *Curr. Opin. Chem. Biol.* **3:**614–622.

**Crystal structure of class B β-lactamases**

*Pseudomonas aeruginosa IMP-1*

**Concha, N. O., C. A. Janson, P. Rowling, S. Pearson, C. A. Cheever, B. P. Clarke, C. Lewis, M. Galleni, J. M. Frère, D. J. Payne, J. H. Bateson, and S. S. Abdel-Meguid.** 2000. Crystal structure of the IMP-1 metallo β-lactamase from *Pseudomonas aeruginosa* and its complex with a mercaptocarboxylate inhibitor: binding determinants of a potent, broad-spectrum inhibitor. *Biochemistry* **39:**4288–4298.

*Bacteroides fragilis Zn(II) β-lactamase*

**Carfi, A., Duée, E., R. Paul-Soto, M. Galleni, J. M. Frère, and O. Dideberg.** 1998. X-ray structure of the ZnII β-lactamase from *Bacteroides fragilis* in an orthorhombic crystal form. *Acta Crystallogr.* **D54:**47–57.

**Concha, N. O., B. A. Rasmussen, K. Bush, and O. Herzberg.** 1996. Crystal structure of the wide-spectrum binuclear zinc β-lactamase from *Bacteroides fragilis. Structure* **4:**823–836.

*Stenotrophomonas maltophilia L1 metallo-β-lactamase*

**Spencer, J., A. R. Clarke, and T. R. Walsh.** 2001. Novel mechanism of hydrolysis of therapeutic β-lactamase by *Stenotrophomonas maltophilia* L1 metallo-β-lactamase. *J. Biol. Chem.* **276:**33638–33644.

**Ullah, J. H., T. R. Walsh, I. A. Taylor, D. C. Emery, C. S. Verma, S. J. Gamblin, and J. Spencer.** 1998. The crystal structure of the L1 metallo-β-lactamase from *Stenotrophomonas maltophilia* at 1.7 Å resolution. *J. Mol. Biol.* **284:**125–136.

*Bacillus cereus Zn(II) β-lactamase*

**Carfi, A., E. Duée, M. Galeni, J. M. Frère, and O. Dideberg.** 1998. 1.85 Å resolution structure of the ZnII β-lactamase from *Bacillus cereus. Acta Crystallogr.* **D54:**313–323.

**Carfi, A., S. Pares, E. Duée, M. Galleni, C. Duez, J. M. Frère, and O. Dideberg.** 1995. The 3-D structure of a zinc metallo-β-lactamase reveals a new type of protein fold. *EMBO J.* **14:**4914–4921.

**Fabiane, S. M., M. K. Sohi, T. Wan, D. J. Payne, J. H. Bateson, T. Mitchell, and B. Sutton.** 1998. Crystal structure of the zinc-dependent β-lactamase from *Bacillus cereus* at 1.9 Å resolution: binuclear active site with features of a mononuclear enzyme. *Biochemistry* **37:**12404–12411.

**Metallo-β-lactamases: catalytic mechanism**

**Bounaga, S., A. P. Laws, M. Galleni, and M. I. Page.** 1998. The mechanism of catalysis and the inhibition of the *Bacillus cereus* zinc-dependent β-lactamase. *Biochem. J.* **331:**703–711.

**Cricco, J. A., E. G. Orellano, R. R. Rasia, E. A. Ceccarelli, and A. J. Vila.** 1999. Metallo-β-lactamases: does it take two to tango? *Coord. Chem. Rev.* **190–192:**519–535.

**Crowder, M. W., Z. Wang, S. L. Franklin, E. P. Zovinka, and S. J. Benkovic.** 1996. Characterization of the metal-binding sites of the β-lactamase from *Bacteroides fragilis. Biochemistry* **35:**12126–12132.

**Franceschini, N. A., B. Caravelli, J. D. Docquier, M. Galleni, J. M. Frère, G. Amicosante, and G. M. Rossolini.** 2000. Purification and biochemical characterization of the VIM-1 metallo-β-lactamase. *Antimicrob. Agents Chemother.* **44:**3003–3007.

**Hernandez Valladares, M., A. Felici, G. Weber, H. W. Adolph, M. Zeppezauer, G. M. Rossolini, G. Amicosante, J. M. Frère, and M. Galleni.** 1997. Zn(II) dependence of the *Aeromonas hydrophila* AE036 metallo-β-lactamase activity and stability. *Biochemistry* **36:**11534–11541.

**Kaminskaia, N. V., B. Spingler, and S. J. Lippard.** 2000. Hydrolysis of β-lactam antibiotics catalyzed by dinuclear zinc(II) complexes: functional mimics of metallo-β-lactamases. *J. Am. Chem. Soc.* **122:**6411–6422.

**McManus-Munoz, S., and M. W. Crowder.** 1999. Kinetic mechanism of metallo-β-lactamase L1 from *Stenotrophomonas maltophilia. Biochemistry* **38:**1547–1553.

**Orellano, E. G., J. E. Giardini, J. A. Cricco, E. A. Ceccarelli, and A. J. Vila.** 1998. Spectroscopic characterization of a binuclear metal site in *Bacillus cereus* β-lactamase II. *Biochemistry* **37:**10173–10180.

**Paul-Soto, R., R. Bauer, J. M. Frère, M. Galleni, W. Meyer-Klauckel, H. Nolting, G. M. Rossolini, D. de Seny, M. Hernandez-Valladares, M. Zeppezauer, and H. W. Adolph.** 1999. Mono- and binuclear $Zn^{2+}$-β-lactamase. Role of the conserved cysteine in the catalytic mechanism *J. Biol. Chem.* **274:**13242–13249.

**Sutton, B. J., P. J. Artyniuk, A. E. Cordero-Barbosa, C. Little, D. C. Phillips, and S. G. Welley.** 1987. An X-ray crystallographic study of β-lactamase I from *Bacillus cereus* at 0.35 nm resolution. *Biochem. J.* **248:**181–188.

**Wang, Z., W. Fast, and S. J. Benkovic.** 1998. Direct observation of an enzyme-bound intermediate in the catalytic cycle of the metallo-β-lactamase from *Bacteroides fragilis. J. Am. Chem. Soc.* **120:**10788–10789.

**Wang, Z., W. Fast, and S. J. Benkovic.** 1999. On the mechanism of the metallo-β-lactamase from *Bacteroides fragilis. Biochemistry* **38:**10013–10023.

chapter 8

Historical Perspective on the Development of Penicillins and β-Lactamase Inhibitors

Classification of the Penicillins

*Natural Penicillins*

*Antistaphylococcal β-Lactamase-Resistant Penicillins*

*Aminopenicillins*

*Antipseudomonal Penicillins*

Penicillin Preparations Commercially Available in the United States

REFERENCES AND FURTHER READING

# Inhibitors of Peptidoglycan Biosynthesis
## Penicillins

Derivatives of penicillanic acid comprise two groups of therapeutic agents of considerable importance in medicine: the antibacterial **penicillins** and the **β-lactamase inhibitors** (Fig. 8.1).

### Historical Perspective on the Development of Penicillins and β-Lactamase Inhibitors

The historic observation of the antibacterial properties of a culture of the fungus *Penicillium notatum* marked the starting point for penicillin antibiotics. By 1940 a research team at the Sir William Dunn School of Pathology of Oxford University (Oxford, United Kingdom), led by Florey, Chain, and Abraham, isolated penicillin for the first time in a reasonably pure form. The team discovered its clinical importance in the treatment of bacterial infections of human patients.

One of the most important landmarks in the story of β-lactam compounds was the X-ray crystallographic determination of penicillin G, previously named penicillin (see Fig. 8.4), by Hodgkin and Roger-Low in May 1945. They established the fused nature of the bicyclic β-lactam-thiazolidine nucleus and the relative stereochemistry of the substituents of the penicillin molecule.

As mentioned in chapter 7, the penicillins as well as other β-lactam antibiotics exert their antibacterial activity because they disrupt bacterial cell wall synthesis. They specifically bind to proteins located in the cytoplasmic membrane, penicillin-binding proteins (PBPs), inhibiting their activity as peptidoglycan-DD-transpeptidases that catalyze the cross-linking reaction of D-alanyl peptides on peptidoglycan strands of the growing cell wall in the final step of the biosynthetic sequence. Other PBPs exhibit DD-carboxypeptidase activity (see Fig. 7.9).

In 1940, Abraham and Chain described the first bacterial enzyme that destroyed penicillin; they termed it penicillinase, since the β-lactam struc-

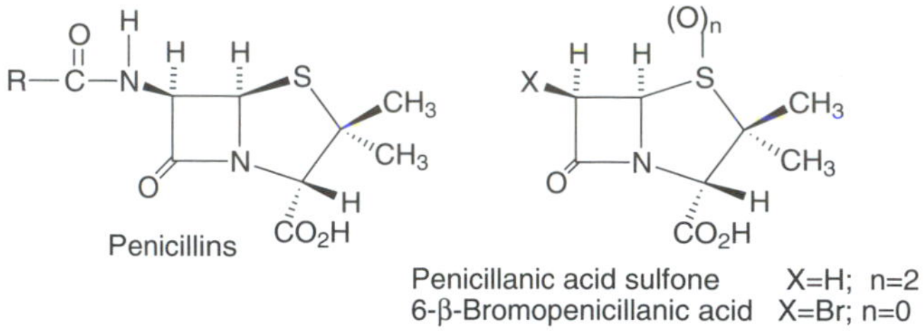

Penicillins

Penicillanic acid sulfone X=H; n=2
6-β-Bromopenicillanic acid X=Br; n=0

**Figure 8.1** Chemical structures of penicillanic acid compounds with antibiotic activity (penicillins) and a β-lactamase inhibitory activity (penicillanic acid sulfone and 6-β-bromopenicillanic acid).

ture of penicillin was not then known. Since enzymes of this type also attack other β-lactam antibiotics such as cephalosporins, carbapenems, and monobactams, they are now called β-lactamases and represent the defense mechanism of bacteria against the action of β-lactam antibiotics; they are therefore clinically important. This finding implies that the genes coding for β-lactamases pre-date the clinical use of β-lactam antibiotics and that β-lactamases clearly destroyed the β-lactam compounds in nature. Penicillin G is a typical example of compounds that inhibit the action of DD-transpeptidases, but it is also a substrate for β-lactamases (Fig. 8.2).

The propensity of β-lactamase genes to be rapidly transmitted on plasmids throughout the bacteria, the heterogeneity of β-lactamases, and the ease with which mutations of the original β-lactamases (e.g., TEM-1, TEM-2, and SHV-1 mutations) were produced threatened the usefulness of penicillins and other β-lactam antibiotics, as well as of β-lactamase inhibitors. Attempts to overcome this problem have developed in two directions: (i) the synthesis of β-lactam antibiotics stable to the hydrolytic action of these enzymes, and (ii) the search for inhibitors of these enzymes.

The use of β-lactam compounds as β-lactamase inhibitors was reported in 1976, when workers at Beecham Pharmaceuticals (currently GlaxoSmithKline) in the United Kingdom isolated clavulanic acid from *Streptomyces clavuligerus* and determined its structure (Fig. 8.3). Clavulanic acid is a typical example of a very poor inhibitor of DD-transpeptidases, but it acts as an irreversible inhibitor of β-lactamases (Fig. 8.3).

As a result of this discovery, a new approach to the treatment of infection caused by bacteria that produce β-lactamases emerged: the association of a β-lactam antibiotic susceptible to the action of β-lactamases with a β-lactamase inhibitor. Therefore, this kind of association expanded the spectrum of action of the β-lactam antibiotic when used in combination.

Since penicillins are bactericidal antibiotics with a high therapeutic index, they are frequently the drugs of choice for infections caused by susceptible organisms in nonallergic patients.

## Classification of the Penicillins

Penicillins are conveniently divided on the basis of antibacterial activity into the following subclasses (or subgroups):

Natural penicillins
  Penicillin G or benzylpenicillin
  Penicillin V or phenoxymethylpenicillin
Antistaphylococcal β-lactamase-resistant penicillins (penicillinase-resistant penicillins)
  Methicillin

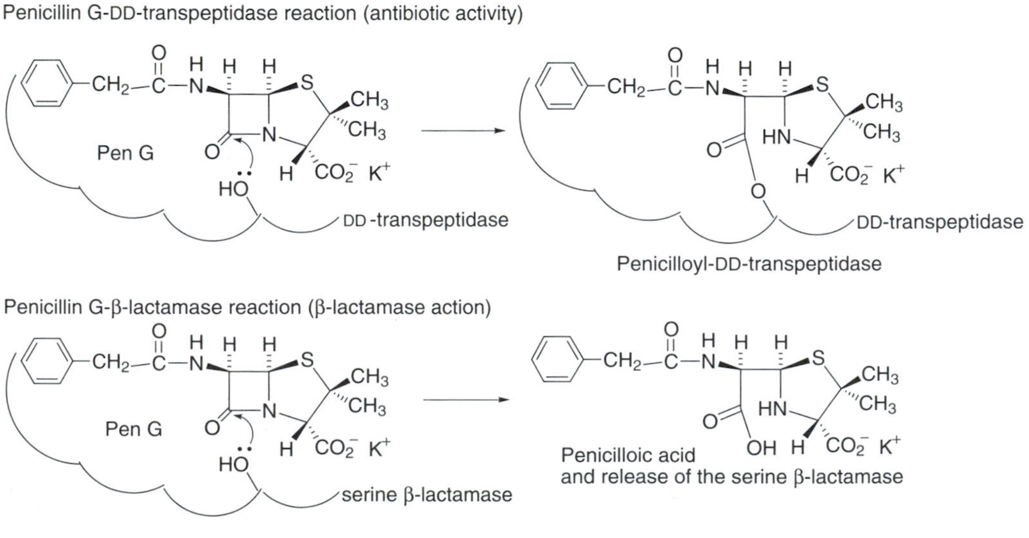

Penicillin G-DD-transpeptidase reaction (antibiotic activity)

Pen G — DD-transpeptidase — Penicilloyl-DD-transpeptidase

Penicillin G-β-lactamase reaction (β-lactamase action)

Pen G — serine β-lactamase — Penicilloic acid and release of the serine β-lactamase

**Figure 8.2** Reactions in which penicillin G inhibits a DD-transpeptidase enzyme and the β-lactam ring is opened by the action of a serine β-lactamase, yielding a penicilloic acid derivative and liberating β-lactamase, which is ready to catalyze a new reaction.

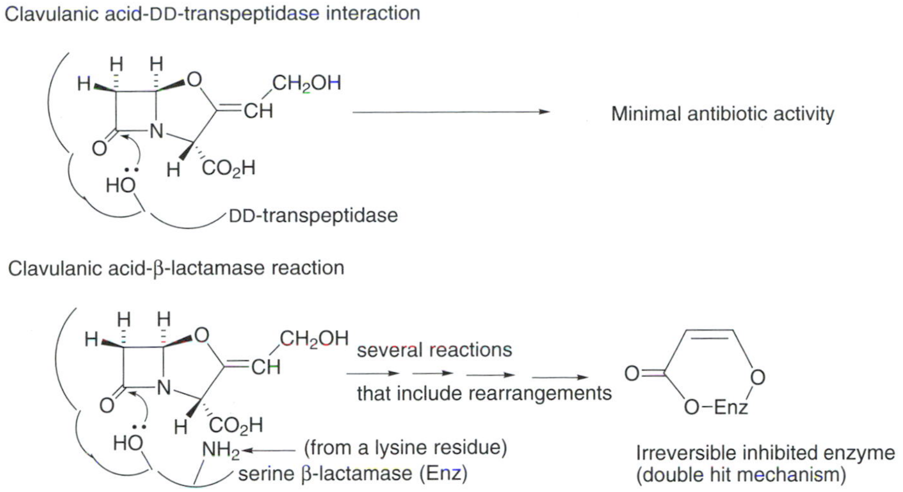

Clavulanic acid-DD-transpeptidase interaction

Minimal antibiotic activity

DD-transpeptidase

Clavulanic acid-β-lactamase reaction

several reactions that include rearrangements

(from a lysine residue)

serine β-lactamase (Enz)

Irreversible inhibited enzyme (double hit mechanism)

**Figure 8.3** Lack of antibiotic activity of clavulanic acid and irreversible inhibition of a serine β-lactamase through a double-hit mechanism. See the text and Fig. 11.1 for a detailed explanation of this last mechanism of reaction. The irreversibly inhibited enzyme structure formed by the so-called "suicide inhibitors" is also shown.

Nafcillin
Isoxazolyl penicillins
Oxacillin
Cloxacillin
Dicloxacillin
Aminopenicillins
Ampicillin
Amoxicillin
Bacampicillin
Antipseudomonal penicillins
Carboxypenicillins
Indanyl carbenicillin
Ticarcillin
Ureidopenicillins (also known as extended-spectrum penicillins)
Mezlocillin
Piperacillin

## Natural Penicillins

Penicillin G (benzylpenicillin) and penicillin V (phenoxymethylpenicillin) (Fig. 8.4) are the only penicillins used in their natural forms as obtained from the fermentation of *Penicillium chrysogenum*. The other penicillins in clinical use are derived from 6-β-aminopenicillanic acid using standard synthetic procedures.

Penicillins G and V are highly active against sensitive strains of gram-positive cocci (the exception being ***Streptococcus** faecalis* and β-lactamase-producing staphylococci). Therefore, they are ineffective against most strains of *Staphylococcus aureus*. Penicillin G is 5 to 10 times more active than penicillin V against *Neisseria* species and against certain anaerobes sensitive to penicillins.

### PHARMACOLOGICAL PROPERTIES

It is outside the scope of this book to address the pharmacological and toxicological properties of antibacterial agents. However, it is considered convenient to describe the absorption of penicillin G, its long-acting parenteral salts procaine penicillin G and benzathine penicillin G, and penicillin V.

About one-third of an orally administered dose of penicillin G is absorbed from the gastrointestinal tract, primarily from the duodenum. Gastric juice, with its pH of 2, rapidly destroys penicillin G. Penicillin V is more stable in acidic media, and therefore is better absorbed from the gastrointestinal tract. When penicillin G is administered intramuscularly, maximum levels in blood are achieved within 15 to 30 min. These levels decline rapidly, and penicillin G is rapidly excreted by the kidneys.

There are two parenteral preparations of penicillin G that slowly release penicillin G from the site of intramuscular injection—procaine penicillin G and benzathine penicillin G. Such insoluble salts of penicillin G release penicillin G slowly but result in persistent concentrations of antibiotic in the blood. Procaine penicillin G yields peak levels in blood at 2 to 4 h, and these levels decline to negligible by 24 to 48 h. Benzathine penicillin G is absorbed from the injection site more slowly, and very low levels of antibiotic are detectable in serum for 3 to 4 weeks after administration (the structures of these salts are shown in Fig. 8.5). Since these two salts of penicillin G result in low antibiotic levels in blood, they must be employed only in the treatment of infections caused by organisms that are *very sensitive* to penicillin G, such as group A beta-

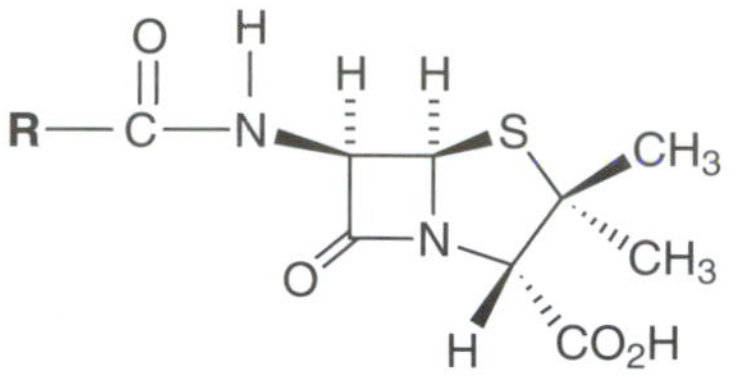

| Name | R | Routes of administration | Sensitivity to β-lactamases |
|---|---|---|---|
| Benzylpenicillin or penicillin G | 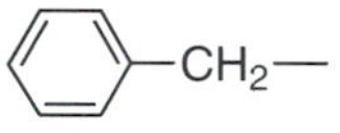 $C_6H_5$–$CH_2$— | IM, IV<br>oral | Sensitive |
| Phenoxymethylpenicillin or penicillin V | 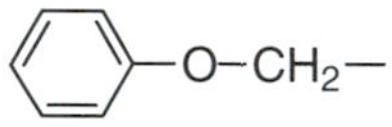 $C_6H_5$–O–$CH_2$— | oral | Sensitive |

**Figure 8.4** Structures and some properties of natural penicillins.

hemolytic streptococci. Some medical specialists prescribe these long-acting parenteral forms as prophylaxis against rheumatic fever disease. In some texts, such as *Mandell, Douglas, and Bennett's Principles and Practice of Infectious Diseases* (see the reference at the end of this chapter), these long-acting derivatives of penicillin G are called repository penicillins, and it is remarked that these salts are only for intramuscular use and cannot be used intravenously, subcutaneously, or topically.

Penicillin V is available only for oral use as a sodium or potassium salt in suspension or tablets. Because of the instability of penicillin G in acid, oral penicillin G is not used. Usually amoxicillin is the penicillin of choice for oral administration.

CLINICAL USES

Penicillin G is preferred for infections caused by the most frequently encountered gram-positive bacteria (except staphylococci and, in some situations, enterococci) and susceptible gram-negative cocci. It is also the drug of choice for infections caused by certain gram-negative bacilli.

### Antistaphylococcal β-Lactamase-Resistant Penicillins

The antistaphylococcal β-lactamase-resistant semisynthetic penicillins were developed to be resistant to β-lactamase-mediated hydrolysis. The structural modification in the side chain of these penicillins provides enough steric hindrance to make them very poor substrates for most β-lactamases. The antistaphylococcal β-lactamase-resistant penicillins have a narrow spectrum of action and are used only for the treatment of infections due to penicillinase-producing staphylococci (*S. aureus*).

The β-lactamase-resistant penicillins available for parenteral use are methicillin, oxacillin, and nafcillin

**Figure 8.5** Chemical structures of procaine penicillin G and benzathine penicillin G.

Procaine penicillin G

Benzathine penicillin G

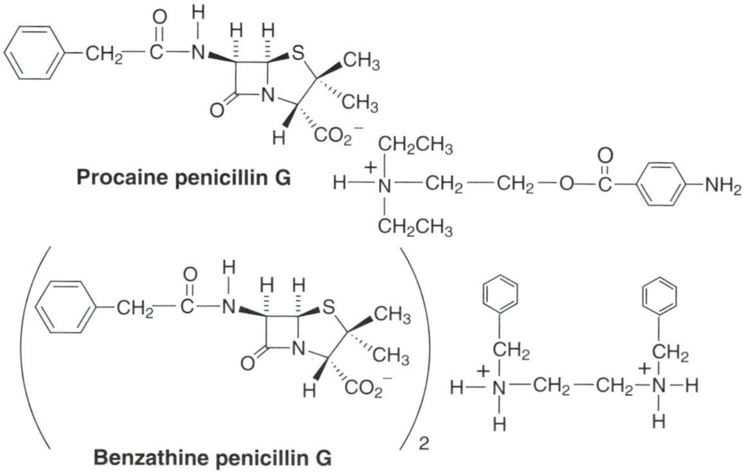

(Fig. 8.6). Cloxacillin and dicloxacillin are the only members of this group available for oral use.

METHICILLIN

Methicillin (2,6-dimethoxyphenylpenicillin) was the first of several antistaphylococcal β-lactamase-resistant penicillins developed. Methicillin is acid labile and therefore can be administered only parenterally. Methicillin sodium is a water-soluble salt for parenteral use. It is active against most strains of *S. aureus*. Methicillin has been in clinical use longer than the other members of this group, but is no longer widely used because it has been associated with interstitial nephritis more often than have oxacillin and nafcillin. Oxacillin and nafcillin are now both widely used for parenteral administration.

## Acquired Resistance

Over the past 20 years, strains of methicillin-resistant *S. aureus* (MRSA) have become prevalent worldwide. MRSA strains contain an additional high-molecular-weight PBP, PBP2a, which has a very low binding affinity for methicillin and for all subclasses of the β-lactamase-resistant penicillins and cephalosporins. The β-lactamase-resistant penicillins are generally more resistant to staphylococcal β-lactamases than are cephalosporins. Often, such MRSA strains are also resistant to aminoglycosides, tetracyclines, erythromycin, and clindamycin. Vancomycin and teicoplanin are considered to be the antibiotics of choice to treat infections caused by MRSA.

About 40 to 60% of strains of *S. epidermidis* are also resistant to β-lactamase-resistant penicillins because of their PBP2a production.

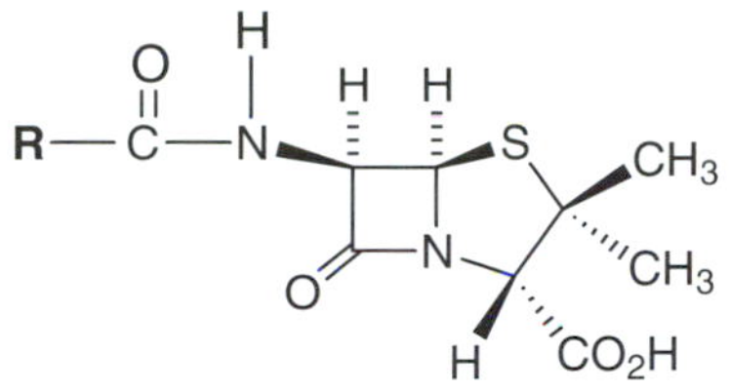


**Figure 8.6** Structures and some properties of penicillinase-resistant penicillins. Abbreviations: IM, intramuscular; IV, intravenous.

| Name | R | | Routes of administration | Sensitivity to β-lactamases |
|---|---|---|---|---|
| Methicillin | 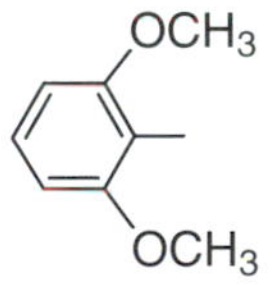 | | IM, IV | Resistant |
| Nafcillin | 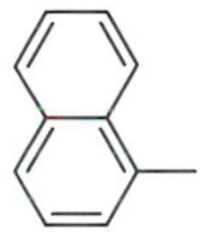 | | IM, IV, oral | Resistant |
| **ISOXAZOLYL PENICILLINS** | | | | |
| 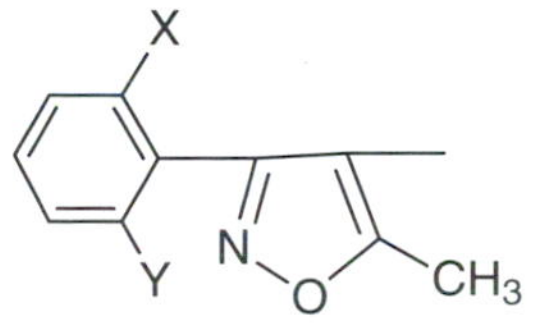 | | | | |
| | X | Y | | |
| Oxacillin | H | H | IM, IV, oral | Resistant |
| Cloxacillin | Cl | H | oral | Resistant |
| Dicloxacillin | Cl | H | oral | Resistant |

NAFCILLIN AND ISOXAZOLYL PENICILLINS
Nafcillin and the isoxazolyl penicillins also are used only for treatment of infection due to β-lactamase-producing staphylococci. These penicillins are not as potent as penicillin G against penicillinase-nonproducing strains of staphylococci and are not useful against gram-negative bacteria.

## Aminopenicillins

The aminopenicillins include ampicillin, its ethoxycarbonyloxyethyl ester (bacampicillin), and amoxicillin. With few exceptions, the antibacterial spectra of ampicillin and amoxicillin are similar. The aminopenicillins have a broader antibacterial spectrum than do penicillin G and the antistaphylococcal β-lactamase-resistant penicillins because they have a better capacity to penetrate the porins of gram-negative bacteria. The aminopenicillins are inactivated by β-lactamases (Fig. 8.7).

Most strains of *Haemophilus influenzae* are susceptible to aminopenicillins, but resistant β-lactamase-producing strains are prevalent. In contrast to penicillin G, the aminopenicillins are active against several aerobic gram-negative enteric bacilli, including strains of *E. coli, Proteus mirabilis, Salmonella,* and *Shigella.* However, resistant β-lactamase-producing strains of these organisms are prevalent. Other members of the *Enterobacteriaceae* (e.g., *Klebsiella, Enterobacter, Serratia,* indole-positive *Proteus,* and *Providencia*), *Pseudomonas aeruginosa,* and most *Acinetobacter* species are resistant to the aminopenicillins.

AMPICILLIN
Ampicillin was the first aminopenicillin introduced into clinical use. The clinical uses of ampicillin as well as other antibacterials are very well presented in the book *The Use of Antibiotics* by Kucers et al.; readers seeking comprehensive information on these aspects should consult this excellent source (see the references at the end of this chapter).

BACAMPICILLIN
Several ampicillin prodrugs have been developed, but only one, bacampicillin, is currently marketed in the United States. Bacampicillin (Fig. 8.8) is an orally administered prodrug that is rapidly hydrolyzed, after absorption from the gastrointestinal tract, by esterases present in the serum. This penicillin yields peak levels in serum twofold higher than those attained after oral administration of equivalent doses of ampicillin itself.

**Figure 8.7** Structures and some properties of aminopenicillins. Abbreviations: IM, intramuscular; IV, intravenous.

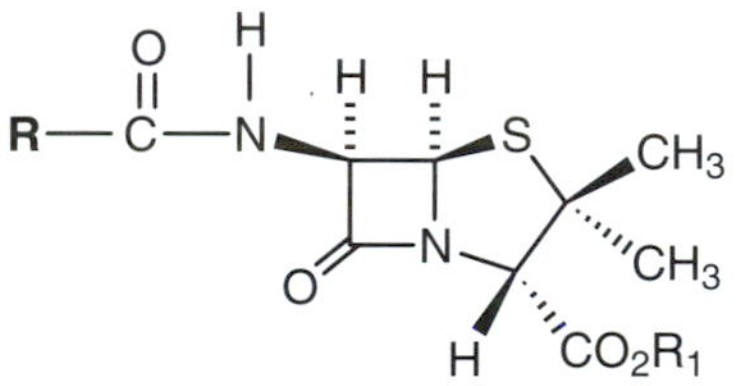

| Name | R | $R_1$ | Routes of administration | Sensitivity to β-lactamases |
|---|---|---|---|---|
| Ampicillin | 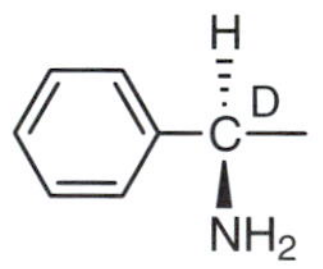 H, D, $NH_2$ | -H | IM, IV, oral | Sensitive |
| Bacampicillin | 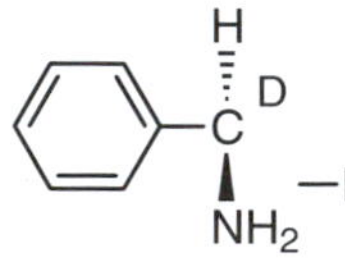 H, D, $NH_2$ | $-HC(CH_3)-O-C(=O)-O-CH_2CH_3$ | Oral | Sensitive |
| Amoxicillin | 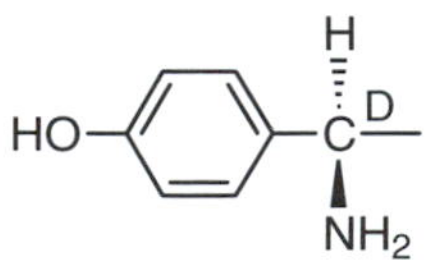 HO–, H, D, $NH_2$ | -H | Oral | Sensitive |

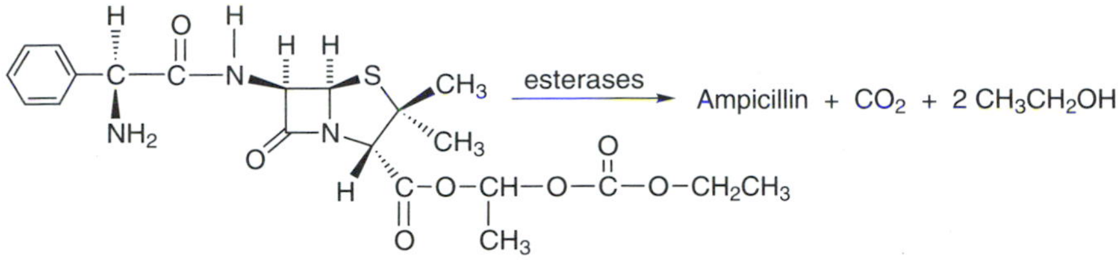

esterases → Ampicillin + $CO_2$ + 2 $CH_3CH_2OH$

**Figure 8.8** Products of cleavage of the double ester of bacampicillin by nonspecific esterases.

Therefore, the clinical uses of bacampicillin are the same as the clinical use of oral ampicillin.

## Antipseudomonal Penicillins

The antipseudomonal penicillins are semisynthetic derivatives of penicillanic acid (Fig. 8.9) and are categorized into two subgroups: the **carboxypenicillins,** which include carbenicillin indanyl ester (an ester for oral administration) and ticarcillin, and the **ureidopenicillins,** which include mezlocillin and piperacillin. In the United States, carbenicillin is currently available only as its indanyl ester, and azlocillin is no longer marketed.

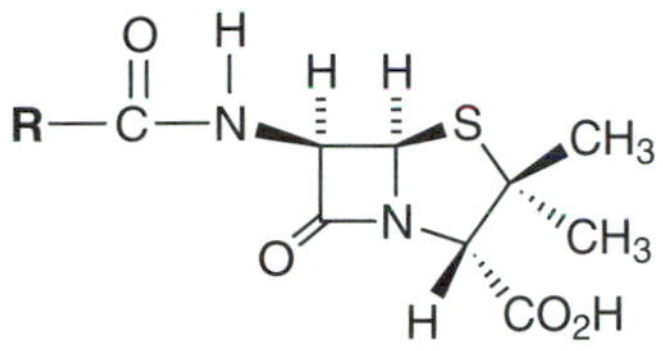

**Figure 8.9** Structures and some properties of antipseudomonal penicillins. IM, intramuscular; IV, intravenous.

| Name | R | Routes of administration | Sensitivity to β-lactamases |
|---|---|---|---|
| **Carboxypenicillins** | | | |
| Indanyl carbenicillin | 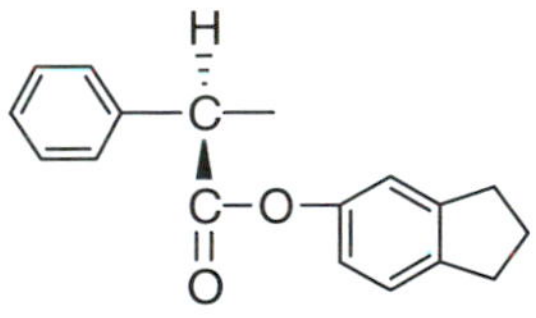 | Oral | Sensitive |
| Ticarcillin | 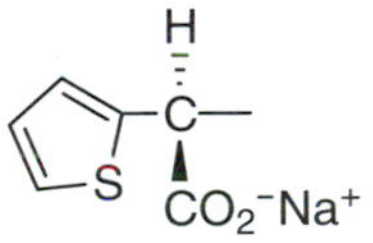 | IM, IV | Sensitive |
| **Ureidopenicillins** | | | |
| Mezlocillin | 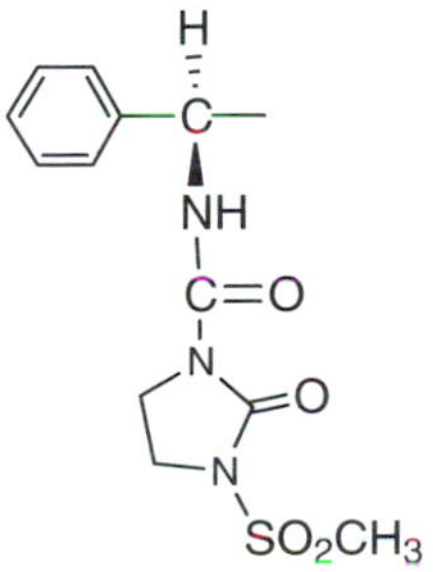 | IM, IV | Sensitive |
| Piperacillin | 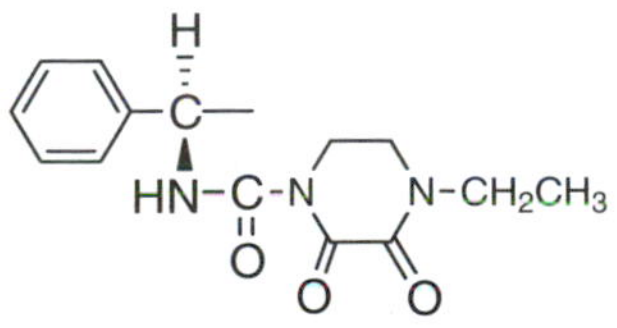 | IM, IV | Sensitive |

Antipseudomonal penicillins retain most of the antibacterial activity of the aminopenicillins but have added activity against gram-negative bacilli. These antibiotics are active against *P. aeruginosa* and indole-positive *Proteus* strains.

## INDANYL CARBENICILLIN

Carbenicillin is unstable in acid, due to a decarboxylation reaction that yields penicillin G. It is well known that orally administered acid is absorbed rather inefficiently by the gastrointestinal tract but that esters are well absorbed. However, the simple alkyl or aryl esters of penicillins are hydrolyzed only slowly by esterases in the human body, and so a special type of ester was needed, such as ethoxycarbonyloxyethyl ester in bacampicillin or an indanyl ester in carbenicillin. Indanyl carbenicillin (with a carboxyl side chain) is well absorbed from the gastrointestinal tract and is hydrolyzed by esterases to carbeniclin, the active drug. Clinically, there are several shortcomings to the use of carbenicillin. Its potency against *P. aeruginosa* is relatively poor, necessitating doses as high as 30 g/day, which carries a risk of toxicity. Carbenicillin itself was marketed as the disodium salt; however, the daily intake of sodium ions ($Na^+$) can be as much as 3.2 g, a large sodium load for cardiac patients or those on a low-sodium regimen. The bioisosteric thienyl analog, ticarcillin, which has a similar spectrum, is somewhat more potent, thus permitting equivalent efficacy at lower doses (15 to 18 g/day). The sodium loads, however, are still more than 2 g per day. Quinolones have replaced indanyl carbenicillin for the treatment of urinary tract infections and prostatitis.

## MEZLOCILLIN

Mezlocillin (Fig 8.9) is an acylureidopenicillin that is more active in vitro than carbenicillin or ticarcillin against enteric gram-negative bacilli such as *E. coli, Klebsiella* spp. (only about 50% of strains are sensitive), *Enterobacter* spp., *Citrobacter* spp., *Acinetobacter* spp., *Serratia* spp., and *Bacteroides fragilis*. Its activity against *P. aeruginosa* is comparable to that of ticarcillin. Mezlocillin is acid labile and therefore must be given parenterally.

## PIPERACILLIN

Piperacillin is a piperazine derivative of ampicillin with increased activity against enterobacteria. It is probably the most active antipseudomonal penicillin available. In the treatment of moderate to severe *Pseudomonas* infection, an aminoglycoside of the antipseudomonal group (gentamicin, tobramycin, or amikacin) is often administered along with an antipseudomonal penicillin (mezlocillin or piperacillin). Such a combination is synergistic against many strains of *Pseudomonas.*

**Table 8.1** Generic and common trade names of penicillins, the preparations available, and manufacturers in the United States

| Generic name | Trade name (preparation)[a] | Manufacturer |
|---|---|---|
| Natural penicillins | | |
| Penicillin G potassium | Pfizerpen (injectable) | Pfizer |
| Penicillin G benzathine | Permapen Isojet (injectable for i.m. use only) | Pfizer |
| Penicillin G benzathine | Bicillin L-A (injectable for i.m. use only) | Monarch |
| Penicillin G benzathine and penicillin G procaine (equal amounts) | Bicillin C-R Injection (for i.m. use only) | Monarch |
| Penicillin G benzathine (900,000 units) and penicillin G procaine (300,000 units) | Bicillin 900/300 (injectable for i.m. use only) | Monarch |
| Penicillin V potassium | Pen-Veetids (tablets) | Apothecon (Bristol-Myers Squibb) |
| β-Lactamase-resistant penicillins | | |
| Dicloxacillin | Dynapen (capsules) | Apothecon (Bristol-Myers Squibb) |
| Aminopenicillins | | |
| Amoxicillin | Amoxil (capsules, tablets, chewable tablets) | GlaxoSmithKline |
| | Amoxil pediatrics (drops, powder for oral suspension) | GlaxoSmithKline |
| | Trimox (capsules, tablets) | Bristol-Myers Squibb |
| Ampicillin | Principen (capsules) | Bristol-Myers Squibb |
| Bacampicillin | Spectrobid (oral preparation) | Roerig |
| Antipseudomonal penicillins | | |
| Carbenicillin indanyl sodium | Geocilin (tablets) | Pfizer |
| Piperacillin sodium | Pipracil (injectable for i.v. and i.m. use) | Lederle |

[a] i.m., intramuscular; i.v., intravenous.

Piperacillin is given parenterally and is excreted predominantly by the renal route.

## Penicillin Preparations Commercially Available in the United States

A list of generic and common trade names and preparations, as well as the pharmaceutical laboratory manufacturers, for antibiotics and other antibacterial agents is presented at the end of each chapter for the remainder of this volume. Table 8.1 lists the generic and trade names of penicillins. For a list of antibiotics and synthetic antibacterial agents on the market in other countries, the reader should consult Parfitt (1999).

### REFERENCES AND FURTHER READING

#### Penicillins

**American Medical Association.** 1995. *Drug Evaluations Annual,* p. 1377–1411. American Medical Association, Chicago, Ill.

**Chambers, H. F.** 2000. Penicillins, p. 261–274. *In* G. L. Mandell, J. E. Bennet, and R. Dolin (ed.), *Principles and Practice of Infectious Diseases,* 5th ed. Churchill Livingstone, Inc., Philadelphia, Pa.

**Kucers, A., S. M. Crowe, M. L. Grayson, and J. F. Hoy.** 1997. *The Use of Antibiotics. A Clinical Review of Antibacterial, Antifungal and Antiviral Drugs,* 5th ed. Butterworth-Heinemann, Oxford, United Kingdom.

**Petri, W. A.** 2001. Penicillins, cephalosporins, and other β-lactam antibiotics, p. 1189–1218. *In* J. G. Hardman and L. E. Limbird (ed.), *Goodman & Gilman's The Pharmacological Basis of Therapeutics,* 10th ed. McGraw-Hill Book Co., New York, N.Y.

**Scholar, E. M., and W. B. Pratt.** 2000. *The Antimicrobial Drugs,* 2nd ed. Oxford University Press, New York, N.Y.

**Sutherland, R.** 1997. β-Lactams: penicillins, p. 256–305. *In* F. O'Grady, H. P. Lambert, R. G. Finch, and D. Greenwood (ed.), *Antibiotic and Chemotherapy,* 7th ed. Churchill Livingstone, Inc., New York, N.Y.

#### Antibacterials Marketed around the World

**Parfitt, K.** 1999. *Martindale. The Complete Drug Reference,* 32nd ed. Pharmaceutical Press, London, United Kingdom. Pages 112 through 270 cover the antibacterial agents used in the treatment and prophylaxis of bacterial infections, with particular reference to preparations BP1998, USO23, and proprietary preparations in many countries.

chapter 9



# Inhibitors of Peptidoglycan Biosynthesis
## Cephalosporins

### Introduction and History of Development

The cephalosporins are a subgroup of β-lactam antibiotics, whose bicyclic system, called the cephem nucleus, consists of a four-member β-lactam ring fused through the nitrogen and adjacent tetrahedral carbon atom to a second heterocycle forming a six-member dihydrothiazine ring.

In 1945, examining the microbial flora of the seawater near a sewage outlet in Cagliari, Sardinia, Italy, Giuseppe Brotzu isolated a fungus of the *Cephalosporium* genus similar to *Cephalosporium acremonium*. Cultures of this fungus produced material which inhibited a variety of gram-positive and gram-negative bacteria. Brotzu published an account of this investigation in 1948. Also in 1948, he sent the results of his investigation and a culture of this fungus to Howard Florey at the Sir William Dunn School of Pathology at Oxford University. In 1953, E. P. Abraham and his coworkers at Oxford isolated a crystalline substance that they called cephalosporin C. Cephalosporin C was active against *Staphylococcus aureus*, *Salmonella enterica* serovar Typhi, and *E. coli*. The structure of cephalosporin C (Fig. 9.1) was determined by a series of chemical degradation studies performed by Abraham and Newton and subsequent X-ray crystallography performed by Hodgkin and Maslem of Oxford University.

The clinical use of cephalosporin C was limited by its generally weak antibacterial activity. With the penicillins as a precedent, the 7-acylamino group was the first target for variation in the search for improved acitivity. Initially, enzymatic cleavage of the aminoadipoyl side chain was attempted, but cephalosporin C proved impervious to such treatment. Morin and coworkers at Eli Lilly Laboratories succeeded in developing a chemical method to convert cephalosporin C efficiently into 7-aminocephalosporanic acid (7-ACA). As soon as 7-ACA became available, numerous novel cephalosporins were prepared by a reacylation reaction involving a similar method to the one developed for the acylation of 6-β-aminopenicillanic acid (6-APA). The nucleus of 7-ACA has been modified with different side

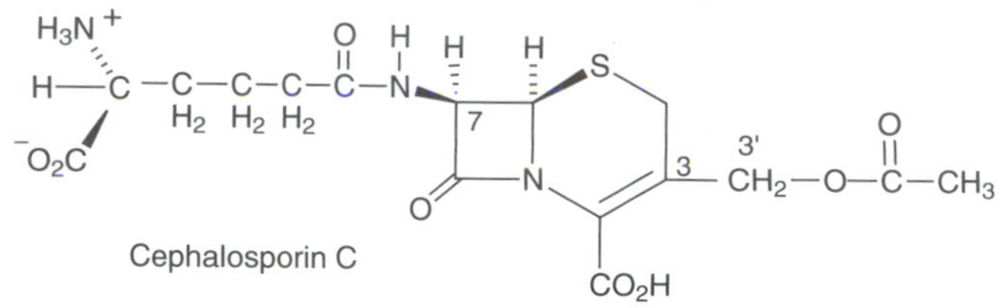

Cephalosporin C

**Figure 9.1** Chemical structure of cephalosporin C.

chains. Modifications ($R_1$) at position 7 of the β-lactam ring are associated with changes in antibacterial activity and stability to β-lactamases. Substitution ($R_2$) at position 3 of the dihydrothiazine ring affects the metabolism and pharmacokinetic properties of the antibiotics to a greater extent than it affects antibacterial activity.

Chemical reacylation of 7-ACA provided cephalothin, which became clinically available in the early 1960s. Later, other cephalosporins, such as cephaloridine, cephapirin, cephacetrile, cefazolin, cephalexin, and cephradine, were developed. All these semisynthetic cephalosporins are classified as **first-generation agents** because they have similar in vitro antimicrobial spectra.

## Classification

Several classifications of the cephalosporins by using microbiologic profiles, pharmacokinetic parameters, and β-lactamase stability have been proposed. In this volume, classification into four generations according to spectrum of activity is adopted (Table 9.1).

The first-generation compounds have a relatively narrow spectrum of activity, being effective primarily against gram-positive cocci. The second-generation cephalosporins have variable activity against gram-positive cocci but increased activity against gram-negative bacteria. The third-generation cephalosporins have an extended spectrum of activity against gram-negative bacilli and most members of the *Enterobacteriaceae*, including some strains that are resistant to second-generation cephalosporins. The fourth generation is the latest group of cephalosporins to be marketed. Cefepime is the only cephalosporin of this group currently approved in the United States. Cefpirome is available in the United Kingdom and in some other European countries.

The main characteristic of fourth-generation cephalosporins is their stability to class 1b β-lactamases of the functional classification of Bush-Jacoby-Medeiros and their enhanced ability to penetrate the porins in the outer membrane of gram-negative bacteria. This last property has been attributed to the presence of a positively charged quaternary ammonium ion at position 3 of the cephem nucleus. The cephalosporins in this generation have activity against gram-positive cocci and a broad array of gram-negative bacteria, including *P. aeruginosa* and many *Enterobacteriaceae* with inducible chromosomal β-lactamases.

Cephalosporin C is produced from a mutant of *Acremonium chrysogenum* (formerly *Cephalosporium acremonium*). The production of cephalosporins derived from 7-ACA is outlined in Fig. 9.2.

**Table 9.1** Classification of cephalosporins by generations

| Class | Generic name | Clinically useful antibacterial spectrum |
|---|---|---|
| First generation | | |
| Parenteral | Cefazolin, cephalothin, cephapirin, cephradine | Gram positive: *S. aureus* (except for methicillin-resistant strains), *S. epidermidis, Streptococcus* species (except enterococci); gram-negative: most strains of *E. coli, K. pneumoniae*, and *P. mirabilis;* anaerobes: most anaerobes are susceptible in vitro except *B. fragilis* |
| Oral | Cefadroxil, cephalexin | |
| Second generation | | |
| Parenteral | Cefamandole, cefmetazole, cefonicid, ceforanide, cefotetan, cefoxitin, cefuroxime | Less effective than first-generation cephalosporins against gram-positive bacteria, but expanded activity against gram-negative bacteria including cephalothin-resistant *E. coli, Klebsiella* spp., and *P. mirabilis;* cefamandole is particularly active against *H. influenzae, Enterobacter* spp., and indole-positive *Proteus* spp.; cefoxitin and cefotetan have enhanced activity against *B. fragilis* and *S. marcescens* |
| Oral | Cefaclor, cefprozil, cefuroxime axetil, loracarbef | |
| Third generation | | |
| Parenteral | Cefoperazone, cefotaxime, ceftazidime, ceftizoxime, ceftriaxone, moxalactam | Increased activity against the gram-negative organisms cited above, *Providencia stuartii, P. aeruginosa*, and *B. fragilis;* cefotaxime and ceftriaxone have especially high activity against β-lactamase producing strains of *H. influenzae* and *N. gonorrhoeae* |
| Oral | Cefdinir, cefixime, cefpodoxime proxetil, ceftibuten | |
| Fourth generation | | |
| Parenteral | Cefepime, cefpirome | Comparable to the third-generation compounds but more stable against β-lactamases |

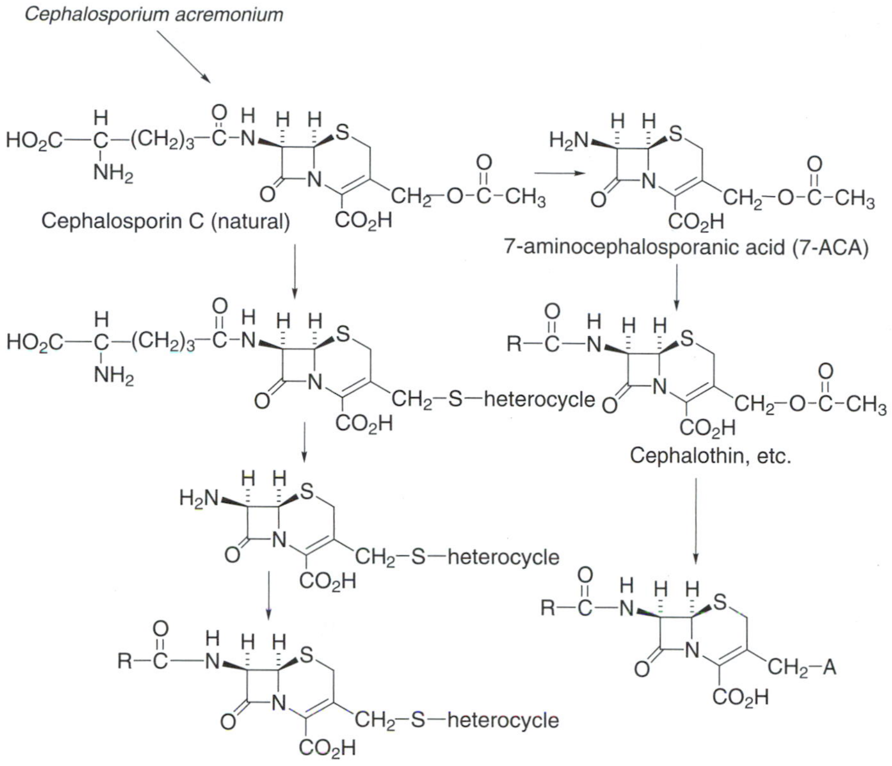

**Figure 9.2** Semisynthesis of cephalosporins through the intermediate 7-ACA.

## Chemical Transformation of a Penicillin Nucleus to a Cephalosporin Nucleus

The pencillin β-sulfoxides have provided a means for chemical conversion of the penicillin structure, by ring expansion, into a cephalosporin structure with the 3-methyl group. In 1963, Morin and coworkers at Eli Lilly Laboratories (Indianapolis, Ind.) reported that heating of penicillin V β-sulfoxide methyl ester in toluene with a trace of acid, followed by removal of the phenoxyacetyl side chain via the imino chloride-imino ether intermediates and hydrolysis of the carboxyl ester, gave 7-aminodeacetoxycephalosporanic acid (7-ADCA) (Fig. 9.3). This process has been studied intensively and patented. Easily removable groups used for protecting the carboxyl functionality include silyl ester, which is easily hydrolyzed. This conversion, starting with cheaper natural penicillins, is currently the basis of the primary industrial route for 7-ADCA production. 7-ADCA is thus the starting compound for 3-methylcephalosporins: cefadroxil, cephalexin, cephradine, ceftizoxime, and ceftibuten, which all have the 3-methyl group as a common feature.

## 3-Substituent Modification

The 3'-acetoxy group of the cephalosporins can be easily displaced with nitrogen and sulfur nucleophiles; for example, pyridine readily displaces the acetoxy group to give the corresponding pyridinium salt (quaternary ammonium salt). Displacement by azide ion followed by reduction results in the amine group; alternatively, the azide group can be modified further, for example, to be cyclized to 1,2,3-triazole or tetrazole structures.

For additional discussions of chemical modification, see the references at the end of this chapter.

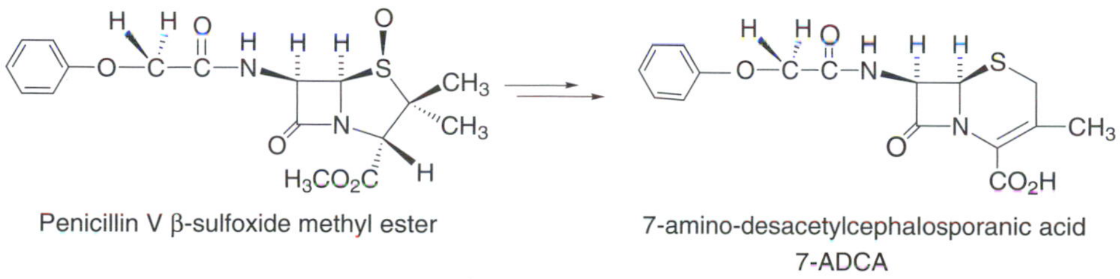

**Figure 9.3** Chemical conversion of the penicillin V β-sulfoxide methyl ester into 7-amino desacetylcephalosporanic acid as originally developed by Morin and coworkers at Eli Lilly Laboratories.

## Deacylation by Esterases of the 3'-Acetoxymethyl Group

The 3'-acetoxymethylcephalosporins are metabolically unstable because the cephalosporins that carry this group are readily deacetylated by serum esterases to give the less active deacetylcephalosporins. This is the case for cephalothin, where the intermediate, 3-hydroxymethylcephalothin, may cyclize to the inactive lactone (Fig. 9.4). To overcome these problems, researchers turned to chemical modification of the 3-substituent to produce pharmacokinetically stable cephalosporins.

## Chemical Reactivity of the β-Lactam Ring in Cephalosporins

Much of the chemical reactivity of the β-lactam antibiotics is associated with the β-lactam ring. It is more reactive than the amide functional group in peptides, where the amide is stabilized by resonance. The reactivity of the β-lactam carbonyl group is a hybrid between a ketone functional group and a tertiary amide.

When cephalosporins are attacked at the β-lactam carbonyl group by a serine β-lactamase (serine hydroxyl group) or by nucleophiles such as alkoxides or hydroxylamine, the presence of the dihydrothiazine ring means that the products formed are not equivalent to those obtained with penicillins. Rather, studies indicate involvement of the 3'-substituent as a leaving group, as shown in Fig. 9.5.

## Departure of a Leaving Group from C-3' of Cephalosporin Molecules

It is well established that the mechanism by which most β-lactamases inactivate β-lactam antibiotics involves acylation of a serine residue at the active site of the enzyme. In the case of a cephalosporin, opening of the β-lactam ring correlates with elimination of a potential leaving group other than the acetoxy group at the 3' position. The experimental circumstances in which opening of the β-lactam ring in cephalosporins can lead the removal of a 3'-substituent during the enzymatic reaction are currently being exploited in fields other than antibiotic research. The term "dual action" has been used to describe the situation when the leaving group possess intrinsic antibacterial activity, including the development of cephalosporin 3'-quinolone esters. This dual mode of action was conceived as an opportunity to expand the antibacterial spectrum to include organisms which are resistant to the third-generation cephalosporins. The quinolones should provide a broad class of antibacterials that are well suited to the role of the second agent for several reasons, the major one being that the antibacterial spectra of cephalosporins and quinolones are complementary, with quinolones being active against β-lactam-resistant strains and cephalosporins being more active against streptococci. A series of cephalosporins has been prepared in which quinolones are linked to the 3' position through an ester bond. This approach has not been as successful as was expected, and so far no new agents that use the reactivity of the cephalosporin moiety to release the fluoroquinolone portion have reached the market.

The concept of dual-action antibacterials was soon realized for monoclonal antibody–β-lactamase-targeted enzyme catalysis. The leaving group could be an anticancer agent, in which case the enzymatic acylation of the cephalosporin moiety by a serine β-lactamase releases the anticancer agent. The use of enzyme-activatable prodrugs in conjunction with antibody-enzyme fusion proteins is a promising strategy for enhancing the antitumor efficacy of antibodies and minimizing the toxicity of chemotherapeutics. Since the concept was first described in 1987, the use of antibody-enzyme conjugates directed at tumor-associated antigens to achieve site-specific activation of prodrugs to potentially cytotoxic agents, termed **antibody-directed enzyme prodrug therapy (ADEPT)**, also named **antibody direct catalysis (ADC)**, has attracted considerable interest.

**Figure 9.4** Hydrolysis of the 3'-acetoxy group of cephalothin by serum esterases.

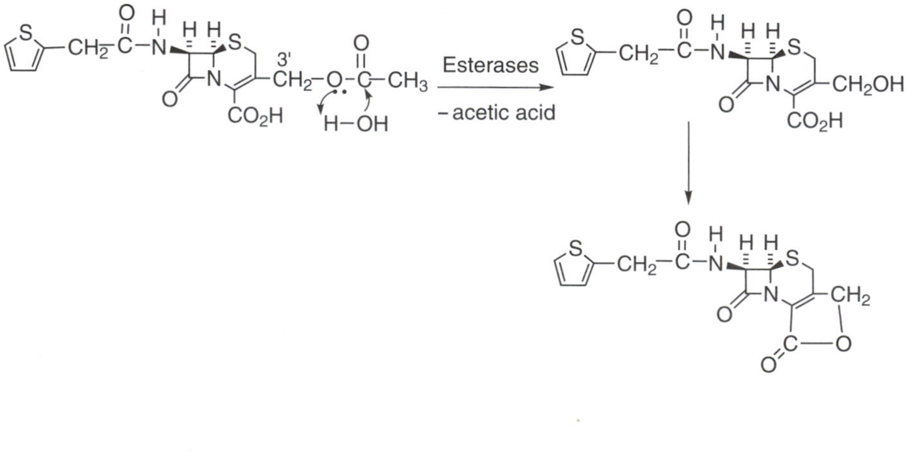

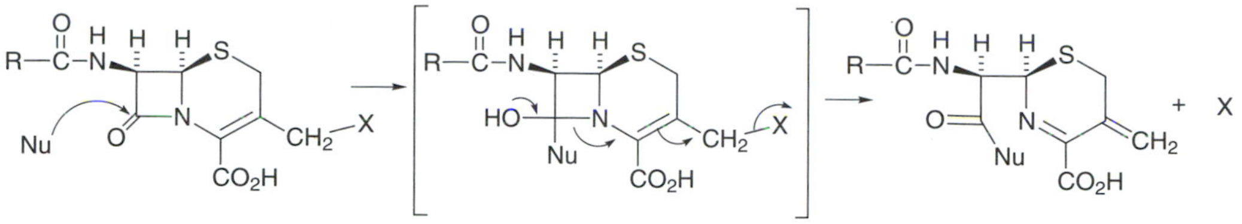

**Figure 9.5** Participation of a leaving group (X) in the opening of the β-lactam ring of cephalosporins. Nu can be the hydroxyl group of serine β-lactamases or a chemical nucleophile such as an alkoxide ion or hydroxylamine.

A novel approach under development involves administration of a β-lactamase covalently bound to a monoclonal antibody which localizes on the targeted tumor cell surface. Subsequent administration of a synthetic cephalosporin compound, the anticancer prodrug, which carries a masked cytotoxic entity and has been specifically designed as a substrate of the enzyme, will result in cleavage of the cephalosporin by the β-lactamase, allowing the specific enzyme-catalyzed release of the cytotoxic agent at the site of the tumor, as depicted in Fig. 9.6.

β-Lactamase is an attractive enzyme for prodrug activation because of its high catalytic efficiency and broad structure specificity. Furthermore, there are no known similar activities, competing substrates, or inhibition endogenous to humans. Cephalosporins have proven to be highly versatile substrate triggers in the construction of enzyme-activatable prodrugs such as vinca alkaloids, retinoic acid, nitrogen mustard drugs, a carboplatin analog, doxorubicin, and taxol.

## First-Generation Cephalosporins

The first-generation cephalosporins (Fig. 9.7) are very active against gram-positive cocci and have moderate activity against community-acquired *Moraxella catarrhalis, E. coli, Proteus mirabilis* (indole negative), and *K. pneumoniae.* The antibacterial activity of these agents for other members of the *Enterobacteriaceae* is unpredictable and should not be assumed. The first-generation cephalosporins are inactive against methicillin-resistant *Staphylococcus aureus* and *Enterococcus* spp. Cefazolin usually is preferred to other first-generation cephalosporins due to its superior pharmacokinetic properties, which result in a higher and more sustained concentration in serum. Cephalothin and cephapirin are now unavailable in the United States.

The first-generation cephalosporins are used to treat *S. aureus* and nonenterococcal streptococcal infections when it is necessary to avoid the use of penicillin. Most commonly, these infections include skin and soft tissue infections and streptococcal pharyngitis. Cefazolin frequently is the drug of choice for prophylaxis during some surgical procedures because of its spectrum of action, favorable toxicity profile, relatively long half-life, modest cost, and proven efficacy.

**Figure 9.6** Schematic drawing of the cephalosporin-platinum complex prodrugs for the ADEPT method. mAb, monoclonal antibody. Reprinted from S. Hanessian and J. Wang, *Can. J. Chem.* **71:**896–906, 1993, with permission from the publisher.

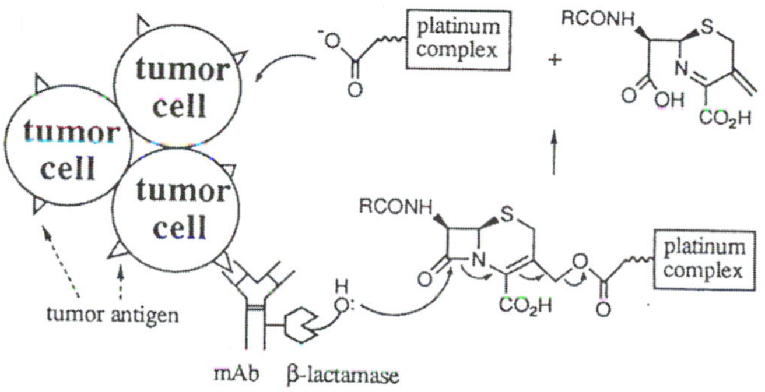

## Second-Generation Cephalosporins

The second-generation cephalosporins should be considered in three groups: the true cephalosporins, the cephamycins, and the carbacephems (Fig. 9.8) In comparison with the first-generation agents, the cephalosporins of this group include a range of parenteral and oral antibiotics that provide significantly improved activity against *H. influenzae, M. catarrhalis, Neisseria meningitidis,* and *N. gonorrhoeae,* as well as enhanced activity against staphylococci and nonenterococcal streptococci. The cephamycins have inferior activity against staphylococci but are active against *Bacteroides* spp., particularly *B. fragilis.* Loracarbef is an orally administered carbacephem. It has comparable activity to this group of cephalosporins against *S. aureus, S. pneumoniae,* and streptococci but has a modestly increased activity against *H. influenzae* and *M. catarrhalis.*

For the sake of clarity, the structure of the orally administered second-generation cefuroxime axetil is not

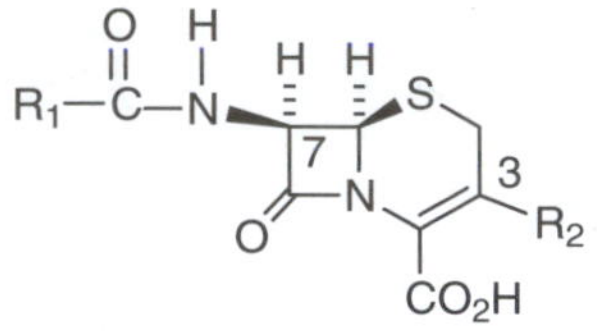

| Name | Routes of administration | R1 | R2 |
|---|---|---|---|
| Cephalothin | IM, IV | 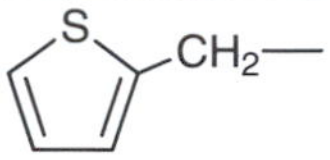 (thienyl)—$CH_2$— | 3' —$CH_2$—O—C(=O)—$CH_3$ |
| Cefazolin | IM, IV | 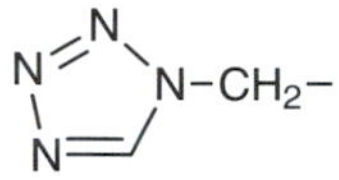 (tetrazolyl)N—$CH_2$— | 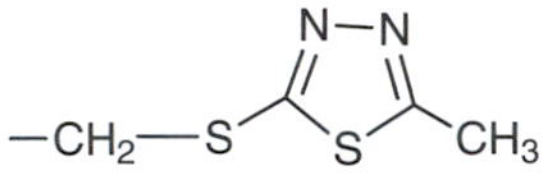 —$CH_2$—S—(thiadiazolyl)—$CH_3$ |
| Cephapirin | IM, IV | 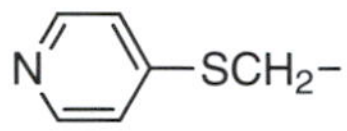 (pyridyl)—$SCH_2$— | —$CH_2$—O—C(=O)—$CH_3$ |
| Cephradine | Oral, IM, IV | 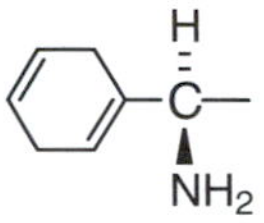 (cyclohexadienyl)—C(H)($NH_2$)— | —$CH_3$ |
| Cephalexin | Oral | 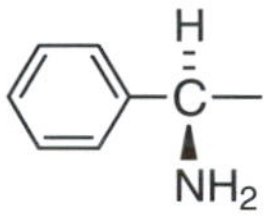 (phenyl)—C(H)($NH_2$)— | —$CH_3$ |
| Cefadroxil | Oral | 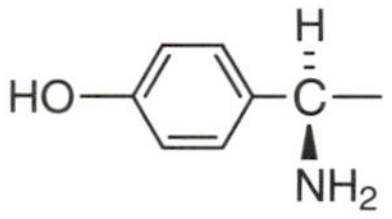 HO—(phenyl)—C(H)($NH_2$)— | —$CH_3$ |

**Figure 9.7** Structures and routes of administration of first-generation cephalosporins.

included in Fig. 9.8. Cefuroxime axetil is the acetoxyethyl ester [—$CH(CH)_3OC(O)CH_3$] of cefuroxime. Between 30 and 50% of an oral dose is absorbed from the gastrointestinal tract and is rapidly hydrolyzed by esterases to the parent compound cefuroxime.

## Cephamycins

The cephamycins are chemically related to cephalosporins, differing primarily by possessing a 7α-methoxy group ($R_3$), which enhances their stability to certain β-lactamases. The cephamycins were first reported by Nagarajan et al. (1971). Cephamycin C is structurally related to cephalosporin C. The aminoadipoyl group at C-7 is the same as that in cephalosporin C; however, cephamycins have a 7α-methoxy group and a 3-carbamoylmethyl group. Unlike penicillins G and V and cephalosporin C, the cephamycins are obtained from *Streptomyces* species, which are bacteria rather than fungi. Exchanging the 7α-adipoyl group of cephamycin C with a thienyl moiety produces cefoxitin, a cephamycin that is commercially available.

The clinically available second-generation cephamycins include cefoxitin, cefotetan, and cefmetazole. Their enhanced activity against gram-negative bacilli and rods results from the presence of the methoxy group at position 7. However, this structural modification accounts for the poor binding affinity of these drugs to the PBPs of gram-positive cocci. Therefore, the cephamycins have reduced activity against staphylococci and streptococci.

## Clinical Uses of Specific Second-Generation Cephalosporins

It is beyond the scope of this book to discuss in detail the pharmacology, absorption, metabolism, excretion, adverse reactions, toxicology, and clinical uses of any

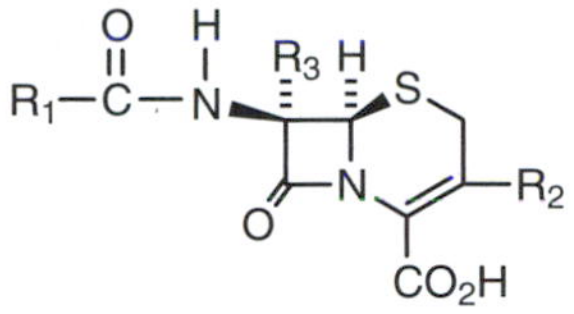

$R_1$–C(=O)–NH– (cephem nucleus with $R_3$, H, S, N, $R_2$, $CO_2H$)

| Generic name | Routes of administration | $R_1$ | $R_2$ | $R_3$ |
|---|---|---|---|---|
| **Cephalosporins** | | | | |
| Cefamandole nafate | IM, IV | 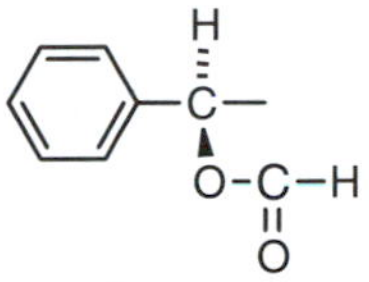 $C_6H_5$–CH(O–C(=O)–H)– | 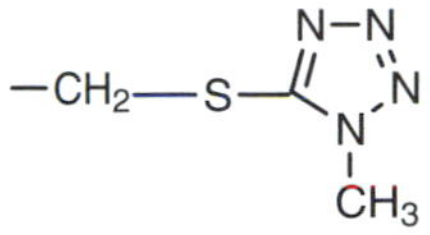 $-CH_2-S-$(1-methyltetrazol-5-yl; N–N, N, N, $CH_3$) | –H |
| Cefuroxime | IM, IV | 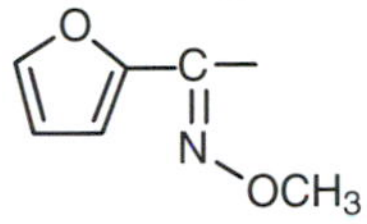 (2-furyl)–C(=N–$OCH_3$)– | $-CH_2-O-C(=O)-NH_2$ | –H |
| Cefuroxime axetil | Oral | | | |
| $-CO_2-CH(CH_3)-O-C(=O)-CH_3$ instead of $-CO_2H$ | | | | |
| Cefaclor | Oral | 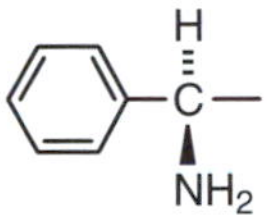 $C_6H_5$–CH($NH_2$)– | –Cl | –H |
| Cefprozil | Oral | 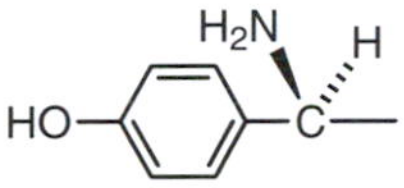 HO–$C_6H_4$–CH($NH_2$)– | 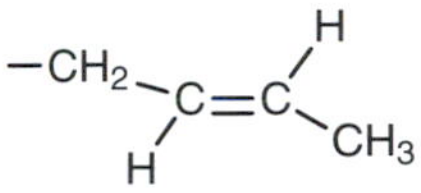 $-CH_2-CH=CH-CH_3$ | –H |
| Cefonicid | IM, IV | 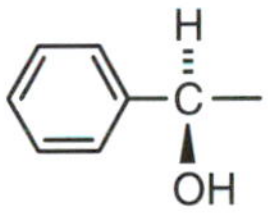 $C_6H_5$–CH(OH)– | 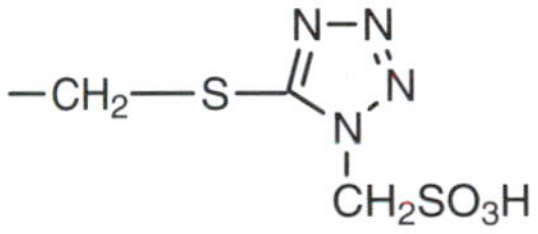 $-CH_2-S-$(tetrazol-5-yl; N–N, N, N, $CH_2SO_3H$) | –H |
| **Cephamycins** | | | | |
| Cefoxitin | IM, IV | 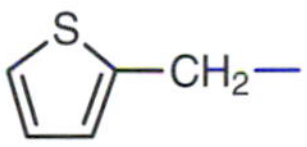 (2-thienyl)–$CH_2$– | $-CH_2-O-C(=O)-NH_2$ | $-OCH_3$ |
| Cefmetazole | IV | $N{\equiv}C-CH_2-S-CH_2-$ | 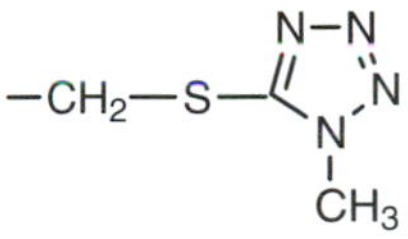 $-CH_2-S-$(1-methyltetrazol-5-yl; N–N, N, N, $CH_3$) | $-OCH_3$ |
| Cefotetan | IM, IV | 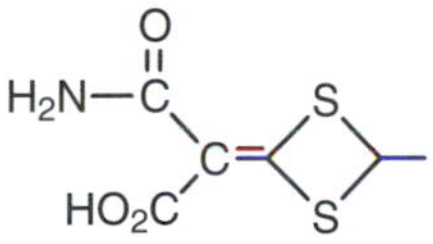 $H_2N-C(=O)$, $HO_2C$ >C=(1,3-dithietan-2-yl)– | 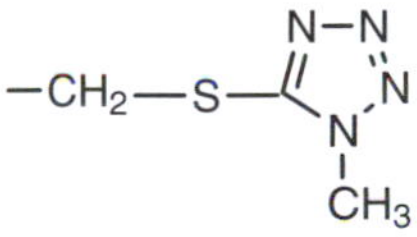 $-CH_2-S-$(1-methyltetrazol-5-yl; N–N, N, N, $CH_3$) | $-OCH_3$ |
| **Carbacephem** | | | | |
| Loracarbef | Oral | 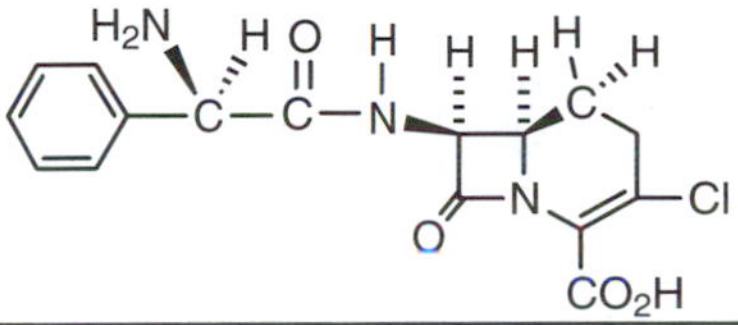 $C_6H_5$–CH($NH_2$)–C(=O)–NH–(carbacephem nucleus: H, H, H, H, H, N, O, Cl, $CO_2H$) | | |

**Figure 9.8** Structures and routes of administration of second-generation cephalosporins.

antibacterial agents. There are excellent reviews and texts to cover these topics.

## Modifications of the Cephem Nucleus at Position 3

The C-3 substituent influences the pharmacokinetic properties as well as the intrinsic antibacterial activity. Changes at the 3' position are thought to affect pharmacologic activity. In the second-generation cephalosporins, several functional groups or atoms have been introduced by different chemical modifications of 7-ACA or 7-ADCA. This is the case for cefprozil, which has a methylene-vinyl methyl group; cefaclor and loracarbef, which have a chlorine atom directly joined to the carbon atom at position 3; cefamandole, cefmetazole, and cefotetan, which have a methylene-methylthiotetrazole (MTZ) group; and cefuroxime, which has a methylene carbamate.

## Modifications of the Side Chain of Position 7 of the Cephem Nucleus

It is well established that the in vitro antibacterial activity of any particular cephalosporin is the result of a combination of its affinity for the molecular target, its ability to penetrate different biological barriers, and its ability to resist the attack of destructive enzymes. The nature and complexity of the biochemical target(s) for β-lactam antibiotics are fairly well established, and the mechanism of penetration is also understood to some extent. In cephalosporins, structural variations have been made in the substituents of the side chain and the C-3 position of the cephem nucleus. These chemical modifications have profound effects on the antibacterial activity and the pharmacokinetic and pharmacodynamic properties of these drugs.

One of the main deficiencies of the first-generation cephalosporins was their lack of resistance to β-lactamases. Compounds with improved β-lactamase resistance have one or more of the following characteristics. The first is a syn-oxime, found in cefuroxime as well as in other third- and fourth-generation cephalosporins. A marked stereochemical preference has been found in the syn isomer, where the methoxy (or alcoxy) and amido groups lie on the same side (*cis*) as each other. The anti-isomer is considerably less active (Fig. 9.8 and 9.9), with a formamido group (as in cefamandole) and a methoxy substituent on the side chain of the 7α position (as in cefoxitin, cefotetan, and cefmetazole). However, these various substituents have different effects, and increased resistance to one kind of β-lactamase does not indicate resistance to all kinds.

It is worth mentioning here that when a new chiral center, such as the quaternary carbon atom of the side chain, is created by the introduction of an amino or hydroxy group, the stereoisomer with the D-configuration (*R*) according to the Cahn-Ingold-Prelog nomenclature is preferred for the phenylglycine- and mandelic acid-derived side chains; in both cases, the other enantiomeric configuration for this carbon atom, the L-configuration (or *S*), is considerably less active.

Introduction of the aminothiazole group into the third- and fourth-generation cephalosporins imparts high potency against gram-negative bacteria.

## Modifications of the Cephem Nucleus: Carbacephems and Oxacephems

### LORACARBEF

Loracarbef is a semisynthetic carbacephem antibiotic that is a member of the second-generation cephalosporins. Structurally, it is a 1-carbacephem that is identical to the cephalosporin cefaclor, except that a methylene group replaces the sulfur atom at position 1 of the dihydrothiazine ring, resulting in a tetrahydropyridine ring. Although loracarbef is not a true cephalosporin, its antibacterial activity is closely related to that of cephalosporin antibiotics.

### MOXALACTAM

Moxalactam is a totally synthetic drug. Structurally it is not a cephalosporin since the dihydrothiazine sulfur atom has been isosterically replaced by an oxygen atom, so that it has a dihydroxazine ring instead of the dihydrothiazine ring common to cephalosporins and cephamycins. Thus, it is the only oxacephem used clinically and is generally viewed as a cephalosporin-type antibiotic. Moxalactam (Fig. 9.9) is a third-generation cephalosporin. The oxygen-for-sulfur replacement has increased its intrinsic activity severalfold. The 7α-methoxy group present in cephamycins, which is known to confer β-lactamase stability to the molecule, was incorporated into moxalactam. The effectiveness against gram-negative bacteria that is so notable in carbenicillin and ticarcillin (which is attributed to the α-COOH group on the acyl side chain) is also present in moxalactam. As in amoxicillin, the *p*-OH group on the benzene ring is incorporated in moxalactam to increase the level of drug in blood and to increase its half-life. Finally, the 1-methyl-tetrazolyl group at position 3, which has been useful in several third-generation cephalosporins, was also incorporated into the structure of this molecule. The net result is a highly effective agent that has been approved by the Food and Drug Administration (FDA) for clinical use in the United States.

# Meaning of the Term "Generation"

The cephalosporins are classified into four "generations" based on their activity against gram-negative bac-

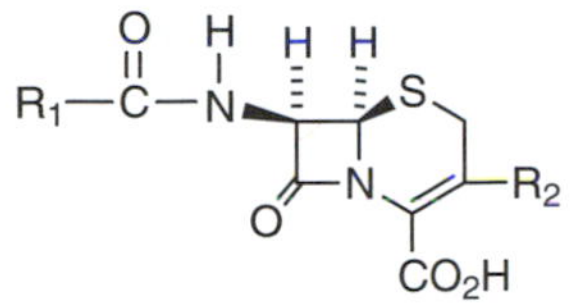

| Generic name | Routes of administration | $R_1$ | $R_2$ |
|---|---|---|---|
| Cefotaxime | IM, IV | 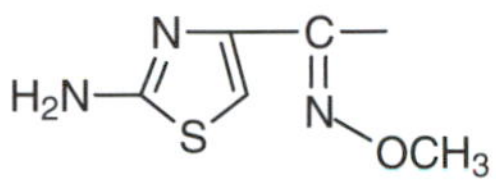 $H_2N$–thiazolyl–C(=N–$OCH_3$)– | 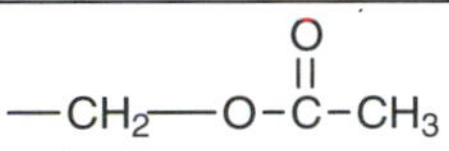 –$CH_2$–O–C(=O)–$CH_3$ |
| Ceftizoxime | IM, IV | 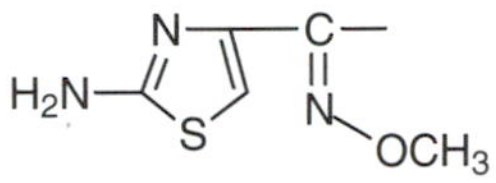 $H_2N$–thiazolyl–C(=N–$OCH_3$)– | –H |
| Ceftriaxone | IM, IV | 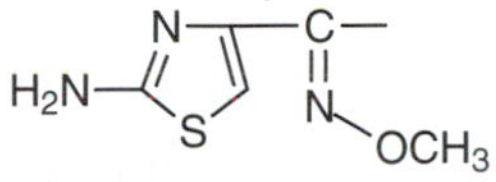 $H_2N$–thiazolyl–C(=N–$OCH_3$)– | 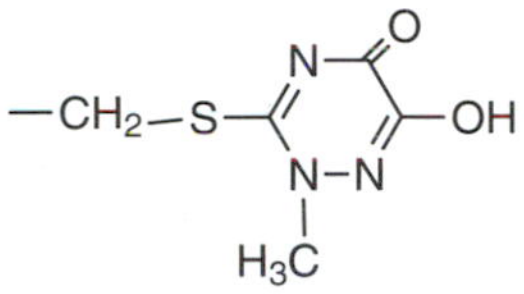 –$CH_2$–S–(triazinyl: N, =O, –OH, N–N, $H_3C$) |
| Ceftazidime | IM, IV | 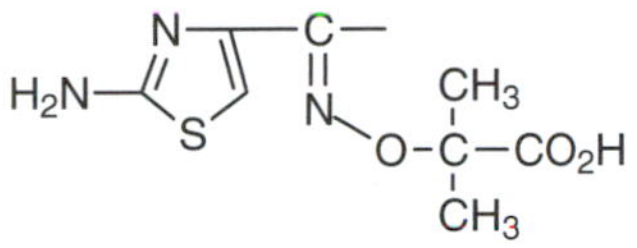 $H_2N$–thiazolyl–C(=N–O–C($CH_3$)($CH_3$)–$CO_2H$)– | 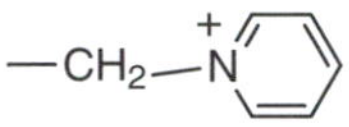 –$CH_2$–$N^+$(pyridinium) |
| Cefixime | Oral | 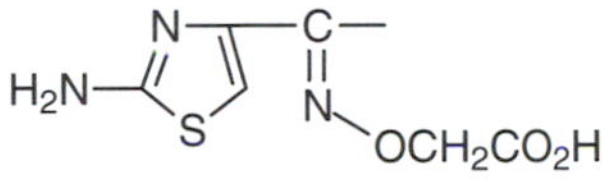 $H_2N$–thiazolyl–C(=N–$OCH_2CO_2H$)– | –HC=$CH_2$ |
| Ceftibuten | Oral | 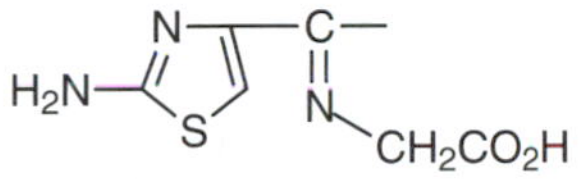 $H_2N$–thiazolyl–C(=N–$CH_2CO_2H$)– | –H |
| Cefpodoxime proxetil (an ester) -$CO_2CH(CH_3)OC(O)CH(CH_3)_2$ | Oral | 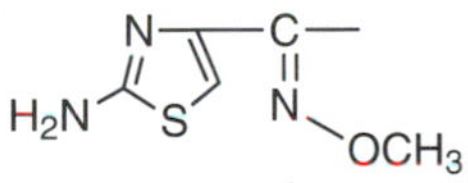 $H_2N$–thiazolyl–C(=N–$OCH_3$)– | –$CH_2$–$OCH_3$ |
| Cefdinir | Oral | 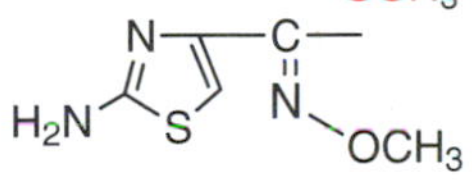 $H_2N$–thiazolyl–C(=N–$OCH_3$)– | 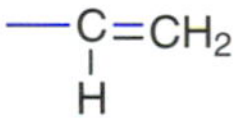 –C(H)=$CH_2$ |
| Cefoperazone | IM, IV | 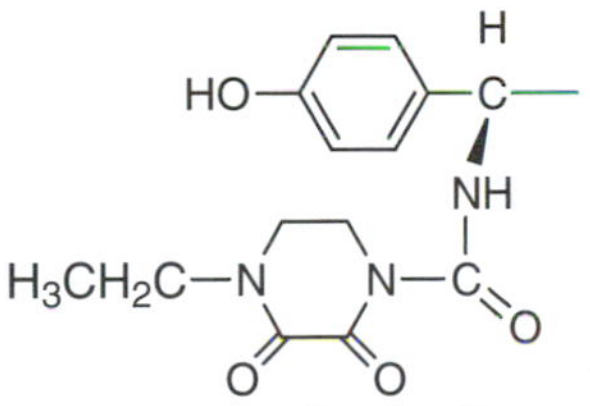 HO–phenyl–C(H)(–NH–C(=O)–N(piperazinedione, O, O)–N–$CH_2CH_3$)– ($H_3CH_2C$–N) | 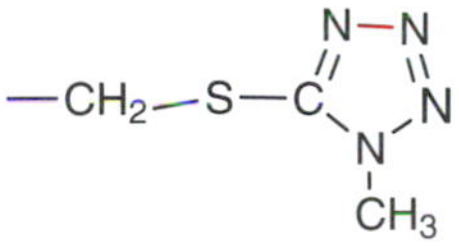 –$CH_2$–S–C(tetrazolyl: N–N, N, N–$CH_3$) |
| Cefditoren pivoxil (an ester) --$CO_2CH_2OC(O)C(CH_3)_3$ | Oral | 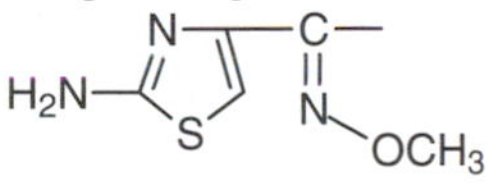 $H_2N$–thiazolyl–C(=N–$OCH_3$)– | 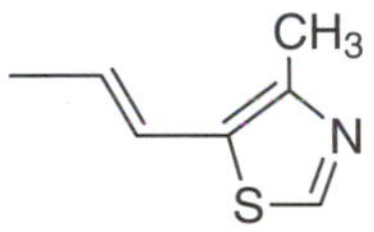 –CH=CH–(thiazolyl, $CH_3$, N, S) |
| **Oxacephem** | | | |
| Moxalactam | IM, IV | 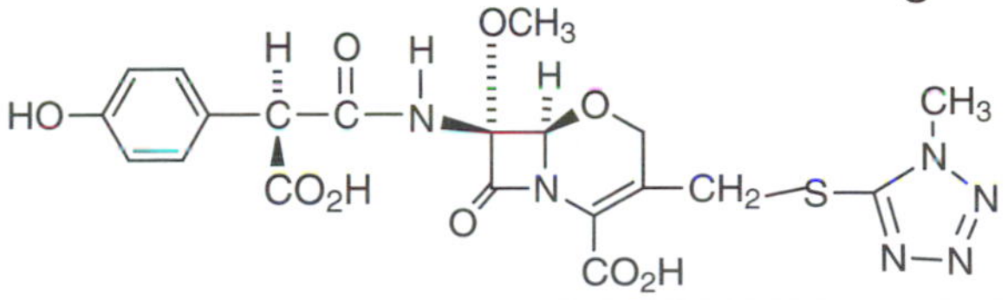 HO–phenyl–C(H)($CO_2H$)–C(=O)–N(H)–[$OCH_3$, H, O; N; $CO_2H$]–$CH_2$–S–(tetrazolyl: N–$CH_3$, N, N, N–N) | |

**Figure 9.9** Structures and routes of administration of third-generation cephalosporins.

teria. The first-generation cephalosporins have a spectrum that includes *E. coli, K. pneumoniae, P. mirabilis,* and most gram-positive cocci, although not enterococci or methicillin-resistant enterococci. The second-generation cephalosporins have broader in vitro activity against gram-negative bacteria. Cefuroxime, cefuroxime axetil, cefamandole, and cefaclor have more activity than the first-generation drugs against gram-negative bacilli and have good activity against most gram-positive bacteria. Cefonicid and ceforanide are less active against some of these organisms. Cefuroxime axetil is more active than cefaclor against *M. catarrhalis*. Cefoxitin, cefotetan, and cefmetazole are active against *B. fragilis*.

## Third-Generation Cephalosporins

The third-generation cephalosporins are more active than first- and second-generation drugs against gram-negative organisms. They are highly active against most strains of facultative enteric gram-negative bacilli and against *H. influenzae* and *N. gonorrhoeae*, including β-lactamase-producing strains. Ceftazidime and cefoperazone are active against most strains of *P. aeruginosa*.

These drugs were introduced primarily to overcome the shortcomings of the earlier cephalosporins. Thus, they have a broader antibacterial spectrum, increased resistance to β-lactamase inactivation, a better penetrability through the porins, particularly of gram-negative bacteria such as *P. aeruginosa*, and are less toxic, or some combination of these features. They have additional advantages in clinical situations that are not satisfactorily addressed by the earlier cephalosporins.

The third-generation cephalosporins are active against facultative gram-negative bacilli, including *E. coli, Klebsiella, P. mirabilis,* indole-positive *Proteus* spp., *Citrobacter* spp., *Providencia* spp., and *Serratia marcescens*. In addition, they have potent antimicrobial activity against *Streptococcus pneumoniae* (including strains with relative penicillin resistance), *S. pyogenes*, and other streptococci; except for ceftazidime, they are effective against *H. influenzae, N. meningitidis, N. gonorrhoeae,* and *M. catarrhalis*. Ceftazidime and cefoperazone are the only two cephalosporins of this generation with significant activity against *P. aeruginosa*. If these drugs are used to treat severe *Pseudomonas* infection outside the urinary tract, they should be used in conjunction with an aminoglycoside antibiotic active against this species.

Ceftizoxime is a third-generation cephalosporin that contains an aminothiazolyl-acetyl side chain with a syn-methoximino group (Fig. 9.9). The aminothiazolyl group enhances its antibacterial activity, particularly against members of the *Enterobacteriaceae*, and the syn-methoximino group impart stability against hydrolysis by many β-lactamases.

Cefotaxime was the first third-generation cephalosporin to be marketed in the United States. It is metabolized to desacetyl-cefotaxime, which, although less potent than the parent compound, has a longer half-life and may act synergistically with cefotaxime to allow effective dosing at 8-h intervals in patients with moderate infections. Cefotaxime reliably enters the cerebrospinal fluid.

Ceftazidime is an aminothiazolyl cephalosporin similar to cefotaxime, ceftizoxime, and ceftriaxone. However, it differs from these agents because it contains a 2-carboxy-2-oxypropaneimino group (Fig. 9.9) rather than a methoximino group. The presence of this group increases the activity of ceftazidime against *P. aeruginosa*. Ceftazidime penetrates into the cerebrospinal fluid and is the drug of choice in the treatment of meningitis due to *P. aeruginosa*. It is a weak inducer of β-lactamases.

Although ceftazidime is stable to the hydrolytic activity of most β-lactamases, resistance has emerged during therapy with this antibiotic, particularly among *Pseudomonas* and *Enterobacter* species, due to selection of the derepressed phenotype. Therefore, third-generation cephalosporins should be indicated with caution to treat infections produced by these organisms. Plasmid-mediated transferable β-lactamases, called **extended-spectrum β-lactamases** (**ESBL**) (e.g., mutants of TEM-1 and TEM-2 β-lactamases), confer resistance to ceftazidime. Ceftazidime has been used in combination with aminoglycoside antibiotics for presumptive and definitive therapy of bacterial infections in neutropenic patients.

Cefoperazone differs structurally from other third-generation cephalosporins. Similar to the penicillin piperacillin, cefoperazone contains a piperazine group in the side chain at position 7 with a β orientation at the cephem nucleus. This group enhances its antipseudomonal activity but decreases its stability to certain β-lactamases. Like cefamandole and moxalactam, it contains a methylthiotetrazole substituent at position 3 in the cephem nucleus that prevents metabolism of this antibiotic.

Ceftriaxone possesses antibacterial activity similar to that of cefotaxime. Ceftriaxone is 90% protein bound and has a half-life in serum of 8 h. Therefore, many serious infections can be treated with once-daily dosing, although meningitis should be treated every 12 h.

Cefpodoxime proxetil is an ester of cefpodoxime with increased oral absorption.

Cefditoren pivoxil (Spectracef [TAP Pharmaceutical Products]) is a new oral third-generation cephalosporin, approved by the FDA in August 2001, for treament of acute exacerbations of chronic bronchitis, pharyngitis, tonsilitis, and uncomplicated skin and soft tissue infections in adults and children aged 12 years or older. The drug has been used in Japan for 7 years. Cefditoren pivoxil is similar to cefdinir and cefpodoxime proxetil in its antibacterial activity. It is highly active against *S. pyogenes,* penicillin-susceptible pneumococci, and methicillin-susceptible *S. aureus.* Like all broad-spectrum oral cephalosporins, it is also active against *H. influenzae, M. catarrhalis, N. gonorrhoeae,* and many enteric gram-negative bacilli but is not active against methicillin-resistant *S. aureus* and *S. epidermidis, P. aeruginosa, Enterococcus* spp., many anaerobes, or the pathogens that cause atypical pneumonia (*Legionella, Mycoplasma,* and *Chlamydia*).

For additional information on the clinical uses of third-generation cephalosporins, the reader should consult the references quoted at the end of this chapter.

## Fourth-Generation Cephalosporins

Cefepime and cefpirome are the only two fourth-generation cephalosporins that have been approved for clinical use. They have been classified as fourth-generation cephalosporins because their potencies against members of the *Enterobacteriaceae* are higher than those of the earlier broad-spectrum cephalosporins. In addition, both antibiotics remain effective against β-lactamase-overproducing gram-negative strains resistant to other expanded-spectrum cephalosporins.

Cefepime hydrochloride (Maxipime [Bristol-Myers Squibb]) was approved by the FDA on 19 January 1996 for parenteral treatment of urinary tract infections and skin infections due to susceptible pathogens and for moderate to severe pneumonia caused by *S. pneumoniae* (pneumococci), *P. aeruginosa, Enterobacter* spp., or *K. pneumoniae.*

Cefpirome is not available in the United States but is available in the United Kingdom under the proprietary name of Cefrom (Hoechst-Roussel). It is used to treat patients with complicated urinary tract infections, community-acquired or nosocomial pneumonia, skin and soft tissue infections, and septicemia (usually due to gram-negative organisms).

The fourth-generation cephalosporins are characterized by a quaternary ammonium substituent at C-3' of the cephem nucleus; cefepime has a *N*-methylpyrrolidine group, and cefpirome has a cyclopentane-pyridinium group. These quaternary ammonium groups provide one positive charge in the molecule, and since there is one carboxy group at position 4, these cephalosporins are in a zwitterionic form (one positive and one negative charge in the same molecule, like the amino acids) (Fig. 9.10).

Permeability plays a key part in β-lactam antibiotic activity in gram-negative cells, since the β-lactam molecules have to penetrate the outer membrane before they reach their target at the surface of the bacterial inner membrane. Cefepime and cefpirome possess increased penetration through the porin channels, which is more rapid than that of ceftazidime. This has been attributed

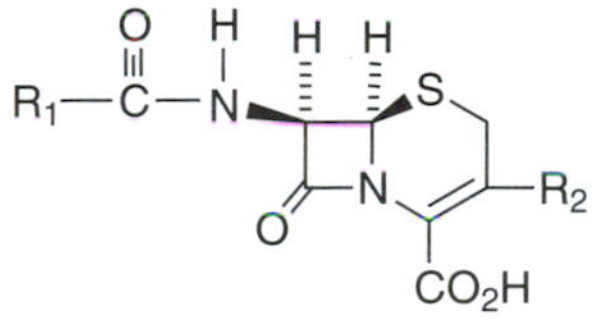

| Generic name | Routes of administration | $R_1$ | $R_2$ |
|---|---|---|---|
| Cefepime | IM, IV | 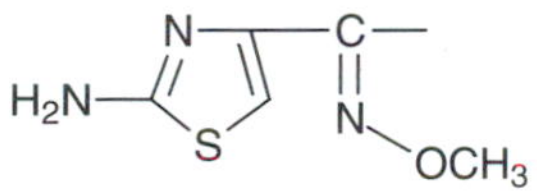 | 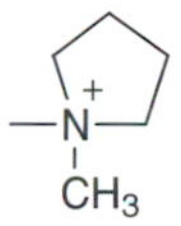 |
| Cefpirome | IM, IV | 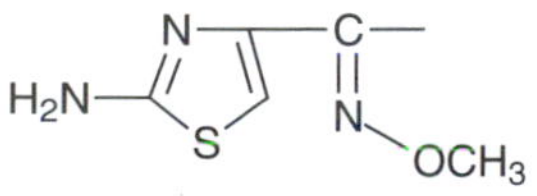 | 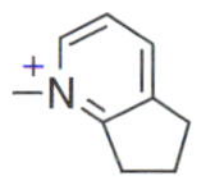 |

**Figure 9.10** Structures and routes of administration of fourth-generation cephalosporins.

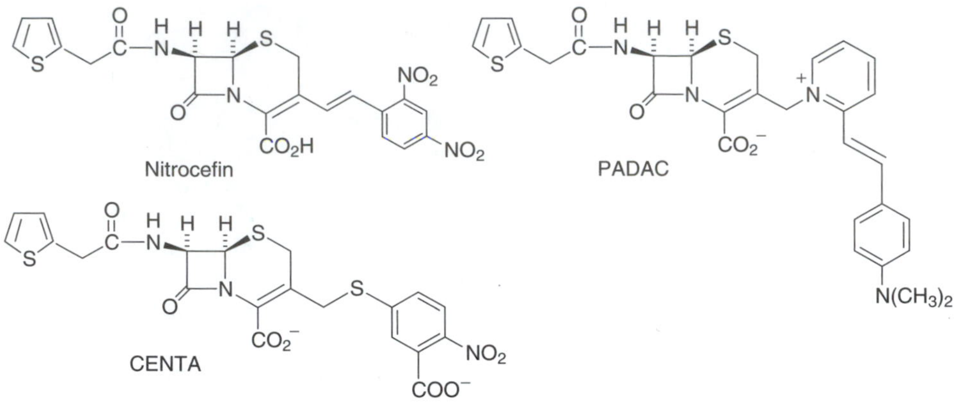

**Figure 9.11** Chemical structures of the chromogenic cephalosporins nitrocefin, PADAC, and CENTA.

**Table 9.2** Generic and common trade names of cephalosporins, the preparations available, and manufacturers in the United States

| Generic name | Trade name (preparation) | Manufacturer |
|---|---|---|
| First generation | | |
| Parenteral | | |
| Cefazolin sodium | Ancef (injectable) | GlaxoSmithKline |
| | Kefzol (injectable) | Eli Lilly |
| Oral | | |
| Cefadroxil monohydrate | Duricef (capsules, tablets) | Bristol-Myers Squibb |
| Cefdinir | Omnicef (capsules, oral suspension) | Abbott |
| Cephalexin | Keflex (pulvules, oral suspension) | Dista |
| Cephalexin USP | Apothecon (capsules) | |
| Cephalexin hydrochloride USP | Keftab (tablets) | Pharma |
| Cephradine | Velocef (capsules) | Apothecon (Bristol-Myers Squibb) |
| Second generation | | |
| Parenteral | | |
| Cefamandole nafate | Mandol (injectable) | Eli Lilly |
| Cefotetan | Cefotan (injectable) | AstraZeneca |
| Cefoxitin | Mexofin (injectable) | Merck |
| Cefuroxime | Zinacef (injectable) | GlaxoSmithKline |
| Oral | | |
| Cefaclor | Ceclor (pulvules, oral suspension) | Eli Lilly |
| | Ceclor (extended-release tablets) | Dura |
| | Cefaclor USP (capsules) | Apothecon |
| Cefprozil | Cefzil (tablets, oral suspension) | Bristol-Myers Squibb |
| Cefuroxime axetil | Ceftin (tablets, oral suspension) | GlaxoSmithKline |
| Loracarbef | Lorabid (oral suspension, pulvules) | Eli Lilly, Monarch |
| Third generation | | |
| Parenteral | | |
| Cefoperazone | Cefobid (injectable) | Pfizer |
| Cefotaxime sodium | Claforan (injectable) | Aventis |
| Ceftazidime L-arginine | Ceptaz (injectable) | GlaxoSmithKline |
| Ceftazidime sodium | Fortaz (injectable) | GlaxoSmithKline |
| Ceftazidime (inner salt) | Tazicef (injectable) | GlaxoSmithKline |
| | Tazidime (injectable) | Eli Lilly |
| Ceftizoxime sodium | Cefizox (injectable) | Fugisawa |
| Ceftriaxone | Rocephin (injectable) | Roche |
| Oral | | |
| Cefixime | Suprax (tablets) | Lederle |
| Cefpodoxime proxetil | Vantin (tablets, oral suspension) | Pharmacia, Upjohn |
| Ceftibuten dihydrate | Cedax (capsules, oral suspension) | D. J. Pharma |
| Cefditoren pivoxil | Spectracef (tablets) | TAP |
| Fourth generation | | |
| Parenteral | | |
| Cefepime | Maxipime (injectable) | Dura |

to the zwitterionic form (net charge of 0), whereas ceftazidime has a net charge of –1 because of the additional carboxylate group in the oximino substituent in the side chain of the cephem nucleus. Nikaido et al. have argued that the differential rate is a consequence of the repulsive effect of anions in the periplasm. However, the exact contribution of the outer membrane permeability to the activity of the fourth-generation cephalosporins is still under discussion.

These fourth-generation cephalosporins have broad antibacterial activity against gram-negative bacteria, including *P. aeruginosa,* and organisms that produce group C (class 1) AmpC β-lactamases.

Cefepime and cefpirome have high affinity for the PBPs of gram-positive bacteria and thus have excellent antibacterial activity against methicillin-susceptible *S. aureus, S. pneumoniae,* and other streptococci. However, they are not active against methicillin-resistant strains.

## Chromogenic Cephalosporins

The chromogenic cephalosporins nitrocefin, PADAC, and CENTA have no clinical uses but instead are used routinely as diagnostic agents for the detection of β-lactamase-producing strains of bacteria. Nitrocefin, PADAC, and CENTA (Fig. 9.11) are used to detect β-lactamases because these compounds undergo a color change when the β-lactam ring is hydrolyzed by β-lactamases. Nitrocefin is light yellow in solution ($\lambda_{max}$ = 386 nm). After treatment with β-lactamases, the color changes to a deep red ($\lambda_{max}$ = 482 nm). PADAC is purple, but treatment with β-lactamases releases the 3'-pyridinium group, with concomitant loss of the purple color.

Recently, Bebrone et al. (2001) reported that CENTA is readily hydrolyzed by β-lactamases of all classes except for the *Aeromonas hydrophila* metalloenzyme. These authors remark that although CENTA cannot practically be used for the detection of β-lactamase-producing strains on agar plates, it should be quite useful for kinetic studies and for detection of the enzymes in crude extracts and chromatographic fractions.

## Cephalosporin Preparations Commercially Available in the United States

Table 9.2 lists the cephalosporins marketed in the United States.

### REFERENCES AND FURTHER READING

#### General

**American Medical Association.** 1995. *Drug Evaluations Annual,* p. 1413–1480. American Medical Association, Chicago, Ill.

**Karchmer, A. W.** 2000. Cephalosporins, p. 274–291. *In* G. L. Mandell, J. E. Bennet, and R. Dolin (ed.), *Principles and Practice of Infectious Diseases,* 5th ed. Churchill Livingstone, Inc., Philadelphia, Pa.

**Kucers, A., S. M. Crowe, M. L. Grayson, and J. F. Hoy.** 1997. *The Use of Antibiotics. A Clinical Review of Antibacterial, Antifungal and Antiviral Drugs,* 5th ed. Butterworth-Heinemann, Oxford, United Kingdom.

**Morin, R. B., B. G. Jackson, R. A. Mueller, E. R. Lavagnino, W. B. Scanlon, and S. L. Andrews.** 1969. Chemistry of cephalosporin antibiotics. XV. Transformation of penicillin sulfoxide. A synthesis of cephalosporin compounds. *J. Am. Chem. Soc.* **91:**1401–1407.

**Nagarajan, R., L. D. Boeck, M. Gorman, R. L. Hamill, C. E. Higgens, M. M. Hoehn, W. M. Stark, and J. G. Whitney.** 1971. β-Lactam antibiotics from *Streptomyces. J. Am. Chem. Soc.* **93:**2308–2310.

**Petri, W. A.** 2001. Penicillins, cephalosporins, and other β-lactam antibiotics, p. 1189–1218. *In* J. G. Hardman and L. E. Limbird (ed.), *Goodman & Gilman's The Pharmacological Basis of Therapeutics,* 10th ed. McGraw-Hill, New York N.Y.

**Scholar, E. M., and W. B. Pratt.** 2000. *The Antimicrobial Drugs,* 2nd ed. Oxford University Press, New York, N.Y.

**Wise, R.** 1997. β-Lactams: cephalosporins, p. 202–255. *In* F. O'Grady, H. P. Lambert, R. G. Finch, and D. Greenwood (ed.), *Antibiotic and Chemotherapy,* 7th ed. Churchill Livingstone, Inc., New York, N.Y.

#### Cefepime

**Jauregui, L., and C. Black.** 1995. Clinical efficacy of cefepime in the therapy of bacterial infections. *Drugs Today* **31:** 241–271.

**Pechère, J. C.** 1995. Antibacterial activity of cefepime. *Drugs Today* **31:**227–233.

**Sanders, C. C.** 1995. Cefepime, the next generation: the reasons and rationale behind its development. *Drugs Today* **31:**221–225.

#### Cefuroxime Axetil

**Scott, L. J., D. Ormrod, and K. L. Goa.** 2001. Cefuroxime axetil: an updated review of its use in the management of bacterial infections. *Drugs* **61:**1455–1500.

#### Cefditoren Pivoxil

**Anonymous.** 2002. Cefditoren. A new oral cephalosporin. *Med. Lett.* **44:**5–6.

**Darkes, M. J., and G. L. Plossker.** 2002. Cefditoren pivoxil. *Drugs* **62:**319–336.

**TAP Pharmaceutical Products.** 2001. Prescribing information. http://www.tap.com.products/spectracef.

#### Antibody-Directed Enzyme-Prodrug Therapy (ADEPT)

**Hakimelahi, G. H., T. W. Ly, S. F. Yu, M. Zakerinia, A. Khalafi-Nezhad, M. N. Soltani, M. N. Gorgani, A. R. Chadegani, and A. A. Moosavi-Movahedi.** 2001. Design and synthesis of a cephalosporin retinoic acid prodrug activated by a monoclonal antibody-β-lactamase conjugate. *Bioorg. Med. Chem.* **9:**2139–2147.

**Hanessian, S., and J. Wang.** 1993. Design and synthesis of a cephalosporin-carboplatinum prodrug activatable by a β-lactamase. *Can. J. Chem.* **71:**896–906.

**Hay, M. P., and W. A. Denny.** 1996. Antibody-directed enzyme-prodrug therapy (ADEPT). *Drug Future* **21:**917–931.

**Jungheim, L. N., and T. A. Sheperd.** 1994. Design of antitumor prodrugs: substrates for antibody targeted enzymes. *Chem. Rev.* **94:**1553–1566.

**Melton, R. G., R. J. Knox, and T. A. Connors.** 1996. Antibody-directed enzyme-prodrug therapy (ADEPT). *Drug Future* **21:**167–181.

### Cephalosporin-3'-Quinolone with a Dual Mode of Action

**Albrecht, H. A., G. Beskid, K. K. Chan, J. G. Christenson, R. Cleeland, K. H. Deitcher, N. H. Georopapadakou, D. D. Keith, D. L. Pruess, J. Sepinwall, A. C. Specian, R. L. Then, M. Weigele, K. F. West, and R. Yang.** 1990. Cephalosporin 3'-quinolone esters with a dual mode of action. *J. Med. Chem.* **33:**77–86.

### Chemical Synthesis and Modifications of Cephalosporins

**Newall, C. E., and P. D. Hallam.** 1990. β-Lactam antibiotics: penicillins and cephalosporins, p. 609–653. *In* C. Hansch, P. G. Sammes, and J. B. Taylor (ed.), *Comprehensive Medicinal Chemistry,* vol. 2. Pergamon Press, Oxford, United Kingdom.

**Roberts, J.** 1993. Cephalosporins, p. 259–313. *In* M. Howe-Grant (ed.), *Chemotherapeutics and Disease Control.* John Wiley & Sons, Inc., New York, N.Y.

### Chromogenic Cephalosporins

**Bebrone, C., C. Moali, F. Mahy, S. Rival, J. D. Docquier, G. M. Rossolini, J. Fastrez, R. F. Pratt, J.-M. Frère, and M. Galleni.** 2001. CENTA as a chromogenic substrate for studying β-lactamases. *Antimicrob. Agents Chemother.* **45:**1868–1871.

chapter 10



# Inhibitors of Peptidoglycan Biosynthesis
## Carbapenems and Monobactams

### Carbapenems

The carbapenems are a subgroup of β-lactam antibiotics with a common carbapenem nucleus.

In the period up to 1970, most β-lactam research was concerned with the penicillin and cephalosporin subgroups of β-lactam antibiotics. Since then, new bicyclic and monocyclic β-lactam structures have been described. Carbapenems are broad-spectrum antibiotics with activity against gram-positive and gram-negative bacteria. It has been suggested that this activity is due to the combined effects of good penetration into gram-negative bacteria via a specific carbapenem uptake pathway involving the OprD porin channel, good stability to hydrolysis by most β-lactamases, and strong binding to essential penicillin-binding proteins (PBPs).

A member of a new group containing a 1-carbapenem ring system was reported in 1976. This compound was named thienamycin and was isolated from fermentation broths of the soil microorganism *Streptomyces cattleya*. It was a novel β-lactam antibiotic of particular interest because of its exceptional antibacterial potency and spectrum, including activity against *Pseudomonas* and β-lactamase-producing species. Thienamycin (Fig. 10.1) is a zwitterionic compound. Broad infrared absorption at ca. 1,580 $cm^{-1}$ is characteristic of a carboxylate anion, and a sharper band at 1,765 $cm^{-1}$ is reminiscent of the β-lactam carbonyl absorption of cephalosporins and cephamycins. The elemental composition $C_{11}H_{16}N_2O_4S$ was deduced from field desorption mass spectra of the antibiotic ($MH^+$, 273) and from high-resolution mass spectra of its *N*-acetylmethyl ester derivative. In 1978, the relative configuration of the three chiral centers C-5, C-6, and C-8 was determined by X-ray crystallographic analysis (Albers-Schönberg et al., 1978). Because of its instability, the monocrystals were obtained by crystallization of the *N*-acetylmethyl ester. The absolute stereochemistry at the three chiral centers was determined to be 5*R*, 6*S*, and 8*R* (according to the Cahn-Ingold-Prelog rules). Thienamycin has an aminoethylthio substituent

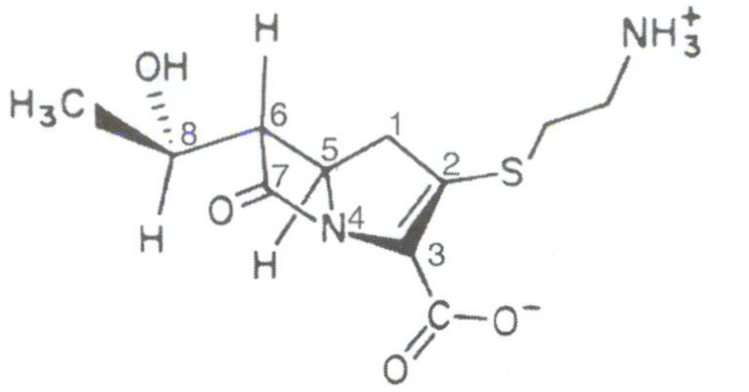

**Figure 10.1** Chemical structure of thienamycin.

at the C-2 and a 1-hydroxyethyl substituent at C-6 of the carbapenem nucleus. It should be noted that it has a *trans*-substitued configuration at C-5 and C-6 of the carbapenem nucleus. Carbapenems are characterized by the presence of a bicyclic ring system which possesses a double bond at C-2 and C-3 in the five-member ring.

Earlier studies at Merck Sharp & Dohme Research Laboratories (Nutley, N.J.) on thienamycin were aimed at improving or maintaining its antibacterial properties while imparting chemical stability to the nucleus. One of the many derivatives, produced by changes at the amino group as well as the hydroxy group, was *N*-formimidoylthienamycin. This structural change fulfilled the objective of producing a chemically stable compound, and the new antibiotic was more potent than the parent compound, thienamycin. This derivative is known as imipenem (Fig. 10.2). Like other β-lactam antibiotics, imipenem and thienamycin interfere with the biosynthesis of the peptidoglycan polymer. Imipenem has greatest affinity for PBP2, as demonstrated in *Escherichia coli* and *Pseudomonas aeruginosa*, although it also binds well to PBP1, PBP4, PBP5, and PBP6.

Subsequently, detailed metabolic studies established that imipenem is extensively metabolized by dehydropeptidase-1 (DHP-1), a dipeptidase enzyme located in the brush border of the kidneys. Later, it was found that imipenem could be protected from the deleterious action of DHP-1 by combining it with cilastatin (Fig. 10.2), an efficient inhibitor of this enzyme. At present, imipenem is used in the clinical setting in combination with cilastatin at a 1:1 ratio. Indeed, the dihydropeptide part of the imipenem molecule is not present in penicillins, cephalosporins, and cephamycins.

## Imipenem

Imipenem is the *N*-formimidoyl derivative of thienamycin. It was the first carbapenem marketed in the United States (although meropenem was already available elsewhere). Imipenem has excellent in vitro activity against aerobic gram-positive species, including staphylococci, streptococci, and pneumococci, but does not inhibit methicillin-resistant *Staphylococcus aureus* or coagulase-negative staphylococci. Among the enterococci (group D streptococci), *Enterococcus faecalis* is usually susceptible but *E. faecium* isolates are usually resistant.

Among gram-negative bacteria, the imipenem spectrum includes most members of the *Enterobacteriaceae*, against which its activity is comparable to that of aztreonam and the most potent third-generation cephalosporins, and strains that are resistant to antipseudomonal penicillins, aminoglycoside, and third-generation cephalosporins.

Imipenem has excellent activity against meningococci, gonococci, and *Haemophilus influenzae*, including β-lactamase-producing strains. It is very active against *Acinetobacter* strains, and its in vitro potency against *P. aeruginosa* equals or exceeds that of ceftazidime. *Burkholderia cepacia* and *Stenotrophomonas maltophilia* are resistant to this drug. Except for *Clostridium difficile*, imipenem is highly active against most clinically important anaerobic species; its potency is comparable to that of clindamycin and metronidazole.

### CLINICAL USES

Imipenem plus cilastatin has been effective in the treatment of a wide variety of serious nosocomial and com-

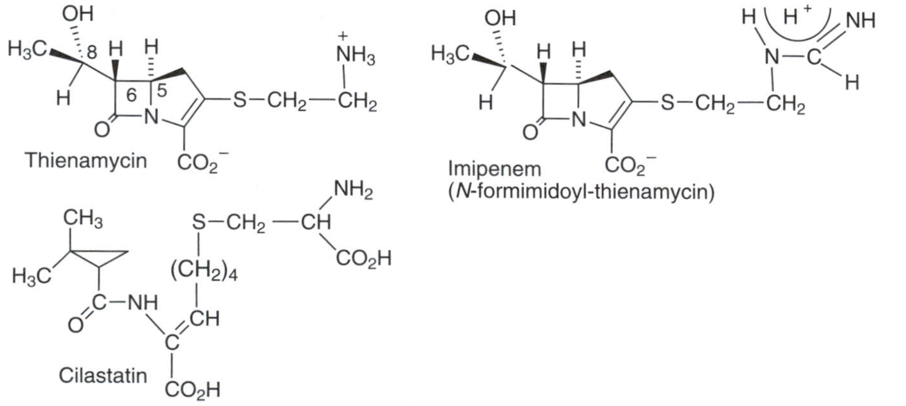

**Figure 10.2** Chemical structures of thienamycin, imipenem (*N*-formimidoylthienamycin), and cilastatin. The carbapenem antibiotics are represented in the zwitterionic form. In imipenem there are two basic functionalities, a secondary amine and the imino group; for that reason, the protons are placed covering these functionalities.

munity-acquired bacterial infections caused by gram-positive cocci, gram-negative aerobic bacilli, and anaerobes, including strains resistant to other antibacterial agents.

Imipenem, ceftazidime, and piperacillin associated with an aminoglycoside are frequently preferred for treatment of febrile neutropenic cancer patients and of infections caused by *P. aeruginosa*. For additional information on clinical uses, the reader is referred to specialized publications, some of which are listed at the end of this chapter.

## Meropenem

Meropenem (Fig. 10.3) is a new β-lactam antibiotic belonging to the carbapenem subgroup. It possesses a 1-β-methyl group on the carbapenem nucleus and a substituted 2' side chain.

The discovery of the excellent activity of the imipenem-cilastatin combination against a wide variety of aerobic and anaerobic microorganisms led to extensive work on the isolation of other naturally occurring carbapenems by pharmaceutical laboratories worldwide as well as on the production of synthetic compounds. Among these, attention was focused on 1-β-methylcarbapenems, owing to their high stability against DHP-1 without loss of antimicrobial activity.

Meropenem exhibits potent antibacterial activity against a wide range of gram-positive and gram-negative bacteria, and its stability against renal DHP-1 is significantly higher than that of the corresponding desmethylmeropenem compound. On 24 June 1996, the Food and Drug Administration conferred drug status on meropenem for use as an injectable drug. Currently, the drug is indicated in the United States to treat intra-abdominal infections in adults and children and bacterial meningitis in children aged 3 months and older. Outside the United States, meropenem intravenous is licensed for use in a wide range of infections.

In vitro, meropenem, like imipenem, is active against most clinically important gram-positive and gram-negative aerobes and anaerobes. Meropenem is slightly more active against gram-negative organisms, while imipenem is slightly more active against gram-positive bacteria. Meropenem, like imipenem, is highly active in vitro against *Listeria monocytogenes* and has moderate activity against β-lactamase-resistant pneumococci. Both imipenem and meropenem are inactive against methicillin-resistant staphylococci, *E. faecium*, or *S. maltophilia*. As with imipenem, emergence of meropenem-resistant *P. aeruginosa* has been reported.

### CLINICAL USES

Meropenem, like imipenem, could be used for the treatment of hospital-acquired infections that may be resistant to other antibiotics. It is indicated for treatment of intra-abdominal infections and bacterial meningitis.

## Ertapenem

Ertapenem (Invanz [Merck]) (Fig. 10.3) is a new β-lactam antibiotic belonging to the carbapenem subgroup. It possesses a 1-β-methyl group on the carbapenem nucleus. It was approved by the FDA in November 2001 for once-daily intravenous treatment of adult patients with the following moderate to severe infections caused by susceptible strains of the designated microorganisms:

- complicated intra-abdominal infections due to *E. coli*, *Clostridium clostridioforme*, *Eubacterium lentum*, *Peptostreptococcus* spp., *Bacteroides fragilis*, *Bacteroides distasonis*, *Bacteroides ovatum*, *Bacteroides thetaiotaomicron*, and *Bacteroides uniformis*
- complicated skin and skin structure infections due to *S. aureus* (methicillin-susceptible strains only), *Streptococcus pyogenes*, *E. coli*, or *Peptostreptococcus* species
- community-acquired pneumonia due to *Streptococcus pneumoniae* (penicillin-susceptible strains only), including cases with concurrent bacteremia, *H. influenzae* (beta-lactamase-negative strains only), or *Moraxella catarrhalis*

**Figure 10.3** Chemical structures of meropenem and ertapenem.

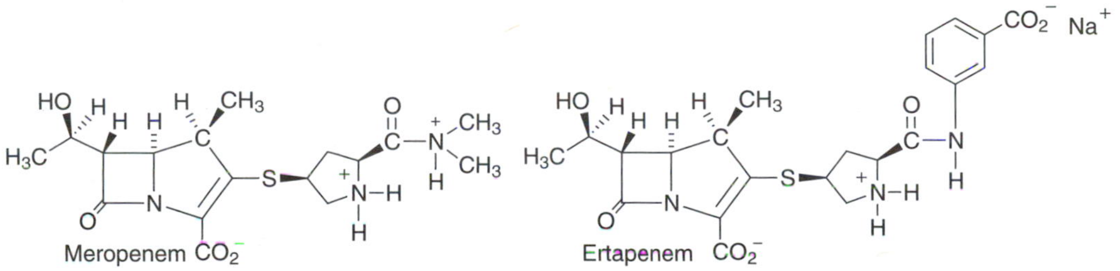

**Table 10.1** Generic and common trade names of carbapenems, the preparations available, and manufacturers in the United States

| Generic name | Trade name (preparation) | Manufacturer |
|---|---|---|
| Imipenem-cilastatin | Primaxin (for intramuscular and intravenous use) | Merck Sharp & Dohme |
| Meropenem | Merrem (for intravenous use only) | AstraZeneca |
| Ertapenem | Invanz | Merck & Co. |

- complicated urinary tract infections including pyelonephritis due to *E. coli*, including cases with concurrent bacteremia, or *Klebsiella pneumoniae*
- acute pelvic infections including postpartum endomyometritis, septic abortion, and postsurgical gynecologic infections due to *Streptococcus agalactiae, E. coli, B. fragilis, Porphyromonas asaccharolytica, Peptostreptococcus* spp., or *Prevotella bivia*

### Carbapenem Preparations Commercially Available in the United States

Table 10.1 lists the carbapenems marketed in the United States.

## Monobactams

The term "monobactam" was introduced by Sykes et al. in 1982 to denote monocyclic bacterially produced β-lactam antibiotics. Unlike the penicillins, cephalosporins, and carbapenems, which are bicyclic compounds, the monobactams are monocyclic β-lactam antibiotics.

The first monobactams to be discovered were naturally occurring compounds isolated from bacteria (e.g., *Gluconobacter, Acetobacter,* and *Chromobacterium* spp.), but they exhibited poor antibacterial activity. These naturally occurring monobactams are characterized by the 2-oxoazetidine-1-sulfonic acid moiety with an acyl side chain at the 3 position with the β-orientation and, for most monobactams, a 3-α-methoxy group (Fig. 10.4).

**Figure 10.4** General chemical structure of naturally occurring monobactams.

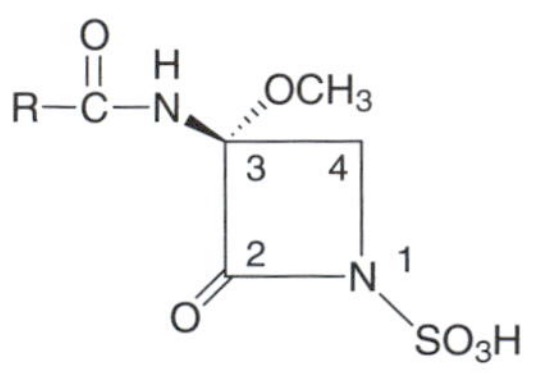

### Aztreonam

Aztreonam is the only member of the monobactam subgroup of β-lactam antibiotics to be approved for clinical uses in the United States. It is a totally synthetic monobactam. It was rationally designed because of the successful structure-activity relationships developed for third-generation cephalosporins, such as the aminothiazol-2-carboxy-2-oxypropaneimino group of ceftazidime. Adding a 4-α-methyl group as the substituent enhances the stability of the ring to β-lactamase catalytic hydrolysis (Fig. 10.5).

Aztreonam exhibits potent and specific activity against a wide range of β-lactamase-producing and nonproducing aerobic gram-negative bacteria, including *P. aeruginosa*, but it causes minimal inhibition of gram-positive bacteria and anaerobes. The exceptional activity of aztreonam appears to be its high and somewhat exclusive affinity for essential PBP3 of gram-negative bacteria. As a consequence of this interaction, bacterial cell division is inhibited, leading to cell death.

#### ANTIBACTERIAL ACTIVITY

The antibacterial spectrum of aztreonam is unique among β-lactam antibiotics. Aztreonam exhibits little or no activity against gram-positive bacteria, such as staphylococci and streptococci, or against anaerobic bacteria. The potent antibacterial activity of aztreonam is directed specifically against aerobic gram-negative bacteria. This activity varies from high against organisms such as *Neisseria* and *Haemophilus* spp. to intermediate against *P. aeruginosa* to poor against *Acinetobacter* spp.

#### CLINICAL USES

Aztreonam has been used for the treatment of a variety of infections such as cystitis, pyelonephritis, lower respiratory tract infections including pneumonia and bronchitis, septicemia, skin and skin structure infections, infections of postoperative wounds or ulcers and burns, intra-abdominal infections including peritonitis, and gynecological infections including endometritis and pelvis cellulitis caused by gram-negative bacteria.

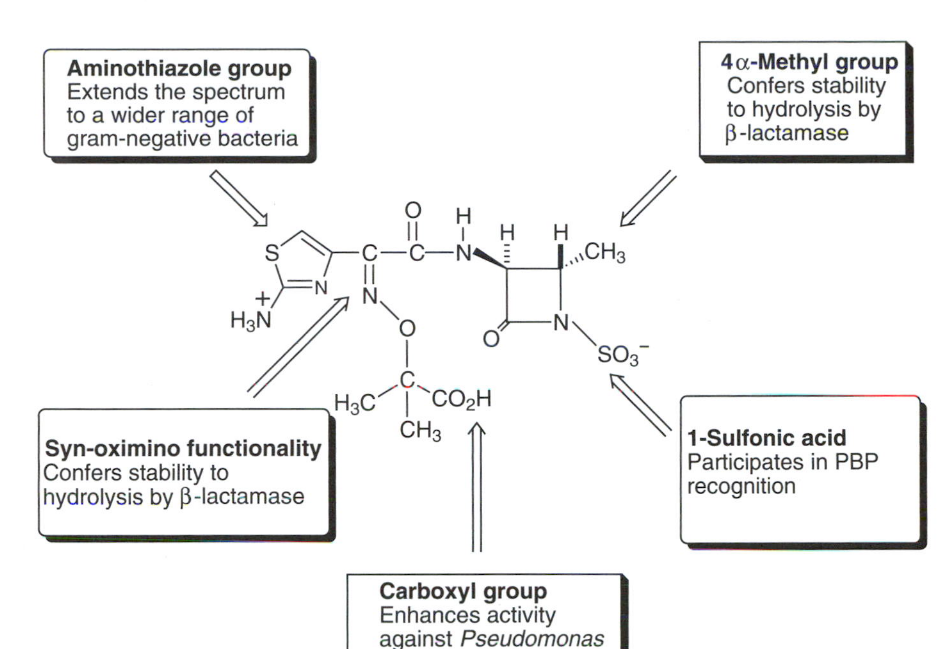


**Figure 10.5** Structure-activity relationships in the different groups or functionalities of the aztreonam molecule.

## Monobactam Preparation Commercially Available in the United States

Aztreonam (Azactam, manufactured by Dura [Bristol-Myers Squibb] for injection) is the only monobactam marketed in the United States.


## REFERENCES AND FURTHER READING

### General

**American Medical Association.** 1995. *Drug Evaluation Annual 1995,* p. 1488–1494. American Medical Association, Chicago, Ill.

**Bush, K.** 1997. Other β-lactams, p. 306–327. *In* F. O'Grady, H. P. Lambert, R. G. Finch, and D. Greenwood (ed.), *Antibiotic and Chemotherapy,* 7th ed. Churchill Livingstone, Inc., New York, N.Y.

**Chambers, H. F.** 2000. Other β-lactam antibiotics, p. 291–293. *In* G. L. Mandell, J. E. Bennet, and R. Dolin, (ed.), *Principles and Practice of Infectious Diseases,* 5th ed. Churchill Livingstone, Inc., Philadelphia, Pa.

**Kucers, A., S. M. Crowe, M. L. Grayson, and J. F. Hoy.** 1997. *The Use of Antibiotics. A Clinical Review of Antibacterial, Antifungal and Antiviral Drugs,* 5th ed. Butterworth-Heinemann, Oxford, United Kingdom.

**Scholar, E. M., and W. B. Pratt.** 2000. *The Antimicrobial Drugs,* 2nd ed. Oxford University Press, New York, N.Y.

### Chemistry of Carbapenems

**Albers-Schönberg, G., B. H. Arison, O. D. Hensens, J. Hirshfield, K. Hoogsteen, E. A. Kaczka, R. E. Rhodes, J. S. Kahan, F. M. Kahan, R. W. Ratcliffe, E. Walton, L. J. Ruswinkle, R. B. Morin, and B. G. Christensen.** 1978. Structure and absolute configuration of thienamycin. *J. Am. Chem. Soc.* **100:**6491–6499.

**Southgate, R., and N. F. Osborne.** 1993. Carbapenems and penems, p. 232–258. *In* M. Howe-Grant (ed.), *Chemotherapeutics and Disease Control.* John Wiley & Sons, Inc., New York, N.Y.

### Ertapenem

**Merck & Co.** 2001. *Invanz. Prescribing Information.* Merck & Co., Rahway, N.J.

### Chemistry of Monobactams

**Cimarusti, C. M., and R. B. Sykes.** 1984. Monocyclic β-lactam antibiotics. *Med. Res. Rev.* **4:**1–24.

**Floyd, D. M., A. W. Fritz, and C. M. Cimarusti.** 1982. Monobactams. Stereospecific synthesis of (*S*)-3-amino-2-oxoazetidine-1-sulfonic acid. *J. Org. Chem.* **47:**176–178.

**Lindener, K. R., D. P. Bonner, and W. H. Koster.** 1993. Monobactams, p. 338–360. *In* M. Howe-Grant (ed.), *Chemotherapeutics and Disease Control.* John Wiley & Sons, Inc., New York, N.Y.

chapter 11



# β-Lactam Compounds as β-Lactamase Inhibitors

As previously stated, the first description of an enzyme of bacterial origin that destroyed penicillins (which were later determined to be penicillins G and V) was reported in 1940 by Edward Abraham and Ernst Chain, who carried out their investigations in the Sir William Dunn School of Pathology of the University of Oxford.

The number, diversity, and substrate specificity of enzymes that are now called β-lactamases have increased considerably since this discovery over 60 years ago. Therefore, it has been necessary to develop new semisynthetic antibiotics resistant to β-lactamases. In that way, chemical modifications were made that gave rise, for example, to syn-oximinocephalosporins, the so-called extended-spectrum cephalosporins. Since 1976, investigations in many pharmaceutical and academic laboratories were also seeking natural or synthetic compounds that could act as irreversible inhibitors of those enzymes (for a description of reversible inhibition mechanisms, see appendix C).

An understanding of the mechanism of action (against substrates) and the inactivation (by inhibitors) of native and mutant β-lactamases, together with the identification of positions where amino acid substitutions occurred (primary structure of the enzyme) and the folding properties and flexibility of these enzymes (tertiary structure of the enzyme), is important for the design of new effective β-lactam antibiotics as well as inhibitors of these enzymes.

## Natural Irreversible Inhibitors

### Clavulanic Acid

At the SmithKline Beecham pharmaceutical laboratory in the United Kingdom, a screening program for substances that were produced by microorganisms and that presented β-lactamase inhibitory activity was implemented. This program was successful, and clavulanic acid was isolated for the first time in 1976 (Fig. 11.1). Clavulanic acid is a β-lactam

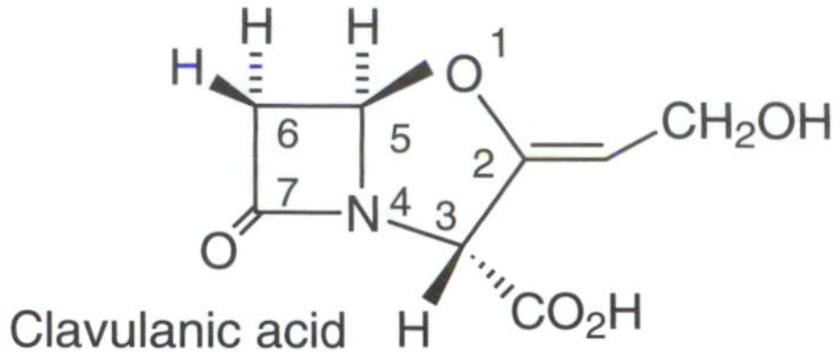

**Figure 11.1** Structure of clavulanic acid.

compound isolated from the fermentation of *Streptomyces clavuligerus*. This metabolite presented very little antibiotic activity but was shown to have potent properties as an irreversible inhibitor of a wide variety of β-lactamases. This discovery was a landmark in the investigations of β-lactam compounds, and its importance lay in the possibility of developing combinations, in certain proportions, with a β-lactam antibiotic sensitive to destruction by β-lactamases.

The chemical structure of this new bicyclic β-lactam derivative consists of a β-lactam ring fused to an oxazolidine. Clavulanic acid lacks the side chain bound to the C-6 atom that is characteristic of β-lactam antibiotics, and it also has an exocyclic substituent attached to the C-2 atom (a hydroxyethylidine group) instead of the gem-dimethyl group present in all penicillins. The stereochemistry at C-3 and C-5 atoms is the same as in penicillins.

Table 11.1 lists the characteristics of a selected group of β-lactamases, along with the activities of several β-lactam antibiotics against them, expressed relative to the value for benzylpenicillin hydrolysis (set at 100), and the 50% inhibitory concentrations of clavulanic acid.

## MECHANISTIC ASPECTS OF THE CLAVULANIC ACID INHIBITION OF CLASS A β-LACTAMASES

Studies of the mechanism of inactivation of class A β-lactamases have been carried out in many laboratories. Various pathways have been proposed for the inhibition of class A β-lactamases by clavulanic acid (in some publications it is designated clavulanate, because the commercial form of the inhibitor is the lithium or sodium salt). An inactivation mechanism for TEM-2 β-lactamase by clavulanic acid has been proposed (see Chen and Herzberg, 1992, and references therein). First, the active site serine 70 residue reacts with the carbonyl carbon atom of the inhibitor molecule (structure 1) to form a tetrahedral intermediate (not represented in Fig. 11.2),

**Table 11.1** Antibiotic activities and inhibition of selected β-lactamases by clavulanic acid[a]

| β-Lactamase | Activity of[b]: Ampicillin | Cephaloridine | Cefotaxime | Ceftazidime | Imipenem | Clavulanic acid inhibition (μM)[c] | Ambler's classification | Bush-Jacoby-Medeiros classification |
|---|---|---|---|---|---|---|---|---|
| *E. cloacae* P99 | 1.3 | 6.700 | >7 | <0.7 | <0.7 | >10 | C | 1 |
| *E. coli* Amp | 9.1 | 290 | 0.37 | ND[e] | <0.03 | 190 | C | 1 |
| *S. aureus* PC1 | 180 | 1.1 | ND | ND | ND | 0.03 | A | 2a |
| TEM-1 | 110 | 140 | 0.07 | 0.01 | <0.01 | 0.09 | A | 2b |
| TEM-3 | 110 | 120 | 170 | 8.3 | 0.01 | 0.03 | A | 2be |
| TEM-10 | 130 | 77 | 1.6 | 68 | <0.02 | 0.03 | A | 2be |
| SHV-1 | 150 | 48 | 0.18 | 0.02 | <0.01 | 0.03 | A | 2b |
| SHV-4 | 200 | 320 | 120 | 52 | <0.01 | 0.03 | A | 2be |
| *Klebsiella* K1 | 61 | 36 | 1.8 | 0.01 | <0.01 | 0.007 | A | 2be |
| TEM-31 (IRT-1) | 250 | 13 | <1 | <1 | <1 | 9.4 | A | 2br |
| PSE-4 | 88 | 40 | 0.02 | 0.02 | 0.01 | 0.15 | A | 2c |
| OXA-10 | 270 | 32 | 1 | 0.12 | 0.05 | 0.81 | D | 2d |
| *P. vulgaris* | 100 | 3,000 | 13 | 0.17 | ND | 0.35 | A | 2e |
| *S. marcescens* Sme-1 | 1,300 | 1,200 | 18 | ND | 310 | 0.28 | A | 2f |
| *B. fragilis* CcrA[d] | 98 | 22 | 51 | 68 | 100 | >500 | B | 3 |
| *P. paucimobilis* | 62 | 3.9 | ND | <0.1 | <0.1 | 19 | ND | 4 |

[a] Reprinted from F. O'Grady, H. P. Lambert, R. G. Finch, and D. Greenwood (ed.), *Antibiotic and Chemotherapy*, 7th ed. (Churchill Livingstone, Inc., New York, N.Y., 1997), with permission from the publisher.

[b] Values are expressed relative to benzylpenicillin hydrolysis (set at 100).

[c] Values are the $IC_{50}$ of clavulanic acid.

[d] Inhibited by EDTA.

[e] ND, not determined.

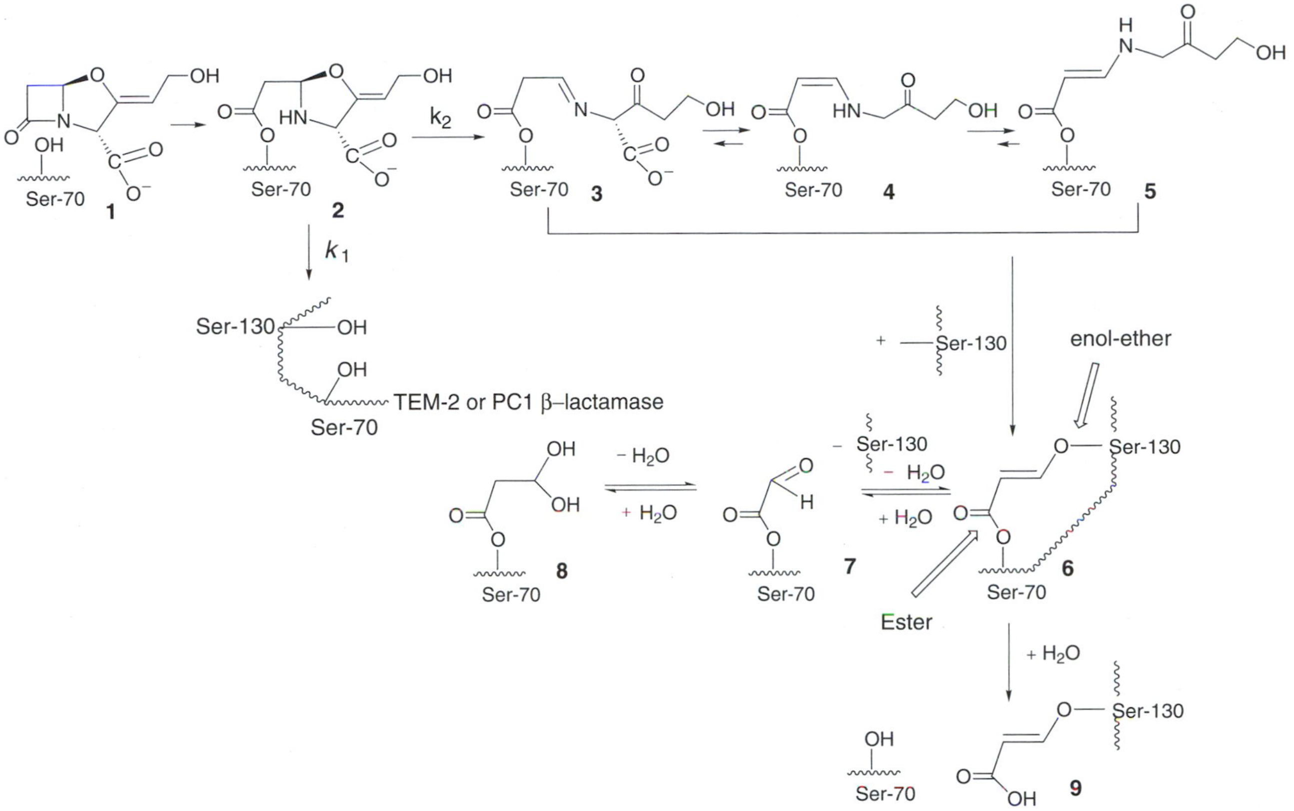

**Figure 11.2** Proposed mechanism for the inactivation of the TEM-2 or PC1 β-lactamase by clavulanic acid. For clarity, the residues Ser70 and Ser130 are represented separately, but they belong to the same molecule. Amino acid numbering is that of Ambler et al. Reprinted from C. C. H. Chen and O. Herzberg, *J. Mol. Biol.* **224:**1103–1113, 1992, with permission from the publisher.

and then the C—N bond in the β-lactam ring is cleaved, resulting in the opening of the β-lactam ring and the formation of the acyl-enzyme complex (structure 2). All mechanisms proposed share these two steps.

Note that the difference between a substrate and an inhibitor depends on the relative rates of deacylation by hydrolysis of this acyl-enzyme complex ($k_1$ in Fig. 11.2), which leads to the free enzyme, and the ring-opening step ($k_2$ in Fig. 11.2), which leads to the imine. For β-lactamase inhibitors, $k_2 >> k_1$.

Opening of the oxazolidine ring gives rise to the imino-keto derivative (step 3). Imtiaz et al. (1993) suggested that the ring opening is accelerated by protonation of an oxonium intermediate mediated by a water molecule bound to the guanidinium group of Arg-244. It has been proposed that intermediate 3 is tautomerized to the *cis*- and *trans*-enamines (not shown in Fig. 11.2) and that this step is followed by a decarboxylation reaction of these β-keto acids leading to their corresponding decarboxylated products (structures 4 and 5). These transient forms undergo an intramolecular nucleophilic displacement of the imine or enamine moiety by the hydroxyl group of Ser130 to form a cross-linked vinyl ether. This is known as a "double-hit" mechanism. Brown et al. (1996) showed that the reaction leading to irreversible inhibition involved cross-linking to a Ser130 residue in the active site. In this way, the enzyme is irreversibly inhibited by the formation of two covalent bonds. One bond is formed between an oxygen atom of the Ser70 at the active site and the residual structure of the inhibitor forming an ester functionality, and the other is formed between an oxygen atom of Ser130 and a three-carbon atom fragment of the original clavulanic acid molecule, forming an enol-ether functional group. Before these findings, it was suggested that an ϵ-amino group of a lysine residue was the nucleophile that reacts with the transient structures 5 or 6, giving rise to an enzyme linked via two covalent bonds (to an oxygen atom from Ser70 and a nitrogen atom from a lysine residue) to the same residual fragment of clavulanic acid.

Note that this double-hit mechanism is similar to that proposed by Navia et al. (1989), on the basis of the crystal structure of the porcine pancreatic elastase-inhibitor complex, for the inactivation of porcine pancreatic elastase (a serine protease) by *tert*-butyl-7α-chlorocephalosporinate sulfone.

The TEM-2 and PC1 β-lactamases, like perhaps all class A β-lactamases, are bifunctional in the presence of clavulanic acid. Approximately 115 molecules of clavulanic acid per enzyme molecule are destroyed before the β-lactamase is irreversibly inactivated (this is the **turnover number**).

The mechanism of β-lactamase inactivation by clavulanic acid has been studied extensively. Crystallographic studies performed on the clavulanic acid-PC1 covalent adduct revealed the acyl-enzyme complex to consist of a clavulanic acid fragment covalently bound to the active Ser70, as either the *cis* or *trans* decarboxylated enamines (Chen and Herzberg, 1992). Electrospray ionization mass spectrometry studies by Brown et al. (1996) provided direct evidence for the complex between clavulanic acid and TEM-2 β-lactamase. Clavulanic acid fragments with masses of 70 and 88 Da were attached to the peptide fragment containing Ser70, and a second 70-Da mass fragment, identified as a β-linked acrylate, was localized to a peptide containing Ser130.

TEM mutants of the TEM-1 and TEM-2 enzymes showed six positions prevalent for substitutions: Met69, Trp165, Met182, Arg244, Arg275, and Asn276. However, three of these sites are of major importance for resistance to clavulanic acid. These sites are at positions 69, 244, and 276. Additional information regarding the amino acid substitution pattern for the 88 mutants tabulated so far by Bush and Jacoby can be found through the Internet (http://www.lahey.org/studies/webt.htm), and a detailed account of the many replacements of specific amino acids on TEM β-lactamases can be found in the excellent publication by Knox et al. (1995) as well as in some recent publications discussing this aspect of extended-spectrum and inhibitor-resistant β-lactamase mutants.

## Semisynthetic Irreversible Inhibitors

### Sulbactam

In 1978, at Pfizer Laboratories in the United States, the penicillanic acid sulfone (sulbactam) (Fig. 11.3) was obtained by semisynthesis. Sulbactam has very good β-lactamase-inhibitory activity, mainly against class A β-lactamases according to Ambler's classification. Like clavulanic acid, sulbactam has very little antibiotic activity but high β-lactamase affinity. This high affinity allows it to be used in combination with penicillins or cephalosporins susceptible to β-lactamases.

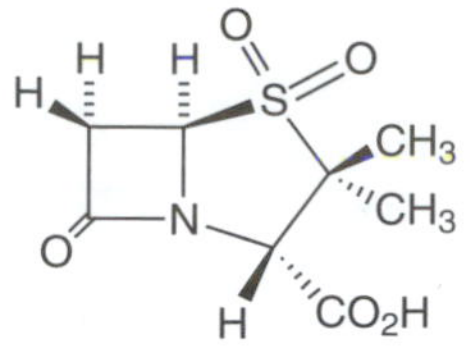

Penicillanic acid sulfone (sulbactam)

**Figure 11.3** Chemical structure of penicillanic acid sulfone (sulbactam).

MECHANISTIC ASPECTS OF THE SULBACTAM INHIBITION OF CLASS A β-LACTAMASES

Knowles et al. (1981), working at Harvard University, proposed a mechanism for the inactivation of *Escherichia coli* TEM-1 (RTEM) β-lactamase by sulbactam. These scientists identified three different processes. First, sulbactam is a substrate in the sense that the enzyme catalyzes the hydrolytic opening of the β-lactam ring. Second, it is an enzyme inhibitor; at pH 8, about 10 molecules of inhibitor are consumed per enzyme molecule. Thus, interaction of the enzyme with sulbactam gives rise to irreversible inhibition.

In Fig. 11.4, this process is schematically represented. Initially, the Ser70 hydroxyl group (nucleophile) in the β-lactamase attacks the β-lactam carbonyl group to form the tetrahedral intermediate A, which gives rise to the formation of the penicilloyl-enzyme B via the opening of the thiazolidine ring: here, the sulfone group acts as a strong electron-withdrawing group. If the acyl-enzyme B follows the normal deacylation course, the imine C is formed and is spontaneously hydrolyzed to give the malonic semialdehyde and penicillamine sulfinate. Intermediate B can also give rise to a second reaction that leads to a transient enzyme inhibition (resulting in enamine structure D). Intermediate B undergoes another reaction leading to the irreversibly inactivated enzyme (structure E). It was found that approximately 7,000 sulbactam molecules are hydrolyzed before the enzyme is completely inactivated. In structure E, the β-lactamase has remained covalently bound, through an oxygen atom and an X atom (see below), to a three-carbon fragment of the inhibitor's original structure that, during the inhibition process, undergoes transformations that lead to the destruction of its original structure. Knowles et al. (1981) suggested in their original report that X is an ϵ-amino group of a lysine residue; Imtiaz et al. (1994) postulated that the nucleophile X has to be

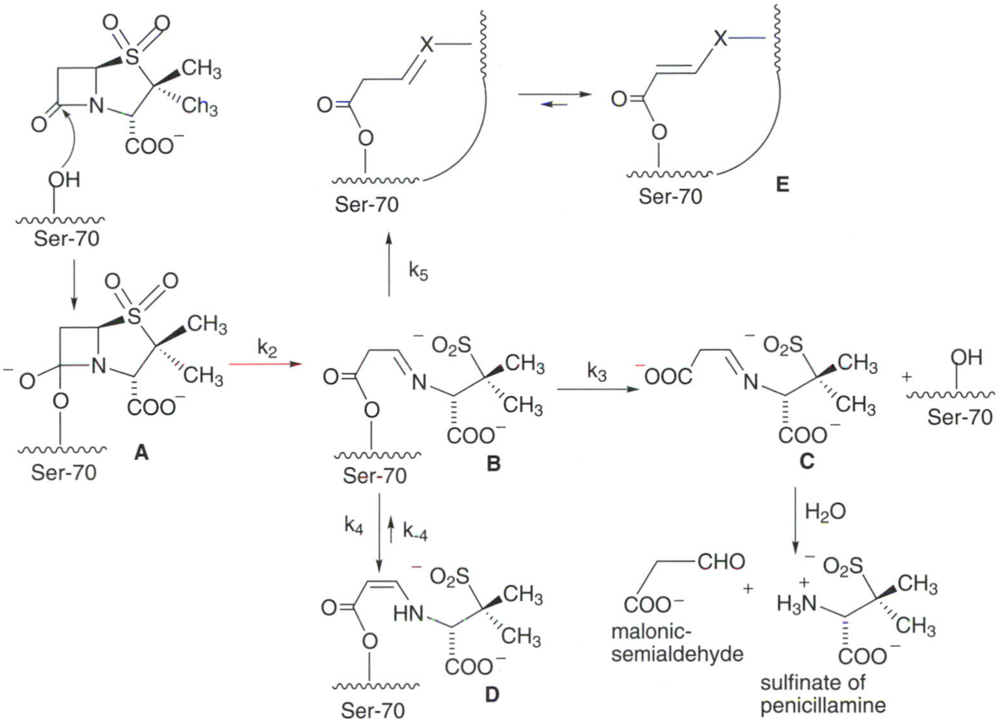

**Figure 11.4** Proposed mechanism for the inactivation of the TEM-1 β-lactamase of *E. coli* by sulbactam. See the text for an explanation and the identity of X. Reprinted from J. R. Knowles, *Acc. Chem. Res.* **18:**97–104, 1985, with permission from the publisher.

Lys273 or Ser130. Yang et al. (2000) demonstrated that with tazobactam, X is an enol-ether from Ser130.

## Tazobactam

In 1987, Micetich et al. reported the semisynthesis of tazobactam (Fig. 11.5), an inhibitor with good affinity to most class A and some class D β-lactamases and with moderate affinity to class C β-lactamases, particularly from *Morganella morganii.* Sulbactam and tazobactam are prepared synthetically from 6-β-aminopenicillanic acid (6-APA).

Clavulanic acid, sulbactam, and tazobactam are mechanism-based inhibitors that are also called suicide substrates because they are recognized as a substrate by β-lactamase, which is then covalently bound to them, subsequently producing chemical modifications (rearrangements and bond cleavages) in the inhibitor so that these modified structures are irreversibly bound to the enzyme.

### MECHANISTIC ASPECTS OF THE TAZOBACTAM INHIBITION OF CLASS A β-LACTAMASES

A mechanism of tazobactam inactivation of β-lactamases from all major structural classes, based on UV spectrometric assays, was proposed by Bush et al. (1983). The proposed mechanism for the interaction between tazobactam and class A β-lactamases is similar to the mechanism proposed for sulbactam. However, Yang et al. (2000), working at Wyeth-Ayerst Research, proposed a mechanism for the inhibition of the class A β-lactamases PC1 and TEM-1 by tazobactam based on an observation of reaction products by electrospray ionization mass spectrometry (Fig. 11.6). In this scheme, initial acylation of Ser70 by tazobactam and opening of the lactam ring are followed by three different events: (i) the rapid decomposition of tazobactam to form propiolylated enzyme 11 (compound numbers are those in

**Figure 11.5** Chemical structure of 2-β-triazolylmethyl penicillanic acid sulfone (tazobactam).

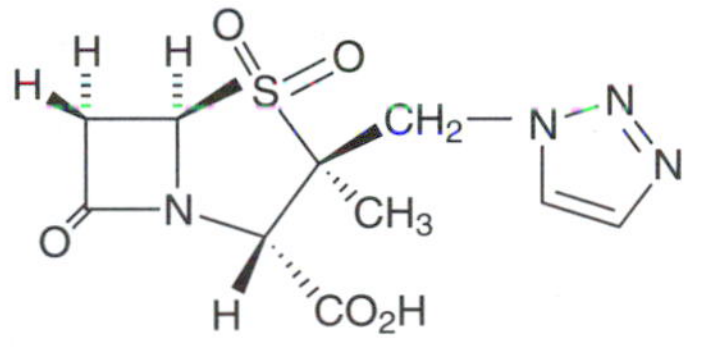

2-β-triazolylmethyl penicillanic acid sulfone (tazobactam)

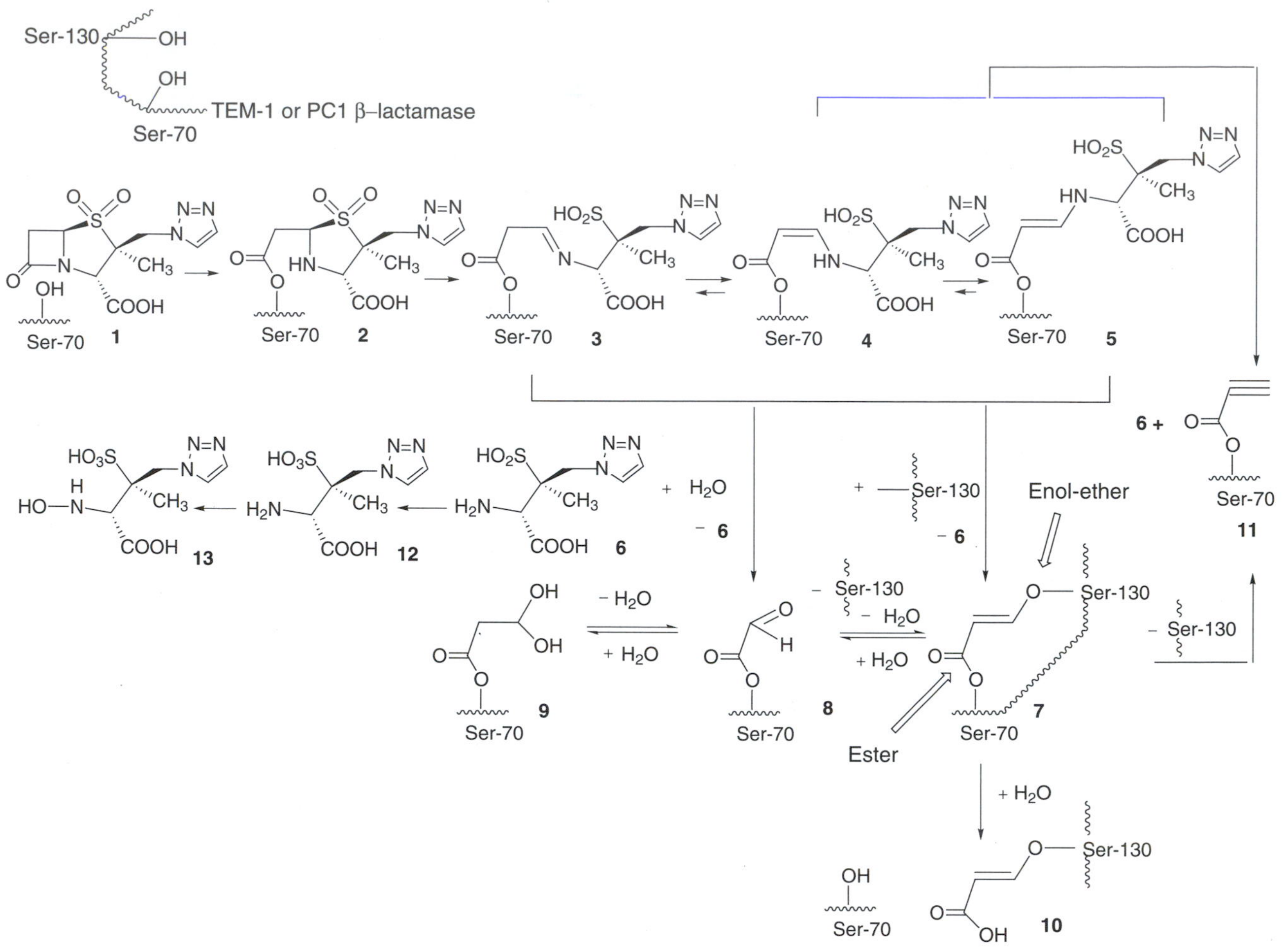

Figure 11.6 Mechanism of inactivation of TEM-1 or PC1 β-lactamase by tazobactam. For clarity, the residues Ser70 and Ser130 are represented separately, but they belong to the same molecule. Amino acid numbering is according to Ambler et al. Reprinted from Y. Yang, K. Janota, K. Tabei, N. Huang, M. M. Siegel, Y.-I. Lin, B. A. Rasmussen, and D. M. Shlaes, *J. Biol. Chem.* **275:**26674–26682, 2000, with permission from the publisher.

Fig. 11.6) that could result from decomposition of intermediates 4, 5, and 7; (ii) an intramolecular nucleophilic displacement of the imine (3) or enamine (4 [*cis*] or 5 [*trans*]) moieties by Ser130 to form a cross-linked vinyl ether (structure 7); and (iii) hydrolysis of the imine or enamine to form a cross-linked aldehyde 8 with concomitant release of a sulfinic acid derivative 6 in these reactions. In turn, intermediate 6 undergoes an oxidation reaction, giving the sulfonic acid derivatives 12 and 13. Intermediate 7 can give a rearrangement reaction to form propiolated enzyme 11 and/or hydrolysis to give compounds 8 and 10. The aldehyde-enzyme compound 8 is in equilibrium with its hydrated form, 9.

The authors concluded that although clavams and tazobactam (a penicillanic acid sulfone) are structurally distinct, the inactivation processes of class A β-lactamases by clavulanic acid and tazobactam are very similar, especially after the formation of the acyl-imine inhibitor-enzyme complex.

## Clavulanic Acid, Sulbactam, and Tazobactam β-Lactamase-Inhibitory Activity

Clavulanic acid, sulbactam, and tazobactam inhibit exocellular β-lactamases encoded by plasmids from *Staphylococcus* spp. and group 2 periplasmic β-lactamases (except for some TEM mutants) of gram-negative bacteria. These β-lactamases include the TEM-1 β-lactamase, which is plasmid encoded and is one of the most common β-lactamases found in bacteria. It is present in *Haemophilus influenzae, Neisseria gonorrhoeae, Salmo-*

*nella* and *Shigella* spp., *Klebsiella pneumoniae, E. coli,* and *S. marcescens.*

Clavulanic acid, sulbactam, and tazobactam do not inhibit chromosomally encoded group 1 β-lactamases (according to the Bush-Jacoby-Medeiros classification), found in *Enterobacter, Morganella, Citrobacter, Pseudomonas,* and *Acinetobacter* spp.

## Combinations of a β-Lactam Antibiotic and a Serine β-Lactamase Inhibitor

In clinical settings, clavulanic acid, sulbactam, and tazobactam are used in combination with a penicillin prone to β-lactamase degradation. As of October 2001, only four β-lactam antibiotic–β-lactamase inhibitor combinations have been approved in the United States and the United Kingdom: amoxicillin-potassium clavulanate, ticarcillin-potassium clavulanate, ampicillin-sulbactam, and piperacillin-tazobactam. In other countries, cefoperazone-sulbactam and sultamicillin (a prodrug which combines an antibiotic and an inhibitor) are also in use.

In these combinations, the β-lactamase inhibitor possesses greater β-lactamase affinity; therefore, since there is no free β-lactamase (either exocellular in gram-positive bacteria or within the periplasm in gram-negative bacteria), the β-lactam antibiotic cannot be hydrolyzed. In this way, the antibacterial range of the β-lactam antibiotic is broadened to include β-lactamase-producing bacterial strains.

### Amoxicillin-Potassium Clavulanate

The mixture of amoxicillin trihydrate and potassium clavulanate is made in 2:1 and 4:1 ratios for oral administration. In mixtures with clavulanic acid, amoxicillin has an enlarged range of antimicrobial action, including β-lactamase-producing strains of *H. ducreyi, N. gonorrhoeae, S. aureus* (except for methicillin-resistant strains), and *M. catarrhalis.* Many β-lactamase-producing strains of *E. coli, Klebsiella, Proteus,* and *Citrobacter diversus* are also susceptible. *Bacteroides fragilis* and *Legionella pneumophila* are also sensitive in in vitro assays; however, this combination is inactive against *Pseudomonas aeruginosa, Serratia,* and several strains of *Enterobacter, Morganella,* and *Providencia.*

#### THERAPEUTIC USES

The amoxicillin-potassium clavulanate mixture is a useful alternative to erythromycin-sulfisoxazole, trimethoprim-sulfamethoxazole, and cefaclor for the treatment of acute otitis media and acute sinusitis produced by amoxicillin-resistant strains of *H. influenzae* and *M. catarrhalis.* It is also effective against mild and moderate infections of the lower respiratory tract, such as exacerbations of chronic bronchitis caused by β-lactamase-producing strains of *H. influenzae* and *M. catarrhalis.* Numerous studies have demonstrated the effectiveness of this clavulanate mixture in primary and recurrent urinary infections caused by susceptible strains of *E. coli* and other bacteria, including many amoxicillin-resistant strains. It has been used with good results in infections of the skin and soft tissues caused by susceptible pathogens, including β-lactamase-producing strains of *S. aureus.* It is currently considered an alternative to penicillinase-resistant penicillins (e.g., dicloxacillin), erythromycin, and first-generation oral cephalosporins (e.g., cephalexin) for the treatment of skin infections caused by *Staphylococcus.*

### Ticarcillin-Potassium Clavulanate

The mixture of disodium ticarcillin and potassium clavulanate is marketed in 30:1 and 15:1 ratios for parenteral administration. Ticarcillin is a broad-spectrum antipseudomonas penicillin that is active in in vitro assays against most of the gram-positive bacteria (except for *Enterococcus* and β-lactamase-producing *Staphylococcus* species) and gram-negative cocci. The ticarcillin-potassium clavulanate mixture does not alter the susceptibility of the strains already sensitive to ticarcillin itself; it extends the activity in vitro to include many strains of β-lactamase-producing enterobacteria, including *E. coli, K. pneumoniae, K. oxytoca, C. diversus, C. amalonecticus, Proteus vulgaris, P. rettgeri, Providencia stuarti,* some *S. marcescens* and *Enterobacter* strains, the *Bacteroides fragilis* group, *Haemophilus* and *Neisseria* species, *M. catarrhalis,* and *Staphylococcus* spp. (except for methicillin-resistant strains). However, the in vitro susceptibility of *P. aeruginosa, Acinetobacter calcoaceticus,* some strains of *S. marcescens,* and strains of *Enterobacter* does not increase due to the presence of clavulanic acid.

#### THERAPEUTIC USES

The ticarcillin-potassium clavulanate mixture is prescribed for serious infections of the lower respiratory tract, the urinary tract, bones and articulations, and skin and soft tissues and for septicemia produced either by susceptible β-lactamase-producing strains of several gram-negative bacilli and *S. aureus* or by ticarcillin-susceptible microorganisms. Due to the broad antibacterial spectrum against gram-positive and both aerobic and anaerobic gram-negative bacteria, including many β-lactamase-producing species, this mixture is useful for the treatment of mixed infections such as intra-abdominal and gynecological infections.

### Sodium Ampicillin-Sodium Sulbactam

The mixture of sodium ampicillin and sodium sulbactam is marketed in a 2:1 ratio for parenteral administration. This mixture extends the in vitro activity of ampicillin to include β-lactamase-producing strains of *S. aureus* and *S. epidermidis* (but not the methicillin-resistant *Staphylococcus* strains), *N. gonorrhoeae, M. catarrhalis, H. influenzae, H. ducreyi, E. coli, Proteus, Klebsiella, Enterobacter aerogenes, A. calcoaceticus,* and many anaerobic bacteria (including those of the *B. fragilis* group). It is inactive against *P. aeruginosa, E. cloacae,* most *Serratia* species, and some other members of the *Enterobacteriaceae.*

THERAPEUTIC USES

The ampicillin-sulbactam mixture is suitable for intra-abdominal, gynecological, skin, and soft tissue infections caused by susceptible bacteria. Its broad antibacterial spectrum, including gram-positive and gram-negative, aerobic and anaerobic β-lactamase-producing bacteria, makes it a useful mixture to treat these types of infections of polymicrobial origin. The ampicillin-sulbactam mixture has similar effectiveness to clindamycin plus gentamicin for intra-abdominal infections, to metronidazole plus gentamicin for infections of pelvic soft tissues and acute inflammatory pelvic illnesses, and to clindamycin plus tobramycin for infections of soft tissues. It is also effective for the treatment of infections in bone, articulations, lower respiratory tract and urinary tract and noncomplicated gonorrheas, as well as for prophylaxis during gastrointestinal and obstetric-gynecological surgery. It is effective in treating children's acute epiglottitis and has similar action to ampicillin plus chloramphenicol against bacterial meningitis in babies and children.

### Sultamicillin

Sultamicillin is a double ester in which ampicillin and sulbactam are bound through a methylene group (Fig. 11.7). It is not commercially available in the United States or the United Kingdom. This double ester is a prodrug. After oral administration and absorption, it is hydrolyzed by nonspecific esterases, releasing the antibiotic ampicillin and the β-lactamase inhibitor sulbactam in a 1:1 ratio.

**Figure 11.7** Chemical structure of sultamicillin.

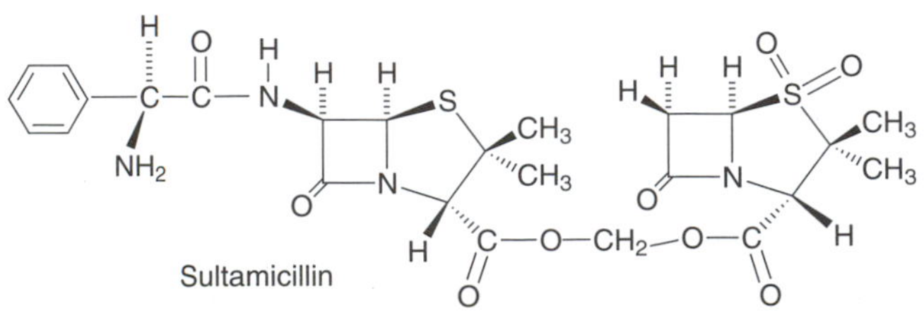

Sultamicillin

### Piperacillin-Sodium Tazobactam

Tazonam is a mixture of piperacillin and tazobactam. Its commercial form is produced in an 8:1 ratio for parenteral administration. Tazobactam reduces the MIC and enlarges the bactericidal activity of piperacillin mainly against bacteria producing penicillinases and oxyimino-cephalosporinases (enzymes that destroy oxyimino-cephalosporins), but it is only moderately active against bacteria that produce other cephalosporinases. Tazobactam increases antibiotic activity of piperacillin against β-lactamase-producing strains of *S. aureus, H. influenzae,* and *M. catarrhalis,* and it diminishes the MIC of piperacillin for many enterobacteria and bacteroids and, sometimes, *Enterobacter* and *C. freundii.* It does not have good activity against *Pseudomonas* and *Acinetobacter* species.

Tazobactam inhibits chromosomal and plasmidic β-lactamases of gram-positive and gram-negative bacteria. It also inhibits the SHV-3 and SHV-4 enzymes produced by strains of *K. pneumoniae* that are resistant to cefotaxime, ceftazidime, aztreonam, and cefoxitin.

THERAPEUTIC USES

The piperacillin-tazobactam mixture is effective in treating intra-abdominal infections. Clinical trials demonstrated that 89% of the patients affected by appendicitis, peritonitis, colecistitis, colangitis, diverticulitis, surgical wounds and abscesses responded to treatment with this mixture. Most of the intra-abdominal infections are polymicrobial, produced by either aerobic or anaerobic bacteria; among them are *B. fragilis* and other species of *Bacteroides,* including some resistant β-lactamase-producing strains of group 2be (Bush-Jacoby-Medeiros classification). The excellent activity of the piperacillin-tazobactam mixture against *E. coli, Klebsiella, Enterobacter, Proteus,* and *Pseudomonas* is beneficial since these bacteria generally cause intra-abdominal infections.

A second and important indication for the piperacillin-tazobactam mixture involves infections in the inferior respiratory tract (pneumonias, bronchitis, bronchopneumonias, and unspecified infections of the inferior respiratory tract). The excellent diffusibility of the piperacillin-tazobactam mixture to the lung parenchyma, the pleura, and the bronchoalveolar secretions and its activity against the bacteria that produce acute pneumopathies make it the favorite treatment for this pathology.

The mixture is also effective in the treatment of pneumonia and bronchopneumonia in elderly, alcoholic, diabetic, renal chronic and bronchoenphysematous patients undergoing chronic treatment with corticoids. In this group of immunocompromised patients, the mixture has very good activity against *S. pneumoniae, M. catarrhalis, S. aureus,* and several species of *Klebsiella* and *Haemophilus*. It is also exceptionally active against *A. calcoaceticus,* a pathogenic gram-negative bacterium.

A third therapeutic indication for piperacillin-tazobactam involves infections of the skin and soft tissues caused by susceptible bacterial species, including those that usually appear in the postoperative wounds, ulcers on the feet of diabetic patients, septic wounds, necrotic infections in soft tissues, and burns. In these cases, the activity of piperacillin-tazobactam against non-methicillin-resistant strains of *S. aureus* and *S. epidermidis* is particularly important.

A fourth indication for piperacillin-tazobactam comprises gynecological infections, including endometritis, salpingitis, pelvic cellulitis of the vaginal sac and tubo-ovarian abscesses. These infections are caused by gram-negative bacteria (e.g., *E. coli* and *Klebsiella* and *Enterobacter* species); enterococci can also be present in these infections.

Urinary tract infections (pyelonephritis, cystitis, cystopyelitis, prostatitis, and urosepsis), especially those caused by *E. coli, P. mirabilis, P. aeruginosa, S. faecalis,* and *Enterobacter* spp., which are the most complicated ones, can be treated with piperacillin-tazobactam.

Septicemia is a serious, potentially fatal infection characterized by the presence of pathogenic bacteria and their toxins in the blood. The use of broad-spectrum antibiotics is generally necessary for treatment. Because of its effectiveness, piperacillin-tazobactam, which turned out to be efficient in 91% of bacteremia cases, seems to be particularly useful for the treatment of septicemia. Patients with osteomelitis, septic arthritis, infected fractures, and other bone and articulation infections caused by sensitive microorganisms can be treated with this mixture.

Due to its broad spectrum of activity, piperacillin-tazobactam is an excellent antibacterial agent for treating infections caused by bacteria that have not yet been identified. It is also used to treat known or suspected polymicrobial infections, such as intra-abdominal infections, soft tissue necrotic infections, and gynecological infections.

Finally, the broad antibacterial spectrum of this mixture makes it most useful for the treatment of infections in immunodepressed or neutropenic patients. In these cases, where the capacity to combat infections is clearly diminished, it is essential to use broad-spectrum bactericidal antibiotics that offer a special protection against gram-negative, gram-positive, and anaerobic bacteria. The piperacillin-tazobactam mixture has excellent activity in neutropenic patients with urinary infections, pneumonia,

**Table 11.2** Activity of drug combinations against some common pathogenic bacteria[a]

| | Activity ($MIC_{90}$, mg/liter)[b] of: | | | |
|---|---|---|---|---|
| **Bacterium** | **Amoxicillin + clavulanate** | **Ticarcillin + clavulanate** | **Ampicillin + sulbactam** | **Piperacillin + tazobactam** |
| *S. aureus* (MSSA)[c] | 0.5 | 4.0 | 0.5 | 2.0 |
| *S. pneumoniae* | | | | |
| Penicillin susceptible | 0.12 | 2.0 | 0.13 | ≤0.06 |
| Penicillin resistant | 1.0 | 128.0 | 4.0 | 4.0 |
| *H. influenzae* | 0.5 | ≤1.0 | ≤1.0 | ≤1.0 |
| *M. catarrhalis* | 0.12 | ≤1.0 | ≤1.0 | ≤1.0 |
| *E. coli* | 8.0 | 32.0 | 64.0 | 16.0 |
| *Klebsiella* spp. | 16.0 | 16.0 | 64.0 | 16.0 |
| *Enterobacter* spp. | 128.0 | 128.0 | 64.0 | 128.0 |
| *P. mirabilis* | 2.0 | 2.0 | 8.0 | 2.0 |
| *P. vulgaris/Providencia* | 8.0 | 32.0 | 32.0 | 2.0 |
| *S. marcescens* | 128.0 | 32.0 | 128.0 | 32.0 |
| *Citrobacter* spp. | 128.0 | >256.0 | 64.0 | 128.0 |
| *P. aeruginosa* | >64.0 | 128.0 | >256.0 | 64.0 |
| *B. fragilis* | 4.0 | 8.0 | 4.0 | 4.0 |

[a] Reprinted from H. P. Lambert, R. G. Finch, and D. Greenwood (ed.), *Antibiotic and Chemotherapy*, 7th ed. (Churchill Livingstone, Inc., New York, N.Y., 1997), with permission from the publisher.

[b] $MIC_{90}$, MIC for 90% of organisms.

[c] MSSA, methicillin-susceptible *S. aureus*.

**Table 11.3** Generic and common trade names of combinations of a β-lactam antibiotic plus a β-lactamase inhibitor, the preparations available, and manufacturers in the United States

| Generic name | Trade name (preparation)[a] | Manufacturer |
|---|---|---|
| Parenteral | | |
| Ampicillin sodium + sulbactam sodium | Unasyn (vial dry powder) | Pfizer |
| Piperacillin sodium + tazobactam sodium | Zosyn (injectable for i.v. administration) | Lederle |
| Ticarcillin disodium + potassium clavulanate | Timentin (injectable for i.v. administration) | GlaxoSmithKline |
| Oral | | |
| Amoxicillin + potassium clavulanate | Augmentin (tablets and oral suspension) | GlaxoSmithKline |

[a] i.v., intravenous.

sinusitis, pharyngitis, septicemia, cellulitis, perianal infections, periodontitis, and fever of unknown origin caused by gram-negative bacteria, especially *E. coli, Klebsiella* species, *P. aeruginosa,* and *Enterobacter* species.

## Cefoperazone-Sulbactam

Cefoperazone-sulbactam is a combination of the third-generation cephalosporin cefoperazone and the β-lactamase inhibitor sulbactam. It has not been authorized for marketing in the United States or the United Kingdom. Sulbactam effectively inhibits a wide variety of β-lactamases produced by common aerobic and anaerobic gram-positive bacteria. Among aerobic bacteria, the cefoperazone-sulbactam combination shows important synergic in vitro effects against β-lactamase-producing strains of *E. coli, Klebsiella, P. vulgaris, Citrobacter,* and *Staphylococcus*. It also has increased activity against β-lactamase-producing strains that are resistant to cefoperazone, particularly *S. epidermidis, E. coli, K. pneumoniae,* and *E. cloacae.*

## Comparative Studies of the Activity of the Combinations of a β-Lactam Antibiotic and a β-Lactamase Inhibitor

In Table 11.2, the MICs for 90% of bacteria of four commercial combinations approved in the United States and the United Kingdom to treat important clinic pathogenic bacteria are tabulated for comparison.

## Combinations of a β-Lactam Antibiotic and a β-Lactamase Inhibitor Commercially Available in the United States

Table 11.3 lists the combinations of β-lactam antibiotic plus β-lactamase inhibitor that are marketed in the United States.

## REFERENCES AND FURTHER READING

### β-Lactamase Inhibitors

**American Medical Association.** 1995. *Drug Evaluations Annual,* p. 1484–1492. American Medical Association, Chicago, Ill.

**Brown, A. G., M. J. Pearson, and R. Southgate.** 1989. Other β-lactam agents. Clavulanic acid and analogues, p. 662–671. *In* C. Hansch (ed.), *Comprehensive Medicinal Chemistry,* vol. 2. Pergamon Press, Oxford, United Kingdom.

**Bush, K.** 1997. Other β-lactams: β-lactamase inhibitors, p. 320–327. *In* F. O'Grady, H. P. Lambert, R. G. Finch, and D. Greenwood (ed.), *Antibiotic and Chemotherapy,* 7th ed. Churchill Livingstone, Inc., New York, N.Y.

**Cartwright, S. J., and S. G. Waley.** 1983. β-Lactamase inhibitors. *Med. Res. Rev.* **3**:341–382.

**Cherry, P. C., and C. E. Newall.** 1982. Clavulanic acid, p. 362–399. *In* R. B. Morin and M. Gorman (ed.), *Chemistry and Biology of β-Lactam Antibiotics,* vol. 2. Academic Press, Inc., New York, N.Y.

**Cole, M.** 1981. Inhibitors of bacterial β-lactamases. *Drugs Future* **6**:697–727.

**Coulton, S., and I. François.** 1994. β-Lactamases: targets for drug design. *Prog. Med. Chem.* **31**:297–349.

**Knox, J. R.** 1995. Extended-spectrum and inhibitor-resistant TEM-type β-lactamases: mutations, specificity, and three-dimensional structure. *Antimicrob. Agents Chemother.* **39**: 2593–2601.

**Maiti, S. N., O. A. Phillips, R. G. Micetich, and D. M. Livermore.** 1998. β-Lactamase inhibitors: agents to overcome bacterial resistance. *Curr. Med. Chem.* **5**:441–456.

**Mascaretti, O. A., C. E. Boschetti, G. O. Danelon, E. G. Mata, and O. A. Roveri.** 1996. β-Lactam compounds. Inhibitors of transpeptidases, β-lactamases and elastases: a review. *Curr. Med. Chem.* **1**:441–470.

**Mascaretti, O. A., O. A. Roveri, and G. O. Danelon.** 1993. Recent advances in the chemistry and biochemistry of β-lactams as β-lactamase inhibitors, p. 677–749. *In* G. Lukacs (ed.), *Recent Progress in the Chemical Synthesis of Antibiotics and Related Microbial Products,* vol. 2. Springer-Verlag KG, Berlin, Germany.

**Page, M. G. P.** 2000. β-Lactamase inhibitors. *Drug Resist. Update* **3**:109–125.

**Payne, D. J., R. Cramp, D. J. Winstanley, and D. J. C. Knowles.** 1994. Comparative activities of clavulanic acid, sulbactam and tazobactam against clinically important β-lactamases. *Antimicrob. Agents Chemother.* **38**:767–772.

**Pratt, R. F.** 1992. β-Lactamase inhibition, p. 229–271. *In* M. I. Page (ed.), *The Chemistry of β-Lactams*. Blackie Academic & Professional, London, United Kingdom.

**Pratt, R. F.** 1989. β-Lactamase inhibitors, p. 178–205. *In* M. Sandler and H. J. Smith (ed.), *Design of Enzyme Inhibitors as Drugs*. Oxford Science Publications, Oxford, United Kingdom.

**Rotschafer, J. C., and B. E. Ostergaard.** 1995. Combinations β-lactam and β-lactamase-inhibitor products: antimicrobial activity and efficiency of enzyme inhibition. *Am. J. Health Syst. Pharm.* **52**(Suppl. 2)**:**15–22.

**Stam, J. G.** 1993. β-Lactamase inhibitors, p. 314–338. *In* M. Howe Grant (ed.), *Chemotherapeutics and Disease Control*. John Wiley & Sons, Inc., New York, N.Y.

**Sutherland, R.** 1995. β-Lactams/β-lactamase inhibitor combinations: development, antibacterial activity and clinical applications. *Infection* **23:**191–200.

**Therrien, C., and R. C. Levesque.** 2000. Molecular basis of antibiotic resistance and β-lactamase inhibition by mechanism-based inactivators: perspectives and future directions. *FEMS Microbiol. Rev.* **24:**251–262.

**Wise, R.** 1982. β-Lactamase inhibitors. *J. Antimicrob. Chemother.* **9**(Suppl. B)**:**31–40.

### Clavulanic Acid

**Brown, R. P. A., R. T. Aplin, and C. J. Schofield.** 1996. Inhibition of TEM-2 β-lactamase from *Escherichia coli* by clavulanic acid: observation of intermediate by electrospray ionization mass spectrometry. *Biochemistry* **35:**12421–12432.

**Chen, C. C. H., and O. Herzberg.** 1992. Inhibition of β-lactamase by clavulanate. Trapped intermediates in crystallographic studies. *J. Mol. Biol.* **224:**1103–1113.

**Imtiaz, U., E. Billings, J. A. Knox, E. K. Manavathu, S. A. Lerner, and S. Mobashery.** 1993. Inactivation of class A β-lactamases by clavulanic acid: the role of arginine-244 in a proposed nonconcerted sequence of events. *J. Am. Chem. Soc.* **115:**4435–4442.

### Penicillanic Acid Sulfone (Sulbactam)

**Fisher, J., R. L. Charnas, S. M. Bradley, and J. R. Knowles.** 1981. Inactivation of the RTEM β-lactamase from *Escherichia coli*. Interaction of penam sulfones with enzyme. *Biochemistry* **20:**2726–2731.

**Imtiaz, U., E. M. Billings, J. R. Knox, and S. Mobashery.** 1994. A structure-based analysis of the inhibition of class A β-lactamase by sulbactam. *Biochemistry* **33:**5728–5738.

**Knowles, J. R.** 1995. Penicillin resistance: the chemistry of β-lactamase inhibition. *Acc. Chem. Res.* **18:**97–104.

### Tazobactam

**Bush, K., C. Macalintal, B. A. Rasmussen, V. J. Lee, and Y. Yang.** 1993. Kinetic interactions of tazobactam with β-lactamases from all major structural classes. *Antimicrob. Agents Chemother.* **37:**851–858.

**Micetich, R. G., S. N. Maiti, P. Sperak, T. W. Hall, S. Yamabe, N. Ishida, M. Tanaka, T. Yamazaki, A. Nakai, and K. Ogawa.** 1987. Synthesis and β-lactamase inhibitory properties of 2β-[(1,2,3-triazol-1-yl) methyl]-2α-methylperam-3α-carboxylic acid 1,1-dioxide and related triazolyl derivatives. *J. Med. Chem.* **30:**1469–1474.

**Yang, Y., K. Janota, K. Tabei, N. Huang, M. M. Siegel, Y.-I. Lin, B. A. Rasmussen, and D. M. Shlaes.** 2000. Mechanism of inhibition of the class A β-lactamases PC1 and TEM-1 by tazobactam. *J. Biol. Chem.* **275:**26674–26682.

### Piperacillin-Tazobactam

**Anonymous.** 1994. Piperacillin/tazobactam. *Med. Lett.* **36:** 7–10.

**Sanders, W. E., and C. C. Sanders.** 1996. Piperacillin/tazobactam: a critical review of the evolving clinical literature. *Clin. Infect. Dis.* **22:**107–123.

chapter 12



# Bacterial Resistance to β-Lactam Antibiotics and β-Lactam Inhibitors of β-Lactamases

Most bacteria are harmless and are an integral part of the world and our lives, but a few are pathogenic to humans. Because bacteria have short generation times (from minutes to hours), they can respond rapidly to changes in their environments. The expanding problem of resistance to β-lactam antibiotics and β-lactam inhibitors of β-lactamases illustrates the genetic adaptability of bacterial populations. The evolution of bacterial resistance to the action of these drugs occurred over a relatively short period. Bacteria become resistant to antibiotics and β-lactam compounds acting as β-lactamase inhibitors either through **mutations** or by acquisition of **specific resistance genes** from other bacteria.

As previously mentioned, penicillin G was the first β-lactam antibiotic and was introduced into medical practice in the 1940s. Since then, many different β-lactams, including penicillins, cephalosporins, cephamycins, carbapenems, monobactams, and β-lactamase inhibitors, have been developed. The β-lactam antibiotics and the combinations of a β-lactam antibiotic and a β-lactamase inhibitor approved for clinical use by the Food and Drug Administration are listed chronologically in Table 12.1.

The use of antibacterial drugs for prophylactic or therapeutic purposes in humans and for veterinary and agricultural purposes has provided selective pressure favoring the survival and spread of resistant organisms. Bacteria have evolved their mechanism of resistance in response to the selective pressure exerted by the presence of these antibacterial agents in clinical usage, although some authors thought that some of the mechanisms of bacterial resistance, like the production of β-lactamases, existed before the introduction of penicillins into clinical use in 1940.

Indeed, the most common mechanism of bacterial resistance to β-lactam compounds is the production of β-lactamases. The genes encoding β-lactamases are found on the bacterial chromosome or on plasmids and transposable elements. The presence of resistance genes on plasmids and transposons allows the genes to be transferred to distinctly related bacteria

**Table 12.1** Dates when β-lactam antibiotics and the combinations of a β-lactam antibiotic with a β-lactamase inhibitor were approved for use in the United States

| Antibiotic | Approval date |
|---|---|
| Penicillin G | May 1946 |
| Methicillin | October 1960 |
| Oxacillin | January 1962 |
| Ampicillin | December 1963 |
| Nafcillin | January 1964 |
| Cephalothin | July 1964 |
| Carbenicillin | August 1970 |
| Ticarcillin | October 1976 |
| Cefamandole | September 1978 |
| Cefoxitin | October 1978 |
| Cefotaxime | March 1981 |
| Piperacillin | December 1981 |
| Cefoperazone | November 1982 |
| Ceftizoxime | September 1983 |
| Cefuroxime | October 1983 |
| Amoxicillin-clavulanate | August 1984 |
| Ceftriaxone | December 1984 |
| Ticarcillin-clavulanate | April 1985 |
| Ceftazidime | July 1985 |
| Imipenem-cilastatin | November 1985 |
| Cefotetan | December 1985 |
| Ampicillin-sulbactam | December 1986 |
| Aztreonam | December 1986 |
| Piperacillin-tazobactam | October 1993 |
| Ceftibuten | December 1995 |
| Cefepime | January 1996 |
| Cefuroxime axetil | March 1996 |
| Meropenem | June 1996 |
| Cefprozil | September 1996 |
| Ticarcillin-clavulanate | December 1997 |
| Cefpodoxime axetil | December 1997 |
| Cefdinir | July 1999 |

by either conjugation, transduction, or transformation (see chapter 4). This means that these genes are mobile within a bacterial community and can spread widely.

Combinations of a β-lactamase inhibitor (clavulanic acid, sulbactam, or tazobactam) and a penicillin antibiotic (amoxicillin, ampicillin, piperacillin, or ticarcillin) remain one of the most successful strategies for the treatment of bacterial infections, particularly those due to bacteria possessing class A β-lactamases. Unfortunately, the number of class A enzymes having low affinity for the current β-lactamase inhibitors is increasing rapidly. The inhibitor-resistant TEM (IRT) and SHV class A β-lactamases have point mutations in critical amino acids important for catalysis. For instance, a recent report (Bermudes et al., 1999) detailed the molecular characterization of TEM-59 (IRT-17) in a clinical isolate of *Klebsiella oxytoca* resistant to amoxicillin and ticarcillin without substantial potentiation by clavulanic acid. These authors found that the resistance was carried by a ca. 50-kb conjugative plasmid that encoded a single β-lactamase with an isoelectric point (pI) of 5.6. Nucleotide sequence analysis revealed identity between the $bla_{\text{TEM-1}}$ gene of this strain of *K. oxytoca* and the $bla_{\text{TEM-2}}$ gene, except for a single A-to-G change at position 590, leading to an amino acid change from Ser130 to Gly. Over the last 5 years, SHV extended-spectrum β-lactamases (ESBLs) have arisen through single-amino-acid substitutions in the parenteral enzyme SHV-1. For instance, Kurokawa et al. (2000) recently reported a new SHV-derived ESBL (SHV-24) that hydrolyzes ceftazidime through a single-amino-acid change from Asp179 to Gly in the enzyme. It is remarkable how a single amino acid change produces such a dramatic effect on the function of the β-lactamase.

## Extended-Spectrum β-Lactamases

ESBLs are a group of enzymes that have the common property of providing resistance to extended-spectrum β-lactam antibiotics such as oxyiminocephalosporins (e.g., cefotaxime, ceftazidime, ceftriaxone, cefepime, and cefpirome), as well to aztreonam, an oxyiminomonobactam. Cephamycins or 7α-methoxycephalosporins (cefoxitin and cefotetan) and carbapenems (imipenem and meropenem) are stable to most of these ESBLs.

ESBLs were first identified in European centers in 1988 in at least three widely separated geographical areas and shortly after this were identified in outbreaks of infection due to ceftazidime-resistant organisms in the United States. Because of the more intensive antibiotic use in hospital than in the community, higher rates of resistance are noted in hospital pathogens, especially in the intensive care unit. Genetic methods for assessing bacterial resistance have been developed and have become part of the standard testing protocols in clinical microbiology laboratories.

These groups of β-lactamases are derived from common plasmid-encoded β-lactamases by one or more amino acid substitutions that expand the spectrum of action of the enzyme. Among the most prevalent types of ESBLs are members of the TEM and SHV families, derived from TEM-1 or TEM-2 and SHV-1 β-lactamases or from OXA-10, which belong to Ambler's class A and class D, respectively. A few class A ESBLs encoded by chromosomal genes have also been found.

An excellent review by Bradford (2001) has recently discussed the characterization, epidemiology, and detection of ESBLs.

## TEM β-Lactamases

The TEM-1 β-lactamase (Color Plate 12.1 [see color insert]) is an Ambler class A enzyme which corresponds to group 2b of the Bush-Jacoby-Medeiros functional classification. TEM-1 is encoded by the $bla_{TEM-1}$ gene, which is present on the Tn*3* transposon.

Jacoby and Bush have tabulated the amino acid sequences for TEM, SHV, and OXA inhibitor-resistant ESBLs; they can be found at http://www.lahey.org/studies/webt.htm. The last update I found was reported in May 2001. At the time, the listing comprised 96 TEM and 37 SHV inhibitor-resistant ESBLs and 12 OXA (OXA-2, OXA-10, OXA-11, and OXA-14 to OXA-20) ESBLs.

Strains of *Escherichia coli* isolated in Athens, Greece, were the first to be identified as producing TEM-1 β-lactamase; this occurred soon after aminopenicillins (ampicillin and amoxicillin) had been introduced into clinical practice. In 1992, TEM-derived β-lactamases that exhibited resistance to β-lactamase inhibitors were first described simultaneously in France and the United Kingdom. The inhibitor-resistant β-lactamases are collectively designated inhibitor-resistant TEM (IRT) or simply inhibitor resistant (IR).

The TEM-type ESBLs are the most widespread plasmid-encoded serine β-lactamases. The gene encoding the TEM-1 enzyme has undergone very high selective pressure, because of the large number of broad-spectrum (or extended-spectrum) β-lactam antibiotics and β-lactamase inhibitors introduced into clinical use since the 1980s. It is therefore not surprising to see the recent emergence of bacterial strains producing ESBLs and IRTs, which are TEM- and SHV-derived mutant enzymes produced by different clinical bacteria because of the frequent use of clavulanic acid, sulbactam, tazobactam, extended-spectrum cephalosporins, and aztreonam in hospitals and in general practice. This implies that the use of extended-spectrum β-lactam antibiotics or the coadministration of a penicillin plus a β-lactamase inhibitor, thought to be suitable for treating frequently isolated bacterial strains, could become increasingly less successful. The TEM β-lactamase variants are the most frequently encountered enzymes of this type, and more than 90 variants have been characterized so far.

Crystallographic studies carried out with TEM-1, variants of TEM-1, and other class A β-lactamases, as well as TEM-1 from *E. coli* covalently bound to penicillin G or to 6-α-(hydroxymethyl)penicillanate, were reported in 1992 by Strynadka et al. and in 1996 by Maveyraud et al., respectively. These studies provided a large base of structural information useful in elucidating the behavior of mutated enzymes either isolated naturally or produced by site-directed mutagenesis.

### THE STRUCTURE-FUNCTION RELATIONSHIP IN TEM β-LACTAMASES

Excellent reviews by Knox (1995), Matagne and Frère (1995), Matagne et al. (1998), and Du Bois et al. (1995) have recently discussed the structure-function relationships in TEM ESBLs; for this reason, this topic is not included here.

### INHIBITOR-RESISTANT β-LACTAMASES

As discussed in previous chapters, the resistance of clinical bacterial isolates to the action of β-lactam antibiotics, whether they are penicillins, cephalosporins, cephamycins, carbapenems, or monobactams, is most frequently due to the production of β-lactamases, which hydrolyze the β-lactam bond of the molecule and thus inactivate the antibiotic. We also have seen that bacteria use other resistance mechanisms to counteract the effects of antibiotics, notably the production of modified target sites as seen in the alteration of penicillin-binding protein 2 (PBP2a) of methicillin-resistant staphylococci or reduced permeability through the porins of gram-negative bacteria, especially in *Pseudomonas aeruginosa*, associated with efflux mechanisms to pump out the antibiotics.

We have just seen that one approach to the problem posed by β-lactamase-producing bacteria is the coadministration of a β-lactamase-sensitive penicillin antibiotic and a mechanism-based inhibitor or inactivator. The most common and effective β-lactamase inhibitors (e.g., clavulanic acid, sulbactam, and tazobactam) are structurally similar to the β-lactam antibiotics in that they retain the β-lactam ring and its reactive carbonyl group for enzymatic recognition. Effective inhibitors frequently contain or generate an additional electrophilic site for simultaneous or subsequent attack by enzymatic nucleophiles. The subsequent reactions leading to the blockage of the active site of the β-lactamase minimize or inhibit substrate turnover and allow the coadministered antibiotic to inactivate its target (PBPs). Indeed, for a β-lactamase inhibitor to be effective, it must not only have good intrinsic inhibitory activity but also, for gram-negative bacteria, be able to penetrate the outer membrane to reach its target in the periplasm. The three β-lactamase inhibitors on the market (clavulanic acid, sulbactam, and tazobactam) have no antibacterial activity by themselves but are all effective against class A β-lactamases while having little or no activity against class C and B β-lactamases.

The discovery of clavulanic acid in 1976 and its development as a β-lactamase inhibitor marked a triumph for humankind over bacteria. However, there are clear and persistent signs that the victory was certainly not total and that the "struggle" against pathogenic bacteria will continue. Unfortunately, the bacteria have been exceptionally adaptive. The bacterial genome continually produces novel ways of overcoming the action of antibacterial agents, forcing the scientific community to continually develop novel strategies for combating bacterial infection.

The IRT β-lactamases were first discovered in a mutant of TEM-1 β-lactamase in 1992, simultaneously in France and the United Kingdom. For several years, β-lactamase inactivators were free from selection for resistance, but more than 23 TEM mutants and at least one SHV mutant have evolved to show less sensitivity to inactivation by one or more β-lactamase inhibitors.

Who will be the winner of the battle, bacteria or humans? It is a challenge to molecular biologists, microbiologists, geneticists, and medicinal and bioorganic chemists to have new ideas for development of new antibacterial agents directed to novel targets in the bacteria. What the bacteria will do next is unpredictable.

## INACTIVATION OF NATIVE TEM β-LACTAMASES AND ACTION OF MUTANT TEM β-LACTAMASES

The action and inactivation of β-lactamases are closely related. There is a bewildering variety of TEM β-lactamase mutants (at least 96 have been tabulated as of May 2001). This is now an exciting time in the study of β-lactamases, because although these enzymes have been known for more than 60 years, it is only in the last 10 to 15 years that we started to glimpse their primary and tertiary structures and their functions at the molecular level. The crystallographic and electrospray ionization mass spectrometry approaches to investigate the unique reactivity of the hydroxyl group of active-site Ser70 and Ser130 of class A β-lactamases have been pursued. The kinetics of inactivation of many class A enzymes by clavulanic acid, sulbactam, and tazobactam have been studied.

The worldwide onset and spread of bacterial resistance to extended-spectrum β-lactam antibiotics and β-lactamase inhibitors as a result of the production of β-lactamase mutants is a cause for concern. The TEM ESBLs and the IRT β-lactamases differ from the parent enzymes TEM-1 and TEM-2 by one, two, or three amino acid substitutions at different locations. Table 12.2 lists the amino acid substitutions in the TEM ESBLs and IRT β-lactamases.

### SHV-Type β-Lactamases

The group of plasmid-mediated SHV β-lactamases includes SHV-1 and at least 23 variants, most of which possess extended-spectrum activity against the extended-spectrum cephalosporins such as cefotaxime and ceftazidime. The name SHV was given by M. Matthew and colleagues and means "sulfhydryl variable," in reference to the enzyme interaction with *p*-chloromercuribenzoate (PCMB), an agent that binds to sulfhydryl groups. This is a particular trait of SHV-1 since other common plasmidic β-lactamases, of the TEM, OXA, and PSE types, are unaffected by PCMB.

SHV enzymes belong to Ambler class A or the Bush-Jacoby-Medeiros group 2be β-lactamases. From the clinical point of view, the plasmid-encoded class A β-lactamases are the most significant. β-Lactamase genes (*bla*), such as $bla_{TEM-1}$, $bla_{TEM-2}$, and $bla_{SHV-1}$, are found in gram-negative bacteria. It is of interest that the early members of the SHV β-lactamases were found particularly in strains of *Klebsiella pneumoniae*. Later, *E. coli* was also found to be a host for variants of this family. The genes encoding these β-lactamases are, however, frequently found on self-transmissible plasmids. They have now been found in other members of the *Enterobacteriaceae,* including important nosocomial pathogens such as *Citrobacter diversus, K. oxytoca,* and *Morganella morganii.* In 1999 an SHV-derived ESBL was found in *Pseudomonas aeruginosa.* This finding indicates that the $bla_{SHV}$ genes have mobilized to a wide range of gram-negative pathogens.

As mentioned earlier, to combat the increasing problem of antibiotic-resistant bacteria, particularly in the hospital setting, extended-spectrum cephalosporins were introduced into clinical practice in the early 1980s. Cefotaxime was the first of these antibiotics to be used in clinical practice. Ceftazidime and ceftriaxone are other examples of this subgroup of cephalosporins. By the mid-1980s, resistance to extended-spectrum cephalosporins had appeared in clinically significant gram-negative bacteria. The first extended-spectrum SHV enzyme was described in 1983 in clinical isolates of *K. pneumoniae* and *Serratia marcescens.* This new enzyme was named SHV-2. A single-amino-acid substitution had altered the spectrum of activity of the SHV-1 β-lactamase to encompass extended-spectrum cephalosporins. Glycine at position 238 in SHV-1 is replaced by serine in SHV-2. SHV-1 is a narrow-spectrum β-lactamase with activity against penicillins (ampicillin, amoxicillin, penicillin G, ticarcillin, mezlocillin, and piperacillin) and early-generation cephalosporins, such as cephaloridine, cephalothin, cephalexin, and cefazolin. The three-dimensional structure of SHV-1 was established by X-ray crystallography (Kuzin et al., 1999).

## THE STRUCTURE-FUNCTION RELATIONSHIP IN SHV β-LACTAMASES

The amino acid substitutions and isoelectric points in SHV ESBLs and IR β-lactamases are summarized in Table 12.3.

**Table 12.2** Amino acid substitutions in the TEM ESBLs and inhibitor-resistant β-lactamases[a]

| β-Lactamase | Alternative name(s) | Amino acid[b] at position: | | | | | | | | | | | | | | | | | | | | | | | | | | | pI |
|---|---|---|---|---|---|---|---|---|---|---|---|---|---|---|---|---|---|---|---|---|---|---|---|---|---|---|---|---|---|
| | | 6 | 21 | 39 | 42 | 51 | 69 | 92 | 104 | 115 | 127 | 130 | 153 | 164 | 165 | 182 | 196 | 204 | 218 | 237 | 238 | 240 | 244 | 262 | 265 | 268 | 275 | 276 | |
| TEM-1 | RTEM-1 | Q | L | Q | A | L | M | G | E | D | I | S | H | R | W | M | G | R | G | A | G | E | R | V | T | S | R | N | 5.4 |
| TEM-2 | | | | K | | | | | | | | | | | | | | | | | | | | | | | | | 5.6 |
| TEM-3 | CTX-1, TEM-14 | | | K | | | | | K | | | | | | | | | | | | S | | | | | | | | 6.3 |
| TEM-4 | | | F | | | | | | K | | | | | | | | | | | | S | | | | M | | | | 5.9 |
| TEM-5 | CAZ-1 | | | | | | | | | | | | | S | | | | | | T | | K | | | | | | | 5.55 |
| TEM-6 | | | | | | | | | K | | | | | H | | | | | | | | | | | | | | | 5.9 |
| TEM-7 | | | | K | | | | | | | | | | S | | | | | | | | | | | | | | | 5.4 |
| TEM-8 | CAZ-2 | | | K | | | | | K | | | | | S | | | | | | | S | | | | | | | | 5.9 |
| TEM-9 | RHH-1 | | F | | | | | | K | | | | | S | | | | | | | | | | | M | | | | 5.5 |
| TEM-10 | MGH-1, TEM-E3, TEM-23 | | | | | | | | | | | | | S | | | | | | | | K | | | | | | | 5.6 |
| TEM-11 | CAZ-lo | | | K | | | | | | | | | | H | | | | | | | | | | | | | | | 5.6 |
| TEM-12 | YOU-2, CAZ-3, TEM-E2 | | | | | | | | | | | | | S | | | | | | | | | | | | | | | 5.25 |
| TEM-13 | | | | | | | | | K | | | | | | | | | | | | | | | | M | | | | 5.6 |
| TEM-15 | | | | | | | | | K | | | | | | | | | | | | S | | | | | | | | 6.0 |
| TEM-16 | CAZ-7 | | | K | | | | | K | | | | | H | | | | | | | | | | | | | | | 6.3 |
| TEM-17 | | | | | | | | | K | | | | | | | | | | | | | | | | | | | | 5.5 |
| TEM-18 | | | | K | | | | | K | | | | | | | | | | | | | | | | | | | | 6.3 |
| TEM-19 | | | | | | | | | | | | | | | | | | | | | S | | | | | | | | 5.4 |
| TEM-20 | | | | | | | | | | | | | | | | T | | | | | S | | | | | | | | 5.4 |
| TEM-21 | | | | K | | | | | K | | | | R | | | | | | | | S | | | | | | | | 6.4 |
| TEM-22 | | | | K | | | | | K | | | | | | | | | | | G | S | | | | | | | | 6.3 |

(continued)

**Table 12.2** Amino acid substitutions in the TEM ESBLs and inhibitor-resistant β-lactamases (*continued*)

| β-Lactamase | Alternative name(s) | Amino acid[b] at position: | | | | | | | | | | | | | | | | | | | | | | | | | | | pI |
|---|---|---|---|---|---|---|---|---|---|---|---|---|---|---|---|---|---|---|---|---|---|---|---|---|---|---|---|---|---|
| | | 6 | 21 | 39 | 42 | 51 | 69 | 92 | 104 | 115 | 127 | 130 | 153 | 164 | 165 | 182 | 196 | 204 | 218 | 237 | 238 | 240 | 244 | 262 | 265 | 268 | 275 | 276 | |
| TEM-24 | CAZ-6 | | | K | | | | | K | | | | | S | | | | | | T | | K | | | | | | | 6.5 |
| TEM-25 | CTX-2 | | F | | | | | | | | | | | | | | | | | | S | | | | M | | | | 5.3 |
| TEM-26 | YOU-1 | | | | | | | | K | | | | | S | | | | | | | | | | | | | | | 5.6 |
| TEM-27 | | | | | | | | | | | | | | H | | | | | | | | K | | | M | | | | 5.9 |
| TEM-28 | | | | | | | | | | | | | | H | | | | | | | | K | | | | | | | 6.1 |
| TEM-29 | | | | | | | | | | | | | | H | | | | | | | | | | | | | | | 5.42 |
| TEM-30 | IRT-2, TRI-2, E-GUER | | | | | | | | | | | | | | | | | | | | | | S | | | | | | 5.2 |
| TEM-31 | IRT-1, TRI-1, E-SAL | | | | | | | | | | | | | | | | | | | | | | C | | | | | | 5.2 |
| TEM-32 | IRT-3 | | | | | | I | | | | | | | | | T | | | | | | | | | | | | | 5.4 |
| TEM-33 | IRT-5 | | | | | | L | | | | | | | | | | | | | | | | | | | | | | 5.4 |
| TEM-34 | IRT-6 | | | | | | V | | | | | | | | | | | | | | | | | | | | | | 5.4 |
| TEM-35 | IRT-4 | | | | | | L | | | | | | | | | | | | | | | | | | | | | D | 5.2 |
| TEM-36 | IRT-7 | | | | | | V | | | | | | | | | | | | | | | | | | | | | D | 5.2 |
| TEM-37 | | | | | | | I | | | | | | | | | | | | | | | | | | | | | D | 5.2 |
| TEM-38 | IRT-9 | | | | | | V | | | | | | | | | | | | | | | | | | | | L | | 5.2 |
| TEM-39 | IRT-10 | | | | | | L | | | | | | | | R | | | | | | | | | | | | | D | 5.4 |
| TEM-40 | IRT-167, IRT-11 | | | | | | I | | | | | | | | | | | | | | | | | | | | | | 5.4 |
| TEM-41[c] | IRT-12 | | | | | | | | | | | | | | | | | | | | | | T | | | | | | 5.2 |
| TEM-42 | | | | K | V | | | | | | | | | | | | | | | | S | K | | | M | | | | 5.8 |
| TEM-43 | | | | | | | | | K | | | | | H | | T | | | | | | | | | | | | | 6.1 |
| TEM-44 | IRT$_2$-2, IT-13 | | | K | | | | | | | | | | | | | | | | | | | S | | | | | | 5.4 |
| TEM-45 | IRT-14 | | | | | | L | | | | | | | | | | | | | | | | | | | | Q | | 5.2 |
| TEM-46 | CAZ-9 | | | K | | | | | K | | | | | S | | | | | | | | K | | | | | | | 6.5 |
| TEM-47 | | | | | | | | | | | | | | | | | | | | | S | K | | | M | | | | 6.0 |
| TEM-48 | | | F | | | | | | | | | | | | | | | | | | S | K | | | M | | | | 6.0 |
| TEM-49 | | | F | | | | | | | | | | | | | | | | | | S | K | | | M | G | | | 6.0 |
| TEM-50 | CMT-1 | | | | | | L | | K | | | | | | | | | | | | S | | | | | | | D | 5.6 |
| TEM-51 | IRT-15 | | | | | | | | | | | | | | | | | | | | | | H | | | | | | 5.2 |

| | | | | | | | | | | | | | | | | | | | | | | | | | | | | | |
|---|---|---|---|---|---|---|---|---|---|---|---|---|---|---|---|---|---|---|---|---|---|---|---|---|---|---|---|---|---|
| TEM-52 | | | | | | | | | K | | | | | | | T | | | | | S | | | | | | | | 6.0 |
| TEM-53 | | | F | | | | | | | | | | | S | | | | | | | | | | | | | | | |
| TEM-54 | | | | | | | | | | | | | | | | | | | | | | | L | | | | | | |
| TEM-55 | | | | | | | | | | | | | | | | | | | E | | | | | | | | | | 5.2 |
| TEM-56 | | | | K | | | | | K | | | | R | | | | | | | | | | | | | | | | 6.4 |
| TEM-57 | | | | | | | | D | | | | | | | | | | | | | | | | | | | | | 5.2 |
| TEM-58 | | | | | | | | | | | | | | | | | | | | | | | S | I | | | | | 5.2 |
| TEM-59 | IRT-17 | | | K | | | | | | | | G | | | | | | | | | | | | | | | | | 5.6 |
| TEM-60 | | | | K | | P | | | K | | | | | S | | | | | | | | | | | | | | | 6.4 |
| TEM-61 | CAZ-hi | | | K | | | | | | | | | | H | | | | | | | | K | | | | | | | 6.5 |
| TEM-62 | Reserved | | | | | | | | | | | | | | | | | | | | | | | | | | | | |
| TEM-63 | TEM-64 | | F | | | | | | K | | | | | S | | T | | | | | | | | | | | | | 5.6 |
| TEM-65 | IRT-16 | | | K | | | | | | | | | | | | | | | | | | | C | | | | | | 5.4 |
| TEM-66 | | | | K | | | | D | K | | | | | | | | | | | | S | | | | | | | | 6.0 |
| TEM-67 | Not yet released | | | | | | | | | | | | | | | | | | | | | | | | | | | | |
| TEM-68 | | | | | | | | | | | | | | | | | | | | | S | K | | | M | | L | | 5.7 |
| TEM-69 | Withdrawn | | | | | | | | | | | | | | | | | | | | | | | | | | | | |
| TEM-70 | | | | | | | | | | | | | | | | | | Q | | | | | | | | | | | 5.2 |
| TEM-71 | Not yet released | | | | | | | | | | | | | | | | | | | | | | | | | | | | |
| TEM-72 | | | | K | | | | | | | | | | | | T | | | | | S | K | | | | | | | 5.9 |
| TEM-73 | IRT-18 | | F | | | | | | | | | | | | | | | | | | | | C | | M | | | | 5.2 |
| TEM-74 | IRT-19 | | F | | | | | | | | | | | | | | | | | | | | S | | M | | | | 5.2 |
| TEM-75 | Not yet released | | | | | | | | | | | | | | | | | | | | | | | | | | | | |
| TEM-76 | IRT-20 | | | | | | | | | | | G | | | | | | | | | | | | | | | | | |
| TEM-77 | IRT-21 | | | | | | L | | | | | | | | | | | | | | | | S | | | | | | |
| TEM-78 | IRT-22 | | | | | | V | | | | | | | | R | | | | | | | | | | | | | D | |
| TEM-79 | IRT-23 | | | | | | | | | | | | | | | | | | | | | | G | | | | | | |

*(continued)*

**Table 12.2** Amino acid substitutions in the TEM ESBLs and inhibitor-resistant β-lactamases *(continued)*

| β-Lactamase | Alternative name(s) | Amino acid[b] at position: | | | | | | | | | | | | | | | | | | | | | | | | | | | pI |
|---|---|---|---|---|---|---|---|---|---|---|---|---|---|---|---|---|---|---|---|---|---|---|---|---|---|---|---|---|---|
| | | 6 | 21 | 39 | 42 | 51 | 69 | 92 | 104 | 115 | 127 | 130 | 153 | 164 | 165 | 182 | 196 | 204 | 218 | 237 | 238 | 240 | 244 | 262 | 265 | 268 | 275 | 276 | |
| TEM-80 | Not yet released | | | | | | | | | | | | | | | | | | | | | | | | | | | | |
| TEM-81 | IRT-23 | | | | | | L | | | | V | | | | | | | | | | | | | | | | | | |
| TEM-82 | IRT-24 | | | | | | V | | | | | | | | | | | | | | | | | | | | Q | | |
| TEM-83 | IRT-25 | | | | | | I | | | | | | | | C | | | | | | | | | | | | Q | | |
| TEM-84 | IRT-26 | | | | | | | | | | | | | | | | | | | | | | | | | | | D | |
| TEM-85 | | | F | | | | | | | | | | | S | | | | | | | | K | | | M | | | | |
| TEM-86 | | | F | | | | | | | | | | | S | | | | | | T | | K | | | M | | | | |
| TEM-87 | | K | | | | | | | K | | | | | C | | T | | | | | | | | | | | | | |
| TEM-88 | | | | | | | | | K | | | | | | | T | D | | | | S | | | | | | | | 5.6 |
| TEM-89 | | | | K | | | | | K | | | G | | | | | | | | | S | | | | | | | | |
| TEM-90 | TLE-1 | | | | | | | | | G | | | | | | | | | | | | | | | | | | | 5.55 |
| TEM-91 | | | | | | | | | | | | | | C | | T | | | | | | K | | | | | | | |
| TEM-92 | | K | | | | | | | K | | | | | | | T | | | | | S | | | | | | | | 6.0 |
| TEM-93 | | | | | | | | | | | | | | | | T | | | | | S | K | | | | | | | |
| TEM-94 | Not yet released | | | | | | | | | | | | | | | | | | | | | | | | | | | | |
| TEM-95 | Not yet released | | | | | | | | | | | | | | | | | | | | | | | | | | | | |
| TEM-96 | Not yet released | | | | | | | | | | | | | | | | | | | | | | | | | | | | |

[a] Reprinted with permission from Lahey Clinic.

[b] Abbreviations: A, alanine; C, cysteine; D, aspartic acid; E, glutamic acid; F, phenylalanine; G, glycine; H, histidine; I, isoleucine; K, lysine; L, leucine; M, methionine; N, asparagine; P, proline; Q, glutamine; R, arginine; S, serine; T, threonine; V, valine; W, tryptophan. DBL numbering per Couture et al., *Mol. Microbiol.* 6:1693–1705, 1992.

[c] Withdrawn.

**Table 12.3** Amino acid substitutions in the SHV ESBLs and IRT β-lactamases[a]

| β-Lactamase | Alternative name | Amino acid[b] at position: | | | | | | | | | | | | | | | | | | | | | | | | | pI |
|---|---|---|---|---|---|---|---|---|---|---|---|---|---|---|---|---|---|---|---|---|---|---|---|---|---|---|---|
| | | 7 | 8 | 18 | 35 | 43 | 48 | 54 | 75 | 80 | 89 | 122 | 129 | 130 | 140 | 156 | 158 | 173 | 179 | 187 | 188 | 192 | 193 | 205 | 238 | 240 | |
| SHV-1 | | F | I | | L | R | E | G | V | V | E | L | M | S | A | G | N | N | D | A | A | K | L | R | G | E | 7.6 |
| SHV-2 | | | | | | | | | | | | | | | | | | | | | | | | | S | | 7.6 |
| SHV-2A | | | | | Q | | | | | | | | | | | | | | | | | | | | S | | 7.6 |
| SHV-3 | | | | | | | | | | | | | | | | | | | | | | | | L | S | | 7.0 |
| SHV-4 | | | | | | | | | | | | | | | | | | | | | | | | L | S | K | 7.8 |
| SHV-5 | | | | | | | | | | | | | | | | | | | | | | | | | S | K | 8.2 |
| SHV-6 | | | | | | | | | | | | | | | | | | | A | | | | | | | | 7.6 |
| SHV-7 | | | F | | | S | | | | | | | | | | | | | | | | | | | S | K | 7.6 |
| SHV-8 | | | | | | | | | | | | | | | | | | | N | | | | | | | | 7.6 |
| SHV-9 | SHV-5a | | | | | | | Del | | | | | | | R | | | | | | | N | V | | S | K | 8.2 |
| SHV-10 | | | | | | | | Del | | | | | | G | R | | | | | | | N | V | | S | K | 8.2 |
| SHV-11 | SHV-1-2a | | | | Q | | | | | | | | | | | | | | | | | | | | | | 7.6 |
| SHV-12 | SHV-5-2a | | | | Q | | | | | | | | | | | | | | | | | | | | S | K | 8.2 |
| SHV-13 | | | | | Q | | | | | | | | | | | | | | | | | | | | A | | 7.6 |
| SHV-14 | | | F | | | S | | | | | | | | | | | | | | | | | | | | | 7.0 |
| SHV-15 | | | | | Q | | K | | M | M | K | | | | | | | | | | | | | | S | K | |
| SHV-16 | 163DRWET167 insertion | | | | | | | | | | | | | | | | | | | | | | | | | | 7.6 |
| SHV-17 | Withdrawn | | | | | | | | | | | | | | | | | | | | | | | | | | |
| SHV-18 | | | F | | | S | | | | | | | | | | | | | | | | | | | A | K | 7.8 |
| SHV-19 | | | | | | | | | | | | | | | | | | F | | | | | | | | | 7.6 |
| SHV-20 | | | | | | | | | | | | | | | | | | F | | | | | | | S | | 7.6 |
| SHV-21 | | | | | | | | | | | | F | | | | | | F | | | | | | | S | | 7.6 |
| SHV-22 | | | | | | | | | | | | | | | | | K | | | | | | | | D | K | 7.6 |
| SHV-23 | Not yet released | | | | | | | | | | | | | | | | | | | | | | | | | | |
| SHV-24 | | | | | | | | | | | | | | | | | | | | G | | | | | | | |
| SHV-25 | | | | A | Q | | | | | | | | V | | | | | | | | | | | | | | 7.5 |
| SHV-26 | | | | | | | | | | | | | | | | | | | | | T | | | | | | 7.6 |
| SHV-27 | | | | | | | | | | | | | | | | D | | | | | | | | | | | |
| SHV-28 | Not yet released | | | | | | | | | | | | | | | | | | | | | | | | | | |
| SHV-29 | Not yet released | | | | | | | | | | | | | | | | | | | | | | | | | | |

(continued)

**Table 12.3** Amino acid substitutions in the SHV ESBLs and IRT β-lactamases *(continued)*

| β-Lactamase | Alternative name | Amino acid[b] at position: | | | | | | | | | | | | | | | | | | | | | | | | pI |
|---|---|---|---|---|---|---|---|---|---|---|---|---|---|---|---|---|---|---|---|---|---|---|---|---|---|---|
| | | 7 | 8 | 18 | 35 | 43 | 48 | 54 | 75 | 80 | 89 | 122 | 129 | 130 | 140 | 156 | 158 | 173 | 179 | 187 | 188 | 192 | 193 | 205 | 238 | 240 | |
| SHV-30 | Not yet released | | | | | | | | | | | | | | | | | | | | | | | | | | |
| SHV-31 | Not yet released | | | | | | | | | | | | | | | | | | | | | | | | | | |
| SHV-32 | Not yet released | | | | | | | | | | | | | | | | | | | | | | | | | | |
| SHV-33 | Not yet released | | | | | | | | | | | | | | | | | | | | | | | | | | |
| SHV-34 | Not yet released | | | | | | | | | | | | | | | | | | | | | | | | | | |
| SHV-35 | Not yet released | | | | | | | | | | | | | | | | | | | | | | | | | | |
| SHV-36 | Not yet released | | | | | | | | | | | | | | | | | | | | | | | | | | |
| SHV-37 | Not yet released | | | | | | | | | | | | | | | | | | | | | | | | | | |

[a] Reprinted with permission from Lahey Clinic.

[b] DBL numbering per Couture et al., *Mol. Microbiol.* **6**:1693–1705, 1992.

Excellent reviews by Tzouvelekis and Bonomo (1999), Heritage et al. (1999), and Du Bois et al. (1995) have recently discussed the structure-function relationship in SHV β-lactamases; for this reason, this topic is not included here.

## OXA-Type β-Lactamases

The OXA-type (oxacillin-hydrolyzing) enzymes are widespread and have been found mostly in members of the *Enterobacteriaceae* and *P. aeruginosa*. They usually confer resistance to amino- and ureidopenicillins and possess high-level hydrolytic activity against oxacillin, cloxacillin, and methicillin. Their activities are weakly inhibited by clavulanic acid.

OXA-type β-lactamases belong to Ambler class D or to group 2d of the Bush-Jacoby-Medeiros functional classification. The class D β-lactamase group was classified as a new group of serine β-lactamases in the late 1980s, as a result of acquisition of sequencing data. Recently, the crystal structure of the OXA-10 β-lactamase was reported at high resolution, simultaneously by the groups of Mobashery and Samama (2000) and Strynadka et al. (2000).

Five OXA-type ESBLs which are poorly inhibited by clavulanic acid have been described for *P. aeruginosa:* OXA-11, OXA-14, OXA-16, and OXA-17, which are derived from OXA-10, and OXA-15, which is derived from OXA-2. OXA-18 has a broad substrate profile, hydrolyzing amoxicillin, ticarcillin, cephalothin, ceftazidime, cefotaxime, and aztreonam but not imipenem, meropenem, or cephamycins (cefoxitin).

THE STRUCTURE-FUNCTION RELATIONSHIP IN OXA β-LACTAMASES

The amino acid substitutions and isoelectric points for OXA ESBLs are summarized in Table 12.4.

## Altered Penicillin-Binding Proteins

Resistance to β-lactam antibiotics can also develop through the production of altered PBPs. When penicillin G and penicillin V were introduced into clinical practice in the 1940s, over 95% of *Staphylococcus aureus* isolates were susceptible; by 1950, half were resistant. By 1960, many hospitals had experienced outbreaks of infections due to multiresistant *S. aureus*. Methicillin was introduced in 1960 to treat staphylococcal infections. This compound has a bulky *o,o*'-dimethoxybenzene acyl group that sterically hinders the attack on the β-lactam ring, allowing the antibiotic to be effective even in the presence of β-lactamase-producing *S. aureus* strains (see chapter 8). However, in 1961, shortly after methicillin was introduced, methicillin-resistant *S. aureus* (MRSA) containing the *mec* determinant were isolated. These strains had acquired a chromosomal gene (*mecA*) encoding a new high-molecular-mass PBP, PBP2a (or PBP2'), that has a low affinity for β-lactam antibiotics. PBP2a is

**Table 12.4** Amino acid substitution in the OXA extended-spectrum β-lactamases[a]

| β-Lactamase | Alternative name | Amino acid[b] at position: | | | | | | | | | | | | | | | | | | pI |
|---|---|---|---|---|---|---|---|---|---|---|---|---|---|---|---|---|---|---|---|---|
| | | 10 | 20 | 48 | 58 | 67 | 76 | 110 | 127 | 131 | 144 | 149 | 164 | 167 | 184 | 208 | 240 | 258 | 272 | |
| OXA-10 | PSE-2 | I | G | | D | | N | T | A | | Y | | W | G | Y | | E | S | E | 6.1 |
| OXA-11 | | | | | | | | | | | S | | | D | | | | | | 6.4 |
| OXA-14 | | | | | | | | | | | | | | D | | | | | | 6.2 |
| OXA-16 | | | | | | | | | T | | | | | D | | | | | | 6.2 |
| OXA-17 | | | | | | | S | | | | | | | | | | | | | 6.1 |
| OXA-19 | | T | S | | N | | | S | | | | | | D | F | | G | N | A | 7.5 |
| OXA-28 | | T | S | | N | | | S | | | | | G | | F | | G | N | A | 8.1 |
| OXA-2 | | | | | | | | | | | | D | | | | | | | | 7.7 |
| OXA-15 | | | | | | | | | | | | G | | | | | | | | 8.0 |
| OXA-18 | See reference for sequence data | | | | | | | | | | | | | | | | | | | 5.5 |
| OXA-1 | | | | A | | A | | | | R | | | | | | D | | | | |
| OXA-31 | | | | V | | P | | | | G | | | | | | L | | | | |

[a] Reprinted with permission from Lahey Clinic.

[b] DBL numbering per Couture et al., *Mol. Microbiol.* 6:1693–1705, 1992. For abbreviations, see footnote *b* in Table 12.2.

an inducible 76-kDa PBP that determines methicillin resistance.

Both susceptible and resistant strains of *S. aureus* produce four major PBPs, PBP1 through PBP4, with approximate molecular masses of 85, 81, 75, and 45 kDa, respectively. Whereas susceptible *S. aureus* strains employ these PBPs to catalyze the transpeptidation reaction that cross-links the peptidoglycan linear polymers of the bacterial cell wall, it is generally thought that PBP2a in MRSA takes over the enzymatic functions of PBP1 through PBP4 in peptidoglycan synthesis when they are inactivated. Profound differences become apparent only when the normal PBPs of the cell are inactivated by methicillin and PBP2a alone is left functional. *MRSA is resistant to all the currently marketed β-lactam antibiotics.* Whereas more than 60% of peptidoglycan is cross-linked in normal growing cells, cross-linking decreases to 15% in the presence of MRSA, suggesting that PBP2a restricts the formation of the cross-linked polymer.

Approximately 30 kb of additional chromosomal DNA *mec,* not found in susceptible strains of staphylococci, is present in MRSA. *mec* is always found near the *pur-nov-his* gene cluster on the *S. aureus* chromosome (Fig. 12.1). *mec* contains *mecA,* the structural gene for PBP2a; *mecI* and *mecR1,* regulatory elements controlling *mecA* transcription; and 20 to 45 kb of *mec*-associated DNA.

**Figure 12.1** Molecular organization of the approximately 50-kb *mec* region and its chromosomal location relative to *fem* factors and *pur-nov-his.* IS indicates IS*431* elements flanking the tobramycin resistance plasmid pUB110. Tn*554* is a transposon containing e*rmA,* encoding inducible erythromycin resistance. Reprinted from H. F. Chambers, *Clin. Microbiol. Rev.* **10:**781–791, with permission from the American Society for Microbiology.

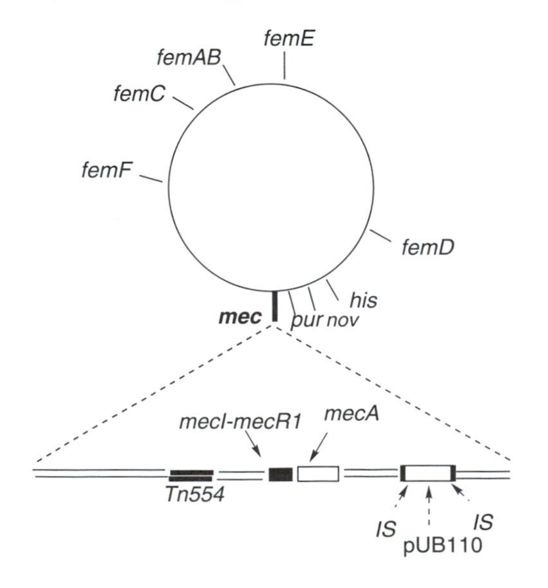


*mecA,* the gene encoding PBP2a, has been cloned and sequenced. Its flanking sequences (*mec* DNA) make up a chromosomal DNA locus that is unique to methicillin-resistant bacteria; no equivalent locus exists in methicillin-susceptible strains. The origin of *mec* DNA is not known, but evidence supports the horizontal transfer of *mec* DNA between different staphylococcal species and of the *mecA* gene between different gram-positive genera. Therefore, the spread is clonal.

The methicillin resistance determinant *mec* gives staphylococci an intrinsic resistance to all β-lactam antibiotics. MRSA has become one of the most important nosocomial pathogens worldwide and poses serious infection control problems because of its tendency to accumulate additional unrelated resistance determinants and incorporate them into its genome. The adaptability and quick response of MRSA strains to antibiotic selection has led, in less than 40 years, to the evolution of strains that are multiresistant to many antibiotics, including in many cases glycopeptides such as vancomycin and teicoplanin. Glycopeptide intermediate-resistant *S. aureus,* first called vancomycin intermediate-resistant *S. aureus,* strains have now been documented in Europe, Japan, and the United States. Alternative treatments are sometimes necessary due to intolerance or treatment failures associated with glycopeptides. Recently, the antibiotics quinupristin-dalfopristin (a mixture of streptogramins) have been introduced for the treatment of MRSA infections in patients intolerant of or failing prior therapy.

## *fem* Factors

Transposon insertional mutagenesis producing susceptible mutants from methicillin-resistant strains has led to the identification of chromosomal genes necessary for full expression of resistance. *fem* factors (for "factors essential for methicillin resistance") are present in both susceptible and resistant strains. Six *fem* genes, *femA, femB, femC, femD, femE,* and *femF,* which map to numerous sites throughout the staphylococcal genome have been characterized.

Figure 12.2 shows the sites of peptidoglycan precursor synthesis in which the block occurs in *fem* mutants. Except for *femE,* which is of unknown function, *fem* mutants produce altered peptidoglycan composition.

## Inducible AmpC β-Lactamases

AmpC β-lactamases are active-site serine β-lactamases. These enzymes belong to class C of Ambler's structural classification or to group 1 of the Bush-Jacoby-Medeiros functional classification. They are encoded by chromosomally located *bla* genes in members of the *Enterobac-*

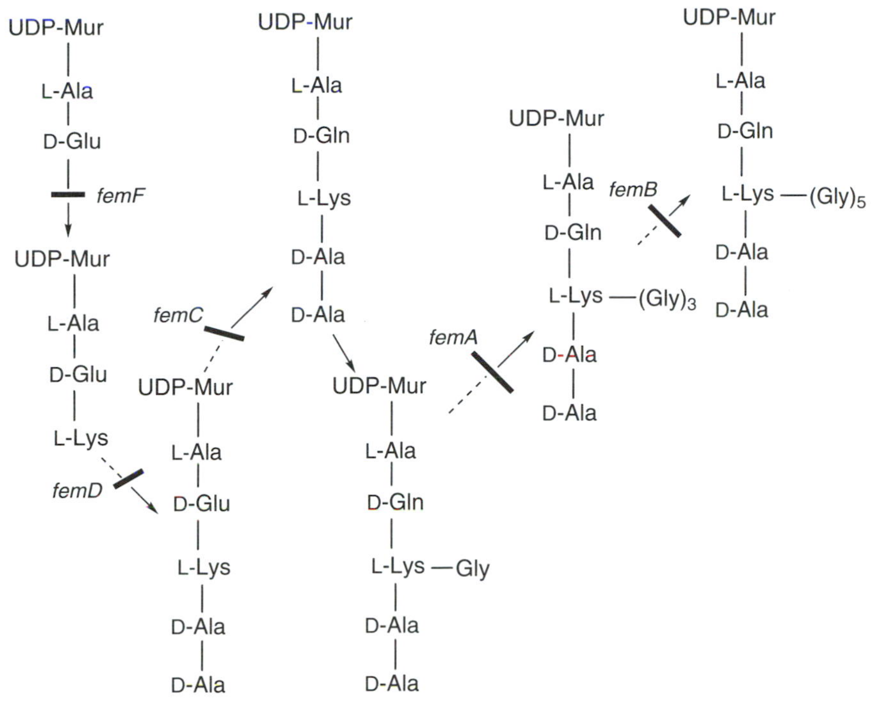

**Figure 12.2** Sites of peptidoglycan precursor synthesis at which blocks occur in *fem* mutants. UDP-Mur, uridine diphosphomuramyl peptide precursors; NAG-NAM, *N*-acetylglucosamine-*N*-acetylmuramic acid disaccharide (not shown in figure).

*teriaceae,* principally *Enterobacter, Serratia, Citrobacter,* and *Providencia* species and *Escherichia coli.* Unlike many class A enzymes, all class C enzymes are relatively resistant to inhibitor compounds such as clavulanic acid, sulbactam, and tazobactam. They are described as primarily cephalosporinases, although they hydrolyze most β-lactam antibiotics.

In many gram-negative organisms, including *Enterobacter* spp., *Citrobacter freundii, S. marcescens, M. morganii,* and *P. aeruginosa,* the expression of chromosomal *ampC* genes is generally low but inducible in response to β-lactam antibiotics. However, not all gram-negative species carry an inducible *ampC* gene. Expression of *ampC* in *E. coli* is constitutively low, not inducible, and only small amounts of the enzyme are produced. The noninducibility of *ampC* expression in *E. coli* results from the absence of the *ampR* gene, which encodes a DNA-binding protein required for induction.

Class C β-lactamases were previously called chromosomally mediated enzymes. However, in 1989, clinically derived strains of *K. pneumoniae* which exhibited transferable resistance to a broad range of β-lactam antibiotics including cephamycins were isolated. In July 2002, R. A. Powers and B. K. Shoichet of Northwestern University (Evanston, Ill.) reported the identification of an AmpC-binding site on the enzyme by using experimental X-ray crystal structures of AmpC complexed with four boromic acid inhibitors and a high-resolution (1.72 Å) native apoenzyme structure and computational approaches. Data about AmpC-type enzymes can be found at http://www.rochester.edu/College/BIO/labs/HallLab/AmpC_Phylo.html.

On interaction of a β-lactam antibiotic with PBPs, peptidoglycan synthesis is perturbed and increased amounts of degradation products of the peptidoglycan are formed. These degradation products accumulate in the periplasm. They are transported to the cytoplasm, where they can act as an inducer, converting the transcriptional regulator protein AmpR into an activator of β-lactamase expression. AmpG, a transmembrane protein, is required for high-level expression of β-lactamase. Jacobs et al. suggested in 1994 that this protein acts as a permease for the muropeptide *N*-acetylglucosaminyl-1, 6-anhydro-*N*-acetylmuramyl-L-alanyl-D-glutamylmesodiaminopimelic acid (Gluc-NAc-anhMurNAc tripeptide). Most β-lactam antibiotics are poor inducers, particularly third- and fourth-generation cephalosporins, but ampicillin, amoxicillin, and cephalexin are moderately good inducers and cefoxitin, imipenem, and meropenem are strong inducers. Figure 12.3 is a diagram of the relative induction of β-lactamase synthesis in a *C. freundii* mutant as a function of concentrations of various β-lactam antibiotics.

The working model for *ampC* induction by β-lactam antibiotics involves the binding of the drug to PBPs, thus disrupting cell wall synthesis. Interference with these cell wall-synthesizing enzymes results in the release of Glc-NAc-anhMurNAc peptides into the periplasmic space (Fig. 12.4). These liberated peptides enter the cytoplasm

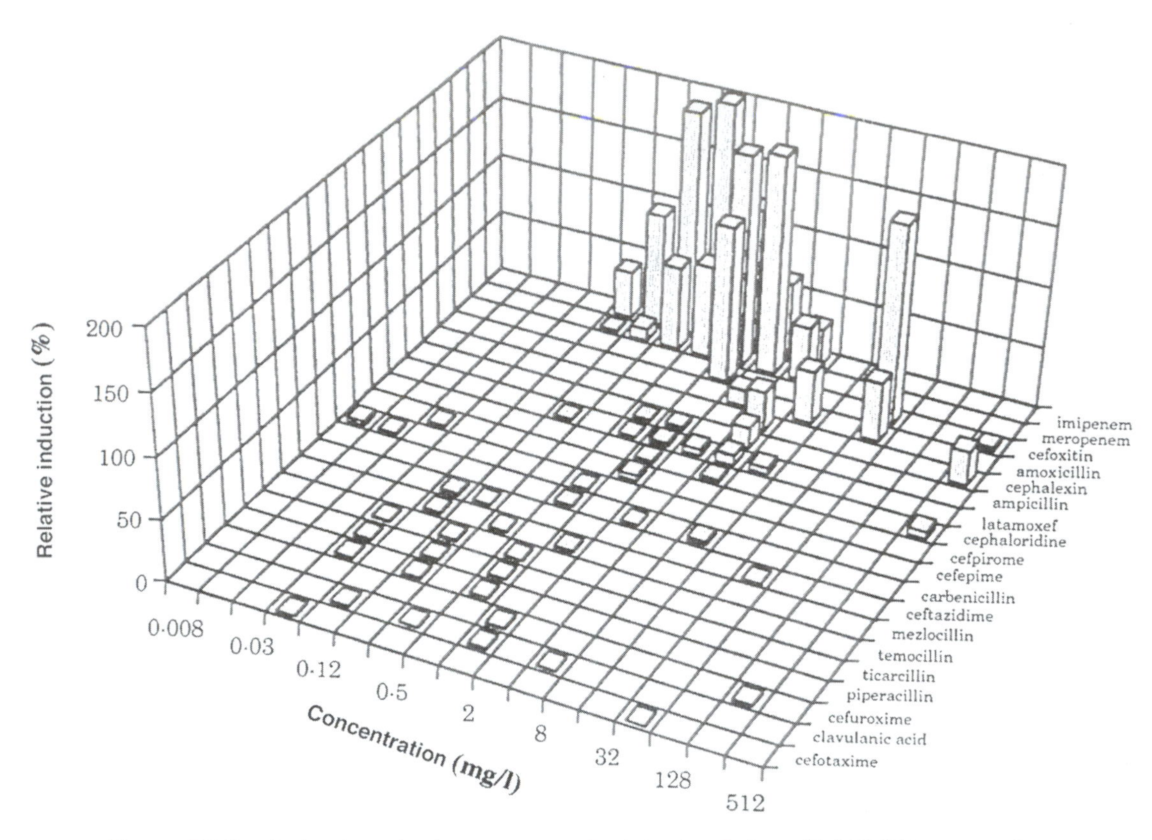


**Figure 12.3** Induction of β-lactamase synthesis in the *C. freundii* 382 010 mutant by β-lactam antibiotics. Reprinted from P. Stapleton, K. Shannon, and I. Phillips, *J. Antimicrob. Chemother.* **36:**483–496, 1995, with permission from the publisher.

**Figure 12.4** Hypothetical model for β-lactamase induction in gram-negative bacteria. BL, β-lactam; PG, peptidoglycan; G, AmpG; R, repressor form of AmpR; A, activator form of AmpR; D, AmpD; E, AmpE. For details, see the text. Reprinted from P. M. Bennett and I. Chopra, *Antimicrob. Agents Chemother.* **37:**153–159, 1993, with permission from the American Society for Microbiology.

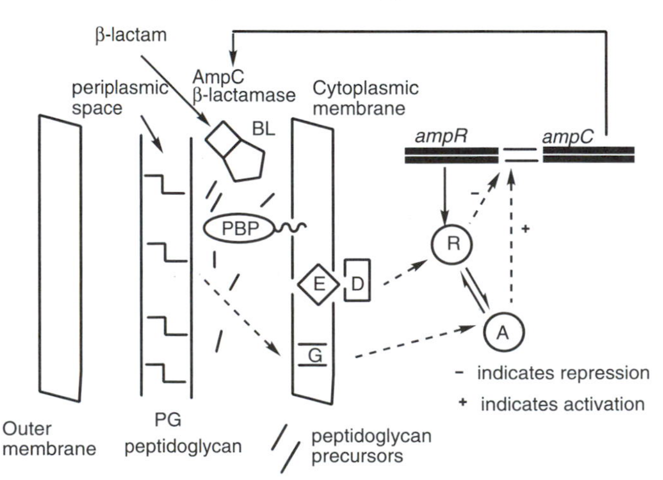


via the AmpG protein where the *N*-acetylglucosamine residue is cleaved off. These anhMurNAc peptides, which are now in excess, interact with AmpR, resulting in induction of *ampC* expression. It has been suggested that the anhMurNAc tripeptide acts as an antirepressor. In the uninduced state, the UDP-MurNAc pentapeptide is bound to AmpR, keeping AmpR in a conformational state incapable of inducing *ampC*. As cell wall synthesis is disrupted, the concentration of anhMurNAc tripeptide increases in the cytoplasm. The change in equilibrium of the anhMurNAc tripeptide and pentapeptide allows the displacement of UDP-MurNAc pentapeptide from AmpR and replacement with anhMurNAc tripeptide. It is possible that other anhMurNAc peptides could also act as antirepressors.

## Research and Developments in Novel β-Lactam Antibiotics and β-Lactam Inhibitors of Serine and Metallo-β-Lactamases

Global antibacterial resistance is an increasing public health problem. The pharmaceutical industries are reacting to the problem by discovering novel antibacterial agents to overcome the emergence of bacterial resistance to antibiotics and β-lactamase inhibitors.

This section summarizes the status of the development of structural modifications of existing groups and subgroups of antibacterial agents to make them less susceptible to degradation by β-lactamases, to increase penetrability through the outer membrane of gram-negative bacteria, or to have an increased affinity for mutated PBPs. It also covers the research and development of novel β-lactam inhibitors of serine and metallo-β-lactamases. Chapter 27 describes the research and development of new antibacterials with novel modes of action that have unique targets such that cross-resistance does not really occur.

This discussion starts with an overview of what has been done in the four years up to 2000 in the pharmaceutical industry as well as in academic laboratories to modify existing molecules of β-lactam antibiotics that act as cell wall inhibitors.

## New Developments in β-Lactam Antibiotics

### NOVEL ANTI-MRSA ANTIBIOTICS

The search for new anti-MRSA β-lactam antibiotics appears to be focused on developing agents that inhibit PBP2a, which gives rise to methicillin resistance in staphylococci and penicillin-resistant pneumococci. The need for new anti-MRSA antibiotics is clear. One effective strategy to overcome bacterial resistance is to improve the potency of known antibacterial agents through synthesis of analogues (structural modification).

#### Structural Modification of Cephalosporins

J. Singh and coworkers at Bristol-Myers Squibb Pharmaceutical Research Institute described a practical synthesis of an anti-MRSA cephalosporin, BMS-247243, a double zwitterion compound (Fig. 12.5). The 2,5-dichlorophenylthioacetamido C-7 group was discovered in the late 1970s by researchers at Eli Lilly to be a lipophilic side chain that conferred excellent activity to the cephem class of antibiotics against gram-positive bacteria. The C-3 thiopyridinium group was substituted at nitrogen with a 4-methylmorpholino-4-yl-propyl group that, according to the authors, was designed to confer aqueous solubility, as required for the intravenous formulation. The synthesis and anti-MRSA biological evaluation of other novel cephalosporin derivatives similar to this one were recently reported by S. D'Andrea and coworkers from the same laboratory.

#### Structural Modifications of Carbapenems

*Novel Anti-MRSA 1β-Methylcarbapenem Derivatives.* Because of their broad spectrum and high activity against many pathogenic bacteria, including *P. aeruginosa,* the carbapenem antibiotics imipenem and meropenem have opened a new field for the use of β-lactam antibiotics. However, these antibiotics do not show high activity against MRSA. MRSA is a major pathogen and is increasingly being disseminated in many medical institutions worldwide. In the search for new carbapenem compounds with activity against MRSA, several pharmaceutical companies are currently developing novel 1β-methylcarbapenem derivatives as well as tricyclic carbapenems (trinems).

Two syntheses of the very promising anti-MRSA 1β-methylcarbapenem (L-786,392) containing a **releasable side chain** were simultaneously reported by M. Jensen and coworkers and G. Humphrey and coworkers at Merck Research Laboratories. L-786,392 has a naphthosultam side chain substituent at position C-2, attached through a methylene linker (Fig. 12.6). This group was designed to provide high binding affinity for the hydrophobic pocket of PBP2a. This structural arrangement was also designed so that on opening of the β-lactam ring by acylation of the specific PBP, the side chain was released into solution. Consequently, red blood cells are not labeled with the potential immunogenic naphthosultam side chain. The researchers therefore hoped that hemolytic anemia, i.e., lysis of the marked red blood cells by the immune system, would not pose a problem for this compound. In this way, the pharmacophore is present to support the initial interaction with the PBP2a and the immunodominant epitope is expelled when the β-lactam ring is opened. L-786,392 proved to be effective against MRSA, methicillin-resistant coagulase-negative staphylococci, and resistant *E. faecium.* It also has good activity against vancomycin intermediate-resistant *S. aureus* strains.

Sumita and coworkers at the research laboratories of Sumimoto Pharmaceuticals in Japan synthesized the

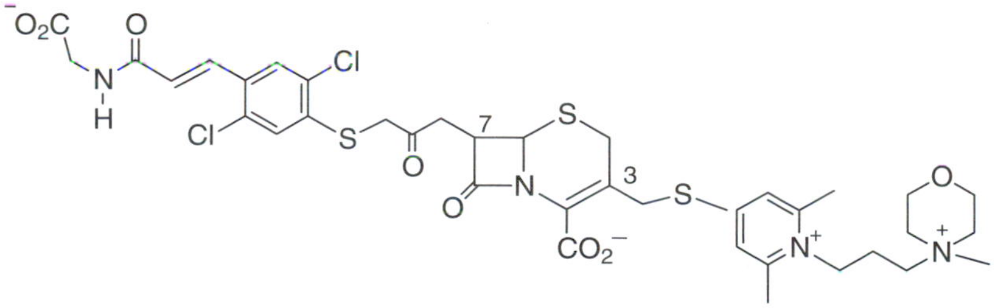

**Figure 12.5** Chemical structure of the cephalosporin BMS-247243.

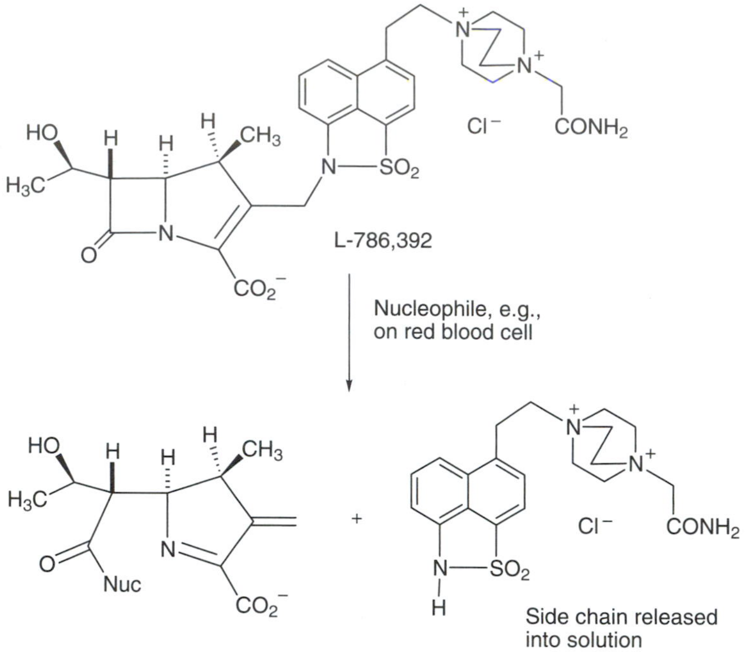

**Figure 12.6** Chemical structure of the β-methylcarbapenem L-786,392 (synthesized at Merck) and illustration of the releasable-hapten hypothesis. Nuc, nucleophile that could be the hydroxyl group of the serine PBPs.

novel 1β-methylcarbapenem compound SM-17466, which is active against MRSA (Fig. 12.7). The introduction of a pyridinothiazole group into C-2 of the carbapenem nucleus gave the molecule high activity against MRSA, and the 1β-methyl group and quaternary ammonium group provide good stability against renal dehydropeptidase I.

The in vitro and in vivo antibacterial activities of SM-17466 were evaluated against a wide range of clinical bacterial isolates and compared with the activities of meropenem, imipenem, vancomycin, and arbekacin (an aminoglycoside antibiotic). SM-17466 has a broad spectrum of action against gram-positive bacteria, showing especially potent activity against methicillin-resistant staphylococci. The MICs of SM-17466, meropenem, imipenem, vancomycin, and arbekacin at which 90% of clinical isolates of MRSA were inhibited ($MIC_{90}$s) were 3.13, 50, 100, 1.56, and 3.13 μg/ml, respectively.

*Novel Anti-MRSA Tricyclic Carbapenems (Trinems).* Tricyclic carbapenems (trinems) are a novel class of semisynthetic antibiotics, discovered by Glaxo Wellcome Research Centre in Italy. The general chemical structure of trinem 1a (Fig. 12.8) is characterized by a tricyclic ring system in which ring C may be five, six, or seven membered and may contain heteroatoms. When ring C is six membered, as in structure 1b (Fig. 12.8), C-4 is generally substituted. Glaxo's C-4 trinem, sanfetrinem, has a methoxy group in this position and a hydroxyethyl side chain at position 10. Sanfetrinem and its biolabile ester sanfetrinem cilexetil, developed by the same pharmaceutical laboratory, are now undergoing phase II clinical studies, particularly for the treatment of infections caused by penicillin-resistant strains.

**Figure 12.7** Chemical structure of SM-17466.

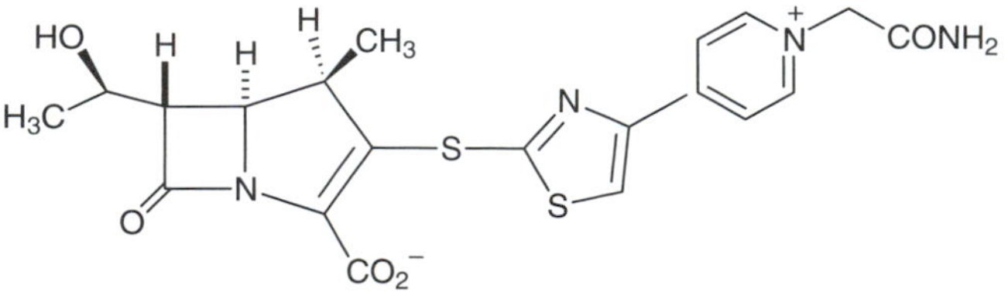

In 2000, Kanno and Kowamoto of the Medicinal Chemistry Research Laboratories of Sankyo (Tokyo, Japan) synthesized several novel trinem derivatives (Fig. 12.9) to explore their activity against MRSA. The antibacterial activities of trinem derivatives 1a to 1d were compared with that of vancomycin (Table 12.5). The MICs of 1a and 1b indicated that these compounds, with the basic amine moiety, had good antibacterial activity against both gram-positive and gram-negative bacteria. Interestingly, the introduction of an aromatic moiety (1c and 1d) remarkably increased the activity against gram-positive bacteria, including MRSA. Particularly, 1d proved to have the same high activity against MRSA as did vancomycin. These workers have recently stated that 1c showed therapeutically valuable activity against gram-negative bacteria and that it could be further developed as a broad-spectrum anti-MRSA agent.

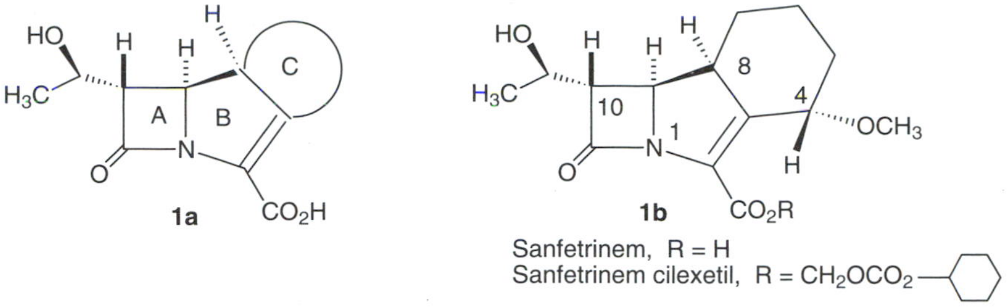

**Figure 12.8** Structure 1a is the general chemical structure of trinems. Structure 1b is the chemical structure of sanfetrinem and the ester sanfetrinem cilexetil.

### SIDEROPHORES (MICROBIAL IRON CHELATORS) LINKED TO β-LACTAM ANTIBIOTICS

Assimilation of iron is essential for microbial growth. As a strategy to increase the penetration of antibacterials through the outer membrane of gram-negative bacterial pathogens, facilitated transport through siderophore receptors has been frequently exploited. The β-lactam conjugates which incorporate an iron-binding moiety work by first binding iron; they can then be recognized by bacteria as if they were naturally occurring siderophore-iron(III) complexes and assimilated into the bacterial cell by *tonB*-dependent pathways. Most of the modification to cephalosporins involves incorporation of a catechol or a hydroxamic acid at the C-7 or C-3 position of the cephem nucleus to form the conjugate.

LB-10517 (Fig. 12.10), a new catechol-cephalosporin using the iron uptake pathway, has good activity against *P. aeruginosa* ($MIC_{90}$, 2.0 μg/ml) and is potent against imipenem- and ofloxacin (a fluoroquinolone)-resistant strains. BMS-180680, a new catechol-containing monobactam (Fig. 12.11), was reported in 1997 by a group of workers at Bristol-Myers Squibb Pharmaceutical Research Institute. The in vitro activities of BMS-180680 were compared to those of aztreonam, ceftazidime, imipenem, piperacillin-tazobactam, ciprofloxacin, amikacin, and trimethoprim-sulfamethoxazole. BMS-180680 was often the most active compound against many species of the family *Enterobacteriaceae*. It was inactive against most strains of *P. fluorescens*.

Readers desiring additional information on these topics should consult the references at the end of this chapter.

**Figure 12.9** Chemical structures of novel trinems synthesized at Sankyo Laboratories.

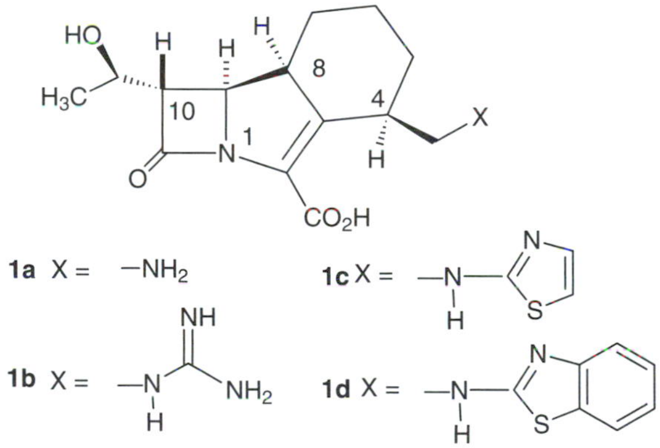

## New Developments in Inhibitors of Serine and Metallo-β-Lactamases

### CURRENT RESEARCH AND DEVELOPMENTS IN SERINE β-LACTAMASE INHIBITORS

As discussed in chapter 11, commercially available serine β-lactamase inhibitors are all effective against class A β-lactamases but have little or no activity against class C

**Table 12.5** Antibacterial activity of trinem derivatives and vancomycin[a]

| Strain | MIC (μg/ml) of: | | | | |
|---|---|---|---|---|---|
| | 1a | 1b | 1c | 1d | VCM[b] |
| *S. aureus* 209P | 0.056 | <0.01 | 0.02 | <0.012 | 0.2 |
| *S. aureus* 56R | 0.2 | 0.05 | 0.05 | 0.025 | 0.78 |
| *S. aureus* 535 (MRSA) | 25 | 6.2 | 3.1 | 1.56 | 1.56 |
| *E. faecalis* 681 | 6.2 | 3.1 | 1.5 | 0.78 | 0.78 |
| *E. coli* NIHJ | 0.1 | 0.1 | 0.8 | 12.5 | >100 |
| *K. pneumoniae* 806 | 0.1 | 0.2 | 0.8 | 25 | >100 |
| *S. marcescens* 1184 | 0.2 | 0.2 | 1.5 | 12.5 | >100 |
| *P. aeruginosa* 1001 | 6.2 | 50 | >100 | >100 | >100 |

[a] Reprinted from O. Kanno and I. Kawamoto, *Tetrahedron* 56:5639–5648, 2000, with permission from the publisher.

[b] VCM, vancomycin.

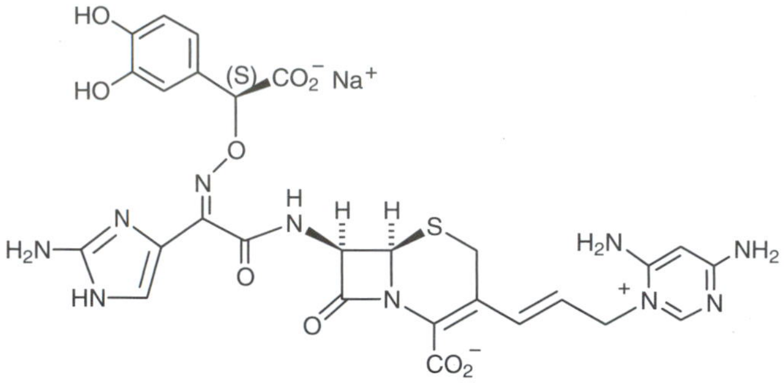

**Figure 12.10** Chemical structure of LB-10517.

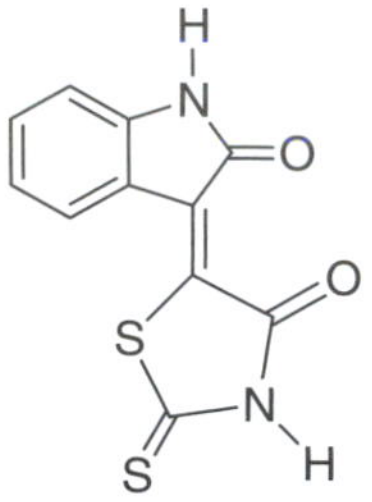

**Figure 12.12** Chemical structure of a rhodanine synthesized at Johnson Pharmaceutical.

β-lactamases. Since bacteria producing class C β-lactamases are increasing in prevalence among organisms causing nosocomial infections, there is a need to develop an inhibitor of class C β-lactamases.

A series of rhodanines (non-β-lactam inhibitors of class C β-lactamases) was recently developed by E. Grant and coworkers at the R. W. Johnson Pharmaceutical Research Institute (Raritan, N.J.). These rhodanines were evaluated against β-lactamase P99 isolated from *E. cloacae* and TEM-1 isolated from *E. coli,* which are prototypical class C and class A β-lactamases, respectively. One of these rhodanines (Fig. 12.12) had a 50% inhibitory concentration ($IC_{50}$) of 2.64 μM against the class C enzyme P99 and an $IC_{50}$ of 8.7 μM against the class A enzyme TEM-1.

Recently, J. Buynak and coworkers, of the Department of Chemistry of Southern Methodist University (Dallas, Tex.), reported the synthesis and evaluation of a series of 3-substituted 7-alkylidenecephalosporin sulfones as inhibitors of class A and class C β-lactamases. The same group, in 1999, also reported the synthesis and evaluation of 6-alkylidene-2'β-substituted penam sulfones as inhibitors of class A and class C β-lactamases. The most active compounds cited in these reports are shown in Figure 12.13. The penam derivative also contains a catechol functionality in the 2'-side chain. The researchers think that this iron-chelating functionality allows the inhibitor to enter the cell via an iron transport pathway. Such a modified inhibitor molecule typically shows improved profiles against gram-negative strains, including *P. aeruginosa* and *S. marcescens.*

**Figure 12.11** Chemical structure of BMS-180680.

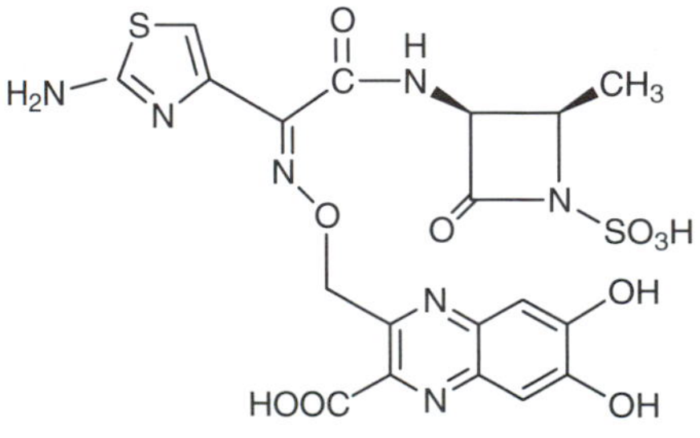

Table 12.6 lists the inhibitory activities of these compounds, relative to tazobactam, against a class C enzyme, P99 from *E. cloacae,* two class A enzymes, TEM-1 and PC1 from *S. aureus,* and a class C enzyme, GC1 from *E. cloacae.* Generally, the 7-alkylidenecephalosporin sulfones and the 6-alkylidenepenam sulfones are better inhibitors of the class C enzyme than is tazobactam. Buynak and coworkers remarked that the cephalosporins that incorporate a vinylogous electron-withdrawing group exhibit biological activity in the order 1a > 1b > 1c > 1d. Compound 2 shows in vitro synergy with the antibiotic piperacillin in the treatment of penicillin-resistant organisms.

P. Bitha and coworkers from Wyeth-Ayerst Research Laboratory (Pearl River, N.Y.) reported in two simultaneous publications in 1999 the synthesis of five 6-(1-hydroxyalkyl)penam sulfone derivatives and two 6-(hydroxymethyl)penams (in one paper) and the synthesis of a 6β-(1-hydroxyethyl)penem sulfone and four 6β-hydroxymethylpenam sulfones (in the other paper). These workers performed a structure-activity relationship (SAR) study to determine the effect of the substituents and the stereochemical requirements for the biological activity of the 6α- and 6β-(1-hydroxyalkyl) groups of penam sulfone derivatives. From these studies, they found that only compound 1 (Fig. 12.14) improved the $IC_{50}$ of sulbactam against both TEM-1 and AmpC β-lactamases.

In 1999, J. Marchand-Brynaert, C. Beauve, and coworkers at the Department of Organic Chemistry of the Université Catholique de Louvain in Belgium described the synthesis and biochemical activity of 1,3-substituted azetidin-2-ones as serine and metallo-β-lactamase and serine protease inhibitors. They found that these novel monobactams were inactive against class A,

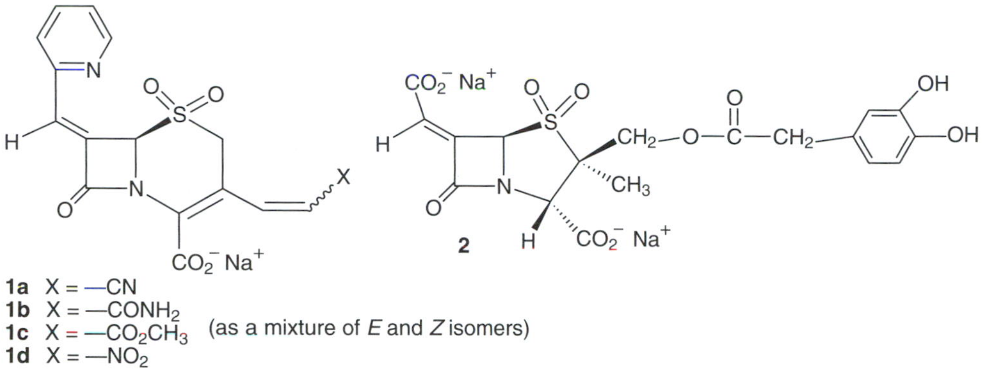

**1a** X = —CN
**1b** X = —$CONH_2$
**1c** X = —$CO_2CH_3$ (as a mixture of *E* and *Z* isomers)
**1d** X = —$NO_2$

**Figure 12.13** Chemical structures of 3-substituted 7-(alkylidene)cephalosporin sulfones and 6-alkylidene-2'β-substituted penam sulfones.

C, D, and B β-lactamases but were weak inhibitors of porcine pancreatic elastase, a serine protease.

A recent review (Mascaretti et al., 1999) described the chemistry and biochemistry of β-lactams as inhibitors of serine β-lactamases as of June 1999. Therefore, for this reason, the earlier work done in the synthesis and SAR studies of β-lactamase inhibitors are not included here. The reader should consult this reference and earlier reviews by the same authors.

## CURRENT RESEARCH AND DEVELOPMENTS IN METALLO-β-LACTAMASE INHIBITORS

The metallo-β-lactamases have been deemed the most problematic of all the β-lactamases studied yet, due to their high turnover ($k_{cat}$) and specificity ($k_{cat}/K_m$) exhibited against β-lactam antibiotics. Moreover, evidence is mounting that metallo-β-lactamase genes are being passed to other bacterial strains by plasmid- and integron-borne transfer pathways. Consequently, metallo-β-lactamases present a serious threat to the clinical utility of β-lactam antibiotics. There is a critical need to develop potent inhibitors of metallo-β-lactamases to overcome metallo-β-lactamase-mediated resistant bacteria. Analysis of the three-dimensional structure of the metallo-β-lactamase of *B. fragilis* suggested that the thiolate group is important for enzymatic activity. Therefore, the structure of the active site has been a topic of interest with respect to the design of inhibitors against metallo-β-lactamases.

Recent reports of metallo-β-lactamase inhibitors have appeared. In a collaborative effort between chemists and biologists at the Dyson Perrins Laboratory (Oxford, United Kingdom), the Laboratoire d'Enzymologie, Université de Liège (Liège, Belgium), and the Università dell'Aquila (L'Aquila, Italy), the synthesis and inhibitory activity of α-amido trifluoromethylketones and trifluoromethyl alcohols (Fig. 12.15) against metallo-β-lactamases from *S. maltophilia* ULA-511, *Aeromonas hydrophila* AE036, *Bacillus cereus* 569H, and *P. aeruginosa* 101 were reported in 1996. The authors mentioned that trifluoromethylketones significantly inhibited the enzymes from *S. maltophilia* and *A. hydrophila* but that they were less active against the enzymes from *B. cereus* and *P. aeruginosa*. The C-2 epimeric trifluoromethyl

**Table 12.6** Inhibition of representative serine β-lactamases by compounds 1a to 1d and 2 relative to tazobactam[a]

| | $IC_{50}$ (μM) of compound for: | | | |
|---|---|---|---|---|
| **Compound** | **P99** | **TEM-1** | **PC1** | **GC1** |
| Tazobactam | 49.8 | 0.32 | 2.8 | 3.4 |
| 1a | 0.01 | 0.014 | 0.72 | 0.012 |
| 1b | 0.026 | 0.09 | 0.10 | 0.01 |
| 1c | 0.20 | 0.02 | 0.30 | 0.30 |
| 1d | 0.02 | 0.07 | 0.20 | 0.10 |
| 2 | 0.026 | 0.06 | 0.7 | ND[b] |

[a] Adapted from J. D. Buynak, V. R. Doppalapudi, A. S. Rao, S. D. Nidamarthy, and G. Adam, *Bioorg. Med. Chem. Lett.* 10:847–851, 2000, with permission from the publisher.

[b] ND, not determined.

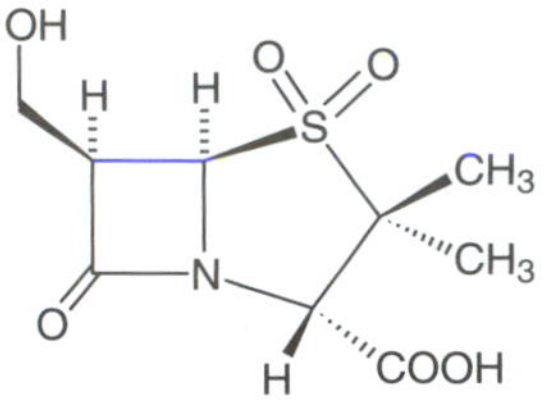

6β-hydroxymethylpenam sulfone 1

**Figure 12.14** Chemical structure of compound 1 synthesized by Bitha et al. at Wyeth-Ayerst.

alcohols 1b and 2b also had significant inhibitory activity, and 2b was the most potent inhibitor of the *B. cereus* and *P. aeruginosa* enzymes. These compounds are proposed to bind to one of the zinc atoms in the active center, as observed with the metalloprotease carboxypeptidase A.

In 1999, in another collaborative project by researchers in Oxford; Liège; L'Aquila; and Siena, Italy, synthesis of the amino acid-derived hydroxamates 1 to 3 (Fig. 12.16) and their evaluation against metallo-β-lactamases from *B. cereus* and *A. hydrophila* were reported. None of these compounds were active against the *B. cereus* enzyme, but they inhibited the *A. hydrophila* enzyme.

J. Marchand-Brynaert and coworkers and J. Bateson, D. Payne, and coworkers at SmithKline Beecham Pharmaceuticals, United Kingdom, reported in 1997 the discovery of a series of mercaptoacetic acid thiol ester derivatives (thiodepsipeptides) that are irreversible inhibitors of metallo-β-lactamases from *B. cereus, S. maltophilia,* and *A. hydrophila* but not from *B. fragilis.* These compounds were named SB214751, SB214752, SB213079, and SB216968 (Fig. 12.17). Compound SB216968 was found to be particularly active against the *A. hydrophila* CphA metallo-β-lactamase. Thiol esters inhibit the *B. cereus* enzyme via hydrolytic release of mercaptoacetic acid, which subsequently covalently modifies a cysteine residue at the active site.

In 1997, D. Payne and coworkers from SmithKline Beecham reported the inhibition of metallo-β-lactamases by a series of thiol ester derivatives of mercaptophenylacetic acid. In 1999, a group of researchers led by M. L. Greenlee at Merck Research Laboratories reported the synthesis and SAR studies of thioester and thiol inhibitors of IMP-1 metallo-β-lactamase. These researchers found that thiodepsipeptides (structure 1) are inhibitors of the IMP-1 metallo-β-lactamase and that introduction of more hydrophobic C-terminal substituents (Fig. 12.18) greatly enhances its activity. Furthermore, the simpler thioesters (structure 2, where $R_3$ = $CH_3$ or phenyl) are even more active inhibitors of IMP-1. Thioesters 1 and 2 are substrates for the IMP-1 enzyme and, as such, are hydrolyzed to thiols 3, which are themselves potent inhibitors of IMP-1. The *R*-stereochemistry of the α-mercapto-acid chiral center is important for the inhibitory activity of these compounds. The novel compounds were assayed for activity against the IMP-1 and CcrA (*B. fragilis*) metallo-β-lactamases. It was found that the thiodepsipeptides were generally inhibitors of IMP-1 but were poorly active against the CcrA enzyme. The two most active compounds of these series of thiodepsipeptides were found to be 1z and 2f, with $IC_{50}$s against IMP-1 of 0.025 and 0.0004 μM, respectively. In the thiol group, compound 3b was the most active ($IC_{50}$, 0.023 μM).

Thiols are well-known inhibitors of metalloproteases. Recently, M. Page and coworkers have synthesized several thiol derivatives as potential inhibitors of the class B metallo-β-lactamases. Figure 12.19 shows the general chemical structures of these novel thiol derivatives. These compounds were indeed competitive inhibitors of the *B. cereus* class B β-lactamase. M. Goto (1997) reported that mercaptoacetic acid and 2-mercaptopropionic acid are competitive inhibitors of the metallo-β-lactamase IMP-1 from *S. marcescens.* Two recent publications by H. Dyson and coworkers at the Department of Molecular Biology of the Scripps Research Institute in the United States reported the nuclear magnetic resonance spectroscopy characterization and the dynamics of the metallo-β-lactamase from *B. fragilis* in the presence and absence of the thiol SB225666 obtained from SmithKline Beecham (Fig. 12.19). The rationale behind these studies was that understanding the struc-

**Figure 12.15** Chemical structures of the 3*S* and 3*R* enantiomers of four α-amido trifluoromethylketones and racemic trifluoromethyl alcohols.

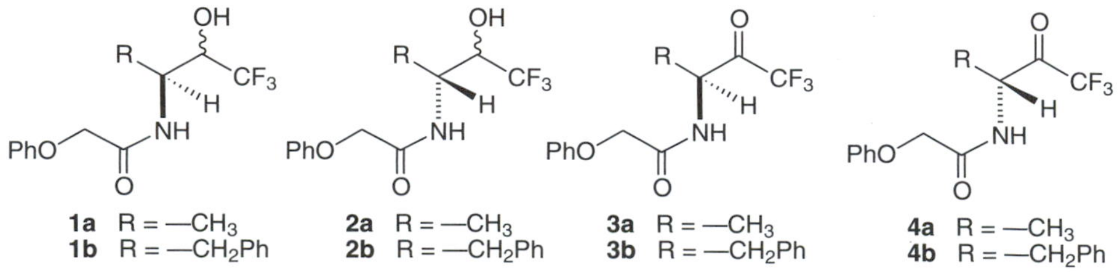

**1a** R = —$CH_3$
**1b** R = —$CH_2Ph$

**2a** R = —$CH_3$
**2b** R = —$CH_2Ph$

**3a** R = —$CH_3$
**3b** R = —$CH_2Ph$

**4a** R = —$CH_3$
**4b** R = —$CH_2Ph$

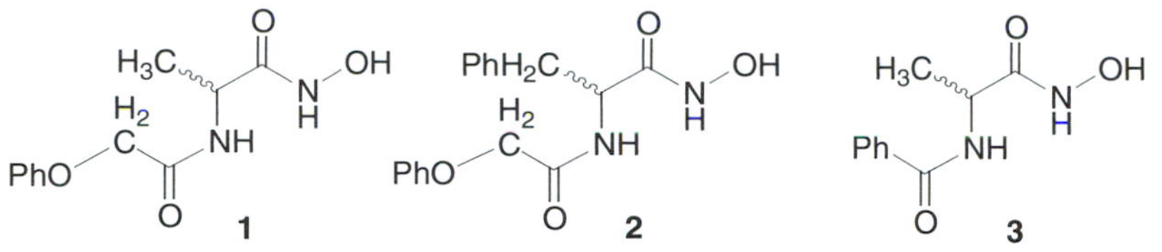

**Figure 12.16** Chemical structures of amino acid-derived hydroxamates synthesized by Walter et al. (1999).

ture and dynamics of the enzymes that mediate antibiotic resistance of pathogenic bacteria, in this case of metallo-β-lactamases, will allow a rational design of specific inhibitors. The authors of these publications suggested that SB225666, which contains a free sulfhydryl group, is probably bound to the protein through a disulfide-like interaction with one of the cysteine side chains near the zinc-binding site.

Another important piece of work was the determination of the crystal structure of the native IMP-1 metallo-β-lactamase from *P. aeruginosa* and its complex with a mercaptocarboxylate (thiol) inhibitor by N. Concha and coworkers in 2000. Binding of the inhibitor in the active site induces a conformational change. The authors stated that in an attempt to understand which critical binding interactions need to be established by a potent inhibitor, they determined the crystal structure of a complex of the IMP-1 metallo-β-lactamase with a mercaptocarboxylate inhibitor. This mercaptocarboxylate inhibits the IMP-1, *B. fragilis* and L1 enzymes with $IC_{50}$ between 100 and 500 nM (or 0.1 and 0.5 μM) and is slightly less effective against the L1 enzyme. The authors of the publication also pointed out that the IMP-1 metallo-β-lactamase native structure highlights the structural diversity in this family of proteins.

In 1998, J. Toney and coworkers at Merck Research Laboratories in the United States reported that the use of screening techniques for the search of inhibitors of

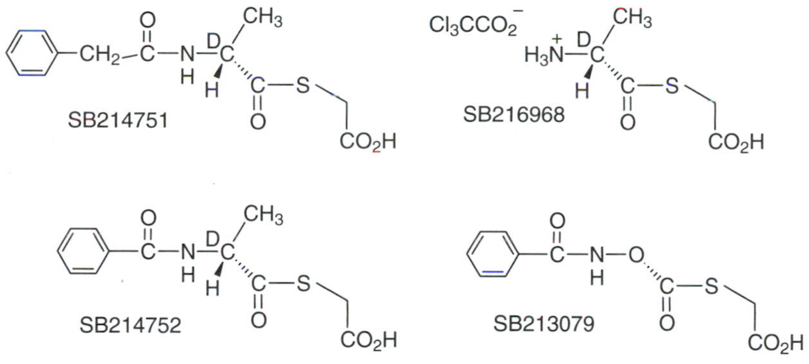

**Figure 12.17** Chemical structures of four derivatives of mercaptoacetic acid thiol ester.

**Figure 12.18** Chemical structures of some of the thioesters and thiols developed by the Merck group.

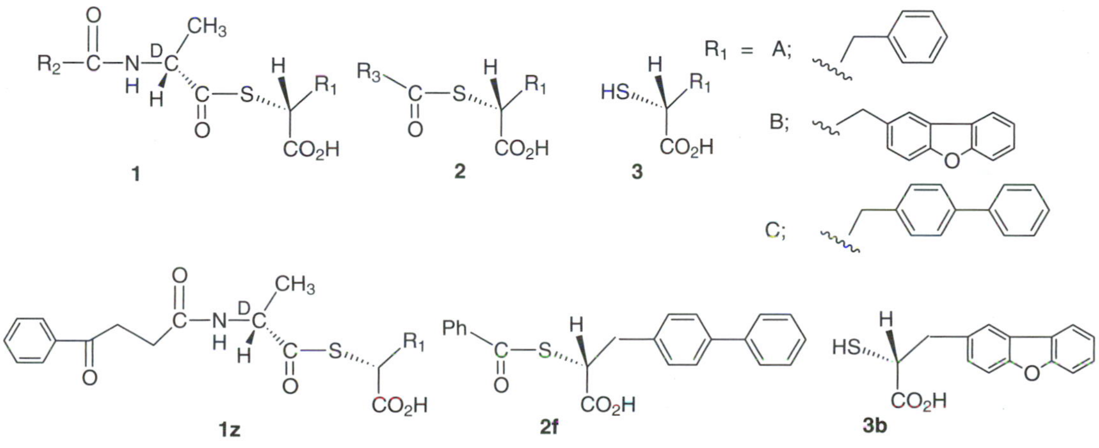

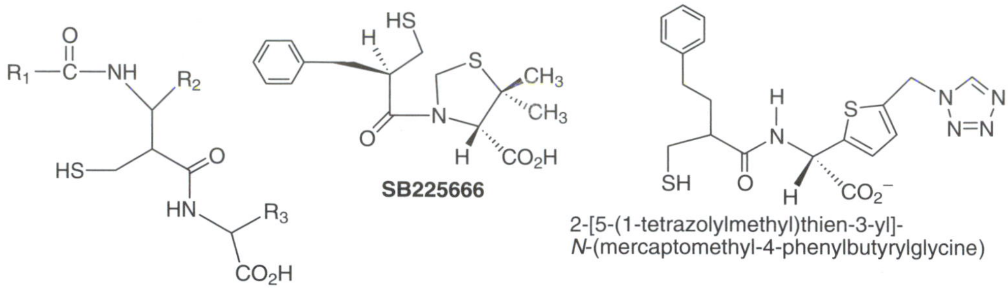

SB225666

2-[5-(1-tetrazolylmethyl)thien-3-yl]-
*N*-(mercaptomethyl-4-phenylbutyrylglycine)

**Figure 12.19** General structure of thiols developed by M. Page and coworkers (1998). The chemical structures of two thiols synthesized at SmithKline Beecham are shown.

metallo-β-lactamases led to the identification of biphenyl tetrazoles (BPTs), and a series of biphenyl tetrazoles linked to various heterocyclic aromatic rings, as potent competitive inhibitors. In their initial publication, these authors reported a SAR study for these compounds by using *B. fragilis* metallo-β-lactamase and the mammalian dehydropeptidase I, a metalloenzyme that is capable of hydrolyzing imipenem. From this study, these workers identified the three most active substituted BPTs, whose chemical structures are represented in Fig. 12.20. The activities against *B. fragilis* metallo-β-lactamase were as follows: L-158,817, $IC_{50}$ = 3.5 μM; L-159,061, $IC_{50}$ = 1.9 μM; and L-159,906, $IC_{50}$ = 0.4 μM. They selected L-159,061 for cocrystallization with the *B. fragilis* metallo-β-lactamase. This inhibitor was chosen for crystallographic studies to define the mode of BPT binding. Fig. 12.21 is a schematic representation of the coordination environment of the zinc sites in the *B. fragilis* metallo-β-lactamase, showing the bridging water (Wat1) and the water coordinated to Zn2 (Wat2).

The X-ray crystal structure of the *B. fragilis* metallo-β-lactamase bound to the BPT L-159,061 (Fig. 12.22) shows that the tetrazole moiety of the inhibitor interacts directly with one of the zinc atoms in the active site, replacing a metal-bound water molecule. For a more detailed description of additional interactions, the reader should consult the original publication.

## FUTURE RESEARCH AND DEVELOPMENTS IN INHIBITORS OF METALLO-β-LACTAMASES

The increase in β-lactam antibiotic resistance due to the production and rapid spread of resistance encoded by plasmids or transposons in pathogenic bacteria has made these enzymes an attractive target for drug development. The IMP-1 enzyme is encoded by both plasmids and integrons, and these highly mobile genetic elements are thought to be responsible for the rapid spread of IMP-1 in clinical isolates from Japan.

## CONTRIBUTION FROM MY LABORATORY TO DEVELOPMENT OF IMP-1 METALLO-β-LACTAMASE INHIBITORS

It was realized by D. J. Payne et al. in 1997 that one approach to overcoming the threat of metallo-β-lactamase would be the discovery or design of inhibitors of these enzymes, to be used in combination with β-lactam antibiotics.

With a view to developing potent inhibitors of metallo-β-lactamases, the design, synthesis, and evaluation of IMP-1 inhibitors were initiated. The choice of this particular enzyme as the target was motivated by its impeding emergence in other pathogenic organisms via the facile interspecies transfer of the plasmid-borne $bla_{IMP}$ gene.

Knowledge of the high-resolution structure of a target metallo-β-lactamase significantly facilitates the design of

**Figure 12.20** Chemical structures of L-158,817, L-159,061, and L-159,906.

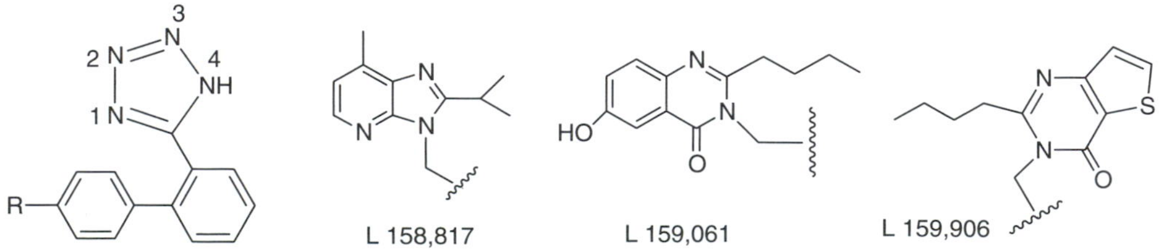

L 158,817 L 159,061 L 159,906

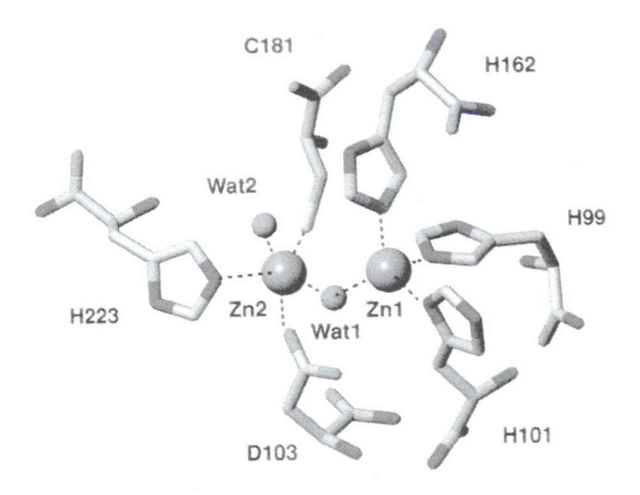


**Figure 12.21** Coordination environment of the zinc sites in the *B. fragilis* metallo-β-lactamase, showing the bridging water (Wat1) and the water coordinated to Zn2 (Wat2). Reprinted from S. D. B. Scrofani, J. Chung, J. J. A. Huntley, S. J. Benkovic, P. E. Wright, and H. J. Dyson, *Biochemistry* **38:**14507–14514, 1999, with permission from the publisher.

potent and selective inhibitors. The crystal structure of the native IMP-1 metallo-β-lactamase and of the lactamase complexed with a mercaptocarboxylate inhibitor provided useful information about the binding interaction of inhibitor with IMP-1 as well as the environment of the two zinc ions. In these 3.1-Å (native) and 2.0-Å (complex) resolution structures, a cysteine residue is linked to Zn2 in the active site. These structures provided us a structural basis for the binding specificity and the preparation of a series of potent and selective inhibitors of IMP-1.

Some of the reported non-β-lactam-based inhibitors of these enzymes have been presented already and include trifluoromethyl alcohols and ketones, amino acid-derived hydroxamates, thiols, thioester derivatives, cysteinyl peptides, biphenyl tetrazoles, mercaptocarboxylates, 1β-methylcarbapenem derivatives, 2,3-disubstituted succinic acid derivatives, and *N*-sulfonyl hydrazones. Some of these inhibitors were found by screening of natural products, others were found by screening libraries of compounds, and yet others were found by computational approaches involving "molecular docking," i.e., placing putative ligands in conformations appropriate for interaction with three-dimensional structures (atomic coordinates for enzymes can be obtained from X-ray crystallography, nuclear magnetic resonance spectroscopy, and homology modeling).

It is generally acknowledged that the unique stereospecificity seen in biological systems is due to a **complementary three-dimensional interaction** between an inhibitor (or inactivator) and an asymmetric active site of the enzyme. These interactions can best be characterized by the crystal structure of the enzyme-inhibitor complex. The X-ray diffraction method remains the

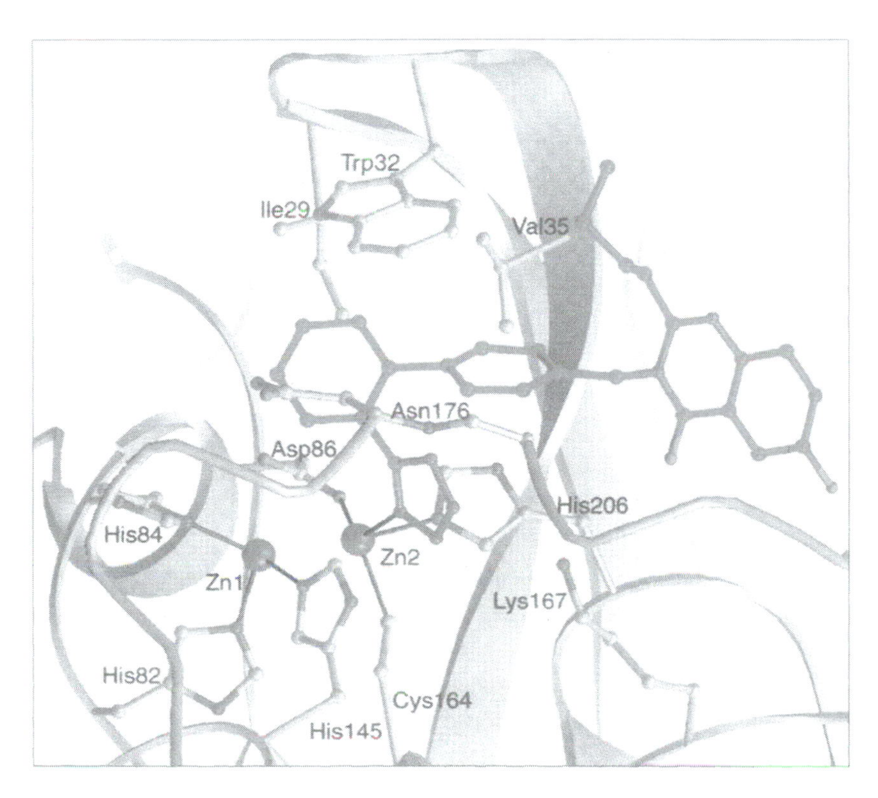


**Figure 12.22** X-ray crystal structure of L-159,061 bound in the active site of the *B. fragilis* metallo-β-lactamase. A view of the active site is shown. Reprinted from J. H. Toney, P. M. Fitzgerald, N. Grover-Sharma, et al., *Chem. Biol.* **5:**185–196, 1998, with permission from the publisher.

most accurate approach to describe the spatial molecular organization. When the characterization of the enzyme (protein) is lacking, an indirect approach is used. The receptor-mapping method involves deducing the structure and other properties of the enzyme by studying the properties of small ligand molecules. As it is well known, the β-lactam amide portion of the penam nucleus is essentially planar, although the C-5 carbon can be displaced up to 0.4 Å toward the β face of the molecule (i.e., toward the sulfur atom) (see Fig. 4.2). The β-lactam ring has little conformational flexibility. On the other hand, the thiazolidine ring of the penam nucleus is flexible and has been observed in two principal conformations. In penicillanic acid sulfide derivatives, the C-2 carbon can be flipped "up" so that the 2β and 3α substituents (-$CH_3$ and -$CO_2H$, respectively) are roughly axial. On the other hand, in the α- and β-sulfoxides and sulfones, the C-2 carbon is flipped "down" so that the 2β and 3α substituents are more nearly equatorial. Obviously, when the 2β and 3α groups are axial (equatorial), the 2α and 3β groups are equatorial (axial).

## Design and Synthesis of New IMP-1 Metallo-β-Lactamase Inhibitors: Substrate Analog Inhibitor Design

Creating molecules with specific properties has been a cherished goal of chemists for generations. Substrate-based and structure-based design of the substrate specificity of metallo-β-lactamase inhibitors is important for developing new drugs. Most of these inhibitors that we have designed emulate the substrate; we have designed these inhibitors as β-bicyclic lactam-derived β-lactam mimetics due to their close relationship to the enzyme-bound β-lactam substrates (penicillin, cephalosporin, carbapenems, clavulanic acid, and sulbactam).

As a logical extension of the principles of substrate-based design of inhibitor, if we can explain how a metallo-β-lactamase works in terms of mono- and binuclear $Zn^{2+}$, we should be able to design inhibitors whose properties are predictable from what we know about these enzymes. The central assumption of substrate-based design of metallo-β-lactamase inhibitors is that good inhibitors must possess significant structural and chemical complementarity with the enzyme.

To design novel metallo-β-lactamase inhibitors, we set up three criteria: (i) what they structurally mimic; (ii) where they might bind, the transition metal ion Zn(II) or the amino acids or water in their environment; and (iii) how they inhibit. Functional groups in the penam nucleus influencing both their efficacy and their selectivity were considered.

Our goal was to design and synthesize molecules whose properties mimic those of substrate molecules but which have a higher affinity than the substrate for the enzyme. We will shortly publish our progress in the design of metallo-β-lactamase inhibitors aimed at inhibitory activity.

## Specific Targeting of Metallo-β-Lactamases

β-Lactam-based inhibitors containing several types of chemical functionalities that serve as the central reactive group were designed by investigating the topology of the active site of mono- and binuclear Zn(II) metallo-β-lactamases. As mentioned above, the binuclear Zn(II) metallo-β-lactamases are characterized by a water (hydroxide) bridge between Zn1 and Zn2 that validated the hypothesis that they act as the nucleophile to attack the carbonyl carbon of β-lactams.

## Functional Groups

We have an ongoing interest in the design and synthesis of metallo-β-lactamases based in the scaffold of the penam nucleus binding to functional groups that might act as nucleophile, electrophile, or transition state analogues, because such compounds are expected to have therapeutic potential for counteracting some of the bacterial resistance mechanisms that have recently emerged due to the action of class B β-lactamases. The importance of these enzymes has been demonstrated in some outbreaks in Japan, Malaysia, Italy, and Greece, as well as other countries, in patients infected by bacteria that produced IMP-1 and VIM-1 enzymes and variants.

Several compounds have been synthesized in my laboratory and tested against IMP-1 metallo-β-lactamase. They have proved to be very active and will be reported in due course.

## REFERENCES AND FURTHER READING

### General

**Bermudes, H., F. Jude, E. B. Chaibi, C. Arpin, C. Bebear, R. Labia, and C. Quentin.** 1999. Molecular characterization of TEM-59 (IRT-17), a novel inhibitor-resistant TEM-derived β-lactamase in a clinical isolate of *Klebsiella oxytoca. Antimicrob. Agents Chemother.* **43:**1657–1661.

**Kurokawa, H., T. Yagi, N. Shibata, K. Shibayama, K. Kamachi, and Y. Arakawa.** 2000. A new SHV-derived extended-spectrum β-lactamase (SHV-24) that hydrolyzes ceftazidime through a single-amino-acid substitution (D179G) in the Ω-loop. *Antimicrob. Agents Chemother.* **44:**1725–1727.

### Extended-Spectrum β-Lactamases

**Bradford, P. A.** 2001. Extended-spectrum β-lactamases in the 21st century: characterization, epidemiology, and detection of this important resistance threat. *Clin. Microbiol. Rev.* **14:**933–951.

**Du Bois, S. K., M. S. Marriot, and S. G. B. Aymes.** 1995. TEM and SHV extended-spectrum β-lactamases: relationship between selection, structure and function. *J. Antimicrob. Chemother.* 35:7–22.

**Jacoby, G. A.** 1994. Genetics of extended-spectrum beta-lactamases. *Eur. J. Clin. Microbiol. Infect. Dis.* **13**(Suppl. 1):2–11.

**Jacoby, G. A.** 1998. Epidemiology of extended-spectrum β-lactamases. *Clin. Infect. Dis.* **27**:81–83.

**Jacoby, G. A., and A. A. Medeiros.** 1991. More extended-spectrum β-lactamases. *Antimcrob. Agents Chemother.* **35**:1697–1704.

**Knox, J. R.** 1995. Extended-spectrum and inhibitor-resistant TEM-type β-lactamases: mutations, specificity, and three-dimensional structure. *Antimicrob. Agents Chemother.* **39**:2593–2601.

**Matagne, A., and J. M. Frère.** 1995. Contribution of mutant analysis to the understanding of enzyme catalysis: the case of class A β-lactamases. *Biochim. Biophys. Acta* **1246**:109–127.

**Matagne, A., J. Lamotte-Brasseur, and J. M. Frère.** 1998. Catalytic properties of class A β-lactamases: efficiency and diversity. *Biochem. J.* **330**:581–598.

**Patzkill, T.** 1998. β-Lactamases are changing their activity spectrums. *ASM News* **64**:90–95.

**Philippon, A., G. Arlet, and P. H. Lagrange.** 1994. Origin and impact of plasmid-mediated extended-spectrum beta-lactamases. *Eur. J. Clin. Microbiol. Infect. Dis.* **13**(Suppl. 1):17–29.

**Philippon, A., R. Labia, and G. Jacoby.** 1989. Extended-spectrum β-lactamases. *Antimicrob. Agents Chemother.* **33**:1131–1136.

### TEM β-Lactamases

**Brown, R. P. A., R. T. Aplin, and C. J. Schofield.** 1996. Inhibition of TEM-2 β-lactamase from *Escherichia coli* by clavulanic acid: observation of intermediates by electrospray ionization mass spectrometry. *Biochemistry* **35**:12421–12432.

**Bush, K., and G. Jacoby.** 1997. Nomenclature of TEM β-lactamases. *J. Antimicrob. Chemother.* **39**:1–3.

**Goussard, S., and P. Courvalin.** 1999. Updated sequence information for TEM β-lactamase genes. *Antimicrob. Agents Chemother.* **43**:367–370.

**Maveyraud, L., I. Massova, C. Birck, K. Miyashita, J. P. Samama, and S. Mobashery.** 1996. Crystal structuire of 6α-(hydroxymethyl)penicillanate complexed to the TEM-1 β-lactamase from *Escherichia coli:* evidence on the mechanism of action of a novel inhibitor designed by a computer-aided process. *J. Am. Chem. Soc.* **118**:7435–7440.

**Ness, S., R. Martin, A. M. Kindler, M. Paetzel, M. Gold, S. E. Jensen, J. B. Jones, and N. C. J. Strynadka.** 2000. Structure-based design guides the improved efficacy of deacylation transition state analogue inhibitors of TEM-1 β-lactamase. *Biochemistry* **39**:5312–5321.

**Strynadka, N. C. J., H. Adachi, S. E. Jensen, K. Johns, A. Sielecki, C. Betzel, K. Sutoh, and M. N. G. James.** 1992. Molecular structure of the acyl-enzyme intermediate in β-lactam hydrolysis at 1.7 Å resolution. *Nature* **359**:700–705.

**Strynadka, N. C. J., R. Martin, S. E. Jensen, M. Gold, and J. B. Jones.** 1996. Structure-based design of a potent transition state analogue for TEM-1 β-lactamase. *Nat. Struct. Biol.* **3**:688–695.

### Inhibitor-Resistant TEM β-Lactamases

**Blazquez, J., M. R. Baquero, R. Canton, I. Alos, and F. Baquero.** 1993. Characterization of a new TEM-type β-lactamase resistant to clavulanate, sulbactam, and tazobactam in a clinical isolate of *Escherichia coli. Antimicrob. Agents Chemother.* **37**:2059–2063.

**Chaibi, E. B., D. Sirot, G. Paul, and R. Labia.** 1999. Inhibitor-resistant TEM β-lactamases: phenotypic, genetic and biochemical characteristics. *J. Antimicrob. Chemother.* **43**:447–458.

**Nicolas-Chanoine, M. H.** 1997. Inhibitor-resistant β-lactamases. *J. Antimicrob. Chemother.* **40**:1–3.

### SHV-Type β-Lactamases

**Du Bois, S. K., M. S. Marriot, and S. G. B. Amyes.** 1995. TEM- and SHV-derived extended-spectrum β-lactamases: relationship between selection, structure and function. *J. Antimicrob. Chemother.* **35**:7–22.

**Heritage, J., F. H. M'Zali, D. Gascoyne-Binzi, and P. M. Hawkey.** 1999. Evolution and spread of SHV extended-spectrum β-lactamases in gram-negative bacteria. *J. Antimicrob. Chemother.* **44**:309–318.

**Kuzin, A. P., M. Nukaga, Y. Nukaga, A. M. Hujer, R. A. Bonomo, and J. R. Knox.** 1999. Structure of SHV-1 β-lactamase. *Biochemistry* **38**:5720–5727.

**Randegger, C. C., A. Keller, M. Irla, A. Wada, and H. Hächler.** 2000. Contribution of natural amino acid substitutions in SHV extended-spectrum β-lactamases to resistance against various β-lactams. *Antimicrob. Agents Chemother.* **44**:2759–2763.

**Tzouvelekis, L. S., and R. A. Bonomo.** 1999. SHV-type β-lactamases. *Curr. Pharm. Des.* **5**:847–864.

### OXA-Type β-Lactamases

**Golemi, D., L. Maveyraud, S. Vakulenko, S. Tranier, A. Ishiwata, L. P. Kotra, J. P. Samama, and S. Mobashery.** 2000. The first structural and mechanistic insights for class D β-lactamases: evidence for a catalytic process for turnover of β-lactam antibiotics. *J. Am. Chem. Soc.* **122**:6132–6133.

**Maveyraud, L., D. Golemi, L. P. Kotra, S. Tranier, S. Vakulenko, S. Mobashery, and J. P. Samama.** 2000. Insights into class D β-lactamases are revealed by the crystal structure of the OXA 10 enzyme from *Pseudomonas aeruginosa. Structure* **8**:1289–1298.

**Naas, T., and P. Nordmann.** 1999. OXA-type β-lactamases. *Curr. Pharm. Des.* **5**:865–879.

**Paetzel, M., F. Danel, L. de Castro, S. C. Mosimann, M. G. P. Page, and N. C. J. Strynadka.** 2000. Crystal structure of the class D β-lactamase OXA-10. *Nat. Struct. Biol.* **7**:918–925.

### Altered Penicillin-Binding Proteins

**Georgopapadakou, N. H.** 1993. Penicillin-binding proteins and bacterial resistance to β-lactams. *Antimicrob. Agents Chemother.* **37**:2045–2053.

**Spratt, B. G.** 1994. Resistance to antibiotics mediated by target alterations. *Science* **264**:388–393.

### Methicillin-Resistant *Staphylococcus aureus*

**Archer, G. L., and D. M. Niemeyer.** 1994. Origin and evolution of DNA associated with resistance to methicillin in staphylococci. *Trends Microbiol.* **2:**343–346.

**Berger-Bächi, B.** 1994. Expression of resistance to methicillin. *Trends Microbiol.* **2:**389–393.

**Berger-Bächi, B.** 1997. Resistance not mediated by β-lactamases (methicillin resistance), p. 158–174. *In* K. B. Crossley and G. L. Archer (ed.), *The Staphylococci in Human Disease.* Churchill Livingstone, Inc., New York, N.Y.

**Chambers, H. F.** 1997. Methicillin resistance in staphylococci: molecular and biochemical basis and clinical implications. *Clin. Microbiol. Rev.* **10:**781–791.

**de Lencastre, H., L. M. Boudewijn de Jonge, P. R. Matthews, and A. Tomasz.** 1994. Molecular aspects of methicillin resistance in *Staphylococcus aureus. J. Antimicrob. Chemother.* **33:**7–24.

**Drew, R. H., J. R. Perfect, L. Srinath, E. Kurkimilis, M. Dowzicky, and G. H. Talbot.** 2000. Treatment of methicillin-resistant *Staphylococcus aureus* infections with quinupristin-dalfopristin in patients intolerant of or failing prior therapy. *J. Antimicrob. Chemother.* **46:**775–784.

**Ehlert, K.** 1999. Methicillin-resistance in *Staphylococcus aureus*—molecular basis, novel targets and antibiotic therapy. *Curr. Pharm. Des.* **5:**45–55.

**Grubb, W. B.** 1998. Genetics of MRSA. *Rev. Med. Microbiol.* **9:**153–162.

**Hackbarth, C. J., and H. F. Chambers.** 1989. Methicillin-resistant staphylococci: genetics and mechanisms of resistance. *Antimicrob. Agents Chemother.* **33:**991–994.

**Komatsuzawa, H., K. Ohta, H. Labischinski, M. Sugal, and H. Suginaka.** 1999. Characterization of *fmtA,* a gene that modulates the expression of methicillin resistance in *Staphylococcus aureus. Antimicrob. Agents Chemother.* **43:**2121–2125.

**Labischinski, H., K. Ehlert, and B. Berger-Bächi.** 1998. The targeting of factors necessary for expression of methicillin resistance in staphylococci. *J. Antimicrob. Chemother.* **41:**581–584.

**Lu, W. P., Y. Sun, M. D. Bauer, S. Paule, P. M. Koenings, and W. G. Kraft.** 1999. Penicillin-binding protein 2a from methicillin-resistant *Staphylococcus aureus:* kinetic characterization of its interactions with β-lactams using electrospray mass spectrometry. *Biochemistry* **38:**6537–6546.

**Ubukata, K., R. Nonoguchi, R. Matsuhashi, and M. Konno.** 1989. Expression and inducibility in *Staphylococcus aureus* of the *mecA* gene, which encodes a methicillin-resistant *S. aureus*-specific penicillin-binding protein. *J. Bacteriol.* **171:**2882–2885.

### Inducible AmpC β-Lactamases

**Bennett, P. M., and I. Chopra.** 1993. Molecular basis of β-lactamase induction in bacteria. *Antimicrob. Agents Chemother.* **37:**153–158.

**Dietz, H., D. Pfeifle, and B. Wiedemann.** 1996. Location of *N*-acetylmuramyl-L-alanyl-D-glutamyl-mesodiaminopimelic acid, presumed signal molecule for β-lactamase induction in the bacterial cell. *Antimicrob. Agents Chemother.* **40:**2173–2177.

**Hanson, N. D., and C. C. Sanders.** 1999. Regulation of inducible AmpC beta-lactamase expression among Enterobacteriaceae. *Curr. Pharm. Des.* **5:**881–894.

**Jacobs, C., J. M. Frère, and S. Normark.** 1997. Cytosolic intermediates for cell wall biosynthesis and degradation control inducible β-lactam resistance in Gram-negative bacteria. *Cell* **88:**823–832.

**Normark, S., E. Bartowsky, J. Erickson, C. Jacobs, F. Lindberg, S. Lindquist, K. Weston-Hafer, and M. Wikström.** 1994. Mechamism of chromosomal β-lactamase induction in Gram-negative bacteria, p. 485–503. *In* J. M. Ghuysen and R. Hakenbeck (ed.), *Bacterial Cell Wall.* Elsevier, Amsterdam, The Netherlands.

**Philippon, A., G. Arlet, and G. A. Jacoby.** 2002, Plasmid-determined AmpC-type β-lactamases. *Antimicrob. Agents Chemother.* **46:**1–11.

**Powers, R. A., and B. K. Shoichet.** 2002. Structure-based approach for binding site identification on AmpC β-lactamase. *J. Med. Chem.* **45:**3222–3234.

**Sanders, C. C.** 1987. Chromosomal cephalosporinases responsible for multiple resistance to newer β-lactam antibiotics. *Annu. Rev. Microbiol.* **41:**573–593.

**Sanders, C. C., P. A. Bradford, A. T. Ehrhardt, K. Bush, K. D. Young, T. A. Henderson, and W. E. Sanders.** 1997. Penicillin-binding proteins and induction of AmpC β-lactamase. *Antimicrob. Agents Chemother.* **41:**2013–2015.

**Stapleton, P., K. Shannon, and I. Phillips.** 1995. The ability of β-lactam antibiotics to select mutants with derepressed β-lactamase synthesis from *Citrobacter freundii. J. Antimicrob. Chemother.* **36:**483–496.

**Weber, D. A., and C. C. Sanders.** 1990. Diverse potential of β-lactamase inhibitors to induce class I enzymes. *Antimicrob. Agents Chemother.* **34:**156–158.

**Wiedemann, B., H. Dietz, and D. Pfeifle.** 1998. Induction of β-lactamase in *Enterobacter cloacae. Clin. Infect. Dis.* **27**(Suppl. 1)**:**S42–S47.

### Novel Anti-MRSA Antibiotics

**D'Andrea, S. V. D., D. Bonner, J. J. Bronson, J. Clark, K. Denbleyker, J. Fung-Tomc, S. E. Hoeft, T. W. Hudyma, J. D. Matiskella, R. F. Miller, P. F. Misco, M. Pucci, R. Sterzycki, Y. Tsai, Y. Ueda, J. A. Wichtowski, J. Singh, T. P. Kissick, J. T. North, A. Pullockaran, M. Humora, B. Boyhan, T. Vu, A. Fritz, J. Heikes, R. Fox, J. D. Godfrey, R. Perrone, M. Kaplan, D. Kronenthal, and R. H. Mueller.** 2000. Synthesis and anti-MRSA activity of novel cephalosporin derivatives. *Tetrahedron* **56:**5687–5698.

**Humphrey, G. R., R. A. Miller, P. J. Pye, K. Rossen, R. A. Reamer, A. Maliakal, S. S. Ceglia, E. J. J. Grabowski, R. P. Volante, and P. J. Reider.** 1999. Efficient and practical synthesis of a potent anti MRSA β-methylcarbapenem containing a releasable side chain. *J. Am. Chem. Soc.* **121:**11261–11266.

**Jensen, M. S., C. Yang, Y. Hsiao, N. Rivera, K. M. Welles, J. Y. L. Chung, N. Yasuda, D. L. Hughes, and P. J. Reider.** 2000. Synthesis of an anti-methicillin resistant *Staphylococcus aureus* (MRSA) carbapenem via stannatrane-mediated Stille coupling. *Org. Lett.* **2**:1081–1084.

**Rosen, H., R. Hadju, L. Silver, H. Kropp, K. Dorso, J. Kohler, J. G. Sundelof, J. Huber, G. G. Hammond, J. J. Jackson, C. J. Gill, R. Thompson, B. A. Pelak, J. H. Epstein-Toney, G. Lankas, R. R. Wilkening, K. J. Wildonger, T. A. Blizzard, F. P. DiNinno, R. W. Ratcliffe, J. V. Heck, J. W. Kozarich, and M. L. Hammond.** 1999. Reduced immunotoxicity and preservation of antibacterial activity in a releasable side-chain carbapenem antibiotic. *Science* **283**:703–706.

**Singh, J., O. K. Kim, T. P. Kissick, K. J. Natalie, B. Zhang, G. A. Crispin, D. M. Springer, J. A. Wichtowski, Y. Zhang, J. Goodrich, Y. Ueda, B. Y. Luh, B. D. Burke, M. Brown, A. P. Dutka, B. Zheng, D. M. Hsieh, M. J. Humora, J. T. North, A. J. Pullockaran, J. Livshits, S. Swaminathan, Z. Gao, P. Schierling, P. Ermann, R. K. Perrone, M. C. Lai, J. Z. Gougoutas, J. D. DiMarco, J. J. Bronson, J. E. Heikes, J. A. Grosso, D. R. Kronenthal, T. W. Denzel, and R. H. Mueller.** 2000. A practical synthesis of an anti-methicillin-resistant *Staphylococcus aureus* cephalosporin BMS-247243. *Organic Process Res. Dev.* **4**:488–497.

### Novel Tricyclic Carbapenems (Trinems)

**Babini, G. S., M. Yuan, and D. M. Livermore.** 1998. Interactions of β-lactamases with sanfetrinem (GV 104326) compared to those with imipenem and with oral β-lactams. *Antimicrob. Agents Chemother.* **42**:1168–1175.

**Biondi, S., G. Gaviraghi, and T. Rossi.** 1996. Synthesis and biological activity of novel tricyclic β-lactams. *Bioorg. Med. Chem. Lett.* **6**:525–528.

**Biondi, S., A. Pecunioso, F. Busi, S. A. Contini, D. Donati, M. Maffeis, D. A. Pizzi, L. Rossi, T. Rossi, and F. M. Sabbatini.** 2000. Highly diastereoselective synthesis of 4-*N*-methylformamidino trinem (GV 129606), a potent antibacterial agent. *Tetrahedron* **56**:5649–5655.

**Hanessian, S., and B. Reddy.** 1999. Total synthesis of tricyclic β-lactams. *Tetrahedron* **55**:3427–3443.

**Iso, Y., and Y. Nishitani.** 1998. Stereocontrolled synthesis of 1β-aminoalkylcarbapenems and tricyclic carbapenems (trinems). *Heterocycles* **48**:2287–2308.

**Kanno, O., and I. Kawamoto.** 2000. Stereoselective synthesis of novel anti-MRSA tricyclic carbapenems (trinems). *Tetrahedron* **56**:5639–5648.

### Siderophores Linked to β-Lactam Antibiotics

**Kline, T., M. Fromhold, T. E. McKennon, S. Cai, J. Treiberg, N. Ilhe, D. Sherman, W. Schan, M. J. Hickey, P. Warrener, P. R. Witte, L. L. Brody, L. Goltry, L. M. Barker, S. U. Anderson, S. K. Tanaka, R. M. Shawar, L. Y. Nguyen, M. Langhorne, A. Bigelow, L. Embuscado, and E. Naeemi.** 2000. Antimicrobial effects of novel siderophores linked to β-lactam antibiotics. *Bioorg. Med. Chem.* **8**:73–93.

**Roosenberg, J. M., Y. M. Lin, Y. Lu, and M. J. Miller.** 2000. Studies and syntheses of siderophores, microbial iron chelators, and analogs as potential drug delivery agents. *Curr. Med. Chem.* **7**:159–197.

Catechol-monobactam BMS-180680

**Fung-Tomc, J., K. Bush, B. Minassian, B. Kolek, R. Flamm, E. Gradelski, and D. Bonner.** 1997. Antibacterial activity of BMS-180680, a new catechol-containing monobactam. *Antimicrob. Agents Chemother.* **41**:1010–1016.

Catechol-cephalosporin (LB-10517)

**Song, H. K., J. I. Oh, M. Y. Kim, Y. Z. Kim, I. C. Kim, and J. H. Kwak.** 1996. In-vitro and in-vivo antibacterial activity of LB10517, a novel catechol-substituted cephalosporin with a broad antibacterial spectrum. *J. Antimicrob. Chemother.* **39**:711–726.

Catechol and hydroxamate carbacephalosporins

**Ghosh, A., M. Ghosh, C. Niu, F. Malouin, U. Moellmann, and M. J. Miller.** 1996. Iron transport-mediated drug delivery using mixed-ligand siderophore-β-lactam conjugates. *Chem. Biol.* **3**:1011–1019.

### TonB Protein

**White, D.** 2000. *The Physiology and Biochemistry of Prokaryotes*, 2nd ed., p. 22. Oxford University Press, New York, N.Y. An explanation of uptake of iron siderophores.

### Novel Inhibitors of Serine β-Lactamases

**Beauve, C., M. Bouchet, R. Touillaux, J. Fastrez, and J. Marchand-Brynaert.** 1999. Synthesis, reactivity and biochemical evaluation of 1,3-substituted azetidin-2-ones as enzyme inhibitors. *Tetrahedron* **55**:13301–13320.

**Bitha, P., Z. Li, G. D. Francisco, B. A. Rasmussen, and Y. I. Lin.** 1999. 6-(1-Hydroxyalkylpenam sulfone derivatives as inhibitors of class A and class C β-lactamases. I. *Bioorg. Med. Chem. Lett.* **9**:991–996.

**Bitha, P., Z. Li, G. D. Francisco, Y. Yang, P. J. Petersen, E. Lenoy, and Y. I. Lin.** 1999. 6-(1-Hydroxyalkyl)penam sulfone derivatives as inhibitors of class A and class C β-lactamases. I. *Bioorg. Med. Chem. Lett.* **9**:997–1002.

**Buynak, J. D., V. R. Doppalapudi, A. S. Rao, S. D. Nidamarthy, and G. Adam.** 2000. The synthesis and evaluation of 2-substituted-7-(alkylidene)cephalosporin sulfones as β-lactamase inhibitors. *Bioorg. Med. Chem. Lett.* **10**:847–851.

**Buynak, J. D., V. R. Doppalapudi, M. Frotan, R. Kumar, and A. Chambers.** 2000. Catalytic approaches to the synthesis of β-lactamase inhibitors. *Tetrahedron* **56**:5709–5718.

**Buynak, J. D., V. R. Doppalapudi, and G. Adam.** 2000. The synthesis and evaluation of 3-substituted-7-(alkylidene)cephalosporin sulfones as β-lactamase inhibitors. *Bioorg. Med. Chem. Lett.* **10**:853–857.

**Buynak, J. D., A. S. Rao, V. R. Doppalapudi, G. Adam, P. J. Petersen, and S. D. Nidamarthy.** 1999. The synthesis and evaluation of 6-alkylidene-2"β-substituted penam sulfones as β-lactamase inhhiitors. *Bioorg. Med. Chem. Lett.* **9**:1997–2002.

**Grant, E. B., D. Guiadeen, E. Z. Baum, B. D. Foleno, H. Jin, D. A. Montenegro, E. A. Nelson, K. Bush, and D. J. Hlasta.** 2000. The synthesis and SAR of rhodanines as novel class C β-lactamase inhibitors. *Bioorg. Med. Chem. Lett.* **10:**2179–2182.

**Heinze-Krauss, I., P. Angehrn, R. L. Charnas, K. Gubernator, E. M. Gutknecht, C. Hubschwerlen, M. Kania, C. Oefner, M. G. P. Page, S. Sogabe, J. L. Specklin, and F. Winkler.** 1998. Structure-based design of β-lactamase inhibitors. 1. Synthesis and evaluation of bridged monobactams. *J. Med. Chem.* **41:**3961–3971.

**Hubschwerlen, C., P. Angehrn, K. Gubernator, M. G. P. Page, and J. L. Specklin.** 1998. Structure-based design of β-lactamase inhibitors. 2. Synthesis and evaluation of bridged sulfactams and oxamaxins. *J. Med. Chem.* **41:**3972–3975.

**Mascaretti, O. A., G. O. Danelon, E. L. Setti, M. Laborde, and E. G. Mata.** 1999. Recent advances in the chemistry of β-lactam compounds as selected active-site serine β-lactamase inhibitors. *Curr. Pharm. Des.* **5:**939–953.

### Novel Inhibitors of Metallo-β-Lactamases

#### Trifluoromethyl Alcohol and Ketone Inhibitors

**Walter, M. W., A. Felici, M. Galleni, R. P. Soto, R. M. Adlington, J. E. Baldwin, J. M. Frère, M. Gololobov, and C. J. Schofield.** 1996. Trifluoromethyl alcohol and ketone inhibitors of metallo-β-lactamases. *Bioorganic Med. Chem. Lett.* **6:** 2455–2458.

#### Amino acid-derived hydroxamates

**Walter, M. W., M. Hernandez Valladares, R. M. Adlington, G. Amicosante, J. E. Baldwin, J. M. Frère, M. Galleni, G. M. Rossolini, and C. J. Schofield.** 1999. Hydroxamate inhibitors of *Aeromonas hydrophila* AE036 metallo-β-lactamase. *Bioorg. Chem.* **27:**35–41.

#### Mercaptoacetic acid thiol ester

**Concha, N. O., C. A. Janson, P. Rowling, C. A. Cheever, B. P. Clarke, C. Lewis, M. Galeni, J. M. Frère, D. J. Payne, J. H. Bateson, and S. S. Abdel-Meguid.** 2000. Crystal structue of the IMP-1 metallo-β-lactamase from *Pseudomonas aeruginosa* and its complex with a mercaptocarboxylate inhibitor: binding determinants of a potent broad-spectrum inhibitor. *Biochemistry* **39:**4288–4298.

**Goto, M., T. Takahashi, F. Yamashita, A. Koreeda, H. Mori, M. Ohta, and Y. Arakawa.** 1997. Inhibition of the metallo-β-lactamase produced from *Serratia marcescens* by thiol compounds. *Biol. Pharm. Bull.* **20:**1136–1140.

**Greenlee, M. L., J. B. Laub, J. M. Balkovec, M. L. Hammond, G. G. Hammond, D. L. Pompliano, and J. H. Epstein-Toney.** 1999. Synthesis and SAR of thioester and thiol inhibitors of IMP-1 metallo-β-lactamase. *Biorg. Med. Chem. Lett.* **9:** 2549–2554.

**Huntley, J. J. A., S. D. B. Scrofani, M. J. Osborne, P. E. Wright, and H. J. Dyson.** 2000. Dynamics of the metallo-β-lactamase from *Bacteroides fragilis* in the presence and absence of a tight-binding inhibitor. *Biochemistry* **39:**13356–13364.

**Page, M. I., and A. P. Laws.** 1998. The mechanism of catalysis and the inhibition of β-lactamases. *J. Chem. Soc. Chem. Commun.* **1998:**1609–1617.

**Payne, D. J., J. H. Bateson, B. C. Gasson, D. Proctor, T. Khushi, T. H. Farmer, D. A. Tolson, D. Bell, P. W. Skett, A. C. Marshall, R. Reid, L. Ghosez, Y. Combret, and J. Marchand-Brynaert.** 1997. Inhibition of metallo-β-lactamases by a series of mercaptoacetic acid thiol ester derivatives. *Antimicrob. Agents Chemother.* **41:**135–140.

**Payne, D. J., J. H. Bateson, B.C. Gasson, T. Khushi, D. Proctor, S. C. Pearson, and R. Reid.** 1997. Inhibition of metallo-β-lactamases by a series of thiol ester derivatives of mercaptophenylacetic acid. *FEMS Microbiol. Lett.* **157:** 171–175.

**Scrofani, S. D. B., J. Chung, J. J. A. Huntley, S. J. Benkovic, P. E. Wright, and H. J. Dyson.** 1999. NMR characterization of the metallo-β-lactamase from *Bacteroides fragilis* and its interaction with a tight-binding inhibitor: role of an active-site loop. *Biochemistry* **38:**14507–14514.

#### Biphenyl tetrazoles

**Toney, J. H., P. M. Fitzgerald, N. Grover-Sharma, S. H. Olson, W. J. May, J. G. Sundelof, D. E. Vanderwall, K. A. Cleary, S. K. Grant, J. K. Wu, J. W. Kozarich, D. L. Pompliano, and G. G. Hammond.** 1998. Antibiotic sensitization using biphenyl tetrazoles as potent inhibitors of *Bacteroides fragilis* metallo-β-lactamase. *Chem. Biol.* **5:**185–196.

**Toney, J. H., K. A. Cleary, G. G. Hammond, X. Yuan, W. J. May, S. M. Hutchins, W. T. Ashton, and D. E. Vanderwall.** 1999. Structure-activity relationships of biphenyl tetrazoles as metallo-β-lactamase inhibitors. *Bioorg. Med. Chem. Lett.* **9:**2741–2746.

chapter 13



# Inhibitors of Peptidoglycan Biosynthesis
## Fosfomycin and D-Cycloserine

### Overview of Biosynthesis of the Bacterial Peptidoglycan Unit

The structure and biosynthesis of the peptidoglycan unit have special significance relative to the action of the antibiotics fosfomycin, D-cycloserine, and bacitracin and the glycopeptides vancomycin and teicoplanin, as well as the β-lactam antibiotics. The biosynthesis of peptidoglycan was first worked out with *Staphylococcus aureus*. Although bacteria show variations in peptidoglycan structure, the biosynthetic sequence in *S. aureus* serves to illustrate the general features of the process. The biosynthesis of peptidoglycan may be conveniently divided into stages inhibited by fosfomycin and those inhibited by D-cycloserine antibiotics (Fig. 13.1).

### Fosfomycin

Fosfomycin [2-*R*-(3-methyl-2-oxiranyl)phosphonic acid] (Fig. 13.2) is active against both gram-positive and gram-negative bacteria. It was isolated in 1969 and introduced into clinical use in the early 1970s in Europe. It has been extensively used clinically in many countries, such as Spain, Italy, France, and Japan. In July 1997, the Food and Drug Administration approved fosfomycin tromethamine (a sodium salt of fosfomycin trometamol) for single-dose oral treatment of uncomplicated urinary tract infection in women in the United States.

Fosfomycin was originally discovered as a fermentation product of *Streptomyces fradiae, S. viridochromogenes,* and *S. wedmorensis,* but it is now produced synthetically. It enters the cells of susceptible bacteria via an L-α-glycerylphosphate active transport system.

#### Mechanism of Action

Fosfomycin acts as a phosphoenolpyruvate analogue and irreversibly inhibits the enol-pyruvyl transferase that catalyzes the transfer of phosphoenolpyru-

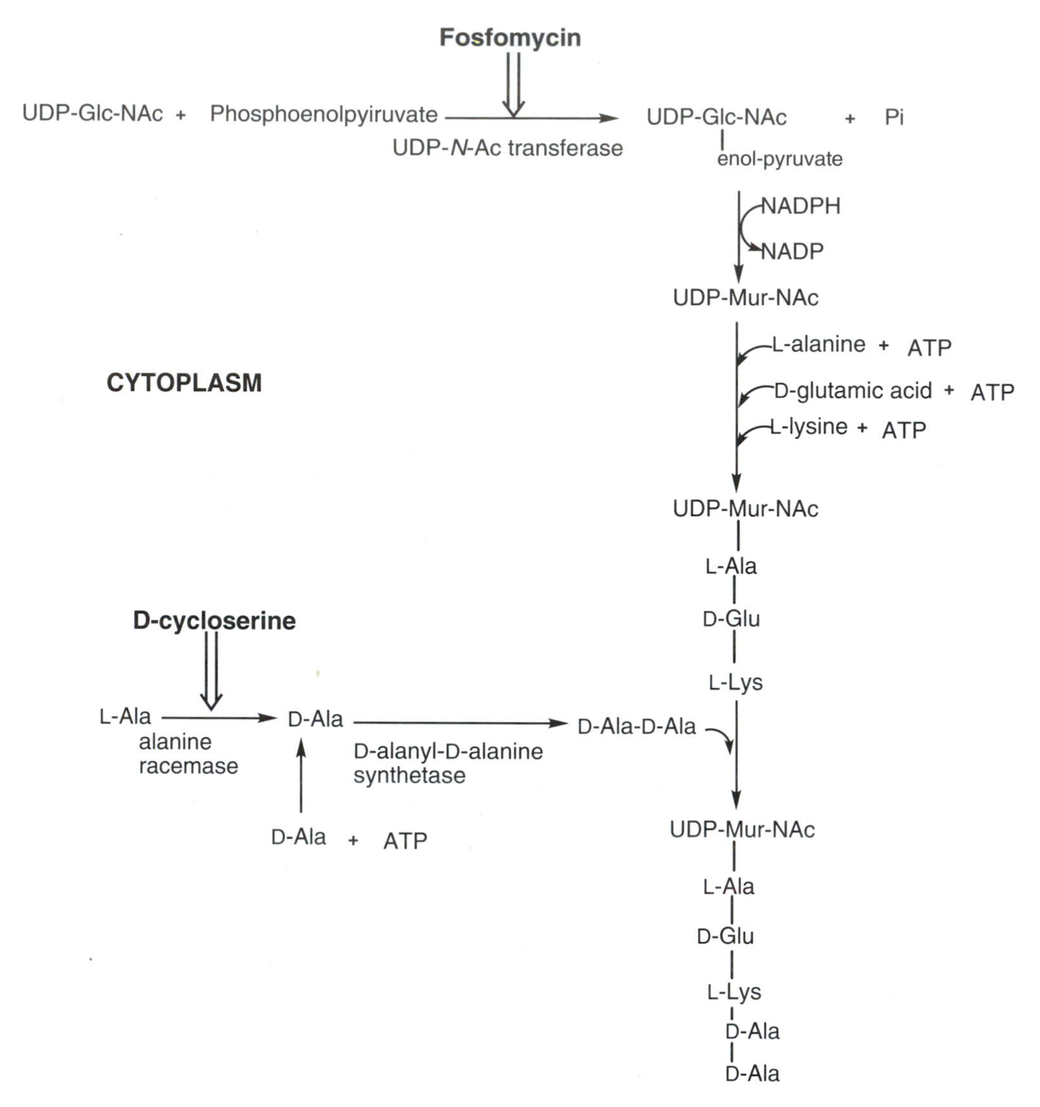


**Figure 13.1** Schematic representation of the reactions in the cytoplasm during peptidoglycan synthesis in *S. aureus*. The sites of inhibition by fosfomycin and D-cycloserine are indicated by large arrows.

vate in the formation of UDP–N-acetylglucosamine (UDP-NAG) enolpyruvate. Mechanistically, the oxygen atom in the oxirane ring is protonated and the carbon atom at position 2 undergoes a nucleophilic attack by a sulfhydryl group of a cysteine residue of the enzyme active site, forming a covalent inactive product (Fig. 13.3).

## Spectrum of Antibacterial Activity

Fosfomycin is moderately active in vitro against *Escherichia coli* and many other common pathogens that cause uncomplicated urinary tract infections.

## Clinical Uses

Fosfomycin has been used in Europe to treat gastrointestinal and urinary tract infections. The fosfomycin tromethamine salt has been particularly effective for single-dose treatment of urinary tract infections in women and has eradicated bacteriuria in over 90% of treated women.

## Bacterial Resistance

In multiple-dose regimens, resistance to fosfomycin emerges rapidly; however, cross-resistance with other antibacterial agents has not been common. Resistance to fosfomycin occurs by three mechanisms; two of them are encoded on the chromosome, whereas the third is of plasmid origin. Chromosomal mutants either lack a functional uptake system for L-α-glycerophosphate or glucose-6-phosphate or have a phosphoenolpyruvate-UDPNAG enolpyruvyl transferase (pyruvyl transferase)

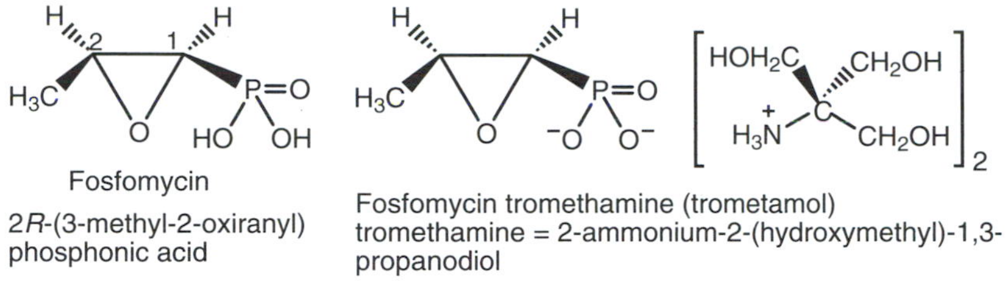

Fosfomycin
2*R*-(3-methyl-2-oxiranyl) phosphonic acid

Fosfomycin tromethamine (trometamol)
tromethamine = 2-ammonium-2-(hydroxymethyl)-1,3-propanodiol

**Figure 13.2** Chemical structures of fosfomycin and fosfomycin tromethamine.

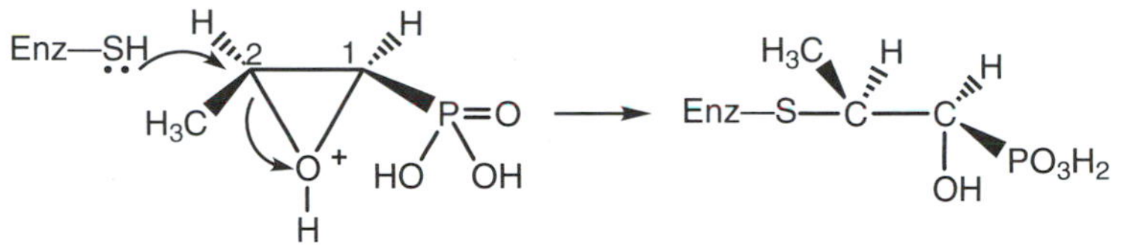

**Figure 13.3** Suggested mechanism of action of fosfomycin at the molecular level. Enz, enzyme.

which discriminates between phosphoenolpyruvic acid and fosfomycin. This mutation of the enzyme decreases the pyruvyl transferase affinity of fosfomycin and makes these strains resistant. Plasmid-encoded resistance to fosfomycin was first described in 1980. The resistant bacteria actively take up the antibiotic and have a fully sensitive pyruvyl transferase enzyme. These strains harbor plasmid genes which confer fosfomycin resistance. One of these, *fosA,* has been found in various members of the *Enterobacteriaceae,* in *Pseudomonas* spp., and in *Acinetobacter* spp. The biochemical mechanism of fosfomycin resistance mediated by the FosA protein has been established. Fosfomycin is inactivated by the opening of its epoxide ring followed by formation of an adduct between its C-1 atom and the sulfhydryl group of the cysteine of the tripeptide glutathione (Fig. 13.4). *E. coli* strains that contain FosA protein but cannot synthesize glutathione (such as *gsh* mutants) are fosfomycin sensitive. The *fosA* gene has not been found so far in gram-positive bacteria. The biochemical mechanism of fosfomycin resistance conferred by the *fosB* gene remains undetermined.

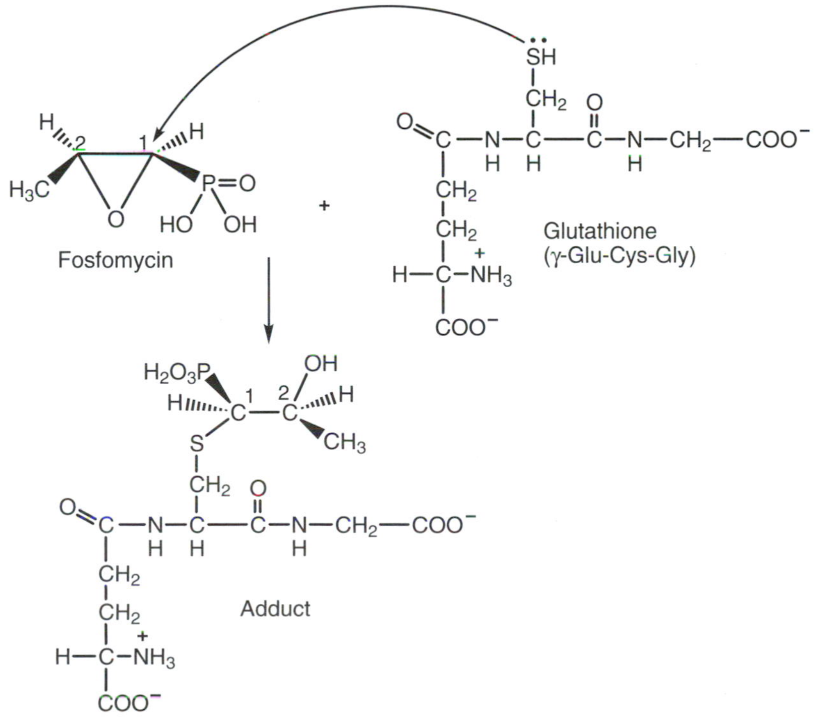

Fosfomycin

Glutathione
(γ-Glu-Cys-Gly)

Adduct

**Figure 13.4** Mechanism of fosfomycin resistance mediated by the FosA protein.

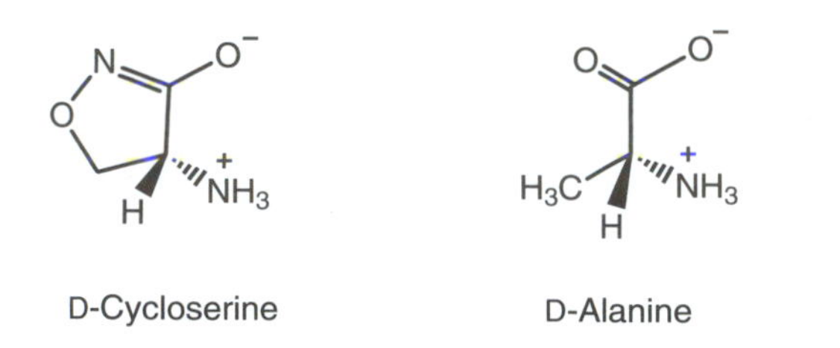


**Figure 13.5** Chemical structures of D-cycloserine and D-alanine in the zwitterionic form.

## Fosfomycin Preparation Commercially Available in the United States

Fosfomycin tromethamine (Monurol, a sachet for oral administration), marketed by Forest, is the only form of fosfomycin marketed in the United States.

# D-Cycloserine

D-Cycloserine (D-4-amino-*S*-isoxalidone) is a structural analog of D-alanine (Fig. 13.5). It was isolated in 1955 from cultures of *Streptomyces orchidaceus* and *S. garyphalus*. Currently it is produced synthetically.

## Mechanism of Action

D-Cycloserine, being a structural analog of D-alanine, acts as a "suicide substrate" for the enzyme racemase and irreversibly forms a covalent complex. That is, the D-cycloserine occupies the substrate site on the alanine racemase, inactivating it in an irreversible manner. Originally, it was supposed that D-cycloserine was a competitive inhibitor of both the alanine racemase and D-alanyl-D-alanine synthetase (Fig. 13.1).

## Spectrum of Antibacterial Activity

*Mycobacterium tuberculosis* is sensitive to D-cycloserine, and the strains which become resistant to streptomycin, isoniazid, and rifampin usually remain sensitive to D-cycloserine.

## Clinical Uses

D-Cycloserine is now used only for the treatment of patients with tuberculosis whose *M. tuberculosis* strains are resistant to several first-line drugs (multiresistant tuberculosis). Under such circumstances, D-cycloserine should always be combined with at least one drug (preferably two) to which the *M. tuberculosis* strain is still sensitive.

## Bacterial Resistance

Resistance in *M. tuberculosis* is rare and develops only slowly in patients treated with D-cycloserine alone.

## REFERENCES AND FURTHER READING


### Fosfomycin Tromethamine Salt

**Reeves, D. S.** 1994. Fosfomycin trometamol. *J. Antimicrob. Chemother.* **34:**853–858.

#### Mechanism of action

**Kahan, F. M., J. S. Kahan, P. J. Cassidy, and H. Kropp.** 1974. The mechanism of action of fosfomycin (phosphonomycin). *Ann. N. Y. Acad. Sci.* **235:**364–386.

#### Resistance

**Arca, P., G. Reguera, and C. Hardisson.** 1997. Plasmid-encoded fosfomycin resistance in bacteria isolated from the urinary tract in a multicentre survey. *J. Antimicrob. Chemother.* **40:**393–399.

**Arca, P., C. Hardisson, and J. E. Suárez.** 1990. Purification of a glutathione *S*-transferase that mediates fosfomycin resistance in bacteria. *Antimicrob. Agents Chemother.* **34:**844–848.

**Arca, P., M. Rico, A. F. Braña, C. J. Villar, C. Hardisson, and J. E. Suárez.** 1988. Formation of an adduct between fosfomycin and glutathione: a new mechanism of antibiotic resistance in bacteria. *Antimicrob. Agents Chemother.* **32:** 1552–1556.

#### Clinical uses

**Anonymous.** 1997. Fosfomycin for urinary tract infections. *Med. Lett. Drugs Ther.* **39:**66–68.

### D-Cycloserine

#### Mechanism of Action

**Wang, E., and C. H. Walsh.** 1978. Suicide substrates for the allanine racemase of *Escherichia coli*. *Biochemistry* **17:**1313.

#### Clinical uses

**Kucers, A., S. M. Crowe, M. L. Grayson, and J. F. Hoy.** 1997. *The Use of Antibiotics,* 5th ed., p. 542–543. Butterworth-Heineman, Oxford, United Kingdom.

chapter 14



# Inhibitors of Peptidoglycan Biosynthesis
## Bacitracin and Glycopeptides

Figure 13.1 shows the early steps in peptidoglycan biosynthesis, which are cytoplasmic. The succeeding steps occur on the cytoplasmic membrane (Fig. 14.1). Normally, undecaprenylpyrophosphate is converted by a pyrophosphatase to the corresponding undecaprenylphosphate and phosphate; the undecaprenylphosphate thus becomes available for reaction with another molecule of UDP-Mur*N*Ac-pentapeptide. Formation of a complex between the lipid pyrophosphate and bacitracin blocks this process and so eventually halts the synthesis of peptidoglycan.

The action of vancomycin and teicoplanin depends on their ability to bind specifically to the terminal D-alanyl-D-alanine group on the peptide side chain of the membrane-bound intermediates in peptidoglycan synthesis (see Fig. 14.3). The formation of the complex between vancomycin or teicoplanin and D-alanyl-D-alanine blocks the enzyme transglycosylase, which is involved in the incorporation of the disaccharide-peptide into the growing peptidoglycan chain.

## Summary of the Last Steps of Peptidoglycan Biosynthesis

As described in chapter 1, bacteria resist hypotonic shock and cell lysis because they are protected by their peptidoglycan-containing cell wall. This rigid polymer consists of a repeating disaccharide unit of *N*-acetylglucosamine-($\beta$1→4)-*N*-acetylmuramic acid to which a pentapeptide is linked through the muramyl lactyl ether carboxylate (Fig. 1.10 and 1.11). This pentapeptide is composed of L-Ala–$\gamma$-D-Glu–(a dibasic amino acid, either L-Lys or *meta*-diaminopimelic acid)–D-Ala–D-Ala. The free amine group of the dibasic amino acid forms an amide bond with the penultimate D-Ala of another pentapeptide, either directly or through a peptide linker such as Gly-5.

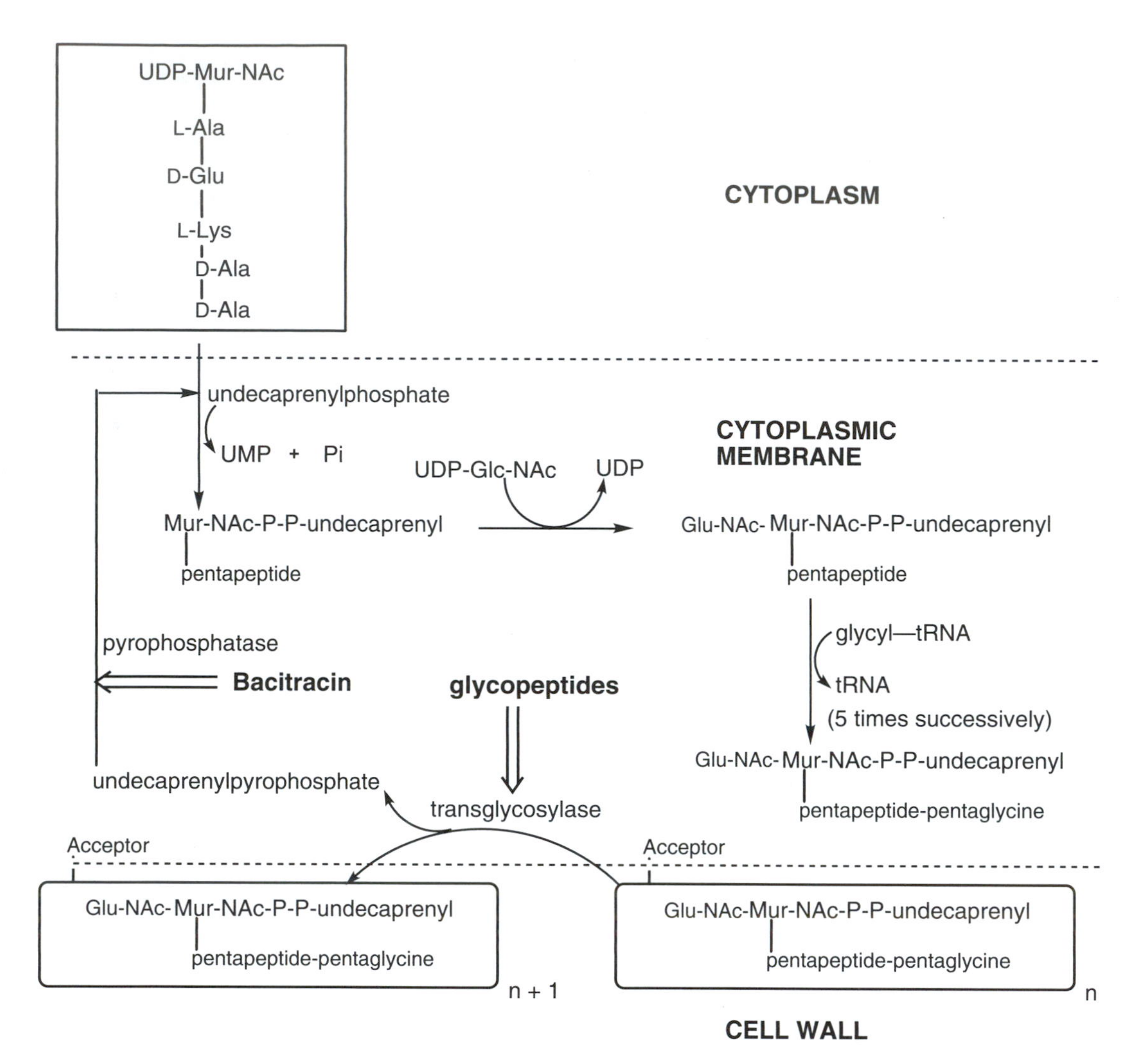


**Figure 14.1** Schematic representation of the reactions at the cytoplasmic membrane during peptidoglycan synthesis in *S. aureus*. The sites of inhibition by bacitracin and the glycopeptides (vancomycin and teicoplanin) are indicated by large hollow arrows.

## Bacitracin

Bacitracin is a mixture of at least nine water-soluble cyclic peptides. The principal component (60 to 80%) is bacitracin A. The structure of bacitracin A is shown in Fig. 14.2. Bacitracin was isolated in 1943 from a strain of a *Bacillus* sp., which was originally classified as *Bacillus subtilis* but now is known as *B. licheniformis*.

### Mechanism of Action

Bacitracin complexes with the membrane-bound undecaprenylpyrophosphate or C-55 isoprenylpyrophosphate. Binding of bacitracin prevents the enzymatic dephosphorylation of the lipid carrier molecule to its monophosphate form, a reaction which occurs during the second stage of peptidoglycan biosynthesis. Therefore, bacitracin interacts with the substrate rather than

**Figure 14.2** Chemical structure of bacitracin A.

D-Phe — L-His — D-Asp — L-Asn — L-Lys — D-Orn — L-Ile — D-Phe

L-Lys — L-Ile — D-Glu — L-Leu (thiazoline ring with N and S) — CH($+NH_3$) — HC($CH_3$) — $CH_2$ — $CH_3$

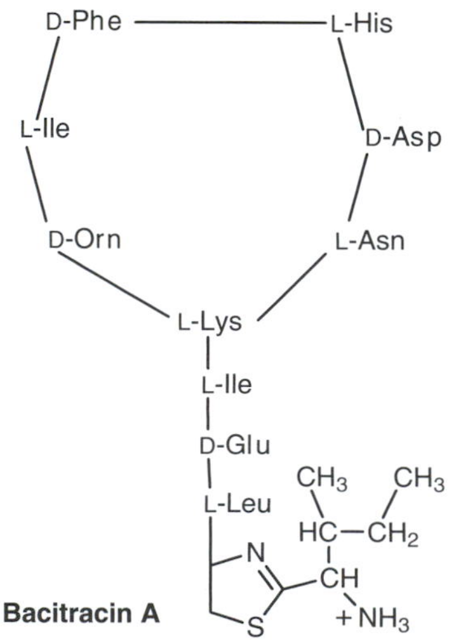

Bacitracin A

with the dephosphorylating enzyme. There is a metal requirement for the binding of bacitracin to the substrate, and zinc is especially active.

### Antibacterial Activity

Bacitracin is highly active against most gram-positive bacteria, particularly *Staphylococcus aureus* and *Streptococcus pyogenes*. *Clostridium difficile* is usually sensitive, and most *C. difficile* strains are synergistically inhibited by the combination of bacitracin and rifampin. Pathogenic neisseriae (meningococci and gonococci) are also sensitive, but gram-negative bacilli are resistant.

### Clinical Uses

Bacitracin received drug certification in 1949. From that time until 1960, it was used systematically, mainly for the treatment of severe staphylococcal infections. Because of its nephrotoxicity, it is no longer administered parenterally. It is now mainly restricted for topical use. Zinc bacitracin is the form used in human medicine; it contains 2 to 10% zinc on a dry-weight basis.

### Bacterial Resistance

Acquired bacterial resistance to bacitracin is rare.

### Bacitracin Preparations Commercially Available in the United States

Table 14.1 lists the bacitracin preparations commercially available in the United States.

## Glycopeptides

Vancomycin and teicoplanin (Fig. 14.3) are natural antibiotics and are members of a large class called the glycopeptides, which are effective in the treatment of gram-positive bacterial infections. Vancomycin and teicoplanin are widely used to treat severe staphylococcal infections, including those caused by methicillin-resistant *S. aureus* (MRSA). The glycopeptide group of antibiotics encompasses hundreds of natural products obtained from various species of soil actinomycetes and almost 1,000 semisynthetic analogues; however, vancomycin and teicoplanin are the only ones used clinically in antibacterial chemotherapy.

In the mid-1950s, scientists at Eli Lilly Laboratories (Indianapolis, Ind.) isolated vancomycin from a fermentation broth of *Amycolatopsis orientalis* (previously called *Nocardia orientalis* and *Streptomyces orientalis*). It became available for clinical use after being approved by the Food and Drug Administration in 1958. Teicoplanin was isolated in 1975 at Lepetit Research Laboratories (Varese, Italy) from a fermentation broth of *Actinoplanes teichomyceticus*. Teicoplanin is a complex of five related molecules with similar antibiotic potency. It is the most significant glycopeptide antibiotic discovered since vancomycin was marketed. It is widely used in Europe under the proprietary name of Targocid (Aventis), although it is an experimental drug in the United States.

### Chemical Structures of Vancomycin and Teicoplanins

The chemical structure of vancomycin and the five teicoplanins consists of a peptide backbone containing seven amino acids, numbered 1 to 7 from the N to the C terminus. Residues 4 and 5 are *p*-hydroxyphenylglycines, residues 2 and 6 are tyrosines, and residue 7 is 3,5-dihydroxyphenylglycine. In vancomycin, residue 1 is *N*-methylleucine and residue 3 is asparagine. In the teicoplanins, residues 1 and 3 are *p*-hydroxyphenylglycine and 3,5-dihydroxyphenylglycine, respectively. In both vancomycin and teicoplanin, the stereochemistry of the α-C chiral center of amino acids 1 to 7 is *R, R, S, R, R, S, S*.

The amino acid backbones of vancomycin and teicoplanin are highly cross-linked and substituted. The aromatic portions of residues 2 and 6 are each cross-linked to residue 4 through an oxygen atom, and the aromatic rings of residues 5 and 7 are cross-linked through a carbon-carbon bond. In addition, residues 1 and 3 of teicoplanin are cross-linked through an oxygen atom. The aromatic rings of residues 2 and 6 have a chlorine substituent. Residues 2 and 6 of vancomycin also have a β-hydroxy group, whereas in teicoplanin only residue 6 contains a β-hydroxy group. Lastly, sugar molecules are attached to the peptide backbone. In vancomycin the disaccharide D-glucosylvancosamine is linked through a glycosidic bond to residue 4. In teicoplanin, three monosaccharides, *N*-acetyl-D-glucosamine, *N*-acetyl-D-mannose, and D-mannose, are attached to residues 4, 6, and 7 respectively.

Vancomycin has a molecular mass of 1,448 Da. The exclusion limit of 500 to 600 Da of the porin proteins

**Table 14.1** Generic and common trade names of bacitracin, the preparations available, and manufacturers in the United States

| Generic name | Trade name (preparation) | Manufacturer |
|---|---|---|
| Bacitracin zinc plus polymyxin B sulfate | Polysporin (ointment) | Pfizer |
| Bacitracin zinc plus polymyxin B sulfate and pramoxine HCl | Betadine brand plus (ointment) | Purdue Frederick |
| Bacitracin zinc plus neomycin and polymyxin sulfate | Neosporin (ophthalmic ointment) | Monarch |

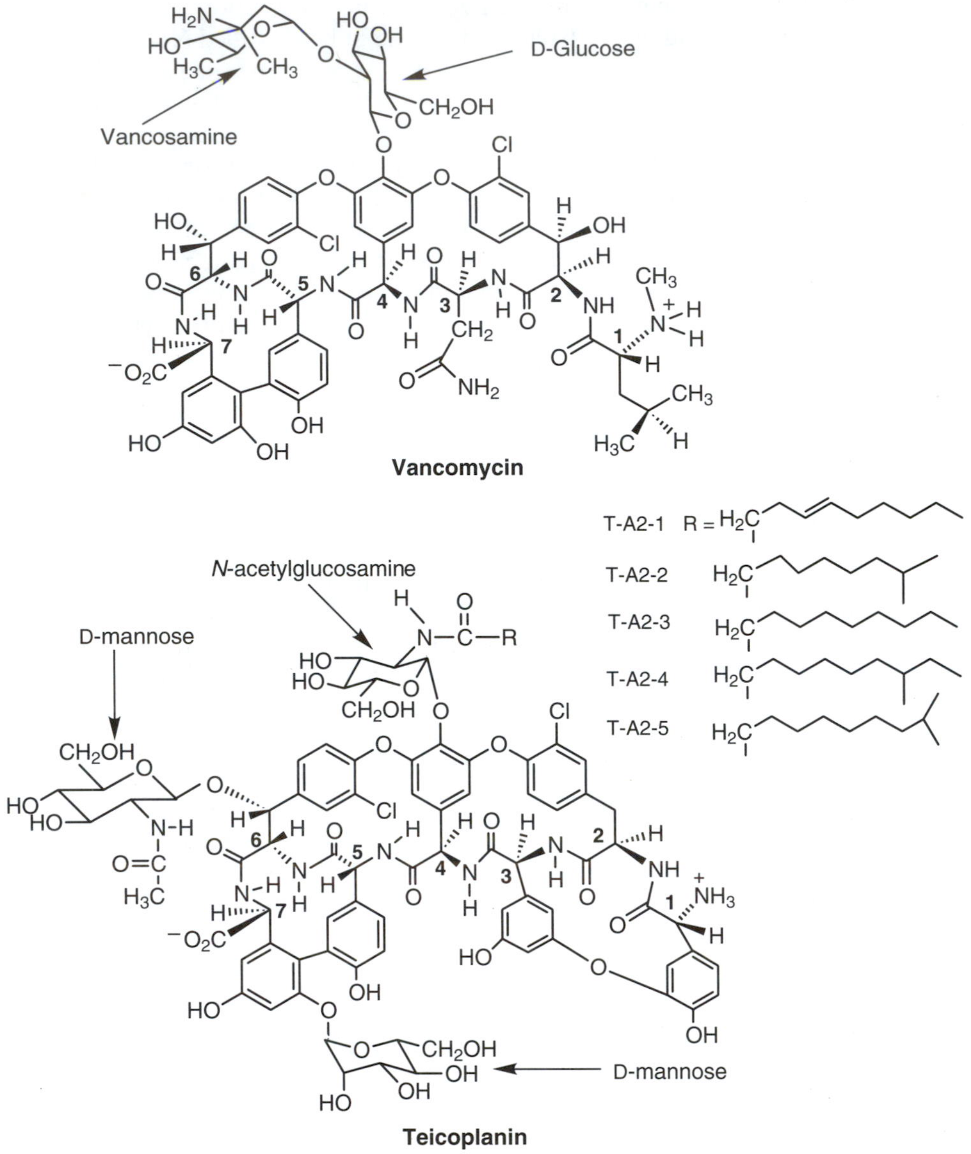

**Figure 14.3** Chemical structures of vancomycin and of five teicoplanins.

(permeability barrier) of the gram-negative bacterial outer membrane prevents the transport of glycopeptides to the periplasmic space, and so only gram-positive bacteria such as enterococci, streptococci, and staphylococci are susceptible to the antibiotic activity of glycopeptides.

## Target of Glycopeptide Antibiotics

### MECHANISM OF ACTION OF VANCOMYCIN

A precise structural picture for the binding of vancomycin to peptidoglycan has been determined by nuclear magnetic resonance spectroscopy and X-ray crystallographic techniques. Vancomycin binds reversibly to the D-Ala–D-Ala dipeptide segment of the muramyl pentapeptide present in peptidoglycan monomers which are exposed at the external cell surface of the cytoplasmic membrane. This binding is mediated through five well-defined hydrogen bonds, as shown in Fig. 14.4. Glycopeptide antibiotics such as vancomycin and teicoplanin do not penetrate into the cytoplasm. Thus, their interaction with the D-Ala–D-Ala dipeptide can take place only after translocation of the undecaprenolate monomer to the outer surface of the cytoplasmic membrane.

**Transglycosylase enzymes** act on the growing peptidoglycan strand to extend the sugar chains by incorporating a new monomeric unit from *N*-acetylglucosamine-(β1→4)-*N*-acetylmuramyl pentapetide pyrophosphoryl undecaprenol (lipid II). Glycopeptide antibiotics, such as vancomycin and teicoplanin, tie up the dipeptide substrate (D-Ala–D-Ala), thereby preventing it from interacting with the transglycosylase (Fig. 14.5) and transpeptidase (Fig. 7.9) enzymes. The binding of the D-Ala–D-Ala carboxy-terminal dipeptide by the bulky mole-

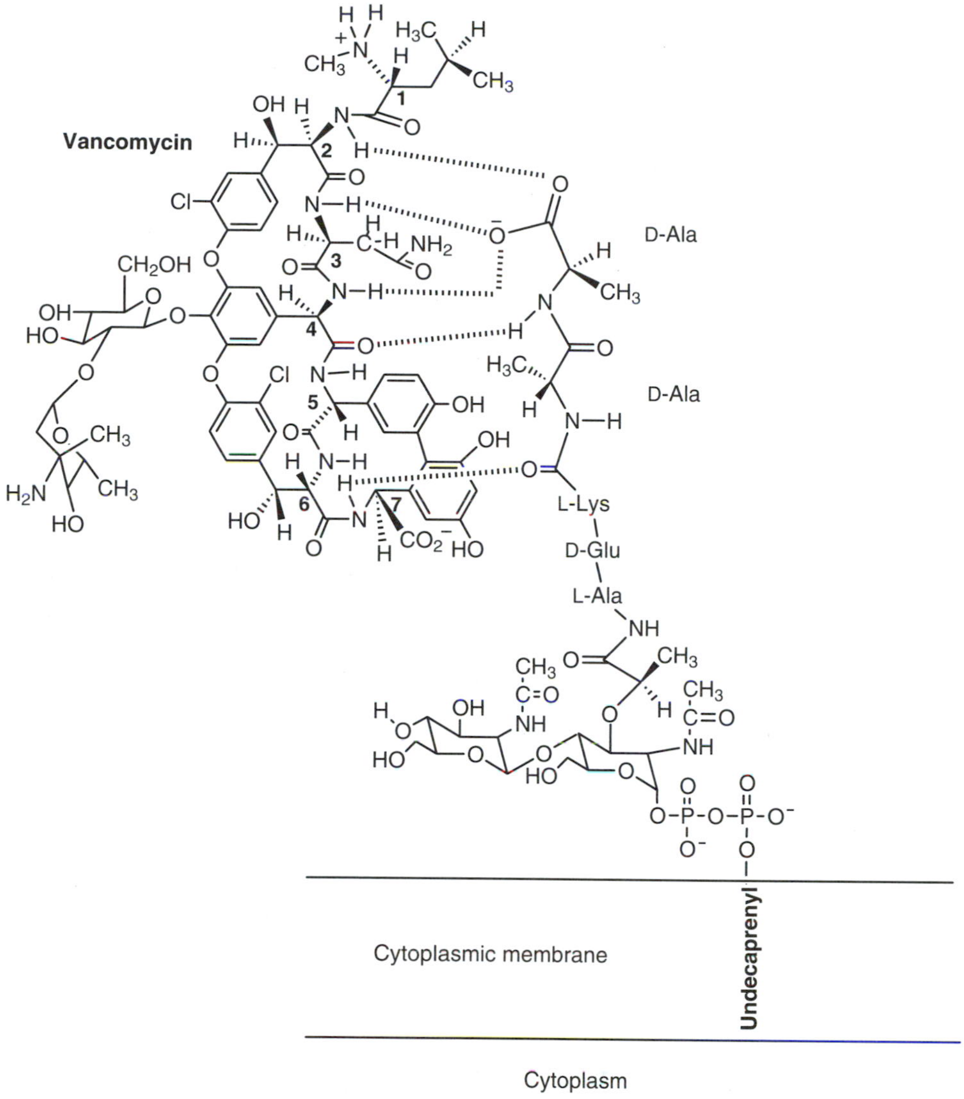

**Figure 14.4** Chemical representation of the 1:1 complex between vancomycin and the dipeptide D-Ala–D-Ala from the muramyl pentapeptide pyrophosphoryl undecaprenol (lipid II).

cule of vancomycin or teicoplanin is thought to physically occlude the approach of transglycosylase and transpeptidase and thus to inhibit their crucial enzymatic activities. The net effect is to retard the growth of the cell wall peptidoglycan, thus leading to accumulation of peptidoglycan precursors. The failure to execute a cross-linking lowers the interstrand covalent connectivity. This causes a reduction in the mechanical strength of the peptidoglycan cell wall and renders the bacteria susceptible to lysis by changes in osmotic pressure.

## DIMERIZATION OF GLYCOPEPTIDES

In 1989, Waltho and Williams at the University of Cambridge (Cambridge, United Kingdom) discovered by using proton nuclear magnetic resonance spectroscopy that the glycopeptide antibiotic ristocetin A forms a dimer in aqueous solution. They were also able to elucidate the relative orientations of the two monomer units of this dimer and the hydrogen-bonding network formed between them. The overall structure of the dimer was later independently confirmed by X-ray crystallography by M. Schäfer (1996) and by P. J. Loll and coworkers (1997). It is well established that ristocetin A, vancomycin, and a number of other glycopeptides dimerize in solution to form head-to-tail complexes. In contrast, teicoplanin does not dimerize in solution.

The dimeric structure of vancomycin is held together by four hydrogen bonds between the two amide backbones. Two units of the D-Ala–D-Ala ligand can dock in the two binding pockets of the vancomycin dimer through five hydrogen bonds each, as depicted in Fig. 14.6. Dimerization results in an enhanced antibacterial activity of vancomycin and other glycopeptides through cooperative binding effects. By dimerizing, vancomycin increases the binding affinity for the nascent bacterial peptide units terminating in D-Ala–D-Ala.

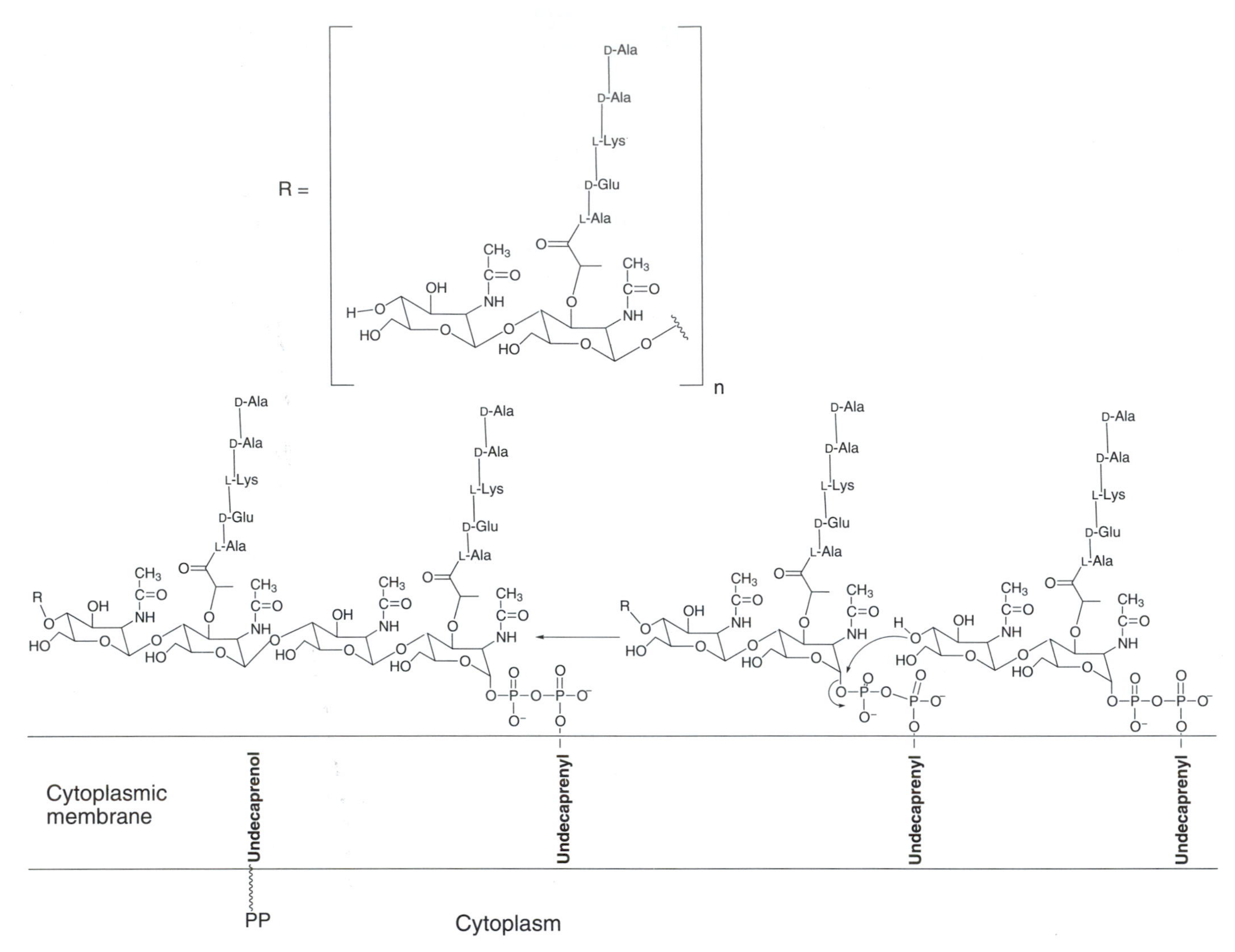


**Figure 14.5** Reaction catalyzed by the transglycosylases at the outer surface of the cytoplasmic membrane. Reprinted from R. C. Goldman and D. Gange, *Curr. Med. Chem.* **7**:801–820, 2000, with permission from the publisher.

While teicoplanin does not dimerize in solution, it does exhibit enhanced activity. The long hydrocarbon chain attached to one of its carbohydrate units is inserted into the bacterial phospholipid layer. This action locates the teicoplanin molecule at its site of action and facilitates the formation of an "intramolecular" complex at the outside of the cytoplasmic membrane, which increases the in vivo effectiveness of the antibiotic.

## Antibacterial Activity and Clinical Uses of Vancomycin

Vancomycin is active against most species of gram-positive cocci and bacilli, such as *S. aureus,* including MRSA strains, *S. epidermidis,* including multiple-drug-resistant strains, *Streptococcus pneumoniae,* viridans group *Streptococcus, Streptococcus bovis, Clostridium* species, diphtheroids, *Listeria monocytogenes, Actinomyces* spp., and *Lactobacillus* spp. *Neisseria gonorrhoeae* strains are usually susceptible, but gram-negative bacilli and mycobacteria are resistant. Most enterococcal isolates remain susceptible to vancomycin, but vancomycin-resistant enterococcal isolates have emerged as a significant problem since the late 1980s.

In an effort to slow the spread of vancomycin-resistant bacteria, the American Medical Association has recommended that the use of intravenous vancomycin be restricted to serious infections caused by susceptible gram-positive bacteria when other antibiotics are ineffective or not tolerated. Vancomycin is the antibiotic of

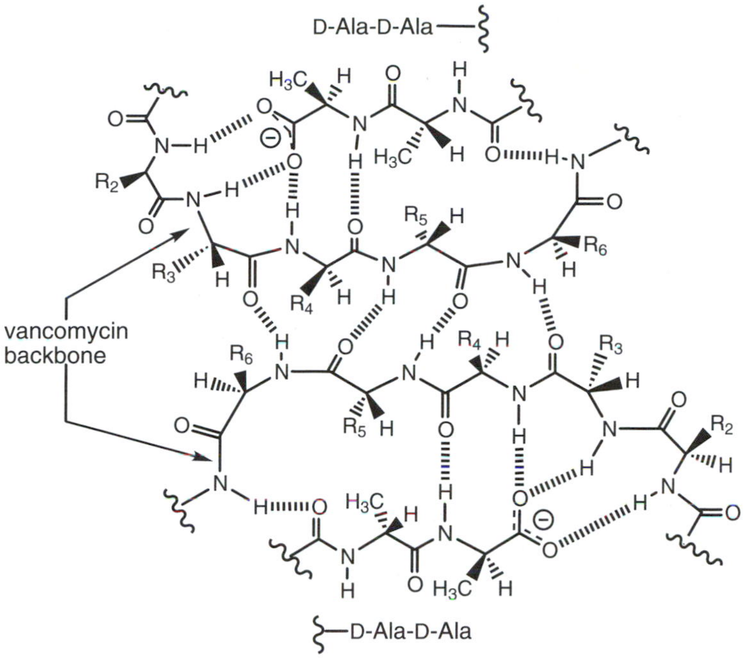

**Figure 14.6** The hydrogen-bonding network in the vancomycin dimer. The peptide backbones of two molecules of vancomycin dimerize, forming four hydrogen bonds. In the structure shown in the figure, two units of the D-Ala–D-Ala ligand dock in the two binding pockets of the dimer through five hydrogen bonds each. Reprinted from D. A. Beauregard, A. J. Maguire, D. H. Williams, and P. E. Reynolds, *Antimicrob. Agents Chemother.* **41:**2418–2423, 1997, with permission from the American Society for Microbiology.

choice for serious infections caused by MRSA and coagulase-negative staphylococci, including methicillin-resistant *S. epidermidis*. These infections include septicemia, endocarditis, osteomyelitis, pneumonia, lung abscesses, soft tissue infections, wound infections, and meningitis. The combination of vancomycin with gentamicin or streptomycin (for susceptible strains) is the regimen of choice for enterococcal endocarditis in penicillin-allergic patients.

## Antibacterial Activity and Clinical Uses of Teicoplanin

The in vitro antimicrobial spectrum of teicoplanin is qualitatively equivalent to that of vancomycin, but there are a few notable differences. For examples, teicoplanin is more active than vancomycin against all streptococci and enterococci but is less active against *S. heamolyticus*. For additional discussion on the use of vancomycin and teicoplanin, see *The Use of Antibiotics*, by Kucers et al. (1997).

## Bacterial Resistance

Vancomycin and teicoplanin are considered the last lines of defense against a variety of serious infections caused by gram-positive organisms such as enterococci, MRSA, and *Clostridium difficile*. Due to the widespread use of vancomycin in the last decade for the treatment of infections caused by gram-positive bacteria, clinically significant resistance has arisen in enterococcal strains. The first report of vancomycin-resistant enterococci (VRE) appeared in 1988, and over the last 15 years the frequency has increased, causing grave concern for both medical practitioners and their patients. Despite the nonpathogenic nature of enterococci to healthy individuals, VRE have become recognized as important opportunistic human pathogens and can be lethal if contracted by immunodeficient patients, such as AIDS patients and organ transplant recipients. Although acquired resistance to vancomycin thus far has been confined to the enterococci, there is concern that it will spread to *Staphylococcus* spp., including MRSA strains.

### MECHANISM OF VANCOMYCIN RESISTANCE

The genetic basis and enzymology of vancomycin resistance have been elucidated by major contributions from the groups of Patrice Courvalin at the Institut Pasteur (Paris, France) and Christopher Walsh at Harvard Medical School (Boston, Mass.).

#### Phenotypic Description

To date, five clinical phenotypes of vancomycin resistance have been recognized and are designated VanA, VanB, VanC, VanD, and VanE. VanA, VanB, VanD, and VanE are acquired resistance phenotypes, and the VanC phenotype is intrinsic to *Enterococcus gallinarum* and *E. casseliflavus* (Table 14.2).

In VanA, VanB, and VanD resistance to vancomycin, the terminal dipeptide D-Ala–D-Ala from the pentapeptide precursor is replaced with D-alanine–D-lactate (D-Ala–D-Lac). The replacement of D-alanine by D-lactic acid leads to a drastic drop in the binding affinity of vancomycin for the peptidoglycan precursors, and hence to

**Table 14.2** Phenotypic characteristics of glycopeptide-resistant enterococci[a]

| Characteristic | Phenotype | | | | |
|---|---|---|---|---|---|
| | VanA | VanB | VanC | VanD | VanE |
| Vancomycin MIC (μg/ml) | 64–1,000 | 4–1,024 | 2–32 | 128 | 16 |
| Teicoplanin MIC (μg/ml) | 16–512 | ≤0.5 | ≤0.5 | 4 | 0.5 |
| Most frequent enterococcal species | *E. faecium*, *E. faecalis* | *E. faecium*, *E. faecalis* | *E. gallinarum*, *E. casseliflavus*, *E. flavescens* | *E. faecium* | *E. faecalis* |
| Genetic determinant | Acquired | Acquired | Intrinsic | Acquired | Acquired |
| Transferable | Yes | Yes | No | No | No |

[a] Reprinted from Y. Cetinkaya, P. Falk, and C. G. Mayhall, *Clin. Microbiol. Rev.* **13**:686-707, 2000, with permission from the American Society for Microbiology, and adapted from H. S. Gold and R. C. Moellering, Jr., *N. Engl. J. Med.* 335:1445–1453, with permission from the Massachusetts Medical Society. Original version © 1996 Massachusetts Medical Society (all rights reserved).

high levels of resistance. In VanC and VanE vancomycin resistance, the final D-alanine is replaced by a D-serine (D-Ala–D-Ser), leading to low-level resistance.

## VanA Resistance

VanA is the phenotype reported most frequently and is associated with high-level resistance to vancomycin and cross-resistance to teicoplanin. These bacteria have developed a strategy for reprogramming peptidoglycan biosynthesis. The molecular mechanism of VanA resistance has been well characterized. The transposon Tn*1546* (a Tn*3* family transposon) in *Enterococcus faecium* contains five genes necessary for resistance, *vanR*, *vanS*, *vanH*, *vanA*, and *vanX*, which encode two regulatory proteins (VanS and VanR) and three enzymes (VanH, VanA, and VanX). These genes enable the bacteria to produce the depsipeptide D-Ala–D-Lac in the cytoplasm phase of peptidoglycan biosynthesis (Fig. 14.7).

VanH, VanA, and VanX act together to modify the peptidoglycan structure (Fig. 14.8). The α-ketoreductase, VanH, reduces pyruvate to D-lactate, while the depsipeptide ligase, VanA, couples the lactacte (D-Lac) to D-Ala, to form D-Ala–D-Lac. In a synergistic manner, the zinc-containing dipeptidase, VanX, hydrolyzes D-Ala–D-Ala but not D-Ala–D-Lac. The new D-Ala–D-Lac dipeptide accumulates and becomes incorporated into the growing peptidoglycan structure. The *vanS* and *vanR* genes encode a two-component signal transduction pathway that activates the transcription of *vanH*, *vanA*, and *vanX* (Fig. 14.8). As mentioned above, D-Ala–D-Lac is incorporated in place of D-Ala–D-Ala in peptidoglycan intermediates to yield a muramyl tetrapeptide ester terminating in D-lactate rather than a muramyl pentapeptide with the usual amide linkage terminating in D-alanine. The replacement of the terminal D-alanine with D-lactate removes one of the hydrogen atoms that participates in binding to vancomycin from the peptidoglycan ligand. Thus, there are only four hydrogen bonds, rather than five, between the binding pocket of vancomycin and the ligand (Fig. 14.9). The affinity of vancomycin for D-Ala–D-Lac is 1,000-fold lower than for D-Ala–D-Ala, accounting for the difference in MIC

**Figure 14.7** Schematic representation of the genes required for a high level of resistance to vancomycin in the VanA phenotype found within transposon Tn*1546*. The *vanR*, *vanS*, *vanH*, *vanA*, and *vanX* genes are essential for high-level resistance; the *vanY* and *vanZ* genes are nonessential. *ORF1* and *ORF2* encode proteins required for transposition. Reprinted from M. Arthur and P. Courvalin, *Antimicrob. Agents Chemother.* **37**:1563–1571, 1993, with permission from the American Society for Microbiology.

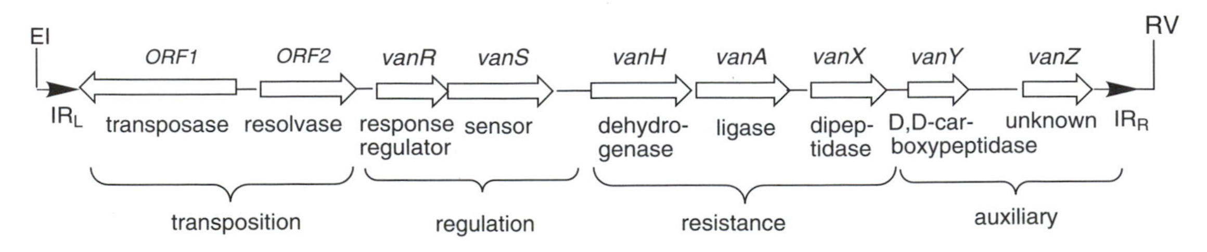

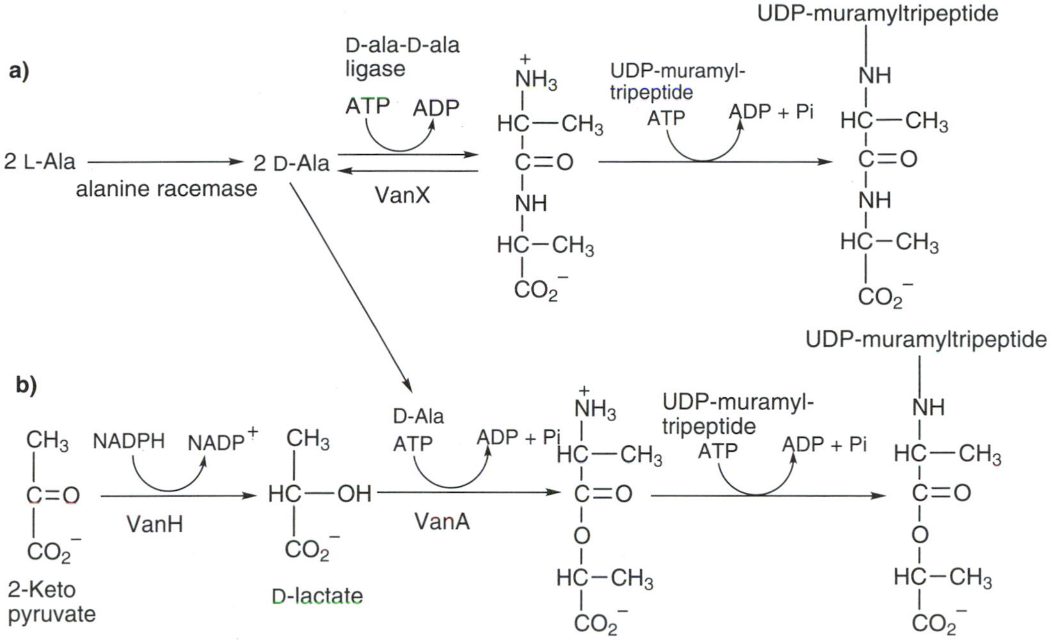

**Figure 14.8** Pathway for the assembly of cytoplasmic peptidoglycan precursors. (a) Synthesis of UDP-muramyl pentapeptide in enterococci susceptible to glycopeptides. (b) Incorporation of D-lactate at the C-terminal position of peptidoglycan precursors of enterococci resistant to glycopeptides. There is an amide functional group in panel a and an ester functionality in the equivalent position in panel b.

**Figure 14.9** Illustration of vancomycin interacting with the peptidoglycan termini of vancomycin-sensitive bacteria (a) and vancomycin-resistant bacteria of the VanA phenotype (b). Adapted from V. L. Healy, E. S. Park, and C. T. Walsh, *Chem. Biol.* **5:**197–207, 1998, with permission from the publisher.

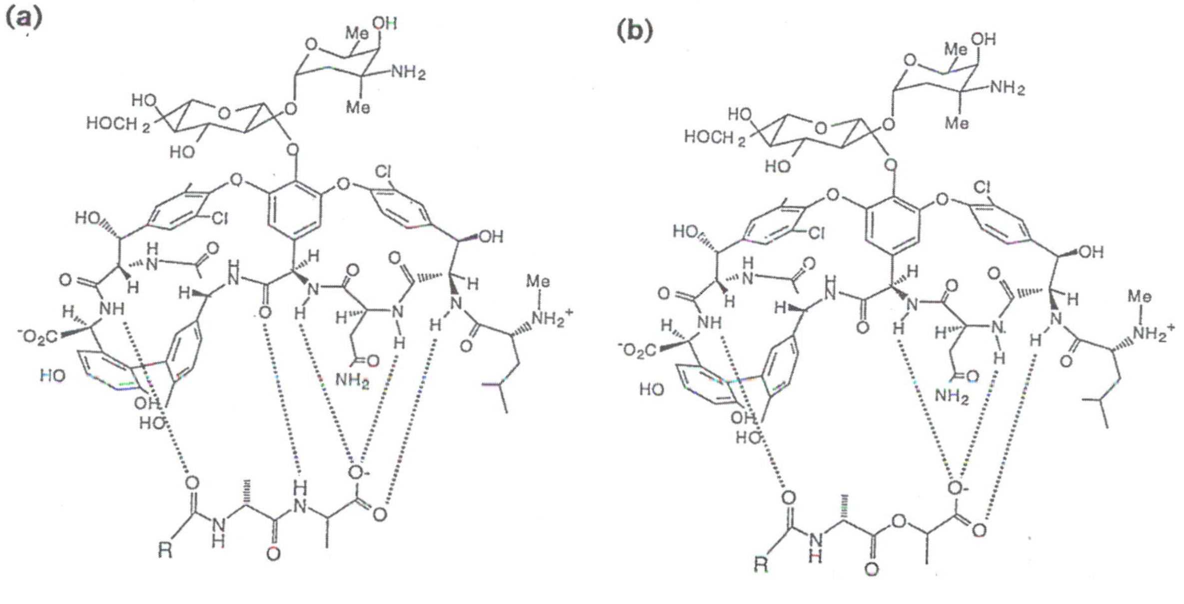

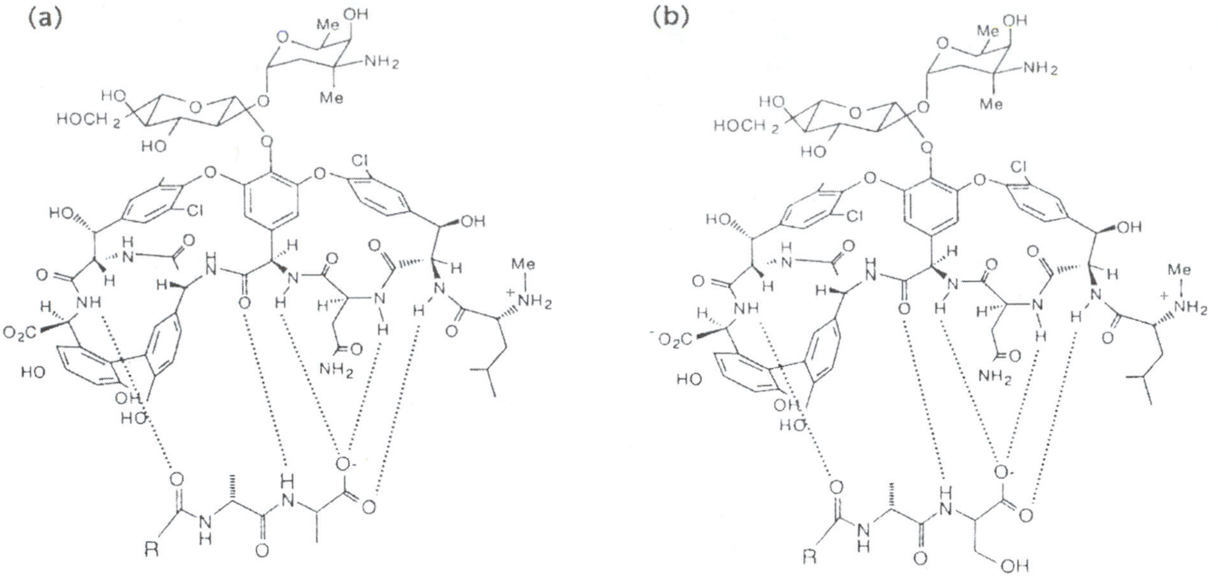

**Figure 14.10** Illustration of vancomycin interacting with the peptidoglycan termini of vancomycin-sensitive bacteria (a) and mildly vancomycin-resistant bacteria of the VanC type (b). Adapted from V. L. Healy, E. S. Park, and C. T. Walsh, *Chem. Biol.* **5:**197–207, 1998, with permission from the publisher.

**Figure 14.11** Chemical structures of vancomycin, LY264826, and its semisynthetic derivative LY333328. Reprinted from T. I. Nicas et al., *Antimicrob. Agents Chemother.* **39:**2585–2597, 1995, with permission from the American Society for Microbiology.

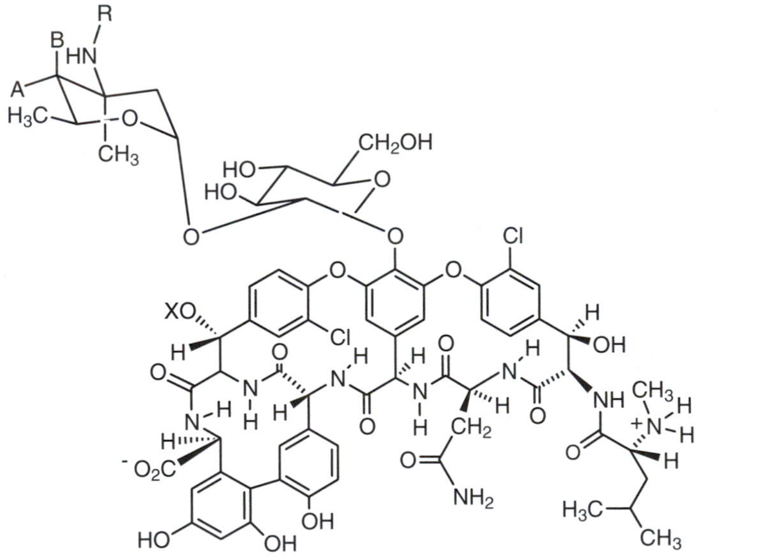

| | A | B | X | R |
|---|---|---|---|---|
| Vancomycin | H | OH | H | H |
| LY 264826 | OH | H | 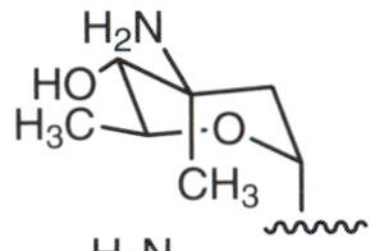 | H |
| LY 333328 | OH | H | 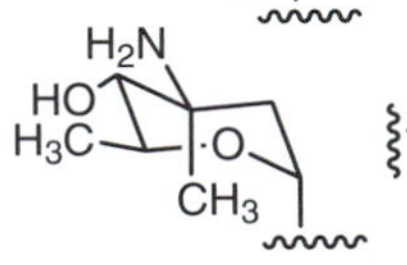 | 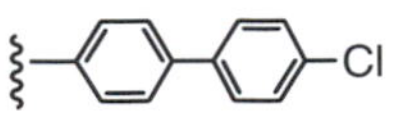 |

observed in resistant and sensitive strains of *E. faecium* and *E. faecalis*.

### VanC Resistance

VanC resistance is characterized by low-level vancomycin resistance (MICs, 4 to 32 μg/ml) but susceptibility to teicoplanin. The genes responsible for the VanC phenotype are an intrinsic property of most isolates of *E. gallinarum, E. casseliflavus,* and *E. flavescens.* Glycopeptide-resistant enterococci of the VanC type synthesize UDP-muramyl-pentapeptide (D-Ser) for cell wall assembly and prevent the synthesis of peptidoglycan precursors ending in D-Ala. The *vanC* cluster of *E. gallinarum* consist of five genes, *vanC1, vanXY-c, vanT, vanRc,* and *vanSc.* Three genes are sufficient for resistance: *vanC1* encodes a ligase that synthesizes the dipeptide D-Ala–D-Ser, *vanXYc* encodes a D,D-dipeptidase-carboxypeptidase that hydrolyzes D-Ala–D-Ala and removes D-Ala from UDP-MurNAc-pentapeptide (D-Ala), and *vanT* encodes a membrane-bound serine racemase that provides D-Ser for the synthetic pathway. These genes are expressed constitutively. If D-Ala is replicated by D-Ser, it is presumed to weaken the binding of vancomycin to the pentapeptide (Fig. 14.10).

## Novel Glycopeptide Antibiotics under Development

The development of glycopeptide antibiotics with activity against vancomycin- and teicoplanin-resistant organisms is of utmost importance because of the recent emergence of low-level vancomycin resistance in *S. aureus* and the prevelance of VRE in immunocompromised patients. While this section does not include all current investigational drugs, it highlights some of the recent highly promising analogs of vancomycin and teicoplanin.

### VANCOMYCIN ANALOGS

The chemical structure of oritavancin (LY333328), the most active compound reported by scientists at Eli Lilly Laboratories, is shown in Fig. 14.11. This compound is a

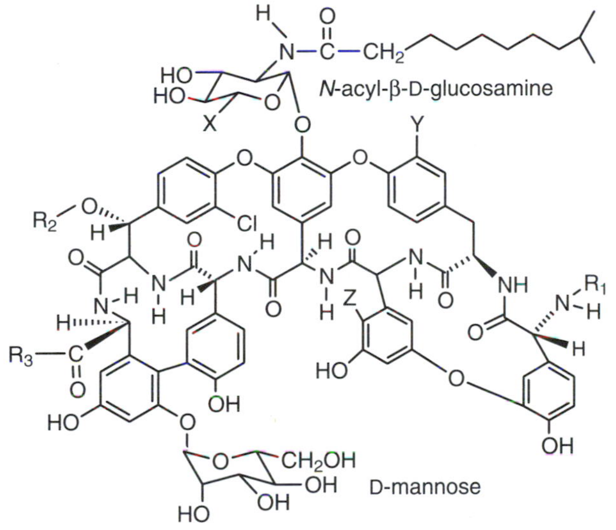

| | X | Y | Z | $R_1$ | $R_2$ | $R_3$ |
|---|---|---|---|---|---|---|
| Teicoplanin | $-CH_2OH$ | Cl | H | H | 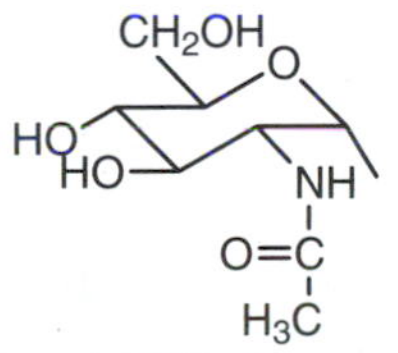 *N*-acetyl-β-D-glucosamine | -OH |
| MDL 62,476 | -COOH | H | Cl | $CH_3$ | H | -OH |
| MDL 63,246 | $-CH_2OH$ | H | Cl | $CH_3$ | H | 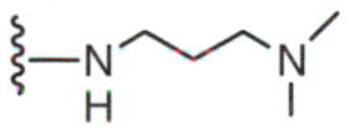 –N(H)–⁄⁄–N⁄ |

**Figure 14.12** Chemical structures of teicoplanin and the related semisynthetic compounds MDL62,476 and MDL63,246. Commercial teicoplanin is a complex of five structurally related natural products that differ in the fatty acid acylating the glucosamine sugar. The most abundant component is shown. Reprinted from T. I. Nicas, M. L. Zeckel, and D. K. Braun, *Trends Microbiol.* **5:**240–250, 1997, with permission from the publisher.

semisynthetic derivative of the naturally occurring compound LY264826, containing a *p*-chlorobiphenyl group attached to the amino functionality of the epivancosamine. Compound LY264826 (Fig. 14.11) has the same peptide backbone as vancomycin but differs in that 4-epivancosamine replaces the vancosamine sugar. In addition, it has an additional epivancosamine sugar attached to residue 6.

LY333328 shows promise for the treatment of VRE as well as other resistant gram-positive organisms such as MRSA. Scientists at the University of Cambridge and at the Institut Pasteur in Paris have found that LY333328 is moderately active against enterococci with the VanA and VanB phenotypes. Scientists at Eli Lilly suggested that LY333328 is able to compensate for a poor binding affinity to D-Ala–D-Lac by using a combination of two other effects. LY333328 is able to form homodimers, which increases its ability to bind to peptidoglycan. Also, it contains an aromatic side chain that can act as a membrane anchor, helping to localize the compound to the outside of the cytoplasmic membrane; this could account for the activity observed against VRE.

TEICOPLANIN ANALOGS

Scientists at Lepetit Research Laboratories have performed extensive structure-activity studies on teicoplanin and its analogs. To identify a compound with improved activity against coagulase-negative staphylococci, they have investigated many naturally occurring teicoplanin analogs and a host of semisynthetic compounds. As a result of these studies, naturally occurring glycopeptide MDL62,476 and the semisynthetic compound MDL63,246 were identified as highly promising candidates (the chemical structures are shown in Fig. 14.12). MDL63,246 was found to be more active in vitro against *S. aureus* and coagulase-negative staphylococci than were teicoplanin, vancomycin, and MDL62,476. MDL63,246 also shows excellent activity in vitro against streptococci, teicoplanin-susceptible enterococci, and some VanA enterococcal isolates. It was more active than teicoplanin and vancomycin against acute staphylococcal, streptococcal, and enterococcal septicemia in immunocompetent and neutropenic mice. According to the authors of the publication of this work (Goldstein et al., 1995), the in vivo activity of MDL63,246 correlates both with its in vitro antibacterial activity and with its long life in rodents.

## Glycopeptide Preparation Commercially Available in the United States

Vancomycin hydrochloride (Vancocin oral, available as capsules, pulvules, and oral solution, and Vancocin parenteral, available as vials for intravenous administration), marketed by Eli Lilly, is the only glycopeptide marketed in the United States.

## REFERENCES AND FURTHER READING


### Clinical Uses of Bacitracin

**Kucers, A., S. M. Crowe, M. L. Grayson, and J. F. Hoy.** 1997. *The Use of Antibiotics*, 5th ed., p. 542–543. Butterworth-Heinemann, Oxford, United Kingdom.

### Vancomycin

**Cooper, M. A., M. T. Fiorini, C. Abell, and D. H. Williams.** 2000. Binding of vancomycin group antibiotics to D-alanine and D-lactate presenting self-assembled monolayers. *Bioorg. Med. Chem.* **8:**2609–2616.

**Kaplan, J., B. D. Korty, P. H. Axelsen, and P. J. Loll.** 2001. The role of sugar residues in molecular recognition by vancomycin. *J. Med. Chem.* **44:**1837–1840.

**Nicolau, K. C., C. N. C. Body, S. Bräse, and N. Winssinger.** 1999. Chemistry, biology, and medicine of the glycopeptide antibiotics. *Angew. Chem. Int. Ed.* **38:**2096–2152.

**Williams, D., and B. Bardsley.** 1999. The vancomycin group of antibiotics and the fight against resistant bacteria. *Angew. Chem. Int. Ed.* **38:**1172–1193.

### Teicoplanin

**Hunt, A. H., R. M. Molloy, J. L. Occolowitz, G. G. Marconi, and M. Debono.** 1984. Structure of the major glycopeptide of the teicoplanin complex. *J. Am. Chem. Soc.* **106:**4891–4895.

### Mechanism of Action of Glycopeptide Antibiotics

**Nagarajan, R.** 1991. Antibacterial activities and modes of action of vancomycin and related glycopeptides. *Antimicrob. Agents Chemother.* **35:**605–609.

### Dimerization Mode for Vancomycin and Some Other Glycopeptide Antibiotics

**Beauregard, D. A., A. J. Maguire, D. H. Williams, and P. E. Reynolds.** 1997. Semiquantitation of cooperativity in binding of vancomycin-group antibiotics to vancomycin-susceptible and -resistant organisms. *Antimicrob. Agents Chemother.* **41:**2418–2423.

**Groves, P., M. S. Searle, J. P. Waltho, and D. H. Williams.** 1995. Asymmetry in the structure of glycopeptide antibiotic dimers: NMR studies of the ristocetin A complex with a bacterial cell wall analogue. *J. Am. Chem. Soc.* **117:**7958–7964.

**Loll, P. J., A. E. Bevivino, B. D. Korty, and P. H. Axelsen.** 1997. Simultaneous recognition of a carboxylate-containing ligand and an intramolecular surrogate ligand in the crystal structure of an asymmetric vancomycin dimmer. *J. Am. Chem. Soc.* **119:**1516–1522.

**Loll, P. J., R. Miller, C. M. Weeks, and P. H. Axelsen.** 1998. A ligand-mediated dimerization mode for vancomycin. *Chem. Biol.* **5:**293–298.

**Schäfer, M., T. R. Schneider, and G. M. Sheldrick.** 1996. Crystal structure of vancomycin. *Structure* **4:**1509–1515.

**Waltho, J. P., and D. H. Williams.** 1989. Aspects of molecular recognition: solvent exclusion and dimerization of the antibiotic ristocetin when bound to a model bacterial cell-wall precursor. *J. Am. Chem. Soc.* **111:**2475–2480.

### Glycopeptide Resistance

**Aráoz, R., E. Anhalt, L. René, M.-A. Badet-Denisot, P. Courvalin, and B. Badet.** 2000. Mechanism-based inactivation of VanX, a D-alanyl-D-alanine dipeptidase necessary for vancomycin resistance. *Biochemistry* **39:**15971–15979.

**Arias, C. A., P. Courvalin, and P. E. Reynolds.** 2000. *vanC* cluster of vancomycin-resistant *Enterococcus gallinarum* BM4174. *Antimicrob. Agents Chemother.* **44:**1660–1666.

**Arthur, M., and P. Courvalin.** 1993. Genetics and mechanisms of glycopeptide resistance in enterococci. *Antimicrob. Agents Chemother.* **37:**1563–1571.

**Arthur, M., F. Depardieu, L. Cabanié, P. Reynolds, and P. Courvalin.** 1998. Requirement of the VanY and VanX D,D-peptidases for glycopeptide resistance in *Enterococci. Mol. Microbiol.* **30:**819–830.

**Arthur, M., C. Molinas, T. D. H. Bugg, G. D. Wright, C. T. Walsh, and P. Courvalin.** 1992. Evidence for in vivo incorporation of D-lactate into peptidoglycan precursors of vancomycin resistant enterococci. *Antimicrob. Agents Chemother.* **36:** 867–869.

**Arthur, M., C. Molinas, F. Depardieu, and P. Courvalin.** 1993. Characterization of Tn*1546*, a Tn*3*-related transposon conferring glycopeptide resistance by synthesis of depsipeptide peptidoglycan precursors in *Enterococcus faecium* BM4147. *J. Bacteriol.* **175:**117–127.

**Arthur, M., and R. Quintiliani, Jr.** 2001. Regulation of VanA- and VanB-type glycopeptide resistance in enterococci. *Antimicrob. Agents Chemother.* **45:**375–381.

**Arthur, M., P. Reynolds, and P. Courvalin.** 1996. Glycopeptide resistance in enterococci. *Trends Microbiol.* **4:**401–407.

**Baptista, M., F. Depardieu, P. Reynolds, P. Courvalin, and M. Arthur.** 1997. Mutations leading to increased levels of resistance to glycopeptide antibiotics in VanB-type enterococci. *Mol. Microbiol.* **25:**93–105.

**Baptista, M., P. Rodrigues, F. Depardieu, P. Courvalin, and M. Arthur.** 1999. Single-cell analysis of glycopeptide resistance gene expression in teicoplanin-resistant mutants of a VanB-type *Enterococcus faecalis. Mol. Microbiol.* **32:**17–38.

**Boyce, J. M., S. M. Opal, J. W. Chow, M. J. Zervos, G. Potter-Bynoe, C. E. Sherman, R. L. C. Romulo, S. Fortna, and A. A. Medeiros.** 1994. Outbreak of multidrug-resistant *Enterococcus faecium* with transferable *vanB* class vancomycin resistance. *J. Clin. Microbiol.* **32:**1148–1153.

**Bugg, T. D. H., G. D. Wright, S. Dutka-Malen, M. Arthur, P. Courvalin, and C. T. Walsh.** 1991. Molecular basis for vancomycin resistance in *Enterococcus faecium* BM4147: biosynthesis of a depsipeptide peptidoglycan precursor by vancomycin resistance proteins VanH and VanA. *Biochemistry* **30:**10408–10415.

**Casadewall, B., and P. Courvalin.** 1999. Characterization of the *vanD* glycopeptide resistance gene cluster from *Enterococcus faecium* BM4339. *J. Bacteriol.* **181:**3644–3648.

**Cetinkaya, Y., P. Falk, and C. G. Mayhall.** 2000. Vancomycin-resistant enterococci. *Clin. Microbiol. Rev.* **13:**686–707.

**Cooper, M. A., and D. H. Williams.** 1999. Binding of glycopeptide antibiotics to a model of a vancomycin-resistant bacterium. *Chem. Biol.* **6:**891–899.

**de Jonge, B. L. M., S. Handwerger, and D. Gage.** 1996. Altered peptidoglycan composition in vancomycin-resistant *Enterococcus faecalis. Antimicrob. Agents Chemother.* **40:**863–869.

**Evers, S., and P. Courvalin.** 1996. Regulation of VanB-type vancomycin resistance gene expression by the VanSB-VanRB two-component regulatory system in *Enterococcus faecalis* V585. *J. Bacteriol.* **178:**1302–1309.

**Fan, C., P. C. Moews, C. T. Walsh, and J. R. Knox.** 1994. Vancomycin resistance: structure of D-alanine-D-alanine ligase at 2.3 Å resolution. *Science* **266:**439–443.

**Fines, M., B. Perichon, P. Reynolds, D. F. Sahm, and P. Courvalin.** 1999. VanE, a new type of acquired glycopeptide resistance in *Enterococcus faecalis* BM4405. *Antimicrob. Agents Chemother.* **43:**2161–2164.

**French, G. L.** 1998. Enterococci and vancomycin resistance. *Clin. Infect. Dis.* **27**(Suppl. 1)**:**S75–S83.

**Geisel, R., F. J. Schmitz, L. Thomas, G. Berns, O. Zetsche, B. Ulrich, A. C. Fluit, H. Labischinsky, and W. Witte.** 1999. Emergence of heterogeneous intermediate vancomycin resistance in *Staphylococcus aureus* isolates in the Düsseldorf area. *J. Antimicrob. Chemother.* **43:**846–848.

**Hakenbeck, R.** 1994. Resistance to glycopeptide antibiotics, p. 535–545. *In* J. M. Ghuysen and R. Hakenbeck (ed.), *Bacterial Cell Wall.* Elsevier, Amsterdam, The Netherlands.

**Healy, V. L., L. S. Mullins, X. Li, S. E. Hall, F. M. Raushel, and C. T. Walsh.** 2000. D-Ala-D-X ligases: evaluation of D-alanyl phosphate intermediate by MIX, PIX and rapid quench studies. *Chem. Biol.* **7:**505–514.

**Healy, V. L., I. S. Park, and C. T. Walsh.** 1998. Active-site mutants of the VanC2 D-alanyl-D-serine ligase, characteristic of one vancomycin-resistant bacterial phenotype, reverts towards wild-type D-alanyl-D-alanine ligase. *Chem. Biol.* **5:**197–207.

**Lessard, I. A. D., and C. T. Walsh.** 1999. Mutational analysis of active-site residues of the enterococcal D-Ala-D-Ala dipeptidase VanX and comparison with *Escherichia coli* D-Ala-D-Ala ligase and D-Ala-D-Ala carboxypeptidase VanY. *Chem. Biol.* **6:**177–187.

**Palepou, M. F. I., A. M. Adebiyi, C. H. Tremlett, L. B. Jensen, and N. Woodford.** 1998. Molecular analysis of diverse elements mediating VanA glycopeptide resistance in enterococci. *J. Antimicrob. Chemother.* **42:**605–612.

**Park, I. S., C. H. Lin, and C. T. Walsh.** 1996. Gain of D-alanyl-D-lactate or D-lactyl-D-alanine synthetase activities in three active-site mutants of the *Escherichia coli* D-alanyl-D-alanine ligase B. *Biochemistry* **34:**10464–10471.

**Sahm, D. F., L. Free, and S. Handwerger.** 1995. Inducible and constitutive expression of *vanC1*-encoded resistance to vancomycin in *Enterococcus gallinarum*. *Antimicrob. Agents Chemother.* **39:**1480–1484.

**Shlaes, D. M., and L. B. Rice.** 1994. Bacterial resistance to the cyclic glycopeptides. *Trends Microbiol.* **2:**383–388.

**Walsh, C. T., S. L. Fisher, I. S. Park, M. Prahalad, and Z. Wu.** 1996. Bacterial resistance to vancomycin: five genes and one missing hydrogen bond tell the story. *Chem. Biol.* **3:**21–28.

**Woodford, N., A. P. Johnson, D. Morrison, and D. C. E. Speller.** 1995. Current perspectives on glycopeptide resistance. *Clin. Microbiol. Rev.* **8:**585–615.

**Wright, G. D., and C. T. Walsh.** 1992. D-Alanyl-D-alanine ligases and the molecular mechanism of vancomycin resistance. *Acc. Chem. Res.* **25:**468–473.

### Clinical Uses of Vancomycin and Teicoplanin

**Kucers, A., S. M. Crowe, M. L. Grayson, and J. F. Hoy.** 1997. *The Use of Antibiotics,* 5th ed., p. 762–801. Butterworth-Heinemann, Oxford, United Kingdom.

### Novel Glycopeptides

**Arthur, M., F. Depardieu, P. Reynolds, and P. Courvalin.** 1999. Moderate-level resistance to glycopeptide LY333328 mediated by genes of the *vanA* and *vanB* clusters in enterococci. *Antimicrob. Agents Chemother.* **43:**1875–1880.

**Goldstein, B. P., G. Candiani, T. M. Arain, G. Romano, I. Ciciliato, M. Berti, M. Abbondi, R. Scotti, M. Mainini, F. Ripamonti, A. Resconi, and M. Denaro.** 1995. Antimicrobial activity of MDL 63,246, a new semisynthetic glycopeptide antibiotic. *Antimicrob. Agents Chemother.* **39:**1580–1588.

**Nicas, T. I., M. L. Zeckel, and D. K. Braun.** 1997. Beyond vancomycin: new therapies to meet the challenge of glycopeptide resistance. *Trends Microbiol.* **5:**240–250.

**Schwalbe, R. S., A. C. McIntosh, S. Qaiyumi, J. A. Johnson, R. J. Johnson, K. M. Furness, W. J. Holloway, and L. Steele-Moore.** 1996. In vitro activity of LY333328, an investigational glycopeptide antibiotic, against enterococci and staphylococci. *Antimicrob. Agents Chemother.* **40:**2416–2419.

# chapter 15



# Antibiotics That Affect Membrane Permeability
## Cyclic Peptides

## Polymyxins

The polymyxins are a group of cyclic, polycationic peptide antibiotics with a fatty acid chain attached to the peptide through an amide linkage. They are produced by fermentation of strains of *Bacillus polymyxa*. Polymyxins B and E (colistin) are the least toxic and are the only polymyxins used clinically. These antibiotics contain a 7-amino-acid ring attached to a 3-amino-acid tail, to which is attached a fatty acyl group. Polymyxin B is a mixture containing mostly polymyxins $B_1$ and $B_2$. They were first isolated in 1947. Initially, only the sulfate form of polymyxin B was commercially available. Colistin, which became available for clinical use in 1959, was isolated in 1949 in Japan from *B. polymyxa* subsp. *colistinus*. Initially it was thought to be a new antibiotic, but in 1963, S. Wilkinson demonstrated that it was identical to polymyxin E.

Polymyxin $B_1$ (Fig. 15.1) contains several L-α,γ-diaminobutyric acid (Dab), L-threonine, and D-phenylalanine residues and 3-methyl octanoic acid as a fatty acid residue.

### Mechanism of Action

Both polymyxins B and E (colistin) are cationic surface-active compounds at physiological pH. It has been suggested that the fatty acid part of the polymyxin molecule penetrates into the hydrophobic region of the outer membrane and the ammonium groups interact with the lipopolysaccharides and phospholipids, competitively displacing divalent cations (calcium and magnesium) from the negatively charged phospholipid group of the membrane lipids. This displacement disrupts membrane organization and increases the permeability of the membrane.

### Antibacterial Activity

Polymyxins B and E are active almost exclusively against aerobic gram-negative bacilli. In particular, they exhibit quite good activity against

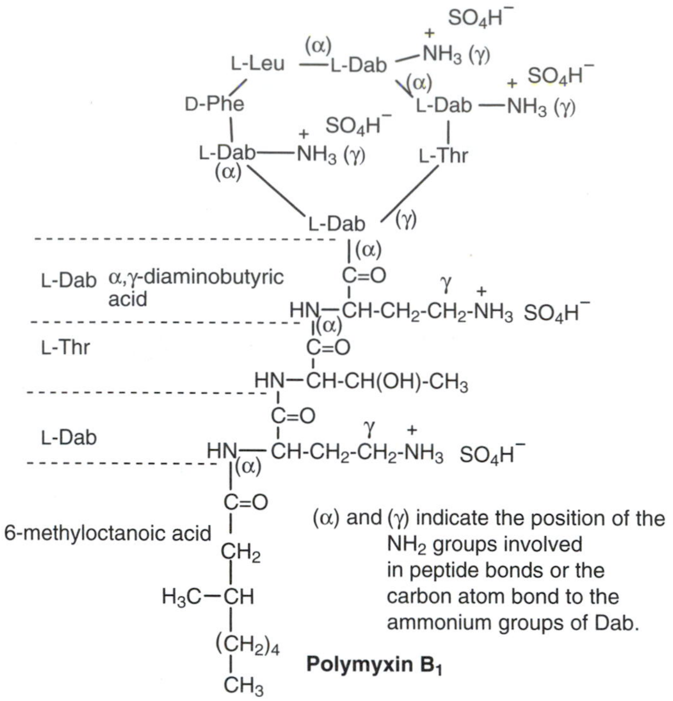

**Figure 15.1** Chemical structure of polymyxin $B_1$ sulfate.

**Table 15.1** Generic and common trade names of polymyxins, the preparations available, and manufacturers in the United States

| Generic name | Trade name (preparation) | Manufacturer |
|---|---|---|
| Polymyxin B sulfate plus bacitracin zinc | Polysporin (ointment) | Pfizer |
| Polymyxin B sulfate plus bacitracin zinc and pramoxine HCl | Betadine Plus (ointment) | Purdue Frederick |
| Polymyxin B sulfate plus neomycin and bacitracin zinc | Neosporin (ophthalmic ointment) | Monarch |
| Polymyxin B sulfate plus neomycin sulfate | Neosporin (solution for irrigation) | Monarch |
| Polymyxin B sulfate plus neomycin | Cortisporin (ophthalmic suspension) | Monarch |

*Pseudomonas aeruginosa.* They are inactive against gram-positive bacteria and most obligate anaerobes.

## Clinical Uses

Both the neurotoxicity and the nephrotoxicity of polymyxins B and E reflect their rather nonspecific interaction with bacterial and mammalian cell membranes. In the past, these polymyxins were often used for the treatment of *P. aeruginosa* infections. Nowadays, because of the availability of effective and less toxic drugs such as gentamicin, tobramycin, amikacin, ticarcillin, piperacillin, and ceftazidime, polymyxins are not the antibiotics of choice to treat infections caused by this bacterium. Polymyxins B and E have now been relegated to topical preparations for dermatological, otic, or ophthalmic infections; these preparations usually contain another antibiotic such bacitracin, neomycin, oxytetracycline, or trimethoprim.

## Polymyxin Preparations Commercially Available in the United States

Table 15.1 lists the polymyxins marketed in the United States.

### REFERENCES AND FURTHER READING


**Clinical Uses of Polymyxins B and E**

**Kucers, A., S. M. Crowe, M. L. Grayson, and J. F. Hoy.** 1997. *The Use of Antibiotics,* 5th ed., p. 667–675. Butterworth-Heinemann, Oxford, United Kingdom.

**Wilkinson, S.** 1963. Identity of colistin and polymyxin E. *Lancet* i:922.

chapter 16



# Antibiotic Inhibitors of Bacterial Protein Synthesis

## Protein Synthesis: an Overview

Before starting a discussion of antibiotics that inhibit bacterial protein synthesis, it is appropriate to mention the biochemical background of protein synthesis. Readers interested in a more comprehensive overview should consult the references cited at the end of this chapter.

In the last step of the gene expression pathway, genomic information encoded in mRNA is translated into protein by the **ribosome.** An individual bacterial ribosome is a complex ribonucleoprotein particle which is designated a 70S ribosome since it sediments in centrifugation with a velocity of 70 Svedberg units. The complete 70S ribosome is formed by association of the 30S and 50S subunits through a network of intermolecular bridges. In prokaryotes, the small subunit (30S particle) is composed of 16S rRNA (of ~1,500 nucleotides) and 21 proteins while the large subunit (50S particle) consists of two RNAs, the 5S rRNA (of about 120 nucleotides) and the 23S rRNA (of ~2,900 nucleotides), together with about 31 proteins. Both subunits, 30S and 50S, are targets of several clinically relevant antibiotics.

The intersubunit space of the ribosome is occupied by the tRNAs, whose anticodons form base pairs with mRNA codons in the 30S subunit, whereas their 3' CCA ends, which carry the growing polypeptide chain and the incoming amino acid, reach into the 50S subunit, the location of the peptidyltransferase center, where peptide bond formation is catalyzed (Fig. 16.1).

## The Ribosome at Atomic Resolution

Structures of 70S ribosome complexes containing mRNA and tRNA have been solved by X-ray crystallography at up to 7.8-Å resolution. Together with the recently published atomic-resolution crystal structures for the large ribosomal subunit from *Haloarcula marismortui* and the small ribosomal

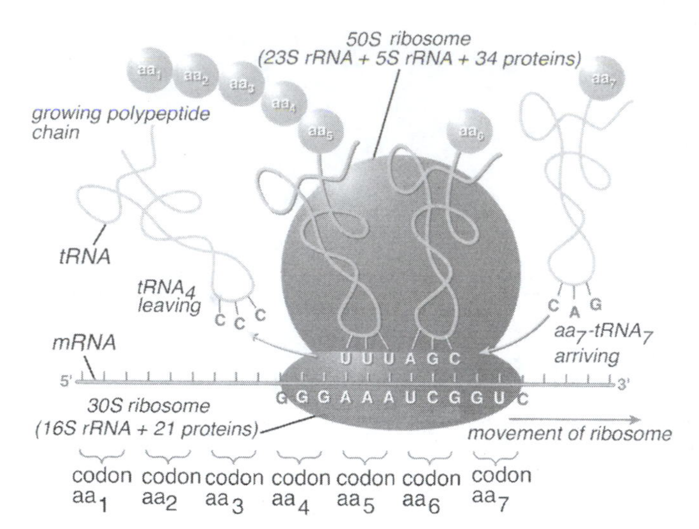


**Figure 16.1** The ribosome, showing the tRNA-mRNA complex between the 30S and 50S ribosomal subunits, and the A and P sites. The growing polypeptide chain extends through the tunnel of the large subunit. Reprinted from T. K. Ritter and X. H. Wong, *Angew. Chem. Int. Ed.* **40:**3508–3533, 2001, with permission from the publisher.

subunit from *Thermus thermophilus,* these structures have altered the way in which protein synthesis is conceptualized.

## X-Ray Crystal Structure of the 70S Ribosome

In September 1999, Jamie Cate, Harry Noller, and their colleagues from the Center for Molecular Biology of RNA at the University of California at Santa Cruz described for the first time the cocrystallization of complete *T. thermophilus* 70S ribosomes in functional complexes with mRNA and A-, P-, and E-site tRNAs and the solution of their crystal structures at resolutions of up to 7.8 Å. In May 2001, the same group of scientists published the three-dimensional structure of the 70S ribosome containing mRNA and tRNAs bound to the P and E sites at 5.5-Å resolution. They found that features of the 50S subunit that were disordered in the high-resolution *H. marismortui* structure were ordered in the 70S *T. thermophilus* structure, providing a nearly complete view of the 50S subunit. The three tRNAs are closely juxtaposed with certain intersubunit bridges.

## X-Ray Crystal Structure of the 30S Ribosomal Subunit

In August 1999, the Ramakrishnan group from the MRC Laboratory of Molecular Biology (Cambridge, United Kingdom) published the first electron density map of the 30S ribosomal subunit obtained by X-ray diffraction at 5.5 Å. The 30S subunit comes from the hyperthermophilic bacterium, *T. thermophilus.* Several proteins with known three-dimensional structures and many regions of double-stranded rRNA have been located in the highly complex electron density map of the 30S ribosomal subunit. In another publication in September 2000, the same group reported the same crystal structure refined at 3-Å resolution. In May 2001, the same group published the crystal structure of the 30S ribosomal subunit complexed with mRNA and cognate tRNA in the A site, in both the presence and absence of the antibiotic paromomycin, at 3.1- and 3.3-Å resolution, respectively.

In December 1999, scientists from the Max Planck Institute (Hamburg and Berlin, Germany) in a collaborative project with scientists from the Weizmann Institute (Rehovot, Israel) reported the electron density map of the 30S ribosomal subunit from *T. thermophilus* at 4.5-Å resolution. In September 2000 the same group published an improved structure solved at 3.3-Å resolution. The structure now includes 1,457 nucleotides and most of the 19 proteins belonging to this small ribosomal subunit.

### SMALL-SUBUNIT FUNCTION

The primary function of the small ribosomal subunit is mediation of the interaction between mRNA and tRNA by monitoring base pairing between the **codon** on mRNA and the **anticodon** on tRNA. There are three tRNA-binding sites, designated A (aminoacyl), P (peptidyl), and E (exit), after their respective tRNA substrates. The anticodon stem-loops for the A- and P-site tRNAs bind to the 30S subunit, where they are base-paired with adjacent codons on mRNA. The decoding of mRNA into protein requires the correct recognition of each A-site codon by the anticodon of the corresponding aminoacyl-tRNA.

## X-Ray Crystal Structure of the 50S Ribosomal Subunit

In 1998, Nenad Ban and her group at Yale University reported the first electron density map of the 50S ribosomal subunit from *H. marismortui.* The crystal structure was solved at 9-Å resolution. It showed features recognizable as duplex RNA. A year later, extension of the phasing of the map to 5-Å resolution made it possible to locate several proteins and nucleic acid sequences. In 2000, the same group of scientists published an atomic structure of the *H. marismortui* 50S ribosomal subunit at 2.4-Å resolution. The model includes 2,711 of the 2,923 nucleotides of 23S rRNA, all 122 nucleotides of the 5S rRNA, and structures for 27 of its 31 proteins.

In December 1999, scientists from Imperial College (London, United Kingdom) and the Max Planck Institute (Berlin, Germany) published the cryoelectron microscopy-determined structure of the *Escherichia coli* 50S ribosomal subunit at 7.5-Å resolution.

In November 2001, Ada Yonat and colleagues from the Weizmann Institute and the Max Planck Insitute at Hamburg and Berlin, published the 3.1-Å structure of the 50S subunit from *Deinococcus radiodurans,* a gram-positive mesophile, suitable for binding of antibiotics and functionally relevant ligands. In a recent publication, this group determined the structure of the 50S subunit of *D. radiodurans* complexed with each of the following antibiotics: chloramphenicol, clindamycin, and the macrolides erythromycin, clarithromycin and roxithromycin (Schlünzen et al., 2001). Analysis of the X-ray structures obtained in that study allowed these scientists to propose structural principles for the action of the antibiotics examined. Their work reveals the functional interactions involved in the binding of these antibiotics to the peptidyltransferase cavity of the 50S subunit. They have found that none of the antibiotics examined show any direct interaction with ribosomal proteins. Chloramphenicol targets mainly the A site, where it interferes directly with substrate binding. Clindamycin interferes with the A-site and P-site substrate binding and physically hinders the path of the growing peptide chain. Macrolides bind at the entrance to the tunnel, where they sterically block the progression of the nascent peptide.

#### LARGE-SUBUNIT FUNCTION

The function of the large subunit includes the activity that catalyzes peptide bond formation with peptidyl transferase and the binding site for the G-protein (GTP-binding protein) and EF-Tu factors that assist in the initiation, elongation, and termination phases of protein synthesis.

#### Location of the Peptidyltransferase Site

As mentioned above, peptide bond formation is catalyzed in the peptidyltransferase center of the large ribosomal subunit. The region of the structure of the 50S subunit defining the catalytic core is adjacent to the entrance of a tunnel along which nascent protein progress until they emerge from the ribosome. This tunnel is referred to as the peptidyltransferase-associated region. The location of the peptidyltransferase site has been determined crystallographically by solving structures of the large subunit complexed with an A-site substrate analogue and with a peptidyltransferase state analogue. This work was published in 2000. Previously, the location of the peptidyltransferase site was determined at low resolution by electron microscopy.

## tRNA Structure

tRNA molecules are the interpreters of the genetic code. An overview of the structure of tRNA will provide a useful outline for understanding how tRNAs serve as interpreters in the translation of mRNA nucleotide sequences into the amino acid sequence of a polypeptide chain.

The yeast alanine-tRNA was the first nucleic acid to be completely sequenced. It contains 76 nucleotide residues, 10 of which have modified bases. Most tRNAs have a guanylate (pG) residue at the 5' end, and all have the trinucleotide sequence CCA (3') at the 3' end. In the secondary structure (Fig. 1.30), the hydrogen-bonding pattern of all tRNAs forms a cloverleaf structure with four arms; the long tRNAs have a short fifth or variable arm. In three dimensions, the conformation of a tRNA has the form of a twisted L (Color Plate 16.1 [see color insert]). The drawing shows the anticodon loop that contains the 3-base sequence that binds to the complementary codon in mRNA. tRNA and mRNA molecules interact through base pairing of anticodons and codons. The anticodon of a tRNA molecule therefore determines where the amino acid attached to its acceptor stem will be added to a growing polypeptide chain. Much of the base pairing between the codon and anticodon is governed by the rules of Watson-Crick base pairing: A pairs with U, G pairs with C, and the strands in the base pairs are antiparallel.

## Protein Synthesis

The final step in gene expression is protein synthesis, a very rapid process. In *E. coli,* protein synthesis occurs at a rate of about 300 to 450 residues per min. Protein synthesis in either prokaryotic or eukaryotic cells can be conveniently divided into three stages: **initiation, elongation,** and **termination.** These fundamental processes are accompanied by the activation of amino acids before their incorporation into the polypeptide chain and the postsynthetic processing of the polypeptide chain.

### Activation of Amino Acids

Activation takes place in the cytosol, not on the ribosome. Prior to the first stage of protein synthesis, each of the 20 natural amino acids (their structures are shown in appendix A) is esterified to the corresponding tRNA molecule via the carboxylate group of each amino acid. This activation occurs at the expense of ATP energy, using $Mg^{2+}$-dependent activating enzymes called **aminoacyl-tRNA synthetases.** The overall reaction for amino acid activation is essentially irreversible. The aminoacyl-tRNA synthetases have been divided into two classes based on substantial differences in primary and tertiary structure and their respective reaction mechanisms (Table 16.1).

The activation reaction occurs in two steps in the enzyme active site. In the first step, an enzyme-bound intermediate, aminoacyl adenylate (aminoacyl-AMP),

**Table 16.1** Two classes of aminoacyl-tRNA synthetases[a,b]

| Class I | Class II |
|---|---|
| Arg | Ala |
| Cys | Asn |
| Gln | Asp |
| Glu | Gly |
| Ile | His |
| Leu | Lys |
| Met | Phe |
| Trp | Pro |
| Tyr | Ser |
| Val | Thr |

[a] Reprinted from D. L. Nelson and M. M. Cox, *Lehninger Principles of Biochemistry*, 4th ed. (Worth Publishers, New York, N.Y., 2000), with permission from the publisher.

[b] All amino acids represent aminoacyl-tRNA synthetases.

forms when the carboxylate group of the amino acid reacts with the α-phosphoryl group of ATP to form an acyl-phosphate, with displacement of pyrophosphate. In the second step, the aminoacyl group is transferred from enzyme-bound aminoacyl-AMP to its corresponding specific tRNA. The course of this second step depends on the class to which the enzyme belongs (Fig. 16.2).

## Initiation Step in Prokaryotes

In initiation, the first amino acid is incorporated via *N*-formylmethionyl-$tRNA_f^{Met}$ (fMet-$tRNA_f^{Met}$) (Fig. 16.3), being positioned in the A site. After protein synthesis is under way, the protein is deformylated and the N-terminal methionine can be removed. Since fMet-$tRNA_f^{Met}$ already has an amide bond, it can be incorporated only as the N-terminal residue of a polypeptide.

The tRNA that recognizes the initiation codon, $tRNA_f^{Met}$, differs from the tRNA that carries internal Met residues, $tRNA^{Met}$, although both recognize the same AUG codon. Nonetheless, the amino acid residue incorporated in response to the (5')AUG initiation codon is *N*-formylmethionine (fMet) in eubacteria. It arrives at the ribosome as fMet-$tRNA_f^{Met}$, which is formed in two successive reactions. First, methionine is attached to $tRNA_f^{Met}$ by the Met-tRNA synthetase. Next, a transformylase transfers a formyl group to the amino group of the Met residue from $N^{10}$-formyltetrahydrofolate.

Color Plate 16.2 (see color insert) shows the initiation process in prokaryotes. Formation of the initiation complex depends on the action of three initiation factors: IF-1, IF-2, and IF-3. The initiator fMet-$tRNA_f^{Met}$ binds to a free 30S subunit. One of the key roles of the initiation factors is to position fMet-$tRNA_f^{Met}$ at the P site in the ternary complex; mRNA then attaches to the 30S subunits. Next, the 50S subunit binds to the 30S preinitiation complex. Finally, IF-1 and IF-3 are liberated and the GTP bound to IF-2 is hydrolyzed, liberating GDP–IF-2 and $P_i$. Completion of the steps shown in Color Plate 16.2 produces a functional 70S ribosome called the **initiation complex.** The initiation complex is now ready to initiate the elongation step.

## Elongation of the Polypeptide Chain

The second stage of protein synthesis is elongation. Every amino acid addition to a growing polypeptide chain is the result of a three-step cycle. The steps in this cycle involve positioning the correct aminoacyl-tRNA in the A site of the ribosome (step 1), forming the peptide bond (step 2), and shifting (translocating) the mRNA by one codon relative to the ribosome (step 3). These three steps are repeated as many times as there are amino acid residues to be added.

### ELONGATION STEP 1: AMINOACYL-tRNA DOCKING IN THE A SITE

At the beginning of an elongation cycle, the peptidyl site (the P or donor site) is filled with either fmet-$tRNA_f^{Met}$ or peptidyl-tRNA and the aminoacyl site (the A or acceptor site) is empty (Color Plate 16.3 [see color insert]). A set of three soluble cytosolic proteins called elongation factors, EF-Tu, EF-Ts, and EF-G, are needed. The appropriate incoming aminoacyl-tRNA is first bound to GTP-bound EF-Tu, forming a ternary complex that fits into the A site of a ribosome. On GTP hydrolysis, an EF-Tu–GDP complex is released from the 70S ribosome. Elongation factor EF-Ts catalyzes the exchange of bound GDP for GTP in EF-Tu (Color Plate 16.4 [see color insert]). Once the EF-Tu–GTP complex is re-formed, it binds to a new aminoacyl-tRNA molecule. Note that one GTP molecule is hydrolyzed for every aminoacyl-tRNA docked at the A site. Kjeldgaard and Nyborg (1992) determined the X-ray structure of the 393-residue EF-Tu from *E. coli* in complex with GDP. The mRNA codon in the ribosomal A site is recognized by aminoacyl-tRNA in a ternary complex with EF-Tu and GTP. In 2002, Holger Stark and coworkers reported the 13-Å-resolution three-dimensional reconstruction determined by cryoelectron microscopy of the kirromycin-stalled codon recognition complex.

### ELONGATION STEP 2: FORMATION OF PEPTIDE BONDS

Aminoacyl-tRNA binding to the A site initiates the second step of the elongation cycle, the formation of the **peptide bond.** This reaction is catalyzed by the peptidyltransferase center, composed entirely of 23S rRNA

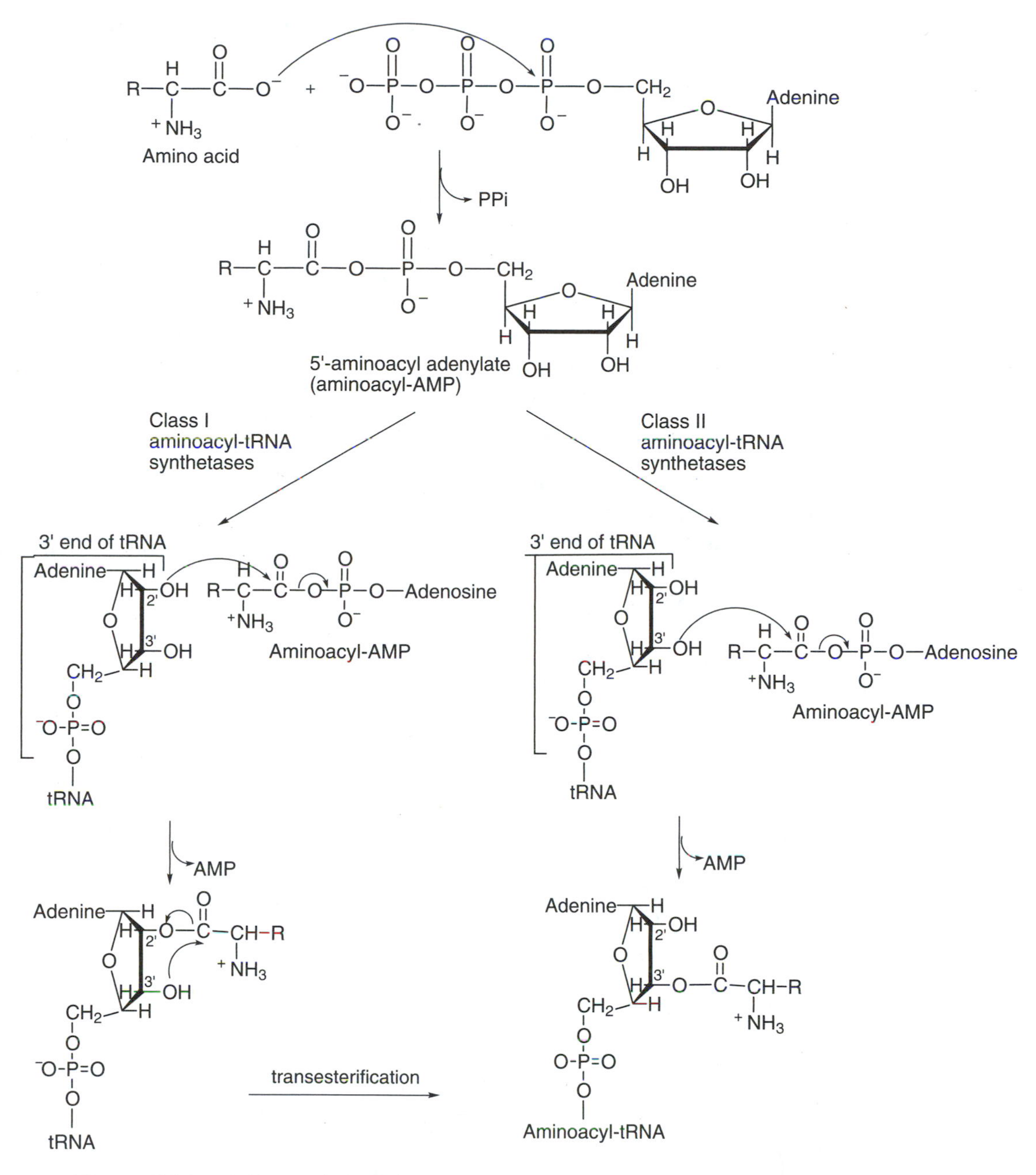


**Figure 16.2** Aminoacylation of tRNA by aminoacyl-tRNA synthetases. The first step is formation of aminoacyl adenylate, which remains bound to the active site. In the second step, the aminoacyl group is transferred to the tRNA. The mechanism of this step is somewhat different for the two classes of aminoacyl-tRNA synthetases. For class I enzymes, the aminoacyl group is initially transferred to the 2' hydroxyl of the 3'-terminal adenylate residue and then moved to the 3' hydroxyl by a transesterification reaction. For class II enzymes, the aminoacyl group is tranferred directly to the 3' hydroxyl of the terminal adenylate, as shown.

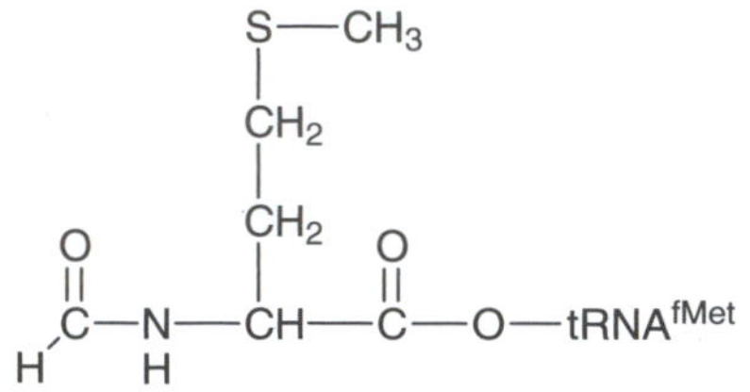

*N*-Formylmethionyl-$tRNA^{fMet}$

**Figure 16.3** Chemical structure of fMet-$tRNA_f^{Met}$.

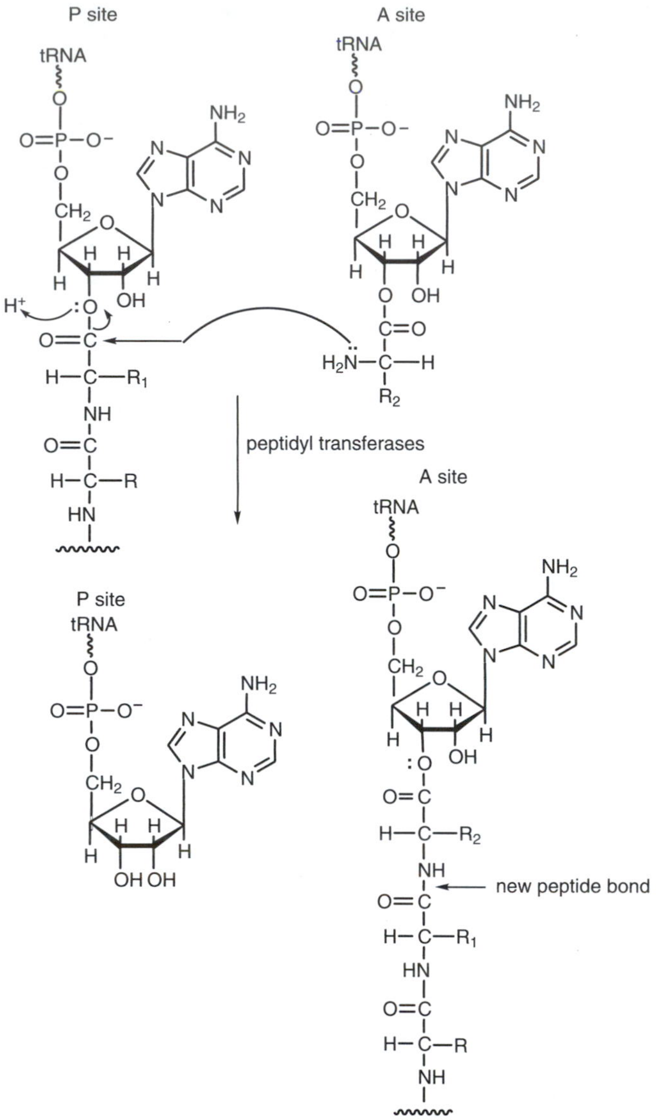

residues in the 50S subunit. Recent evidence suggests that 23S rRNA participates in the peptidyltransferase function. It has been hypothesized that the N-3 of adenylate residue 2451 in 23S RNA acts as a general base to facilitate nucleophilic attack by the α-amino group of the aminoacyl-tRNA bound in the A site. Reaction of the α-carboxyl carbon of fMet-tRNA forms the ester or, if it is not the first peptide bond, the nascent peptidyl-tRNA. A peptide bond is formed, leaving the tRNA in the P site (Fig. 16.4). This reaction produces a dipeptidyl-tRNA in the A site for the first peptide bond formed and a polypeptidyl-tRNA subsequently. The nascent dipeptide (or polypeptide) chain is thereby lengthened at its C terminus by one amino acid and transferred to the A-site tRNA.

P site

A site

peptidyl transferases

new peptide bond

**Figure 16.4** Formation of a peptide bond by an aminoacyl group transfer reaction. The reaction is catalyzed by the peptidyltransferase center and lengthens the peptide chain by one amino acid residue. Note the transfer of the nascent peptide to the tRNA in the A site.

ELONGATION STEP 3: TRANSLOCATION

The third step in the chain elongation cycle is called **translocation.** In prokaryotes, translocation requires the participation of the elongation factor EF-G. Only the EF-G–GTP complex is active in translocation. After the peptide bond has formed (in the second step of the elongation cycle), the newly created peptidyl-tRNA is partially in the A site and partially in the P site, as shown in Color Plate 16.5 (see color insert). The deaminoacylated tRNA has been displaced somewhat from the P site and occupies a position in the E site. Binding of EF-G–GTP to the ribosome releases the deaminoacylated tRNA from the E site and completes the translocation of the peptidyl-tRNA from the A site to the P site.

On GTP hydrolysis, EF-G–GDP leaves the ribosome, free to begin a cycle of reformation. The ribosome moves one triplet codon toward the 3' end of mRNA, using the energy provided by hydrolysis of GTP bound to EF-G, and hence leaving the A site open for the incoming (third) aminoacyl-tRNA.

## Termination of Protein Synthesis

Protein synthesis stops when the ribosome reaches one of the three special **nonsense codons**—UAA, UAG, and UGA. In *E. coli,* the termination codons are the only codons that have no corresponding aminoacyl-tRNA and are recognized by protein release factors. In bacteria, once a terminating codon occupies the ribosomal A site, three release factors, designated RF-1, RF-2, and RF-3, participate in the termination of protein synthesis. RF-1 recognizes the termination codons UAA and UAG, whereas RF-2 recognizes UAA and UGA. RF-3 forms a heterodimer with either RF-1 or RF-2; it also binds to GTP. The binding of the heterodimers to the mRNA at the A site alters the activity of the peptidyltransferase center, causing it to hydrolyze the ester of the peptidyl-tRNA.

As a result of this hydrolysis, the last tRNA and the free polypeptide chain are released from the ribosome. Simultaneously, the release factors leave the ribosome, along with GDP plus $P_i$. The 70S ribosome dissociates from the mRNA and separates into its 30S and 50S subunits. The initiation factors IF-1 and IF-3 are ready for the next round of protein synthesis.

## Newly Synthesized Polypeptide Chains Undergo Processing and Folding

### Processing

As mentioned earlier, *N*-formylmethionine is the first amino acid incorporated during the initiation step in prokaryotic protein synthesis, and as a result, all newly synthesized polypeptides carry transiently a formylated N terminus. Peptide deformylase (PDF) catalyzes the subsequent removal of the N-terminal formyl group from those polypeptides, many of which undergo further N-terminal processing by methionine aminopeptidase (MAP). MAP, a metalloproteinase, hydrolyzes the initial methionine once the *N*-formyl group has been removed (see Fig. 16.5). PDF is a unique metallopeptidase, which utilizes a ferrous ion ($Fe^{2+}$) to catalyze the amide bond hydrolysis. It is present in all eubacteria.

MAP is now known to have a particular substrate specificity, and the removal or maintenance of the methionine depends on the nature of the second amino acid of the polypeptide chain.

### Folding

As the newly synthesized polypeptide emerges from the ribosome, it folds into its characteristic secondary and tertiary structures. This folding is possible because protein conformation is a direct function of the amino acid sequence (the primary structure) (see appendix A). The protein assumes its native conformation with the formation of appropriate hydrogens bonds and van der Waals, ionic, and hydrophobic interactions. Folded proteins occupy a low-energy well that makes the native structure much more stable that alternative conformations.

#### PROTEIN FOLDING IS ASSISTED BY CHAPERONES

Studies of protein folding have led to the observation that large protein chains use **molecular chaperones** to adopt their correct conformation. Chaperones are proteins that increase the rate of correct folding of some proteins by binding newly synthesized polyketide before they are completely folded. Chaperones appear to inhibit incorrect folding and assembly by forming stable complexes with surfaces on polyketide chains that are exposed only during synthesis, folding, and assembly. Even in the presence of chaperones, protein folding is spontaneous.

In 2002, Bitto and McKay reported the crystallographic structure of SurA, a molecular chaperone that facilitates folding of the outer membrane proteins. In *Escherichia coli, surA* (for "survival") was first identified as a gene whose disruption impaired cell survival in stationary phase. The SurA protein was subsequently shown to be involved in the process of folding and assembly of the outer membrane porins. SurA specifically facilitates the conversion of apparently unfolded monomers to folded monomers, which then assemble into unstable trimers, which in turn are converted to stable trimers. SurA therefore functions as a molecular chaperone that facilitates the correct folding of outer membrane proteins in gram-negative bacteria. The crystallographic structure of the *E. coli* SurA protein has been solved and refined to 3.0-Å resolution.

## Antibiotics That Inhibit Bacterial Protein Synthesis

Although the mechanism of protein synthesis is similar in prokaryotes and eukaryotes, prokaryotic ribosomes differ substantially from those in eukaryotes. This explains the effectiveness of several clinically important antibiotics.

There are many clinically useful antibiotic classes that interact with rRNA and inhibit protein synthesis. Examples include aminoglycosides, tetracyclines, macrolides, ketolides, lincosamides, streptogramins (quinupristin and dalfopristin), chloramphenicol, oxazolidinones, and mupirocin. The structurally diverse nature of these inhibitors suggests that a multiplicity of unique rRNA-binding sites are available for effective small molecule drug interactions. Aminoglycoside-RNA and tetracycline-RNA interactions have been well defined for 16S rRNA (Color Plate 16.6 [see color insert]). Table 16.2 summarizes the modes of action of all clinically important antibacterial agents that inhibit bacterial protein synthesis.

*N*-formylmethionyl-polypeptide —PDF ($H_2O$ → Formate)→ *N*-methionyl-polypeptide —MAP ($H_2O$ → Methionine)→ *N*-polypeptide

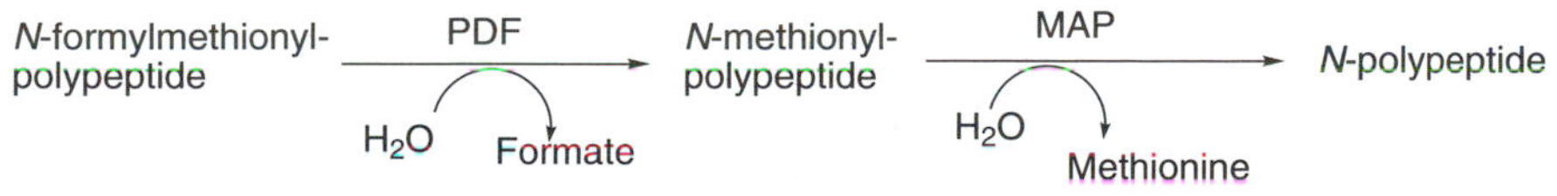

**Figure 16.5** Reaction catalyzed by PDF and MAP.

**Table 16.2** Antibiotics and synthetic antibacterial agents that act by inhibition of bacterial protein synthesis

| Antibiotics | Mechanism of action |
|---|---|
| Inhibitors of the 30S ribosomal subunit | |
| Aminoglycoside group | Aminoglycosides bind conserved sequences within the 16 rRNA of the 30S ribosomal subunit, thereby inducing codon misreading as well as the inhibition of translocation |
| Tetracyclines | Tetracycline binds to two sites, both composed entirely of RNA, one of which is close enough to the A site to account for the well-known drug activity as an inhibitor to the A-site tRNA binding; its second binding site is close to helix 27 (H27), where a conformational switch is known that affects the fidelity of decoding |
| Inhibitors of the 50S ribosomal subunit | |
| Macrolides and ketolides | Bind at the entrance to the tunnel, where they sterically block the progression of the nascent peptide |
| Lincosamides | Interfere with the A-site and P-site substrate binding and physically hinder the path of the growing peptide chain |
| Quinupristin (streptogramin B) | Thought to be similar to that of the 14-member ring macrolides, which sterically hinder the growth of the nascent peptide during early rounds of translation |
| Dalfopristin (streptogramin A) | Related to inhibition of peptide bond formation, either due to weakening of the binding of the peptidyl- and aminoacyl-tRNA to the 50S subunit of the ribosome or due to direct interference with the peptidyl transfer reaction |
| Chloramphenicol | Targets mainly the A site, where it interferes directly with substrate binding |
| Inhibitors of the formation of the 70S ternary complex | |
| Oxazolidinones | The oxazolidinone linezolid inhibits the binding of fMet-$tRNA_f^{Met}$ to 70S ribosomes |
| Inhibitors of the isoleucyl-tRNA synthetase | |
| Mupirocin | Inhibits isoleucyl-tRNA synthetase |

Puromycin (Fig. 16.6) is not clinically useful since it also blocks protein synthesis in eukaryotes and therefore is poisonous to humans. However, it is an effective research tool because its structure closely resembles that of the 3' end of an aminoacyl-tRNA molecule. Puromycin binds to the A site on the ribosome, and the growing peptide is transferred to it instead of to an aminoacyl-tRNA. The puromycin-peptide complex is then released from the ribosome, halting the elongation step.

In the following chapters, the mechanism of action of antibiotics that act by inhibition of bacterial protein synthesis is covered in detail.

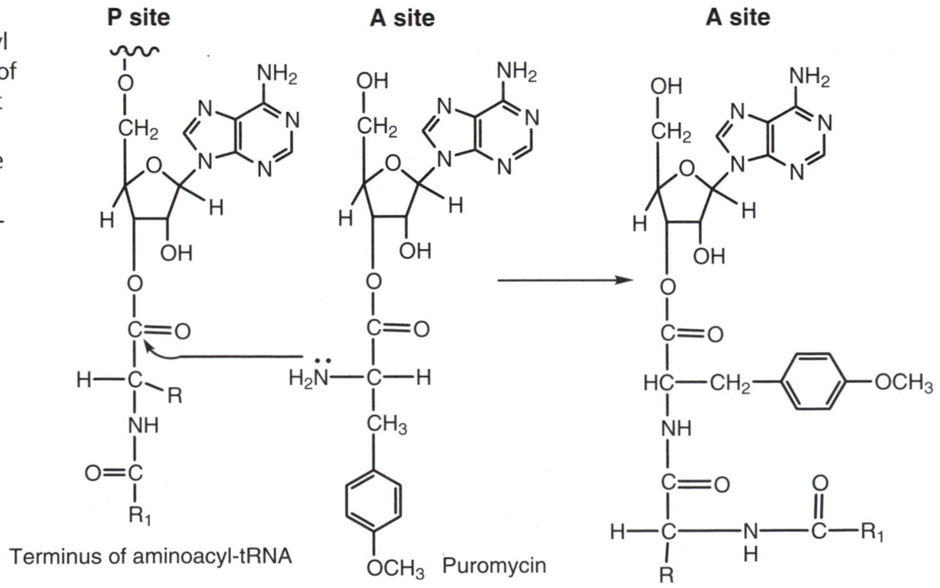

**Figure 16.6** Structural analogy between puromycin and the aminoacyl terminus of tRNA, showing formation of a peptide bond between puromycin at the A site of a ribosome and the nascent peptide bound to the tRNA at the P site. The product of the reaction is bound weakly at the A site and dissociates from the ribosome, terminating protein synthesis and producing an incomplete, inactive peptide.

## REFERENCES AND FURTHER READING


### General

**Horton, H. R., L. A. Moran, R. S. Ochs, J. D. Rawn, and K. G. Scrimgeour.** 2002. *Principles of Biochemistry,* 3rd ed. Prentice-Hall, Inc., Upper Saddle River, N.J.

**Nelson, D. L., and M. M. Cox.** 2000. *Lehninger Principles of Biochemistry,* 4th ed. Worth Publishers, New York, N.Y.

**Prescott, L. M., J. P. Harley, and D. A. Klein.** 1999. *Microbiology,* 4th ed. McGraw-Hill, Boston, Mass.

**Ritter, T. K., and C. H. Wong.** 2001. Carbohydrate-based antibiotics: a new approach to tackling the problem of resistance. *Angew. Chem. Int. Ed.* **40:**3508–3533.

### The Ribosome at Atomic Resolution

**Moore, P. B.** 2001. The ribosome at atomic resolution. *Biochemistry* **40:**3243–3250.

Crystal structure of the 70S ribosome

**Cate, J. H., M. M. Yusupov, G. Z. Yusupova, T. N. Earnest, and H. F. Noller.** 1999. X-ray crystal structures of 70S ribosome functional complexes. *Science* **285:**2095–2104.

**Yusupov, M. M., G. Z. Yusupova, A. Baucom, K. Lieberman, T. N. Earnest, J. H. Cate, and H. F. Noller.** 2001. Crystal structure of the ribosome at 5.5 Å resolution. *Science* **292:**883–896.

Crystal structure of the 30S ribosomal subunit

**Carter, A. P., W. M. Clemons, D. E. Brodersen, R. J. Morgan-Warren, B. T. Wimberly, and V. Ramakrishnan.** 2000. Functional insights from the structure of the 30S ribosomal subunit and its interactions with antibiotics. *Nature* **407:**340–348.

**Clemons, W. M., J. L. C. May, B. T. Wimberly, J. P. McCutcheon, M. S. Capel, and V. Ramakrishnan.** 1999. Structure of bacterial 30S ribosomal subunit at 5.5 Å resolution. *Nature* **400:**833–840.

**Ogle, J. M., D. E. Brodersen, W. M. Clemons, M. J. Tarry, A. P. Carter, and V. Ramakrishnan.** 2001. Recognition of cognate transfer RNA by the 30S ribosomal subunit. *Science* **292:**897–902.

**Tocilj, A., F. Schlünzen, D. Janell, M. Glühmann, H. A. S. Hansen, J. Harms, A. Bashan, H. Bartels, I. Agmon, F. Franceschi, and A. Yonath.** 1999. The small ribosomal subunit from *Thermus thermophilus* at 4.5 Å resolution: pattern fittings and the identification of a functional site. *Proc. Natl. Acad. Sci. USA* **95:**14252–14257.

**Wimberly, B. T., D. E. Brodersen, W. M. Clemons, R. J. Morgan-Warren, A. P. Carter, C. Vonrhein, T. Hartsch, and V. Ramakrishnan.** 2000. Structure of the 30S ribosomal subunit. *Nature* **407:**327–339.

Crystal structure of the 50S ribosomal subunit

**Ban, N., B. Freeborn, P. Nissen, P. Penczek, R. A. Grassucci, R. Sweet, J. Frank, P. B. Moore, and T. A. Steitz.** 1998. A 9 Å resolution X-ray crystallographic map of the large ribosomal subunit. *Cell* **93:**1105–1115.

**Ban, N., P. Nissen, J. Hansen, M. Capel, P. B. Moore, and T. A. Steitz.** 1999. Placement of protein and RNA structures into a 5 Å resolution map of the 50S ribosomal subunit. *Nature* **400:**841–847.

**Ban, N., P. Nissen, J. Hansen, P. B. Moore, and T. A. Steitz.** 2000. The complete atomic structure of the large ribosomal subunit at 2.4 Å resolution. *Science* **289:**905–920.

**Harms, J., F. Schluenzen, R. Zarivach, A. Bashan, S. Gat, I. Agmon, H. Bartels, F. Franceschi, and A. Yonath.** 2001. High resolution structure of the large ribosomal subunit from a mesophilic eubacterium. *Cell* **107:**679–688.

**Schlünzen, F., R. Zarivach, J. Harms, A. Bashan, A.Tocilj, R. Albrecht, A. Yonath, and F. Franceschi.** 2001. Structural basis for the interaction of antibiotics with the peptidyl transferase centre in eubacteria. *Nature* **413:**814–821.

Cryoelectron microscopy of the 50S ribosomal subunit

**Matadeen, R., A. Patwardhan, B. Gowen, E. V. Orlova, T. Pape, M. Cuff, F. Mueller, R. Brimacombe, and M. van Heel.** 1999. The *Escherichia coli* large ribosomal subunit at 7.5 Å resolution. *Structure* **7:**1575–1583.

### Small-Molecule–RNA Interactions

**Sucheck, S. J., and C. H. Wong.** 2000. RNA as a target for small molecules. *Curr. Opin. Chem. Biol.* **4:**678–586.

### Aminoacyl-tRNA Synthetases

**Berisio, R., F. Schluenzen, J. Harms, A. Bashan, T. Auerbach, D. Baram, and A. Yonath.** 2003. Structural insight into the role of the ribosomal tunnel in cellular regulation. *Nat. Struct. Biol.* **10:**366–370.

**Ruff, M., S. Krishnaswamy, M. Boeglin, A. Poterszman, A. Mitschler, A. Podjarny, B. Rees, J. C. Thierry, and D. Moras.** 1991. Class II aminoacyl transfer RNA synthetases: crystal structure of yeast aspartyl-tRNA synthetase complexed with $tRNA^{Asp}$. *Science* **252:**1682–1689.

**Schlünzen, F., R. Zarivach, J. Harms, A. Bashan, A. Tocilj, R. Albretch, A. Yonath, and F. Franceschi.** 2001. Structural basis for the interaction of antibiotics with the peptidyl transferase centre in eubacteria. *Nature* **413:**814–821.

### Methionyl Aminopeptidase

**Cosper, N. J., V. M. D'Souza, R. A. Scott, and R. C. Holz.** 2001. Structural evidence that the methionyl aminopeptidase from *Escherichia coli* is a mononuclear metalloproteinase. *Biochemistry* **40:**13302–13309.

### Structure of Elongation Factor EF-Tu

**Kjeldgaard, M., and J. Nyborg.** 1992. Refined structure of elongation factor EF-Tu from *Escherichia coli. J. Mol. Biol.* **223:**721–742.

**Stark, H., M. V. Rodnina, H. J. Wieden, F. Zemlin, W. Winthmeyer, and M. van Heel.** 2002. Ribosome interactions of aminoacyl-tRNA and elongation factor Tu in the codon-recognition complex. *Nat. Struct. Biol.* **9:**849–854.

### Molecular Chaperones

**Bitto, E., and D. B. McKay.** 2002. Crystallographic structure of SurA, a molecular chaperone that facilitates folding of outer membrane porins. *Structure* **10:**1489–1498.

chapter 17



# Inhibitors of the 30S Ribosomal Subunit
## Aminoglycosides and Tetracyclines

## Aminoglycosides

The aminoglycoside group of antibiotics are multifunctional hydrophilic carbohydrates that possess two or more amino monosaccharides connected by glycosidic bonds to an aminocyclitol nucleus (i.e., 2-deoxystreptamine, streptamine, or streptidine). These amines are protonated under physiological conditions; hence, the aminoglycosides are considered polycationic compounds. Aminocyclitols are cyclohexane derivatives that have amine and hydroxyl alcoholic functionalities as substituents on the ring. Most aminoglycosides contain a 2-deoxystreptamine cyclitol. However, spectinomycin contains a streptamine cyclitol and streptomycin has the streptidine cyclitol, which contains a bis-guanidino group (Fig. 17.1). Amino monosaccharides are always present in the molecule of aminoglycoside antibiotics; the exception is the pseudosaccharide present in spectinomycin.

The aminoglycoside antibiotics have several features justifying their continued clinical use, including their rapid and potent bactericidal activity, long-lasting postantibiotic effect (defined as the continued suppression of bacterial growth after the microorganism is no longer in contact with the antibiotic), and synergy with other antibiotics. The in vitro antimicrobial spectrum of activity of aminoglycoside antibiotics includes a broad range of aerobic gram-negative bacilli, many staphylococci and certain mycobacteria.

The first aminoglycoside antibiotic, streptomycin, was isolated in 1943 from a strain of *Streptomyces griseus*. In 1949, the neomycins were isolated from *Streptomyces fradiae*. In 1957, Umezawa in Japan discovered the kanamycins. Other aminoglycosides were subsequently isolated from *Streptomyces* and *Micromonospora* species. As a convention to distinguish their origins, the aminoglycosides produced by species of *Streptomyces* have the suffix "-mycin," whereas those produced by *Micromonospora* species have the suffix "-micin."

Currently, four aminoglycoside antibiotics are marketed in the United States: gentamicin sulfate, tobramycin sulfate, amikacin sulfate, and

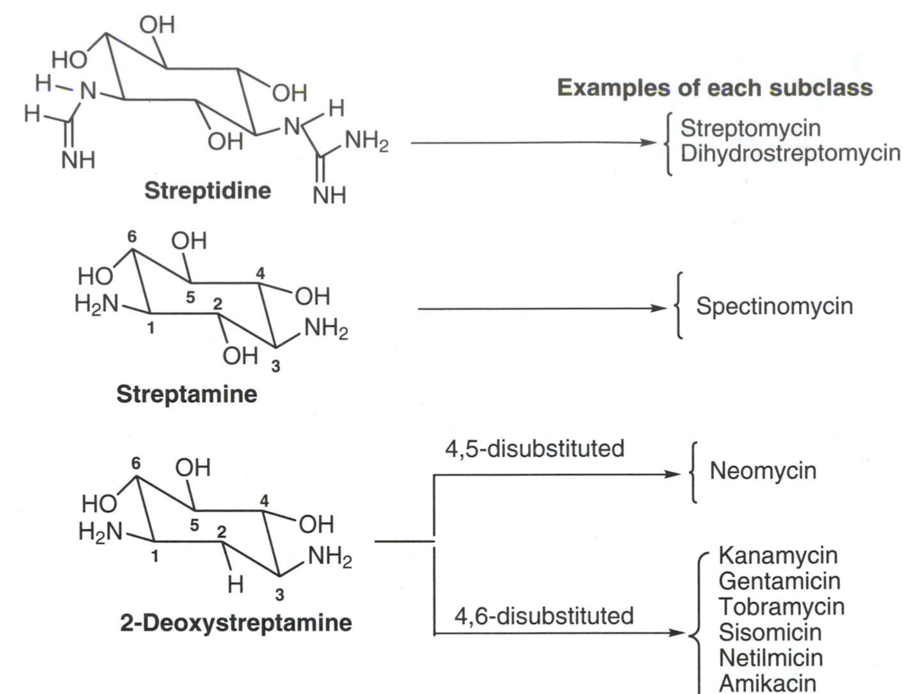


**Figure 17.1** Classification of aminoglycoside antibiotics based on the chemical structure of the aminocyclitol: streptidine, streptamine, or 2-deoxystreptamine, including those in which the amino sugars are linked at the 4- and 5-hydroxyl groups and those substituted at the 4- and 6-hydroxyl positions.

netilmicin sulfate. Streptomycin sulfate is available from Pfizer/Roerig for compassionate use for specific indications but is not commercially available in the United States.

Spectinomycin, neomycin, kanamycin, and tobramycin are derived from different species of *Streptomyces*; gentamicin and sisomicin are derived from *Micromonospora* species; amikacin is produced through chemical modification of kanamycin A, dibekacin is a derivative of kanamycin B, and netilmicin is the $N^1$-ethyl derivative of sisomicin.

## Chemistry

The aminoglycoside antibiotics can be divided into four groups on the basis of the type and substitution pattern of their aminocyclitol molecule.

### DERIVATIVES CONTAINING THE AMINOCYCLITOL STREPTIDINE

#### Streptomycin

Examples of the streptidine class are streptomycin and dihydrostreptomycin. Dihydrostreptomycin is obtained by chemical modifications of streptomycin (Fig. 17.2).

### DERIVATIVES CONTAINING THE AMINOCYCLITOL STREPTAMINE

The only clinically useful example of the streptamine subclass is spectinomycin.

#### Spectinomycin

Spectinomycin is not strictly an aminoglycoside; its chemical structure does contain a streptidine moiety, but there is no aminoglycoside unit (Fig. 17.3). However, this antibiotic is included in the aminoglycoside group. The ketone group at position 3 exists as the hydrate (Fig. 17.3). The molecule possesses nine asymmetric centers.

Spectinomycin was first isolated from a fermentation broth of *Streptomyces spectabilis* in 1961. It is also produced by *S. flavopersicus*. It is supplied as the dihy-

**Figure 17.2** Chemical structures of streptomycin and dihydrostreptomycin.

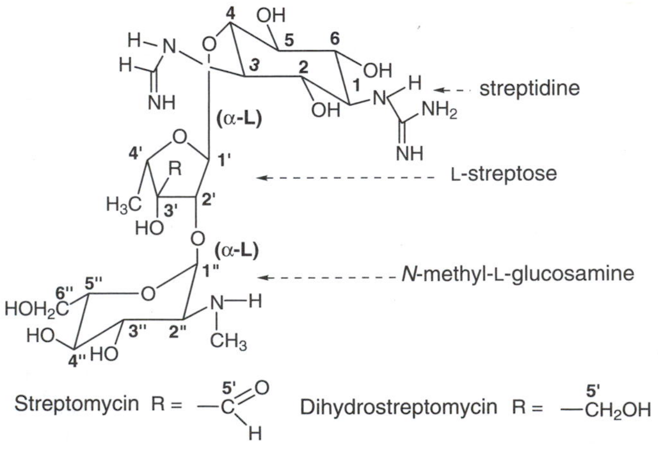

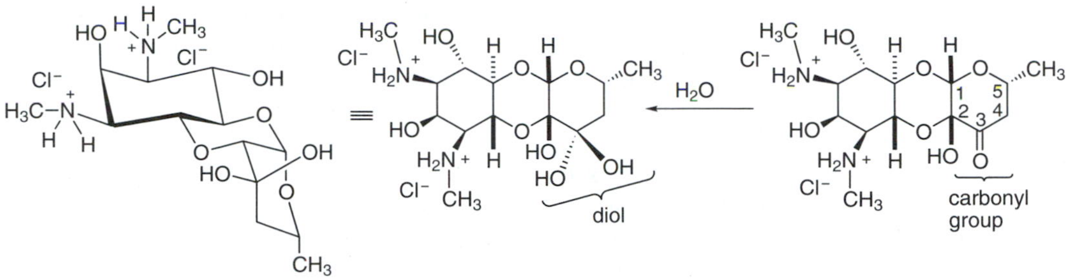

**Figure 17.3** Chemical structure of spectinomycin chlorhydrate drawn conformationally and planar. At the right is the 3-keto form, which is hydrated in aqueous medium, forming a diol group as shown.

drochloride and the sulfate. The clinical use of spectinomycin is restricted to a single-dose treatment of gonorrhea caused by *Neisseria gonorrhoeae* in penicillin-allergic patients or by penicillin-resistant strains of this species.

## DERIVATIVES CONTAINING A 4,5-DISUBSTITUTED 2-DEOXYSTREPTAMINE MOIETY

### Neomycin

Neomycin is the only therapeutically useful aminoglycoside of the 4,5-disubstituted 2-deoxystreptamine class. It was first described by Waksman and Lechevalier in 1949 after being obtained from fermentation products of *S. fradiae* as a mixture of neomycin A, B, and C. Neomycin B is the main component (the structure of neomycin B is given in Fig. 17.4). This mixture is marketed as the sulfate.

Neomycin should not be administered parenterally for systemic infections because this route of administration causes ototoxicity and nephrotoxicity. Therapeutic use is limited to topical and oral administration for local antibacterial effects. Neomycin is used, either alone or in combination with bacitracin or polymyxins, for local treatment of superficial infections by staphylococci and gram-negative bacilli.

### Paromomycin

Paromomycin is produced by the fermentation of certain strains of *Streptomyces rimosus* subsp. *paromomycinus*. It has an antibacterial spectrum similar to that of neomycin. It is not marketed in the United States.

## DERIVATIVES CONTAINING A 4,6-DISUBSTITUTED 2-DEOXYSTREPTAMINE MOIETY

Examples of the 4,6-disubstituted 2-deoxystreptamine class are kanamycins A, B, and C; amikacin; dibekacin; tobramycin; gentamicins C1, C1a, and C2; sisomicin; and netilmicin.

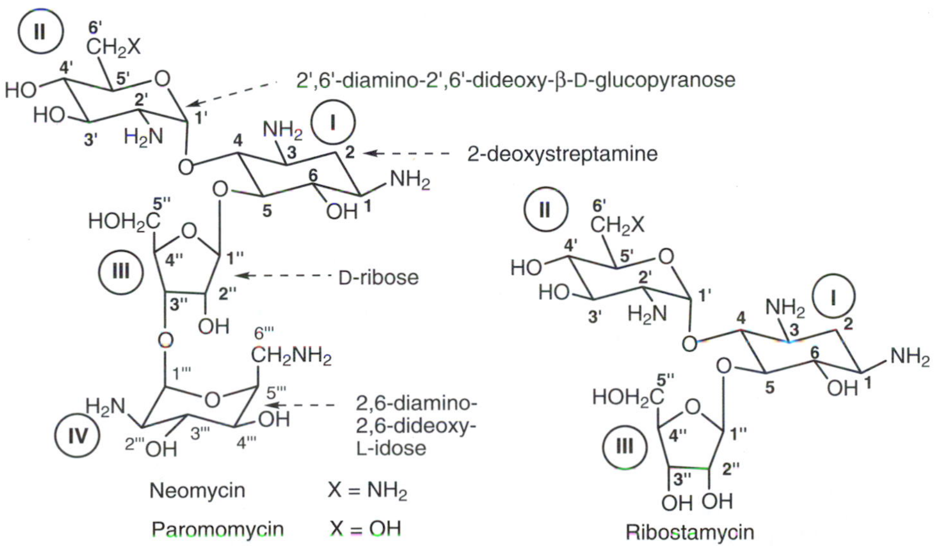

**Figure 17.4** Chemical structures of neomycin B, paromomycin, and ribostamycin, with the atomic and ring-numbering systems denoted in arabic and roman numerals, respectively.

## Kanamycins

Kanamycins A, B, and C are derivatives of 4,6-disubstituted 2-deoxystreptamine (Fig. 17.5). These antibiotics were first obtained in 1957 in Japan from fermentation of *Streptomyces kanamyceticus*. In the United States, kanamycin A is marketed as its sulfate; kanamycins B and C are not available because of high toxicity. Structural modifications of the kanamycins have been extensively investigated, and the most notable outcome of this work was the discovery of amikacin and dibekacin. Kanamycin A is active against enteric bacilli like *Klebsiella* spp., *Enterobacter* spp., *Escherichia coli*, and *Proteus* spp. It is inactive against *Pseudomonas aeruginosa* and *Serratia*. Because it is somewhat more toxic than gentamicin or tobramycin, its use has declined markedly since the 1960s.

## Amikacin

Amikacin is a semisynthetic derivative of kanamycin A. It is a substituted (*S*)-4-amino-2-hydroxybutyryl side chain of the amino group at position 1 of the deoxystreptamine moiety of kanamycin A (Fig. 17.5). The presence of this secondary amine gives amikacin stability against most bacterial plasmid-mediated enzymes (aminoglycoside-modifying enzymes) which are responsible for resistance to aminoglycosides. For this reason, amikacin is active against many gentamicin- and tobramycin-resistant strains of gram-negative bacilli.

Amikacin is used to treat gram-negative infections in hospitalized patients. Some infectious-disease specialists recommend it for the treatment of severe infections due to *P. aeruginosa* infections and infections due to enteric gram-negative bacilli that are suspected to be resistant to gentamicin and tobramycin. Restriction of amikacin use is advocated to minimize the emergence of resistant strains of bacteria.

## Dibekacin

Absence of the hydroxyl groups at positions 3' and 4' of the amino saccharide unit of kanamycin A (Fig. 17.5) renders dibekacin resistant to phosphorylation by 3' phosphorylases [APH(3')] and also resistant to the adenylation catalyzed by 4' adenyltransferases [AAD (4')]. Dibekacin is not commercially available in the United States or the United Kingdom, but it is available in Japan and most European countries. It is used in the treatment of infections due to aerobic gram-negative bacteria resistant to gentamicin, tobramycin, and amikacin.

## Tobramycin

Tobramycin (Fig. 17.5) was isolated from the fermentation broth of *Streptomyces tenebrarius*. It is active against aerobic gram-negative bacilli, including the *Enterobacteriaceae* and *P. aeruginosa*. It is two to five times more active in vitro than gentamicin against *P. aeruginosa*. Strains of *P. aeruginosa* that are sometimes moderately resistant to gentamicin are usually sensitive to tobramycin. Tobramycin is usually less active in vitro than gentamicin against *Serratia marcescens*. Tobramycin is used to treat serious infections caused by aerobic gram-negative bacilli. It is also frequently the

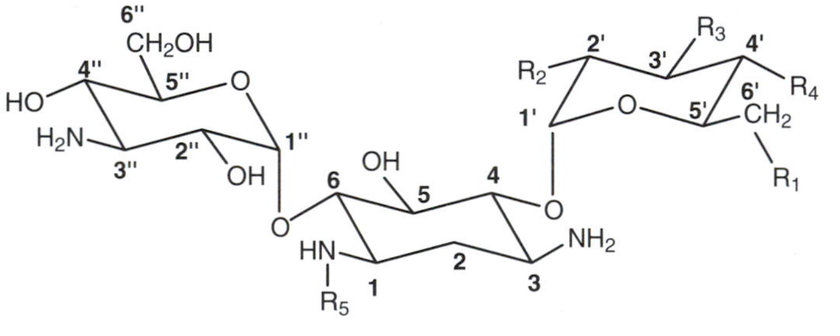

**Figure 17.5** Chemical structures of kanamycins A, B, and C, amikacin, dibekacin, and tobramycin.

| | $R_1$ | $R_2$ | $R_3$ | $R_4$ | $R_5$ |
|---|---|---|---|---|---|
| Kanamycin A | $NH_2$ | OH | OH | OH | H |
| Kanamycin B | $NH_2$ | $NH_2$ | OH | OH | H |
| Kanamycin C | OH | $NH_2$ | OH | OH | H |
| Amikacin | $NH_2$ | OH | OH | OH | (*S*)-$COCHOHCH_2CH_2NH_2$ |
| Dibekacin[a] | $Nh_2$ | $NH_2$ | H | H | H |
| Tobramycin | $NH_2$ | $NH_2$ | H | OH | H |

[a]Available in Europe and Japan, but not in the United Kingdom or the United States.

aminoglycoside of choice for infections caused by *P. aeruginosa* because of its greater in vitro activity against this organism. It usually is given in combination with an antipseudomonal penicillin for severe systemic *P. aeruginosa* infections. It also provides synergy in combination with a β-lactam antibiotic in the treatment of *Staphylococcus aureus* infections.

### Gentamicin

Gentamicin is a mixture of gentamicins C1, C1a, and C2, isolated from the fermentation broth of *Micromonospora purpurea* and related species (Fig. 17.6). Gentamicin was discovered in 1963 by workers at Schering Laboratory.

Gentamicin is used clinically as the sulfate salt. It is used to treat serious infections caused by aerobic gram-negative bacilli, including the lower respiratory tract, intra-abdominal, soft tissue, bone or joint, wound, and complicated urinary tract infections. Gentamicin often is the preferred aminoglycoside for general use in hospitals where the prevalence of bacterial resistance to this antibiotic is low.

The combination of gentamicin and penicillin G (or ampicillin) exhibits synergistic activity against enterococci (e.g., *Enterococcus faecalis*) and viridans group streptococci. Frequently, this combination is the regimen of choice for the treatment of endocarditis caused by these organisms. Gentamicin is also used for the treatment of severe *S. aureus* infections, especially those involving bacteremia. Other uses of gentamicin are discussed in the references.

### Sisomicin

Sisomicin (Fig. 17.7) is produced by fermentation of *Micromonospora inyoensis*. Sisomicin does not offer significant advantages over gentamicin. It has been available in Europe but not in the United States, the United Kingdom, or Australia.

**Figure 17.6** Chemical structures of gentamicins C1, C2, and C1a.

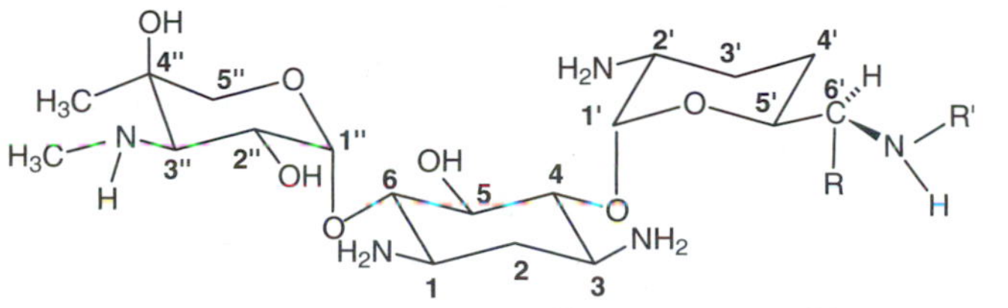

| | R | R' |
|---|---|---|
| Gentamicin $C_1$ | $-CH_3$ | $-CH_3$ |
| Gentamicin $C_2$ | $-CH_3$ | -H |
| Gentamicin $C_{1a}$ | -H | -H |

### Netilmicin

Netilmicin is the $^{1}N$-ethyl derivative of the 2-deoxystreptamine moiety of sisomicin (Fig. 17.7). It is active against staphylococci, including methicillin-resistant and coagulase-negative strains. Streptococci are resistant. It is active against a wide variety of *Enterobacteriaceae* and *Acinetobacter*. It is resistant to many aminoglycoside-modifying enzymes that inactivate gentamicin and is active against some gentamicin-resistant isolates of *E. coli, Klebsiella,* and *Enterobacter*. It is less active than gentamicin against *P. aeruginosa*.

Netilmicin has been used to treat severe gram-negative bacillary infections and complicated urinary tract infections, and it has been used in combination for severe undiagnosed sepsis. It is particularly useful to treat infections caused by gentamicin-resistant organisms.

## Summary of Clinical Uses

The therapeutic use, dosing regimens, and toxicity of aminoglycoside antibiotics are complicated topics and are only briefly considered here. For a comprehensive account, readers should refer to texts covering current therapy of infectious diseases.

Gentamicin, tobramycin, amikacin, and netilmicin can be used interchangeably for the treatment of a large variety of infections by aerobic gram-negative bacteria. However, gentamicin is generally the preferred aminoglycoside antibiotic because of long experience with its use and its relatively low cost. Aminoglycoside antibiotics are potential drugs of choice for the following infectious diseases caused by susceptible organisms.

Gentamicin is used in combination with another antibiotic to achieve synergy against certain pathogens:

- with penicillin, ampicillin, or vancomycin against gentamicin-susceptible enterococci or viridans group streptococci
- with ampicillin against *Listeria monocytogenes*
- with piperacillin against *P. aeruginosa* and susceptible *Enterobacter* spp.

**Figure 17.7** Chemical structures of sisomicin and netilmicin.

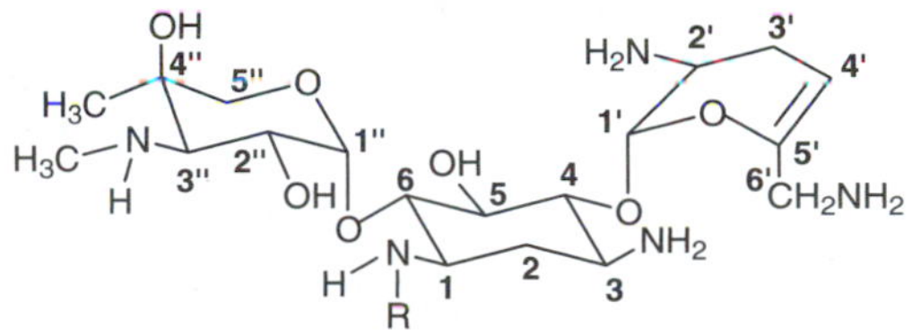

| | |
|---|---|
| Sisomicin | R = -H |
| Netilmicin or *N*-ethyl sisomicin | $R = -CH_2CH_3$ |

- with vancomycin and rifampin against *Streptococcus epidermidis* prosthetic valve infections
- in conjunction with ampicillin (or vancomycin) for endocarditis prophylaxis in high-risk patients before gastrointestinal or genitourinary procedures
- combined with an antistaphylococcal penicillin for endocarditis caused by *S. aureus*
- in association with doxycycline for the treatment of brucellosis

Amikacin has been used in some atypical mycobacterial infections. Amikacin plus clarithromycin is the drug of choice for *Mycobacterium fortuitum* complex infections.

Streptomycin (or gentamicin) is an agent of choice for *Yersinia pestis* infections (plague). It remains a useful agent for *Mycobacterium tuberculosis* therapy.

It should be remembered that penicillin and aminoglycoside antibiotics must never be physically mixed, because both are chemically inactivated to a significant degree on mixing.

## Adverse Effects

All aminoglycoside antibiotics have the potential to produce reversible and irreversible vestibular, cochlear, and renal toxicity. The risk of toxicity must be considered, especially when the drugs are to be used for an extended time. The margin between effective and toxic concentrations is narrow. Levels of the drugs in serum should be measured to ensure the presence of effective concentrations and minimum toxicity. Prolonged use should be restricted to the therapy of life-threating infections and those for which a less toxic agent is contraindicated.

## Transport through Membranes

The aminoglycosides, being cationic under physiological conditions, interact with anionic components on the outer membrane of gram-negative bacteria, such as the polar heads of phospholipids and lipopolysaccharides, and of gram-positive species, such as teichoic acid and lipopolysaccharides. Association of aminoglycosides with these molecules is predominantly electrostatic and involves charge-charge interactions. Ionic binding of this type is nonspecific, energy independent, and reversible. Transport of aminoglycosides across the cytoplasmic membrane involves energy-dependent processes that rely on an electrochemical gradient generated during ATP hydrolysis.

Recently, it has been proposed that the passage of the polycationic aminoglycoside molecules across the outer membrane of gram-negative bacteria is a self-promoted uptake process involving the drug-induced disruption of $Mg^{2+}$ bridges between adjacent lipopolysaccharide molecules. The subsequent transport across the cytoplasmic membrane is dependent on electron transport and is termed energy-dependent phase I (EDP-I).

For additional features not addressed here, readers should consult the references.

## Mechanism of Action

Most aminoglycoside antibiotics cause bacterial cell death (bactericidal action). However, the action of spectinomycin is limited to inhibition of cellular growth and division (a bacteriostatic event). The bactericidal activity of the aminoglycoside antibiotics is derived from their ability to inhibit the translation process. They achieve their effect by binding to phylogenetically conserved sequences within the 16S rRNA of the 30S ribosomal subunit, thereby inducing codon misreading as well as inhibition of translation

Although all the aminoglycosides used clinically bind to the A-site decoding region on bacterial 16S rRNA, causing both misreading of the genetic code and inhibition of translocation, there are differences in the mechanism of action of aminoglycoside derived from the 4,6- and 4,5-disubstituted 2-deoxystreptamine moieties, such as neomycin, gentamicin, kanamycin, and tobramycin, which target the same region of rRNA and interfere with the decoding process, and of the streptidine aminoglycoside, streptomycin, with which they do not overlap because they do not occupy the same binding site.

### THE NEOMYCIN CLASS

Aminoglycosides that contain a 2-desoxystreptamine ring bind to the penultimate stem (helix 44) in 16S rRNA in the region of A1408, A1492, and A1493, which is where A-site tRNAs bind (Fig. 17.8). No ribosomal protein directly contacts this region of rRNA, so that the aminoglycoside-binding site is formed primarily by RNA.

Recent work (Mehta and Champney, 2002) on 30S ribosomal subunit assembly as a target for inhibition by aminoglycosides in *E. coli* showed that neomycin and paromomycin interact with the 16S rRNA of the ribosome within an internal loop of the decoding site. Binding to this region results in a conformational change of the conserved bases within the loop of the A site, which facilitates high-affinity binding between the rRNA of the internal loop and rings I and II of the aminoglycoside antibiotic. The tightly bound antibiotic contributes to codon misreading and mistranslation of mRNA.

Kaul and Pilch (2002) have studied the binding of the neomycin class of aminoglycosides to the A site of 16S rRNA. These scientists used a combination of spectro-

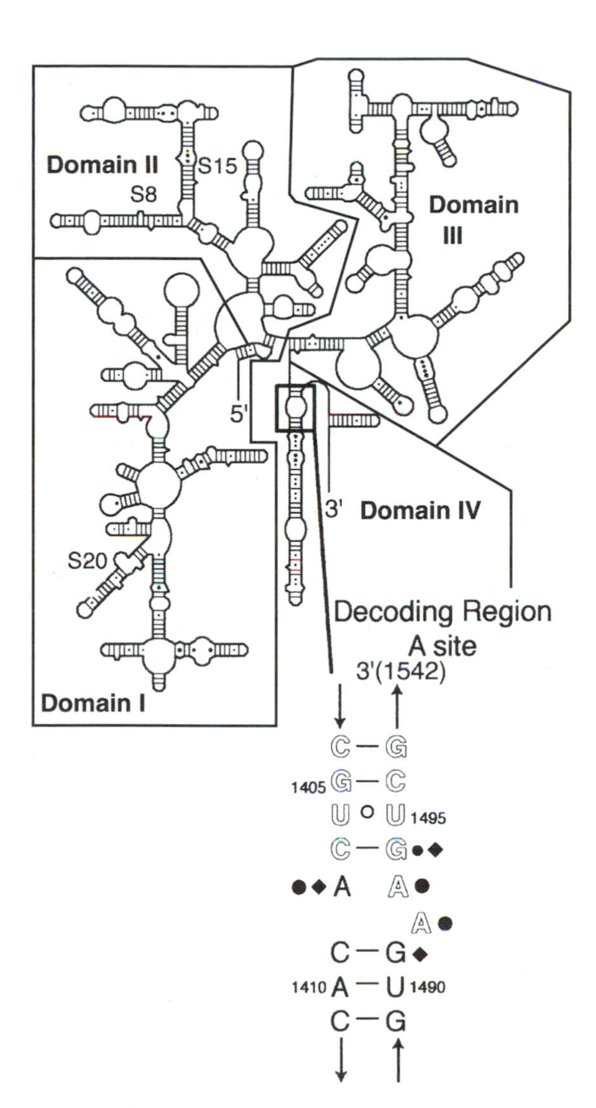


**Figure 17.8** Secondary structure of *E. coli* 16S rRNA . This four-domain structure is 46% base paired, highlighting the decoding region A site. Nucleotides conserved in more than 95% of ribosomal sequences are shown in outline. Adapted from M. O'Connor, M. Bayfield, S. T. Gregory, W.-C. M. Lee, J. S. Lodmell, A. Mankad, J. R. Thompson, A. Vila-Sanjurjo, C. L. Squires, and A. E. Dahlberg, p. 217–227, *in* R. A. Garrett, S. R. Douthwaite, A. Liljas, A. T. Matheson, P. B. Moore, and H. F. Noller (ed.), *The Ribosome: Structure, Function, Antibiotics, and Cellular Interactions* (ASM Press, Washington, D.C., 2000), with permission from the publisher.

scopic and calorimetric techniques to characterize the binding of the aminoglycoside antibiotics neomycin, paromomycin, and ribostamycin to an RNA oligonucleotide that models the A site of *E. coli* 16S rRNA. This model is a 27-mer RNA oligonucleotide that was developed in 1996 by the Puglisi group from the Department of Structural Biology of Stanford University (Stanford,

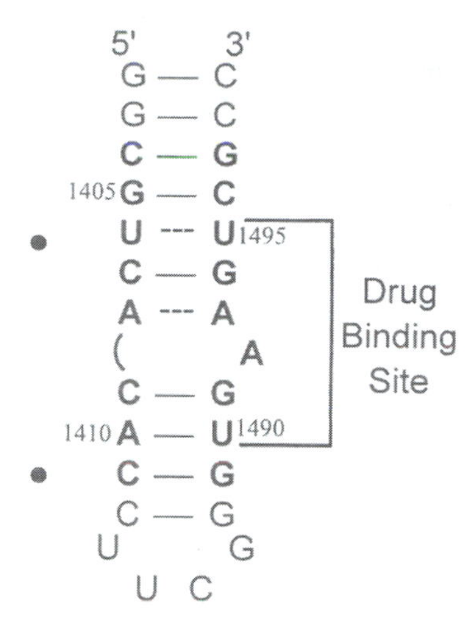


**Figure 17.9** Secondary structure of the 27-mer A-site RNA oligonucleotide, as derived by NMR. Watson-Crick base pairs are denoted by solid lines, while mismatched base pairs are denoted by dashed lines. Bases present in *E. coli* 16S rRNA are depicted in bold type and are numbered as in 16S rRNA. The aminoglycoside-binding site, as revealed by NMR and footprinting studies, is as indicated. Reprinted from M. Kaul and D. S. Pilch, *Biochemistry* **41:**7695–7706, 2002, with permission from the publisher.

Calif.); the sequence and secondary structure are depicted in Fig. 17.9. This model was developed by using footprinting techniques to demonstrate the binding site for the neomycin class of aminoglycosides in the *E. coli* 16S rRNA A site of the 30S ribosomal subunit. Nuclear magnetic resonance studies by the Puglisi group culminated in a solution structure of paromomycin in complex with the A-site RNA oligonucleotide.

The work done by Kaul and Pilch demonstrated that the affinities of the aminoglycosides for the host RNA follow the hierarchy neomycin > paromomycin > ribostamycin. Their results revealed the impact of specific alterations in aminoglycoside structure on the thermodynamics of binding to the A-site model RNA oligonucleotide.

Wong et al. (1998) studied the binding specificity of several aminoglycosides with model RNA sequences derived from the 16S ribosomal A site by using surface plasmon resonance. They found that the neomycin class of aminoglycosides showed specificity for wild-type A-site sequences and that the kanamycins and gentamicins generally showed poor specificity for the same sequences.

Yoshizawa et al. (1998) elucidated the structure of gentamicin C1a bound to an A-site RNA complex. These studies used nuclear magnetic resonance techniques to elucidate the three-dimensional structures of the complexes of RNA and aminoglycosides.

### STREPTOMYCIN

Early experiments indicated that streptomycin made ribosomes error prone by affecting the proofreading step. More recent data indicate interference with both the initial selection and proofreading. Carter et al. (2000) described the crystal structure of the 30S subunit complexed with the antibiotic streptomycin, which interferes with decoding and translocation. Streptomycin is tightly bound to the phosphate backbone of 16S RNA from four different parts of the molecule through both salt bridges and hydogen bonds. It also makes contact with K45 from protein S12. Four nucleotides of 16S RNA (nucleotides 13, 526, 915, and 1490) have been implicated in streptomycin binding. The tight interactions observed for streptomycin indicate that it preferentially stabilizes the *ram* state seen in the crystal structure. The restrictive A site has a low tRNA affinity, whereas the *ram* state has a higher affinity. Therefore, by stabilizing the *ram* state, streptomycin would be expected to increase the initial binding of noncognate tRNAs. Figure 17.10 shows the streptomycin-binding site, its interaction with H27, the 530 loop (H18), H44, and ribosomal protein S12.

## Bacterial Resistance

The emergence of bacterial strains resistant to aminoglycoside antibiotics has somewhat reduced their potential in therapy of bacterial infections. Resistance to the antibacterial action of aminoglycosides is mediated primarily by three mechanisms: (i) a decreased uptake and/or accumulation of the antibiotic in bacteria; (ii) the bacterial expression of plasmid-borne enzymes which modify the chemical structure catalyzing reactions of acetylation, phosphorylation, and or adenylation, thereby inactivating the antibiotic action (this is the most common of the resistance mechanisms and is a significant clinical problem); and (iii) changes in the bacterial ribosomes such that the affinity for the aminoglycosides is significantly decreased (this mechanism of resistance appears to be relatively rare except for streptomycin).

### DECREASE IN DRUG UPTAKE AND ACCUMULATION

Reduced drug uptake is seen primarily in *Pseudomonas* spp. and other nonfermenting gram-negative bacilli and is probably due to reduced membrane permeability.

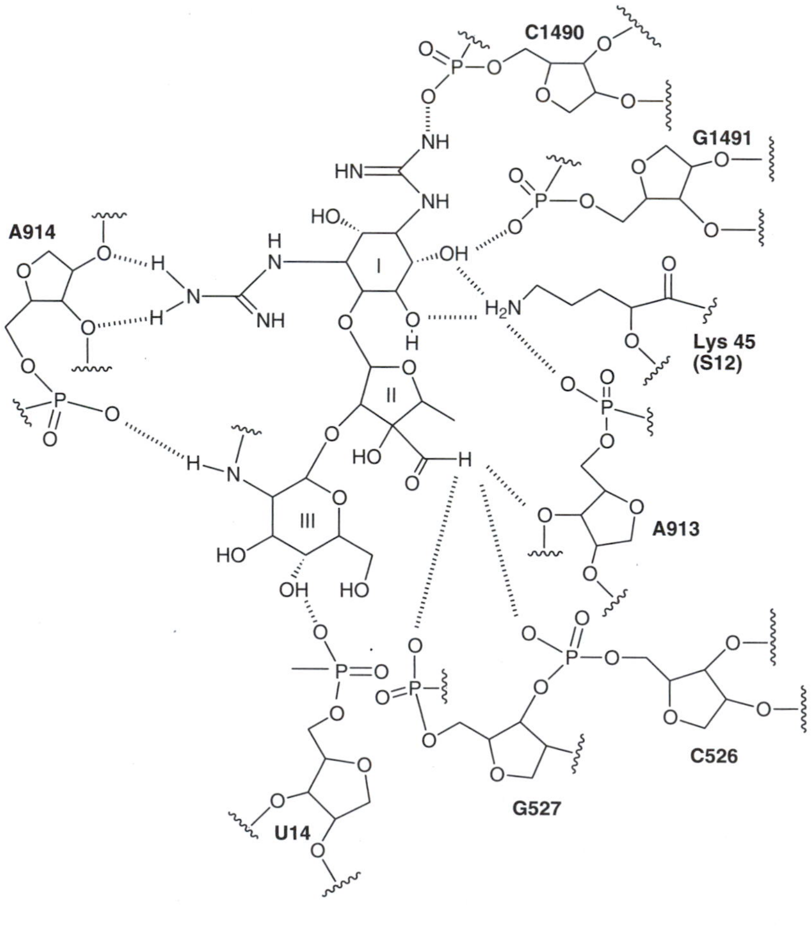

**Figure 17.10** The streptomycin-binding site. For details, see the text. Adapted from A. P. Carter, W. M. Clemons, D. E. Brodersen, R. J. Morgan-Warren, B. T. Wimberly, and V. Ramakrishnan, *Nature* **407:**340–348, 2000, with permission from the publisher.

Active efflux has been demonstrated for kanamycin and neomycin in *E. coli*.

## EXPRESSION OF AMINOGLYCOSIDE-MODIFYING ENZYMES

The presence of aminoglycoside-modifying enzymes is by far the most prevalent and important mechanism of bacterial resistance. Aminoglycoside-modifying enzymes are named according to the type of reaction which they catalyze, i.e., *N*-acylases, *O*-phosphorylases, *O*-nucleotidylases, and adenylases, and can be further subdivided according to the position of the site of modification on the drug.

### Types of Aminoglycoside-Modifying Enzymes

The enzymes modifying aminoglycosides are

- *N*-acetyltransferases, which use the cofactor acetyl coenzyme A (acetyl-CoA) as the acetate donor and affect the amino groups
- *O*-phosphonotransferases, which use nucleoside triphosphates, especially ATP, as the donor of the phosphate group and affect the hydroxyl groups
- *O*-nucleotidyltransferases (mostly *O*-adenyltransferases), which also use nucleoside triphosphates, especially ATP, as the adenylate donor and affect the hydroxyl groups

The chemical transformation of an amino or hydroxyl group in the molecule of a specific aminoglycoside into a derivative that has reduced affinity for the bacterial ribosome, and possibly for which the EDP-II antibiotic uptake fails to occur, results in high-level resistance.

In 1993, Shaw and coworkers from Schering Laboratories proposed a nomenclature system to designate these enzymes and genes. First, each enzyme is designated by class: AAC for acetyltransferases, APH for phosphonotransferases, and ANT for nucleotidyltransferases. The class is followed by a number in parentheses, which describes the regiospecificity of group transfer, and a subsequent roman numeral, which differentiates between enzymes with differing modification profiles. The positions on the cyclitol ring (streptidine and deoxystreptamine) are numbered 1 to 6, those of the 4-substituent ring are numbered from 1' to 6', and those at the 5 or 6 position of deoxystreptamine are numbered 1" to 6". Thus, APH(3')-Ia is an aminoglycoside kinase that transfers a phosphate group to the 3' hydroxyl group of an aminoglycoside molecule of diverse antibiotics. If a different APH(3'), such as APH(3')-II, transfers a phosphate group to the 3'-hydroxyl group, they are distinguished by the roman numeral. *aph(3')-Ia* is a gene that encodes a phosphatidyl transfer protein which modifies aminoglycosides at position 3'. Aminoglycoside resistance genes are generally found on mobile genetic elements, such as integrons, transposons, and plasmids.

Table 17.1 lists the main enzymes which modify the most important clinically used aminoglycosides.

*Aminoglycoside Acetyltransferases.* AACs are acetyl-CoA-dependent acyltransferases that primarily modify amino groups (*N*-acetyltransferases). They can be divided into four groups based on the regiospecificity of the modification: AAC(1), AAC(2'), AAC(3), and AAC(6').

In 1998, Wolf and coworkers presented the first three-dimensional structure of aminoglycoside acetyltransferase when they reported the crystal structure of AAC(3)-Ia from *Serratia marcescens* solved in complex with CoA to 2.3-Å resolution. The protein is derived from a plasmid-located gene that is widespread in *Enterobacteriaceae* and confers resistance to a variety of aminoglycosides including gentamicin.

In 1999, Wybenga-Groot and coworkers reported the crystal structure of the AAC(6')-Ii acetyl-CoA-bound enzyme determined at 2.7-Å resolution. This protein is chromosomally encoded in all *Enterococcus faecium* strains.

In 2002, Vetting and coworkers determined to 1.5- to 1.8-Å resolution the crystal structure of AAC(2')-Ic from *Mycobacterium tuberculosis* in the apoenzyme form and in ternary complexes with CoA and either ribostamycin, tobramycin, or kanamycin A. The AAC(2') enzymes are chromosomally encoded. The aminoglycoside binds the active site of the dimeric structure of AAC(2')-Ic within a pocket adjacent to CoA. The formation of the ternary

**Table 17.1** Typical enzymes modifying clinically used aminoglycosides

| Enzyme | Aminoglycoside(s) modified |
|---|---|
| *N*-Acetyltransferases (AAC) | |
| 2' | Gentamicin, tobramycin |
| 6' | Gentamicin, tobramycin, kanamycin |
| 3 | Gentamicin, tobramycin, kanamycin |
| *O*-Phosphonotransferases (APH) | |
| 2" | Gentamicin, tobramycin, kanamycin |
| 3' (5") | Kanamycin, neomycin |
| 3" | Streptomycin |
| 6 | Streptomycin |
| *O*-Adenyltransferases (ANT) | |
| 2' | Gentamicin, tobramycin |
| 3" | Streptomycin |
| 4' | Kanamycin, neomycin |
| 6 | Streptomycin |
| 9 | Spectinomycin |

complex buries the 2' amino group of the aminoglycoside, as well as the sulfhydryl group of the CoA. The authors have proposed a chemical reaction mechanism for the AAC(2')-Ic catalysis, based on the AAC(2')-Ic ternary complex. The acetylation of aminoglycosides catalyzed by this enzyme is suggested to proceed through a direct nucleophilic attack of the 2' amine to the acetyl-CoA thioester.

*Aminoglycoside Phosphonotransferases.* As mentioned above, APH enzymes catalyze the ATP-dependent phosphorylation of specific aminoglycoside hydroxyl groups. There are several classes of these enzymes which have been classified primarily on the basis of substrate specificity. APH(3')-IIIa catalyzes the transfer of the γ-phosphate of ATP to the 3' hydroxyl group of 4,6-disubstituted deoxystreptamine aminoglycosides and to both 3' and 5' hydroxyls of 4,5-disubstituted deoxystreptamine aminoglycosides, such as neomycin.

In 1997, Hon and coworkers determined the crystal structure of APH(3')-IIIa bound to ADP to 2.2-Å resolution.

*Aminoglycoside Nucleotidyltransferases.* Kanamycin nucleotidyltransferase, as originally isolated from *Staphylococcus aureus,* inactivates the antibiotic kanamycin by catalyzing the transfer of a nucleotidyl group from nucleoside triphosphates such as ATP to the 4'-hydroxyl group of the aminoglycoside. In 1993, Sakon and coworkers reported the determination of the molecular structure of ANT(4')-Ia by X-ray crystallographic analysis to 3.0-Å resolution. The molecular structure of this enzyme was determined both in the absence of substrates and as the ternary complex with kanamycin and the nonhydrolyzable ATP analogue AMPCPP.

### Overview of Modes of Inactivation by Enzymes

Since kanamycin B is subject to inactivation by the three classes of aminoglycoside-modifying enzymes, it is useful to illustrate these modes of inactivation by using this compound as a model (Fig. 17.11).

### Inhibitors of Aminoglycoside-Modifying Enzymes

The emerging structural data now have the potential to be exploited in the design of specific inhibitors of enzyme activity. The challenge is to use this information to synthesize effective and potent inhibitors that will overcome antibiotic resistance produced by these aminoglycoside-modifying enzymes.

## Structure-Activity Relationships among Aminoglycoside Derivatives

As we already have seen, the aminoglycosides vary greatly in structure despite possessing similar antibacterial properties. As mentioned above, enzymatic modifications occur at either the OH or $NH_2$ group, and the likelihood of modification depends on the position of the group in the aminoglycoside molecule. If the antibiotic is chemically altered so that the vulnerable group becomes nonmodifiable or is removed, the aminoglycoside is protected from enzymatic modification. The enzymatic deactivation of the aminoglycosides by resistant bacteria has significantly affected synthetic strategies. Many

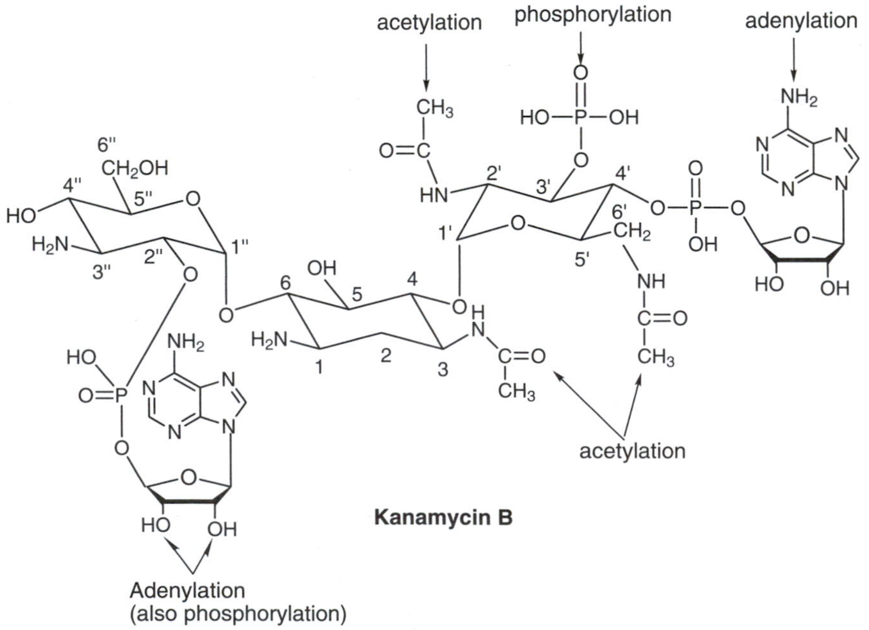

**Figure 17.11** Enzymatic inactivation of kanamycin B. Not all resistant bacteria exhibit every reaction shown.

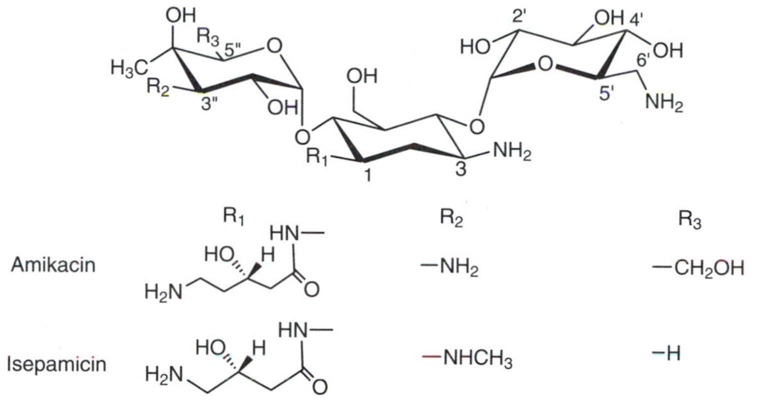

**Figure 17.12** Chemical structure of isepamicin compared to that of amikacin.

efforts have been made to attach substituents to the key amino and hydroxy groups in the hope of overcoming enzymatic deactivation. So far, the semisynthetic aminoglycosides have been essentially limited to chemical transformations or removal of sites susceptible to the action of these bacterial enzymes.

A useful modification of the amino cyclitol unit is acylation or alkylation of the C-1 amino group, as exemplified by amikacin, which is the C-1 *N*-α-hydroxybutyryl derivative of kanamycin A, and netilmicin, which is the $N^1$-ethyl derivative of sisomicin. The C-3' and C-4' positions can be unsubstituted, as in dibekacin (2',4'-dideoxykanamycin B), a derivative of kanamycin B which is stable to both APH(3') and ANT(4') enzymes. The C-4' center can be unsaturated, producing the tetrahydropyranyl derivatives sisomicin and netilmicin. Accordingly, amikacin, dibekacin, and netilmicin often possess activity against bacteria resistant to older, conventional aminoglycosides.

## Recent Advances in the Development of Novel Aminoglycosides

There has been little research activity on the development of novel aminoglycosides by isolation from fermentation and/or chemical semisynthesis. This might be because the aminoglycosides can cause serious ototoxicity and nephrotoxicity which limit their use. Nevertheless, some novel semisynthetic derivatives have recently been developed. For example, isepamicin (Fig. 17.12) is a novel broad-spectrum aminoglycoside with a high level of stability to aminoglycoside-modifying enzymes, especially to AAC(6')-I, a prevalent enzyme that causes resistance to tobramycin, netilmicin, and amikacin.

Another interesting example is the development of arbekacin, a semisynthetic aminoglycoside (Fig. 17.13) with very good activity against a wide variety of known aminoglycoside-resistant bacteria, including methicillin-resistant staphylococci (MRS) with aminoglycoside-modifying enzymes. A remarkable observation is that arbekacin was active against AAC(2')-producing strains despite its susceptibility to the modifying action of this class of enzymes. This finding was attributed to the fact that 2'-*N*-acetylarbekacin, the product of the action of AAC(2') enzymes on arbekacin, has much higher and broader antibiotic activity than known *N*-acetylated aminoglycosides.

## Aminoglycoside Preparations Commercially Available in the United States

Table 17.2 lists the aminoglycosides marketed in the United States.

# Tetracyclines

The tetracyclines are a group of antibiotics with an identical basic skeleton of four linearly fused six-membered rings, named 1,4,4a,5,5a,6,11,12-octahydronaphthacene (rings designated A, B, C, and D) and differing from each other chemically only by substituent variation at positions 5, 6, and 7 (the structures are shown in Fig. 17.14).

## History and Overview

After the successful isolation, structure determination, and clinical use of penicillins G and V, the screening of

**Figure 17.13** Chemical structure of arbekacin.

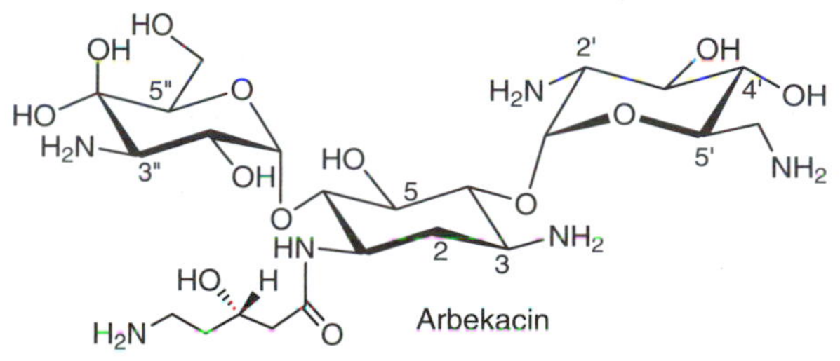

**Table 17.2** Generic and common trade names of aminoglycosides, the preparations available, and manufacturers in the United States

| Generic name | Trade name (preparation) | Manufacturer |
|---|---|---|
| Amikacin sulfate | Amikacin sulfate USP (injection) | Elkins-Sinn |
| Gentamicin sulfate | Gentamicin sulfate (injection) | Elkins-Sinn |
| Kanamycin sulfate | Kantrex (capsules) | Apothecon |
| Neomycin sulfate plus polymyxin B sulfate | Neosporin (cream, ointment) | Pfizer |
| Tobramycin sulfate | Nebcin (vials) | Eli Lilly |
| | Tobramycin sulfate (injection) | Lederle |
| | Tobi (solution for inhalation) | Pathogenesis |
| | TobraDex (ophthalmic ointment) | Alcon |
| | TobraDex (ophthalmic suspension) | Alcon |

naturally occurring sources in search of novel antibacterial substances became a common practice. In 1948, the first tetracycline, chlorotetracycline (formerly called aureomycin) (the structural formula is shown in Fig. 17.14) was obtained from the fermentation of the soil actinomycete *Streptomyces aureofaciens.* This antibiotic was marketed by Lederle Laboratories. Shortly thereafter, a second tetracycline was isolated at Pfizer Inc. This antibiotic, produced by *Streptomyces rimosus,* was named oxytetracycline. Tetracycline was discovered in 1953 and was produced by reductive dechlorination of chlorotetracycline; later it was also obtained by direct fermentation. Demeclocycline was discovered in 1957, as a product of fermentation of a mutant strain of *S. aureofaciens.* Methacycline, doxycycline, and minocycline were obtained by semisynthesis in 1965, 1967, and 1972, respectively.

At present only tetracycline, oxytetracycline, demeclocycline, doxycycline, and minocycline are marketed in the United States.

## Chemical Properties

The tetracyclines are amphoteric compounds that can exist as acid or base salts as well as in the zwitterionic state, depending on the pH of the environment. The tetracycline molecule may be viewed as tribasic acids, with the dimethylammoniun ions (the conjugate acid of the dimethylamino group) being the weakest ($pK_{a3}$, 9.1 to 9.7). The phenol-diketone system accounts for the $pK_{a2}$ range from 7.2 to 7.8, while the tricarbonylmethane moiety generates the strongest acidity ($pKa_1$, 2.8 to 3.3) Minocycline, which has an additional dimethylamino group at C-7, generates a fourth $pK_a$

**Figure 17.14** Chemical structures of seven clinically important tetracyclines (for an explanation, see the text).

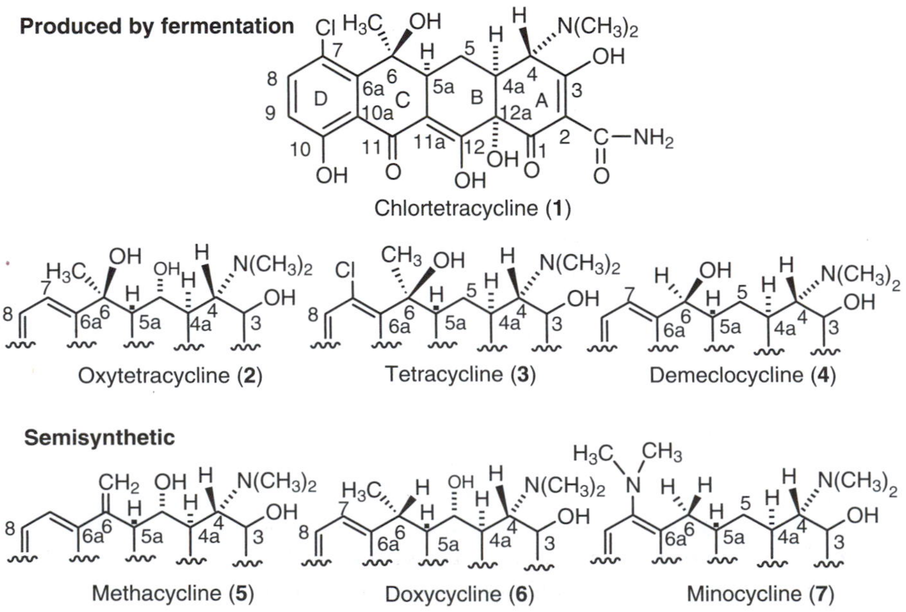

Produced by fermentation

Chlortetracycline (**1**)

Oxytetracycline (**2**) Tetracycline (**3**) Demeclocycline (**4**)

Semisynthetic

Methacycline (**5**) Doxycycline (**6**) Minocycline (**7**)

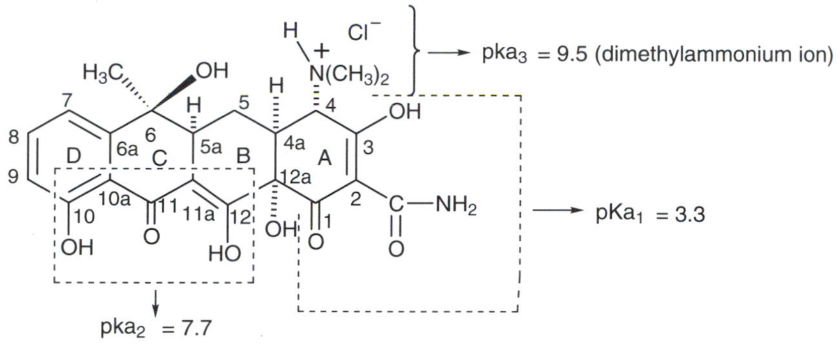

Figure 17.15 Chemical structure of tetracycline hydrochloride.

value, of 5.0. The measured $pK_a$ values of, for example, tetracycline are 3.3, 7.7, and 9.5 (Fig. 17.15).

The tetracyclines are strong chelating agents. The A ring and the keto-enol system at positions C-11 and C-12 are active sites for chelation. This ability to chelate divalent and trivalent metal ions such as calcium results in tooth discoloration when tetracycline is administered to treat bacterial infections. Because of this, the tetracyclines should not be administered to pregnant women or children younger than 8 years.

The tetracycline molecule, as well as those that contain the 6β-hydroxy group, is labile to acid and base degradation. At pH ≤ 2.0, tetracycline eliminates a molecule of water with concomitant aromatization of ring C to form anhydrotetracycline (Fig. 17.16). In basic medium, ring C of tetracycline is opened to form isotetracycline. Also at pH values between 2.0 and 6.0, epimerization of the "natural" C-4α dimethylamine group to the C-4β epimer occurs. Under acidic conditions, a 1:2 equilibrium is established in solution within a day (Fig. 17.17). This instability leads to a decrease in antibacterial activity, to which all clinically used tetracyclines are susceptible. Indeed, these chemical transformations contribute to the chemical instability of the tetracyclines under several reaction conditions.

## Antibacterial Activity and Clinical Uses

The tetracyclines, discovered between 1948 and 1957, were the first broad-spectrum antibiotics introduced into clinical practice. Their spectrum of activity includes gram-positive and gram-negative bacteria, mycobacteria, mycoplasmas, chlamydias, rickettsias, and even some protozoa causing malaria.

Generally, doxycycline and minocycline are more active against *S. aureus* and various streptococci than is tetracycline. These tetracyclines are active against *Streptococcus pneumoniae*. The tetracyclines are active against *Neisseria* spp., including both *N. gonorrhoeae* and *N. meningitidis*, although tetracyline-resistant *N. gonorrhoeae* strains are being identified with increasing frequency.

Among the common gram-negative bacilli, *E. coli*, *Klebsiella* spp., and *Enterobacter* spp. are susceptible to

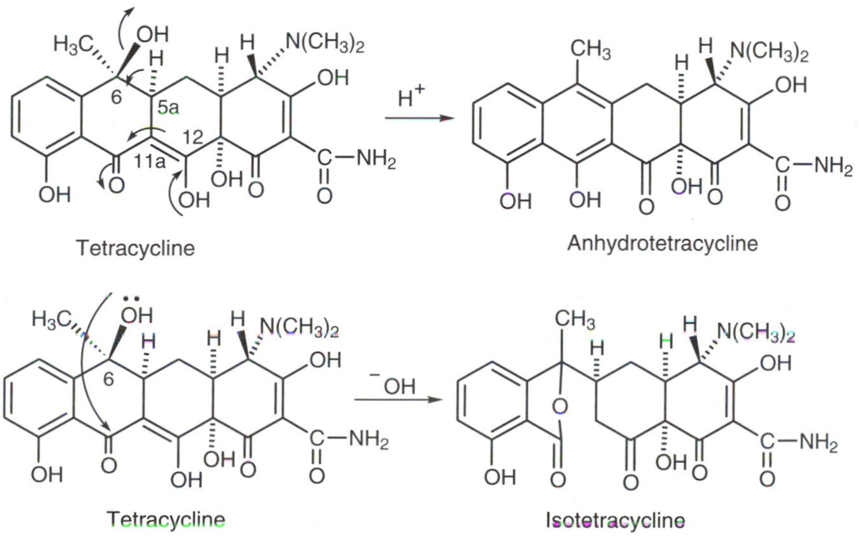

Figure 17.16 Chemical instability of tetracycline under acidic and basic conditions. Note that the 6β-hydroxy group participates in these degradation reactions.

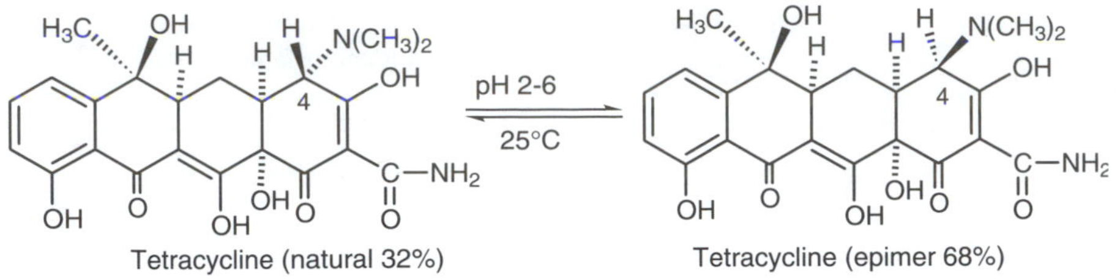

**Figure 17.17** Epimerization of the tetracycline subsituents at C-4 in acidic medium (pH 2 to 6).

Tetracycline (natural 32%) ⇌ (pH 2-6, 25°C) Tetracycline (epimer 68%)

tetracyclines; however, many strains are resistant, especially in hospital environments. *Proteus mirabilis* and *P. aeruginosa* are resistant.

The broad spectrum of action of the tetracyclines is reflected in their effectiveness against several protozoa, including *Entamoeba histolytica* and *Plasmodium falciparum.*

With the advent of highly specific β-lactam antibiotics and the increasing resistance to tetracyclines, the tetracyclines are no longer first-choice antibiotics. They remain important only for particular indications, such as respiratory infections, rickettsial infections, and some gram-negative bacterial infections. Tetracycline is given orally in low doses for the treatment of chronic severe acne vulgaris (an inflammatory lesion of the pilosebaceous follicle). The bacteria isolated from acne lesions are *Propionibacterium acnes, Staphylococcus epidermidis,* and *Pityrosporum ovale,* organisms which are members of the normal flora of the skin.

## Mechanism of Action

At the concentrations achieved in blood during antibacterial therapy, the tetracyclines are bacteriostatic. They bind primarily to the 30S ribosomal subunit, where they inhibit protein synthesis by blocking the binding of aminoacylated tRNA to the A site. They also prevent the binding of both release factors RF-1 and RF-2 during termination, regardless of the stop codon.

Structural information about ribosomal antibiotic-binding sites was first obtained by nuclear magnetic resonance (NMR) spectroscopy of complexes of antibiotic agents with fragments of 16S RNA. In 2000, the Ramakrishnan group from the MRC Laboratory of Molecular Biology (Cambridge, United Kingdom) determined the three-dimensional structure of the 30S ribosomal subunit from *Thermus thermophilus* complexed with tetracycline at 3.3 Å resolution. It was found that tetracycline has a flat fused-ring system with hydrophilic functional groups along one side. The molecule is thus able to make charged interactions with one edge and either hydrophobic or stacking interactions with the other. The scientists who did this work found two binding sites for tetracycline within the small ribosomal subunit. The better occupied site is located near the acceptor site for aminoacylated tRNA between the head and the body of the 30S subunit (the A site), and the less occupied site is at the interface between three RNA domains in the body of the subunit. They also found that there is little or no overall change in the conformation of the 16S RNA on binding of tetracycline to the ribosome.

In 2001, a group of scientists from the Max Planck Institute (Hamburg, Germany), the Weizmann Institute (Rehovot, Israel), and the Universidad de Carabobo (Valencia, Venezuela) determined and refined the 3.2-Å structure of the 30S subunit of *T. thermophilus* and determined and analyzed by X-ray crystallography the structure of the complex with tetracycline. These scientists identified six binding sites for tetracycline in the 30S subunit but did not find common structural traits among the binding sites. Then they investigated which sites could affect translation and to what extent they are involved in the inhibitory action of tetracycline. They found that Tet-1, the site with the highest relative occupancy, interferes with the location where the A site tRNA docks onto the 30S structure. Therefore, they suggest that tetracycline can physically prevent the binding of the tRNA to the A site. They think that this mode of interaction is consistent with the classical model of tetracycline as an inhibitor of A-site occupation and hence offers a clear explanation for the bacteriostatic effect of tetracycline. They think that the presence of five additional binding sites and the biochemical evidence for different locations of tetracycline and the low level of resistance conferred by the ribosomal protection proteins will demand more complex explanations.

Readers interested in a more comprehensive account of these works should consult the references.

## Transport through Membranes

Since the target site for tetracyclines is the 30S ribosomal subunit, the antibiotic molecules must cross either one or two membranes, depending on whether the target is in gram-positive or gram-negative bacteria.

### UPTAKE ACROSS THE GRAM-NEGATIVE OUTER MEMBRANE

Passage of tetracyclines across the outer membrane of *E. coli* involves passive diffusion through outer membrane porins. Since OmpF porins are cationic selective, tetracycline uptake occurs preferentially through this type of porins.

UPTAKE ACROSS THE BACTERIAL CYTOPLASMIC MEMBRANE

Some authors suggested that the passage of tetracycline across the bacterial cytoplasmic membrane occurs by an energy-dependent uptake system that acts in response to a pH gradient.

## Bacterial Resistance

Bacterial resistance to clinically useful tetracyclines is predominantly due to acquired resistance, i.e., when resistant strains emerge from previously sensitive bacterial population by acquisition of resistance genes, usually residing in plasmids and/or transposons. In essence, bacterial resistance results from the selective pressure exerted on bacteria during the administration of tetracyclines for chemotherapy.

Resistance to tetracycline may be mediated by one of three different mechanisms: (i) an energy-dependent efflux of tetracyclines carried out by transmembrane spanning proteins, which results in reduction of the concentration of tetracycline in the cytosol; (ii) ribosomal protection, whereby the tetracyclines no longer bind productively to the bacterial ribosome; or (iii) chemical modification, requiring oxygen and NADPH and catalysis by enzymes.

Efflux and ribosomal protection, mediated by plasmid or chromosomal determinants, are the two major mechanisms of bacterial resistance of clinical significance. Since the chemical alteration mechanism occurs rarely, this discussion focuses on the major mechanisms of tetracycline resistance.

Stuart B. Levy and coworkers had given the following definition of a tetracycline resistance determinant: "a tetracycline resistance determinant can be defined as a naturally occurring, generally contiguous genetic unit which includes all genes (both structural and regulatory) involved in resistance." They also proposed in 1989, and updated in 1999, a nomenclature for tetracycline resistance determinants.

A list of tetracycline resistance determinants is given in Table 17.3.

**Table 17.3** Classification of tetracycline resistance determinants according to their mechanism of resistance[a]

| Mechanism | |
|---|---|
| **Efflux** | **Ribosomal** |
| Tet(A) | Tet(M) |
| Tet(P) | Tet(O) |
| Tet(V) | Tet(Q) |
| Tet(Y) | Tet(S) |
| Tet(Z) | Tet(T) |
| *otraB* | Tet(W) |
| *tcr 3 (tcrC)* | *otraA* |
| Tet30 | *tet* |

[a] Based on information from Levy et al. (1999).

EFFLUX OF TETRACYCLINES

A brief presentation of tetracycline efflux is provided here. Please refer to chapter 5 for a more comprehensive description of this mechanism of resistance.

Levy and coworkers discovered that active efflux is a major mechanism of tetracycline resistance in bacteria. They demonstrated that tetracycline-resistant cells lose the accumulated drug faster than susceptible cells do and that tetracycline enters the bacterial cell by an energy-dependent process. The determinants which confer resistance by removing tetracycline from the cytosol encode transporter proteins located in the cytoplasmic membrane. These proteins mediate energy-dependent efflux of the tetracyclines (Fig. 17.18).

RIBOSOMAL PROTECTION BY SOLUBLE PROTEINS

The ribosomal protection proteins are all polypeptides of approximately 72.5 kDa. These proteins interact with the

**Figure 17.18** Diagrammatic representation of tetracycline accumulation in sensitive (top) and resistant (bottom) bacterial cells. Sensitive cells show a net active uptake, while resistant cells show a net active efflux. T, tetracycline. Reprinted from L. E. Bryan (ed.), *Antimicrobial Drug Resistance* (Academic Press, Inc., New York, N.Y., 1984), with permission from the publisher.

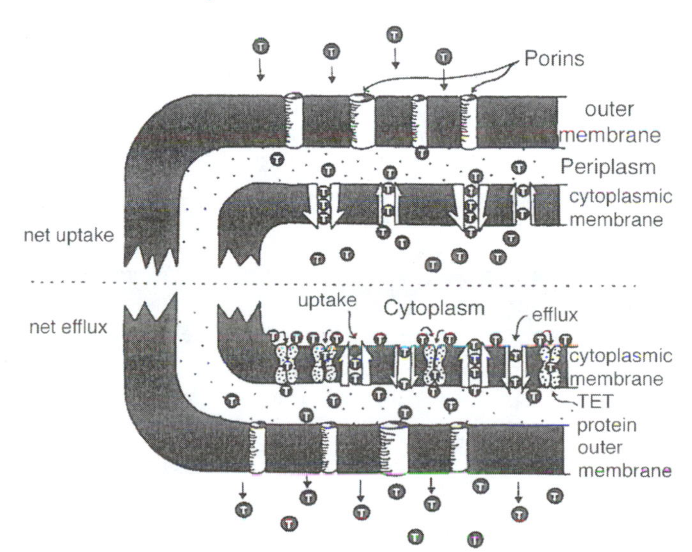

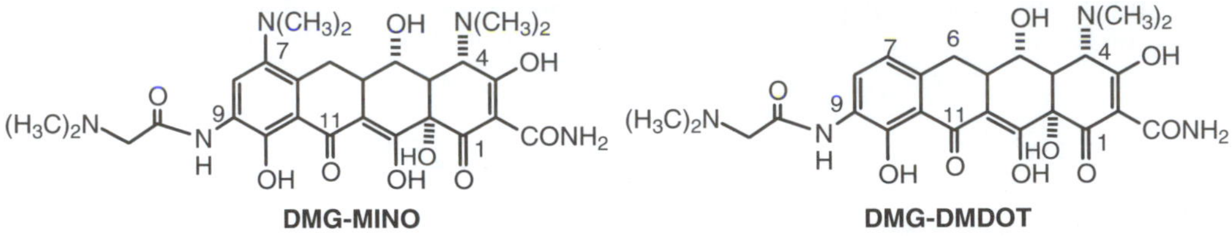

**Figure 17.19** Chemical structure of *N,N*-dimethylglycylamido derivatives of 9-aminominocycline (DMG-MINO) and 9-amino-6-demethyl-6-deoxytetracycline (DMG-DMDOT).

ribosome, making it insensitive to tetracycline inhibition. The exact mode of interaction of these proteins with the ribosomes is not well understood. Several classes of ribosome protection resistance genes have been characterized, and *tet(M)*, *tet(O)*, and *tet(Q)* have been sequenced.

## Development of Novel Semisynthetic Tetracyclines

Since the introduction of the semisynthetic tetracycline minocycline in 1972, numerous natural and semisynthetic tetracyclines have been reported, but no new tetracyclines have been marketed since that time. In 1988, Lederle Laboratories initiated a program to develop new tetracyclines active against tetracycline-resistant bacteria. The first major breakthrough came with the reported synthesis and antibacterial activity of the *N,N*-dimethylglycylamido derivatives of 9-aminominocycline (DMG-MINO) and 9-amino-6-demethyl-6-deoxytetracycline (DMG-DMDOT) (Fig. 17.19). These compounds are called the glycylcyclines.

Both DMG-MINO and DMG-DMDOT showed excellent activity against strains resistant to tetracycline. They bound to the high-affinity tetracycline ribosomal binding site and evaded Tet(M)- and Tet(O)-mediated ribosomal protection and the effect of the efflux pumps of the Tet(A), Tet(B), Tet(C), Tet(D), and Tet(K) proteins, as well as others recently characterized. Thus, modification at position 9 of ring D proved to be effective in overcoming two distinct resistance mechanisms.

## Tetracycline Preparations Commercially Available in the United States

Table 17.4 lists the tetracyclines marketed in the United States.

**Table 17.4** List of generic and common trade names of tetracyclines, the preparations available, and manufacturers in the United States

| Generic name | Trade name (preparation) | Manufacturer |
|---|---|---|
| Demeclocycline hydrochloride | Declomycin (tablets) | Lederle |
| Doxycycline hyclate | Dory (capsules) | Warner Chilcott |
| | Periostat (capsules) | CollaGenex |
| | Vibramycin (capsules) | Pfizer |
| | Vibra-Tabs (coated tablets) | Pfizer |
| Doxycycline monohydrate | Monodox (capsules) | Oclassen |
| | Vibramycin monohydrate (oral suspension) | Pfizer |
| Doxycycline calcium | Vibramycin calcium (oral suspension) | Pfizer |
| Minocycline hydrochloride | Minocin (intravenous, capsules, and oral suspension) | Lederle |
| Oxytetracycline | Terramycin (intramuscular solution) | Pfizer |
| Oxytetracycline plus polymyxin B sulfate | Terramycin (ophthalmic ointment) | Pfizer |
| Oxytetracycline plus hydrocortisone acetate | Terra-Cortril (ophthalmic suspension) | Pfizer |
| Tetracycline chlorhydrate | Achromycin V (capsules) | Lederle |
| | Sumycin (capsules, tablets) | Apothecon |
| | Tetracycline hydrochloride (capsules) | Mylan |

## REFERENCES AND FURTHER READING

### Aminoglycosides


**Structure, functions, and resistance**

**Wright, G. D., A. M. Berghuis, and S. Mobashery.** 1998. Aminoglycoside antibiotics, p. 27–69. *In* B. P. Rosen and S. Mobashery (ed.), *Resolving the Antibiotic Paradox*. Kluwer Academic/Plenum Publishers, New York, N.Y.

**Mechanism of action**

**Carter, A. P., W. M. Clemons, D. E. Brodersen, R. J. Morgan-Warren, B. T. Wimberly, and V. Ramakrishnan.** 2000. Functional insights from the structure of the 30S ribosomal subunit and its interactions with antibiotics. *Nature* **407:**340–348.

**Fourmy, D., M. I. Recht, S. C. Blanchard, and J. D. Puglisi.** 1996. Structure of the A site of *Escherichia coli* 16S ribosomal RNA complexed with an aminoglycoside antibiotic. *Science* **274:**1367–1372.

**Fourmy, D., M. I. Recht, and J. D. Puglisi.** 1998. Binding of neomycin-class aminoglycoside antibiotics to the A-site of 16S rRNA. *J. Mol. Biol.* **277:**347–362.

**Fourmy, D., S. Yoshizawa, and J. D. Puglisi.** 1998. Paromomycin binding induces a local conformational change in the A-site of 16S rRNA. *J. Mol. Biol.* **277:**333–345.

**Kaul, M., and D. S. Pilch.** 2002. Thermodynamics of aminoglycosides-rRNA recognition: the binding of neomycin-class aminoglycosides to the A site of 16S rRNA. *Biochemistry* **41:**7695–7706.

**Kotra, L. P., J. Haddad, and S. Mobashery.** 2000. Aminoglycosides: perspectives on mechanisms of action and resistance and strategies to counter resistance. *Antimicrob. Agents Chemother.* **44:**3249–3256.

**Mehta, R., and W. S. Champney.** 2002. 30S ribosomal subunit assembly is a target for inhibition by aminoglycosides in *Escherichia coli*. *Antimicrob. Agents Chemother.* **46:**1546–1549.

**Moazed, D., and H. F. Noller.** 1987. Interaction of antibiotics with functional sites in 16S ribosomal RNA. *Nature* **327:**389–394.

**Ryu, D. H., A. Litovchick, and R. R. Rando.** 2002. Stereospecificity of aminoglycoside-ribosomal interactions. *Biochemistry* **41:**10499–10509.

**Wang, H., and Y. Tor.** 1998. RNA-aminoglycoside interactions: design, synthesis, and binding of "amino-aminoglycosides" to RNA. *Angew. Chem. Int. Ed.* **37:**109–111.

**Wong, C. H., M. Hendrix, E. S. Priestley, and W. A. Greenberg.** 1998. Specificity of aminoglycoside antibiotics for the A-site of the decoding region of ribosomal RNA. *Chem. Biol.* **5:**397–406.

**Yoshizawa, S., D. Fourmy, and J. D. Puglisi.** 1998. Structural origins of gentamicin antibiotic action. *EMBO J.* **17:**6437–6448.

**Bacterial resistance**

**Davies, J., and G. D. Wright.** 1997. Bacterial resistance to aminoglycoside antibiotics. *Trends Microbiol.* **5:**234–240.

**Mingeot-Leclercq, M. P., Y. Glupczynski, and P. M. Tulkens.** 1999. Aminoglycosides: activity and resistance. *Antimicrob. Agents Chemother.* **43:**727–737.

**Shaw, K. J., P. N. Rather, R. S. Hare, and G. M. Miller.** 1993. Molecular genetics of aminoglycoside resistance genes and familial relationships of the aminoglycoside-modifying enzymes. *Microbiol. Rev.* **57:**138–163.

**Aminoglycoside-modifying enzymes**

**Azucena, E., and S. Mobashery.** 2001. Aminoglycoside-modifying enzymes: mechanism of catalytic processes and inhibition. *Drug Resist. Updates* **4:**106–117.

**Wright, G. D.** 1999. Aminoglycoside-modifing enzymes. *Curr. Opin. Microbiol.* **2:**499–503.

**Bacterial uptake**

**Chung, L., G. Kaloyanides, R. McDaniel, A. McLaughlin, and S. McLaughlin.** 1985. Interaction of gentamicin and spermine with bilayer membranes containing negatively-charged phospholipids. *Biochemistry* **24:**442–452.

**Taber, H. W., J. P. Mueller, P. F. Miller, and A. S. Arrow.** 1987. Bacterial uptake of aminoglycoside antibiotics. *Microbiol. Rev.* **51:**439–457.

**Crystal structure**

**Hon, W. C., G. A. McKay, P. R. Thompson, R. M. Sweet, D. S. C. Yang, G. D. Wright, and A. M. Berghuis.** 1997. Structure of an enzyme required for aminoglycoside antibiotic resistance reveals homology to eukaryotic protein kinases. *Cell* **89:**887–895.

**Pedersen, L. C., M. M. Benning, and H. M. Holden.** 1995. Structural investigation of the antibiotic and ATP-binding sites in kanamycin nucleotidyltransferase. *Biochemistry* **34:** 13305–13311.

**Sakon, J., H. H. Liao, A. M. Kanikula, M. M. Benning, I. Rayment, and H. M. Holden.** 1993. Molecular structure of kanamycin nucleotidyl transferase determined to 3 Å resolution. *Biochemistry* **32:**11977–11984.

**Vetting, M. W., S. S. Hegde, F. Javid-Majd, J. S. Blanchard, and S. L. Roderick.** 2002. Aminoglycoside 2'-*N*-acetyltransferase from *Mycobacterium tuberculosis* in complex with coenzyme A and aminoglycosides substrates. *Nat. Struct. Biol.* **9:** 653–658.

**Wolf, E., A. Vassilev, Y. Makino, A. Sali, Y. Nakatani, and S. K. Burley.** 1998. Crystal structure of a GCN5-related *N*-acetyltransferase: *Serratia marcescens* aminoglycoside 3-*N*-acetyltransferase *Cell* **94:**439–449.

**Wybenga-Groot, L. E., K. Draker, G. D. Wright, and A. M. Berghuis.** 1999. Crystal structure of an aminoglycoside 6'-*N*-acetyltransferase: defining the GCN5-related *N*-acetyltransferase superfamily fold. *Structure* **7:**497–507.

**Toxicity**

**Mingeot-Leclercq, M. P., and P. M. Tulkens.** 1999. Aminoglycosides: nephrotoxicity. *Antimicrob. Agents Chemother.* **43:** 1003–1012.

**Clinical uses**

**American Medical Association.** 1996. *Drug Evaluations Annual 1995,* p. 1537–1558. American Medical Association, Chicago, Ill.

**Dworkin, R. J.** 1999. Aminoglycosides for the treatment of gram-negative infections: therapeutic use, resistance and future outlook. *Drug Resist. Updates* **2:**173–179.

**Kucers, A., S. M. Crowe, M. L. Grayson, and J. F. Hoy.** 1997. *The Use of Antibiotics,* 5th ed., p. 428–541. Butterworth-Heinemann, Oxford, United Kingdom.

**Maurin, M., and D. Raoult.** 2001. Use of aminoglycosides in treatment of infection due to intracellular bacteria. *Antimicrob. Agents Chemother.* **45:**2977–2986.

**Reese, R. E., R. F. Betts, and B. Gumustop.** 2000. *Handbook of Antibiotics,* 3rd ed., p. 415–434. Lippincott Williams & Wilkins, Philadelphia, Pa.

**Isepamicin**

**Jones, R. N.** 1995. Isepamicin (SCH 21420, 1-N-HAPA gentamicin B): microbiological characteristics including antimicrobial potency and spectrum of activity. *J. Chemother.* **7**(Suppl. 2)**:**7–16.

**Miller, G. H., F. J. Sabatelli, L. Naples, R. S. Hare, and K. J. Shaw.** 1995. The changing nature of aminoglycoside resistance mechanisms and the role of isepamicin. A new broad spectrum aminoglycoside. *J. Chemother.* **7**(Suppl. 2)**:**31–44.

**Arbekacin**

**Hotta, K., C. B. Zhu, and T. Ogata.** 1996. Enzymatic 2'-*N*-acetylation of arbekacin and antibiotic activity of its product. *J. Antibiot.* **49:**458–464.

**Lambert, T., G. Gerbaud, M. Galimand, and P. Courvalin.** 1993. Characterization of *Acinetobacter haemolyticus aac*(6')-*Ig* gene encoding an aminoglycoside 6'-*N*-acetyltransferase which modifies amikacin. *Antimicrob. Agents Chemother.* **37:**2093–2100.

## Tetracyclines

**Chopra, I., P. M. Hawkey, and M. Hinton.** 1992. Tetracyclines, molecular and clinical aspects. *J. Antimicrob. Chemother.* **29:**245–277.

**Chopra, I., and M. Roberts.** 2001. Tetracycline antibiotics: mode of action, applications, molecular biology, and epidemiology of bacterial resistance. *Microbiol. Mol. Biol. Rev.* **65:**232–260.

**Hlavka, J. J., G. A. Ellestad, and I. Chopra.** 1993. Tetracyclines, p. 562–577. *In* M. Howe-Grant (ed.), *Chemotherapeutics and Disease Control.* John Wiley & Sons, Inc., New York, N.Y.

**Mechanism of action**

**Brodersen, D. E., W. M. Clemons, A. P. Carter, R. J. Morgan-Warren, B. T. Wimberly, and V. Ramakrishnan.** 2000. The structural basis for the action of the antibiotics tetracycline, pactamycin, and hygromycin B on the 30S ribosomal subunit. *Cell* **103:**1143–1154.

**Pioletti, M., F. Schlünzen, J. Harms, R. Zarivach, M. Glühmann, H. Avila, A. Bashan, H. Bartels, T. Auerbach, C. Jacobi, T. Hartsch, A. Yonat, and F. Franceschi.** 2001. Crystal structures of complexes of the small ribosomal unit with tetracycline, edeine and IF3. *EMBO J.* **20:**1829–1839.

**Nomenclature for tetracycline resistance determinants**

**Levy, S. B., L. M. McMurry, T. M. Barbosa, V. Burdett, P. Courvalin, W. Hillen, M. C. Roberts, J. I. Rood, and D. E. Taylor.** 1999. Nomenclature for new tetracycline resistance determinants. *Antimicrob. Agents Chemother.* **40:**1–5.

**Levy, S. B., L. M. McMurry, V. Burdett, P. Courvalin, W. Hillen, M. C. Roberts, and D. E. Taylor.** 1989. Nomenclature for tetracycline resistance determinants. *Antimicrob. Agents Chemother.* **33:**1373–1374.

**Bacterial resistance**

**Roberts, M. C.** 1994. Epidemiology of tetracycline-resistance determinants. *Trends Microbiol.* **2:**353–357.

**Speer, B. S., N. B. Shoemaker, and A. A. Salyers.** 1992. Bacterial resistance to tetracycline: mechanisms, transfer, and clinical significance. *Clin. Microbiol. Rev.* **5:**387–399.

**Taylor, D. E., and A. Chau.** 1996. Tetracycline resistance mediated by ribosomal protection. *Antimicrob. Agents Chemother.* **33:**1373–1374.

**Active efflux**

**Borges-Walmsley, M. I., and A. R. Walmsley.** 2001. The structure and function of drug pumps. *Trends Microbiol.* **9:**71–79.

**Koronakis, V., A. Sharff, E. Koronakis, B. Luisi, and C. Hughes.** 2000. Crystal structure of the bacterial membrane protein TolC central to multidrug efflux and protein export. *Nature* **405:**914–919.

**Levy, S. B** 1992. Active efflux mechanisms for antimicrobial resistance. *Antimicrob. Agents Chemother.* **36:**695–703.

**Clinical uses**

**American Medical Association.** 1996. *Drug Evaluations Annual, 1995,* p. 1519–1528. American Medical Association, Chicago, Ill.

**Kucers, A., S. M. Grove, M. L. Grayson, and J. F. Hoy.** 1997. *The Use of Antibiotics,* 5th ed., p. 719–762. Butterworth-Heinemann, Oxford, United Kingdom.

**Scholar, E. M., and W. B. Pratt.** 2000. *The Antimicrobial Drugs,* p. 184–197. Oxford University Press, Oxford, United Kingdom.

**Glycylcyclines**

**Bergeron, J., M. Ammirati, D. Danley, L. James, M. Norcia, J. Retsema, C. A. Strick, W. G. Su, J. Sutcliffe, and L. Wondrack.** 1996. Glycylcyclines bind to the high-affinity tetracycline ribosomal binding site and evade Tet(M)- and Tet(O)-mediated ribosomal protection. *Antimicrob. Agents Chemother.* **40:**2226–2228.

**Sum, P. E., V. J. Lee, R. T. Testa, J. J. Hlavka, G. A. Ellestad, J. D. Bloom, Y. Gluzman, and F. P. Tally.** 1994. Glycylcyclines. 1. A new generation of potent antibacterial agents through modifications of 9-aminotetracyclines. *J. Med. Chem.* **37:**184–188.

**Tally, F. P., G. A. Ellestad, and R. T. Testa.** 1995. Glycylcyclines: a new generation of tetracyclines. *J. Antimicrob. Chemother.* **35:**449–452.

chapter 18



# Inhibitors of the 50S Ribosomal Subunit
## Macrolides, Lincosamides, and Streptogramins

## Macrolides

### Structure and Classification

The term **macrolide** is derived from two words, "macro-," meaning large, and "-olide," meaning a lactone. All these compounds are chemically characterized by a macrocyclic lactone (the aglycone) glycosidically linked to one or more amino monosaccharides. The macrolides are classified according to the number of atoms comprising the lactone ring in 12-, 14-, 15-, or 16-member macrolides (Fig. 18.1), each of which has both chemical and biological differentiating characteristics.

The 12-member ring macrolides are not included in Figure 18.1 because they are not represented by drugs used in clinical practice in the United States. The 14-member ring group is composed of compounds of natural origin (e.g., erythromycin and oleandomycin) and semisynthetic derivatives (e.g., roxithromycin, dirithromycin, and clarithromycin). The 15-member ring is represented by azithromycin. This semisynthetic derivative was obtained by a Beckman rearrangement of the 9-oxime of erythromycin A that produces a tertiary amine (a nitrogen atom substituted by a methyl group); therefore, this reaction expands the ring from 14 members to produce a 15-member ring. This macrocycle is now named **azalide.** The 16-member ring group also contains both compounds of natural origin (e.g., josamycin, spiramycin, and midecamycin) and semisynthetic derivatives (e.g., rokitamycin and miokamycin).

The natural macrolide antibiotics are isolated primarily from the genus *Streptomyces.* They are characterized by having antibacterial activity mostly against gram-positive bacteria. The macrolide antibiotics currently used in the United States are erythromycin and the semisynthetic derivatives of erythromycin A, i.e., clarithromycin, azithromycin, and dirithromycin.

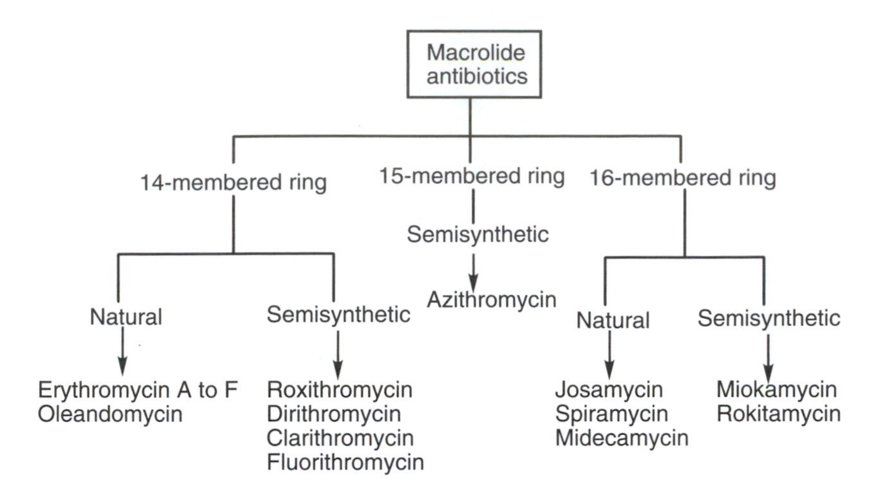


**Figure 18.1** Macrolide antibiotics subgrouped according to the size of the aglycon macrocyclic lactone ring. The 15-membered ring macrolides are named azalide. Taken from A. C. Bryskier and A. Denis, *in* W. Schönfeld and H. A. Kirst (ed.), *Macrolide Antibiotics* (Birkhaüser Verlag, Basel, Switzerland, 2002).

## Historical Perspective on Research to Develop Novel Macrolides

For over 40 years, the macrolide antibiotics have been widely used to treat various infectious conditions. They have been particularly useful as treatment for patients who are allergic to penicillins; they are also effective against pneumococcal, streptococcal, and mycoplasmal infections and hence are clinically useful for the treatment of upper and lower respiractory tract infections. They are also the drugs of choice for *Legionella* and *Chlamydia* infections.

Macrolide research has been through three stages. The first, in the 1950s and 1960s, involved the discovery and development of many macrolides from natural sources; the most important of these was erythromycin A. The macrolide antibiotics are produced as secondary metabolites of soil microorganisms, and the majority have been produced by various strains of *Streptomyces*.

The second stage was the semisynthesis of novel derivatives of erythromycin A to overcome the acid lability of erythromycin A. This led to the development of azithromycin, clarithromycin, and dirithromycin, which have improved pharmacokinetic properties relative to erythromycin, particularly stability to acid and high concentration in tissues.

The last stage started in the late 1990s and still is going on; in this stage, new macrolides have been developed or are under development to overcome erythromycin A resistance, as well as cross-resistance with the new macrolides that has arisen within gram-positive bacteria such as *Streptococcus pneumoniae, S. pyogenes,* and the viridans group of streptococci. So far, the most successful modification of erythromycin A has been the removal of the neutral sugar L-cladinose and subsequent oxidation of the 3-hydroxyl group to form a 3-keto function. This chemical transformation, together with the introduction of a cyclic carbamate at C-11 to C-12 has yielded several ketolides currently undergoing clinical trial. The chemical, biochemical, and microbiological aspects of ketolides are discussed below.

## 14- and 15-Member Ring Macrolides

### ERYTHROMYCIN A

Erythromycins are produced as a complex of six components (A to F) by *Saccharopolyspora erythraea* (formerly *Streptomyces erythraeus*). They were discovered in 1952 at Eli Lilly Laboratories. Only erythromycin A has been developed for clinical use.

The chemical structure of erythromycin A (henceforth called erythromycin) is made up of the aglycone erythronolide A, to which the dimethylamino monosaccharide D-desosamine is linked through a β-glycosidic bond at position 5 and the neutral monosaccharide L-cladinose is linked through an α-bond at position 3 (Fig. 18.2). The complete stereochemistry of erythromycin was established by X-ray analysis of its hydroiodide dihydrate.

Erythromycin is inactivated by gastric acids, and it is thus administered as enteric-coated tablets or as capsules containing enteric-coated pellets that dissolve and are absorbed in the small intestine. The basic dimethylamino group of the D-desosamine moiety present in erythromycin forms salts with acids. Erythromycin stearate is a preparation used for oral administration. Erythromycin ethylsuccinate is an ester of erythromycin suitable for oral administration.

#### Instability of Erythromycin under Acidic Conditions

The erythromycin base itself is orally absorbed; however, as mentioned above, it must be enteric coated since it has poor acid stability due to intramolecular hemiketal and

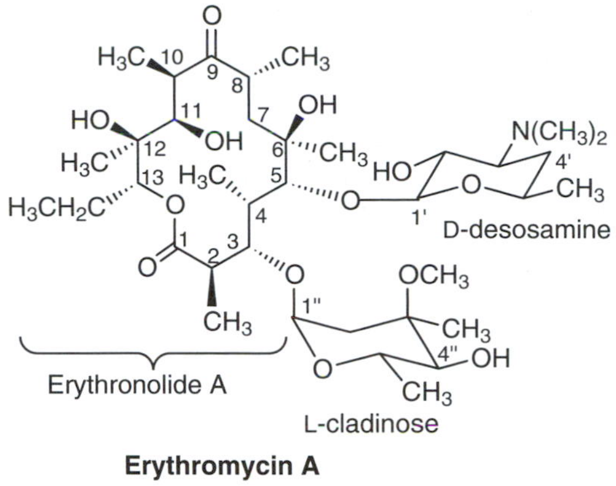

**Figure 18.2** Chemical structure of erythromycin A.

spiroketal formation between the 9-keto group and the 6- and 12-hydroxyl groups to generate erythromycin 8,9-anhydro-6,9-hemiketal and erythromycin 6,9,9,12-spiroketal as shown in Fig. 18.3. These two degradation products have no antibacterial activity.

Since the intramolecular cyclization of erythromycin inactivates its antibiotic activity, much effort has been spent in diminishing or completely blocking this degradation pathway. One approach to preventing this decomposition has utilized water-insoluble, acid-stable salts, esters and/or pharmaceutical formulations to protect erythromycin during its passage through the stomach. In the United States, the stearate salt and the ethylsuccinate ester of erythromycin are commercially available (Fig. 18.4). The dimethylammonium erythromycin stearate salt is formed by stearic acid (octadecanoic acid) and the dimethylamino group at C-3' of the D-desosamine sugar. The stearate is a water-insoluble salt that dissociates in the intestine, and the erythromycin base is absorbed.

By esterifying the 2'-OH group of the D-desosamine sugar with 3-chloroformylpropionate [$ClC(O)CH_2CH_2CO_2CH_2CH_3$], erythromycin ethylsuccinate is formed. Because it is tasteless, it is formulated in flavored pediatric suspensions. This ester derivative is better absorbed than the erythromycin base and is hydrolyzed to the free erythromycin by esterases in the blood. The ethylsuccinate of erythromycin does not have significant antibacterial activity until it is hydrolyzed to the erythromycin base.

## Spectrum of Antimicrobial Activity of Erythromycin

Erythromycin has a relatively broad spectrum and is active in vitro against most gram-positive and some gram-negative bacteria, mycoplasmas, chlamydiae, treponemas, and rickettsias. It is usually active against the following organisms in vitro and in clinical infections.

### *Gram-Positive Aerobic Bacteria*

- Erythromycin is highly active against group A streptococci.
- Penicillin-susceptible *Streptococcus pneumoniae* is susceptible to erythromycin, but many penicillin-resistant strains are also resistant to macrolides.

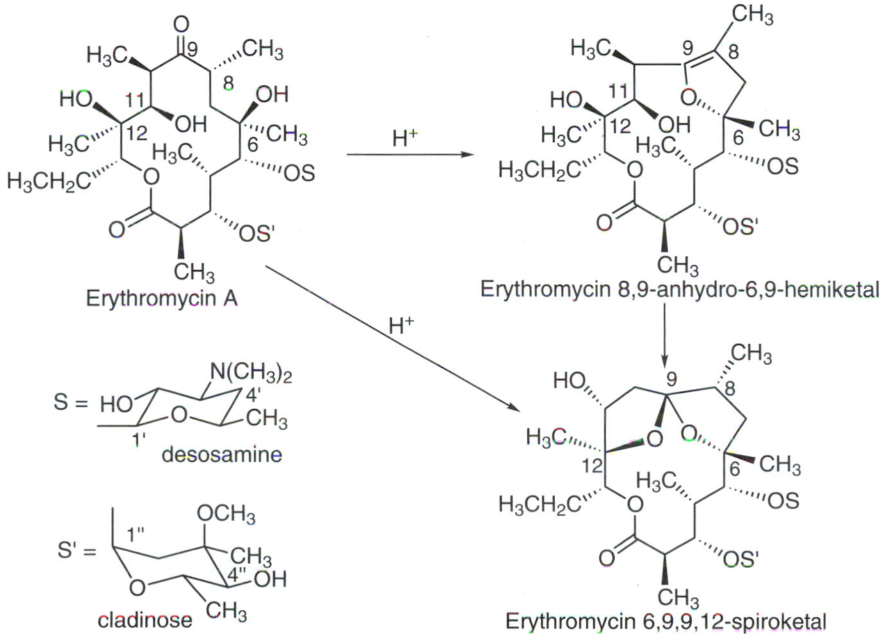

**Figure 18.3** Decomposition of erythromycin A under acidic conditions generates erythromycin 8,9-anhydro-6,9-hemiketal and erythromycin 6,9,9,12-spiroketal.

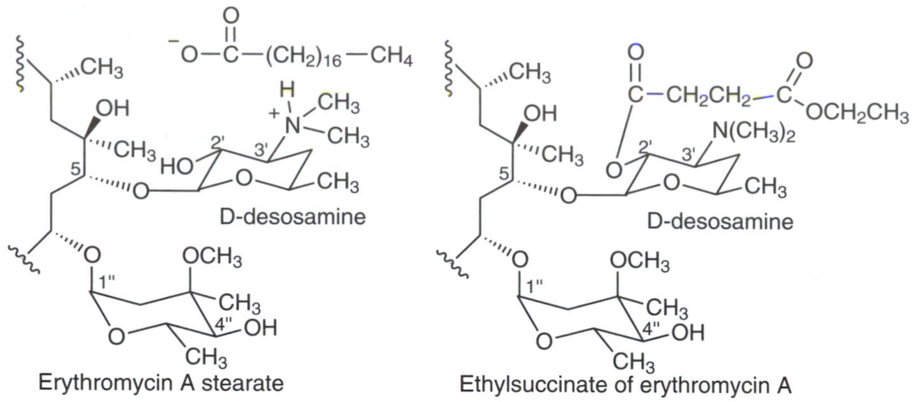

**Figure 18.4** Chemical structures of erythromycin A stearate and the ethylsuccinate ester of erythromycin A.

- Many other streptococci are susceptible (groups B, C, G, etc.), but group D (enterococci) is resistant.
- Methicillin-sensitive *Staphylococcus aureus* is often susceptible, but resistance may emerge during therapy. Methicillin-resistant *S. aureus* (MRSA) strains are resistant to macrolides.
- Erythromycin is the drug of choice for *Corynebacterium diphtheriae* and an alternative agent for *Bacillus anthracis* (the bacterium that causes anthrax).

*Gram-Negative Aerobic Bacteria*

- Erythromycin is the agent of first choice for *Campylobacter jejuni*, *Bordetella pertussis* (the cause of whooping cough), *Legionella* spp., and the agent of bacillary angiomatosis (*Bartonella henselae* or *B. quintana*).
- Erythromycin is an alternative agent for *Haemophilus ducreyi* (the cause of chancroid), *Eikenella corrodens*, and *Moraxella catarrhalis.*
- Erythromycin has weak activity against *H. influenzae*, and many strains are resistant. Erythromycin is not viewed as an alternative agent for this pathogen, although clarithromycin and azithromycin are alternative agents for respiratory isolates.

The most obvious shortcoming in the spectrum of activity of macrolides is that many common gram-negative bacteria such as *Escherichia coli* and *Klebsiella*, *Salmonella*, and *Pseudomonas* species are not susceptible. It was demonstrated that the outer membrane of gram-negative bacteria limits the uptake of macrolides, thus causing an inherent resistance. In essence, the impermeability of the outer membrane prevents the macrolide from entering the cell and reaching its site of action, the 50S bacterial ribosomal subunit.

*Gram-Negative Anaerobic Bacteria.* Erythromycin has activity against some species of gram-negative anaerobes, but *Bacteroides fragilis* strains and *Fusobacterium* spp. usually are resistant.

*Other Pathogens*

1. **Mycoplasmas.** Erythromycin is very active against *Mycoplasma pneumoniae* and is a drug of choice for this pathogen as well as for *Ureaplasma urealyticum.*
2. **Chlamydiae.** Erythromycin is active against *Chlamydia trachomatis* and *C. pneumoniae.*
3. **Spirochetes.** Erythromycin is active against *Borrelia burgdorferi*, the causative agent of Lyme disease.

## Clinical Uses of Erythromycin

Erythromycin has been clinically used for more than 45 years and is considered a drug of choice for treatment of upper and lower respiratory tract infections. However, as mentioned above, it easily loses its antibacterial activity as a result of degradation under acidic conditions, and the degraded products are known to be responsible for undesirable gastrointestinal side effects. Several chemical modifications of erythromycin have been investigated to overcome this acid instability problem. As a result, clarithromycin and azithromycin are widely prescribed due to their efficacy and safety.

Erythromycin is used as a drug of choice in adults and children for treatment of the following conditions:

- respiratory tract infections caused by *Mycoplasma pneumoniae*
- *Legionella* infections (often combined with rifampin)
- *Chlamydia trachomatis* pneumonia, which is more common in children than adults

Erythromycin is used also for the treatment of conjunctivitis of the newborn, pneumonia of infancy, and urogenital infections during pregnancy.

Erythromycin is also used in:

- pertussis (whooping cough) caused by *Bordetella pertussis,* for both therapy and prophylaxis
- *Campylobacter jejuni* infections; *Corynebacterium diphtheriae* infections or carrier states; *Haemophilus ducreyi* (chancroid) genital lesions; infections by *Bartonella henselae,* the agent of bacillary angiomatosis; and *Ureaplasma urealyticum* infections (e.g., urethritis)

Erythromycin has also been used as an alternative in penicillin-allergic patients to treat the following conditions:

- skin and soft tissue infections of mild to moderate degree caused by *Streptococcus pyogenes* and *Staphylococcus aureus*
- upper and lower respiratory tract infections of mild to moderate degree caused by *Streptococcus pyogenes* and *Streptococcus pneumoniae*
- rheumatic fever prophylaxis

Several oral preparations are available: erythromycin base, stearate salt, and the ethylsuccinate ester. Erythromycin should not be administered intramuscularly. Intravenous preparations achieve appreciably higher levels in serum.

OLEANDOMYCIN

Oleandomycin is no longer marketed in the United States. Its triacetate, named troleandomycin, is marketed by Pfizer Laboratories under the trade name of TAO (capsules) and is used occasionally for the treatment of severe asthma in combination with methylprednisolone; however, it is not recommended for the treatmeant of infectious diseases since it offers no advantage over erythromycin.

## Chemical Modification at the Erythronolide A

Coming back to avoid the problem of the instability of erythromycin under acidic conditions, another strategy followed was the chemical modification of the functional groups involved in the intramolecular cyclization, specifically the C-9 ketone, the hydroxyl groups at positions 6 and 12, and the C-8 hydrogen atom.

### Roxithromycin

Roxithromycin, or 9[*O*-(2-methoxyethoxy)methyl]oxime, is a semisynthetic derivative of erythromycin A (Fig. 18.5). It was synthesized for the first time at Roussel-Uclaf in France and introduced into clinical practice in Europe in 1987. Roxithromycin has a pharmacological and antibacterial spectrum similar to that of erythromycin and a sixfold increase in potency over erythromycin after oral administration, which was attributed to the absence of inactivation by internal ketalization. Roxithromycin was marketed under the proprietary name of Rulid by Eli Lilly but currently is not available in the United States.

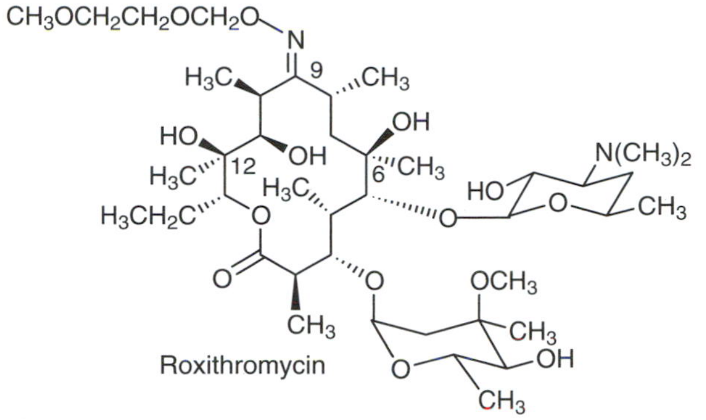

**Figure 18.5** Chemical structure of roxithromycin.

### Dirithromycin

Dirithromycin is a macrolide antibiotic for oral administration that was approved in June 1995 by the Food and Drug Administration (FDA) for use in the United States. It is marketed as Dynabac by Eli Lilly and distributed by Muro. The FDA-approved use of dirithromycin is for a 5-day regimen to treat patients with acute bacterial exacerbations due to *H. influenzae, Moraxella catarrhalis,* or *S. pneumoniae* and to treat patients with uncomplicated skin and skin structure infections caused by *S. aureus* (methicillin-susceptible strains) or *S. pyogenes* and for group A streptococcal pharyngitis.

Dirithromycin is an oxazine derivative of 9-(*S*)-erythromycylamine; it is a prodrug which, after absorption from the gastrointestinal tract, is rapidly converted by nonenzymatic hydrolysis to 9-(*S*)-erythromycylamine, the active compound. Originally, dirithromycin was selected by Boehringer-Ingelheim, but subsequent research into the microbiological, pharmacokinetic, and clinical properties of this macrolide were conducted at Eli Lilly & Co.

Reduction of the oximes and hydrazones of erythromycin A produces 9-(*S*)-erythromycylamine as the principal product, with minor amounts of the 9-(*R*) isomer. Preliminary clinical studies showed that 9-(*S*)-erythromycylamine was poorly absorbed in humans. Then a further modification by cyclocondensation of 9-(*S*)-erythromycylamine with 2-(2-methoxyethoxy)acetaldehyde diethylacetal gave the 9-*N*-11-*O*-oxazine derivative of erythromycin A (Fig. 18.6). On oral administration, dirithromycin is nonenzymatically hydrolyzed to ery-

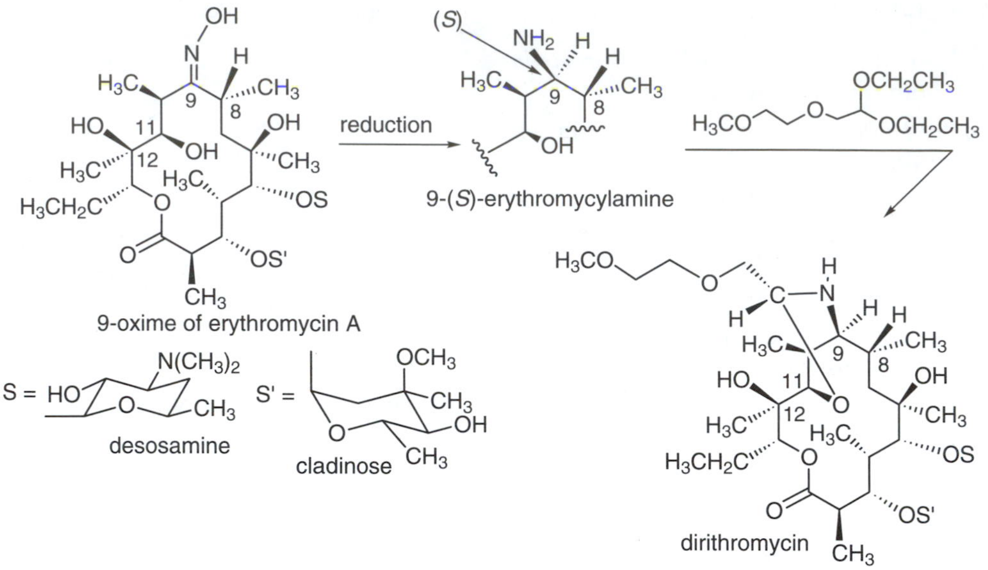

**Figure 18.6** Chemical synthesis of dirithromycin.

thromycylamine, which is microbiologically active. In vitro, dirithromycin is generally similar to erythromycin in its antibacterial activity. It offers the advantage of once-daily dosage.

## Azithromycin

A third approach to solve the problem of the instability of erythromycin under acidic conditions is a chemical transformation of the 9-oxime of erythromycin A to expand the 14-member ring to a 15-member ring by a Beckman rearrangement. This reaction is followed by reduction and N-methylation. These transformations yielded a novel 15-member ring containing a tertiary amine group. The introduction of this basic group converts to azithromycin with a dibasic structure (Fig. 18.7). Synthesis and evaluation of the antibacterial activity were done jointly by Pliva Pharmaceutical in Croatia and Pfizer Laboratories in the United States. The work was published in 1988.

**Figure 18.7** Chemical structure of azithromycin.

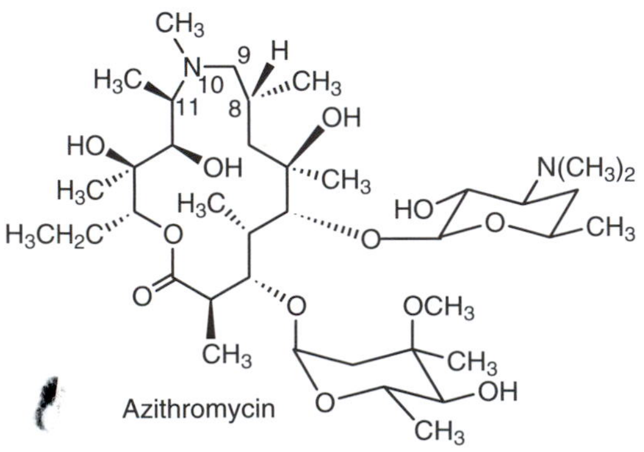

### CLINICAL USES

Azithromycin (Zithromax [Pfizer]) was first approved in the United States in 1992 for administration in capsule form to adults as the first once-daily 5-day oral antibiotic for the treatment of community-acquired respiratory infections and skin infections. Both oral and intravenous formulations of azithromycin for adults are available. For children younger than 16 years, only the oral formulation is approved.

### Approved Uses in Adults

- Acute bacterial exacerbations of chronic obstructive pulmonary disease due to *H. influenzae, M. catarrhalis,* or *S. pneumoniae*
- Community-acquired pneumonia due to *C. pneumoniae, H. influenzae, M. pneumoniae,* or *S. pneumoniae* in patients for whom oral therapy is appropriate
- Pharyngitis and tonsillitis caused by *S. pyogenes* in individuals who cannot tolerate the usual first-line penicillin therapy (penicillin by the intramuscular route is the usual drug of choice in the treatment of *S. pyogenes* infection and the prophylaxis of rheumatic fever; azithromycin is often effective in the eradication of susceptible strains of *S. pyogenes* from the nasopharynx)
- Uncomplicated skin and skin structure infections due to *S. aureus, S. pyogenes,* or *S. agalactiae* (abscesses usually require surgical drainage)

- Urethritis and cervicitis due to *C. trachomatis* or *N. gonorrhoeae*
- Genital ulcer disease due to *H. ducreyi* (the cause of chancroid in men)

### Approved Uses in Children

In December 2001, the FDA approved the use of azithromycin (as an oral suspension) as both a single-dose regimen and a 3-day regimen for the treatment of acute otitis media in pediatric patients. Azithromycin was approved in 1995 as a once-daily, 5-day treatment of juvenile acute otitis media caused by the three most common bacterial pathogens, *S. pneumoniae, H. influenzae,* and *M. catarrhalis.*

Other approved uses are as follows:

- Community-acquired pneumonia due to *C. pneumoniae, H. influenzae, M. pneumoniae,* or *S. pneumoniae* in patients for whom oral therapy is appropriate
- Pharyngitis and tonsillitis caused by *S. pyogenes,* as an alternative to first-line therapy in individuals who cannot tolerate the first-line therapy

### Other Uses

Macrolides are not used exclusively for the treatment of community-acquired respiratory tract infections. Their ability to penetrate eukaryotic cells makes them highly suitable for the treatment of diseases caused by intracellular pathogens, such as nongonococcal urethritis and trachoma. Azithromycin is used for these indications in adults and children. Azithromycin and, to a lesser extent, clarithromycin penetrate polymorphonuclear leukocytes, monocytes, lymphocytes, and alveolar macrophages, and achieve high intracellular concentrations.

## Clarithromycin

Clarithromycin (6-O-methyl erythromycin A) is a semisynthetic macrolide antibiotic chemically derived from erythromycin A. It differs chemically from erythromycin A by having an O-methyl ether substitution at position 6 of the macrolide ring (Fig. 18.8). This change increases the stability of these compounds in gastric acid, improving absorption by the oral route. The spectrum of activity is similar to that of erythromycin, except for the enhanced activity against *H. influenzae* and the activity against atypical mycobacteria, but it has better pharmacokinetic properties, including a twice-daily dose regimen.

Clarithromycin (Biaxin [Abbott]) is available as intermediate-release tablets, extended-release tablets, and granules for oral suspension. It is rapidly absorbed from the gastrointestinal tract after oral administration. It is active in vivo against a variety of aerobic and anaerobic gram-positive and gram-negative microorganisms as well as most *Mycobacterium avium* complex (MAC) microorganisms.

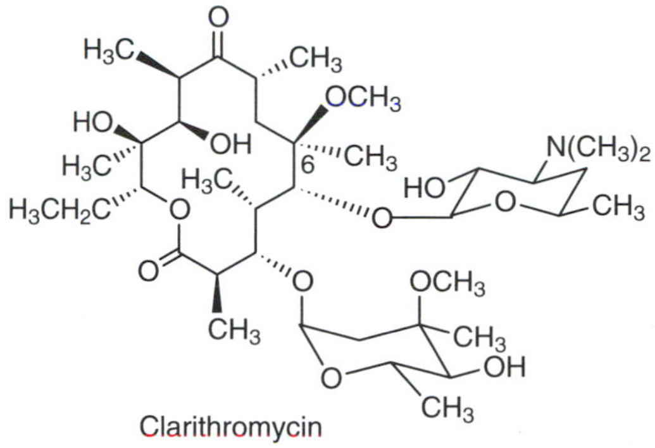

**Figure 18.8** Chemical structure of clarithromycin.

Clarithromycin was approved in 1992 by the FDA for treatment of respiratory, skin, and skin structure infections. It is two- to fourfold more active than erythromycin against most streptococci and staphylococci, including *S. pneumoniae,* but streptococci and staphylococci resistant to erythromycin are resistant to clarithromycin as well. Gram-positive bacteria resistant to erythromycin are resistant to clarythromycin (e.g., MRSA and high-level penicillin-resistant and many intermediate-level penicillin-resistant *S. pneumoniae* strains).

### CLINICAL USES

### Approved Uses in Adults

*Upper Respiratory Tract Infection.* Clarithromycin is indicated for the treatment of the following upper respiratory infections: group A streptococcal pharyngitis (although penicillin is the agent of choice), acute maxillary sinusitis, acute exacerbation of chronic bronchitis, and mild community-acquired pneumonia, especially in young to middle-aged adults.

Mycobacterium avium *Complex Infection.* In October 1995, the FDA approved the use of clarithromycin (Biaxin [Abbott Laboratories]) for the prophylaxis of disseminated MAC infections in human immunodeficiency virus-infected patients. Clarithromycin or azithromycin is currently the first choice for the treatment of MAC bacteremia in AIDS patients.

Helicobacter pylori *Infection.* In April 1996, the FDA approved the combination of clarithromycin with omeprazole, a proton pump inhibitor, for use in the treatment of duodenal ulcers caused by *Helicobacter pylori*. This macrolide was also approved by the FDA in January 1997 for use in combination with ranitidine bismuth citrate for treatment of patients with an active duodenal ulcer associated with *H. pylori* infection. In September 1997, clarithromycin in combination with lansoprazole and amoxicillin was approved for the treatment of patients with *H. pylori* infection and duodenal ulcer disease (active or a 1-year history of duodenal ulcer).

*Skin and Skin Structure Infection.* Uncomplicated skin and skin structure infections due to susceptible *S. pyogenes* or *S. aureus* can be treated with clarithromycin.

### Approved Uses in Children

In July 1996, the FDA approved the use of clarithromycin for the treatment of children with community-acquired pneumonia caused by *M. pneumoniae, C. pneumoniae,* and *S. pneumoniae*. It is also approved for the treatment of otitis media in children. The two latest approvals were granted to Abbott Laboratories in March 2000 and October 2000, for use of clarithromycin in acute exacerbation of chronic bronchitis and acute maxillary sinusitis, and for the addition of *H. influenzae* to previously approved indications for community-acquired pneumonia for Biaxin Filmtab (a recovered tablet pharmaceutical form).

## 16-Member Ring Macrolides

Many 16-member ring macrolide antibiotics have been isolated as fermentation products or obtained by chemical modification of natural macrolides; however, none have so far been approved by the FDA for use in the United States, although a few are in use in Japan and Europe. Generally, these macrolides have a similar antibacterial profile to erythromycin. This section briefly describes the macrolides in clinical use.

### NATURAL MACROLIDES

Josamycin (Fig. 18.9) was obtained from culture broth of *Streptomyces narbonensis* subsp. *josamyceticus*. It has been commercially available in Japan and Europe since 1970. It is recommended for the treatment of lower respiratory tract infections, including those due to *M. pneumoniae*. Its uses are otherwise similar to those of erythromycin.

Spiramycin (Fig. 18.9) was isolated in 1955 from a series of strains of *Streptomyces ambofaciens* in France. This macrolide antibiotic has been used extensively in that country, but its use in the United States and United Kingdom has been limited.

Midecamycin (Fig. 18.10) is a fermentation product of *Streptomyces mycarofaciens*. It has a similar spectrum of activity to erythromycin but is less active. It is available in Japan, Italy, and France.

### SEMISYNTHETIC MACROLIDES

Rokitamycin (Fig. 18.9) is a semisynthetic 16-member ring macrolide; it is unstable in acidic media. It is available in Japan and Italy. Rokitamycin has an identical spectrum of activity to erythromycin but is less active against gram-positive cocci.

Miokamycin (9,3"-di-*O*-acetylmidecamycin) (Fig. 18.10) is a semisynthetic diacetyl derivative of the natural macrolide midecamycin. It is available in Italy, Spain, France, and Japan.

## Intracellular Concentration of Macrolides

All macrolide antibiotics are concentrated, mainly in polymorphonuclear leukocytes and macrophages, and some achieve persistently high intracellular levels. This is

**Figure 18.9** Chemical structures of the 16-member macrolides josamycin, rokitamycin (semisynthetic), and spiramycin I (natural product).

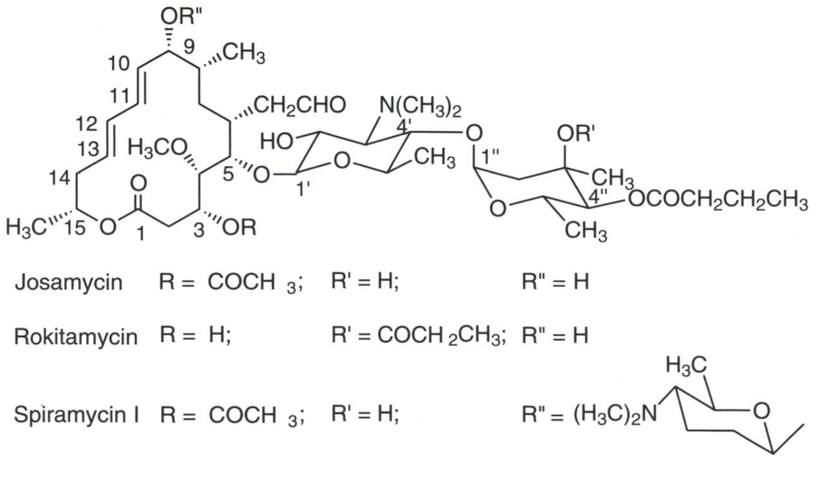

Josamycin R = $COCH_3$; R' = H; R" = H

Rokitamycin R = H; R' = $COCH_2CH_3$; R" = H

Spiramycin I R = $COCH_3$; R' = H; R" = $(H_3C)_2N$–

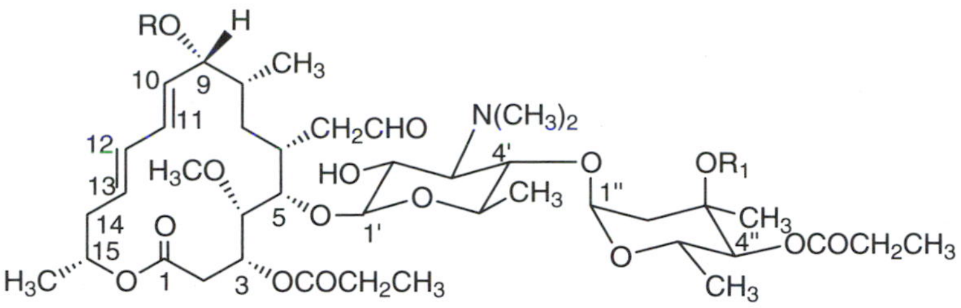

Midecamycin $A_1$ R = $R_1$ = -H

Miokamycin R = $R_1$ = $-COCH_3$

**Figure 18.10** Chemical structures of the macrolides midecamycin (natural) and miokamycin (semisynthetic).

why they are used to treat infections caused by *Chlamydia* spp. and *Legionella* spp.

## Mechanism of Action

Macrolides reversibly bind to the 50S subunit and inhibit the transpeptidation and translocation process, causing premature detachment of incomplete polypeptide chains. Macrolides bind at the entrance of the tunnel, where they sterically block the progression of the nascent peptide. Recent reports describe the three-dimensional modeling of the 50S ribosome complexed with antibiotics, which is essential to understand the molecular mechanisms of antibiotic binding and antibiotic resistance.

An important contribution was made in October 2001 by a group of scientists from the Max Planck Institute (Hamburg, Germany) and from the Weizmann Institute (Rehovot, Israel), who determined the high-resolution X-ray structures of the 50S ribosomal subunit of the eubacterium *Deinococcus radiodurans* complexed individually with the macrolides erythromycin, clarithromycin, and roxithromycin and with clindamycin and chloramphenicol. They found that the antibiotic-binding sites are composed exclusively of segments of 23S rRNA at the peptidyltransferase cavity and do not involve any interactions between the drugs and ribosomal proteins. Their results showed unambiguously that these macrolide antibiotics bind to the same site at the entrance of the tunnel and that their binding contacts clearly differ from those of chloramphenicol but overlap those of clindamycin to a greater extent.

Figure 18.11 shows the interactions of macrolides (erythromycin, clarithromycin, and roxithromycin) with the peptidyltransferase cavity. The reactive groups of the desosamine sugar and the lactone ring mediate all the hydrogen bond interactions of these macrolides with the peptidyltransferase cavity. The 2'-OH group of the desosamine sugar appears to form hydrogen bonds at three positions, $N^6$ and $N^1$ of A2041Dr (A2058Ec) and $N^6$ of A2042Dr (A2059Ec). The hydrogen bonds between the 2'-OH and $N^1$ and $N^6$ of A2041Dr (A2058Ec) explain why this nucleotide is essential for macrolide binding, thus shedding light on the two most common ribosomal resistance mechanisms against macrolides: the $N^6$ dimethylation of A2041Dr (A2058Ec) by the erythromycin resistance family of methylases and rRNA mutations changing the identity of the nucleotide at this position. The dimethylation of the $N^6$ group not only would add a bulky substituent causing steric hindrance for the binding but also would prevent the formation of hydrogen bonds to the 2'-OH group. The mode of interactions proposed by these scientists implies that a mutation at A2041Dr (A2058Ec) disrupts the pattern of hydrogen bonding, thus impairing binding and rendering the bacteria resistant. Mitochondrial and cytoplasmic rRNAs of higher eukaryotes have a guanosine at position 2058Ec. Therefore, the proposed mode of interactions explains the selectivity of macrolides for bacterial ribosomes.

The 2'-OH of the desosamine sugar is located 4 Å away from the $N^6$ of G2040Dr (G2057Ec), a nucleotide forming the last base pair of helix 73. The authors suggested that although an interaction with this group cannot be ruled out, this base pair seems to be engaged mainly in maintaining the right conformation of A2041Dr (A2058Ec).

The reactive groups of the cladinose sugar are not involved in hydrogen bond interactions with the 23S rRNA. Cladinose dispensability was confirmed by structure-activity relationship (SAR) studies showing that the 4"-OH is dispensable for macrolide binding. The ketolides do not have in their structures a cladinose sugar, and they bind more tightly to bacterial ribosomes.

The authors of the publication pointed out that the structures determined do not indicate the presence of a direct interaction between erythromycin and helix 35 (H35) in domain II of the rRNA and the peptidyltransferase loop in domain V, which was previously proposed by Hansen et al. (1999). However, it was mentioned that overall, it is difficult to explain the high affinity of

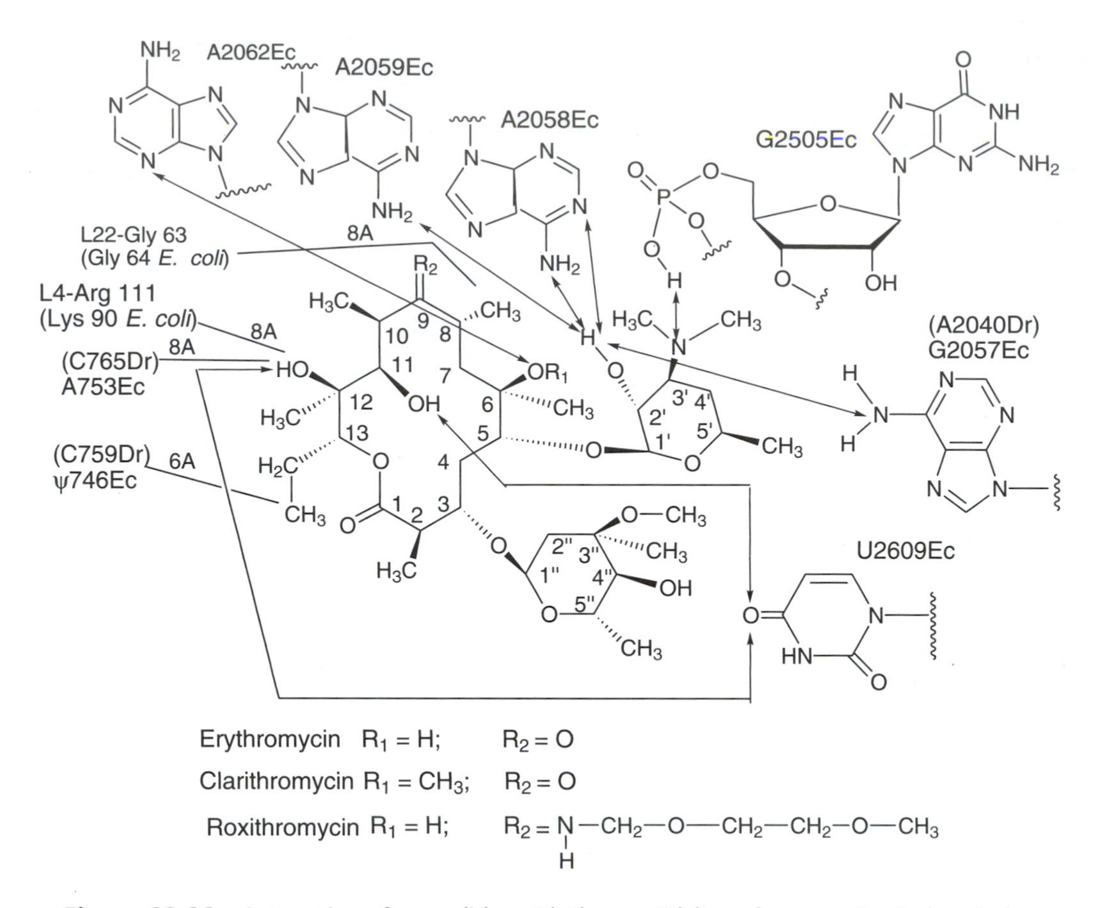


**Figure 18.11** Interaction of macrolides with the peptidyltransferase cavity. A chemical structure diagram of the macrolides (erythromycin, clarithromycin, and roxithromycin) is presented, showing the interactions of the reactive groups of the macrolides with the nucleotides of the peptidyltransferase cavity. Adapted from F. Schlünzen, R. Zarivach, J. Harms, A. Bashan, A. Tocilj, R. Albrecht, A. Yonath, and F. Franceschi, *Nature* **413:**814–821, 1999, with permission from the publisher.

macrolides ($K_d$, $10^{-8}$ M) for the ribosome solely on the basis of the proposed seven hydrogen bonds. The binding of erythromycin and its derivatives may be further stabilized by van der Waals' forces, hydrophobic interactions, and the geometry of the rRNA that tightly surrounds the macrolide molecules.

Regarding the two ribosomal proteins, L4 and L22, it was determined that the shortest distance from erythromycin 12-OH to L4 (R111Dr/K90Ec) is 8 Å whereas the shortest distance from 8-$CH_3$ to L22 (G63Dr/G64Ec) is 9 Å. These distances are longer than one would expect for any meaningful chemical interaction. Therefore, the authors proposed that the resistance of macrolides acquired by mutations in these two proteins is probably the product of an indirect effect produced by a perturbation of the 23S rRNA owing to the mutated proteins.

Overall, the binding sites of the antibiotics suggest that the inhibitory action is not determined solely by their interaction with specific nucleotides. The macrolides, clindamycin, and chloramphenicol could also inhibit peptidyltransferase by interfering with the proper positioning and movement of the tRNAs at the peptidyltransferase cavity. The peptidyltransferase center is the site of interaction of macrolide, lincosamide, and streptogramin B ($MLS_B$) antibiotics. Although this is a chemically diverse group of antibiotics, they have all been involved in interactions with common essential nucleotide moieties in the bacterial 23S rRNA to inhibit ribosomal function. The overlapping of binding sites may explain why clindamycin and the macrolides bind competitively to the ribosome and why most RNA mutations conferring resistance to macrolides also confer resistance to lincosamides.

Like all great endeavors in science, the recently published quest for structural information about the ribosome is the culmination of decades of effort on the part of the entire scientific community concerned with ribosomes.

## Mechanism of Bacterial Resistance

Three different mechanisms account for bacterial resistance to the action of macrolide antibiotics:

- target site modification
- active efflux
- enzymatic inactivation

Target site modification and active efflux are responsible for the majority of macrolide resistance observed in clinical isolates. The first mechanism involves the production of a ribosomal methylase, and the second mechanism involves a family of genes called *mef* and is commonly referred to as efflux resistance.

### TARGET SITE MODIFICATION

Target site modification has resulted in cross-resistance to the macrolides, lincosamides, and streptogramin B, the so-called $MLS_B$ phenotype. Different bacterial species are able to synthesize an enzyme, encoded by a series of structurally related erythromycin-resistant methylase (*erm*) genes, that methylates rRNA. Modifications of the ribosomal target of macrolides are related to the synthesis of adenine-$N^{6,6}$-dimethyltransferases encoded on plasmids transposons. Erm rRNA methylases are a family of highly related proteins that use *S*-adenosylmethionine as a methyl donor to mono- or dimethylate the $N^6$ amino group of an adenosine-2058 residue of bacterial 23S rRNA (Fig. 18.12). The *N*-dimethyltransferases have been found in *Staphylococcus, Streptococcus, Enterococcus,* and *Bacillus* strains. Methyltransferase genes are widely disseminated in both gram-positive and gram-negative bacteria and can be located on plasmids and transposons.

Methylation of the 23S rRNA of the 50S subunit probably leads to a conformational change of the ribosome that yields broad cross-resistance to macrolide-lincosamide-streptogramin B antibiotics.

The *erm* genes can be expressed constitutively or inducibly. **Constitutive resistance** occurs when the methylating enzyme is produced constitutively, and **inducible resistance** occurs when when the enzyme induction is effected by exposure of the organism to both 14-member ring (such as erythromycin and clarithromycin) and 15-member ring (such as azithromycin) but not 16-member ring (such as josamycin and spiramycin) macrolides. Different types of erythromycin resistance methylases, encoded by *erm* genes, are produced by different bacteria. The *erm* genes have been divided into at least 20 different classes on the basis of hybridization studies and sequence comparisons. In 1999, a new nomenclature for $MLS_B$ genes was proposed.

Another type of target site modification resistance to macrolides is that due to site-specific mutations in the 23S rRNA gene. Mankin et al. (1999) showed that erythromycin and the ketolide HMR3647, a member of a new class of macrolide antibiotics (see below), interacts not only with domain V of 23S rRNA but also with rRNA segments in domain II, in the vicinity of the hairpin 35 loop (Fig. 18.13). The authors found that a mutation in the hairpin 35 loop confers resistance to erythromycin and ketolides and that both antibiotics

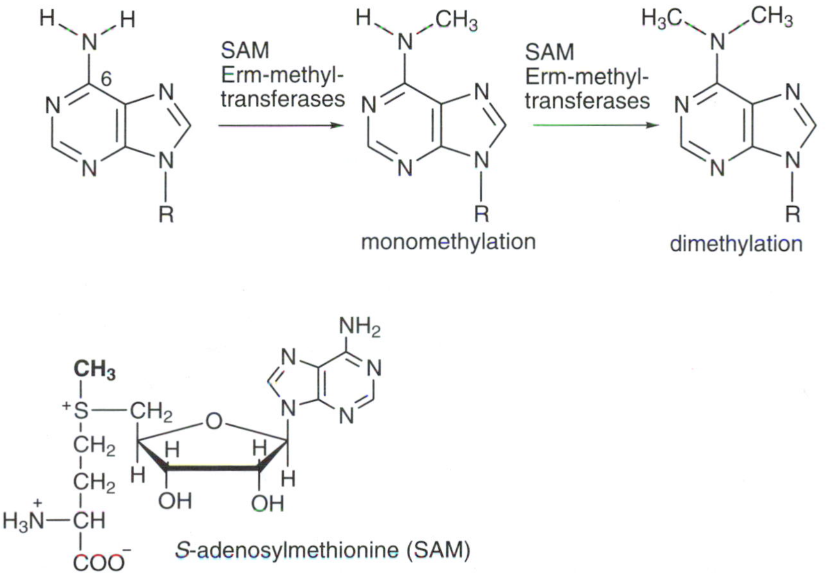

**Figure 18.12** All Erm methyltransferases methylate the same adenine residue, resulting in an $MLS_B$ phenotype.

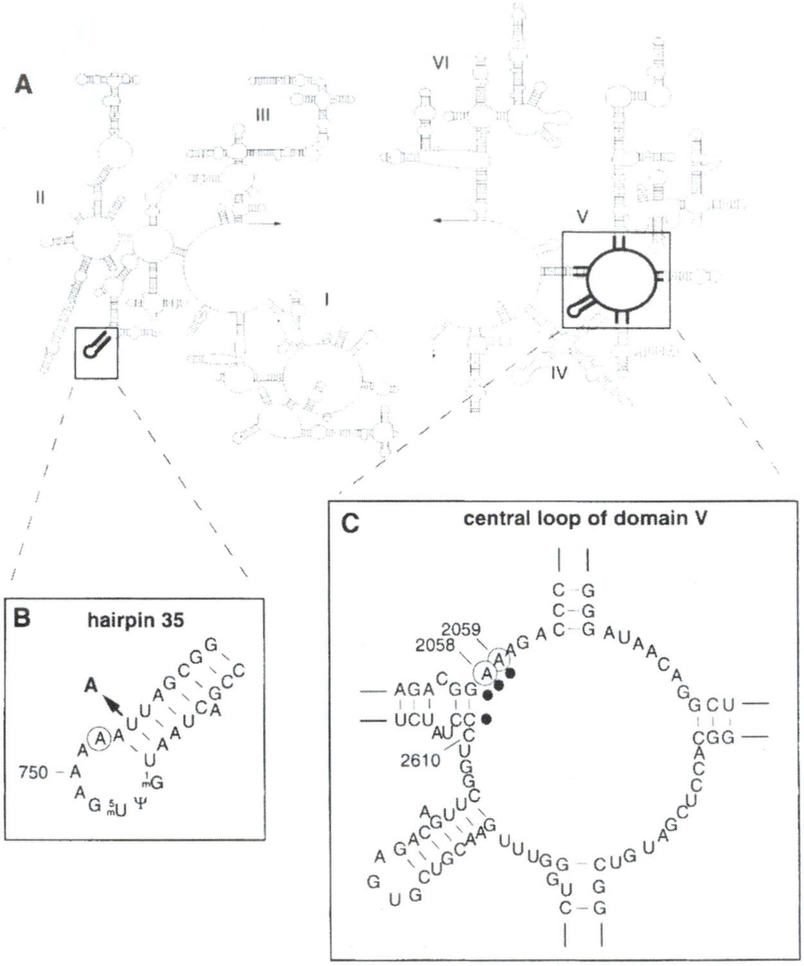

**Figure 18.13** (A) Secondary structure of *E. coli* 23S rRNA. (B and C) Enlargements of the hairpin 35 (B) and the central loop of domain V (C). The positions whose accessibility to modification by dimethyl sulfate is affected by macrolide antibiotics are circled (the position corresponding to ψ746 is protected by HMR3647 in the *Thermus aquaticus* ribosome). The nucleotides in domain V, whose mutations cause resistance to macrolides, are shown by black dots. The newly isolated macrolide resistance mutation in hairpin 35 is shown by an arrow. Posttranscriptional modifications are shown only for hairpin 35. Reprinted from L. Xiong, S. Shah, P. Mauvais, and A. S. Mankin, *Mol. Microbiol.* **31:**633–639, 1999, with permission from the publisher.

affect the accessibility of nucleotides in the hairpin loop to chemical modification. These results indicate that hairpin 35 is located in the vicinity of, and indeed may be part of, the ribosomal peptidyltransferase center. Furthermore, a strong interaction between ketolides and the hairpin 35 region may account for the high activity of these antibiotics against erythromycin-resistant bacteria.

## ACTIVE EFFLUX

The macrolide-specific efflux from resistant cells is effected by a membrane protein encoded by the *mef* gene. This leads to the M phenotype, which is resistant to 14- and 15-member ring macrolides and susceptible to 16-member ring macrolides, ketolides, lincosamides, and streptogramin B. The clarithromycin, azithromycin, or erythromycin MIC can vary from 1 to 32 μg/ml in resistant bacteria. The *mef* genes have been found in a variety of gram-positive bacteria, including corynebacteria, enterococci, micrococci, and a variety of streptococci. The genes of the efflux pump can be either acquired, such as *mef,* or carried intrinsically, such as *acrAB*. A number of different macrolide resistance genes code for transport (efflux) proteins, which pump the macrolides out of the cell or cellular membrane.

The classification of various efflux pumps that expel antibiotics was recently reviewed by Sutcliffe (1999).

### Efflux Pumps in Gram-Positive Organisms

Efflux as a mechanism of macrolide resistance in streptococci has been recently described for *S. pneumoniae*, where it is encoded by *mefE,* and for *S. pyogenes*, where it is encoded by *mefA*. Strains with efflux mechanisms are resistant to macrolides, with an erythromycin MIC generally ranging from 1 to 16 μg/ml, but are uniformly susceptible to clindamycin. Mef is thought to be a membrane protein, which is sufficient for the energy-dependent efflux of 14- and 15-member ring macrolides.

### Efflux Pumps in Gram-Negative Organisms

Macrolides have consistently higher MICs in gram-negative than gram-positive bacteria. The resistance of most gram-negative bacteria to macrolides is currently thought to be due mainly to intrinsic efflux pumps in synergy with slow penetration through the bacterial outer membrane. In gram-negative bacteria, the efflux mechanism involves three components: a pump in the inner cytoplasmic membrane, a porous protein molecule on the outer wall, and a third protein that connects the inner and outer membranes. The major efflux pumps responsible for macrolide resistance in gram-negative species are energy dependent and include AcrAB-TolC in *H. influenzae* and *E. coli* and MexAB-OprM in *P. aeruginosa.* The macrolide efflux resistance mechanisms have been recently reviewed by Zhong and Shortridge (2000) and by Pechère (2001).

ENZYMATIC INACTIVATION

Target site modification and active efflux are the most frequently encountered mechanisms of resistance to macrolide antibiotics. A third mechanism of resistance is due to enzymatic degradation, but the clinical significance of this mechanism has not yet been established.

Between 1986 and 1989, members of the family *Enterobacteriaceae* that were highly resistant to erythromycin because of inactivating enzymes were reported. These strains produce erythromycin esterases (esterases I and II) encoded by the *ereA* and *ereB* (for "erythromycin-resistant esterases") genes, and these enzymes catalyze the cleavage of the lactone ring of 14-membered macrolides (Fig. 18.14). However, the 16-membered macrolides are not efficiently utilized as substrates by these enzymes.

## Trends in Macrolide Research: Novel Macrolides To Overcome Bacterial Resistance

The currently available macrolide antibiotics in the United States, including erythromycin and the acid-stable clarithromycin, azithromycin, and dirithromycin, cover the bacterial species most commonly responsible for upper and lower respiratory tract infections and, depending on the particular macrolide, *Legionella* spp., *Chlamydia* spp., *Mycoplasma* spp., and *H. influenzae.* A remarkable property of macrolides is that they have high intracellular concentrations and activity in various cellular compartments in which intracellular pathogens are found. However, all these macrolide antibiotics, including the 16-membered ring macrolides, have several drawbacks; in particular, they are inactive against $MLS_B$-resistant streptococci and *S. pneumoniae.* Therefore, there is a clear need to overcome pneumococcal resistance.

Structural modification of existing naturally occurring antibiotics remains one of the most effective approaches for overcoming bacterial resistance. Erythromycin A is one of the choices of macrolides for chemical modification of its structure, mainly because the SAR is well known. Indeed, it is necessary for the new structures to conserve the good pharmacokinetic properties and high stability in acid medium (for oral administration) of the macrolides.

KETOLIDES

The ketolides are a new class of macrolides. They are semisynthetic 14-member ring macrolides that differ from erythromycin A in that they have a 3-keto group in place of the α-L-cladinose moiety in the erythronolide A ring. Ketolides derive their name from the 3-keto group (keto) and the lactone ring (olide).

Telithromycin (HMR 3647), developed at Aventis (Romainville, France), was first launched in Germany as Ketex in 2002 as a once-daily oral treatment for respiratory infections including community-acquired pneumonia, acute bacterial exacerbations of chronic bronchitis, acute sinusitis, and tonsillitis/pharyngitis. It shows potent in vitro activity against common respiratory pathogens including *S. pneumoniae, H. influenzae, M. catarrhalis,* and *S. pyogenes,* as well as other atypical pathogens.

Telithromycin is a semisynthetic derivative of the natural macrolide erythromycin (Fig. 18.15). It is prepared by removing the cladinose sugar from the C-3 position of the erythronolide skeleton and oxidizing the remaining

**Figure 18.14** Cleavage of the lactone ring of erythronolide A, catalyzed by esterases.

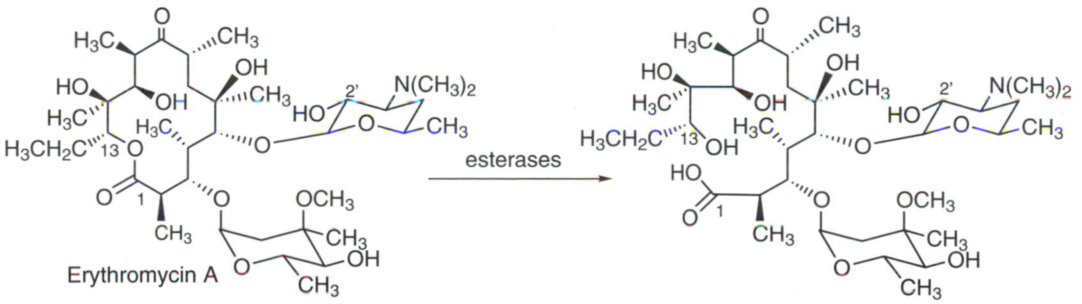

Erythromycin A

esterases

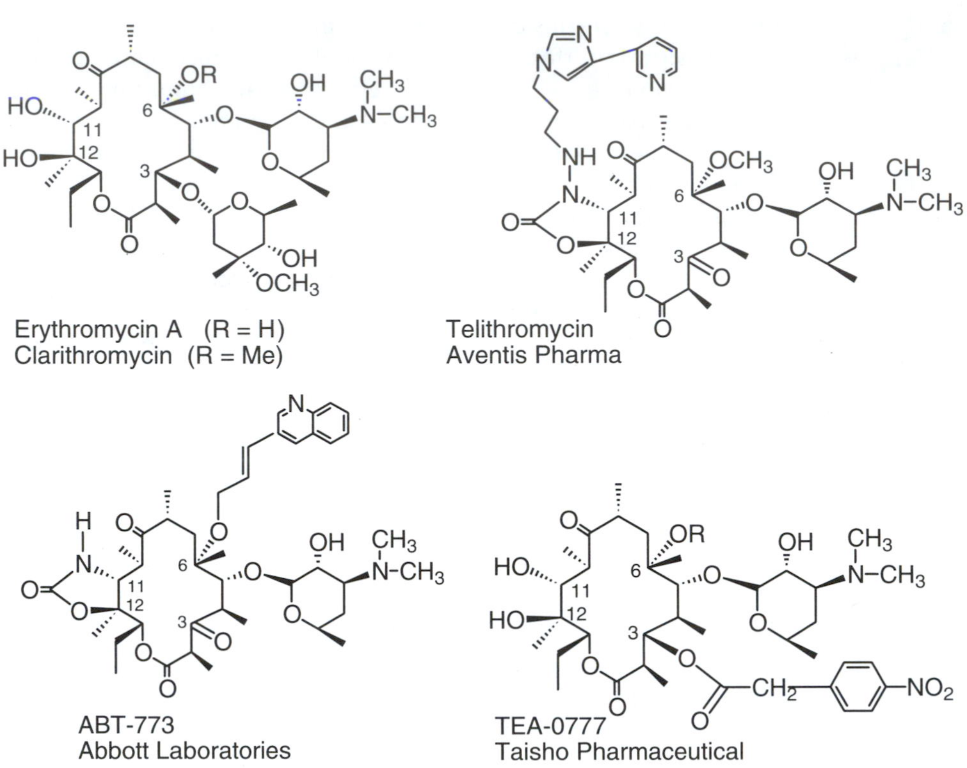

**Figure 18.15** Chemical structures of telithromycin, ABT-773, and TEA-0777. For comparison, the chemical structures of erythromycin A and clarithromycin are also shown. Note that in these structures the $CH_3$ (methyl, Me) and $CH_2CH_3$ (ethyl) groups are represented by a line above or below the plane of the molecule.

hydroxyl group to a keto group. In addition to the C-3 ketone, telithromycin has an aromatic N-substituted carbamate extension at positions C-11 to C-12. This ring has an imidazo-pyridyl group attachment. Telithromycin possesses a 6-$OCH_3$ group (like clarithromycin), avoiding internal hemiketalization with the 3-keto function and giving the ketolide molecule excellent acid stability.

ABT-773, another ketolide (Fig. 18.15), is under development at Abbott Laboratories. It is also a 14-member lactone ring that, in addition to the C-3 ketone, has a quinolylallyl ether at C-6 and a cyclic carbamate group at C-11 to C-12.

This new class of antibiotics was designed to overcome current mechanisms of resistance to erythromycin A within gram-positive cocci, specifically for the treatment of community-acquired respiratory tract infections. Ketolides do not induce $MLS_B$ resistance and are active against erythromycin resistance methylase gene (*erm*)-carrying gram-positive cocci.

## Mechanism of Action

Ketolides prevent bacterial protein synthesis by binding to two domains of the 50S subunit of the bacterial ribosomes. Recently, Namour et al. (2001), working at Aventis Pharmacy, and Ackermann and Rodloff (2003) reported that telithromycin blocked the process in two different ways: (i) by blocking the translation of mRNA, contributed by interactions with domains II and V of the 23S rRNA, and (ii) by interfering with the assembly of precursors of the new ribosomal 50S subunit (r proteins and 5S and 23S rRNA), leading to nucleolytic degradation of the unassembled precursor particles by cellular RNases.

In a recent (2003) publication, Schlünzen, Yonath, and coworkers at the Max Planck Institute (Hamburg, Germany) and the Weizmann Institute (Rehovot, Israel) reported the crystal structure of the 50S ribosomal subunit of *Deinococcus radiodurans* in complex with azithromycin (an azalide macrolide) and the ketolide ABT-773. Their report mentioned that both compounds exert their antibacterial activity by blocking the protein exit tunnel. They mentioned that the interactions of ABT-773 with specific nucleotides of the multibranched loop of domain V of 23S rRNA are rather similar to those of erythromycin A but that additional contacts with domain II and IV led to a tighter binding. They demonstrated for both azithromycin and ABT-773 that the results obtained are consistent with the majority of mutational and footprinting data, thus explaining the different modes of interaction and difference in activity against certain $MLS_B$-resistant phenotypes.

These investigators found that a specific nucleotide of the 23S rRNA, nucleotide 2058, forms tight contacts with erythromycin but less tight contacts with ABT-773, since its position within the tunnel appears displaced by about 3.0 Å compared with the position observed for the macrolides of the erythromycin A class. This position still leads to the same inhibitory mechanism, namely, blockage of the path of the nascent chain through the 50S subunit, because the contacts made with domain II compensate for the loss of contacts with nucleotide 2058; in this way, ABT-773 can bind to mutated or methylated nucleotide 2058, generated by the two dominant resistance mechanisms.

Finally, the investigators conclude that the differences between the binding modes of macrolides, azalides, and ketolides reveal the contributions of the specific chemical modifications of the macrolides, and they explain the enhanced binding properties of the advanced (new) compounds on this basis.

#### ACYLIDES

Introduction of an acyl group instead of L-cladinose into the 3-O position of erythromycin A derivatives led to a novel class of macrolide antibiotics that are named acylides. A recent report from scientists at Taisho Pharmaceutical (Saitama, Japan) and Chiba University (Chiba, Japan) disclosed that the 3-O-nitrophenylacetyl derivative TEA0777 (Fig. 18.15) showed potent activity against erythromycin-susceptible gram-positive pathogens and also against $MLS_B$-resistant *S. aureus* and efflux-resistant *S. pneumoniae.*

### Macrolide Preparations Commercially Available in the United States

Table 18.1 lists the macrolides marketed in the United States.

## Lincosamides

### Historical Perspective and Structure of Lincomycin and Clindamycin

Lincomycin is the major component of lincosamide antibiotics, which were isolated in 1963 from the fermentation of a strain of *Streptomyces lincolnensis* in the Upjohn research laboratories. The structure of lincomycin is shown in Fig. 18.16. Subsequently, lin-

**Table 18.1** Generic and common trade names of macrolides, the preparations available, and manufacturers in the United States

| Generic name | Trade name (preparation) | Manufacturer |
|---|---|---|
| Azithromycin dihydrate | Zithromax (tablets, capsules, oral suspension) | Pfizer |
| | Zithromax (for intravenous infusion) | Pfizer |
| Clarithromycin | Biaxin (tablets, oral suspension) | Abbott |
| | Biaxin XL (Filmtab tablets) | Abbott |
| Dirithromycin | Dynabac (tablets) | Muro |
| Erythromycin | Akne-Mycin (ointment) | Healthpoint |
| | Benzamycin (topical gel) | Dermik |
| | Emgel 2% (topical gel) | GlaxoSmithKline |
| | Ery-Tab (tablets) | Abbott |
| | Erythromycin base (Filmtab tablets) | Abbott |
| | Erythromycin delayed-release (capsules) | Abbott |
| | Ilotycin (ophthalmic ointment) | Dista |
| | PCZ Dispertab (tablets) | Abbott |
| | Eryc delayed-release (capsules) | Warner Chilcott |
| | Theramycin Z (topical solution) | Bioglan |
| Erythromycin ethylsuccinate | E.E.S. 200 (liquid) | Abbott |
| | E.E.S. 400 (Filmtab tablets and liquid) | Abbott |
| | E.E.S. (granules) | Abbott |
| | EryPed 200 and Eryped 400 | Abbott |
| | EryPed (drops and tablets) | Abbott |
| | Pediazole (suspension) | Ross |
| | Erythromycin ethylsuccinate (tablets) | Mylan |
| Erythromycin ethylsuccinate plus sulfisoxazole acetyl | Suspension for oral use | Lederle Standard |
| Erythromycin lactobionate | Erythromycin lactobionate (injection) | Lederle Standard |
| Erythromycin stearate | Erythrocin stearate (Filmtab tablets) | Abbott |
| | Erythromycin stearate (tablets) | Mylan |

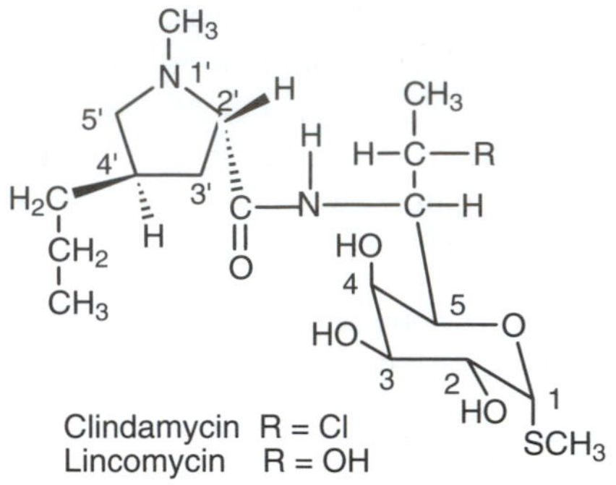

Figure 18.16 Chemical structures of clindamycin and lincomycin.

comycin has been produced by a variety of *Streptomyces* strains and by a strain of *Actinomyces roseolus*. Many chemical modifications of the lincomycin molecule have been developed in an attempt to produce an improved antibiotic. Of these, only clindamycin (Fig. 18.16) was clinically superior to lincomycin. Currently, clindamycin is the only lincosamide antibiotic marketed in the United States and United Kingdom.

Clindamycin, a semisynthetic antibiotic, was introduced in 1966. It is active against a number of clinically important gram-positive bacteria, and its spectrum of activity includes important anaerobic *Bacteroides* species and the malarial parasite *Plasmodium falciparum*. The chemical name for clindamycin is methyl-7(*S*)-chloro-6,7,8-trideoxy-6-(1-methyl-*trans*-4-propyl-L-2-pyrrolidinecarboxamido)-1-thio-L-threo-α-D-galacto-octopyranoside. The chemical structures of both lincomycin and clindamycin contain a thiolated L-threo-α-D-galactopyranoside that is joined to a proline residue through an amide linkage (Fig. 18.16).

## Mechanism of Action

Clindamycin interferes with A-site and P-site substrate binding and physically hinders the path of the growing peptide chain. The 3.1-Å structural model of the *Deinococcus radiodurans* 50S subunit was used by the groups at the Max Planck Institute and the Weizmann Institute as a reference to determine the structure of the 50S-clindamycin complex.

Figure 18.17 illustrates the interaction of clindamycin with the peptidyltransferase cavity. The binding position

**Figure 18.17** Interactions of clindamycin with the peptidyltransferase cavity. (a) Chemical structure diagram of clindamycin showing the interactions (arrows) of its reactive groups with the nucleotides of the peptidyltransferase cavity. Arrows between two chemical moieties indicate that the two groups are less than 4.4 Å apart. (b) Secondary structure of the peptidyltransferase ring of *D. radiodurans,* showing the nucleotides involved in the interaction with clindamycin. Reprinted from F. Schlünzen, R. Zarivach, J. Harms, A. Bashan, A. Tocilj, R. Albrecht, A. Yonath, and F. Franceschi, *Nature* **413:**814–821, 1999, with permission from the publisher.

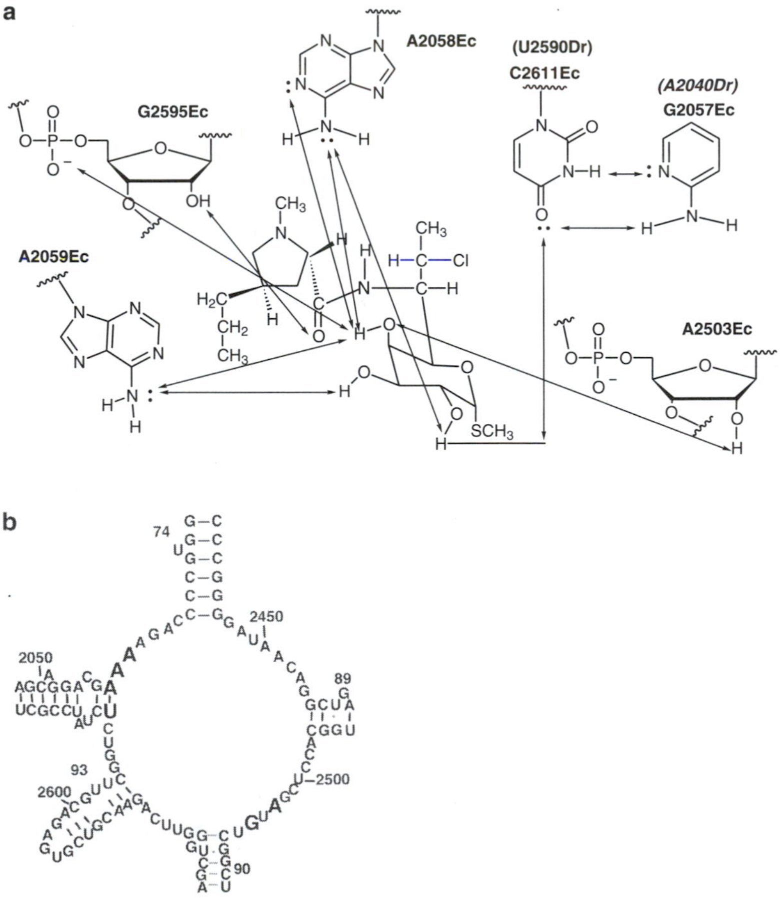

of clindamycin provides a structural explanation for the hybrid nature of its A and P sites. Clindamycin can interfere with the positioning of the aminoacyl group at the A site and the peptidyl group at the P site while also sterically blocking the progression of the nascent peptide toward the tunnel.

In 1992, Stephen Douthwaite published his findings on the interaction of clindamycin and lincomycin with *E. coli* 23S rRNA by comparing chemical footprinting. The results indicated that both antibiotics interact strongly with 23S rRNA bases A2058, A2451, and G2505 and weakly with G2061. The results indicated that the modification patterns differ in that A2059 is bound by clindamycin but not by lincomycin.

### Antimicrobial Activity and Clinical Uses

The antibacterial spectrum of clindamycin includes gram-positive cocci and anaerobes but not aerobic gram-negative organisms. Lincomycin has the same spectrum of activity as clindamycin but is less active both in vitro and in vivo. Clindamycin may be bacteriostatic or bactericidal; the response depends on the bacterial species and the growth conditions.

Except for *Bacteroides* spp., clindamycin is not the antibiotic of choice for any specific pathogen. It is a useful alternative agent for (i) *B. fragilis* infections and/or intra-abdominal and pelvic infections (although clindamycin is sometimes listed as a drug of choice for *Bacteroides* spp., some infectious-disease experts recommend metronidazole, especially for severe *B. fragilis* infections); (ii) mixed aerobic and anaerobic soft tissue infections; (iii) treatment of patients allergic to both penicillins and cephalosporins; (iv) osteomyelitis, since it penetrates bone very well; (v) severe aspiration pneumonia or lung abscesses; and (vi) foot infections in diabetes. It is commonly used in combination with an agent with good activity against aerobic gram-negative rods (e.g., ciprofloxacin).

In March 1998, the FDA authorized the marketing of 2% clindamycin phosphate cream for a 3-day dosing regimen; in July 1999, it approved a suppository form for the treatment of bacterial vaginosis. Readers interested in the medical uses of these antibiotics should refer to a more appropriate source of information.

### Bacterial Resistance

It was mentioned above that resistance to lincosamides is usually due to alteration of the ribosome, following the $N^6$ dimethylation of a specific adenine residue in 23S rRNA, which confers cross-resistance to macrolides, lincosamides, and streptogramin B (i.e., the $MLS_B$ phenotype). As well as this broad spectrum of resistance, resistance specific to lincosamides has been reported. Phosphorylation and nucleotidylation of the hydroxyl group at position 3 of lincosamide molecules have been detected in strains of staphylococci, streptococci, and enterococci. In clinical isolates of *Staphylococcus haemolyticus* and *S. aureus*, *O*-nucleotidyltransferases encoded by two closely related genes named *linA* (for "lincosamide inactivation nucleotidyl") and *linA*', respectively, were characterized. These enzymes inactivate clindamycin by converting it to clindamycin 4-(5'-adenylate) and lincomycin 3-(5'-adenylate) by using ATP, GTP, CTP, or UTP as the nucleotidyl donor and $MgCl_2$ as a cofactor. In 1999, Leclercq and his group in France and scientists from Pharmacia & Upjohn reported a new resistance gene, *linB*, conferring resistance to lincosamides by nucleotidylation in *Enterococcus faecium* HM1025 (Fig. 18.18). This new gene was cloned, and its nucleotide sequence was determined.

### Lincosamide Preparations Commercially Available in the United States

Table 18.2 lists the lincosamides marketed in the United States.

## Streptogramins

### Historical Perspective

Streptogramins are a group of natural and semisynthetic compounds. The natural streptogramins are produced as secondary metabolites from a wide variety of diverse species of soil organisms belonging to the genus *Streptomyces*. Streptogramin was isolated in 1953 from *Streptomyces diastaticus* as a mixture of streptogramins A and B. Two years later, the mixture of virginiamycins A and B was isolated from a strain of *Streptomyces virginiae*. Since then, many other natural streptogramins have been discovered. The particular feature of this group of antibiotics is that they are mixtures of two chemically unrelated compounds.

Streptogramins consist of a mixture of two components: cyclic polyunsaturated macrolactones (group A) and cyclic hexadepsipeptides (group B). The latter are cyclized through an ester bond between the hydroxyl group of an N-terminal threonine and the C-terminal carboxylic acid of a phenylglycine (Fig. 18.19).

Streptogramins A and B act synergistically on the bacterial ribosome to disrupt bacterial protein synthesis. Individually, the A and B compounds are bacteriostatic, whereas in combination they are bactericidal.

### Pristinamycins IA and IIA

Pristinamycin was isolated from *Streptomyces pristinaespiralis*. It has two components: 30 to 40% is pristinamycin IA (a group B streptogramin) and 60 to 70% is

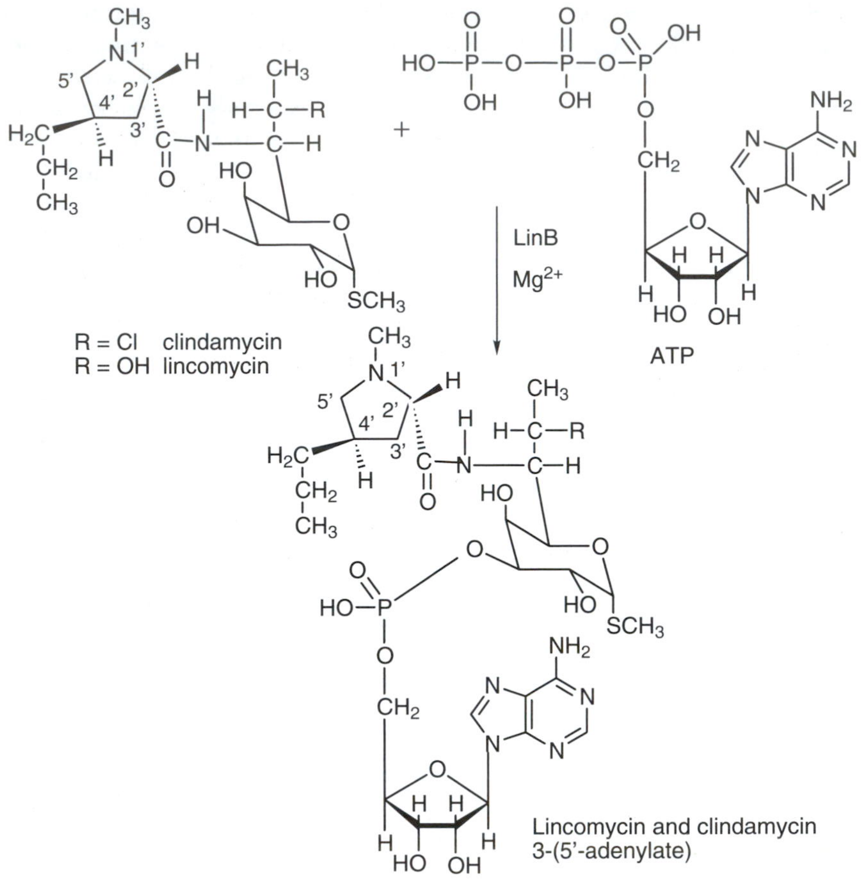

**Figure 18.18** Mechanism whereby clindamycin and lincomycin are converted to lincomycin and clindamycin 3-(5'-adenylate) by LinB in the presence of ATP and $MgCl_2$.

pristinamycin IIA (a group A streptogramin). Chemically, pristinamycin IA is a peptide macrolactone, made up of six amino acid residues: (*S*)-threonine, (*R*)-α-aminobutyric acid, (*S*)-proline, (*S*)-*N*-methyl-4-dimethylaminophenylalanine, (*S*)-oxopipecolic acid, and (*S*)-phenylglycine. The amino function of the threonine residue is acylated with 3-hydroxypicolinic acid. Pristinamycin IIA is a polyunsaturated macrolactone with a dehydroproline ring and an oxazole nucleus incorporated in the macrolactone structure (Fig. 18.19).

Pristinamycin has been used as an oral antistaphylococcal antibiotic in France and other European countries since 1968. It is active against gram-positive bacteria, especially *Staphylococcus* and *Streptococcus* spp., and some gram-negative bacteria such as *Haemophilus, Neisseria, Legionella,* and *Mycoplasma* spp. Pristinamycin I and II are synergistic; the activity of the mixture is 10-fold greater than the sum of the activities of the individual components. A related streptogramin, virginiamycin, has been used as a growth promoter in animals in Europe and the United States, although its use was banned in the European Union in July 1999.

The major drawback of the natural streptogramin compounds is that they are not water soluble. This

**Table 18.2** Generic and common trade names of lincosamides, the preparations available, and manufacturers in the United States

| Generic name | Trade name (preparation) | Manufacturer |
|---|---|---|
| Clindamycin hydrochloride | Cleocin HCl (capsules) | Pharmacia & Upjohn |
| Clindamycin phosphate | Cleocin T phosphate (sterile solution) | Pharmacia & Upjohn |
| | Cleocin T (topical gel and topical lotion) | Pharmacia & Upjohn |
| | Cleocin T (topical solution) | Pharmacia & Upjohn |
| | Cleocin vaginal (ovules) | Pharmacia & Upjohn |

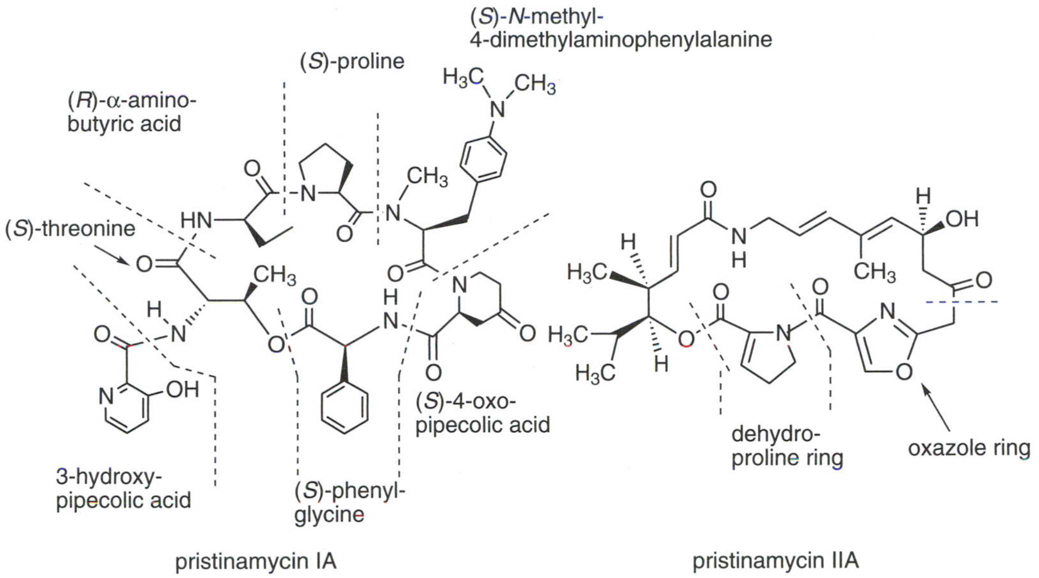

**Figure 18.19** Chemical structures of natural pristinamycin IA and IIA. The components of pristinamycin are indicated (see the text for an explanation). Note that the Cahn-Ingold-Prelog system is used to denote the configuration of the chiral centers, and an ethyl group is represented by a line in the aminobutyric acid moiety. The two rings that are part of the macrolactone of pristinamycin IIA (on the right) are also indicated.

physicochemical characteristic has prompted chemists to work on improving the solubility of each component without affecting the antimicrobial activity.

## Quinupristin and Dalfopristin

Synercid (Rhone-Poulenc Rorer Laboratories) was developed as a water-soluble and injectable semisynthetic streptogramin. The association is made up of the methanesulfonate ($CH_3SO_3^-$) salt of 5(*R*)-[3(-*S*)-quinuclidinyl]thiomethyl pristinamicin I (quinupristin) and of 26(*S*)-diethylaminoethyl sulfonyl pristinamycin $II_B$ (dalfopristin) in a 30:70 (wt/wt) ratio (Fig. 18.20). These compounds display a 5% water solubility at pHs of around 4.5. The mixture is available in a freeze-dried form to be dissolved in sterile isotonic solutions, suitable for intravenous use.

## Antimicrobial Activity and Clinical Uses

In September 1999, the FDA gave accelerated approval to quinupristin plus dalfopristin, marketed in a 30:70 combination ratio as the brand name of Synercid (Rhone-Poulenc Rorer), for the intravenous treatment of bacteremia and life-threatening infection caused by vancomycin-resistant *Enterococcus faecium* (VREF) and for the treatment of complicated skin and skin structure infections caused by *S. aureus* and *S. pyogenes*. It is equally active in vitro against MRSA, methicillin-susceptible *S. aureus*, methicillin-resistant and -susceptible *S. epidermidis* and penicillin-resistant and -susceptible *S. pneumoniae*. It also is active in vitro against *N. meningitidis*, *M. catarrhalis*, *L. pneumophila*, *M. pneumoniae*, and *C. perfringes*. It is inactive against *E. faecalis*. The accelerated approval was based on a demonstrated effect on a surrogate end point that is likely to predict clinical benefit (i.e., the ability of quinupristin-dalfopristin to clear VREF from the bloodstream).

The 1990s have seen an increase in the number of problems associated with multidrug-resistant gram-positive bacteria, including MRSA, methicillin-resistant coagulase-negative staphylococci (MRCNS), penicillin-resistant *S. pneumoniae*, and vancomycin-resistant enterococci. The prevalence of MRSA and MRCNS as major causes of morbidity and mortality in the hospital setting increased during the 1990s. Because glycopeptide antibiotics (vancomycin and teicoplanin) are often the only drugs still active against methicillin-resistant staphylococci, the emergence of staphylococci exhibiting reduced sensitivity to vancomycin was of particular concern. Consequently, there was an urgent need to develop new antibiotics not only to treat infections caused by multidrug-resistant gram-positive cocci but also to reduce the increasing selective pressure exerted by glycopeptides on gram-positive pathogens, mainly in hospitals. Quinupristin-dalfopristin has demonstrated significant in vitro activity against staphylococci including MRSA and MRCNS.

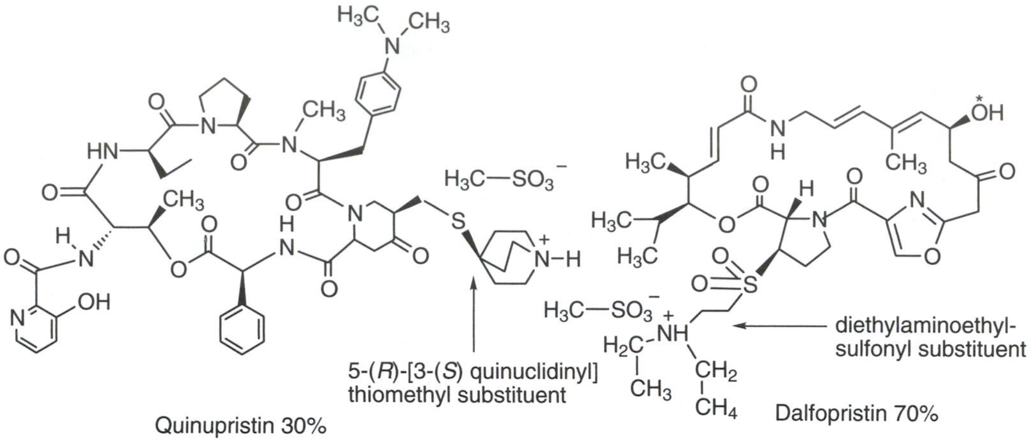

**Figure 18.20** Chemical structures of quinupristin and dalfopristin. An acetyltransferase inactivates dalfopristin (see the text for details). The position of the modified hydroxyl group in dalfopristin is indicated by the asterisk.

## Mechanism of Action

Quinupristin and dalfopristin inhibit bacterial protein synthesis by undergoing strong irreversible binding to the 50S ribosomal subunit. The group A and B streptogramins bind to separate sites on the 50S subunit of the bacterial ribosome. Both group A and B streptogramins apparently act during the early rounds of protein synthesis. The mechanism of action of group A streptogramins is related to inhibition of peptide bond formation either because of weakening of the binding of the peptidyl- and aminoacyl-tRNA to the ribosome or because of direct interference with the peptidyl transfer reaction. The mode of action of the group B streptogramins is thought to be similar to that of 14-member ring macrolides, which sterically hinder the growth of the nascent peptide during early rounds of translation.

Vannuffel et al. (1994) at the University of Louvain (Brussels, Belgium) investigated type B virginiamycin S (or VS) and type A virginiamycin M (or VM). Their results indicate that all rRNA bases, whose reactivity is modified by VM, are located outside the central loop of domain V. In contrast, the protective effect of macrolides and type B streptogramins is restricted to bases in the central loop of domain V. These results confirmed that the binding sites of VM and VS are different; they also confirmed the assumption that type A streptogramins promote a conformational change of ribosomes, which these authors considered responsible for the synergistic effect of the two virginiamycin components.

Cocito et al. (1997) reviewed useful information about the inhibition of protein synthesis by streptogramins and related antibiotics. They remarked that interference with the function of the peptidyltransferase catalytic center of the 50S subunit is fundamental to the antibacterial activity of the $MLS_B$ group of antibiotics. They also mentioned that in the presence of type A streptogramins, ribosomal initiation complexes are assembled in an apparently normal fashion but are functionally inactive, suggesting an inhibition of the elongation phase. Moreover, type B streptogramins inhibit the first two of the three elongation steps, aminoacyl-tRNA binding to the A site and peptidyl transfer from site P, but do not affect the third step, i.e., translocation from the A site to the P site.

## Bacterial Resistance

Since quinupristin (a group B streptogramin) and dalfopristin (a group A streptogramin) are chemically unrelated and have different binding sites, the mechanisms of resistance to these two compounds are different.

In France, resistance to mixtures of group A and B streptogramins was first encountered among staphylococci in 1975. Resistance to class B streptogramins takes place (i) by rRNA methylation catalyzed by a 23S rRNA methylase encoded in the *erm* genes, which also confers resistance to macrolides and lincosamides (the $MLS_B$ phenotype) (Fig. 18.21), or (ii) by an elimination of the hexadepsipeptide ring by specific lyases. This last mechanism has been well studied by Mukhtar and coworkers and was reported in 2001. They established that the enzyme Vgb, encoded by the *vgb* gene from *S. aureus,* inactivates streptogramin B antibiotics by an elimination mechanism rather than by hydrolysis across the ester

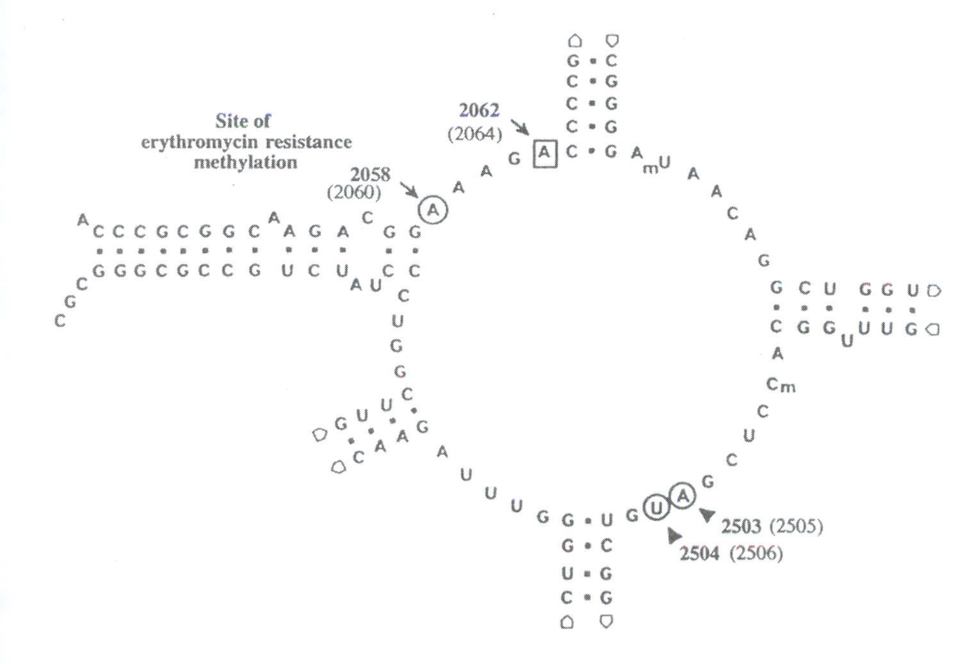


**Figure 18.21** Secondary structure of the peptidyltransferase loop in domain V of 23S rRNA. The mutated position in *S. pneumoniae* BM4455 is indicated by a square. The positions of the other binding sites of streptogramins (A2058, A2503, and U2504) are indicated by circles. Nucleotide sequence and numbering are those of *E. coli* 23S rRNA, and the corresponding *S. pneumoniae* numbering is given in parentheses. Reprinted from F. Depardieu and P. Courvalin, *Antimicrob. Agents Chemother.* **45:**319–323, 2001, with permission from the American Society for Microbiology.

bond, generating an N-terminal dehydrobutyrine group. Therefore, they demonstrated that Vgb is not a hydrolase but a lyase.

Staphylococcal resistance to class A streptogramins is mediated by two mechanisms: (i) active efflux due to ABC transporter protein (VgaA and VgaB) encoded by the genes *vgaA* and *vgaB* or (ii) inactivation coenzyme A-dependent O acetylation of the hydroxyl group of these compounds. This inactivation is catalyzed by streptogramin acetyltransferases encoded by *vatA, vatB,* and *vatC* in staphylococci. All tested staphylococci that are resistant to group A streptogramins carry one or two of the cited genes located on plasmids or on a transposon that might have a chromosomal location. In staphylococci, resistance to the synergistic mixture of the A and B group of streptogramins is always associated with resistance to the group A compound but not necessarily with resistance to the group B compound.

In *E. faecium,* two acetyltransferases encoded by the plasmid-borne *vatD* (previously designated *satA,* for "streptogramin acetyltransferase") and *vatE* (previously designated *satG*) genes inactivate class A streptogramins (in Fig. 18.20, the modified hydroxyl group in dalfopristin is indicated by an asterisk).

In 2002, Sugantino and Roderick reported the crystal structure of VatD, a streptogramin acetyltransferase from a human urinary isolate of *E. faecium* that was determined as an apoenzyme and in complex with either acetyl-coenzyme A or pristinamycin IIA and coenzyme A. These structures illustrate the location and arrangement of residues at the active site and point to His82 as a residue that may act as a general base.

## Streptogramin Preparations Commercially Available in the United States

Quinupristin-dalfopristin (Synercid, available for intravenous infusion), marketed by Aventis, is the only streptogramin antibiotic marketed in the United States.

## REFERENCES AND FURTHER READING

### Macrolides

#### General

**Bryskier, A. J., J. P. Butzler, H. C. Neu, and P. M. Tulkens.** 1993. *Macrolides: Chemistry, Pharmacology, and Clinical Uses.* Arnette Blackwell, Paris, France.

**Bryskier, A. J., and J. P. Butzler.** 1997. Macrolides, p. 377–393. *In* F. O'Grady, H. P. Lambert, R. G. Finch, and D. Greenwood (ed.), *Antibiotic and Chemotherapy.* Churchill Livingstone, Inc., New York, N.Y.

**Kirst, H. A.** 1993. Macrolides, p. 400–444. *In* M. Howe-Grant (ed.), *Chemotherapeutics and Disease Control.* John Wiley & Sons, Inc., New York, N.Y.

#### Mechanism of action

**Ban, N., P. Nissen, J. Hansen, M. Capel, P. B. Moore, and T. A. Steitz.** 1999. Placement of protein and RNA structures into a 5 Å-resolution map of the 50S ribosomal subunit. *Nature* 400:841–847.

**Hansen, L. H., P. Mauvais, and S. Douthwaite.** 1999. The macrolide-ketolide antibiotic binding site is formed by structures in domains II and V of 23S ribosomal RNA. *Mol. Microbiol.* **31:**623–631.

**Mazzei, T., E. Mini, A. Novelli, and P. Periti.** 1993. Chemistry and mode of action of macrolides. *J. Antimicrob. Chemother.* **31**(Suppl. C)**:**1–9.

**Retsema, J., and W. Fu.** 2001. Macrolides: structures and microbial targets. *Int. J. Antimicrob. Agents* **18:**S3–S10.

**Schlünzen, F., R. Zarivach, J. Harms, A. Bashan, A. Tocilj, R. Abrecht, A. Yonath, and F. Franceschi.** 2001. Structural basis for the interaction of antibiotics with the peptidyl transferase centre in eubacteria. *Nature* **413:**814–821.

**Weisblum, B.** 1995. Insights into erythromycin action from studies of its activity as inducer of resistance. *Antimicrob. Agents Chemother.* **39:**797–905.

**Williams, J. D., and A. M. Sefton.** 1993. Comparison of macrolide antibiotics. *J. Antimicrob. Chemother.* **31**(Suppl. C)**:**11–26.

**Mechanism of resistance**

**Clancy, J., J. Petitpas, F. Dib-Hajj, W. Yuan, M. Cronan, A. V. Kamath, J. Bergeron, and J. A. Retsema.** 1996. Molecular cloning and functional analysis of a novel macrolide-resistance determinant, *mefA,* from *Streptococcus pyogenes. Mol. Microbiol.* **22:**867–879.

**Depardieu, F., and P. Courvalin.** 2001. Mutation in 23S rRNA responsible for resistance to 16-membered macrolides and streptogramins in *Streptococcus pneumoniae. Antimicrob. Agents Chemother.* **45:**319–323.

**Douthwaite, S., L. H. Hansen, and P. Mauvais.** 2000. Macrolide-ketolide inhibition of MLS-resistant ribosomes is improved by alternative drug interaction with domain II of 23S rRNA. *Mol. Microbiol.* **36:**183–193.

**Kataja, J., H. Seppälä, M. Skurnik, H. Sarkkinen, and P. Huovinen.** 1998. Different erythromycin resistance mechanisms in group C and group G streptococci. *Antimicrob. Agents Chemother.* **42:**1493–1494.

**Leclercq, R., and P. Courvalin.** 1991. Intrinsic and unusual resistance to macrolide, lincosamide, and streptogramin antibiotics in bacteria. *Antimicrob. Agents Chemother.* **35:**1273–1276.

**Lina, G., A. Quaglia, M. E. Reverdy, R. Leclercq, F. Vandenesch, and J. Etienne.** 1999. Distribution of genes encoding resistance to macrolides, lincosamides, and streptogramins among staphylococci. *Antimicrob. Agents Chemother.* **43:**1062–1066.

**Pechère, J. C.** 2001. Macrolide resistance mechanisms in Gram-positive cocci. *Int. J. Antimicrob. Agents* **18:**S25-S28.

**Portillo, A., F. Ruiz-Larrea, M. Zarazaga, A. Alonso, J. L. Martinez, and C. Torres.** 2000. Macrolide resistance genes in *Enterococcus* spp. *Antimicrob. Agents Chemother.* **44:**967–971.

**Roberts, M. C., J. Sutcliffe, P. Courvalin, L. B. Jensen, J. Rood, and H. Seppala.** 1999. Nomenclature for macrolide and macrolide-lincosamide-streptogramin B resistance determinants. *Antimicrob. Agents Chemother.* **43:**2823–2830.

**Sutcliffe, J., T. Grebe, A. Tait-Kamradt, and L. Wondrack.** 1996. Detection of erythromycin-resistant determinants by PCR. *Antimicrob. Agents Chemother.* **40:**2562–2566.

**Sutcliffe, J. A., J. P. Mueller, and E. A. Utt.** 1999. Antibiotic resistance mechanisms of bacterial pathogens, p. 759–788. *In* A. L. Demain and J. E. Davies (ed.), *Manual of Industrial Microbiology and Biotechnology,* 2nd ed. ASM Press, Washington, D.C.

**Sutcliffe, J., A. Tait-Kamradt, and L. Wondrack.** 1996. *Streptococcus pneumoniae* and *Streptococcus pyogenes* resistant to macrolides but sensitive to clindamycin: a common resistance pattern mediated by an efflux system. *Antimicrob. Agents Chemother.* **40:**1817–1824.

**Weisblum, B.** 2000. Resistance to the macrolide-lincosamide-streptogramin antibiotics, p. 694–710. *In* V. A. Fischetti, R. P. Novick, J. J. Ferretti, D. A. Portnoy, and J. I. Rood (ed.), *Gram-Positive Pathogens.* ASM Press, Washington, D.C.

**Ribosome modification**

**Douthwaite, S., and B. Vester.** 2000. Macrolide resistance conferred by alterations in the ribosome target site, p. 431–439. *In* R. A. Garret, S. R. Douthwaite, A. Liljas, A. T. Matheson, P. B. Moore, and H. E. Noller (ed.), *The Ribosome: Structure, Function, Antibiotics, and Cellular Interactions.* ASM Press, Washington, D.C.

**Leclercq, R., and P. Courvalin.** 1991. Bacterial resistance to macrolide, lincosamide, and streptogramin antibiotics by target modification. *Antimicrob. Agents Chemother.* **35:**1267–1272.

**Skinner, R., E. Cundliffe, and F. J. Schmidt.** 1983. Site of action of a ribosomal RNA methylase responsible for resistance to erythromycin and other antibiotics. *J. Biol. Chem.* **258:**12702–12705.

**Vester, B., and S. Douthwaite.** 2000. Macrolide resistance conferred by base substitutions in 23S rRNA. *Antimicrob. Agents Chemother.* **45:**1–12.

**Weisblum, B.** 1995. Erythromycin resistance by ribosome modification. *Antimicrob. Agents Chemother.* **39:**577–585.

**Xiong, L., S. Shah, P. Mauvais, and A. S. Mankin.** 1999. A ketolide resistance mutation in domain II of 23S rRNA reveals the proximity of hairpin 35 to the peptidyl transferase centre. *Mol. Microbiol.* **31:**633–639.

**Efflux mechanisms**

**Sutcliffe, J.** 1999. Resistance to macrolides mediated by efflux mechanisms. *Curr. Opin. Anti-Infect. Investig. Drugs* **1:**403–412.

**Zhong, P., and V. D. Shortridge.** 2000. The role of efflux in macrolide resistance. *Drug Resist. Updates* **3:**325–329.

**Enzymatic inactivation**

**Noguchi, N., A. Emura, H. Matsuyama, K. O'Hara, M. Sasatsu, and M. Kono.** 1996. Nucleotide sequence and characterization of erythromycin resistance determinant that encode macrolide 2'-phosphotransferase I in *Escherichia coli. Antimicrob. Agents Chemother.* **39:**2359–2362.

Clinical uses

**Kucers, A., S. M. Crowe, M. L. Grayson, and J. F. Hoy.** 1997. *The Use of Antibiotics,* 5th ed., p. 607–666. Butterworth-Heinemann, Oxford, United Kingdom.

Recent developments in macrolides and ketolides

**Agouridas, C., A. Denis, J. M. Auger, Y. Benedetti, A. Bonnefoy, F. Bretin, J. F. Chantot, A. Dussarat, C. Fromentin, S. G. D'Ambrières, S. Lachaud, P. Laurin, O. Le Martret, V. Loyau, and N. Tessot.** 1998. Synthesis and antibacterial activity of ketolides (6-O-methyl-3-oxoerythromycin derivatives): a new class of antibacterials highly potent against macrolide-resistant and -susceptible respiratory pathogens. *J. Med. Chem.* **41:**4080–4100.

**Bonnefoy, A., A. M. Girard, C. Agouridas, and J. F. Chantot.** 1997. Ketolides lack inducibility properties of MLSB resistance phenotype. *J. Antimicrob. Chemother.* **40:**85–90.

**Chu, D. T. W.** 1999. Recent developments in macrolides and ketolides. *Curr. Opin. Microbiol.* **2:**467–474.

**Elliot, R. L., D. Pireh, G. Griesgraber, A. M. Nilius, P. J. Ewing, M. H. Bui, P. M. Raney, R. K. Flamm, K. Kim, R. F. Henry, D. T. W. Chu, J. J. Plattner. and Y. S. Or.** 1998. Anhydrolide macrolides. 1. Synthesis and antibacterial activity of 2,3-anhydro-6-O-methyl 11,12-carbamate erythromycin A analogues. *J. Med. Chem.* **41:**1651–1659.

**Griesgraber, G., M. K. Kramer, R. L. Elliot, A. M. Nilius, P. J. Ewing, P. M. Raney, M. H. Bui, R. K. Flamm, D. T. W. Chu, J. J. Plattner, and Y. S. Or.** 1998. Anhydrolide macrolides. 2. Synthesis and antibacterial activity of 2,3-anhydro-6-O-methyl-11,12-carbazate erythromycin A analogues. *J. Med. Chem.* **41:**1660–1670.

**Or, Y. S., R. F. Clark, S. Wang, D. T. W. Chu, A. M. Nilius, R. K. Flamm, M. Mitten, P. Ewing, J. Alder, and Z. Ma.** 2000. Design, synthesis, and antimicrobial activity of 6-O-substituted ketolides active against resistant respiratory tract pathogens. *J. Med. Chem.* **43:**1043–1049.

Telithromycin

**Ackermann, G., and A. C. Rodloff.** 2003. Drugs of the 21st century: telithromycin (HMR 3647)—the first ketolide. *J. Antimicrob. Chemother.* **51:**497–511.

**Barman Balfour, J. A., and D. P. Figgitt.** 2001. Telithromycin. *Drugs* **61:**815–829.

**Douthwaite, S., and W. S. Champney.** 2001. Structures of ketolides and macrolides determine their mode of interaction with the ribosomal target site. *J. Antimicrob. Chemother.* **48**(Suppl. T1):1–8.

**Felmingham, D., G. Zhanel, and D. Hoban.** 2001. Activity of the ketolide antibacterial telithromycin against typical community-acquired respiratory pathogens. *J. Antimicrob. Chemother.* **48**(Suppl. T1):33–42.

**Fines, M., and R. Leclercq.** 1999. New antibiotics in development in the macrolide, lincosamide and streptogramin group. *Curr. Opin. Anti-Infect. Investig. Drugs* **1:**443–452.

**Graul, A., and J. Castanier.** 1998. HMR-3647. *Drugs Future* **23:**591–597.

**Hammerschlag, M. R., P. M. Roblin, and C. M. Bébéar.** 2001. Activity of telithromycin, a new ketolide antibacterial, against atypical and intracellular respiratory tract pathogens. *J. Antimicrob. Chemother.* **48**(Suppl. T1):25–30.

**Leclercq, R.** 2001. Overcoming antimicrobial resistance: profile of a new ketolide antibacterial, telithromycin. *J. Antimicrob. Chemother.* **48**(Suppl. T1):9–23.

**Namour, F., D. H. Wessels, M. H. Pascual, D. Reynols, E. Sultan, and B. Lenfant.** 2001. Pharmacokinetics of the new ketolide telithromycin (HMR 3647) administered in ascending single and multiple doses. *Antimicrob. Agents Chemother.* **45:**170–275.

**Roberts, M. C.** 1999. Telithromycin. *Curr. Opin. Anti-Infect. Investig. Drugs* **1:**506–513.

ABT-773

**Barry, A. L., P. C. Fuchs, and S. D. Brown.** 2001. In vitro activity of the ketolide ABT-773. *Antimicrob. Agents Chemother.* **45:**2922–2924.

**Schlünzen, F., J. M. Harms, F. Franceschi, H. A. S. Hansen, H. Bartels, R. Zarivach, and A. Yonath.** 2003. Structural basis for the antibiotic activity of ketolides and azolides. *Structure,* **11:**329–338.

TEA0777

**Tanikawa, T., T. Asaka, M. Kashimura, Y. Misawa, K. Suzuki, M. Sato, K. Kameo, S. Morimoto, and A. Nishida.** 2001. Synthesis and antibacterial activity of acylides (3-O-acyl-erythromycin derivatives): a novel class of macrolide antibiotics. *J. Med. Chem.* **44:**4027–4030.

**Lincosamides**

Chemistry

**Bannister, B.** 1993. Lincosamides, p. 390–399. *In* M. Howe-Grant (ed.), *Chemotherapetics and Disease Control.* John Wiley & Sons, Inc., New York, N.Y.

Mechanism of action

**Menninger, J. R., and R. A. Coleman.** 1993. Lincosamide antibiotics stimulate dissociation of peptidyl-tRNA from ribosomes. *Antimicrob. Agents Chemother.* **37:**2027–2029.

**Schlünzen, F., R. Zarivach, J. Harms, A. Bashan, A. Tocilj, R. Abrecht, A. Yonath, and F. Franceschi.** 2001. Structural basis for the interaction of antibiotics with the peptidyl transferase centre in eubacteria. *Nature* **413:**814–821.

Antimicrobial activity and clinical uses

**American Medical Association.** 1996. *Drug Evaluations Annual 1995,* p. 1509–1517. American Medical Association, Chicago, Ill.

**Medical Economics Data Production Co.** 2000. *Physicians' Desk Reference,* 54th ed., p. 2421–2427. Medical Economics Data Production Co., Montvale, N.J.

**Reese, R. E., R. F. Betts, and B. Gumustop.** 2000. *Handbook of Antibiotics,* 3rd ed., p. 435–440. Lippincott Williams & Wilkins, Philadelphia, Pa.

**Scholar, E. M., and W. B. Pratt.** 2000. *The Antimicrobial Drugs,* 2nd ed. Oxford University Press, Oxford, United Kingdom.

Resistance

**Bozgodan, B., B. Berrezouga, M. S. Kuo, D. A. Yurek, K. A. Farley, B. J. Stockman, and R. Leclercq.** 1999. A new resistance gene, *linB,* conferring resistance to lincosamides by nucleotidylation in *Enterococcus faecium* HM1025. *Antimicrob. Agents Chemother.* **43:**925–929.

**Douthwaite, S.** 1992. Interaction of the antibiotics clindamycin and lincomycin with *Escherichia coli* 23S ribosomal RNA. *Nucleic Acids Res.* **20:**4717–4720.

**Lina, G., A. Quaglia, M. E. Reverdy, R. Leclercq, F. Vandenesch, and J. Etienne.** 1999. Distribution of genes encoding resistance to macrolides, lincosamides, and streptogramins among staphylococci. *Antimicrob. Agents Chemother.* **43:**1062–1066.

## Streptogramins

Chemistry, biochemistry, and medicinal chemistry

**Barrière, J. C., N. Berthayd, D. Beyer, S. Dutka-Malen, J. M. Paris, and J. F. Desnottes.** 1998. Recent developments in streptogramin research. *Curr. Pharm. Des.* **4:**155–180.

**Barrière, J. C., and J. M. Paris.** 1993. RP 59500 and related semisynthetic streptogramins. *Drugs Future* **18:**833–845.

**Paris, J. M., J. C. Barrière, C. Smith, and P. E. Bost.** 1990. The chemistry of pristinamycins, p. 183–248. *In* G. Lukacs and M. Ohno (ed.), *Recent Progress in the Chemical Synthesis of Antibiotics.* Springer-Verlag KG, Berlin, Germany.

Spectrum of antimicrobial activity

**Bouanchaud, D. H.** 1997. *In-vitro* and *in-vivo* antibacterial activity of quinupristin/dalfopristin. *J. Antimicrob. Chemother.* **39**(Suppl. A)**:**15–21.

Clinical use

**Reese, R. E., R. F. Betts, and B. Gumustop.** 2000. *Handbook of Antibiotics,* 3rd ed., p. 515–520. Lippincott Williams & Wilkins, Philadelphia, Pa.

Mechanism of action

**Cocito, C., M. Di Giambatista, E. Nyssen, and P. Vannuffel.** 1997. Inhibition of protein synthesis by streptogramins and related antibiotics. *J. Antimicrob. Chemother.* **39**(Suppl. A)**:** 7–13.

**Vannuffel, P., M. Di Giambattista, and C. Cocito.** 1994. Chemical probing of virginiamycin M-promoted conformational change of the peptidyltransferase domain. *Nucleic Acids Res.* **22:**4449–4453.

**Vannuffel, P., M. Di Giambattista, and C. Cocito.** 1992. The role of rRNA bases in the interaction of peptidyltransferase inhibitors with bacterial ribosomes. *J. Biol. Chem.* **267:**16114–16120.

Bacterial resistance

**Baquero, F.** 1997. Gram-positive resistance: challenge for the development of new antibiotics. *J. Antimicrob. Chemother.* **39**(Suppl. A)**:**1–6.

**Depardieu, F., and P. Courvalin.** 2001. Mutation in 23S rRNA responsible for resistance to 16-membered macrolides and streptogramins in *Streptococcus pneumoniae. Antimicrob. Agents Chemother.* **45:**319–323.

**Malbruny, B., A. Canu, B. Bozdogan, B. Fantin, V. Zarrouk, S. Dutka-Malen, C. Feger, and R. Leclercq.** 2002. Resistance to quinupristin-dalfopristin due to mutation of L22 ribosomal protein in *Staphylococcus aureus. Antimicrob. Agents Chemother.* **46:**2200–2207.

**Sugantino, M., and S. L. Roderick.** 2002. Crystal structure of Van(D): an acetyltransferase that inactivates streptogramin group A antibiotics. *Biochemistry* **41:**2209–2216.

**Witte, W.** 1999. Antibiotic resistance in gram-positive bacteria: epidemiological aspects. *J. Antimicrob. Chemother.* **44**(Topic A)**:**1–9.

Streptogramin resistance in staphylococcal isolates

**Allignet, J., S. Aubert, A. Morvan, and N. El Sohl.** 1996. Distribution of genes encoding resistance to streptogramin A and related compounds among staphylococci resistant to these antibiotics. *Antimicrob. Agents Chemother.* **40:**2523–2528.

**Allignet, J., and N. El Sohl.** 1995. Diversity among the gram-positive acetyltransferases inactivating streptogramin A and structurally related compounds and characterization of a new staphylococcal determinant, *vatB. Antimicrob. Agents Chemother.* **39:**2027–2036.

**Allignet, J., N. Liassine, and N. El Sohl.** 1998. Characterization of staphylococcal plasmid related to pUB110 and carrying two novel genes, *vatC* and *vghB,* encoding resistance to streptogramins A and B and similar antibiotics. *Antimicrob. Agents Chemother.* **42:**1794–1798.

**Allignet, J., V. Loncle, C. Simenel, M. Delpierre, and N. El Sohl.** 1993. Sequence of a staphylococcal gene, *vat,* encoding an acetyltransferase inactivating the A-type compounds of virginiamycin-like antibiotics. *Gene* **130:**91–98.

**Allignet, J., V. Loncle, and N. El Sohl.** 1992. Sequence of a staphylococcal plasmid gene, *vga,* encoding a putative ATP-binding protein involved in resistance to virginiamycin A-like antibiotics. *Gene* **117:**45–51.

**Allignet, J., V. Loncle, P. Mazodier, and N. El Sohl.** 1988. Nucleotide sequence of a staphylococcal plasmid gene, *vgb,* encoding a hydrolase inactivating the B component of virginiamycin-like antibiotics. *Plasmid* **20:**271–275.

**Mukhtar, T. A., K. P. Kotera, D. W. Hughes, and G. D. Wright.** 2001. Vgb from *Staphylococcus aureus* inactivates streptogromin B antibiotics by an elimination mechanism, not hydrolysis. *Biochemistry* **40:**8877–8886.

**Werner, G., and W. Witte.** 1999. Characterization of a new enterococcal gene, satG, encoding a putative acetyltransferase conferring resistance to streptogramin A compounds. *Antimicrob. Agents Chemother.* **43:**1813–1814.

Streptogramin resistance in *Enterococcus faecium* isolates

**Bozgodan, B., and R. Leclercq.** 1999. Effects of genes encoding resistance to streptogramins A and B on the activity of quinupristin-dalfopristin against *Enterococcus faecium*. *Antimicrob. Agents Chemother.* **43**:2720–2725.

**Eliopoulos, G. M., C. B. Wennersten, H. S. Gold, T. Schülin, M. Souli, M. G. Farris, S. Cerwinka, H. L. Nadler, M. Dowzicky, G. H. Talbot, and R. C. Moellering.** 1998. Characterization of vancomycin-resistant *Enterococcum faecium* isolates from the United States and their susceptibility in vitro to dalfopristin-quinupristin. *Antimicrob. Agents Chemother.* **42**: 1088–1092.

**Jensen, L. B., A. M. Hammerum, F. K. Aarestrup, A. E. van den Bogaard, and E. E. Stobberingh.** 1998. Occurrence of *satA* and *vgb* genes in streptogramin-resistant *Enterococcus faecium* isolates of animal and human origins in The Netherlands. *Antimicrob. Agents Chemother.* **42**:3330–3331.

**Rende-Fournier, R., R. Leclercq, M. Galimand, J. Duval, and P. Courvalin.** 1993. Identification of the *satA* gene encoding a streptogramin A acetyltransferase in *Enterococcus faecium* BM4145. *Antimicrob. Agents Chemother.* **37**:2119–2125.

**Soltani, M., D. Beighton, J. Philpott-Howard, and N. Woodford.** 2001. Identification of *vat*(E-3), a novel gene encoding resistance to quinupristin-dalfopristin in a strain of *Enterococcus faecium* from a hospital patient in the United Kingdom. *Antimicrob. Agents Chemother.* **45**:645–646. (Erratum, **45**:998.)

chapter 19



# An Inhibitor of the 50S Ribosomal Subunit
# Chloramphenicol

## Structure and Current Status of Clinical Uses

Chloramphenicol has the structure shown in Fig. 19.1. It was originally isolated in 1947 from the fermentation of *Streptomyces venezuelae,* but it is now prepared synthetically. It was the first orally active broad-spectrum antibiotic to be discovered.

The potential of chloramphenicol to cause irreversible fatal aplastic anemia in rare cases has limited its usefulness to the treatment of serious infections in which the anatomic location of the infection, the susceptibility of the causative organism, or individual patient characteristics limit or prevent the use of less toxic antibacterial agents. Chloramphenicol has been available since 1949 under the trade name Chloromycetin (issued to Parke-Davies Co.); it is sold in capsule, suspension, and powder (for injection) forms. Parke-Davies Pharmaceutical Co. no longer markets this drug in the United States.

## Structural Features and Structure-Activity Relationships

The chemical name of chloramphenicol is D-(−)-*threo*-*p*-nitrophenyl-2-dichloroacetamido-1,3-propanediol. The chemical structure of chloramphenicol has functional groups not encountered in other antibiotics, such as a nitro group (-$NO_2$) as substituent in the *para* or 4 position of the benzene ring and the dichloroacetamide group as substituent on the carbon atom at position 2. The two stereogenic centers at carbons 1 and 2 allow for four possible stereoisomers; however, only the naturally occurring D(−)-*threo* configuration, or 1*R*,2*R*, is active.

Structure-activity relationships (SAR) and mechanism-of-action studies indicate that the requirements for chloramphenicol activity are (i) the D-(−)-*threo* configuration (or 1*R*,2*R*), (ii) the 1,3-propanediol moiety, and (iii) a strong withdrawing group on the *para* or 4 position of the benzene ring.

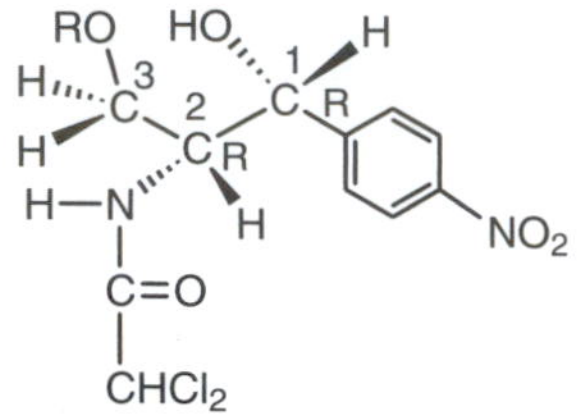

Chloramphenicol $R = H$

Chloramphenicol palmitate $R = -C(=O)-(CH_2)_{12}CH_3$

Chloramphenicol hemisuccinate $R = -C(=O)-CH_3CH_2CO_2H$

**Figure 19.1** Chemical structures of chloramphenicol, chloramphenicol palmitate, and chloramphenicol hemisuccinate.

Attempts to modify the chemical structure of chloramphenicol have generally resulted in a marked loss of activity, except in thiamphenicol, a compound in which a sulfomethyl group ($CH_3SO_2^-$) replaces the nitro group of chloramphenicol. Thiamphenicol has antibacterial activity comparable to that of chloramphenicol itself. After considerable SAR research, it appears that the natural structure of chloramphenicol is the optimum structure for antibacterial activity. Indeed, this is in contrast to what we have seen with other groups of antibiotics, where molecular modifications resulted in improvements of antibacterial activity, extension of the spectrum of activity, reduced toxicity, or improvement in chemical stability or in overcoming bacterial resistance.

## Antimicrobial Activity and Clinical Uses

Chloramphenicol is active against a wide range of gram-positive and gram-negative bacteria. In addition, it is potent against intracellular rickettsial infections, e.g., Rocky Mountain spotted fever (caused by *Rickettsia rickettsii*), the lymphogranuloma-psittacosis group (caused by *Chlamydia trachomatis*), and *Vibrio cholerae* infection. It is particularly active against *Salmonella enterica* serovar Typhi and *Haemophilus influenzae*.

A valuable property of chloramphenicol is that it readily crosses the blood-brain barrier and can therefore be used to treat infections of the central nervous system caused by susceptible organisms. Nowadays, chloramphenicol remains an alternative therapy for the treatment of meningitis caused by *Neisseria meningitidis, Streptococcus pneumoniae,* and *H. influenzae* in penicillin-allergic patients and is also used for the treatment of bacterial brain abscesses. Traditionally, it has been a first choice antibacterial agent against typhoid fever (*S. enterica* serovar Typhi), although resistance is quite widespread nowadays.

Chloramphenicol is poorly soluble in water and tastes extremely bitter. To mask the bitter taste for pediatric patients, esterification of the primary alcohol at C-3 with palmitoyl chloride yields the water-insoluble tasteless prodrug palmitate of chloramphenicol (the chemical structure is given in Fig. 19.1). Its hydrolysis in the duodenum releases chloramphenicol. A freely soluble prodrug is chloramphenicol hemisuccinate (Fig. 19.1). This was formulated to improve solubility in water for intravenous and intramuscular injection. These two prodrugs lack antibacterial activity, but they serve to release chloramphenicol, the active drug.

The main reason for the caution in using chloramphenicol is the possibility of developing fatal aplastic anemia and other blood dyscrasias such as agranulocytosis. Aplastic anemia is rare, occurring in only 1 in 25,000 to 40,000 patients given a course of chloramphenicol therapy, and usually is fatal. The aplasia is not dose related and can become manifest weeks to months after the use of chloramphenicol. The precise mechanism is unknown. Chloramphenicol is therefore reserved for situations where the benefits exceed the risk. These include typhoid fever and bacterial meningitis, which on some occasions are unresponsive to other antibacterial agents either because of resistance or because of the inability of the agents to cross noninflamed meninges.

## Mechanism of Action

Chloramphenicol is known to bind at the peptidyltransferase center of the large ribosomal subunit. It interferes with peptide bond formation by preventing the binding of aminoacyl-tRNA to the ribosomal A site. It may also directly interfere with the transfer of the peptidyl residue from the peptidyl-tRNA to aminoacyl-tRNA.

To elucidate the structural basis of ribosome-chloramphenicol interactions, scientists from the Max Planck Institute (Hamburg and Berlin, Germany) and the Weizmann Institute (Rehovot, Israel) have determined the high-resolution X-ray structure of the 50S ribosomal subunit of the eubacterium *Deinococcus radiodurans* in complex with chloramphenicol (Schlünzen et al., 2001). The result shows that chloramphenicol targets the 50S subunit only at the peptidyltransferase cavity and interacts exclusively with specific nucleotides assigned to a multibranched loop of domain V of the 23S rRNA. Figure 19.2 shows a two-dimensional structure diagram of the interaction of chloramphenicol with the peptidyltransferase cavity. These data confirm the previous mutational and footprint data and also show that the interactions of chloramphenicol do not involve any ribosomal proteins.

The authors suggested that one of the oxygen atoms of the *para*-nitro ($p$-$NO_2$) group of chloramphenicol

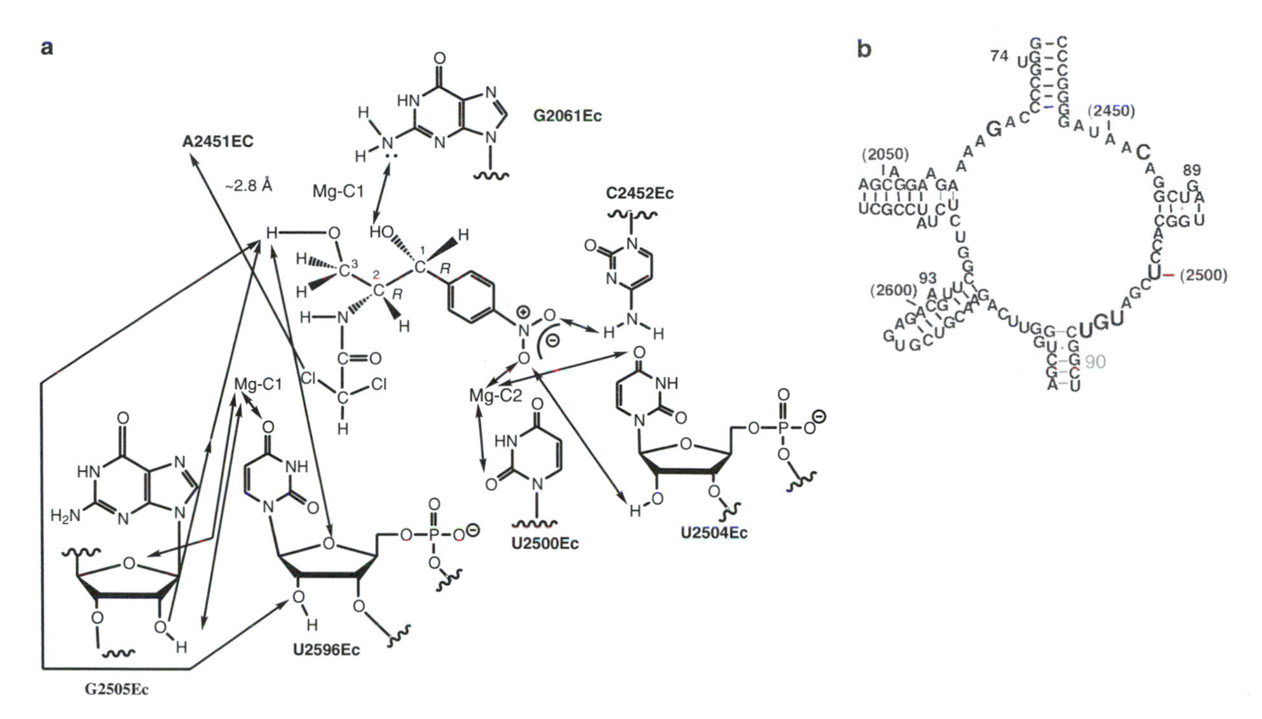


**Figure 19.2** Interactions of chloramphenicol with the peptidyltransferase cavity. (a) Chemical structure diagram of chloramphenicol showing the interactions (arrows) of its reactive groups with the nucleotides of the peptidyltransferase cavity. Arrows between two chemical moieties indicate that the two groups are less than 4.4 Å apart. (b) Secondary structure of the peptidyltransferase ring of *Deinococcus radiodurans,* showing the nucleotides involved in the interaction with chloramphenicol. Reprinted from F. Schlünzen, R. Zarivach, J. Harms, A. Bashan, A. Tocilj, R. Albrecht, A. Yonath, and F. Franceschi, *Nature* **413:**814–821, 1999, with permission from the publisher.

appears to form hydrogen bonds with N-4 of C2431Dr (C2452Ec) (where Ec stands for *Escherichia coli*), which is involved in chloramphenicol resistance. The other oxygen atom of the *p*-$NO_2$ group interacts with O-2' of U2483Dr (U2504Ec). The -OH group of chloramphenicol at C-1 is located within a hydrogen-bonding distance from N-2 of G2044Dr (G2061Ec) of the 23S rRNA. The -OH goup of chloramphenicol at C-3 is fundamental for its activity, and in vivo modification of this group confers resistance to the drug. This group is within a hydrogen-bonding distance from O-4' of U2485Dr (U2506Ec) and O-2' of G2484Dr (G2505Ec). The 4' carbonyl group appears to form a hydrogen bond with the -OH of U2485Dr (U2506Ec) at C-2'.

## Bacterial Resistance

### Enzymatic Inactivation

The most clinically important mechanism of resistance in bacteria is that of O acetylation catalyzed by the enzyme chloramphenicol *O*-acetyltransferase (CAT). Plasmid-borne *cat* genes have been identified among most genera of gram-positive and gram-negative bacteria. One of the most troublesome forms of CAT-mediated resistance is that frequently found in strains of *S. enterica* serovar Typhi, the agent of typhoid fever.

There are three main types of CAT, types I, II, and III. All CAT proteins normally exist in solution as homotrimers. Type III is the most important, and a wealth of information is now available, including the tertiary structure at high resolution and the structural determinants for the binding of each substrate. It will be instructive to present here the postulated general mechanism for the CAT-catalyzed acetylation of chloramphenicol and then consider this reaction in more detail. Fig. 19.3 shows the transformations involved in the conversion in vivo of the two substrates (chloramphenicol and 1-O-acetylchloramphenicol) to their final product, 1,3-di-*O*-acetylchloramphenicol. Acetylation of the primary alcohol at C-3 is catalyzed by CAT with the cofactor acetyl coenzyme A (acetyl-CoA), yielding 3-*O*-acetylchloramphenicol. Next, there is a nonenzymatic reaction (an intramolecular transfer of the acetyl group

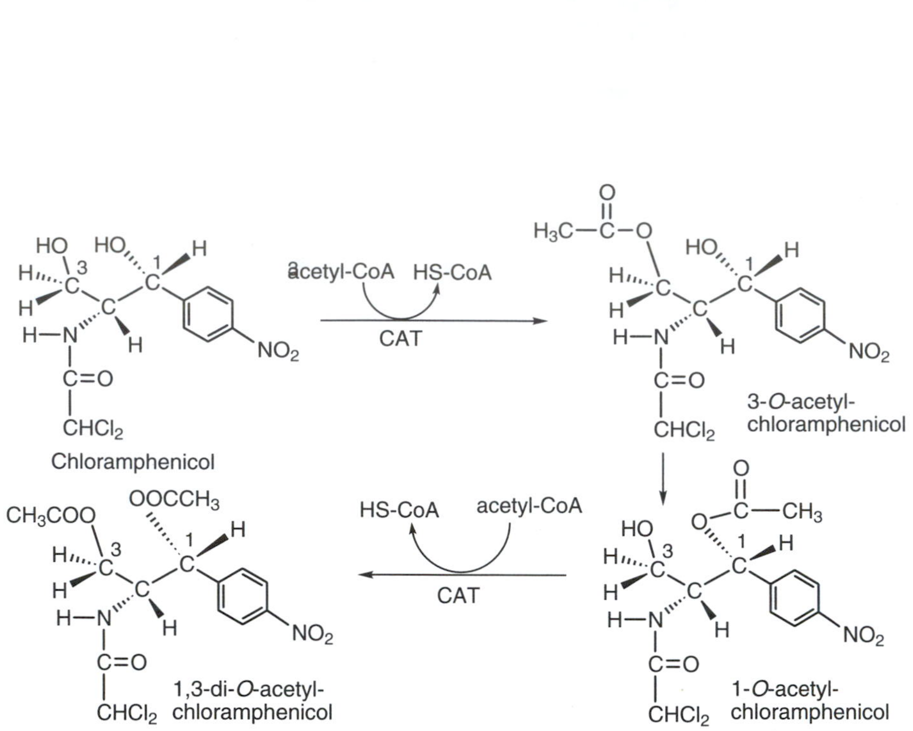

**Figure 19.3** CAT-catalyzed acetylation of chloramphenicol and 1-*O*-acetylchloramphenicol, yielding 1,3-di-*O*-acetylchloramphenicol as the final product.

from C-3 to C-1), yielding a mixture of 1-O- and 3-O-monoacetylated product. The 1-O-acetylchloramphenicol thus formed is available for a second round of enzymatic acetylation at the C-3 position, yielding 1,3-diacetylchloramphenicol. Both the resultant mono- and di-O-acetylated chloramphenicol are inactive. They do not bind to the 50S ribosomal subunit.

The catalytic machinery of type III CAT comprises a number of amino acids which make precise contacts with the substrate molecules of chloramphenicol and 1-O-acetylchloramphenicol. Central to catalysis is His195, which supplies the general base ($N^{\epsilon 2}$ in Figure 19.4) to deprotonate the C-3 hydroxyl of both substrates, producing an oxyanion intermediate which in turn attacks the carbonyl carbon of acetyl-CoA to yield a tetrahedral intermediate (Fig. 19.4). Essential to the stabilization of the latter is a hydrogen bond between its oxygen and the hydroxyl of Ser-148. Collapse of the tetrahedral intermediate yields the 3-O-acetylchloramphenicol and HS-CoA, with the coincident regeneration of the imidazole of His195, ready for another cycle of acetylation.

Although the CAT mechanism for resistance to chloramphenicol is widespread in bacteria, it is not used by the chloramphenicol-producing *Streptomyces* strains to protect themselves against their own toxic product. However, a 3-O phosphoester of chloramphenicol was identified in *Streptomyces venezuelae,* suggesting that the producing organism has a mechanism of chloramphenicol resistance that has not been encountered in other microbial systems. The enzyme responsible for this novel inactivation by phosphorylation, a 3-O-phosphotransferase, was crystallized and its X-ray structure was determined by Izard and Ellis (2000).

## Active Efflux

In 1994, Stuart B. Levy and coworkers reported their studies on active efflux of chloramphenicol in susceptible *E. coli* strains and in multiple-antibiotic-resistant (Mar) mutants; the mechanism was shown to depend on proton motive force.

**Figure 19.4** Mechanism postulated for the inactivation of chloramphenicol catalyzed by CAT with the cofactor acetyl-CoA.

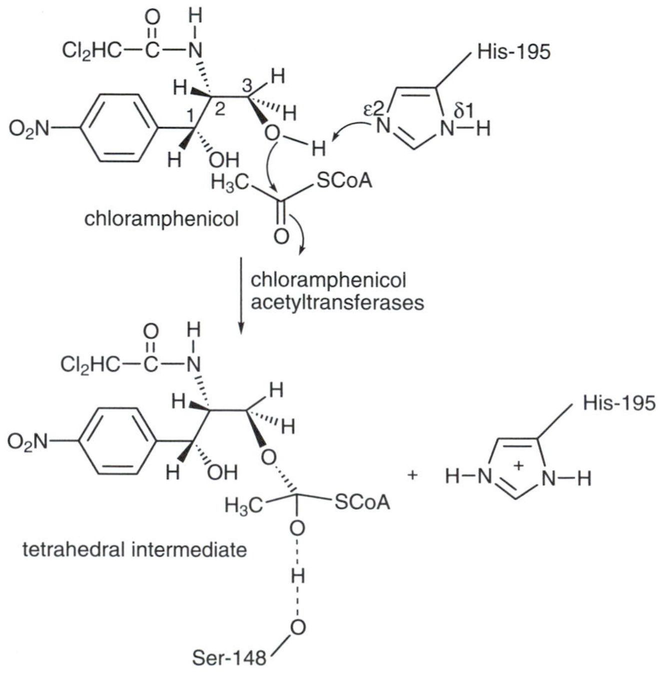

## REFERENCES AND FURTHER READING


### General

**Nagabhushan, T., G. H. Miller, and K. J. Varma.** 1993. Chloramphenicol and analogues, p. 175–192. *In* M. Howe-Grant (ed.), *Chemotherapeutics and Disease Control.* John Wiley & Sons, Inc., New York, N.Y.

### Mechanism of Action

**Schlünzen, F., R. Zarivach, J. Harms, A. Bashan, A. Tocilj, R. Albrecht, A. Yonath, and F. Franceschi.** 2001. Structural basis for the interaction of antibiotics with the peptidyl transferase centre in eubacteria. *Nature* **413:**814–821.

### Clinical Use

**Reese, R. E., R. F. Betts, and B. Gumustop.** 2000. *Handbook of Antibiotics,* p. 441–445. Lippincott Williams & Wilkins, Philadelphia, Pa.

### Bacterial Resistance

#### Chloramphenicol O-acetyltransferases

**Leslie, A. G. W.** 1990. Refined crystal structure of chloramphenicol acetyltransferase at 1.75 Å resolution. *J. Mol. Biol.* **213:**167–186.

**Leslie, A. G. W., P. C. E. Moody, and W. V. Shaw.** 1988. Structure of chloramphenicol acetyltransferase at 1.75 Å resolution. *Proc. Natl. Acad. Sci. USA* **85:**4133–4177.

**Murray, I. A., and W. V. Shaw.** 1997. O-Acetyltransferases for chloramphenicol and other natural products. *Antimicrob. Agents Chemother.* **41:**1–6.

#### Chloramphenicol O-phosphotransferases

**Izard, T., and J. Ellis.** 2000. The crystal structures of chloramphenicol phosphotransferase reveal a novel inactivaction mechanism. *EMBO J.* **19:**2690–2700.

#### Active efflux

**McMurry, L. M., A. M. George, and S. B. Levy.** 1994. Active efflux of chloramphenicol in susceptible *Escherichia coli* strains and in multiple-antibiotic-resistant (Mar) mutants. *Antimicrob. Agents Chemother.* **38:**542–546.

chapter 20



# Inhibitors of the Formation of the First Peptide Bond
## Oxazolidinones

The oxazolidinones are a novel chemical class of synthetic antibacterial agents that target protein synthesis in a wide spectrum of gram-positive pathogens including methicillin-resistant *Staphylococcus aureus* (MRSA), penicillin-resistant *Streptococcus pneumoniae,* and vancomycin-resistant *Enterococus faecium.*

In 1987 the antibacterial activities of the oxazolidinones were first described by scientists at E. I. du Pont de Nemours & Co., Inc. (Wilmington, Del.). It was demonstrated that the oxazolidinone DuP721 inhibited protein synthesis in susceptible bacteria. Oxazolidinones were abandoned for some time after these earlier studies because of their high toxicity. Later, in 1996, new derivatives with superior pharmacological properties were synthesized at Pharmacia & Upjohn (currently Pharmacia Corp.) in Kalamazoo, Mich., and in April 2000 one of the synthesized oxazolidinone antibacterial agents, linezolid, was approved for clinical use. This chapter focuses on linezolid.

## Linezolid

Linezolid (formerly PMN-100766) (Fig. 20.1) is the first oxazolidinone to be approved by the Food and Drug Administration (FDA). It was licensed to Pharmacia & Upjohn, which distributed it under the trade name Zyvox (as tablets and intravenous and oral suspension). The FDA approval covers the treatment of adult patients with vancomycin-resistant *E. faecium* infections, nosocomial pneumonia, complicated and uncomplicated skin and skin structure infections, including those due to MRSA, and nosocomial and community-acquired pneumonia due to *S. aureus* or penicillin-susceptible *S. pneumoniae.* Linezolid was licensed in the United Kingdom in early 2001.

Linezolid is obtained by total synthesis, so it qualifies as an antibacterial agent but not as an antibiotic. Its chemical structure is

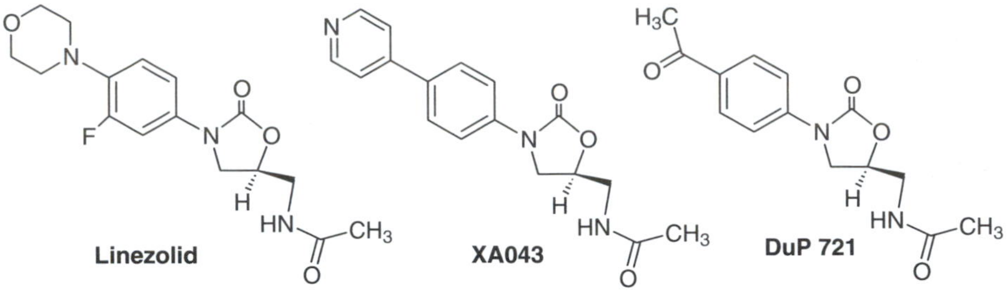

**Figure 20.1** Chemical structures of linezolid, XA043, and DuP721.

3-(fluorophenyl)-2-oxazolidinone that has a morpholin-1-yl group substitution.

### Antibacterial Activity

Linezolid is bacteriostatic against staphylococci and enterococci and often bactericidal against streptococci. In vitro, linezolid is active against both *E. faecium* and *E. faecalis*; as mentioned in chapter 18, quinupristin-dalfopristin is active against vancomycin-resistant *E. faecium* but not *E. faecalis*. Like vancomycin and quinupristin-dalfopristin, linezolid is active against staphylococci, including MRSA and methicillin-resistant *S. epidermidis*. It is also active against penicillin-resistant pneumococci and *S. aureus* with intermediate susceptibility to vancomycin. It does not have clinically useful activity against gram-negative bacteria.

### Clinical Uses

Linezolid, the first of a new class of antibiotics, the oxazolidinones, has been marketed for treatment of infection due to vancomycin-resistant *E. faecium*, nosocomial and community-acquired pneumonia due to *S. aureus* or penicillin-susceptible *S. pneumoniae*, and skin structure infections, including those due to MRSA. It is available for both oral and parenteral use.

## Mechanism of Action of Oxazolidinones

Oxazolidinones are potent inhibitors of bacterial protein biosynthesis. Several hypotheses regarding the mode of inhibition of protein synthesis in sensitive bacteria by oxazolidinone have been proposed. The most recent was published in 2001 in the *Journal of Biological Chemistry* by Patel et al. of Du Pont Pharmaceutical Co. These scientists demonstrated that linezolid, XA043, and DuP721 inhibit the translation of natural mRNA templates but have no significant effect on poly(A)-dependent translation. They found that these oxazolidinones inhibit ribosomal peptidyltransferase activity in the simple reaction of 70S ribosomes. Then they proposed that oxazolidinones inhibit bacterial protein biosynthesis by interfering with the binding of initiation fMet-tRNA to the ribosomal peptidyltransferase P site, which is vacant only prior to the formation of the first peptide bond. Then, as already mentioned, oxazolidinones do not affect the formation of the 30S preinitiation complex but do prevent the formation of the fMet-tRNA–ribosome–mRNA ternary complex.

These scientists mentioned that their findings agree with genetic evidence that oxazolidinone-resistant mutations are associated exclusively with domain V of 23S RNA (peptidyltransferase active site) as well as with their mutagenesis results.

Previous studies by scientists from the Center for Pharmaceutical Biotechnology, Department of Biochemistry and Molecular Biology, University of Chicago, from Odense University, Odense, Denmark, and from the Infectious Diseases Research Department, Pharmacia Corp., also demonstrated that oxazolidinone resistance mutations in 23S rRNA of *Escherichia coli* reveal the central region of domain V to be the primary site of drug action. These authors indicated schematically the location of linezolid resistance mutations in *E. coli* 23S rRNA (Fig. 20.2).

## Bacterial Resistance

Because linezolid is totally synthetic, has recently been introduced into clinical use, and has a mechanism of action different from those of macrolides, lincosamides, quinupristin-dalfopristin, and chloramphenicol, it lacks every clinically significant cross-resistance mechanism. However, a report of characterization of mutations in rRNA and comparison of their occurrence in vancomycin-resistant enterococci was recently published by scientists from Northwestern University, Northwestern Memorial Hospital, and Pharmacia Corp. These researchers assessed the potential for the emergence of resistance during the use of linezolid by testing 10 clinical isolates of vancomycin-resistant enterococci as well as a vancomycin-susceptible control strain of *E. faecalis*. DNA sequencing of the 23S rRNA genes revealed that linezolid resistance in three *E. faecalis* isolates was asso-

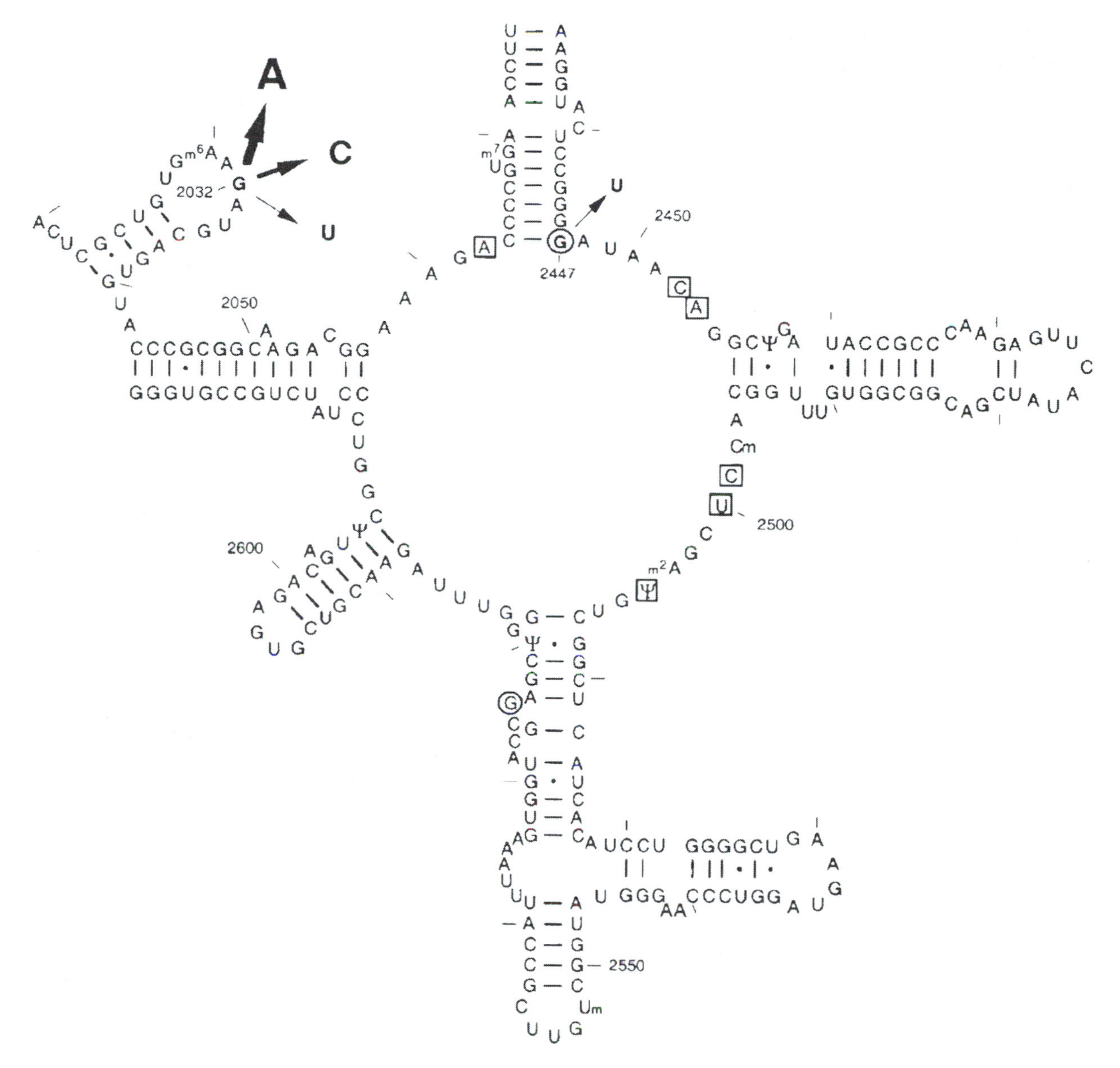


**Figure 20.2** Secondary structure of the segment of *E. coli* 23S rRNA encompassing the central-loop domain V and the neighboring regions. *E. coli* linezolid resistance mutations are marked by arrows. Reprinted from L. Xiong, P. Kloss, S. Douthwaite, N. M. Anderson, S. Swaney, D. L. Shinabarger, and A. S. Mankin, *J. Bacteriol.* **182:**5325–5331, 2000, with permission from the American Society for Microbiology.

ciated with a guanine-to-uracil transversion at bp 2576 while one *E. faecium* isolate for which the MIC was 16 μg/ml contained a guanine-to-adenine transition at bp 2505.

In 2002, Auckland and coworkers reported the first three examples of resistant enterococci (two isolates of *E. faecium* and one of *E. faecalis*) isolated in the United Kingdom; they were obtained from patients who had received linezolid. All the resistant isolates had a G2576U mutation at bp 2576 in the peptidyl transferase region of the 23S rRNA, which is identical to the mutation previously described in a linezolid-resistant mutant of *E. faecalis*.

Vancomycin-resistant *E. faecium* appears capable of developing resistance to linezolid (Gonzales et al., 2001), as does *S. aureus* (Tsiodras et al., 2001).

## Oxazolidinone Preparations Commercially Available in the United States

Linezolid (Zyvox, available for oral use as an oral suspension or in tablet form and available in injectable form), marketed by Pharmacia Corp., is the only oxazolidinone marketed in the United States.

### REFERENCES AND FURTHER READING


#### Oxazolidinones

**Brickner, S. J.** 1996. Oxazolidinone antibacterial agents. *Curr. Pharm. Des.* **2**:175–194.

**Diekema, D. J., and R. N. Jones.** 2001. Oxazolidinone antibiotics. *Lancet* **358**:1975–1982.

**Diekema, D. J., and R. N. Jones.** 2000. Oxazolidinones. A review. *Drugs* **59**:7–16.

**Ford, C. W., J. C. Hamel, D. Stapert, J. K. Moerman, D. K. Hutchinson, M. R. Barbachyn, and G. E. Zurenko.** 1997. Oxazolidinones, new antibacterial agents. *Trends Microbiol.* 5:196–200.

#### Linezolid

**General**

**Anonymous.** 2000. Linezolid. *Med. Lett.* **42**:45–46.

**Perry, C. M., and B. Jarvis.** 2001. Linezolid. A review of its use in the management of serious gram-positive infections. *Drugs* **61**:525–551.

**Antimicrobial activity**

**Brown-Elliot, B. A., S. C. Ward, C. J. Crist, L. B. Mann, R. W. Wilson, and R. J. Wallace.** 2001. In vitro activities of linezolid against multiple *Nocardia* species. *Antimicrob. Agents Chemother.* **45**:1295–1297.

**Cercenado, E., F. Garcia-Garrote, and E. Bouza.** 2001. *In vitro* activity of linezolid against multiply resistant Gram-positive clinical isolates. *J. Antimicrob. Chemother.* **47**:77–81.

**Eliopoulos, G. M., C. B. Wennersten, H. S. Gold, and R. C. Moellering.** 1996. In vitro activities of new oxazolidinone antimicrobial agents against enterococci. *Antimicrob. Agents Chemother.* **40**:1745–1747.

**Patel, R., M. S. Rouse, K. E. Piper, and J. M. Steckelberg.** 2001. Linezolid therapy of vancomycin-resistant *Enterococcus faecium* experimental endocarditis. *Antimicrob. Agents Chemother.* **45**:621–623.

**Rybak, M. J., D. M. Cappelletty, T. Moldovan, J. R. Aeschlimann, and G. W. Kaatz.** 1998. Comparative in vitro activities and postantibiotic effects of the oxazolidinone compounds eperezolid (PNU-100592) and linezolid (PNU-100766) versus vancomycin against *Staphylococcus aureus*, coagulase-negative staphylococci, *Enterococcus faecalis*, and *Enterococcus faecium*. *Antimicrob. Agents Chemother.* **42**:721–724.

**Wallace, R. J., B. A. Brown-Elliott, S. C. Ward, C. J. Crist, L. B. Mann, and R. W. Wilson.** 2001. Activities of linezolid against rapidly growing mycobacteria. *Antimicrob. Agents Chemother.* **45**:764–767.

#### Mechanism of Action of Oxazolidinones

**Aoki, H., L. Ke, S. M. Poppe, T. J. Poel, E. A. Weaver, R. C. Gadwood, R. C. Thomas, D. L. Shinabarger, and M. C. Ganoza.** 2002. Oxazolidinone antibiotics target the P site on *Escherichia coli* ribosomes. *Antimicrob. Agents Chemother.* **46**:1080–1085.

**Kloss, P., L. Xiong, D. L. Shinabarger, and A. S. Mankin.** 1999. Resistance mutations in 23S rRNA identify the site of action of the protein synthesis inhibitor linezolid in the ribosomal peptidyl transferase center. *J. Mol. Biol.* **294**:93–101.

**Lin, A. H., R. W. Murray, T. J. Vidmar, and K. R. Marotti.** 1997. The oxazolidinone eperezolid binds to the 50S ribosomal subunits and competes with binding of chloramphenicol and lincomycin. *Antimicrob. Agents Chemother.* **41**:2127–2131.

**Matassova, N. B., M. V. Rodrina, R. Endermann, H. P. Knoll, U. Pleiss, H. Wild, and W. Wintermeyer.** 1999. Ribosomal RNA is the target for oxazolidinones, a novel class of translational inhibitors. *RNA* **5**:939–946.

**Patel, U., Y. P. Yan, F. W. Hobbs, J. Kaczmarczyk, A. M. Slee, D. L. Pompliano, M. G. Kurilla, and E. V. Bobkova.** 2001. Oxazolidinones mechanism of action: inhibition of the first peptide bond formation. *J. Biol. Chem.* **276**:37199–37205.

**Shinabarger, D. L., K. R. Marotti, R. W. Murray, A. H. Lin, E. P. Melchior, S. M. Swaney, D. S. Dunyak, W. F. Demyan, and J. M. Buysse.** 1997. Mechanism of action of oxazolidinones: effect of linezolid and eperezolid on translation reactions. *Antimicrob. Agents Chemother.* **41**:2132–2136.

**Swaney, S. M., H. Aoki, M. C. Ganoza, and D. L. Shinabarger.** 1998. The oxazolidinone linezolid inhibits initiation of protein synthesis in bacteria. *Antimicrob. Agents Chemother.* **42**:3251–3255.

**Zhou, C. C., S. M. Swaney, D. L. Shinabarger, and B. J. Stockman.** 2002. 1H nuclear magnetic resonance study of oxazolidinone binding to bacterial ribosomes. *Antimicrob. Agents Chemother.* **46**:625–629.

#### Bacterial Resistance

**Auckland, C., L. Teare, F. Cooke, M. E. Kaufmann, M. Warner, G. Jones, K. Bamford, H. Ayles, and A. P. Johnson.** 2002. Linezolid-resistant enterococci: report of the first isolate in the United Kingdom. *J. Antimicrob. Chemother.* **50**:743–746.

**Gonzales, R. D., P. C. Schreckenberger, M. B. Graham, S. Kelkar, K. DenBesten, and J. P. Quinn.** 2001. Infection due to vancomycin-resistant *Enterococcus faecium* resistant to linezolid. *Lancet* **357**:1179–1180.

**Prystowsky, J., F. Siddiqui, J. Cosay, D. L. Shinabarger, J. Millichap, L. R. Peterson, and G. A. Noskin.** 2001. Resistance to linezolid: characterization of mutations in rRNA and comparison of their occurrences in vancomycin-resistant enterococci. *Antimicrob. Agents Chemother.* **45**:2154–2156.

**Tsiodras, S., H. S. Gold, G. Sakoulas, G. M. Eliopoulos, C. Wennersten, L. Venkataraman, R. C. Moellering, and M. J. Ferraro.** 2001. Linezolid resistance in a clinical isolate of *Staphylococcus aureus*. *Lancet* **358:**207–208.

**Xiong, L., P. Kloss, S. Douthwaite, N. M. Andersen, S. Swaney, D. L. Shinabarger, and A. S. Mankin.** 2000. Oxazolidinone resistance mutations in 23S rRNA of *Escherichia coli* reveal the central region of domain V as the primary site of drug action. *J. Bacteriol.* **182:**5325–5331.

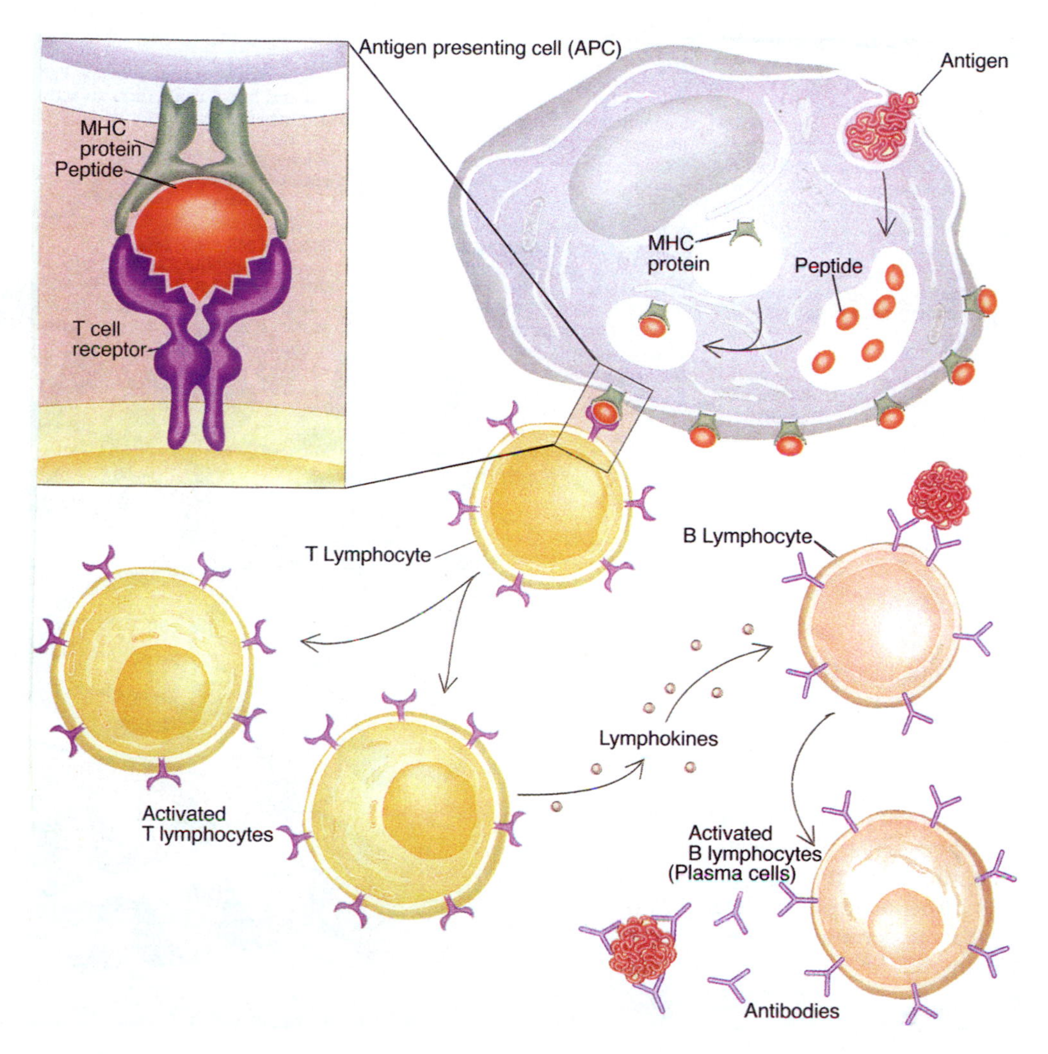


**Color Plate 3.1** A general view of how the specific immune system defends the body against an antigen from a bacterium. Reprinted from *Sci. Am.*, special issue, September 1993, with permission from Carol Donner.

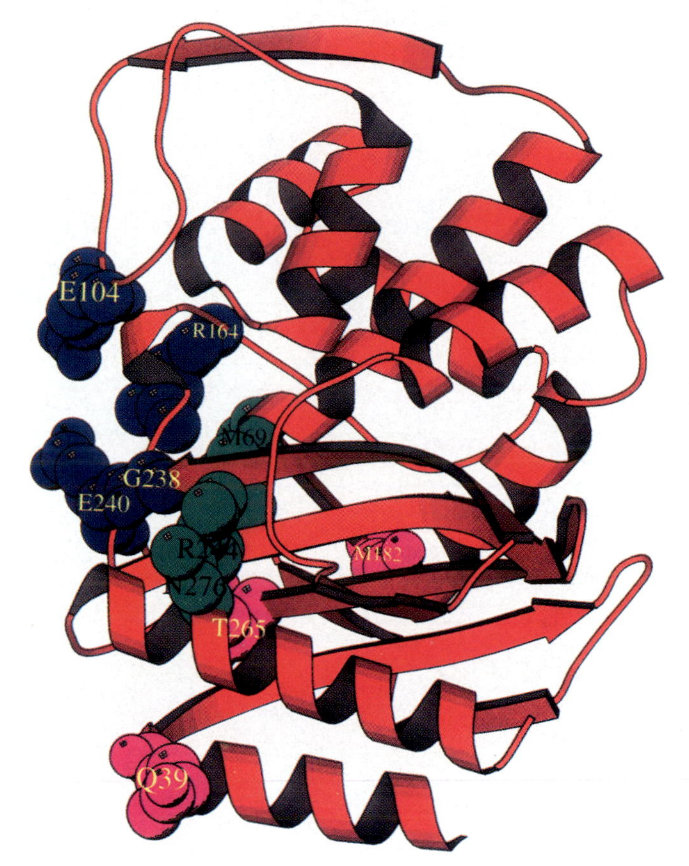


**Color Plate 12.1** Ribbon diagram of TEM-1 β-lactamase showing the positions of amino acid substitutions found in ESBLs. The substitutions that result in increased hydrolysis of the extended-spectrum cephalosporins cefotaxime and ceftazidime are shown in blue. The positions where substitutions decrease the affinity of the enzyme for inhibitors are shown in green. The positions where amino acid substitutions have been identified that do not affect the catalytic activity of the enzyme are shown in magenta. Reprinted from T. Patzkill, *ASM News* **64:**90–95, 1998, with permission from the American Society for Microbiology.

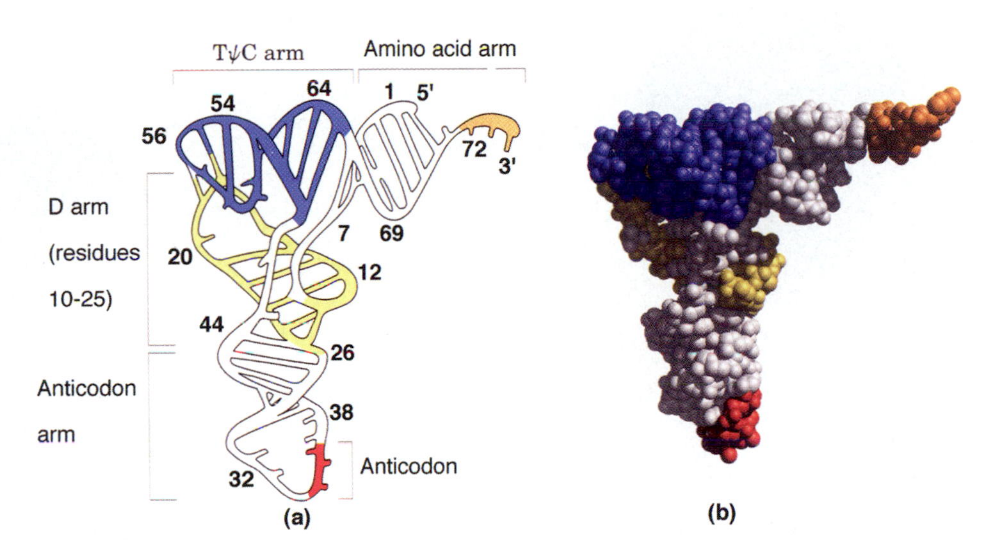


**Color Plate 16.1** Three-dimensional structure of yeast $tRNA^{Phe}$ deduced from X-ray diffraction analysis. The shape resembles a twisted L. (a) Schematic diagram with the various arms identified. (b) Space-filling model. Color coding is the same in both representations. The three bases of the anticodon are shown in red, and the CCA sequence at the 3' end is shown in orange. The TψC and D arms are blue and yellow, respectively. Reprinted from D. L. Nelson and M. M. Cox, *Lehninger Principles of Biochemistry,* 4th ed. (Worth Publishers, New York, N.Y., 2000), with permission from the publisher.

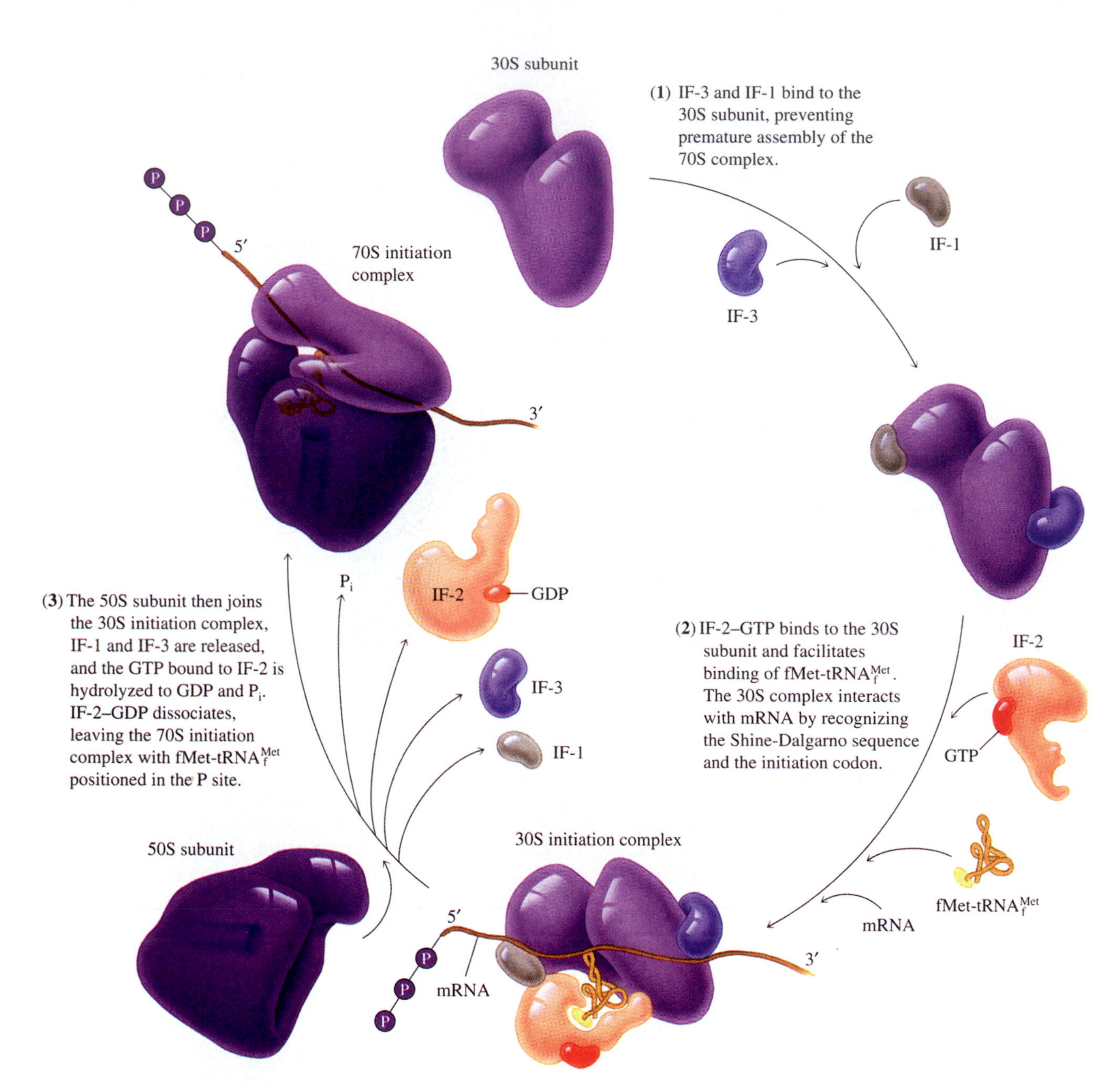
30S subunit
(1) IF-3 and IF-1 bind to the 30S subunit, preventing premature assembly of the 70S complex.
IF-1
IF-3
P
P
P
5′
70S initiation complex
3′
P$_i$
IF-2
GDP
IF-3
IF-1
(3) The 50S subunit then joins the 30S initiation complex, IF-1 and IF-3 are released, and the GTP bound to IF-2 is hydrolyzed to GDP and P$_i$. IF-2–GDP dissociates, leaving the 70S initiation complex with fMet-tRNA$_f^{Met}$ positioned in the P site.
(2) IF-2–GTP binds to the 30S subunit and facilitates binding of fMet-tRNA$_f^{Met}$. The 30S complex interacts with mRNA by recognizing the Shine-Dalgarno sequence and the initiation codon.
IF-2
GTP
50S subunit
30S initiation complex
fMet-tRNA$_f^{Met}$
mRNA
5′
P
P
P
mRNA
3′

**Color Plate 16.2** Steps in the formation of the prokaryotic 70S initiation complex. Reprinted from H. R. Horton, L. A. Moran, R. S. Ochs, J. D. Rawn, and K. G. Scrimgeour, *Principles of Biochemistry,* 3rd ed. (Prentice-Hall, Upper Saddle River, N.J., 2002), with permission from the publisher.

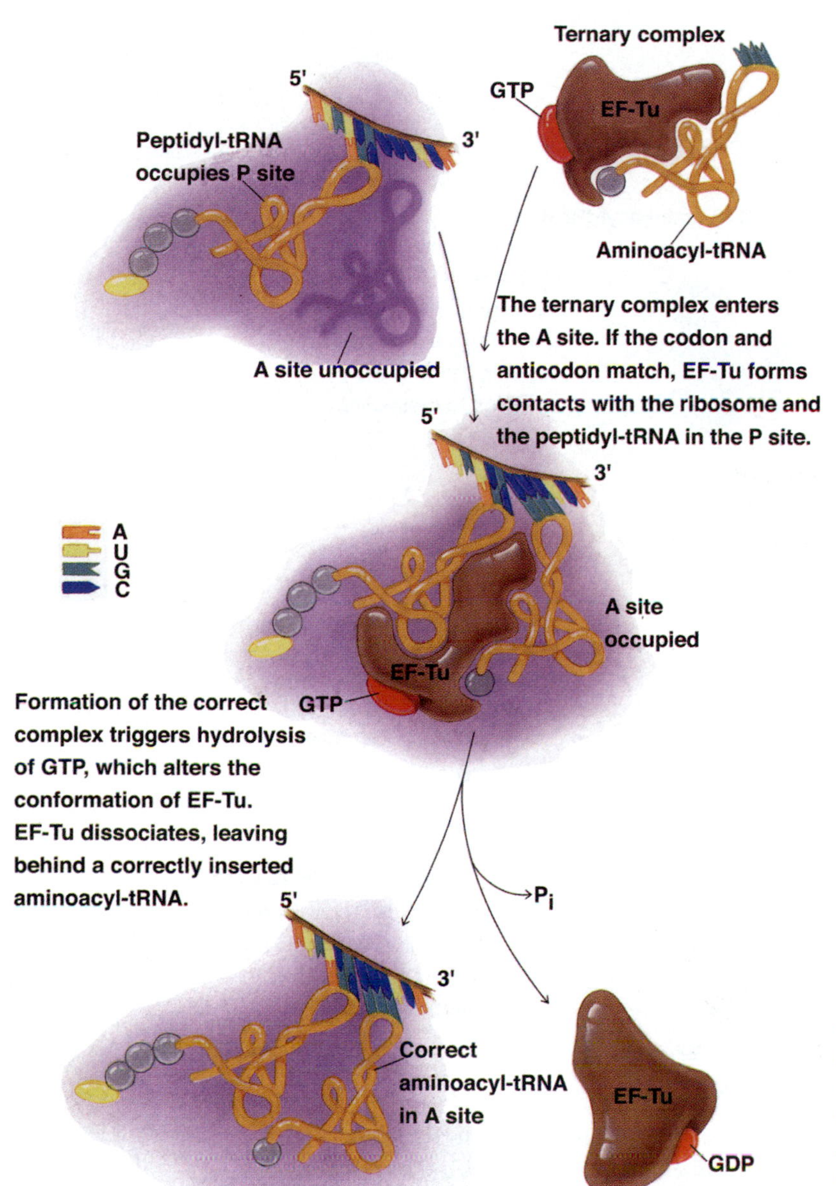

Ternary complex
GTP
EF-Tu
5'
3'
Peptidyl-tRNA
occupies P site
Aminoacyl-tRNA
A site unoccupied
The ternary complex enters
the A site. If the codon and
anticodon match, EF-Tu forms
contacts with the ribosome and
the peptidyl-tRNA in the P site.
5'
3'
A
U
G
C
A site
occupied
EF-Tu
GTP
Formation of the correct
complex triggers hydrolysis
of GTP, which alters the
conformation of EF-Tu.
EF-Tu dissociates, leaving
behind a correctly inserted
aminoacyl-tRNA.
5'
P
3'
Correct
aminoacyl-tRNA
in A site
EF-Tu
GDP

**Color Plate 16.3** Insertion of aminoacyl-tRNA into the A site by GTP–EF-Tu during the chain elongation step in *E. coli.* Reprinted from H. R. Horton, L. A. Moran, R. S. Ochs, J. D. Rawn, and K. G. Scrimgeour, *Principles of Biochemistry,* 3rd ed. (Prentice-Hall, Upper Saddle River, N.J., 2002), with permission from the publisher.

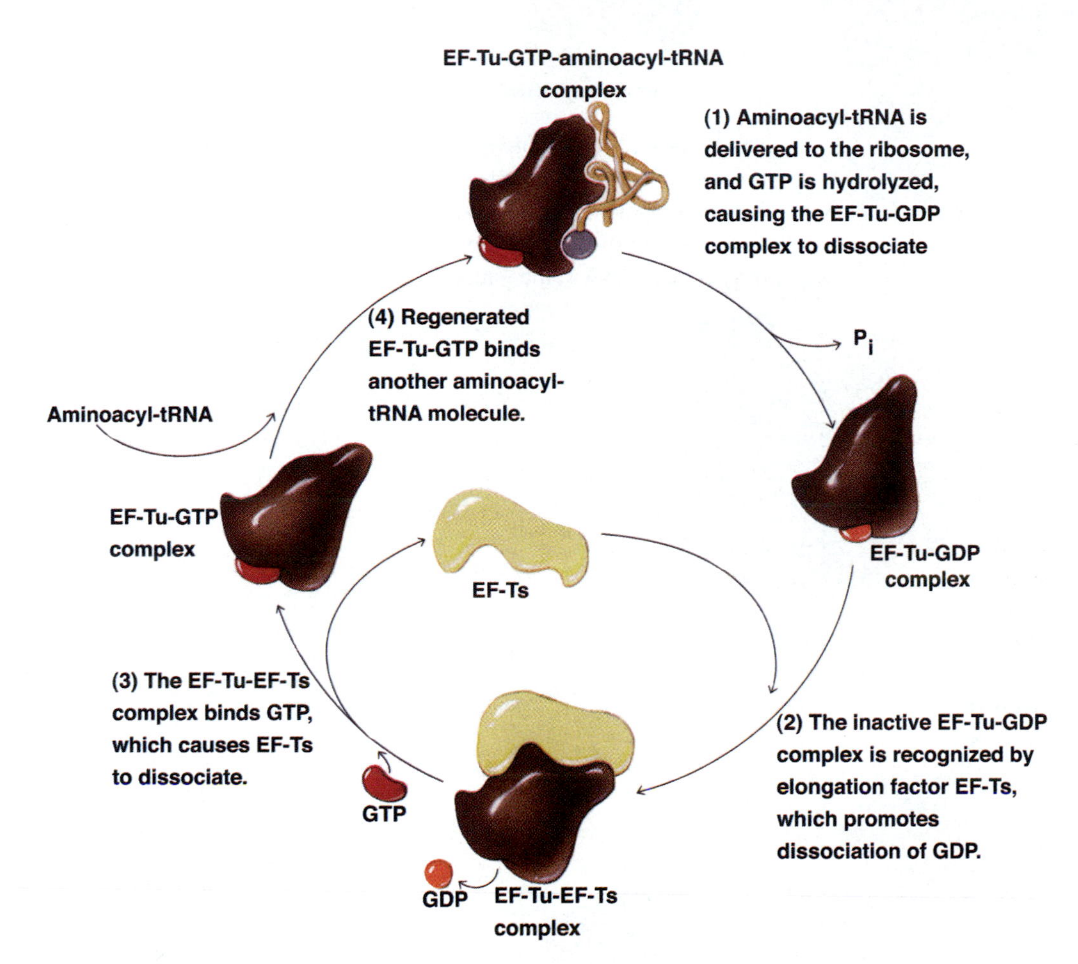
EF-Tu-GTP-aminoacyl-tRNA
complex
(1) Aminoacyl-tRNA is
delivered to the ribosome,
and GTP is hydrolyzed,
causing the EF-Tu-GDP
complex to dissociate
$P_i$
(4) Regenerated
EF-Tu-GTP binds
another aminoacyl-
tRNA molecule.
Aminoacyl-tRNA
EF-Tu-GTP
complex
EF-Ts
EF-Tu-GDP
complex
(3) The EF-Tu-EF-Ts
complex binds GTP,
which causes EF-Ts
to dissociate.
GTP
(2) The inactive EF-Tu-GDP
complex is recognized by
elongation factor EF-Ts,
which promotes
dissociation of GDP.
GDP
EF-Tu-EF-Ts
complex

**Color Plate 16.4** Cycling of EF-Tu–GTP. Reprinted from H. R. Horton, L. A. Moran, R. S. Ochs, J. D. Rawn, and K. G. Scrimgeour, *Principles of Biochemistry,* 3rd ed. (Prentice-Hall, Upper Saddle River, N.J., 2002), with permission from the publisher.

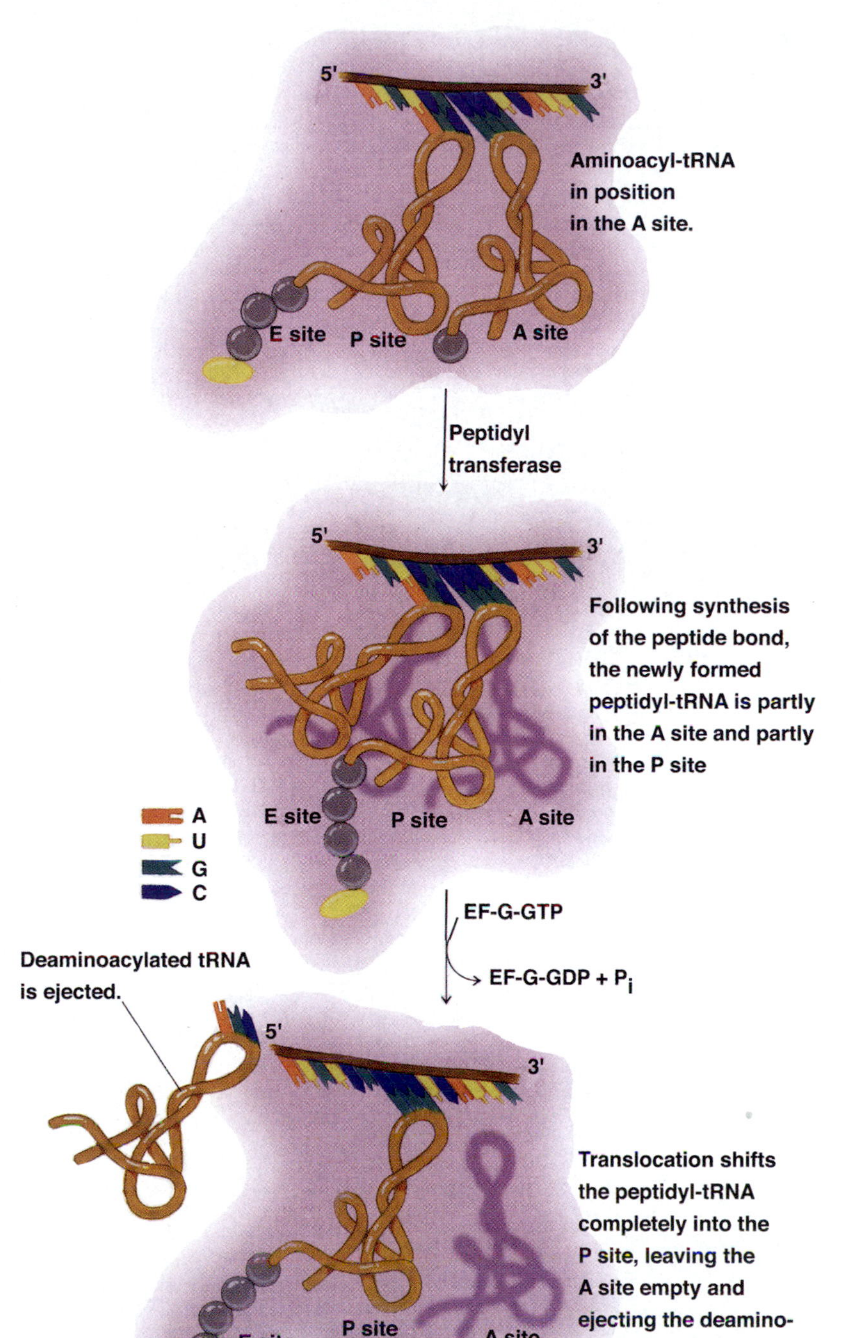
5'
3'
Aminoacyl-tRNA
in position
in the A site.
E site
P site
A site
Peptidyl
transferase
5'
3'
Following synthesis
of the peptide bond,
the newly formed
peptidyl-tRNA is partly
in the A site and partly
in the P site
A
U
G
C
E site
P site
A site
EF-G-GTP
EF-G-GDP + P$_i$
Deaminoacylated tRNA
is ejected.
5'
3'
Translocation shifts
the peptidyl-tRNA
completely into the
P site, leaving the
A site empty and
ejecting the deamino-
acylated tRNA from
the E site.
E site
P site
A site

**Color Plate 16.5** Translocation during protein synthesis in prokaryotes. Reprinted from H. R. Horton, L. A. Moran, R. S. Ochs, J. D. Rawn, and K. G. Scrimgeour, *Principles of Biochemistry,* 3rd ed. (Prentice-Hall, Upper Saddle River, N.J., 2002), with permission from the publisher.

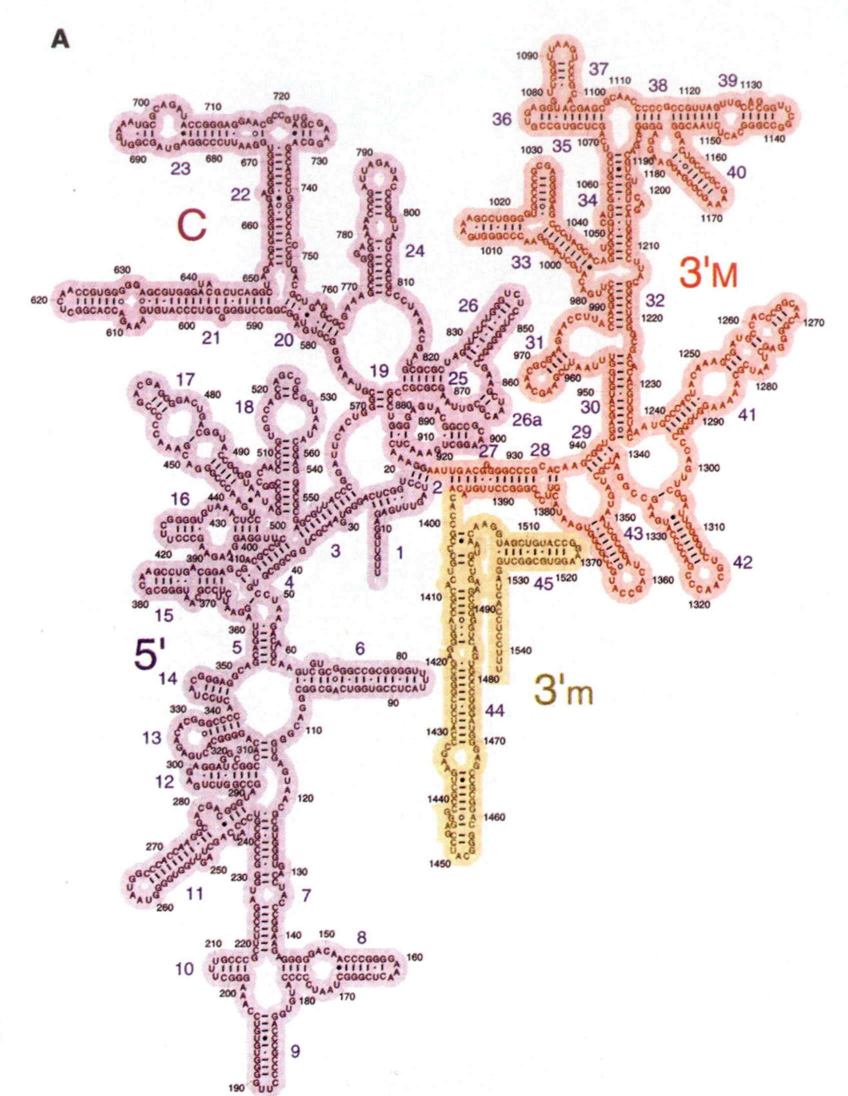


**Color Plate 16.6** Secondary and tertiary structures of 16S, 23S, and 5S rRNA. (A) Secondary structure of *T. thermophilus* 16S rRNA. (B) Secondary structures of *T. thermophilus* 23S and 5S rRNA. (C) Three-dimensional fold of 16S rRNA in the 70S ribosome. (D) Three-dimensional fold of 23S and 5S rRNA. Reprinted from M. M. Yusupov, G. Z. Yusupova, A. Baucom, K. Lieberman, T. N. Earnest, J. H. Cate, and H. F. Noller, *Science* **292:**883–896, 2001, with permission from the publisher.

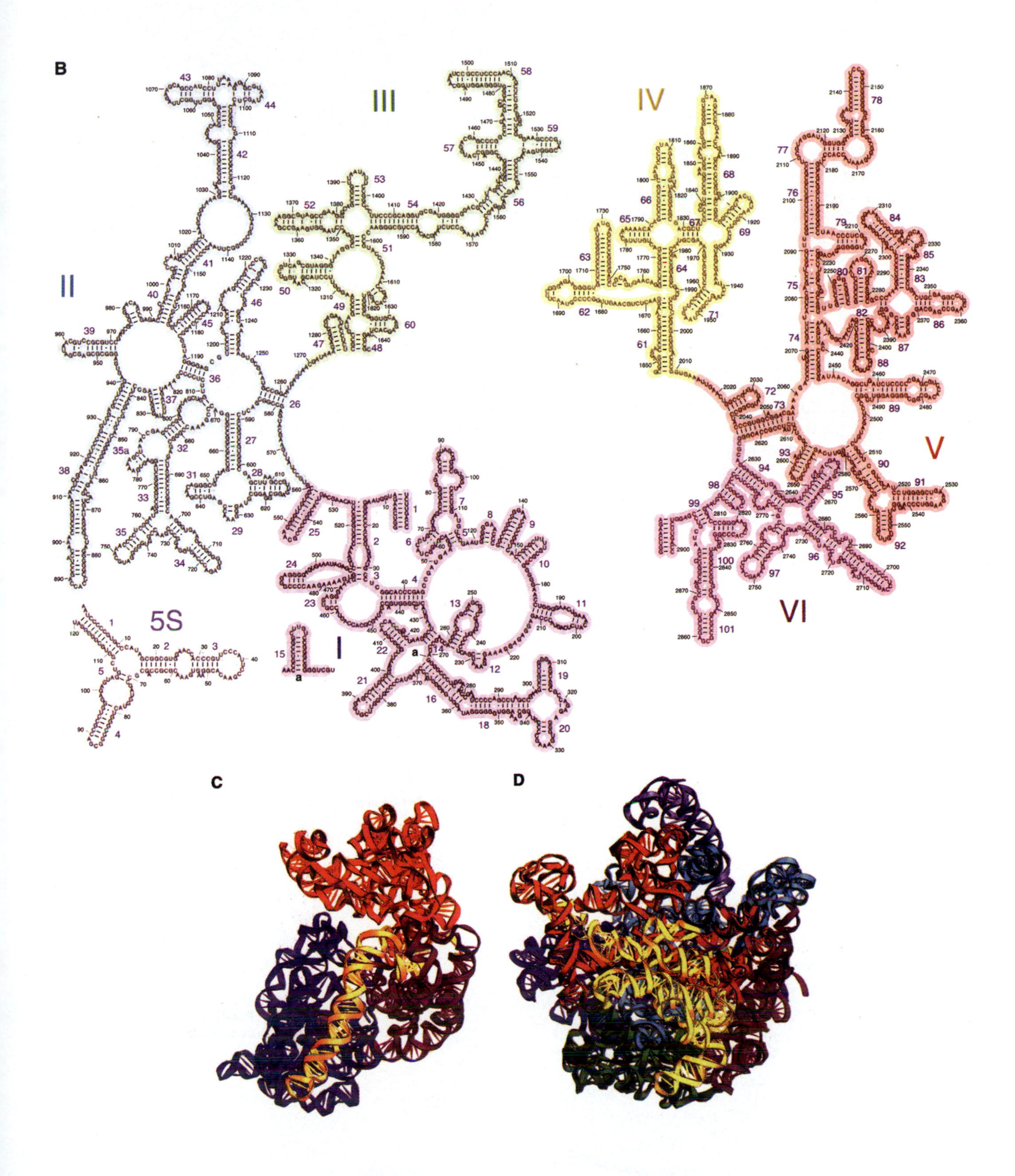

B
II
III
IV
V
VI
I
5S
C
D

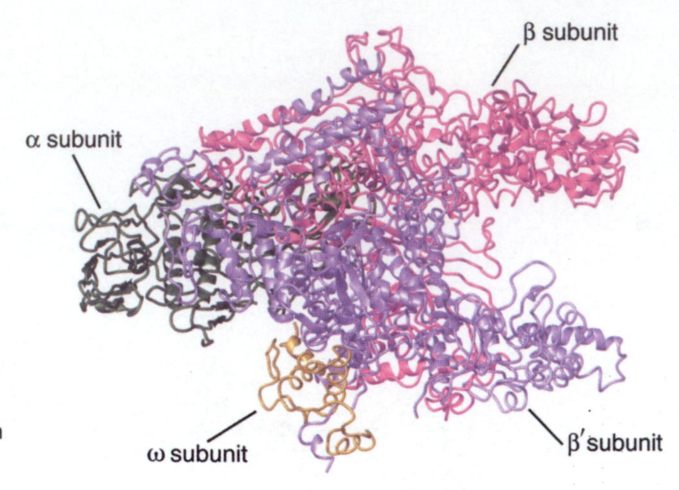


**Color Plate 22.1** Structure of *T. aquaticus* RNA polymerase core enzyme. Reprinted from H. R. Horton, L. A. Moran, R. S. Ochs, J. D. Rawn, and K. G. Scrimgeour, *Principles of Biochemistry,* 3rd ed. (Prentice-Hall, Upper Saddle River, N.J., 2002), with permission from the publisher.

**Color Plate 22.2** Transcription by RNA polymerase in *E. coli.* To synthesize an RNA strand complementary to one of the two DNA strands in a double helix, the DNA is transiently unwound. (a) About 17 bp is unwound at any given time. The transcription bubble moves from left to right as shown, keeping pace with RNA synthesis. The DNA is unwound ahead and rewound behind as RNA is transcribed. Red arrows show the direction in which the DNA and the short RNA-DNA hybrid must rotate to permit this process. (b) Supercoiling of DNA brought about by transcription. Positive supercoils form ahead of the transcription bubble, and negative supercoils form behind. Reprinted from D. L. Nelson and M. M. Cox, *Lehninger Principles of Biochemistry,* 3rd ed. (Worth Publishers, New York, N.Y., 2000), with permission from the publisher.

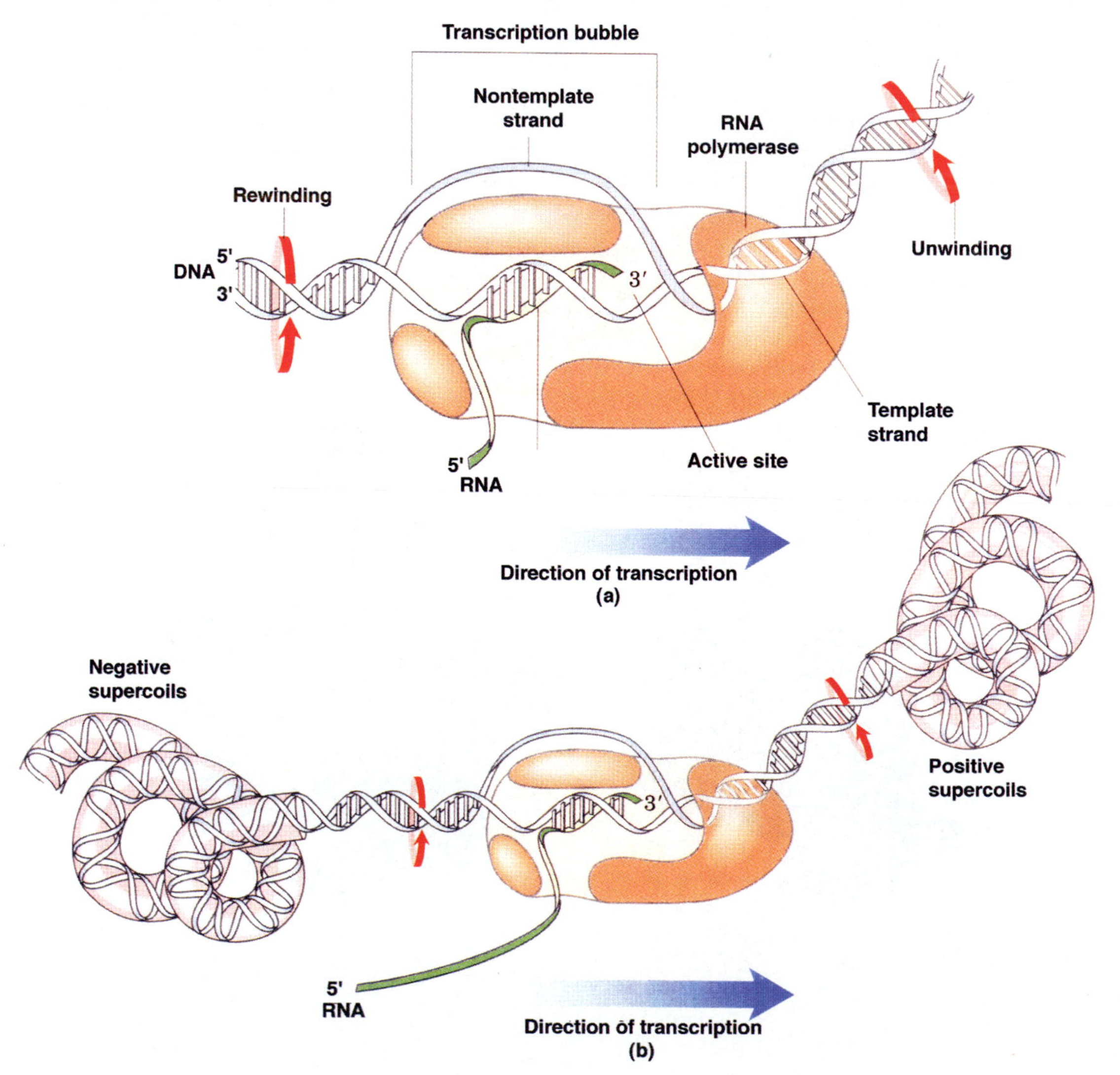

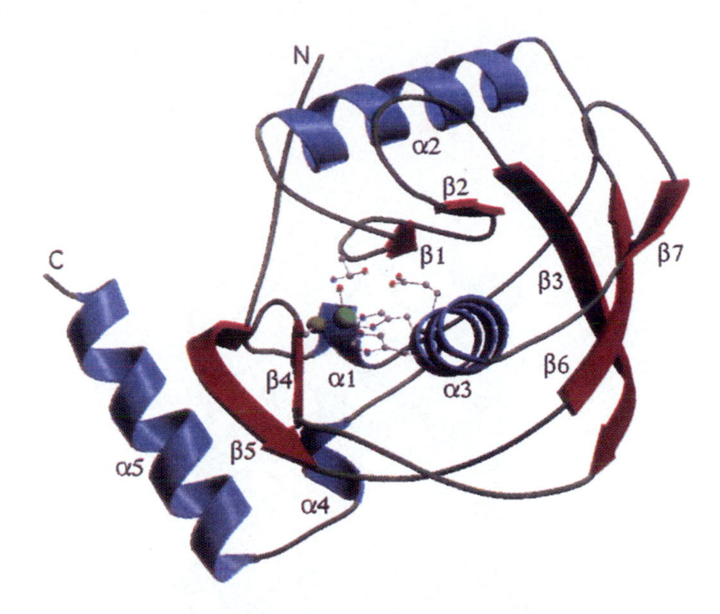


**Color Plate 27.1** Ribbon diagram of the *E. coli* deformylase viewed toward the substrate cleft into the metal center. The secondary structure is color coded with α-helical regions as blue, β-sheet regions as red, and the remainder as green. The metal cation is shown as green. Reprinted from M. K. Chan, W. Gong, P. T. Ravi Rajagopalan, B. Hao, C. M. Tsai, and D. Pei, *Biochemistry* **36:**13904–13909, 1997, with permission from the publisher.

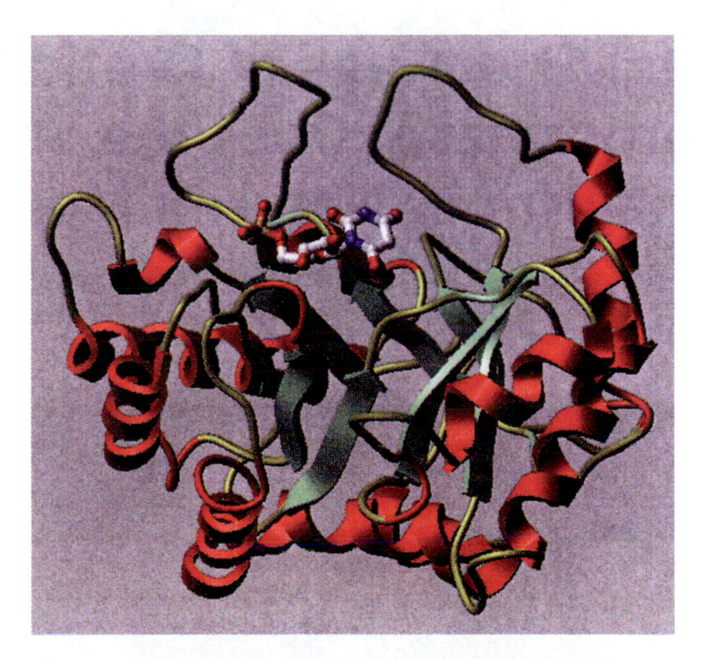

**Color Plate A1** The tertiary structure describes the shape of the fully folded polypeptide chain. The example shown is the inhibitor barbituric acid ribonucleotide bound in the active site of orotidine 5'-monophosphate decarboxylase. The enzyme is composed of β-strands and α-helices. Reprinted from K. N. Houk et al., *ChemBioChem* **2:**113–118, 2001, with permission from the publisher.

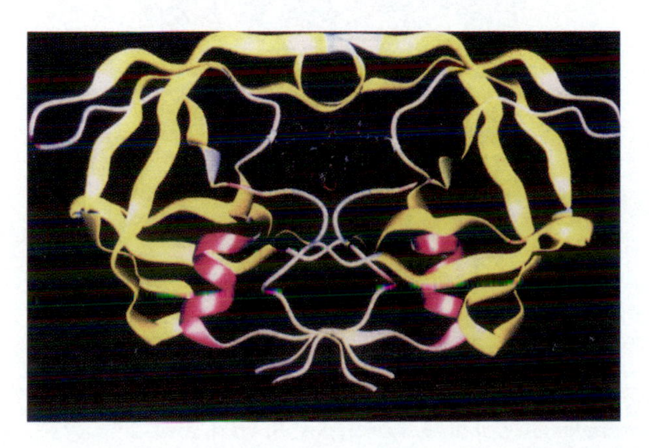

**Color Plate A2** Ribbon diagram of the C2 backbone of HIV-1 protease. Reprinted from E. De Clercq, *J. Med. Chem.* **38:**2491–2517, 1995, with permission from the publisher.

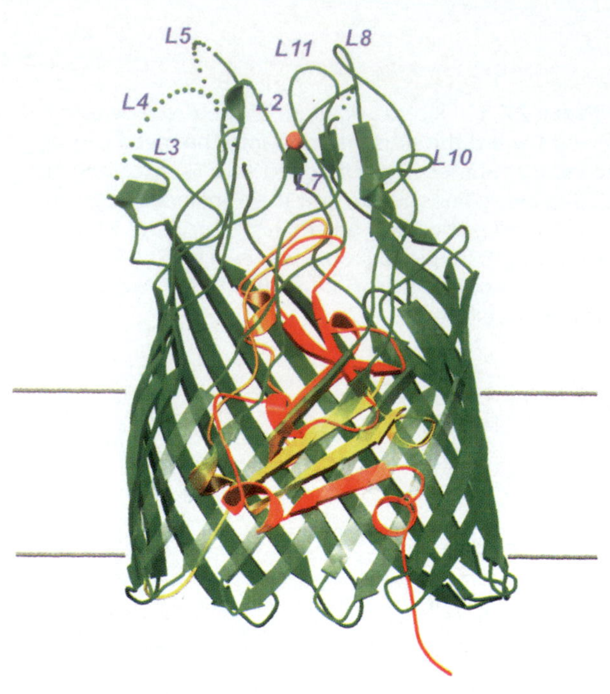


**Color Plate D1** Ribbon diagram of FepA. The extracellular space is located at the top of the illustration, and the periplasmic space is at the bottom. The putative iron position of ferric enterobactin is indicated by the red sphere. Part of the barrel has been rendered transparent to reveal the N-terminal domain located in the channel. Reprinted from S. K. Buchanan, B. S. Smith, L. Venkatramani, D. Xia, L. Esser, M. Palnitker, R. Chakraborty, D. van der Helm, and J. Deisenhofer, *Nat. Struct. Biol.* **6:**56–63, 1999, with permission from the publisher.

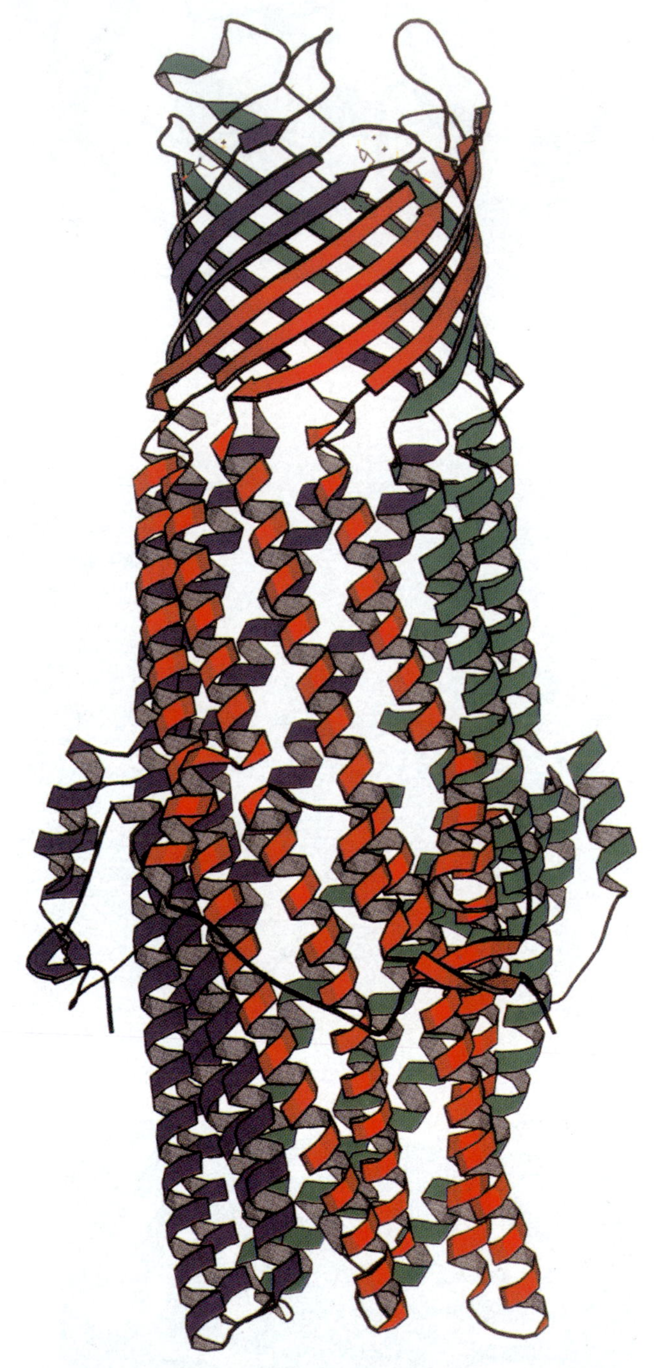

**Color Plate D2** Ribbon diagram of the crystal structure of TolC. A trimeric TolC molecule forms the combined β-barrel/α-helical "tunnel" structure. Each monomer is shown in a different color. Reprinted from V. Koronakis, A. Sharif, E. Koronakis, B. Luisi, and C. Hughes, *Nature* **405:**914–919, 2000, with permission from the publisher.

chapter 21



# An Inhibitor of Isoleucyl-tRNA Synthetase
## Mupirocin

Mupirocin (formerly called pseudomonic acid A) was isolated as the major component of a family of structurally related antibiotics produced by a strain of *Pseudomonas fluorescens*. Mupirocin is used as a topical antibiotic. It was approved in May 1988 by the Food and Drug Administration for the treatment of impetigo and is sold as a 2% ointment of mupirocin calcium. Mupirocin was introduced into clinical practice in the United Kingdom in 1985 (Bactroban; SmithKline Beecham). Mupirocin resistance was initially reported in 1987 in the United Kingdom and has now also been reported in the United States.

The chemical structure of mupirocin (Fig. 21.1), 9-[2(*E*)4-(2*S*,3*R*,4*R*,5*S*)-[5-(2,3-epoxy-5-hydroxy-4-methylhexyl)-3,4-dihydroxytetrahydropyran-2-yl]-3-methylbut-2-enoyloxy]nonanoic acid, was established based on nuclear magnetic resonance spectroscopy, mass spectrometry, and chemical degradative studies of its methyl ester and various derivatives. The chemical structure may be considered as consisting of two parts, a 9-hydroxynonanoic acid moiety and a $C_{17}$ monic acid fragment. Since the epoxide-bearing end of the mupirocin molecule resembles the carbon skeleton of isoleucine, it was expected that mupirocin may compete with this amino acid for the active site of isoleucyl-tRNA synthetase. Later this was proved to be correct.

Although mupirocin is well absorbed after oral and parenteral administration, its concentration in serum is short-lived because of extensive degradation to the antibacterially inactive metabolites, monic acid and 9-hydroxynonanoic acid, carried out by nonspecific esterases (possibly in hepatic or renal tissues).

## Antimicrobial Activity

Mupirocin inhibits the growth of staphylococci (including methicillin-resistant strains) and streptococci (except enterococci) at low concentrations

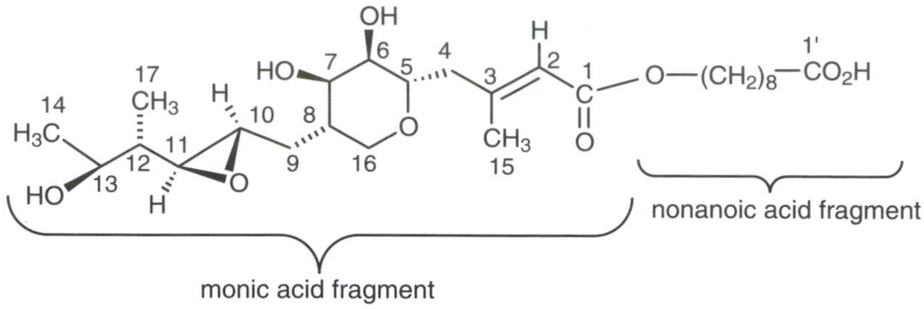

**Figure 21.1** Chemical structure of mupirocin.

and is bactericidal at high concentrations, which are readily achieved by topical application. It is inactive against members of the *Enterobacteriaceae* and *Pseudomonas aeruginosa.*

## Clinical Uses

Mupirocin has no cross-resistance to existing antibacterial agents, is most active at acidic pHs against bacteria included in its spectrum, and penetrates well into superficial layers of the skin and nasal mucosa.

Mupirocin has been successfully used in eradication of the nasal carrier state of *Staphylococcus aureus,* including methicillin-resistant *S. aureus* (MRSA). It has also been used for impetigo. For treatment of this disease, 2% mupirocin in polyethylene glycol base should be applied topically three times a day for 8 days. Impetigo, either bullous or nonbullous, is caused by *S. aureus.* Bullae represent an exfoliative reaction of the epidermis to an exotoxin elaborated by *S. aureus* strains. Patients with folliculitis, furunculosis, impetigo, or other primary skin infections have been treated successfully with mupirocin. Finally, it has been used for burn wounds infected by *S. aureus.* Mupirocin ointment applied to a burn wound twice a day has been effective in eliminating MRSA.

## Mechanism of Action

Isoleucyl-tRNA synthetase (IleRS) joins isoleucine to tRNA at its synthetically active site. The 2.2-Å-resolution crystal structure of *S. aureus* IleRS complexed with $tRNA^{Ile}$ and mupirocin has recently been solved. Mupirocin is an analogue of isoleucine; it competitively binds IleRS. This inhibition decreases or abolishes bacterial protein synthesis by preventing the incorporation of isoleucine into nascent peptides. The epoxide-bearing end of the mupirocin molecule resembles the carbon skeleton of isoleucine and competes for the active site of the IleRS enzyme (Fig. 21.2).

## Bacterial Resistance

Mupirocin resistance in *S. aureus* results from changes in the target enzyme, IleRS. These resistant strains can be divided arbitrarily into two distinct groups, those which exhibit a low level of resistance and those which exhibit a high level of resistance. The definition of mupirocin

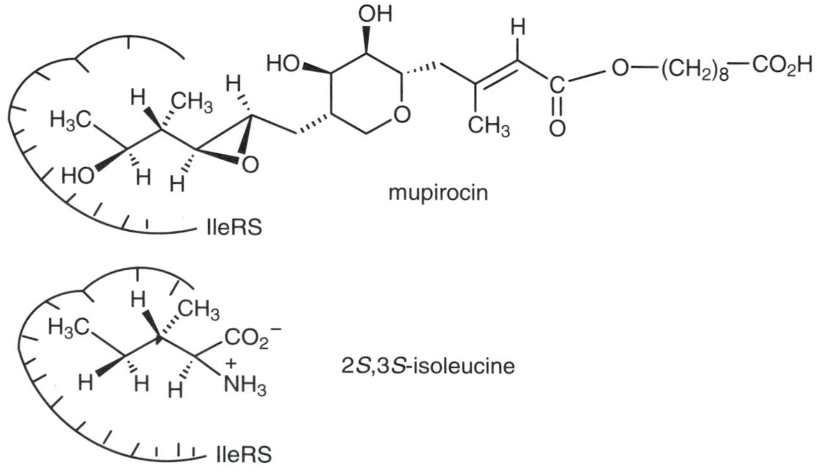

**Figure 21.2** Interaction of mupirocin and 2*S*,3*S*- or (L)-isoleucine with the isoleucine-binding site on IleRS.

resistance has varied, but currently there is an agreement about these two main categories of mupirocin resistance: low-level resistance (MICs ≤ 256 μg/ml) and high-level resistance (MICs ≥ 512 μg/ml). In strains of staphylococci with low-level resistance, the resistance is conferred by the chromosomal *mupA* gene. The MupA phenotype results from a mutation in the staphylococcal native *ileS* gene, which encodes IleRS. However, isolates resistant to high levels of mupirocin contain two biochemically distinct IleRS enzymes, the native mupirocin-sensitive IleRS enzyme and an additional IleRS enzyme that is less sensitive to inhibition by mupirocin. The high level of resistance has been attributed to the presence of the additional plasmid-borne gene *mupA,* which codes for the additional IleRS enzyme. This *mupA* gene has been cloned and sequenced.

## Mupirocin Preparations Commercially Available in the United States

Mupirocin (Bactroban, available as a 2% cream for dermatologic use and as a nasal preparation), marketed by GlaxoSmithKline, is the only IleRS inhibitor marketed in the United States.

### REFERENCES AND FURTHER READING


#### Chemistry

**Chain, E. B., and G. Mellow.** 1977. The structure of pseudomonic acid A, a novel antibiotic produced by *Pseudomonas fluorescens. J. Chem. Soc. Perkin 1* **1997:**294–309.

#### Antibacterial Activity

**Anonymous.** 1988. Mupirocin—a new topical antibiotic. *Med. Lett.* **30:**763–768.

#### Mechanism of Action

**Hughes, J., and G. Mellows.** 1978. On the mode of action of pseudomonic acid: inhibition of protein synthesis in *Staphylococcus aureus. J. Antibiot.* **31:**330–335.

**Hughes, J., and G. Mellows.** 1980. Interaction of pseudomonic acid A with *Escherichia coli* B isoleucyl-tRNA synthetase. *Biochem. J.* **191:**209–219.

#### Bacterial Resistance

**Cookson, B. D.** 1998. The emergence of mupirocin resistance: a challenge to infection control and antibiotic prescribing practice. *J. Antimicrob. Chemother.* **41:**11–18.

**Finlay, J. E., L. A. Miller, and J. A. Poupard.** 1997. Interpretive criteria for testing susceptibility of staphylococci to mupirocin. *Antimicrob. Agents Chemother.* **41:**1137–1139.

**Fujimura, S., A. Watanabe, and D. Beighton.** 2001. Characterization of the *mupA* gene in strains of methicillin-resistant *Staphylococcus aureus* with a low level of resistance to mupirocin. *Antimicrob. Agents Chemother.* **45:**641–641.

**Gilbart, J., C. R. Perry, and B. Slocombe.** 1993. High-level mupirocin-resistance in *Staphylococcus aureus*: evidence for two distinct isoleucyl-tRNA synthetases. *Antimicrob. Agents Chemother.* **37:**32–38.

**Hodgson, J. E., S. P. Curnock, K. G. H. Dyke, R. Morris, D. R. Sylvester, and M. S. Gross.** 1994. Molecular characterization of the gene encoding high-level mupirocin resistance in *Staphylococcus aureus* J2870. *Antimicrob. Agents Chemother.* **38:**1205–1208.

**Leski, T. A., M. Gniadkowski, A. Skoczynska, E. Stefaniuk, K. Trzcinski, and W. Hryniewicz.** 1999. Outbreak of mupirocin-resistant staphylococci in a hospital in Warsaw, Poland, due to plasmid transmission and clonal spread of several strains. *J. Clin. Microbiol.* **37:**2781–2788.

**Ramsey, M. A., S. F. Bradley, C. A. Kauffman, and T. M. Morton.** 1996. Identification of chromosomal location of *mupA* gene, encoding low-level mupirocin resistance in staphylococcal isolates. *Antimicrob. Agents Chemother.* **40:**2820–2923.

**Schmitz, F. J., E. Lindenlauf, B. Hoffmann, A. C. Fluit, J. Verhoef, H. P. Heinz, and M. E. Jones.** 1998. The prevalence of low- and high-level mupirocin resistance in staphylococci from 19 European hospitals. *J. Antimicrob. Chemother.* **42:**489–495.

#### IleRS Structure

**Silvian, L. F., J. Wang, and T. S. Steitz.** 1999. Insights into editing from an Ile-tRNA synthetase structure with $tRNA^{Ile}$ and mupirocin. *Science* **285:**1074–1077.

chapter 22



# Inhibitors of DNA-Dependent RNA Polymerase
## Rifamycins

The clinically useful rifamycins include rifampin (known in some countries as rifampicin), a semisynthetic derivative of rifamycin B. Rifabutin and rifapentine are the other semisynthetic derivatives of rifamycins that are also commercially available in the United States. Rifamycin B is one of a group of structurally related antibiotics produced by a strain of *Amycolatopsis mediterranea* (formerly known as *Streptomyces mediterranei* and *Nocardia mediterranea*). These compounds inhibit bacterial RNA synthesis by binding to the β-subunit of bacterial DNA-dependent RNA polymerases. The principal uses of rifamycin B are in the treatment of tuberculosis and leprosy.

### Overview of the Transcription Process

This section gives an overview of transcription in bacteria and the structure and function of bacterial DNA-dependent RNA polymerase before turning to the rifamycins.

As mentioned in chapter 4, the flow of information from gene to protein requires an RNA intermediate. The transfer of information from DNA to mRNA is called **transcription,** since the DNA base sequence is being copied in an RNA base sequence. The transcription of genetic information from DNA to RNA is carried out through the action of the enzyme **RNA polymerase,** which catalyzes the formation of phosphodiester bonds between ribonucleotides. RNA polymerase requires the presence of DNA, which acts as a template. The last phase of gene expression is **translation** or protein synthesis. As already mentioned, genetic information in the form of an mRNA nucleotide sequence is translated and governs the synthesis of proteins. This pathway was known as the central dogma of molecular biology.

The three major kinds of RNA molecules are derived from information permanently stored in DNA. mRNA encodes the amino acid sequence of polypeptides specified by a gene or a set of genes. tRNA reads the information encoded in the mRNA and transfers the appropriate amino acid to a growing polypeptide chain during protein synthesis. Molecules of rRNA are constituents of ribosomes. RNA and DNA differ only in the hydroxyl group at the 2' position of the pentose monosaccharide (deoxyribose or ribose) and the substitution of the base uracil for thymine. However, unlike DNA, prokaryotic RNA is a single-stranded chain of variable length.

## Structure of Bacterial DNA-Dependent RNA Polymerase

RNA polymerase is an oligomeric protein. The active enzyme of *Escherichia coli* contains different subunits called β', β, σ, α, and ω, which have molecular weights of 155,600, 150,600, 70,300, 36,500, and 11,000, respectively. Five of these subunits combine with a stoichiometry of $\alpha_2\beta\beta'\omega$ to form the core enzyme that participates in many of the transcription reactions.

The structure of core RNA polymerase from the bacterium *Thermus aquaticus* is shown in Color Plate 22.1 (see color insert). The β and β' subunits form a large groove at one end. This is where DNA binds and polymerization takes place. The groove is large enough to accommodate about 16 bp of double-stranded B-DNA.

The β and β' subunits are unrelated despite the similarity of their names. These two subunits make up the active site of the enzyme; the β' subunit contributes to DNA binding, whereas the β subunit contains part of the polymerase active site. The α subunits are the scaffold for assembly of the other subunits; they also interact with many proteins that regulate transcription. The role of the small ω subunit is not well characterized.

RNA polymerases from several different bacteria have been purified and compared. Although all function similarly, their subunit composition varies widely. RNA polymerases from other bacterial species are composed of subunits that differ in size from those of *E. coli* RNA polymerase.

## RNA Synthesis

DNA-dependent RNA polymerase requires a DNA template, all four ribonucleoside 5'-triphosphates (ATP, GTP, CTP, and UDP) as precursors of the nucleotide units of RNA, and $Mg^{2+}$. The chemistry of RNA synthesis has much in common with that of DNA synthesis. RNA polymerase elongates an RNA strand by adding ribonucleotide units to the 3'-hydroxyl end of the strand and builds RNA in the 5' → 3' direction. The 3'-hydroxyl group acts as a nucleophile, attacking the α phosphate of the incoming ribonucleoside triphosphate in a reaction that is driven by the release and subsequent hydrolysis of $PP_i$ (Fig. 22.1).

The overall process for RNA synthesis on a duplex DNA template can be conceptually divided into **initiation, elongation,** and **termination.** In the initiation phase of the reaction, the RNA polymerase binds at a specific site on the DNA called the **promoter.** Here, it unwinds and unpairs a small segment of DNA. Elongation begins by the base pairing of ribonucleoside triphosphate to one strand of the DNA, followed by the stepwise formation of a covalent bond between one base and the next. As elongation proceeds, the DNA unwinds progressively in the direction of synthesis (Color Plate 22.2 [see color insert]). Termination occurs at a sequence recognized by the RNA polymerase as a stop signal. At this point the ternary complex of DNA, RNA, and polymerase breaks up.

## Chemistry of Rifamycins

Rifamycins are characterized by an aliphatic ansa-bridge that connects two nonadjacent positions of a naph-

**Figure 22.1** Reaction catalyzed by DNA-dependent RNA polymerase. For simplicity, B represents a base and hydrogen bonding between bases is designated by a single dashed line.

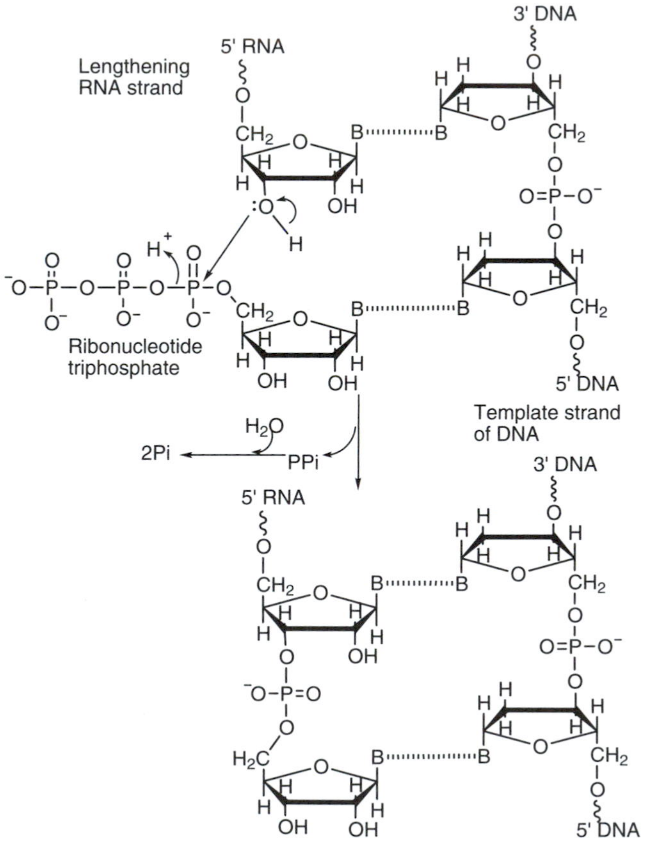

thalenic nucleus. Rifamycin B (Fig. 22.2) was isolated from the broth of fermentation of *A. mediterranei*. Only rifamycin B, which accounts for 10 to 15% of the crude complex, can be isolated as a stable crystalline compound. The addition of sodium diethylbarbiturate to the fermentation medium results in a further improvement in the yield, since with the addition of this additive, rifamycin B is practically the only antibiotic produced under these conditions. Its structure was determined by chemical degradation studies and confirmed by X-ray crystallography. Rifamycin SV (Fig. 22.2) was obtained by chemical modification of other rifamycins but is also produced by fermentation of some strains of *A. mediterranei*.

The rifamycins are biologically active against gram-positive bacteria and mycobacteria, particularly *Mycobacterium tuberculosis*, the agent of tuberculosis. Interest in these antibiotics has been centered on their potent activity against pathogenic mycobacteria. There have been thousands of rifamycin derivatives prepared in an attempt to obtain a broader-spectrum antibiotic with good oral absorption. Rifampin (Fig. 22.2) was the first semisynthetic rifamycin to be approved by the Food and Drug Administration (FDA). Its approval was followed by the approvals of rifabutin in April 1993 and rifapentine in June 1998 (Fig. 22.2); these drugs are more active, have a broader spectrum of biological activity, and are therapeutically more effective than rifamycins B and SV.

## Rifampin

Rifampin, introduced in 1971, has proved to be an effective agent against susceptible strains of *M. tuberculosis* as well as strains resistant to isoniazid or streptomycin. In vitro, it is bactericidal against a wide range of organisms, including mycobacteria. It has been used principally in the management of tuberculous infections at all sites and, more recently, in the treatment of leprosy.

### Mechanism of Action

The mechanism of action of rifampin is based on the inhibition of DNA-dependent RNA polymerase in bacteria and mycobacteria. This results in suppression of the initial chain formation of RNA synthesis. In strains of *E. coli* and mycobacteria susceptible to rifampin, the antibiotic binds to the β subunit of RNA polymerase and leads to abortive initiation of transcription.

### Resistance

Resistance to rifampin develops by alteration of the β subunit of the DNA-dependent RNA polymerase. The

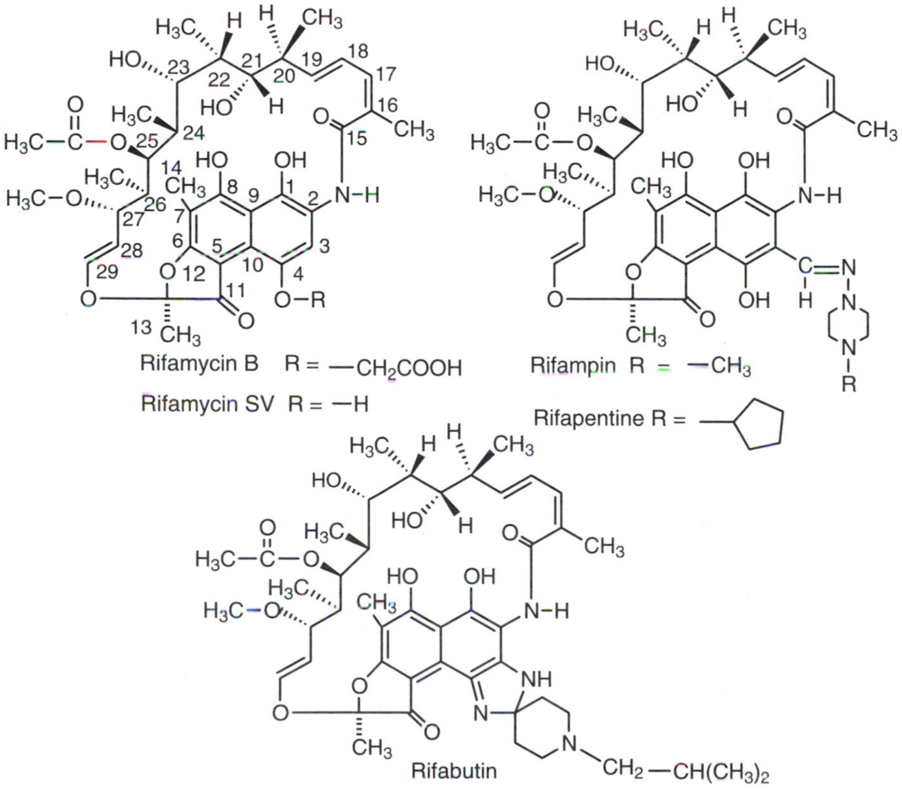

Rifamycin B R = —$CH_2COOH$

Rifamycin SV R = —H

Rifampin R = —$CH_3$

Rifapentine R = (cyclopentyl)

Rifabutin

**Figure 22.2** Chemical structures of rifamycins B and SV, rifampin, and rifapentine.

**Table 22.1** Generic and common trade names of rifamycins and the preparations available in the United States[a]

| Generic name | Trade name (preparation) |
|---|---|
| Rifapentine | Priftin (tablets) |
| Rifampin | Rifadin (capsules) |
| | Rifadin (for intravenous infusion) |
| Rifampin plus isoniazid | Rifamate (capsules) |
| Rifampin plus isoniazid and pyrazinamide | Rifater (tablets) |

[a] All of the rifamycins listed are manufactured by Aventis.

gene that encodes the RNA polymerase β subunit (*rpoB*) was cloned and sequenced. This putative rifampin resistance in *M. tuberculosis* and many *Mycobacterium leprae* isolates is associated with mutations that occur within a 69-bp region of the *rpoB* gene. The types of mutations include single-nucleotide changes and in-frame deletions and insertions.

Because resistance develops rapidly, rifampin should not be used alone; it is normally given in combination with other antituberculosis drugs, such as isoniazid.

## Rifabutin

Rifabutin (Fig. 22.2) is a semisynthetic rifamycin antibiotic, similar to rifampin, which was approved by the FDA in April 1993 for prevention of disseminated *Mycobacterium avium* complex (MAC) disease in patients with advanced human immunodeficiency virus infection. MAC colonization of the respiratory and gastrointestinal tract is common in patients with advanced human immunodeficiency virus infection. The disseminated MAC infection occurs with an increasing frequency as the CD4 lymphocyte count decreases. Treatment of disseminated MAC infection is difficult and requires the use of several drugs.

The mechanism of action of rifabutin against *M. avium* is unknown; in some other susceptible bacteria, this antibiotic inhibits DNA-dependent RNA polymerase. In vitro, rifabutin is about as active as rifampin against MAC organisms and is active against some strains that are resistant to rifampin; strains highly resistant to rifampin, however, may also be resistant to rifabutin. Rifabutin also has a longer half-life than rifampin.

## Rifapentine

Rifapentine (Priftin; Aventis [formerly Hoechst Marion Roussel]) (Fig. 22.2) is a long-acting analog of rifampin. It received an accelerated approval from the FDA in June 1998 for oral use, in conjunction with at least one other drug, in the treatment of pulmonary tuberculosis.

Rifapentine, like other rifamycins, inhibits DNA-dependent RNA polymerase. In addition to its activity against *M. tuberculosis,* it is active in vitro against *M. avium* and *Toxoplasma gondii.* Strains of *M. tuberculosis* resistant to rifamycin are also resistant to rifapentine. Rifapentine has a much longer half-life than rifampin. This is its major advantage over rifampin, since only twice-weekly doses are required for initial treatment of tuberculosis and once-weekly doses are needed during the continuation phase of treatment.

## Rifamycin Preparations Commercially Available in the United States

Table 22.1 lists the rifamycins marketed in the United States.

### REFERENCES AND FURTHER READING


#### RNA Synthesis

**Horton, H. R., L. A. Moran, R. S. Ochs, J. D. Rawn, and K. G. Scrimgeour.** 2002. *Principles of Biochemistry,* 3rd ed. Prentice-Hall, Englewood Cliffs, N.J.

**Nelson, D. L., and M. M. Cox.** 2000. *Lehninger Principles of Biochemistry,* 3rd ed. Worth Publishers, New York, N.Y.

**Zubay, G. L.** 1998. *Biochemistry,* 4th ed. Wm. C. Brown Publishers, Dubuque, Iowa.

#### Rifampin

**Dollery, C.** 1999. *Therapeutic Drugs,* 2nd ed., p. R32–R37. Churchill Livingstone, Edinburgh, United Kingdom.

#### Resistance

**Honore, N., and S. T. Cole.** 1993. Molecular basis of rifampin resistance in *Mycobacterium leprae. Antimicrob. Agents Chemother.* 37:414–418.

**Miller, L. P., J. T. Crawford, and T. M. Shinnick.** 1994. The *rpoB* gene of *Mycobacterium tuberculosis. Antimicrob. Agents Chemother.* 38:805–811.

**Moghazeh, S. L., X. Pan, T. Arain, C. K. Stover, J. M. Musser, and B. N. Kreiswirth.** 1996. Comparative antimycobacterial activities of rifampin, rifapentine, and KRM-1648 against a collection of rifampin-resistant *Mycobacterium tuberculosis* isolates with known *rpoB* mutations. *Antimicrob. Agents Chemother.* **40:**2655–2657.

**Telenti, A., P. Imboden, F. Marchesi, D. Lowrie, S. Cole, M. J. Colston, L. Matter, K. Schopfer, and T. Bodmer.** 1993. Detection of rifampicin-resistant mutations in *Mycobacterium tuberculosis. Lancet* **341:**664–665.

**Valim, A. R., M. L. Rossetti, M. O. Ribeiro, and A. Zaha.** 2000. Mutations in the *rpoB* gene of multidrug-resistant *Mycobacterium tuberculosis* isolates from Brazil. *J. Clin. Microbiol.* **38:**3119–3122.

**Williams, D. L., L. Spring, L. Collins, L. P. Miller, L. B. Heifets, P. R. J. Gangadharam, and T. P. Gillis.** 1998. Contribution of *rpoB* mutations to development of rifamycin cross-resistance in *Mycobacterium tuberculosis. Antimicrob. Agents Chemother.* **42:**1853–1857.

### Resistance in mycobacteria

**Musser, J. M.** 1995. Antimicrobial agent resistance in mycobacteria: molecular genetic insights. *Clin. Microbiol. Rev.* **8:**495–514.

### Rifabutin

**Anonymous.** 1993. Rifabutin. *Med. Lett.* **35:**893–899.

**Dautzenberg, B., P. Castellani, J. L. Pellegrin, D. Vittecoq, C. Truffot-Pernot, N. Pirotta, and D. Sassella.** 1996. Early bactericidal activity of rifabutin versus that of placebo in treatment of disseminated *Mycobacterium avium* complex bacteremia in AIDS patients. *Antimicrob. Agents Chemother.* **40:**1722–1725.

**Dollery, C.** 1999. *Therapeutic Drugs,* 2nd ed., p. R26–R32. Churchill Livingstone, Edinburgh, United Kingdom.

### Rifapentine

**Anonymous.** 1999. Rifapentine. *Med. Lett.* **41:**21–22.

**Araujo, F. G., A. A. Khan, and J. S. Remington.** 1996. Rifapentine is active in vitro and in vivo against *Toxoplasma gondii. Antimicrob. Agents Chemother.* **40:**1335–1337.

chapter 23



# Inhibitors of DNA Gyrase and Topoisomerase IV
Quinolones

The quinolones, also called fluoroquinolones, are derivatives of 1,4-dihydro-4-oxo-3-quinolinecarboxylic acid. This class of antibacterial agents is of synthetic origin. In contrast, the semisynthetic antibiotics are obtained from fermentation of fungus or bacteria and are then modified by chemical or enzymatic reactions.

## Brief History and Overview of the Quinolones

The history of the development of the quinolones as antibacterial agents can be traced to the discovery of an antibacterial by-product formed during the synthesis of the antimalarial agent chloroquine, an isomer of a key intermediate, 7-chloro-1-ethyl-1,4-dihydro-4-oxo-3-quinolinecarboxylic acid (compound 1) (Fig. 23.1). The biological activity of this compound was discovered by a random screening process. Although this compound possessed only modest in vitro antibacterial activity against some gram-negative bacteria, this finding stimulated efforts to synthesize new analogs. In 1962, Lesher et al. described the 7-methyl-1-ethyl-1,4-dihydro-4-oxo-3-quinolinecarboxylic acid (compound 2), also known as nalidixic acid, an 8-aza-4-quinolonecarboxylic acid or 1,8-naphthyridonecarboxylic acid (Fig. 23.1). It was the first commercially available compound of this class and was approved for treatment of urinary tract infections in 1964. The spectrum of activity of nalidixic acid was limited to gram-negative bacteria; it had an unfavorable pharmacokinetic profile, in that its levels in blood were very low, making it ineffective for the treatment of many systemic infections; and resistance developed rapidly. Because of these shortcomings, research and development of new quinolones was renewed to improve activity. Within the next decade, cinoxacin (compound 3), which had improved activity against a limited range of gram-negative bacteria, was synthesized. Parallel developments in Japan yielded 7-piperazine-substituted derivatives of pyridino-pyrimidine, such as pipemidic acid (compound

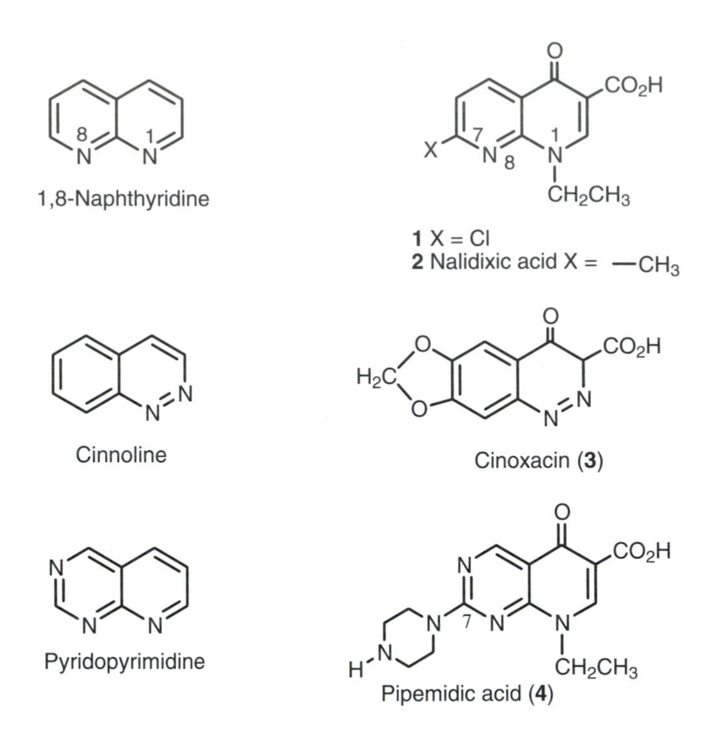


**Figure 23.1** Chemical structures of the first-generation quinolones (structure 1), nalidixic acid (structure 2), cinoxacin (structure 3), and pipemidic acid (structure 4).

4). The introduction of a piperazinyl side chain at position 7, as in pipemidic acid, improved the activity of the drugs against gram-negative bacteria, broadening the spectrum to include *Pseudomonas aeruginosa.* Nalidixic acid was introduced in the United States; it is still in use but only for urinary tract infections. Pipemidic acid has not been approved for use in the United States and is not discussed further in this chapter. By 1977, many other quinolone analogs had been prepared and investigated in terms of antibacterial activity. Compounds 2 to 4 (Fig. 23.1) are now classified as **first-generation quinolones.** It is worth noting that chemically these compounds are naphthyridine, pyridino-pyrimidine, and cinnoline derivatives; however, they are all considered quinoline-type compounds. Thus, one may view naphthyridine drugs as 8-aza-4-quinolones, the pyridopyrimidine ring compounds as 6,8-diaza-4-quinolones, and the cinnoline system as a 2-aza-4-quinolone. Only nalidixic acid, under the trade name of NegGram (tablets and suspension) (Sanofi-Synthelabo) is currently available in the United States.

The first quinolone with a fluorine substituent at position 6 was flumequine, which was patented in 1973 (structure not shown). This compound gave the first indications that activity against gram-positive organisms and pharmacokinetic properties could be improved in this class, but it was not developed as a marketed drug. The real breakthrough came with the combination of these two features in norfloxacin, a 6-fluoroquinolone with a piperazine ring at position 7 (Fig. 23.2). Norfloxacin was patented in 1978. Between then and 1982, many new fluoroquinolones were prepared and patented, several of which are still in use today, including ciprofloxacin (1981) and ofloxacin (1982). These fluoroquinolones are now classified as **second-generation quinolones.**

The second-generation quinolones were introduced into therapy in the mid-1980s as orally administered antibacterial agents with activity that ranges from the *Enterobacteriaceae* and *Pseudomonas* to several gram-positive bacteria. The advantage of these compounds is that they are well absorbed from the gastrointestinal tract, providing adequate levels in blood to allow their use for systemic infections.

The second-generation quinolones are characterized by a fluorine atom on position 6 of the quinoline ring, as well as a carboxyl group at C-3, a keto group at C-4, and a piperazinyl or methylpiperazinyl moiety at C-7 (Fig. 23.2). Levofloxacin is the L isomer (or *S* isomer) of ofloxacin (racemic or 50:50 DL isomers). Figure 23.2 illustrates the chemical structures of all second-generation fluoroquinolones currently marketed in the United States.

A third advance was made in the early 1990s, with the development of the **third-generation quinolones** (Fig. 23.3). All third-generation fluoroquinolones have significantly improved activity against gram-positive bacteria, notably *Streptococcus pneumoniae.* Some of the third-generation quinolones have good activity against anaerobes and atypical pathogens.

There is no agreement about the classification of quinolones among some authors, as exemplified in the book edited by Andriole (2000). Brighty and Gootz classify the quinolones in four generations instead of the three generations used by Ball in the same book.

Grepafloxacin was approved in November 1997 and marketed under the trade name of Raxar (tablets) by Glaxo-Wellcome (now GlaxoSmithKline), but it has recently been withdrawn from the market. Trovafloxacin (Trovan, an oral antibacterial agent) and alatrofloxacin (Trovan-IV, the intravenous formulation of the drug) were withdrawn from the market in April 2001 by Pfizer Inc. because of their liver toxicity. Temofloxacin (Omniflox; Abbott) was approved in 1992 but was withdrawn because of reports of an associated hemolytic uremic syndrome. Norfloxacin

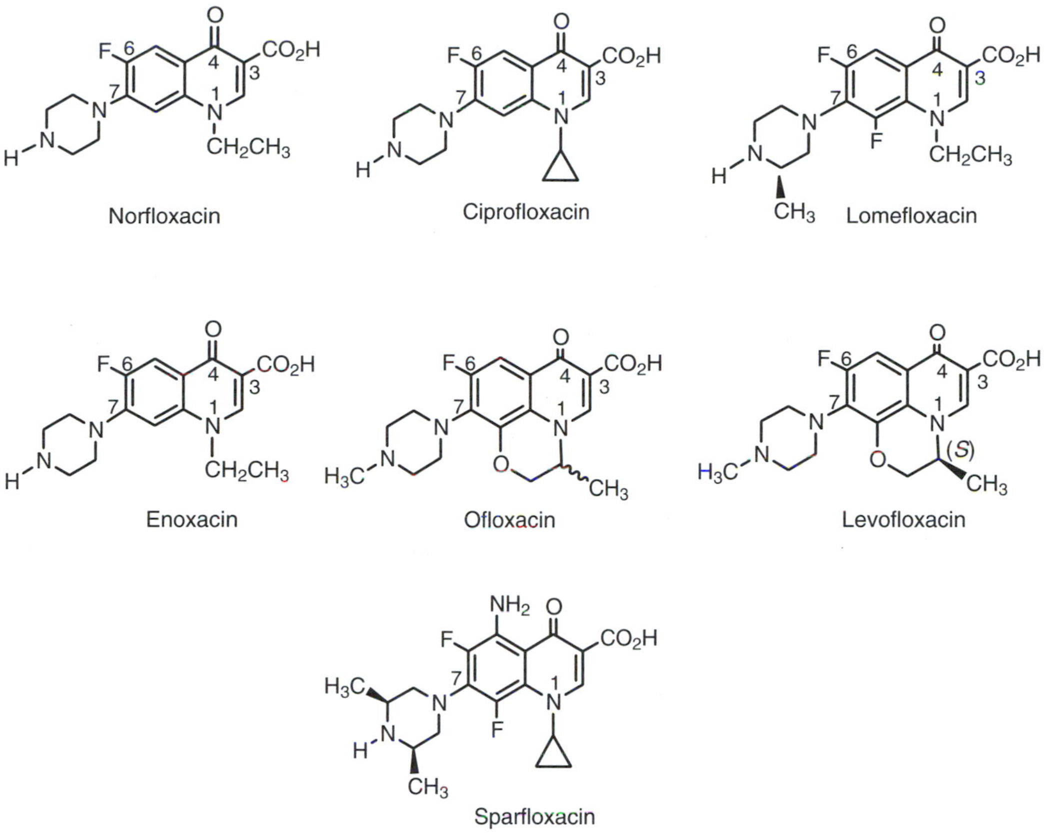

Norfloxacin Ciprofloxacin Lomefloxacin

Enoxacin Ofloxacin Levofloxacin

Sparfloxacin

**Figure 23.2** Chemical structures of second-generation quinolones.

(Noroxin; Merck) is currently not commercially available in the United States.

## Mechanism of Action

In most bacteria, the chromosomes exist as single circles of double-stranded DNA, maintained in a highly supercoiled state. Because of the helical structure of DNA, DNA gyrase and topoisomerase IV are the enzymes responsible for unlinking the parental strands during DNA replication, and their function is essential for removal of the topological constraint to maintain replication fork progression.

Quinolone drugs target bacterial type II topoisomerases; DNA gyrase (also called topoisomerase II) is the primary target in gram-negative bacteria, such as *Escherichia coli*, of nalidixic acid, norfloxacin, and ciprofloxacin, whereas DNA topoisomerase IV is the primary cellular target of third-generation quinolones in gram-positive bacteria, such as *Staphylococcus aureus* and *S. pneumoniae*. These observations suggest that the quinolone chemical structure determines the mode of antibacterial action.

DNA gyrase and topoisomerase IV cleave double-stranded DNA in both strands and then transport another segment of double-stranded DNA through the

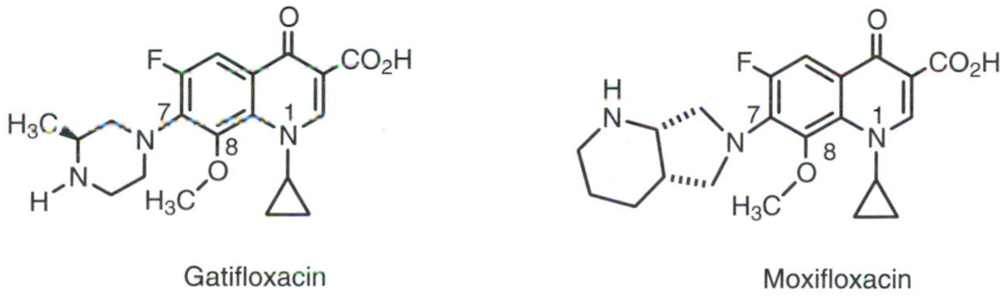

Gatifloxacin Moxifloxacin

**Figure 23.3** Chemical structures of third-generation quinolones.

cleaved DNA segment before religating the DNA. Biochemical studies have revealed two distinct modes of DNA unlinking during DNA replication. DNA gyrase removes the positive supercoils in front of the advancing replication fork, whereas topoisomerase IV decatenates the precatenates behind the replication fork. Thus, the combined efforts of gyrase and topoisomerase IV ensure the completion of DNA unlinking during DNA replication and chromosome segregation. It has been shown that quinolone antibacterial drugs block DNA replication not by depriving the cell of gyrase but instead by trapping a covalent topoisomerase-DNA complex, which leads to the inhibition of DNA replication and the generation of double-strand breaks.

## Supercoiling

**Supercoiling** refers to the coiling of a coil or helix. A telephone cord, for example, is typically a coiled wire. The path taken by the wire between the base of the telephone and the receiver often includes one or more supercoils (Fig. 23.4).

**Figure 23.4** A typical phone cord is coiled like a DNA helix, and the coiled cord can itself coil into a supercoil. Reprinted from D. L. Nelson and M. M. Cox, *Lehninger Principles of Biochemistry,* 3rd ed. (Worth Publishers, New York, N.Y., 2000), with permission from the publisher.

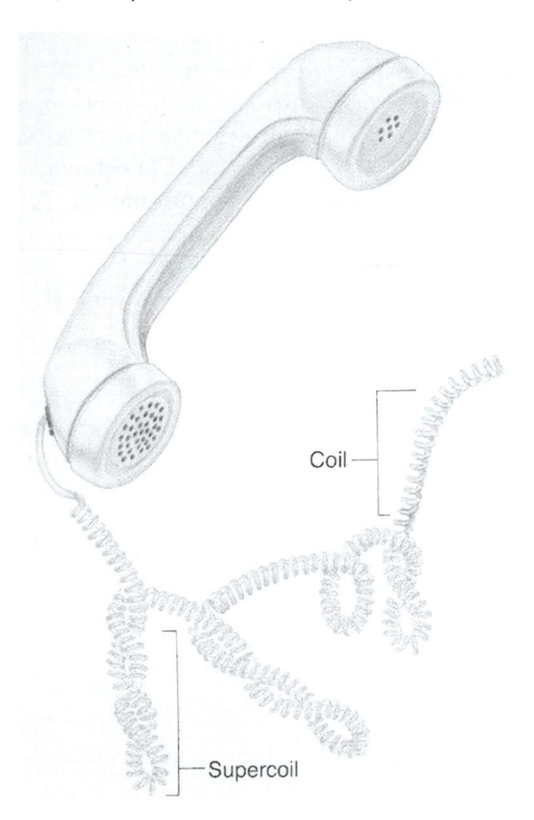


SUPERCOILING OF DNA MOLECULES

Supercoiling is introduced into DNA molecules when the double-stranded helical structure is twisted around its own axis in three-dimensional space. Normally, DNA occurs as a helical, double-stranded molecule in which the two strands are antiparallel and in which nitrogen bases from opposing chains form a complementary structure in which the chains are hydrogen bonded with one another; the classical B-type helix is the structure first solved by Watson and Crick in 1953. Figure 1.28 shows the molecular structure of B-DNA with G-C and A-T base pairing and the antiparallel orientation of the two polynucleotide chains.

The most obvious consequence of DNA supercoiling is the compaction of very large DNA molecules in the relatively small volume of cells.

## Topoisomerases

Topoisomerases are divided into two classes, type I and type II, according to a functional definition which reflects an underlying mechanistic distinction. Type I topoisomerases, represented by *E. coli* topoisomerases I and III, break single strands of duplex DNA, pass another single DNA strand through the break, and then reseal the break. In contrast, type II topoisomerases, represented by *E. coli* DNA gyrase and topoisomerase IV, break both strands of duplex DNA, pass other DNA duplex strands through the break, and reseal both breaks.

DNA gyrase exists as an $A_2B_2$ tetramer encoded by the *gyrA* and *gyrB* genes. Topoisomerase IV exists as a $C_2E_2$ tetramer encoded by the *parC* and *parE* genes. The crystal structures of the 43-kDa N-terminal domain of GyrB (responsible for ATP hydrolysis and for capturing a DNA strand) and a 59-kDa fragment of the 64-kDa N-terminal domain of GyrA (containing the residues for DNA binding and cleavage and the quinolone resistance-determining region [QRDR] [residues 67 to 106]) were determined in 1991 by D. B. Wigley et al. and in 1997 by J. H. Morais Cabral et al., respectively.

Bacterial DNA gyrase plays critical roles in DNA replication, recombination, and transcription, as well as the maintenance of genomic supercoiled density. It can add negative supercoils to DNA, thereby decreasing the linking number. Since supercoiled DNA is under torsional strain, the addition of supercoils requires energy, which is supplied by coupling the reaction to the hydrolysis of ATP. DNA gyrase catalyzes the formation of a transient double-stranded break in one loop of DNA. In the next step, another loop of DNA is passed through the break and then both breaks are resealed. The strand passage process is coupled to the binding and hydrolysis of ATP by the enzymes.

In contrast, *E. coli* topoisomerase I is a monomeric protein with a molecular weight of 100,000. It catalyzes the removal of negative supercoils in DNA. Since supercoiled DNA is under torsional strain, the removal of supercoils is a spontaneous process and no external energy source is needed to drive the topoisomerase I reaction. It might be instructive to present here the proposed mechanism for the reaction of topoisomerase I in *E. coli,* although it is clear that this enzyme is not inhibited by quinolones (Fig. 23.5). In step 1, the hydroxyl group of a tyrosine residue at the active site of the enzyme forms a phosphodiester linkage with a 5'-phosphate group of the DNA chain, breaking the DNA strand. In step 2, the second strand (not shown) passes through the break in the first strand. In step 3, the broken phosphodiester linkage of the first strand is reformed.

## Models of Quinolone Interactions with DNA Gyrase

Shen and Pernet first demonstrated in 1985 that norfloxacin bound neither GyrA not GyrB but did bind double-stranded DNA. Subsequent studies, published by Shen et al. in 1989, showed that DNA gyrase stimulated drug binding to DNA. In 1989, these investigators first proposed the cooperative quinolone–DNA-binding model for DNA gyrase inhibition by quinolones. Figure 23.6 is a schematic representation of the cooperative quinolone–DNA-binding model for DNA gyrase inhibition. Filled and hatched boxes denote the quinolone molecules that self-assemble to form a supermolecule inside the DNA gyrase-induced DNA bubble; the drug binds to the unpaired bases via hydrogen bonds (dotted lines). Details of the proposed mode of self-association

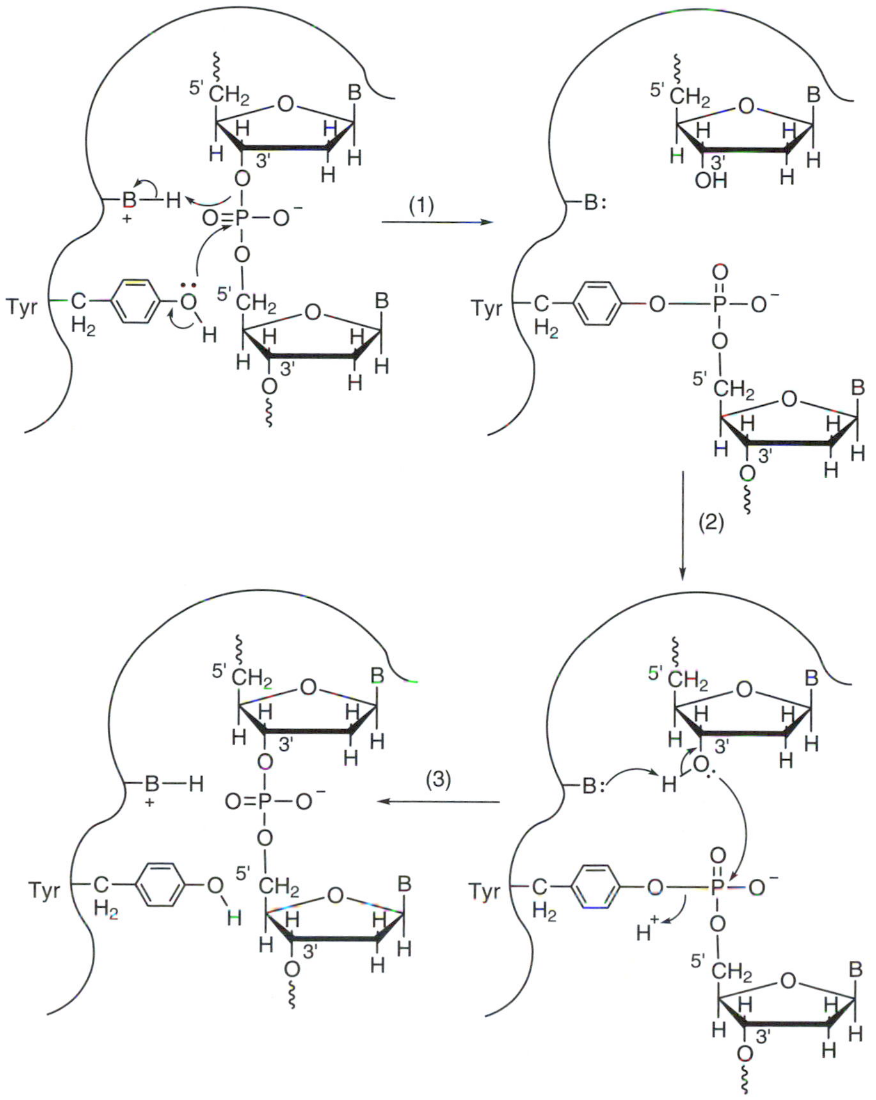

**Figure 23.5** Mechanism of the *E. coli* topoisomerase I reaction. This enzyme is not essential and is not inhibited by quinolones.

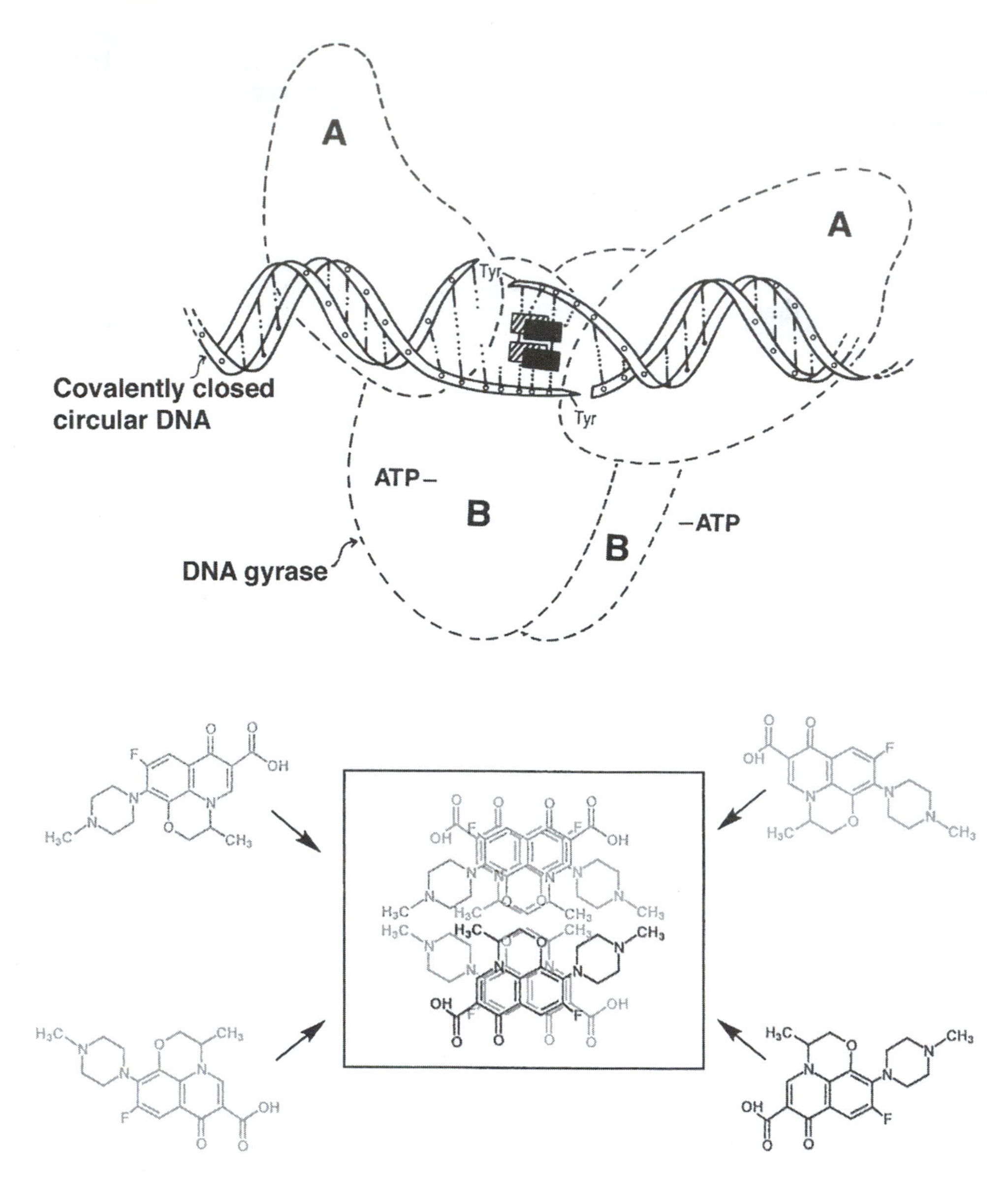


**Figure 23.6** Cooperative quinolone–DNA-binding model for DNA gyrase inhibition proposed by Shen et al. (1989). See the text for details. Reprinted from I. Morrissey, K. Hoshino, K. Sato, A. Yoshida, I. Hayakawa, M. G. Bures, and L. L. Shen, *Antimicrob. Agents Chemother.* **40:**1775–1784, 1996, with permission from the American Society for Microbiology.

are also illustrated below. The binding pocket is thought to be induced during the intermediate gate-opening step of the DNA supercoiling process. DNA gyrase A subunits form covalent bonds between tyrosine 122 and the 5' end of the DNA chain, and the subsequent opening of the DNA chains along the 4-bp staggered cuts results in a locally denatured DNA bubble that is an ideal site for the binding of the drug. Dashed curves outline the shape of the DNA gyrase. Figure 23.7 is another schematic representation of the model described in Fig. 23.6, with the symbols explained in the lower panel.

The cooperative quinolone–DNA-binding model generally agrees with the structure-activity relationships of quinolones. The proposed model suggests three functional domains on the quinolone molecule (Fig. 23.8): the DNA-binding domain, the drug self-association domain, and the drug-enzyme interaction domain. In short, this model suggested that the quinolones bound to a gyrase-DNA complex after strand cleavage, using the exposed single strands in the suggested 4 bp cut as a target and preventing religation. This model has been challenged by the finding of Critchlow and Maxwell in 1996

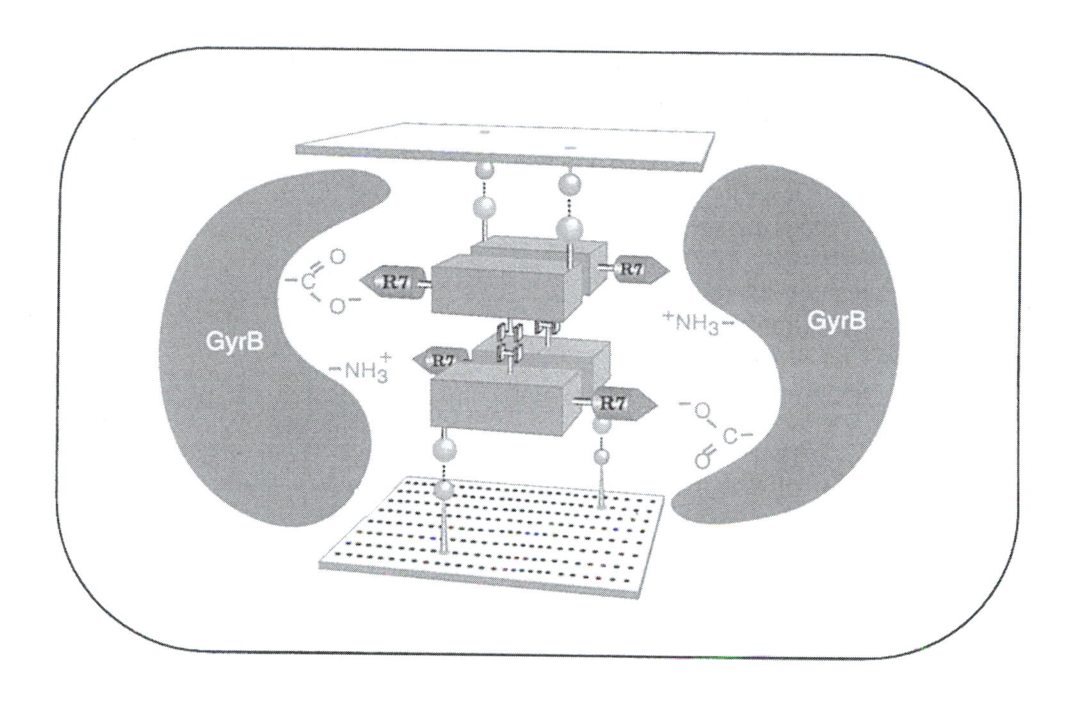


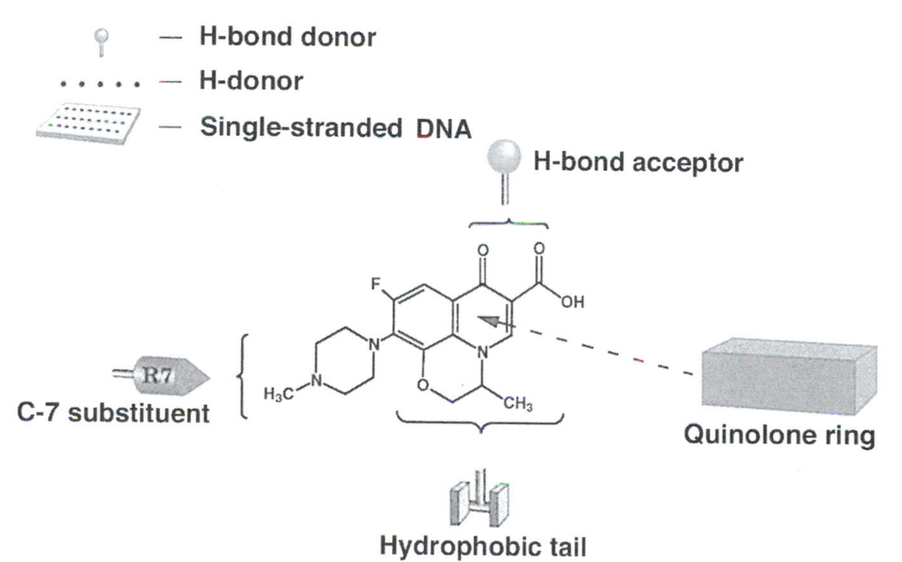


**Figure 23.7** The model represented in Fig. 23.6 is further depicted by a brick-stacking schematic presentation, with the symbols explained in the lower panel. Reprinted from I. Morrissey, K. Hoshino, K. Sato, A. Yoshida, I. Hayakawa, M. G. Bures, and L. L. Shen, *Antimicrob. Agents Chemother.* **40:**1775–1784, 1996, with permission from the American Society for Microbiology.

that DNA cleavage is not required for the binding of [$^3$H]ciprofloxacin to a complex of DNA gyrase bound to DNA.

As mentioned above, bacterial topoisomerase IV is essential for proper chromosome segregation and is a target for quinolone action. Despite the importance of this enzyme to the survival of prokaryotic cells, relatively little is known about the details of its catalytic mechanism or the basis on which quinolones alter its enzymatic function.

## High-Resolution Crystal Structure of a Domain of GyrA

The high-resolution crystal structure of a domain of GyrA has given some insight into the binding pocket and

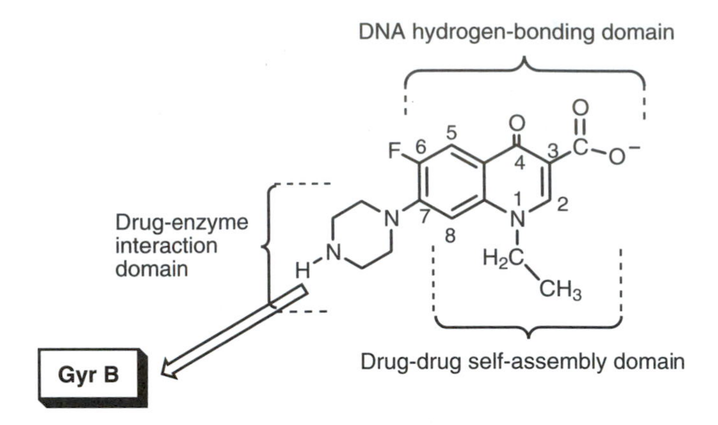


**Figure 23.8** Functional domains of quinolone antibacterial agents.

suggests that two quinolone molecules interact with the GyrA dimer. This binding pocket lies adjacent to the active site for DNA cleavage, and it is likely that drugs interact with DNA bases in that region. Large conformational changes are thought to occur in GyrB during the supercoiling cycle, and it is thought that these lead to the juxtaposition of the quinolone resistance-determining residues from GyrA and GyrB and the assembly of the drug-binding pocket.

## Structure-Activity Relationships

This chapter does not discuss the full details of the structure-activity relationships based on chemical and biological data collected in the quinolone field and described in the literature for over 25 years. There are excellent reviews of this topic, and the reader is directed to the references given at the end of this chapter. Briefly, the basic nucleus of quinolones can be modified at the N-1 position and at the C-6, C-7, and C-8 positions.

- The addition of a fluorine atom at C-6 enhances DNA gyrase-inhibitory activity and provides activity against staphylococci; a second fluorine atom at C-8 results in increased absorption and a longer half-life.
- A piperazine group at C-7 provides better activity against aerobic gram-negative organisms and increased activity against staphylococci and pseudomonads.
- Addition of a methoxy group at C-8 targets DNA gyrase and topoisomerase IV, thereby reducing the resistance of the organisms.

Generally, structure-activity relationship studies of quinolone compounds have been done to identify a position and stereochemistry in the quinolone nucleus or a modification in a substituent that might lead to a broader spectrum of activity, lower toxicity, or improved oral absorption, in addition to other pharmacological or pharmacokinetic aspects. These studies include the following:

- studies of the inhibitory properties of bacterial DNA gyrase and topoisomerase IV
- determinations of MICs for gram-positive and gram-negative organisms (in vitro properties)
- mouse protection models of infection after oral and/or subcutaneous administration (in vivo properties)
- toxicological studies, including 50% lethal dose data and information about side effects
- pharmacokinetic studies with dogs or mice, including oral bioavailability, elimination half-life, and $C_{max}$ (for the most promising candidates)

Current development of a new candidate for synthesis has to consider resistance mechanisms as well as X-ray crystallographic studies of the binding of quinolone drugs to the DNA gyrase-DNA complex. Indeed, from the commercially available molecules we can retrieve a wealth of information about structural aspects of a good antibacterial agent. Certainly, information from X-ray crystallographic studies of enzymes and, even better, of complexes between enzymes and inhibitors and docking of hypothetical molecules has greatly facilitated a rational approach to development of new active drugs.

## Routes of Quinolone Permeation in *Escherichia coli*

At least two factors determine the efficacies of the quinolones against bacteria: the transport of the drug into the cell and the inhibition of the target enzymes, DNA gyrase and topoisomerase IV. In *E. coli,* hydrophilic fluoroquinolones are thought to cross the outer membrane via the OmpF porin. These drugs, particularly the lipophilic ones, also penetrate the lipid bilayer of the inner membrane.

## Clinical Uses of First-, Second-, and Third-Generation Quinolones Available in the United States

Listed at the end of this chapter (see Table 23.3) are the quinolones currently marketed in the United States. Their broad spectrum of activity, along with their excellent bioavailability when given orally, good tissue penetration, and relatively low incidence of adverse effects, accounts for the wide clinical use of these drugs.

The first-generation quinolones are used to treat urinary tract infections. The second-generation quinolones, exemplified by ciprofloxacin, have high activity against gram-negative species and atypical species but only moderate activity against gram-positive bacteria and anaerobes. The third-generation quinolones have improved activity against gram-positive bacteria, especially *S. pneumoniae,* better pharmacokinetic properties, and decreased side effects while the activity against gram-negative and atypical species was preserved and there is good activity against anaerobes, except for gemifloxacin. However, these antibiotics are less active than ciprofloxacin against nonfermentative bacteria.

## First-Generation Quinolones

### NALIDIXIC ACID

The FDA indication for nalidixic acid is for treatment of urinary tract infections. Nalidixic acid was the first oral quinolone approved (in 1962). It is used primarily for urinary tract infections; however, due to the high incidence of side effects, better-tolerated agents should be used.

## Second-Generation Quinolones

This discussion is restricted to the clinical use of the second-generation quinolones. The book by Kucers et al. (1997) and the book edited by Andriole (2000) contain comprehensive accounts of the clinical uses of all quinolones on the market.

### CIPROFLOXACIN

Ciprofloxacin was released in late 1987 and remains a very useful antibacterial agent. Like most other second-generation quinolones, it has a long half-life, allowing twice-daily dosing. It remains active against *P. aeruginosa,* like many quinolones, but is not active against pneumococci, including penicillin-resistant strains and anaerobes, unlike the third-generation agents.

Cipro (available as ciprofloxacin hydrochloride tablets, 5 and 19% oral suspensions, and intravenous infusion) has been developed by Bayer Corp. for oral and parenteral use. Cipro is indicated, according the monograph provided by Bayer Corp., for the treatment of infections caused by susceptible strains of the designated microorganisms under the conditions listed below:

- acute sinusitis caused by *Haemophilus influenzae, S. pneumoniae,* or *Moraxella catarrhalis*
- lower respiratory tract infections caused by *E. coli, Klebsiella pneumoniae, Enterobacter cloacae, Proteus mirabilis, Pseudomonas aeruginosa, H. influenzae, H. parainfluenzae,* or *S. pneumoniae,* as well as acute exacerbations of acute bronchitis caused by *M. catarrhalis* (there is a note in the Bayer monograph that states, "Although effective in clinical trials, ciprofloxacin is not a drug of first choice in the treatment of presumed or confirmed pneumonia secondary to *Streptococcus pneumoniae*")
- urinary tract infections caused by *E. coli, K. pneumoniae, E. cloacae, Serratia marcescens, P. mirabilis, Providencia rettgeri, Morganella morganii, Citrobacter diversus, C. freundii, P. aeruginosa, S. epidermidis, S. saprophyticus,* or *E. faecalis*
- acute uncomplicated cystitis in females as a result of *E. coli* or *S. saprophyticus*
- chronic bacterial prostatitis caused by *E. coli* or *P. mirabilis.*
- complicated intra-abdominal infections (used in combination with metronidazole) caused by *E. coli, P. aeruginosa, P. mirabilis, K. pneumoniae,* or *Bacteroides fragilis*
- skin and skin structure infections caused by *E. coli, K. pneumoniae, E. cloacae, P. mirabilis, P. vulgaris, P. stuartii, M. morganii, C. freundii, P. aeruginosa, S. aureus* (methicillin susceptible), *S. epidermidis,* or *S. pyogenes*
- bone and joint infections caused by *E. cloacae, S. marcescens,* or *P. aeruginosa*
- infectious diarrhea caused by *E. coli* (enterotoxigenic strains), *Campylobacter jejuni, Shigella boydii, S. dysenteriae, S. flexneri,* or *S. sonnei* when antibacterial therapy is indicated
- typhoid fever (enteric fever) caused by *Salmonella enterica* serovar Typhi (Note that the efficacy of ciprofloxacin in the eradication of the chronic typhoid carrier state has not been demonstrated)
- uncomplicated cervical and urethral gonorrhea due to *N. gonorrhoeae*

### ENOXACIN

The FDA indications for enoxacin are as follows:

- cystitis (caused by *Enterobacter* spp., *E. coli, Klebsiella* spp., *Proteus* spp., and *Pseudomonas* spp.)
- gonorrhea (cervical and urethral)
- urinary tract infections

### LEVOFLOXACIN

Levofloxacin is the pure optically active *S*-(–) enantiomer of the racemic drug ofloxacin. The chemical name is (–)-(*S*)-9-fluoro-2,3-dihydro-3-methyl-10-(4-

methyl-1-piperazinyl)-7-oxo-7*H*-pyrido[1,2,3-de]-1,4-benzoxazine-6-carboxylic acid. In 1996, the Food and Drug Administration (FDA) approved the use of levofloxacin hemihydrate for respiratory infections, urinary tract infections, and uncomplicated skin infections; in 2000 the approval was extended to include community-acquired pneumonia due to penicillin-resistant *S. pneumoniae* and complicated skin and skin structure infections; in 2002 the approval was extended to include treatment of nosocomial pneumonia due to methicillin-susceptible *Staphylococcus aureus* (MSSA), *Pseudomonas aeruginosa, Serratia marcescens, Escherichia coli, K. pneumoniae, H. influenzae,* or *S. pneumoniae.*

Levofloxacin (Levaquin; tablets or for injection) is marketed in the United States by Ortho-McNeil Pharmaceutical Inc., a Johnson & Johnson company based in Raritan, N.J. Levaquin tablets or injectable preparations are indicated for the treatment of adults (18 years or older) with mild, moderate, and severe infections caused by susceptible strains of the designated microorganisms under the conditions listed below:

- acute maxillary sinusitis due to *S. pneumoniae, H. influenzae,* or *M. catarrhalis*
- acute bacterial exacerbation of chronic bronchitis due to *S. aureus, S. pneumoniae, H. influenzae, H. parainfluenzae,* or *M. catarrhalis*
- community-acquired pneumonia due to *S. aureus, S. pneumoniae* (including penicillin-resistant strains; MIC of penicillin, 2 μg/ml), *H. influenzae, H. parainfluenzae, K. pneumoniae, M. catarrhalis, Chlamydia pneumoniae, Legionella pneumophila,* or *Mycoplasma pneumoniae*
- complicated skin and skin structure infections due to methicillin-sensitive *S. aureus, E. faecalis, S. pyogenes,* or *P. mirabilis*
- uncomplicated skin and skin structure infections (mild to moderate) including abscesses, cellulitis, furuncles, impetigo, and pyoderma wound infections due to *S. aureus* or *S. pyogenes*
- complicated urinary tract infections (mild to moderate) due to *E. faecalis, E. cloacae, E. coli, K. pneumoniae, P. mirabilis,* or *P. aeruginosa*
- acute pyelonephritis (mild to moderate) caused by *E. coli*
- uncomplicated urinary tract infections (mild to moderate) due to *E. coli, K. pneumoniae,* or *S. saprophyticus*

#### LOMEFLOXACIN

The FDA indications for lomefloxacin are as follows:

- bronchitis
- prophylaxis for urinary tract infections (in transurethral surgical procedures)
- complicated and uncomplicated urinary tract infections

Lomefloxacin is an oral fluoroquinolone that is not frequently used due to the high incidence of photosensitivity and the availability of other alternative fluoroquinolones.

#### SPARFLOXACIN

The FDA indications for sparfloxacin are as follows:

- community-acquired pneumonia
- upper respiratory tract infections (chronic bronchitis and acute bacterial exacerbations of chronic bronchitis)

Sparfloxacin is an oral fluoroquinolone with good activity against penicillin-resistant *S. pneumoniae* but a high incidence of photosensitivity.

### Third-Generation Quinolones

The third-generation quinolones gatifloxacin (Tequin; Bristol-Myers Squibb) and moxifloxacin (Avelox; Bayer) are available for once-daily treatment of patients with community-acquired pneumonia, acute bacterial exacerbations of chronic bronchitis, or acute sinusitis. Gatifloxacin has also received FDA approval for treatment of urinary tract infections and gonorrhea.

#### GATIFLOXACIN

Chemically, gatifloxacin is (±)-1-cyclopropyl-6-fluoro-1,4-dihydro-methoxy-7-(3-methyl-1-piperazinyl)-4-oxo-3-quinoline carboxylic acid (the structure is shown in Fig. 23.3). Several biological effects have been attributed to the methoxy substituent at the C-8 position. First, it appears to impart enhanced activity against DNA gyrase of gram-positive bacteria and hence increased potency against these bacteria. In addition, it increases the bactericidal action of quinolones, particularly against *gyrA* mutants, and also reduces the phototoxicity of the quinolone.

Gatifloxacin (Tequin tablets for oral administration and preparations for injection or intravenous administration) is marketed in the United States by Bristol-Myers Squibb and is indicated for the treatment of infections due to susceptible strains of the microorganisms listed below:

- acute bacterial exacerbations of chronic bronchitis due to *S. pneumoniae, H. influenzae, H. parainfluenzae, M. catarrhalis,* or *S. aureus*

- acute sinusitis due to *S. pneumoniae* or *H. influenzae*
- community-acquired pneumonia due to *S. pneumoniae, H. influenzae, H. parainfluenzae, M. catarrhalis, S. aureus, M. pneumoniae, C. pneumoniae,* or *L. pneumophila*
- uncomplicated urinary tract infections (cystitis) due to *E. coli, K. pneumoniae,* or *P. mirabilis*
- complicated urinary tract infections due to *E. coli, K. pneumoniae,* or *P. mirabilis*
- pyelonephritis due to *E. coli*
- uncomplicated urethral and cervical gonorrhea due to *N. gonorrhoeae*
- acute, uncomplicated rectal infections in women as a result of *N. gonorrhoeae*
- uncomplicated skin and skin structure infections due to MSSA or *Streptococcus pyogenes*

MOXIFLOXACIN

Moxifloxacin (Avelox tablets for oral administration and injection for intravenous administration) is marketed in the United States by Bayer. It has a spectrum of activity similar to those of levofloxacin and gatifloxacin (this includes enhanced activity against *S. pneumoniae*). Its anaerobic activity is moderately good, but there has been only limited clinical experience. The parenteral form was approved in December 2001.

Moxifloxacin is an enantiomerically pure 1-cyclopropyl-7-(2,8-diazabicyclo[4.3.0]nonane)-6-fluoro-9-methoxy-1,4-dihydro-4-oxo-3-quinoline carboxylic acid (Fig. 23.3) which has potent activity against an extensive spectrum of bacteria. It is slightly less active than ciprofloxacin against enterobacteria and much less effective against *P. aeruginosa*. However, it is clearly better than ciprofloxacin in treating infections with atypical microorganisms and especially anaerobic bacteria and aerobic gram-positive cocci, including methicillin-resistant *S. aureus* and penicillin-resistant *S. pneumoniae*.

Avelox is indicated for the treatment of adults (≥18 years of age) with infections caused by susceptible strains of the microorganisms listed below:

- acute bacterial sinusitis caused by *S. pneumoniae, H. influenzae,* or *M. catarrhalis*
- acute bacterial exacerbations of chronic bronchitis caused by *S. pneumoniae, H. influenzae, H. parainfluenzae, K. pneumoniae, S. aureus,* or *M. catarrhalis*
- community-acquired pneumonia (of mild to moderate severity) caused by *S. pneumoniae, H. influenzae, M. pneumoniae, C. pneumoniae,* or *M. catarrhalis*

## Bacterial Resistance

Bacterial resistance to quinolones may result from chromosomal mutations coding for modifications in target subunits of bacterial topoisomerases II and IV or by active efflux via efflux pumps and alterations in the expression of outer membrane proteins, most importantly OmpF. Plasmid-mediated bacterial resistance has not been confirmed to occur.

In gram-positive bacteria in general, and *S. pneumoniae* in particular, resistance to quinolones appears to be due mainly to mutational alterations of the intracellular targets of the drugs, DNA gyrase and topoisomerase IV. Both enzymes are thought to be the primary targets of the drugs in this species, since mutations in the QRDR and the *gyrA* gene, which encodes the A subunit of DNA gyrase, and in the *parC* gene, which encodes the A subunit of topoisomerase IV, confer resistance to single-step mutants. Mutations in *gyrB* and *parE,* which encode the B subunits of DNA gyrase and of topoisomerase IV, respectively, have also been implicated in the quinolone resistance of certain mutants obtained in vitro.

Recall that topoisomerase IV was discovered in 1990 by Kato et al. The enzyme is composed of four subunits, two of each of the ParC and ParE proteins (also called GrlA and GrlB in *S. aureus*), encoded by the *parC* and *parE* genes, respectively.

The quinolones most strongly affected by a single mutation are those for which the mutations occur in their preferred target, for example in gyrase for sparfloxacin, grepafloxacin, gatifloxacin, gemifloxacin, clinafloxacin, and moxifloxacin and in topoisomerase IV for ciprofloxacin, norfloxacin, levofloxacin, and trovafloxacin.

Studies with *E. coli* identified quinolone resistance mutations in a discrete region of GyrA and GyrB termed the QRDR (see Tables 23.1 and 23.2). Recent studies identified similar mutations in the analogous region of ParC.

## DNA Gyrase

In *E. coli,* the most extensively studied gram-negative bacterium, resistance mutations have been found in the amino terminus of GyrA between amino acids 67 and 106, near Tyr-122, which is transiently covalently bound to DNA phosphate groups during the DNA strand-passing reactions of the enzyme (Table 23.1). Therefore, models of quinolone resistance based on alterations in DNA gyrase developed from studies of *E. coli* will probably be applicable to a broad range of gram-negative bacteria. For a complete list of mutations in the GyrA subunit of DNA gyrase and the ParC subunit of topoisomerase IV associated with quinolone resistance and for a comprehensive account of quinolone resistance, see the excellent review by Hooper (1999).

**Table 23.1** Mutations of *E. coli* in the GyrA subunit of DNA gyrase and the ParC subunit of topoisomerase IV associated with quinolone resistance[a]

| GyrA | | | ParC | | |
|---|---|---|---|---|---|
| Amino acid position | Wild-type amino acid | Mutant amino acid | Amino acid position | Wild-type amino acid | Mutant amino acid |
| 67 | Ala | Ser | | | |
| 81 | Gly | Cys, Asp | 78 | Gly | Asp |
| 82 | Asp | Gly | | | |
| 83 | Ser | Leu, Trp, Ala | 80 | Ser | Leu, Ile, Arg |
| 84 | Ala | Pro | | | |
| 87 | Asp | Asn, Val, Gly, Tyr, His | 84 | Glu | Lys, Glu, Val |
| 106 | Gln | His, Arg | | | |

[a] Reprinted from D. C. Hooper, *Drug Resist. Updates* 2:38–55, 1999, with permission from the publisher.

Resistance mutations in GyrB (Table 23.2) generally occur less often than those in GyrA and have been found in *E. coli, S. enterica* serovar Typhimurium, *M. tuberculosis, S. aureus,* and *S. pneumoniae.* The two mutations studied in *E. coli* GyrB, Asp-426–Asn and Lys-476–Glu, are the best known. Both mutations cause resistance to nalidixic acid, but they differ in their effects on susceptibility to fluoroquinolones with a piperazinyl substituent at position 7 of the quinolone ring (norfloxacin, ciprofloxacin, and others).

## Topoisomerase IV

The ParC subunit of topoisomerase IV is homologous to GyrA, and the ParE subunit is homologous to GyrB. The largest body of information concerning the role of topoisomerase IV in quinolone resistance in gram-positive bacteria comes from studies of *S. aureus* and *S. pneumoniae.* Mutations in ParE appear to be uncommon in clinically resistant isolates.

## Resistance Due to Altered Access of Drugs to Target Enzymes

Quinolones must traverse the cell wall and cytoplasmic membrane of gram-positive bacteria and, additionally, the outer membrane of gram-negative bacteria to reach the DNA gyrase and topoisomerase IV in the cytoplasm.

The *S. aureus* NorA protein is a transmembrane multidrug efflux pump that confers low-level resistance to hydrophilic quinolones. The *norA* gene promoter is active in *E. coli.* The best-characterized multidrug efflux pumps in gram-positive pathogens are NorA (*S. aureus*) and PmrA (*S. pneumoniae*). Both pumps are proton motive force-dependent pumps. Quinolone resistance resulting from efflux has been found in a number of gram-negative organisms including *Burkholderia cepacia, Campylobacter jejuni, Citrobacter freundii, E. aerogenes, E. coli, K. pneumoniae, P. vulgaris, P. aeruginosa, S. enterica* serovar Typhimurium, *Shigella dysenteriae, Stenotrophomonas maltophilia, Vibrio parahaemolyticus,* and the anaerobe *B. fragilis.*

For a comprehensive account of efflux-mediated resistance to quinolones in gram-negative bacteria, see the reviews by Hooper (1999) and Poole (2000).

# Quinolone Preparations Commercially Available in the United States

Table 23.3 lists the quinolones marketed in the United States.

**Table 23.2** Mutations of *E. coli* in the GyrB subunit of DNA gyrase and the ParE subunit of topoisomerase IV associated with quinolone resistance[a]

| GyrB | | | ParE | | |
|---|---|---|---|---|---|
| Amino acid position | Wild-type amino acid | Mutant amino acid | Amino acid position | Wild-type amino acid | Mutant amino acid |
| 426 | Asp | Asn | 445 | Leu | His |
| 447 | Lys | Glu | | | |

[a] Reprinted from D. C. Hooper, *Drug Resist. Updates* 2:38–55, 1999, with permission from the publisher.

**Table 23.3** Generic and common trade names of quinolones, the preparations available, and manufacturers in the United States

| Generic name | Trade name (preparation) | Manufacturer |
|---|---|---|
| First generation | | |
| Nalidixic acid | NegGram (tablets, suspension) | Sanofi-Synthelabo |
| Second generation | | |
| Ciprofloxacin | Cipro (intravenous infusion, tablets, oral suspension) | Bayer |
| Enoxacin | Penetrex (tablets) | Aventis |
| Levofloxacin | Levaquin (injection, tablets) | Ortho-McNeil |
| Lomefloxacin | Maxaquin (tablets) | Unimed |
| Ofloxacin | Floxin (intravenous infusion, tablets) | Ortho-McNeil |
| Sparfloxacin | Zagam (tablets) | Bertex |
| Third generation | | |
| Gatifloxacin | Tequim (injection, tablets) | Bristol-Myers Squibb |
| Moxifloxacin | Avelox (tablets) | Bayer |

## REFERENCES AND FURTHER READING

### Books

**Andriole, V. T. (ed.).** 2000. *The Quinolones,* 3rd ed. Academic Press, Inc., San Diego, Calif.

**Hooper, D. C., and E. Rubinstein (ed.).** *Quinolone Antimicrobial Agents,* 3rd ed., in press. ASM Press, Washington, D.C.

### History of the Developments of Quinolones

**Appelbaum, P. C., and P. A. Hunter.** 2000. The fluoroquinolone antibacterials: past, present and future perspectives. *Int. J. Antimicrob. Agents* **16:**5–15.

**Lesher, G. Y., E. D. Forelich, M. D. Gruet, J. H. Bailey, and R. P. Brundage.** 1962. 1,8-Naphthyridone derivatives. A new class of chemotherapeutic agents. *J. Med. Pharm. Chem.* **5:**1063–1068.

### Mechanism of Action

**Anderson, V. E., T. D. Gootz, and N. Osheroff.** 1998. Topoisomerase IV catalysis and the mechanism of quinolone action. *J. Biol. Chem.* **273:**17879–17885.

**Anderson, V. E., R. P. Zaniewski, F. S. Kaczmarek, T. D. Gootz, and N. Osheroff.** 2000. Action of quinolones against *Staphylococcus aureus* topoisomerase IV: basis for DNA cleavage enhancement. *Biochemistry* **39:**2726–2732.

**Drlica, K.** 1999. Mechanism of fluoroquinolone action. *Curr. Opin. Microbiol.* **2:**504–508.

**Drlica, K., and X. Zhao.** 1999. DNA topoisomerase IV as a quinolone target. *Curr. Opin. Anti-Infect. Investig. Drugs* **1:**435–442.

**Heddle, J., and A. Maxwell.** 2002. Quinolone-binding pocket of DNA gyrase: role of GyrB. *Antimicrob. Agents Chemother.* **46:**1805–1815.

**Hiasa, H., M. E. Shea, C. M. Richardson, and M. N. Gwynn.** 2003. *Staphylococcus aureus* gyrase-quinolone-DNA tertiary complexes fail to arrest replication fork progression *in vitro. J. Biol. Chem.* **278:**8861–8868.

**Hooper, D. C.** 1995. Quinolone mode of action. *Drugs* **49**(Suppl. 2)**:**10–15.

**Kampranis, S., and A. Maxwell.** 1998. Conformational changes in DNA gyrase revealed by limited proteolysis. *J. Biol. Chem.* **273:**22606–22614.

**Khodursky, A. B., and N. R. Cozzarelli.** 1998. The mechanism of inhibition of topoisomerase IV by quinolone antibacterials. *J. Biol. Chem.* **273:**27668–27677.

**Marians, K. J., and H. Hiasa.** 1997. Mechanism of quinolone action. *J. Biol. Chem.* **272:**9401–9409.

**Morrissey, I., and J. George.** 1999. Activities of fluoroquinolones against *Streptococcus pneumoniae* type II topoisomerase purified as recombinant proteins. *Antimicrob. Agents Chemother.* **41:**2579–2585.

**Morrissey, I., K. Hoshino, K. Sato, A. Yoshida, I. Hayakawa, M. G. Bures, and L. L. Shen.** 1996. Mechanism of differential activities of ofloxacin enantiomers. *Antimicrob. Agents Chemother.* **40:**1775–1784.

**Noble, C. G., F. M. Barnard, and A. Maxwell.** 2003. Quinolone-DNA interaction: sequence-dependent binding to single-stranded DNA reflects the interaction within the gyrase-DNA complex. *Antimicrob. Agents Chemother.* **47:**854–862.

**Shen, L. L., and A. G. Pernet.** 1985. Mechanism of inhibition of DNA gyrase by analogues of nalidixic acid: the target of the drug is DNA. *Proc. Natl. Acad. Sci. USA* **82:**307–311.

### Supercoiled DNA

**Bowater, R. P.** Supercoiled DNA: structure. Accepted for publication in *Encyclopedia of Life Sciences.* Nature Publishing, London, United Kingdom. [Online.] http://www.els.net.

**Nelson, D. L., and M. M. Cox.** 2000. *Lehninger Principles of Biochemistry,* 3rd ed. Worth Publishers, New York, N.Y.

Topoisomerases

**Adams, D. E., E. M. Shekhtman, E. L. Zechiedrich, M. B. Schmid, and N. R. Cozzarelli.** 1992. The role of topoisomerase IV in partitioning bacterial replicons and the structure of catenated intermediates in DNA replication. *Cell* **71:**277–288.

**Bates, A. D.** Topoisomerases. Accepted for publication in *Encyclopedia of Life Sciences.* Nature Publishing, London, United Kingdom. [Online.] http://www.els.net.

**Drlica, K., and X. Zhao.** 1997. DNA gyrase, topoisomerase IV, and the 4-quinolones. *Microbiol. Mol. Biol. Rev.* **61:**377–392.

**Hooper, D. C.** 1998. Bacterial topoisomerases, anti-topoisomerases, and anti-topoisomerase resistance. *Clin. Infect. Dis.* **27**(Suppl. 1):S54–S63.

**Hoshino, K., A. Kitamura, I. Morrissey, K. Sato, J. Kato, and H. Ikeda.** 1994. Comparison of inhibition of *Escherichia coli* topoisomerase IV by quinolones with DNA gyrase inhibition. *Antimicrob. Agents Chemother.* **38:**2623–2627.

**Kato, J., Y. Nishimura, R. Imamura, H. Niki, S. Hiraga, and H. Suzuki.** 1990. New topoisomerase essential for chromosome segregation in *E. coli. Cell* **63:**393–404.

**Kato, J., H. Suzuki, and H. Ikeda.** 1992. Purification and characterization of DNA topoisomerase IV in *Escherichia coli. J. Biol. Chem.* **267:**25676–25684.

**Levine, C., H. Hiasa, and K. J. Marians.** 1998. DNA gyrase and topoisomerase IV: biochemical activities, physiological roles during chromosome replication, and drug sensitivities. *Biochem. Biophys. Acta* **1400:**29–43.

**Luttinger, A.** 1995. The twisted "life" of DNA in the cell: bacterial topoisomerases. *Mol. Microbiol.* **15:**601–606.

**Wang, J. C.** 1996. DNA topoisomerases. *Annu. Rev. Biochem.* **65:**635–692.

### Mechanism of Quinolone Inhibition of DNA Gyrase

**Critchlow, S. E., and A. Maxwell.** 1996. DNA cleavage is not required for the binding of quinolone drugs to the DNA gyrase-DNA complex. *Biochemistry* **35:**7387–7393.

**Kampranis, S., and A. Maxwell.** 1998. The DNA gyrase-quinolone complex. *J. Biol. Chem.* **273:**22615–22626.

**Morrissey, I., K. Hoshino, K. Sato, A. Yoshida, I. Hayakawa, M. G. Bures, and L. L. Shen.** 1996. Mechanism of differential activities of ofloxacin enantiomers. *Antimicrob. Agents Chemother.* **40:**1775–1784.

**Shen, L. L.** 1993. Quinolone-DNA interaction, p. 77–95. *In* D. C. Hooper and J. S. Wolfson (ed.), *Quinolone Antimicrobial Agents.* American Society for Microbiology, Washington, D.C.

**Shen, L. L., J. Baranowski, and A. G. Pernet.** 1989. Mechanism of inhibition of DNA gyrase by quinolone antibacterials: specificity and cooperativity of drug binding to DNA. *Biochemistry* **28:**3879–3885.

**Shen, L. L., W. E. Kohlbrenner, D. Weigl, and J. Baranowski.** 1989. Mechanism of quinolone inhibition of DNA gyrase. *J. Biol. Chem.* **264:**2973–2978.

**Shen, L. L., L. A. Mitscher, P. N. Sharma, T. J. O'Donnell, D. W. T. Chu, C. S. Cooper, T. Rosen, and A. G. Pernet.** 1989. Mechanism of inhibition of DNA gyrase by quinolone antibacterials: a cooperative drug-DNA binding model. *Biochemistry* **28:**3886–3894.

**Willmott, C. J. R., S. E. Critchlow, I. C. Eperon, and A. Maxwell.** 1994. The complex of DNA gyrase and quinolone drugs with DNA forms a barrier to transcription by RNA polymerase. *Mol. Biol.* **242:**351–363.

### Crystal Structure of the Breakage-Reunion Domain of DNA Gyrase

**Berger, J. M., S. J. Gamblin, S. C. Harrison, and J. C. Wang.** 1996. Structure at 2.7 Å resolution of a 92K yeast DNA topoisomerase II fragment. *Nature* **379:**225–232.

**Morais Cabral, J. H., A. P. Jackson, C. V. Smith, N. Shikotra, A. Maxwell, and R. C. Liddington.** 1997. Crystal structure of the breakage-reunion domain of DNA gyrase. *Nature* **388:**903–906.

**Wigley, D. B., G. J. Davies, E. J. Dodson, A. Maxwell, and G. Dodson.** 1991. Crystal structure of an N-terminal fragment of the DNA gyrase B protein. *Nature* **351:**624–629.

### Structure-Activity Relationships

**Chu, D. T. W., and P. B. Fernandes.** 1989. Structure-activity relationships of the fluoroquinolones. *Antimicrob. Agents Chemother.* **33:**438–442.

**Domagala, J. M.** 1994. Structure-activity and structure-side-effect relationships for the quinolone antibacterials. *J. Antimicrob. Chemother.* **33:**685–706.

**Mitscher, L. A., P. Devasthale, and R. Zavod.** 1993. Structure-activity relationships, p. 3–51. *In* D. C. Hooper and J. S. Wolfson (ed.), *Quinolone Antimicrobial Agents.* American Society for Microbiology, Washington, D.C.

### Permeation in *Escherichia coli*

**Chapman, J. S., and N. H. Georgopapadakou.** 1988. Routes of quinolone permeation in *Escherichia coli. Antimicrob. Agents Chemother.* **32:**131–135.

### Clinical Uses of Quinolones

**Andriole, V. T. (ed.).** 2000. *The Quinolones,* 3rd ed., Academic Press, Inc., San Diego, Calif.

**Bryskier, A.** 1999. Respiratory fluoroquinolones: myth or reality? *Curr. Opin. Anti-Infect. Investig. Drugs* **1:**413–427.

**Hooper, D. C.** 2000. New uses for new and old quinolones and the challenge of resistance. *Clin. Infect. Dis.* **30:**243–254.

**Kucers, A., S. M. Crowe, M. L. Grayson, and J. F. Hoy.** 1997. *The Use of Antibiotics.* Butterworth-Heinemann, Oxford, United Kingdom.

**Piddock, L. J. V.** 1994. New quinolones and gram-positive bacteria. *Antimicrob. Agents Chemother.* **38:**163–169.

### Bacterial Resistance

**Acar, J. F., and F. W. Goldstein.** 1997. Trends in bacterial resistance to fluoroquinolones. *Clin. Infect. Dis.* **24**(Suppl. 1):S67–S73.

**Hooper, D. C.** 1999. Mechanism of fluoroquinolone resistance. *Drug Resist. Updates* **2**:38–55.

**Piddock, L. J. V.** 1995. Mechanisms of resistance to fluoroquinolones: state-of-the-art 1992–1994. *Drugs* **49**(Suppl. 2):29–35.

**Poole, K.** 2000. Efflux-mediated resistance to fluoroquinolones in gram-negative bacteria. *Antimicrob. Agents Chemother.* **44**:2233–2241.

**Sun, L., S. Sreedharan, K. Plummer, and L. M. Fisher.** 1996. NorA plasmid resistance to fluoroquinolones: role of copy number and *norA* frameshift mutations. *Antimicrob. Agents Chemother.* **40**:1665–1669.

chapter 24



# Antibacterial Agents That Cause DNA Damage in Obligate Anaerobic Organisms
## 5-Nitroimidazoles

The members of this class of compounds are 5-nitroimidazoles, of which metronidazole (Fig. 24.1) is the only one that is commercially available in the United States. The other imidazole used clinically is tinidazole, which is available in Europe, including the United Kingdom. This chapter will focus on metronidazole.

Metronidazole was introduced in Europe in 1960 and in the United States in 1963. It has become an extremely important antibacterial agent, especially in the treatment of anaerobic bacterial infections. It causes bacterial DNA damage regardless of the growth phase of the organism and is rapidly bactericidal.

## Brief History of the Development of 5-Nitroimidazoles

The history of the nitroimidazoles began in 1955 in the Rhone-Poulenc research laboratories in France, with the isolation of a mixture of at least three antibiotics from the culture of one *Streptomyces* species. One of these compounds, called azomycin, exhibited activity against protozoans, specifically trichomonads (*Trichomonas vaginalis*). This compound was identified as 2-nitroimidazole. Azomycin was selected as the "lead compound" (lead compounds are compounds possessing sufficient potency in a screening program devised to serve as a model for disease that it justifies significant further investigation). This finding fueled the preparation of many analogues, one of which, 1-(β-hydroxyethyl)-2-methyl-5-nitroimidazole, was found to be the most active derivative against *T. vaginalis* and to have low toxicity. The synthesis of this compound was reported in 1957, and it was named metronidazole. Later, metronidazole was found to be useful for the systemic treatment of urogenital trichomoniasis in humans. It was also shown to be effective in the treatment of human *Giardia lamblia* infections and amebiasis. The observation that metronidazole relieved acute ulcerative

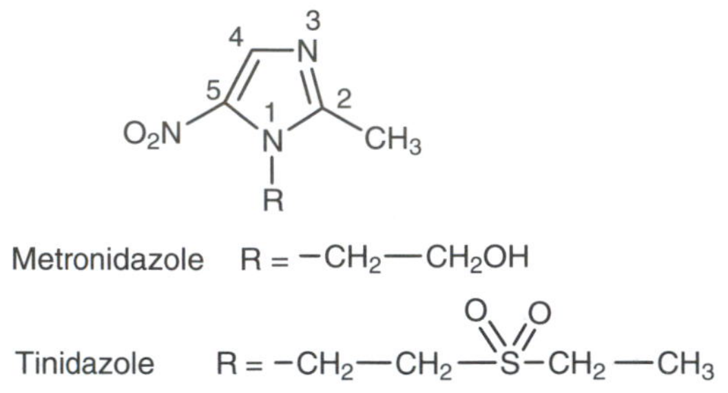

Metronidazole $R = -CH_2-CH_2OH$

Tinidazole $R = -CH_2-CH_2-SO_2-CH_2-CH_3$

**Figure 24.1** Chemical structures of metronidazole and tinidazole.

gingivitis in a patient being treated for trichomonal vaginitis led to studies, culminating in 1962, of its use in anaerobic bacterial infections. Subsequently, it was confirmed that metronidazole was useful for the treatment of Vincent's stomatitis and that it inhibited *Fusobacterium necrophorum*. In 1968, it was approved for use in the prevention of tetanus and gas gangrene. Finally, in 1972 it was shown to be useful for the treatment of infections due to *Bacteroides* spp.; since then, it has been used for a variety of anaerobic infections.

## Metronidazole

Metronidazole has activity against a wide variety of anaerobic prokaryotic and eukaryotic pathogens, including *Helicobacter pylori,* and it is currently the mainstay of therapy for infection with organisms such as *Bacteroides* species, *Clostridium* species, *T. vaginalis, G. lamblia,* and *Entamoeba histolytica.*

### Antimicrobial Activity

Metronidazole is a potent inhibitor of obligate anaerobic protozoa and bacteria.

PROTOZOA

Metronidazole is active against most anaerobic protozoa, including *T. vaginalis, G. lamblia,* and *E. histolytica.* Luminal parasites have two characteristics which distinguish them from other eukaryotes: they live under anaerobic conditions, and they lack mitochondria and enzymes of oxidative phosphorylation. They have an iron-sulfur protein called pyruvate-ferredoxin oxidoreductase, which is involved in metronidazole activation.

GRAM-NEGATIVE ANAEROBIC BACTERIA

Metronidazole is bactericidal for gram-negative anaerobic bacteria, including *B. fragilis,* other *Bacteroides* spp., *Fusobacterium* spp., and *Prevotella* spp. *Veillonella* spp. are also susceptible.

GRAM-POSITIVE ANAEROBIC BACTERIA

The anaerobic cocci such as *Peptococcus* and *Peptostreptococcus* spp. are usually metronidazole sensitive. The anaerobic gram-positive spore-forming bacilli such as *Clostridium perfringes, C. tetani, C. sordellii,* and *C. septicum* are also susceptible. *C. difficile* is quite sensitive. In contrast, the gram-positive asporogenous bacilli such as *Actinomyces, Propiobacterium, Bifidobacterium,* and *Lactobacillus* spp. are usually resistant to metronidazole, except for *Eubacterium* spp., which may be sensitive.

Luminal protozoa and anaerobic bacteria are killed by metronidazole because they share the same fermentation enzyme, pyruvate-ferredoxin oxidoreductase.

### Mechanism of Action

According to published data, the selective activity of 5-nitroimidazoles (metronidazole and tinidazole) against anaerobic organisms is due to the preferential reduction of the 5-nitro group by obligate anaerobes but not by aerobes. This means that the reduction potential is more negative (lower) in anaerobes than in aerobes. In obligate anaerobes, the 5-nitroimidazoles do not have to compete with oxygen.

Any redox system that possesses a reduction potential more negative than that of metronidazole will donate its electrons to metronidazole. The direct donors of electrons in luminal parasites and anaerobic bacteria are thought to be ferredoxin and/or ferredoxin-like Fe-S transport proteins. In anaerobic organisms, the redox potential is –430 to –460 mV, the typical value for ferrredoxin-like Fe-S proteins, whereas metronidazole has a redox potential of –415 mV, making it an efficient electron acceptor. The lowest redox potentials obtainable by aerobic organisms (e.g., *Escherichia coli*) are those of NAD or NADH (–320 mV) and NADP or NADPH (–324 mV), making these organisms intrinsically metronidazole resistant since they are unable to reduce metronidazole.

Before reduction of the 5-nitroimidazoles can occur, the molecule must enter the cell. Reduction by typical anaerobes, including *Clostridium, Bacteroides,* and *Trichomonas* spp., is carried out by the pyruvate-ferredoxin oxidoreductase complex, in which the 5-nitro group of the imidazole acts as an electron sink by capturing one electron to form a nitro radical anion from reduced ferredoxin (Fig. 24.2). In the absence of 5-nitroimidazole, the electron would normally be donated to a proton (hydrogen cation) to form a molecule of hydrogen gas in the hydrogenase reaction ($2H^+ + 2e^- \rightarrow H_2$). In essence, a short-lived reduction product, most probably the protonated nitro radical anion, abstracts electrons (in an oxidation step) from DNA, causing strand breaks,

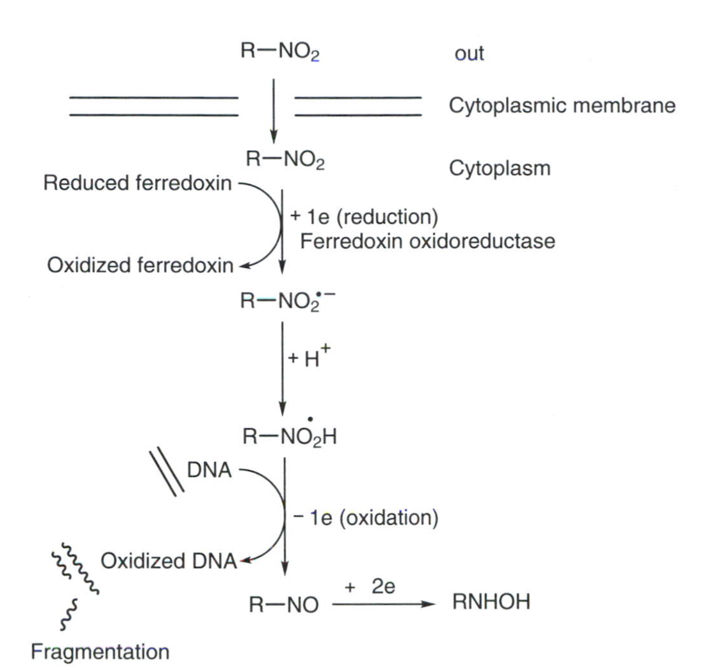


**Figure 24.2** Proposed mechanism for the reduction of 5-nitroimidazoles. Reprinted from D. I. Edwards, *J. Antimicrob. Chemother.* **31:**9–20, 1993, with permission from the publisher.

destabilization of the helix, unwinding, and cell death. The precise nature of the DNA damage is not fully known. In anaerobes, the presence of oxygen molecules would rapidly remove the electron from the nitro radical anion, regenerating the original nitro group and forming a superoxide radical anion:

$$R{-}NO_2^{\bullet-} + O_2 \rightarrow R{-}NO_2 + O_2^{\bullet-}$$

## Clinical Uses

Metronidazole has been the drug of choice for treating trichomoniasis since 1960 and is currently the only drug licensed for this purpose in the United States. It is efficacious against *E. histolytica*, the cause of amebic dysentery and liver abscess. *G. lamblia* (also known as *G. duodenalis*) has been treated with metronidazole since this luminal parasite was recognized as a cause of malabsorption and epigastric pain.

Metronidazole is an important part of combination therapy against *H. pylori*, a major cause of gastritis and peptic ulcer disease. It is particularly useful in treating susceptible anaerobic cerebral infections (usually in combination with antibiotics) and anaerobic meningitis because of its bactericidal activity and good penetration into cerebrospinal fluid and the brain, including abscess contents.

Either alone or in combination with antibiotics, metronidazole is effective as prophylaxis in surgical operations in which the risk of postoperative anaerobic infections is high (e.g., colon surgery and vaginal or abdominal hysterectomy). Readers are referred to recent publications on the clinical uses of metronidazole.

## FDA Indications

The Food and Drug Administration (FDA) indications for metronidazole use are as follows:

- intra-abdominal infections
- skin and skin structure infections
- gynecologic infections (endometritis, endomyometritis, tubo-ovarian abscess, and postsurgical vaginal infections)
- bacterial septicemia
- bone and joint infections
- central nervous system infections (meningitis and brain abscess)
- lower respiratory tract infections (in combination with another agent with activity against microaerophilic streptococci)
- endocarditis (caused by *Bacteroides* species)
- effective colorectal surgery (classified as contaminated or potentially contaminated)

## Acquired Resistance

Understanding of the antimicrobial resistance to metronidazole is based on studies with anaerobic microorganisms such as *Bacteroides, Trichomonas,* and *Clostridium* spp. Although resistance rates of *T. vaginalis* are low, treatment failures due to resistance are significant. The MIC of metronidazole for *T. vaginalis* causing refractory vaginitis is frequently three to eight times the MIC for susceptible strains. Resistant *T. vaginalis* strains have a decrease in the pyruvate-ferredoxin oxidoreductase system.

Usually, metronidazole resistance has been considered the result of insufficient absorption and/or transport of drug to the site of infection. In 1989, the Centers for Disease Control reported that 5% of *T. vaginalis* patients' isolates showed some level of resistance to metronidazole. The prescribed regimen for patients with resistant infections (longer treatment with a higher dosage of drug) is effective for only 80% of the patients with metronidazole-resistant trichomoniasis.

For further information about acquired resistance of *G. intestinalis, Bacteroides,* and *Clostridium* spp., and *H. pylori* to metronidazole, readers are referred to the references at the end of this chapter.

**Table 24.1** Generic and common trade names of metronidazole, the preparations available, and manufacturers in the United States

| Generic name | Trade name (preparation) | Manufacturer |
|---|---|---|
| Oral form | | |
| Metronidazole | Flagyl 375 (capsules) | Searle |
| | Flagyl ER (extended spectrum tablets) | Searle |
| Parenteral form | | |
| Metronidazole hydrochloride | Flagyl I.V. (for intravenous infusion) | SCS Pharmaceuticals |
| Vaginal form | | |
| Metronidazole | MetroGel (vaginal gel) | 3M Pharmaceuticals |
| Topical form | | |
| Metronidazole | Metrocream (topical cream, 0.75%) | Galderma Laboratories |
| | Metrogel (topical gel, 0.75%) | Galderma Laboratories |
| | Metrolotion (topical lotion, 0.75%) | Galderma Laboratories |

## Nitroimidazole Preparations Commercially Available in the United States

Table 24.1 lists the nitroimidazoles marketed in the United States. Preparations are available for oral, parenteral, vaginal, and topical administration.

### REFERENCES AND FURTHER READING


#### General

**Edwards, D. J.** 1987. Nitroimidazoles, p. 404–415. *In* F. O'Grady, H. L. Lambert, R. G. Finch, and D. Greenwood (ed.), *Antibiotics and Chemotherapy,* 7th ed. Churchill Livingstone, Ltd., Edinburgh, United Kingdom.

**Reese, R. E., R. F. Betts, and B. Gumustop.** 2000. *Handbook of Antibiotics.* 3rd ed., p. 521–526. Lippincott Williams & Wilkins, Philadelphia, Pa.

**Scholar, E. M., and W. B. Pratt.** 2000. *The Antimicrobial Drugs,* 2nd ed., p. 422–429. Oxford University Press, Oxford, United Kingdom.

#### Mechanism of Action

**Edwards, D. I.** 1993. Nitroimidazole drugs—action and resistance mechanisms I. Mechanism of action. *J. Antimicrob. Chemother.* **31:**9–20.

**Samuelson, J.** 1999. Why metronidazole is active against both bacteria and parasites. *Antimicrob. Agents Chemother.* **43:**1533–1541.

#### Mechanism of Resistance

**Debets-Ossenkopp, I. J., R. G. Pot, D. J. van Westerloo, A. Goodwin, C. M. J. E. Vandenbroucke-Grauls, D. E. Berg, P. S. Hoffman, and J. G. Kusters.** 1999. Insertion of mini-IS*605* and deletion of adjacent sequences in the nitroreductase (*rdxA*) gene cause metronidazole resistance in *Helicobacter pylori* NCTC11637. *Antimicrob. Agents Chemother.* **43:**2657–2662.

**Edwards, D. I.** 1993. Nitroimidazole drugs—action and resistance mechanisms. II. Mechanism of resistance. *J. Antimicrob. Chemother.* **31:**201–210.

**Haggoud, A., G. Reysset, H. Azeddoug, and M. Sebald.** 1996. Nucleotide sequence analysis of two 5-nitroimidazole resistance determinants from *Bacteroides* strains and of a new insertion sequence upstream of the two genes. *Antimicrob. Agents Chemother.* **38:**1047–1051.

**Kwon, D. H., F. A. K. El-Zaatari, M. Kato, M. S. Osato, R. Reddy, Y. Yamaoka, and D. Y. Graham.** 2000. Analysis of *rdxA* and involvement of additional genes encoding NAD(P)H flavin oxidoreductase (FrxA) and ferredoxin-like protein (FdxB) in metronidazole resistance of *Helicobacter pylori. Antimicrob. Agents Chemother.* **44:**2133–2142.

**Narcisi, E. M., and W. E. Secor.** 1996. In vitro effect of tinidazole and furazolidinone on metronidazole-resistant *Trichomonas vaginalis. Antimicrob. Agents Chemother.* **40:**1121–1125.

**Upcroft, P., and J. A. Upcroft.** 2001. Drug targets and mechanism of resistance in the anaerobic protozoa. *Clin. Microbiol. Rev.* **14:**150–164.

#### Clinical Uses

**Dollery, C.** 1999. *Therapeutic Drugs.* 2nd ed., vol. 2, p. M146–M151. Churchill Livingstone, Ltd., Edinburgh, United Kingdom.

**Kucers, A., S. Crowe, M. L. Grayson, and J. Hoy.** 1997. *The Use of Antibiotics,* 5th ed., p. 936–964. Butterworth-Heinemann, Oxford, United Kingdom.

#### *Trichomonas vaginalis*

**Petrin, D., K. Delgaty, R. Bhatt, and G. Garber.** 1998. Clinical and microbiological aspects of *Trichomonas vaginalis. Clin. Microbiol. Rev.* **11:**300–317.

#### Giardiasis

**Gardner, T. B., and D. R. Hill.** 2001. Treatment of giardiasis. *Clin. Microbiol. Rev.* **14:**114–128.

#### Gram-positive anaerobic cocci

**Murdoch, D. A.** 1998. Gram-positive anaerobic cocci. *Clin. Microbiol. Rev.* **11:**61–120.

chapter 25



# Antibacterial Agents That Cause Damage to DNA
## 5-Nitrofurans

The 5-nitrofurans are a class of totally synthetic antibacterial agents characterized by being derivatives of 5-nitrofuran and by containing the azomethine group (—CH=N—) (Fig. 25.1). Since the initial report of their antibacterial activity, many compounds have been synthesized and tested but only a few have been marketed worldwide for use in humans. However, currently the only 5-nitrofuran derivative in use in the United States is nitrofurantoin, a urinary antiseptic. Nitrofurantoin is still in use in humans because of its broad spectrum of activity, which covers gram-positive and gram-negative bacteria (except *Pseudomonas aeruginosa* and some *Klebsiella* and *Proteus* strains), its relatively low toxicity, and the infrequent development of resistance. Since the introduction of nitrofurantoin for use in urinary tract infections (UTIs), essentially no resistance has occurred, unlike other antibacterial agents. It was suggested that this is due to the ability of the 5-nitrofurans to affect multiple cytoplasmic targets such as inhibiting various enzymes within bacteria and that they may also damage bacterial DNA, leading to DNA strand breakage. Nitrofurans are thought to act at multiple sites in such a way that there is a low probability of developing simultaneous resistance to all these mechanisms of action.

**Urinary antiseptics** are antibacterial agents that concentrate in the urine but do not produce adequate levels in serum. Therefore, these antibacterial agents are useful only in the prevention or therapy of lower UTIs and not for treatment of severe pyelonephritis or associated systemic infections.

The 5-nitroimidazoles and 5-nitrofurans are aromatic compounds that share a 5-nitro group which could be involved in the formation of intermediate reactive free radicals such as nitro radical anions, nitro radicals, and nitroso and hydroxylamine functionalities. The stability of nitro radicals can be explained in terms of resonance of these structures, something that is not present in aliphatic nitro derivatives. The 5-nitroimidazoles such as metronidazole are known to be activated by a bacterial ferredoxin nitroreductase system in susceptible anaerobic organisms. These intermediate free radicals

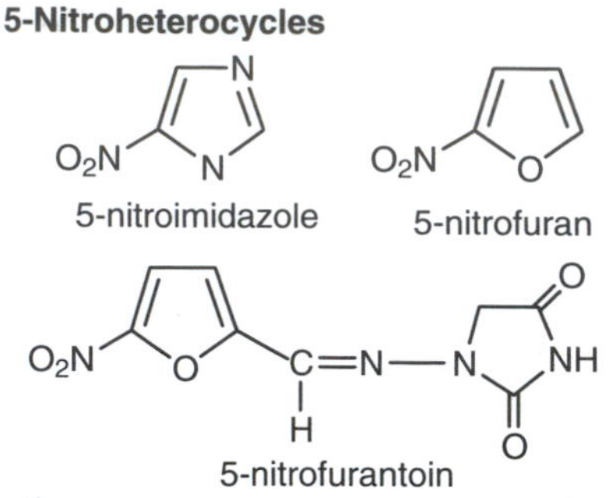

**Figure 25.1** Chemical structures of 5-nitroimidazole, 5-nitrofuran, and 5-nitrofurantoin.

are highly reactive species that are thought to be responsible for the damage to DNA strands that leads to bacterial and protozoal cell death. Some of the differences between luminal protozoa and obligate anaerobic bacteria allow us to understand the selective toxicity of 5-nitroimidazoles. Remember that mitochondria are rod-shaped structures present in almost all eukaryotic cells where oxidative phosphorylation, the citric acid (Krebs) cycle, and fatty acid oxidation take place. The structural organization of mitochondria consists of a double lipoprotein membrane system, with an outer membrane enclosing an inner membrane that, in turn, by a number of invaginations, results in a series of compartments called cristae. Electron transfer, energy generation, and storage processes all occur in the mitochondria. Bacteria have no mitochondria; in bacteria, the mitochondrial functions described above are all carried out within the plasma membrane, where all the needed enzymes are anchored. This relatively poor protection of the enzyme system (compared with that in the host cell) could explain the selective toxicity of these 5-nitro aromatic antibacterial agents. The situation in protozoa is somewhat different. Some species have a few mitochondria; more often, however, some contain simple tubules rather than cristae.

## Nitrofurantoin

Nitrofurantoin is the only 5-nitrofuran derivative available in the United States and the United Kingdom. Its use is restricted to the treatment of acute uncomplicated UTIs (acute cystitis) caused by susceptible strains of *Escherichia coli* or *Staphylococcus saprophyticus*. It is not indicated for the treatment of pyelonephritis or perinephric abscesses, and it lacks the broader tissue distribution of other bactericidal agents approved for treatment of UTIs. It is rapidly excreted into the urine after oral absorption. It is active against most urinary tract pathogens, although *Proteus* spp. and *P. aeruginosa* are usually resistant.

It is of interest that the nausea and vomiting experienced by some patients have been minimized by the development of a macrocrystalline form, whose rate of absorption on oral administration is decreased.

### Spectrum of Antimicrobial Activity

Nitrofurantoin is active against most gram-negative bacilli which commonly cause UTIs. Most strains of *E. coli* are sensitive, but *Enterobacter* and *Klebsiella* spp. are less susceptible. Strains of *Proteus* and *Providencia* spp. vary in their sensitivity, but most are moderately resistant. *Serratia* and *Acinetobacter* spp. and *P. aeruginosa* are usually resistant.

Nitrofurantoin is active against gram-positive cocci which sometimes cause urinary tract infections, such as *E. faecalis, E. faecium, Staphylococcus aureus,* and *Staphylococcus epidermidis*. Vancomycin-resistant *E. faecium* strains are usually nitrofurantoin sensitive. *S. saprophyticus,* an important cause of UTIs in young women, is sensitive to nitrofurantoin.

### Mechanism of Action

The mode of action of nitrofurantoin has not been precisely elucidated.

In the published bibliography by the pharmaceutical laboratory of Procter & Gamble, it is mentioned that "nitrofurantoin is reduced by bacterial flavoproteins to reactive intermediates which in turn inactivate or alter bacterial ribosomal proteins and other macromolecules. As a result of such inactivation, the vital protein synthesis, aerobic energy metabolism, DNA synthesis, RNA synthesis, and cell wall synthesis are inhibited."

### Acquired Resistance

Development of resistance to nitrofurantoin has not been a significant problem since its introduction in 1953. This lack of resistance has been attributed to the broad nature of its mode of action.

### Clinical Uses

As mentioned above, nitrofurantoin is indicated only for the treatment and prophylaxis of UTIs. Its Food and Drug Administration indication is as an antiseptic for uncomplicated UTIs.

## 5-Nitrofuran Preparations Commercially Available in the United States

Macrodantin is the macrocrystalline form (available in capsules), and it appears to be associated with a lower incidence of gastrointestinal tract side effects. Macrobid consists of nitrofurantoin monohydrate (75 mg) and nitrofurantoin macrocrystals (25 mg), with a slow-release formulation, in a capsule equivalent to 100 mg of nitrofurantoin. Both Macrodantin and Macrobid are manufactured by Procter & Gamble Pharmaceuticals.

## REFERENCES AND FURTHER READING

### General

**Hooper, D. C.** 2000. Urinary tract agents: nitrofurantoin and methenamine, p. 423–428. *In* G. L. Mandell, J. E. Bennett, and R. Dolin (ed.), *Principles and Practice of Infectious Diseases,* 5th ed. Churchill Livingstone, Inc., Philadelphia, Pa.

**Scholar, E. M., and W. B. Pratt.** 2000. *The Antimicrobial Drugs,* 2nd ed., p. 245–250. Oxford University Press, Oxford, United Kingdom.

### Mechanism of Action

**McOsker, C. C., and P. M. Fitzpatrick.** 1994. Nitrofurantoin: mechanism of action and implications for resistance development in common uropathogens. *J. Antimicrob. Chemother.* 33(Suppl. A):23–30.

### Clinical Uses

**Kucers, A., S. Crowe, M. L. Grayson, and J. Hoy.** 1997. *The Use of Antibiotics,* 5th ed., p. 922–931. Butterworth-Heinemann, Oxford, United Kingdom.

**Reese, R. E., R. F. Betts, and B. Gumustop.** 2000. *Handbook of Antibiotics,* 3rd ed., p. 564–573. Lippincott Williams & Wilkins, Philadelphia, Pa.

chapter 26



# Compounds That Interfere with Tetrahydrofolic Acid Biosynthesis
## Sulfonamides and Trimethoprim

Two groups of compounds, the sulfonamides and the 2,4-diaminopyrimidines, e.g., trimethoprim (TMP), interfere with the biosynthesis of tetrahydrofolic acid (THF). Their chemistry and development are treated separately, but their mechanism of action relies on the inhibition of two different enzymes, dihydropteroate synthetase (DHPS) and dihydrofolate reductase (DHFR), which are active in the bacterial biosynthesis of THF.

## Sulfonamide Structure

Sulfonamides derived from *p*-aminobenzenesulfonamide are commonly referred to as sulfa drugs. Sulfonamides are synthetic antibacterial agents with an illustrious history. In 1935, they were the first class of antibacterial agents with life-saving potency to be used systematically for the treatment of bacterial infection, initiating the era of the sulfa drugs.

Chemically, the clinically useful sulfonamides are derived from *p*-aminobenzenesulfonamide (Fig. 26.1). Increased activity is associated with monosubstitution of one of the hydrogen atoms of the sulfonamide group attached to position 1 of the benzene ring. Substitution by an aromatic heterocyclic substituent determines the pharmacological properties of the drug, such as absorption, solubility, and gastrointestinal tolerance. A free amino group at position 4 (*para* or *p*) is associated with enhanced activity. Substitution at the amino group results in decreased absorption from the gastrointestinal tract.

## Brief History of the Development of Sulfonamides as Antibacterial Drugs

The history of sulfonamides began in 1932 with the first report by Gerhard Domagk of Bayer Laboratories, who demonstrated that prontosil, a chemical azo sulfonamide dye (Fig. 26.2) was active in vivo against strep-

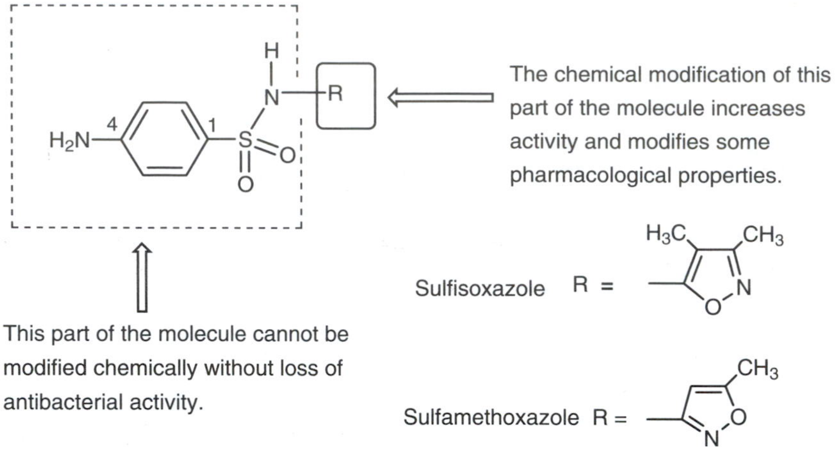

The chemical modification of this part of the molecule increases activity and modifies some pharmacological properties.

This part of the molecule cannot be modified chemically without loss of antibacterial activity.

Sulfisoxazole R =

Sulfamethoxazole R =

**Figure 26.1** General structure of the sulfonamide drugs.

tococcal infections in mice and staphylococcal infections in rabbits but was inactive in vitro. The rationale for screening sulfonamide dyes was based on the known propensity of these dyes to bind wool fibers that are proteins. Analogous attachment of bacterial proteins was expected, but this was not precisely the mode of action of prontosil.

In 1935, Tréfouel and coworkers at the Institut Pasteur (Paris, France) suggested that prontosil undergoes metabolic cleavage of the azo bond to liberate *p*-aminobenzenosulfonamide (sulfanilamide), which may be responsible for the observed antibacterial properties first reported by Domagk. This hypothesis was confirmed in subsequent studies in 1937 by Fuller, who demonstrated the presence of sulfanilamide in blood collected from humans treated with prontosil; sulfanilamide was also isolated from the urine of patients receiving prontosil. These studies established that the therapeutic action of prontosil was indeed due to its metabolism to sulfanilamide, a colorless cleavage product formed by liver metabolism of the administered prontosil. Sulfanilamide became the prototype sulfonamide antibacterial and initiated the research into sulfonamide compounds. Many compounds have been developed since their introduction as antibacterial agents in 1935.

Advances have included increased antibacterial potency, decreased toxicity, and pharmacological properties such as high solubility, low solubility, and prolonged duration of action. The early sulfonamides were produced before the introduction of β-lactam antibiotics initiated the antibiotic era. The more contemporary sulfonamides were produced after 1957, when it had become apparent that penicillins and cephalosporins were not the anticipated cure-all for infectious agents. As expected, precise structure-activity relationships emerged from this research, and a number of sulfa drugs reached the market.

The most useful structural modification of the simple *p*-aminobenzenesulfonamide has been the attachment of aromatic heterocyclic substituents directly to the sulfonamido nitrogen atom. Congeners of this type often offer an enhanced spectrum of activity as well as improved pharmacokinetic properties.

The emergence of strains of bacteria resistant to sulfonamides alone resulted in a steady decrease in their traditional use in uncomplicated urinary tract infection. They are now seldom used alone; however, the combination of sulfamethoxazole with trimethoprim remains clinically relevant in the treatment of many bacterial infections.

## Nomenclature

The generic name of a sulfonamide is built up by adding the prefix "sulfa" (or "sulf" before a vowel) to an abbreviated form of the name of the chemical group attached to the nitrogen atom of the sulfonamide group. For example, sulfisoxazole has an isoxazole ring (Fig. 26.1).

diazo group

liver

[H]

prontosil

*p*-aminobenzenesulfonamide or sulfanilamide

**Figure 26.2** Metabolic transformation in vivo of prontosil into *p*-aminobenzenesulfonamide.

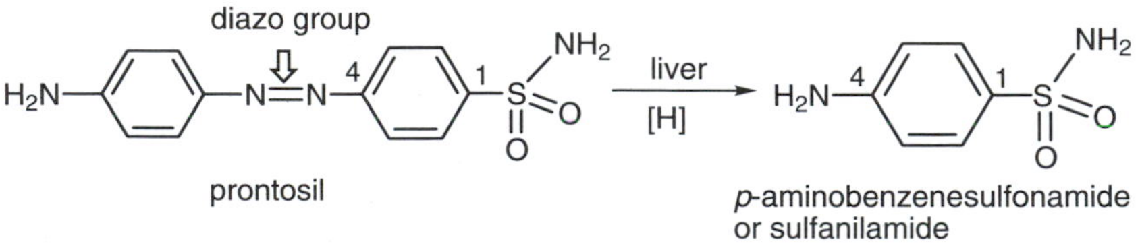

### Sulfamethoxazole

Sulfamethoxazole (SMX) is an intermediate-acting sulfonamide. It is well absorbed from the gastrointestinal tract and is excreted slowly. Because the half-life of SMX (9 to 11 h) is similar to that of TMP (10 to 12 h), it was the sulfonamide of choice for use in combination with TMP.

### Biosynthesis of Sulfonamides in Bacteria

The enzyme DHPS catalyzes the displacement of pyrophosphate from the pteridine substrate. It is well established that the sulfonamides mimic *p*-aminobenzoic acid (PABA), which is the natural substrate required for the biosynthesis of THF. In the absence of sulfonamide competition, the amino group of PABA attaches to the pteridine substrate to produce dihydropteroic acid (Fig. 26.3).

### Mechanism of Action of Sulfonamides

Specifically, the sulfonamides compete with PABA during the formation of dihydropteroate from dihydropteridine pyrophosphate. Sulfonamides have a higher affinity for the bacterial enzyme tetrahydropteroic acid synthetase than does the natural substrate, PABA.

THF is required for growth by both bacterial and human cells. **Folic acid** (Fig. 26.4) is taken into human cells by an active-transport mechanism. Since it does not enter bacterial cells, bacteria must synthesize dihydrofolic acid (DHF) and THF intracellularly de novo. This difference in the biochemistry of the bacterial cell and the human cell is the basis of the selective toxicity of the sulfonamides.

### Inhibitors of Dihydrofolate Reductase

#### Trimethoprim

Trimethoprim, or 2,4-diamino-5-(3,4,5-trimethoxybenzyl)pyrimidine (Fig. 26.5), was synthesized in 1956 by Hitchings at Wellcome Laboratories in the United States. It was first used in 1962 in England. In the 1970s, it was used alone for prophylaxis of urinary tract infections and later for treatment of acute urinary tract infections, as well. Since 1968, TMP and a sulfonamide have been used in combination.

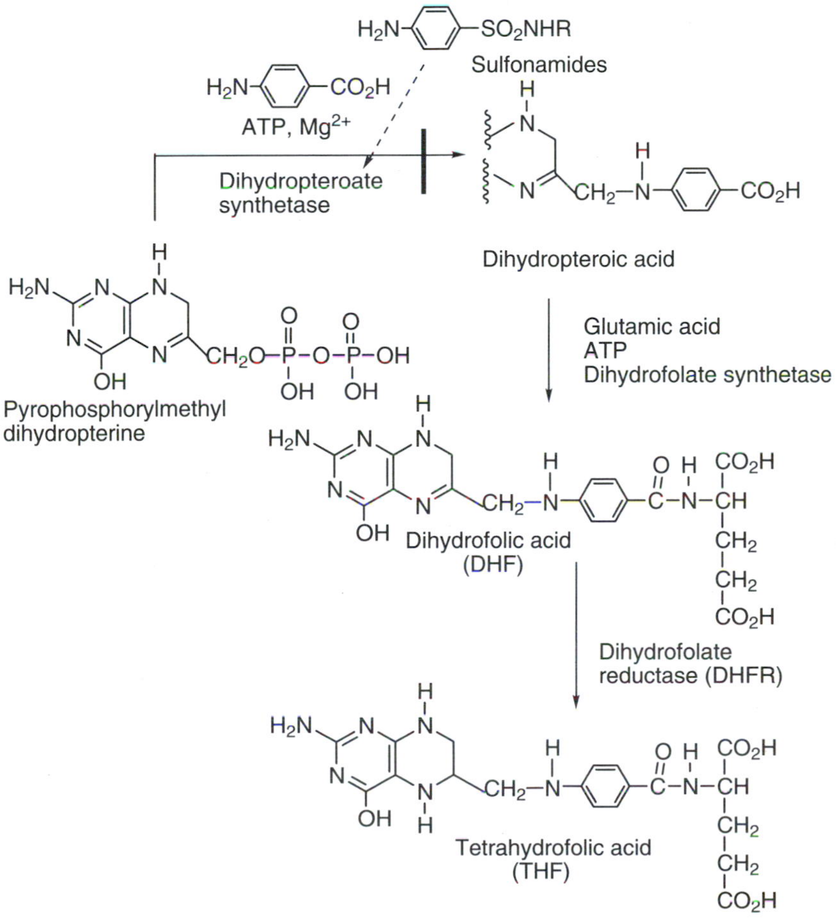

**Figure 26.3** Sulfonamides and THF biosynthesis in bacteria.

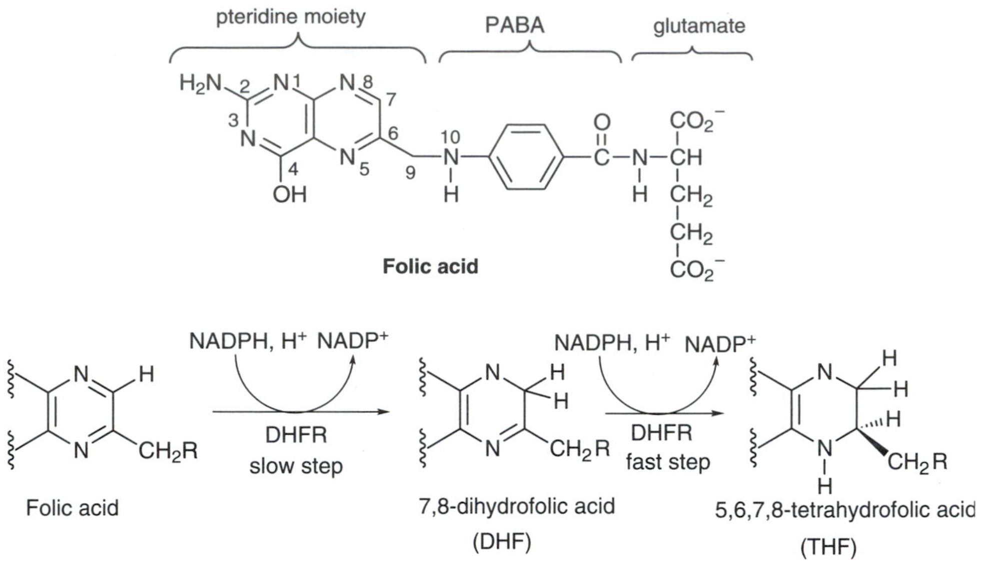

**Figure 26.4** Chemical structure of folic acid and reduction to DHF and THF catalyzed by the human DHFR in the presence of NADPH.

MECHANISM OF ACTION

TMP is a structural analog of the pteridine portion of DHF. It is a competitive inhibitor of DHFR, the enzyme that catalyzes the reduction of DHF to THF in the presence of NADPH (Fig. 26.6). Trimethoprim is 50,000 to 100,000 times more active against bacterial DHFR than against the human enzyme.

Sulfonamides inhibit DHPS, which, as shown in Fig. 26.3, catalyzes the formation of dihydropteroic acid. In the subsequent step of the pathway, TMP inhibits DHFR, which catalyzes the formation of THF from DHF. Although these steps follow one another and cause a sequential blockade, a recent review by Huovinen (2001) indicated that this does not necessarily explain the proposed synergistic effect of the two drugs. Instead, Huovinen pointed out that if a particular pathway is completely inhibited at a particular point, an inhibitor that is present at another point may not be able to increase the degree of inhibition. He also mentioned that a combination of two drugs with slightly different bacterial spectra and different resistance profiles among pathogenic bacteria improves the usefulness of the drug combination, regardless of whether it produces a synergistic effect. The concomitant use of TMP and a sulfonamide acting on the same biosynthetic pathway reduces the rate of developing bacterial resistance with respect to the use of a single pathway blocker.

**Figure 26.5** Chemical structure of TMP.

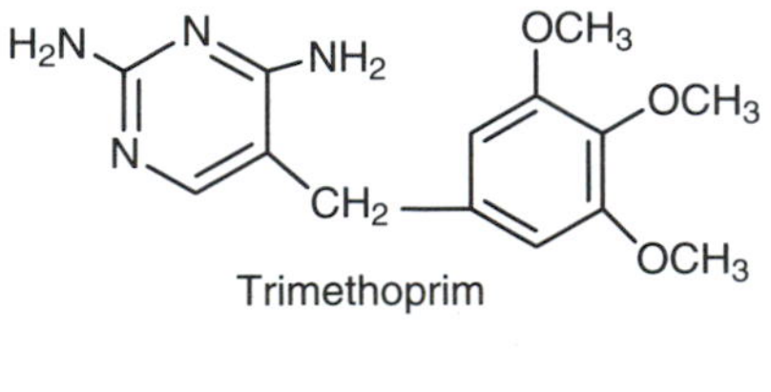

## Trimethoprim-Sulfamethoxazole Combination

TMP-SMX, formerly called co-trimoxazole, became available in the United States in 1973. It is currently available commercially as Bactrim or Septra and in generic preparations. Commercial preparations contain a mixture of TMP and SMX in a fixed ratio of 1:5. Each agent is bacteriostatic when used alone, but together they are bactericidal.

**Figure 26.6** Inhibition of bacterial DHFR by TMP.

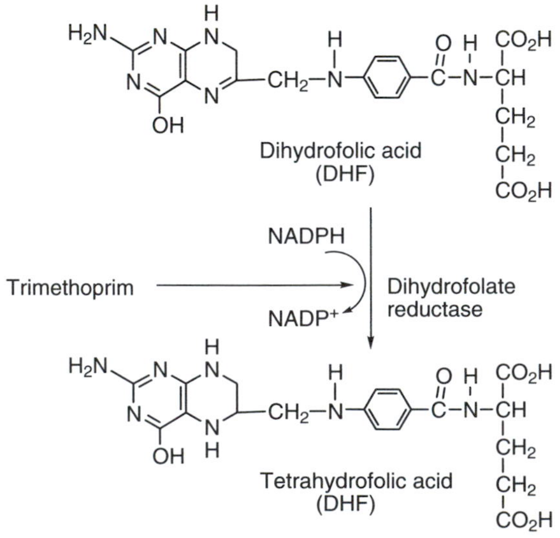

### Spectrum of Antimicrobial Activity

TMP and SMX each cover a wide spectrum of bacteria, including urinary tract pathogens (*Escherichia coli*, other members of the *Enterobacteriaceae*, and *Staphylococcus saprophyticus*) and respiratory tract pathogens (*Streptococcus pneumoniae* and *Haemophilus influenzae*); in combination, they are active against *Moraxella catarrhalis*, skin pathogens (*Staphylococcus aureus*), and enteric pathogens (*E. coli* and *Shigella* species).

## Overview of the Mechanism of Action of Sulfonamides and Trimethoprim

TMP-SMX inhibits thymidylate synthetase and DHFR in sequential steps in the synthesis of THF; from this point, THF is converted to the various one-carbon cofactors required for the synthesis of thymidylate, purine nucleotides, and the amino acids glycine and methionine. This means that bacterial DNA, RNA, and protein synthesis are all affected by these antimicrobials (Fig. 26.7).

### Tetrahydrofolate Cofactors: the Metabolism of $C_1$ Units

THF is required by enzymes that catalyze biochemical transfers of one carbon unit at the oxidation levels of methanol ($CH_3OH$), formaldehyde (HCHO), and formic acid (HCOOH). Thus, the fundamental groups bound to THF are methyl, methylene, or formyl groups. The structures of several one-carbon cofactors of THF and the enzymatic interconversion that occurs among the various substituted forms are shown in Fig. 26.8.

### A Unique Methylation Reaction Produces dTMP from dUMP

Thymidylate (deoxythymidine monophosphate [dTMP]), which is required for DNA formation, is formed from UMP in four steps: UMP is phosphorylated to UDP, which is reduced to dUDP, which is dephosphorylated to dUMP, which is then methylated. The enzyme TS catalyzes the conversion of deoxyuridine monophosphate (dUMP) and $N^5$,$N^{10}$-methylene THF to dTMP and regenerates a molecule of DHF (Fig. 26.9). This is the sole de novo pathway for biosynthesis of dTMP. Because of the central role of TS in the synthesis of an essential DNA precursor and its importance as a chemotherapeutic target, the atomic structure of TS from *Lactobacillus casei* was determined to 3-Å resolution in 1987.

### Precursors of the Purine Ring of Nucleotides

The purine ring of nucleotides is synthesized from amino acids, THF derivatives, and $CO_2$. Cofactors of THF provide the substituents at C-2 and C-8 (Fig. 26.10).

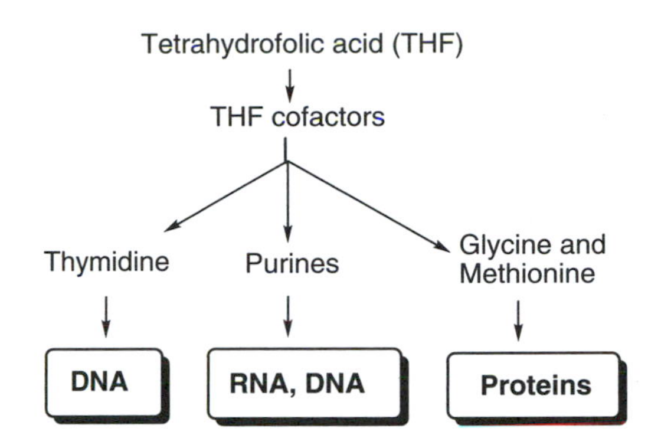


**Figure 26.7** Scheme showing the importance of the essential THF cofactors in the bacterial synthesis of thymidine, purines, glycine, and methionine and how they affect the DNA, RNA, and protein synthesis.

### Biosynthesis of Glycine

The THF derivative donates one carbon unit in the synthesis of glycine from $CO_2$ and $NH_4^+$, in a reaction catalyzed by glycine synthetase (Fig. 26.11).

### Biosynthesis of Methionine

Methionine is regenerated from homocysteine by transfer of the methyl group of $N^5$-methyltetrahydrofolate, as shown in Fig. 26.12.

## Clinical Uses of the Trimethoprim-Sulfamethoxazole Combination

TMP-SMX is a preferred or alternative antibacterial agent for a number of infectious diseases, as outlined below. For a detailed account of clinical uses see Kucers et al. (1997) or another specialized text.

For certain pathogens, TMP-SMX is considered the agent of choice. These include *M. catarrhalis*, *H. influenzae* causing upper respiratory infections and bronchitis, *Y. enterocolitica*, *Aeromonas* spp., *Burkholderia cepacia*, *S. maltophilia*, and *Nocardia* spp.

### Urinary Tract Infections

TMP-SMX is recommended for the treatment of upper or lower urinary tract infections due to susceptible strains of *E. coli*, *Klebsiella* or *Enterobacter* spp., *Proteus mirabilis*, *P. vulgaris*, and *Morganella morganii*. Intravenous administration may be necessary for severe infections (e.g., acute pyelonephritis).

### Respiratory Tract Infections

TMP-SMX is considered the agent of choice for the treatment and prevention of *Pneumocystis carinii* pneu-

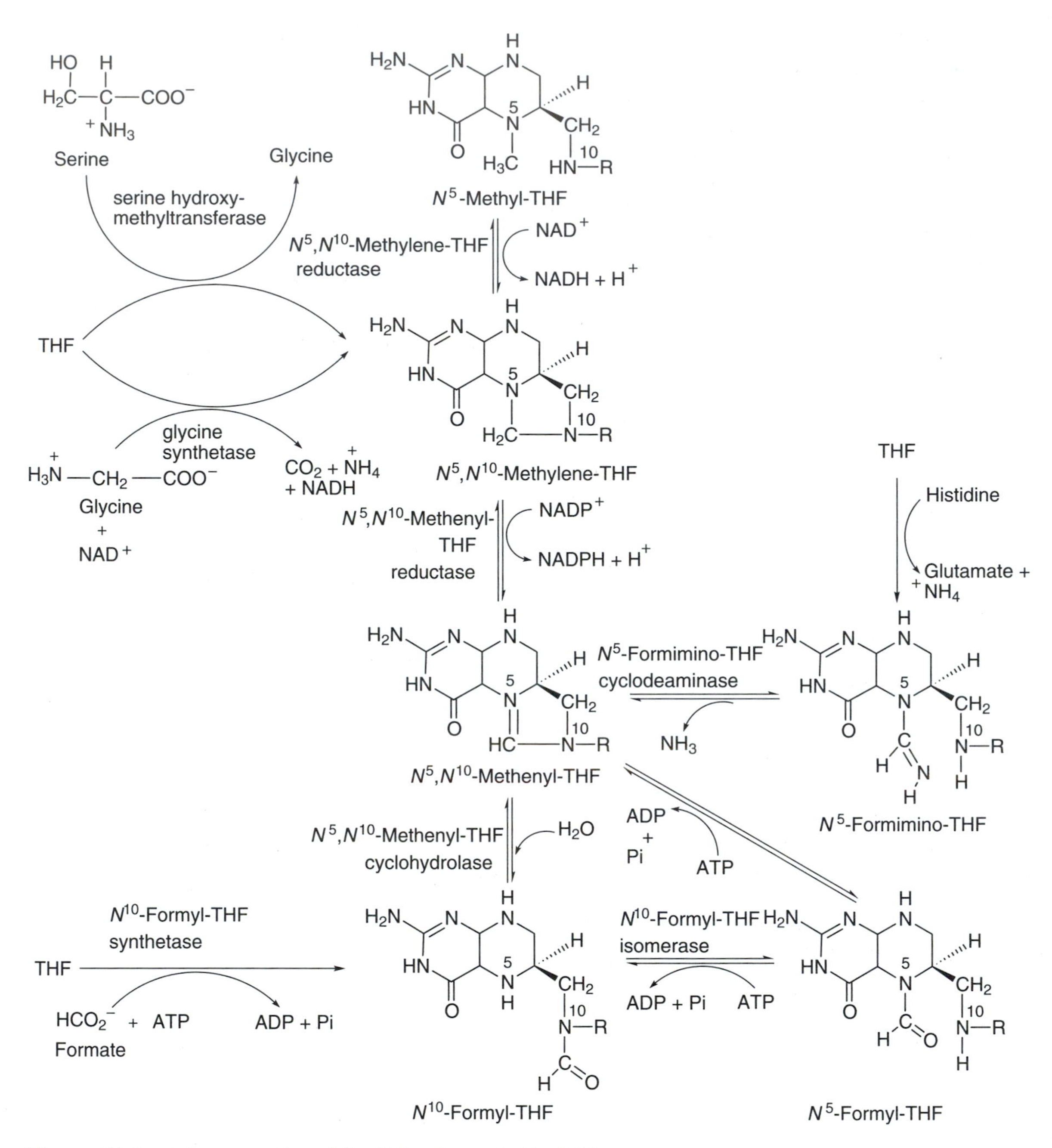


**Figure 26.8** Interconversion of the C-1 units carried by THF.

monia in children and adults. Because TMP-SMX is active against penicillin-susceptible *S. pneumoniae, H. influenzae,* and *M. catarrhalis,* it is a useful agent alone or as part of a rotating antibiotic regimen in the treatment of acute exacerbations of chronic bronchitis. Intravenous TMP-SMX has been used for lower respiratory tract infections (pneumonia) due to susceptible gram-negative pathogens. It has been used as an alternative agent to treat acute otitis media, especially if ampicillin-resistant *H. influenzae* is suspected of being the cause.

## Gastrointestinal Infections

TMP-SMX or a fluoroquinolone is a preferred regimen for shigellosis caused by susceptible strains of *Shigella flexneri* and *S. sonnei.*

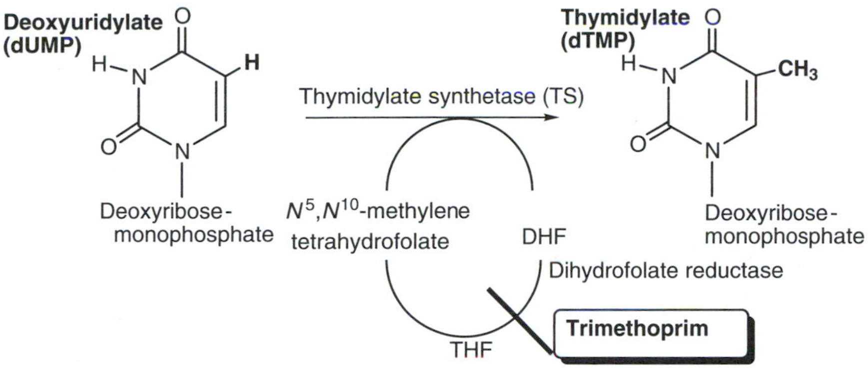

**Figure 26.9** Cycle of reaction in the synthesis of dTMP from dUMP. The reaction for regeneration of THF is blocked by TMP, bringing the cycle to a halt with the folate in the inactive dihydro form.

## Other Infections

Nosocomial infections due to gram-negative bacteria resistant to many antibiotics may be treatable by TMP-SMX (e.g., *Enterobacter, Klebsiella,* and *Proteus* spp.). *B. cepacia* infections that have relapsed or failed to respond to other agents often have responded well to TMP-SMX.

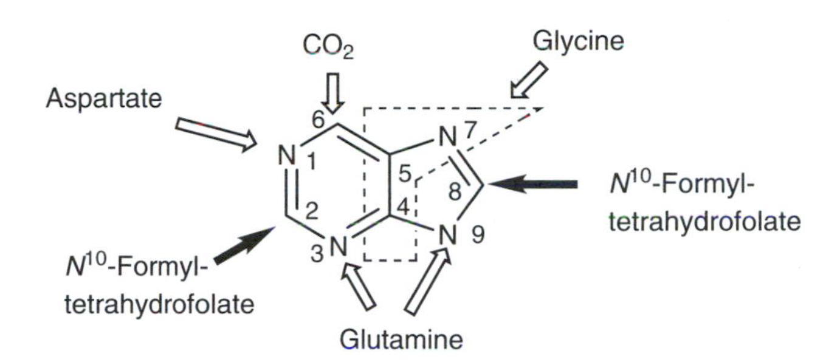


**Figure 26.10** Origins of the atoms in the purine ring.

# Trimethoprim and Sulfonamide Resistance

Sulfonamides were used for the first time in 1932, and TMP was used for the first time in 1962. Since 1968, TMP and sulfonamides have been used in combination. TMP alone has also been used as prophylaxis and treatment for urinary tract infections. Because of its broad spectrum of antibacterial activity, good tolerance, and low cost, the TMP-SMX combination has been used extensively for almost 30 years. As a consequence, virtually all clinically important bacterial species have developed a degree of resistance to TMP and sulfonamides.

In a recent review, Huovinen (2001) noted that bacterial resistance to TMP and to sulfonamides is mediated by the following five main mechanisms:

- the permeability barrier and/or efflux pumps
- naturally insensitive target enzymes
- regulational changes in the target enzymes
- mutational or recombinational changes in the target enzymes
- acquired resistance by drug-resistant target enzymes

## Transport-Related Mechanisms

Resistance mediated by the permeability barrier has been found to act against both sulfonamides and TMP. Moreover, this can also be the case regarding efflux pumps. Huovinen (2001) indicated that it may be difficult to separate the influences of these two mechanisms on resistance levels. *P. aeruginosa* has these types of resistance mechanisms, which explains why the potency of TMP and the sulfonamides against *P. aeruginosa* is limited, with MICs typically in the resistant range. The reason for this **intrinsic resistance** is not well known but has been attributed to a combination of restricted outer membrane permeability and resistance mechanisms such as energy-dependent multidrug efflux.

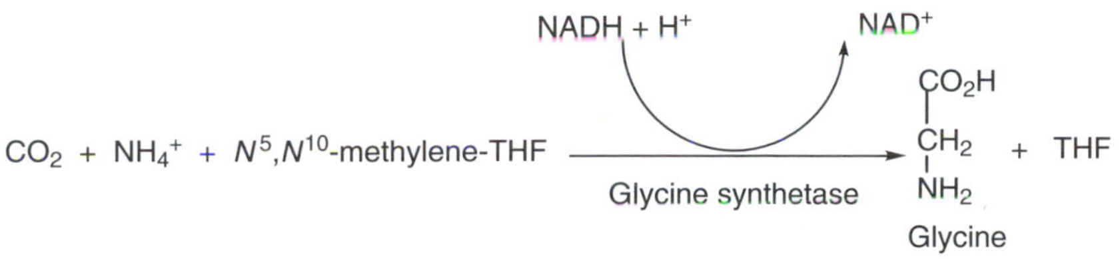

**Figure 26.11** Biosynthesis of glycine in bacteria.

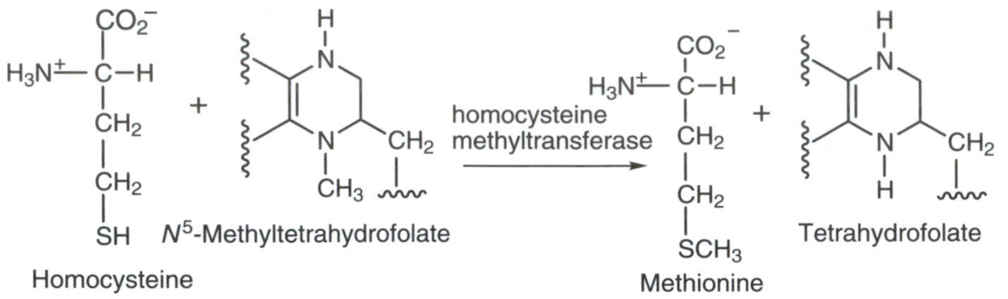

Homocysteine + $N^5$-Methyltetrahydrofolate → (homocysteine methyltransferase) → Methionine + Tetrahydrofolate

**Figure 26.12** Regeneration of methionine from homocysteine by transfer of the methyl group of $N^5$-methyltetrahydrofolate.

## Resistance to Sulfonamides

A number of mechanisms responsible for resistance to the sulfonamides have been elucidated. Organisms may develop resistance by mutation, resulting in either bacterial overproduction of PABA or structural changes in DHPS that produce an enzyme with lowered affinity for sulfonamides. Resistance also may be mediated by plasmids that code for the production of sulfonamide-resistant DHPS.

### OVERPRODUCTION OF PABA

Some *Staphylococcus* and *Neisseria* strains develop sulfonamide resistance through the hyperproduction of PABA. The sulfonamide must be able to compete effectively with the natural substrate PABA for recognition by DHPS. For susceptible bacterial organisms, the sulfonamide can accomplish this task if it is present in sufficient intracellular concentrations to overwhelm the binding of PABA to the enzyme. However, abnormally high levels of PABA can override the antibacterial effects even though the sulfonamide possesses a higher affinity for the enzyme than the natural substrate does.

### CHROMOSOMAL RESISTANCE

Chromosomal mutations in the *dhps* gene can be isolated in the laboratory. An example of this is the nucleotide sequence of the *dhps* gene from *E. coli* C, in which two identical mutants differed from the wild-type strain by a single base pair. In *Neisseria meningitidis*, clinical sulfonamide resistance is linked to the bacterial chromosome. Most of the clinically isolated sulfonamide-resistant strains of *N. meningitidis* contain a *dhps* gene about 10% different from the corresponding gene in sulfonamide-susceptible isolates. It was therefore concluded that the resistance had been introduced by recombination rather than by point mutations.

### PLASMID-BORNE SULFONAMIDE RESISTANCE

The gene encoding DHPS has been cloned, sequenced, and expressed in *E. coli*. The protein has been purified for biochemical characterization and X-ray crystallographic studies. Resistance to sulfonamides in gram-negative enteric bacteria is mediated by the horizontal transfer of foreign *dhps* genes (due to plasmid-encoded DHPS), mediated by *sul* genes, which have diverged considerably from the chromosomal bacterial *dhps* genes. The *sul* gene is part of the integrons. Two such genes, *sul1* and *sul2*, have been sequenced and are found at roughly the same frequency in clinical isolates. These specialized mobile genetic elements facilitate the generation of multiresistance operons. Remarkably, these DHPS enzymes show pronounced insensitivity to sulfonamides but normal binding to the PABA substrate, despite the close structural similarity between substrate and inhibitor.

## Resistance to Trimethoprim

Resistance to TMP is either plasmid borne or chromosome encoded. The most common way in which bacteria resist TMP is by a bypass mechanism. Here the cause of TMP resistance is the production of an additional (plasmid-encoded) DHFR that lacks the capacity to bind TMP. Overproduction of the normal (chromosomally encoded) DHFR is an additional mechanism of resistance.

### PLASMID-ENCODED TRIMETHOPRIM RESISTANCE

Since 1972, about 20 different transferable *dhfr* genes conferring TMP resistance have been characterized. The most prevalent of these genes, *dhfrI* and variants of *dhfrII*, mediate high-level resistance to TMP, with TMP MICs greater than the normal MIC by >1,000-fold; they are most frequently found in gram-negative enteric bacteria.

The nucleotide sequence of a plasmid-borne TMP resistance gene from a commensal fecal *E. coli* isolate revealed a new DHFR gene, *dhfrIV*, which occurred as a gene cassette integrated in a site-specific manner in a class 1 integron.

### CHROMOSOMAL TRIMETHOPRIM RESISTANCE

Chromosomal changes associated with TMP resistance in *E. coli* and *H. influenzae* have been described. In these cases, resistance was due to an altered promoter, resulting in overproduction of the chromosomally encoded DHFR. A single amino acid substitution in the *dhfr* gene and altered chromosomally encoded DHFR have been considered responsible for resistance to TMP among *S. aureus* and *S. pneumoniae*.

A single amino acid substitution, Phe-98 to Tyr-98, in DHFR is the molecular origin of TMP resistance in *S. aureus*. The structural role of Tyr-98 in TMP resistance was determined by obtaining an X-ray crystal structure of the ternary complex of *S. aureus* DHFR with TMP in the presence of NADPH. Critical evidence concerning the resistance mechanism has also been provided by nuclear magnetic resonance spectral analysis of $^{15}N$-labeled TMP in the ternary complex of both wild-type and mutant enzymes. These studies show that the mutations result in the loss of a hydrogen bond between the 4-amino group of TMP and the carbonyl oxygen atom of Leu-5.

Cloning and sequencing of the TMP resistance determinant from *S. pneumoniae* strains indicated that an altered chromosomally encoded DHFR was responsible for resistance. Comparison of DHFR sequences from pneumococcal strains with various susceptibilities to TMP, together with site-directed mutagenesis, revealed that substitution of a leucine residue for isoleucine 100 resulted in TMP resistance. Hydrogen bonding between the carbonyl oxygen atom of isoleucine 100 and the 4-amino group of TMP was proposed to play a critical role in the inhibition of DHFR by trimethoprim.

## Sulfonamides Commercially Available in the United States

Sultrin, marketed as a triple sulfa cream, is a topical antibacterial available for intravaginal administration. Mafenide acetate (Sulfamylon) is a topical antibacterial agent for adjunctive therapy in second- and third-degree burns. Sulfacetamide sodium is indicated in the topical treatment of acne vulgaris. Silver sulfadiazine is a topical antimicrobial drug indicated as an adjuvant for the prevention and treatment of wound sepsis in patients with second- and third-degree burns.

Other sulfonamides, as well as TMP alone and in combination with sulfamethoxazole, are listed in Table 26.1.

**Table 26.1** Generic and common trade names of sulfonamides, trimethoprim, and trimethoprim-sulfonamide combinations, the preparations available, and manufacturers in the United States

| Generic name | Trade name (preparation) | Manufacturer |
|---|---|---|
| Sulfonamides used alone | | |
| Mafenide acetate | Sulfamylon | Bertex |
| Silver sulfadiazine | Silvadene (cream, 1%) | Monarch |
| | Silver sulfadiazine cream | Watson |
| | SSD cream | Par |
| TMP used alone | | |
| TMP sulfate | Proloprim (ophthalmic solution) | Allergan |
| TMP-SMX | Bactrim (for intravenous infusion) | Roche Laboratories |
| | Bactrim (tablets and pediatric suspension) | Roche Laboratories |
| | Septra (for intravenous infusion) | Monarch |
| | Septra (tablets and suspension) | Monarch |
| | TMP + SMX (tablets) | Lederle Standard |
| | TMP + SMX (for injection) | Elkin-Sinn |
| Sulfonamides in other combinations | | |
| Sulfathiazole + sulfacetamide + sulfabenzamide | Sultrin (triple sulfa cream) | Ortho-McNeil |
| Sulfamethizole + oxytetracycline hydrochloride + phenazopyridine hydrochloride | Urobiotic 250 (capsules) | Pfizer |
| Sulfisoxazole + erythromycin ethylsuccinate | Pediazole (for oral suspension) | Ross |
| Sulfacetamide sodium + prednisolone acetate | Blephamide (ophthalmic ointment and suspension) | Allergan |
| Sulfacetamide sodium | Klarol (lotion, 19%) | Dermik |
| Sulfacetamide sodium + sulfur | Sulfacet R (lotion) | Dermik |

## REFERENCES AND FURTHER READING


### History of the Development of Sulfonamides as Antibacterial Drugs

**Fuller, A. T.** 1937. *Lancet* i:194.

**Tréfouel, J., J. Tréfouel, F. Nitti, and D. Bovet.** 1935. *C. R. Soc. Biol.* **120:**756.

### General

**Zinner, S. H., and K. H. Mayer.** 2000. Sulphonamides and trimethoprim, p. 394–404. *In* G. L. Mandell, J. E. Bennett, and R. Dolin, ed., *Principles and Practice of Infectious Diseases,* 5th ed. Churchill Livingstone, Inc., Philadelphia, Pa.

### Clinical Uses

**Kucers, A., S. M. Growe, M. L. Grayson, and J. F. Hoy.** 1997. *The Use of Antibiotics,* 5th ed., p. 806–904. Butterworth-Heinemann, Oxford, United Kingdom.

### Structure and Functions of DHPS and DHFR

**Hampele, I., A. D'Arcy, G. E. Dale, D. Kostrewa, J. Nielsen, C. Oefner, M. G. P. Page, H. J. Schönfeld, D. Stüber, and R. L. Then.** 1997. Structure and function of the dihydropteroate synthase from *Staphylococcus aureus. J. Mol. Biol.* **268:**21–30.

**Hardy, L. W., J. S. Finer-Moore, W. R. Montfort, M. O. Jones, D. V. Sant, and R. M. Stroud.** 1987. Atomic structure of thymidilate synthase: target for rational drug design. *Science* **235:**448–455.

### Bacterial Resistance to Sulfonamides

**Sköld, O.** 2000. Sulfonamide resistance: mechanisms and trends. *Drug Resist. Updates* **3:**178–189.

### Bacterial Resistance to Trimethoprim

**Dale, G. E., C. Broger, A. D'Arcy, P. G. Hartman, R. Hooyt, S. Jolidon, I. Kompis, A. M. Labhardt, H. Langen, H. Locher, M. G. Page, D. Stuber, R. L. Then, B. Wipf, and C. Oefner.** 1997. A single amino acid substitution in *Staphylococcus aureus* dihydrofolate reductase determines trimethoprim resistance. *J. Mol. Biol.* **266:**23–30.

**De Groot, R., M. Sluijter, A. D. De Bruyn, J. Campos, W. H. F. Goessens, A. L. Smith, and P. W. M. Hermans.** 1996. Genetic characterization of trimethoprim resistance in *Haemophilus influenzae. Antimicrob. Agents Chemother.* **40:**2131–2136.

**Huovinen, P.** 1997. Increases in rates of resistance to trimethoprim. *Clin. Infect. Dis.* **24**(Suppl. 1):S63–S66.

**Peter, A. V., M. du Plessis, K. P. Klugman, and S. G. B. Amyes.** 1998. New trimethoprim-resistant dihydrofolate reductase cassette, *dfrXV,* inserted in a class 1 integron. *Antimicrob. Agents Chemother.* **42:**2221–2224.

**Pikis, A., J. A. Donkersloot, W. J. Rodriguez, and J. M. Keith.** 1998. A conservative amino acid mutation in the chromosome-encoded dihydrofolate reductase confers trimethoprim resistance. *J. Infect. Dis.* **178:**700–706.

### Bacterial Resistance to Trimethoprim and Sulfonamides

**Huovinen, P.** 2001. Resistance to trimethoprim-sulfamethoxazole. *Clin. Infect. Dis.* **32:**1608–1614.

**Huovinen, P., L. Sundström, G. Swedberg, and O. Sköld.** 1995. Trimethoprim and sulfonamide resistance. *Antimicrob. Agents Chemother.* **39:**279–289.

**Köhler, T., M. Kok, M. Michea-Hamzehpour, P. Plesiat, N. Gotoh, T. Nishino, L. Kocjancic Curty, and J.-C. Pechere.** 1996. Multidrug efflux in intrinsic resistance to trimethoprim and sulfamethoxazole in *Pseudomonas aeruginosa. Antimicrob. Agents Chemother.* **40:**2288–2290.

chapter 27



# New Antibacterial Drugs in Development That Act on Novel Targets

## Background and Outlook

This book would not be complete without a chapter on the current development of novel antibacterials that are members of new rather than existing classes. So, this chapter presents a survey of some new antibacterial agents that act on novel targets.

The emergence of antibiotic- and synthetic antibacterial agent-resistant bacterial pathogens such as methicillin-resistant *Staphylococcus aureus* (MRSA) and vancomycin-resistant enterococci (VRE) are major concerns in the hospital setting, and multidrug-resistant strains of *Streptococcus pneumoniae, S. pyogenes, Mycobacterium tuberculosis,* and *Neisseria gonorrheae* plague the community. Multiresistance plasmids and transposons, integrons, or mutations can lead to bacteria becoming resistant to almost all existing antibiotics and synthetic antibacterial agents. We have seen that bacterial resistance takes different forms. The bacteria can alter the drug's target, prevent the drug from reaching the target by using mechanisms such as efflux pumps, or destroy or modify the drug so that it cannot bind to its target. Historically, three important cellular functions have been the major targets of antibiotics: cell wall biosynthesis, protein synthesis, and DNA replication.

Although numerous antibiotics and synthetic antibacterial agents are currently on the market, bacterial resistance to conventional antibacterial agents has become a serious problem in infection control and has created an urgent need for the discovery of new agents that operate with distinctly different mechanisms of action from those of current therapies. The pharmaceutical industry, biotechnology companies, and academic centers are responding to the threat of bacterial resistance with renewed efforts to discover new antibacterials in attempts to overcome the problem of global antibacterial resistance. Figure 27.1 summarizes the current status of the discovery and development of some new antibacterial agents with novel modes of action.

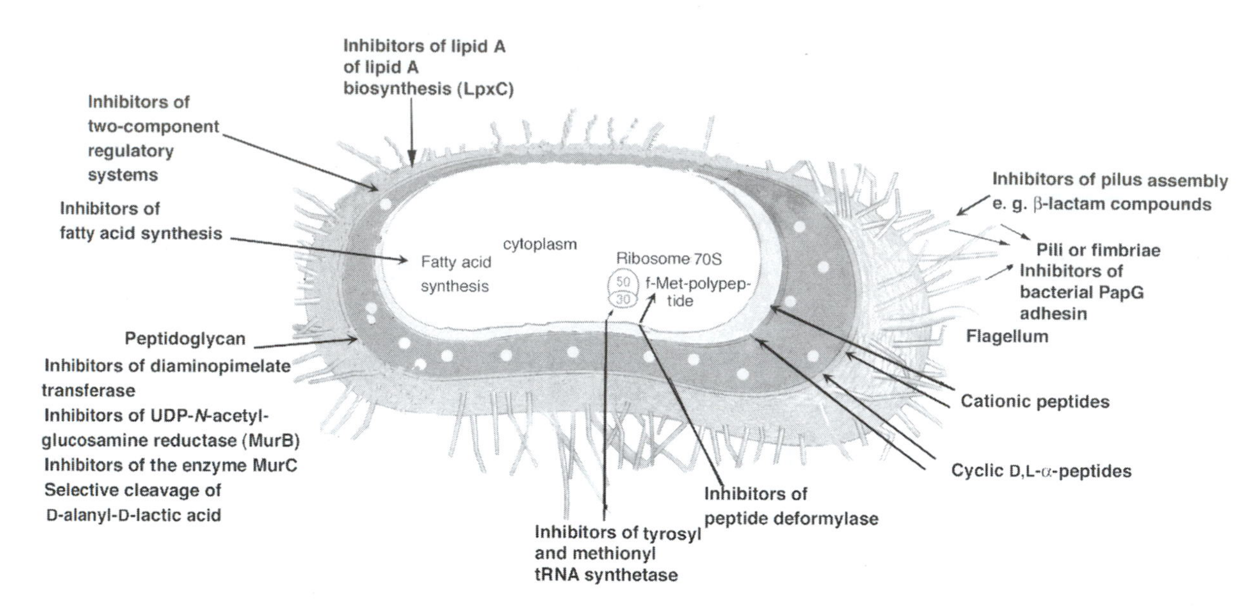


**Figure 27.1** Some of the novel targets for new antibacterial agents currently under development.

# Novel Targets

## Peptide Deformylase

Peptide deformylase (PDF) belongs to a new class of metalloproteases that utilize an $Fe^{2+}$ ion as the catalytic metal ion. PDF is an essential bacterial protease that catalyzes the removal of formyl groups from the N termini of newly synthesized bacterial polypeptides.

In bacteria, ribosome-mediated synthesis of proteins starts with a formylmethionine residue. The amino group of the methionyl moiety carried by the initiator $tRNA^{fMet}$ is N formylated by formyltransferase prior to its incorporation into a polypeptide. Consequently, *N*-formylmethionine is always present at the N terminus of a nascent bacterial polypeptide. However, most mature proteins do not retain the *N*-formyl group or the terminal methionine residue. Following translation, the formyl group is hydrolyzed by PDF, a reaction necessary for further processing at the N terminus by methionine aminopeptidase (see Fig. 16.5).

In *Escherichia coli*, PDF is encoded by the *def* gene, homologues of which are present in all sequenced pathogenic bacterial genomes and are essential to bacterial growth. Since protein synthesis in the eukaryotic cytoplasm does not involve the formyl group and PDF is apparently absent in eukaryotes, including humans, it has been proposed that specific deformylase inhibitors could selectively block the growth of bacterial cells with minimal toxicity to eukaryotic hosts, which provides a sound basis for antibacterial therapy.

Although PDF was discovered in 1968, it was difficult to isolate and characterize the enzyme due to its instability. Subsequent studies have shown that the instability is due to the oxidation of an $Fe^{2+}$ ion at the enzyme active site. The ferrous ion has been replaced by $Zn^{2+}$, $Ni^{2+}$ and $Co^{2+}$. While $Zn^{2+}$ reduces the activity by over two orders of magnitude, the $Ni^{2+}$ and $Co^{2+}$ forms of PDF retain almost all the catalytic activity of the native enzyme. X-ray crystallographic and nuclear magnetic resonance studies of the $Fe^{2+}$, $Zn^{2+}$, $Co^{2+}$ and $Ni^{2+}$ PDF forms revealed virtually identical three-dimensional structures. The metal ion is tetrahedrally coordinated by the side chains of two histidines of the conserved HEXXH motif (H = histidine, E = glutamic acid, X = an undetermined or nonstandard amino acid) and a conserved cysteine from an EGCLS motif (G = glycine, C = cysteine, L = leucine, and S = serine). The fourth coordination position is occupied by a water molecule that presumably hydrolyzes the amide bond.

In 1997, Michael K. Chan, Dehua Pei, and coworkers reported the crystallization and 2.9-Å X-ray structure of the zinc-containing *E. coli* PDF (Color Plate 27.1 [see color insert]). PDF from *E. coli* is a monomeric protein of 168 residues which shares the HEXXH and EGCLS fingerprint motifs with PDF sequences from other eubacteria; this suggests a common architecture of the catalytic region in these proteins.

Analysis of PDF gene sequences suggested that the enzyme family could be divided into two classes: class I from gram-negative organisms such as *E. coli* and class

II from gram-positive organisms. Class I PDF from the gram-negative organism *E. coli* has been well characterized in terms of both enzymology and structure. In 2002, Baldwin et al. reported the X-ray structure of class II PDF from the gram-positive bacterium *S. aureus* at 1.9-Å resolution. Comparison of this structure with the prototype class I enzyme from *E. coli* reveals minor structural variation arising from the insertions as well as specific active-site differences that may impact the design of pathogen-specific PDF inhibitors.

ACTIVE SITE OF PEPTIDE DEFORMYLASE

Structural studies of PDFs reveal that the catalytic metal cation ($Fe^{2+}$) in the native enzyme or the $Ni^{2+}$, $Zn^{2+}$, or $Co^{2+}$ analog is tetracoordinated to the imidazole groups of two histidines from the conserved HEXXH motif, one thiol ligand of a cysteine from the conserved EGCLS motif, and a water molecule. The water molecule is also hydrogen bonded to the amide side chain of a glutamine and is therefore activated for nucleophilic attack (Fig. 27.2).

## Enzyme Inhibitors as Drugs

Enzymes are frequent targets for drug design. In considering the topic of enzyme inhibitors as drugs, we should remind ourselves that it is one thing for a compound to inhibit an enzyme and quite another for that compound to become a marketed pharmaceutical product. For an enzyme inhibitor to become a practical drug, several criteria must be met.

- The biochemical pathway that is inhibited must be related to a disease state in such a way that inhibition of that pathway in a patient is therapeutic.
- The enzyme inhibitor must be specific so that unwanted inhibition of other pathways or receptors does not occur at therapeutic doses.
- The compound must have the pharmacokinetic characteristics of a practical drug; i.e., it must be absorbed, must penetrate to the site of action, and must have a reasonably predictable dose-response relationship and duration of action.
- The compound must have an acceptable toxicological profile in animals, and the results of clinical studies in humans must demonstrate an appropriate balance between benefits and risk in therapeutic use.
- The compound must be approved by regulatory agencies such as the Food and Drug Administration in the United States.

## Amide Hydrolysis: Peptidases (Proteases)

Peptidases (sometimes called proteases when protein hydrolysis is involved) are a family of enzymes whose function is to catalyze the hydrolysis of amide bonds in peptides. Some proteases hydrolyze the carboxy terminus (**carboxypeptidases**), some hydrolyze the amino terminus (**aminopeptidases**), and some hydrolyze interior peptide bonds, depending on what the amino acid side chain groups ($R_2$ and $R_3$) are. The ones that cleave interior peptide bonds are called **endopeptidases,** and the ones that cleave terminal amide bonds are called **exopeptidases** (Fig. 27.3).

Most proteases are sequence specific; the size and the hydrophobicity and hydrophilicity of enzyme active sites define which amino acid side chains of polypeptide substrates are bound. The standard nomenclature used to designate substrate/inhibitor residues (e.g., P3, P2, P1, P1', P2', P3') that bind to corresponding enzyme subsites (e.g., S3, S2, S1, S1', S2', S3') is shown in Fig. 27.4.

Recently, it has been convincingly demonstrated for a wide range of proteases that metalloproteases and serine, aspartic, and cysteine proteases universally bind their inhibitors or substrates in extended or β-strand

**Figure 27.2** Schematic representation of the active site of PDF.

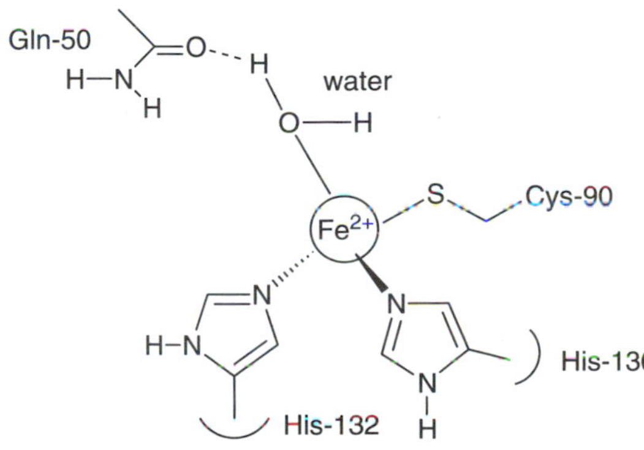

**Figure 27.3** Classification of peptidases according to the site of amide bond cleavage.

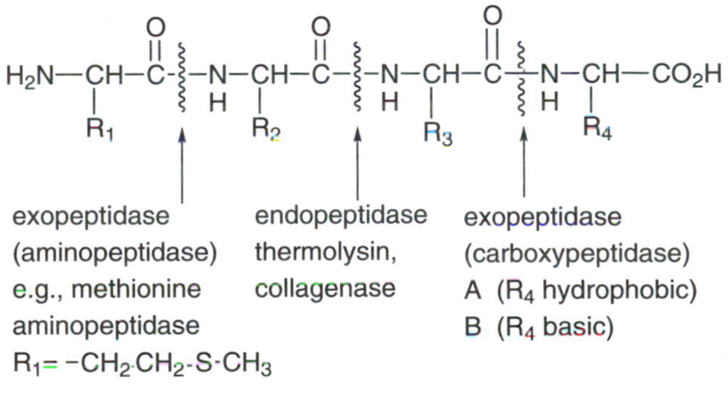

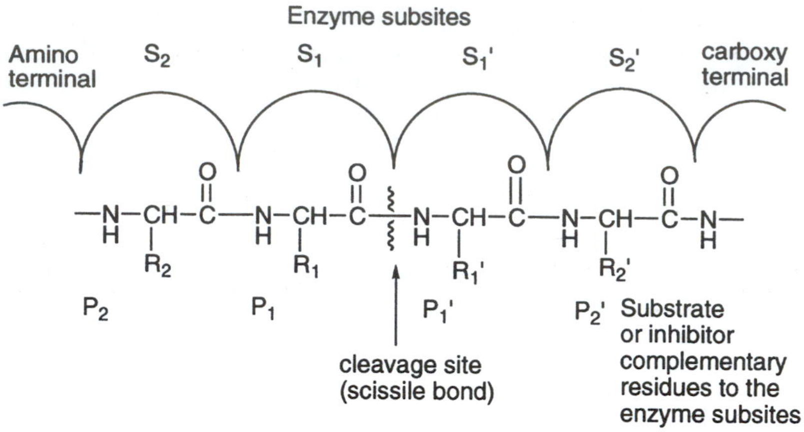

**Figure 27.4** Standard nomenclature for substrate residues and their corresponding binding sites. Reprinted from I. Schechter and A. Berger, *Biochem. Biophys. Res. Commun.* **27:**157–162, 1967, with permission from the publisher.

conformations; that is, the peptide backbone or equivalent is drawn out into a linear arrangement.

## PROTEASE INHIBITORS

The four classes of protease enzymes (serine, aspartic, and cysteine proteases, and metalloproteases) selectively catalyze the hydrolysis of polypeptide bonds. Clues to how specific proteases selectively recognize small molecules most often come from studies of peptide substrates for proteases. Protease inhibitors have been traditionally developed by natural-product screening for lead compounds with subsequent optimization or by empirical substrate-based methods involving truncating polypeptide substrates to short peptides, replacing the cleavable amide bond by a noncleavable isostere, and optimizing inhibitor potency through trial-and-error structural modifications.

This substrate-based drug design has been substantially improved in recent years by incorporation of mechanism-based drug design strategies; more recently, it was further enhanced by the availability of three-dimensional structural information determined by X-ray crystallography and nuclear magnetic resonance spectroscopy. Today the design of protease inhibitors involves a combination of all these traditional drug discovery approaches, supplemented by combinatorial chemistry. Advances in molecular biology have had a dramatic impact on the drug discovery process. The ability to produce significant quantities of pure protein has facilitated the determination of the structures of many of these biologically relevant targets. Systematically docking compounds from chemical databases to molecular targets of known structure has become a powerful tool in the discovery of enzyme inhibitors.

A key mechanistic feature of many metalloprotease inhibitors is the presence of a transition-state isostere that simulates to some extent the transition state of amide bond hydrolysis. As drugs, metalloprotease inhibitors need to be designed to adopt an extended conformation and to form hydrogen bonds with and/or interact with the enzyme through hydrophobic effects. The peptide backbones of metalloproteases are susceptible to degradation, and these inhibitors also suffer from delivery problems such as low membrane permeability and very low oral bioavailability.

The design principles for several different structural classes of PDF inhibitors have now become apparent through a number of methods including structure-based design and combinatorial chemistry.

### Discovery and Development of Peptide Deformylase Inhibitors

Since PDF is a metallohydrolase, compounds containing a metal ion-chelating group are potential inhibitors of the enzyme. A variety of functionalities and backbones have been exploited.

The requirement for a molecule to be an effective inhibitor of PDF is the possession of a functional group (e.g., H-phosphonate, hydroxamic acid, or thiol [sulfhydryl]) capable of coordinating (bidentate or monodentate) to the active-site $Fe^{2+}$ ion and one or more side chains which undergo effective van der Waals interactions with the enzyme subsites.

### Inhibitors of Peptide Deformylase Based on Phosphorus Derivatives

Inhibitors based on phosphonoamidates, phosphonates, and phosphinic acid zinc-binding groups have long been used as transition-state isosteres to replace scissile amide bonds in the development of metalloprotease inhibitors.

In a preliminary communication in 1998, Pei and coworkers reported the synthesis of (*S*)-2-*O*-(H-phosphonoxy)-L-caproyl-L-leucyl-*p*-nitroanilide (PCLNA) (Fig. 27.5). PCLNA acts as a competitive inhibitor of

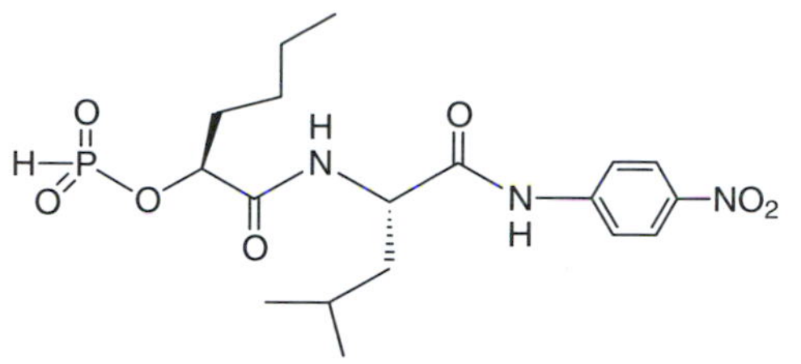

(*S*)-2-*O*-(H-phosphonoxy)-L-caproyl-L-leucyl-*p*-nitroanilide

**Figure 27.5** Structure of PCLNA.

both the zinc and iron forms of PDF, with $K_i$ values of 76 and 37 μM, respectively. The following year, these workers reported the three-dimensional crystal structures of the cobalt and the zinc PDF from *E. coli* complexed to the transition state analogue PCLNA (Hao et al., 1999).

## Proposed Mechanism of Deformylation by *E. coli* Peptide Deformylase

On the basis of the structures of the native PDF and the PDF-PCLNA complex, the same group of scientists proposed a mechanism for the deformylation by PDF as shown in Figure 27.6.

## Thiol Inhibitors of Peptide Deformylase

Thiols are attractive as binding groups for $Fe^{2+}$ or $Zn^{2+}$ ions, and inhibition of PDF with thiols has been explored. Among the earliest examples of thiol PDF inhibitors are dipeptide analogues, such as 2-thiomethyl-Nle-Arg-$OCH_3$ (called TNR) (Meinonel et al., 1999) (Fig. 27.7).

TNR behaves as a competitive inhibitor of PDF. It exhibits a binding constant in the micromolar range. Nuclear magnetic resonance spectroscopy has provided

**Figure 27.6** Proposed mechanism of PDF-mediated deformylation based on the native (top left) and complexed (bottom right) structures. Reprinted from B. Hao, W. Gong, P. T. Ravi Rajagopalan, Y. Zhou, D. Pei, and M. K. Chan, *Biochemistry* **38:**4712–4719, 1999, with permission from the publisher.

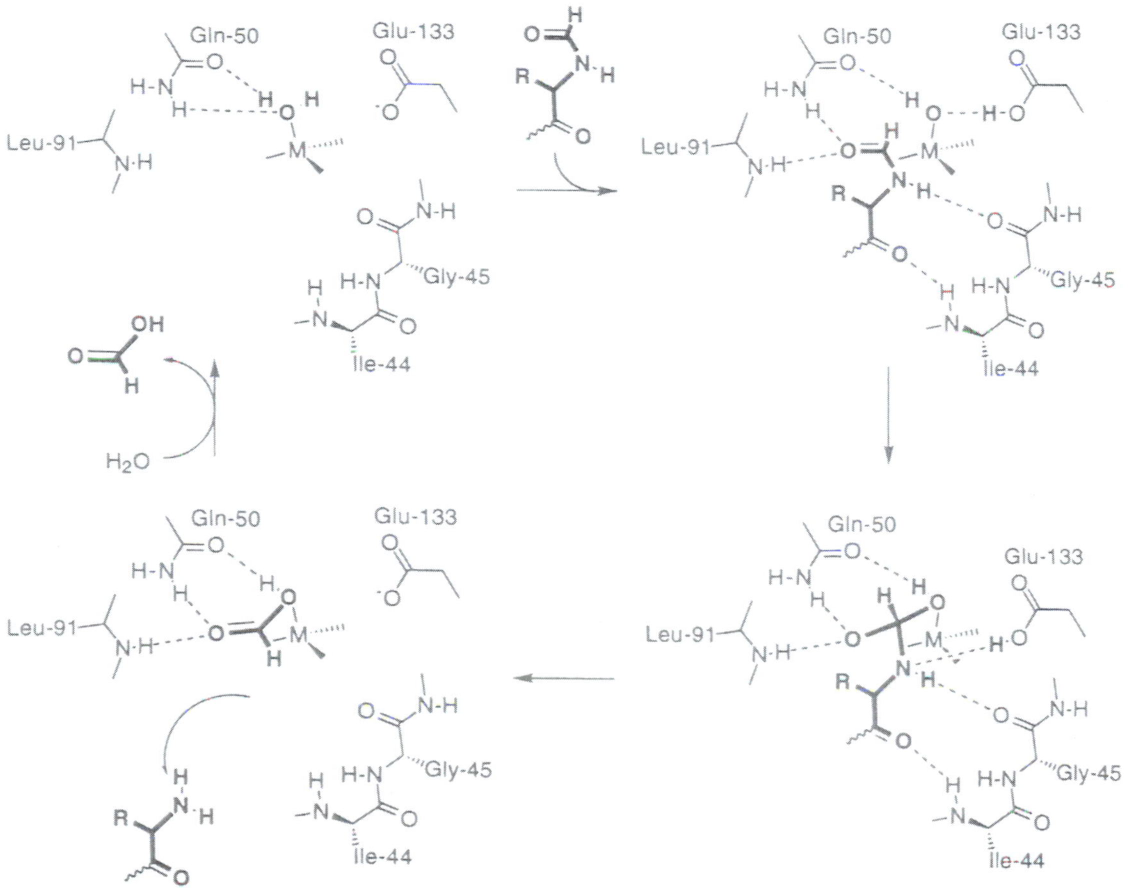

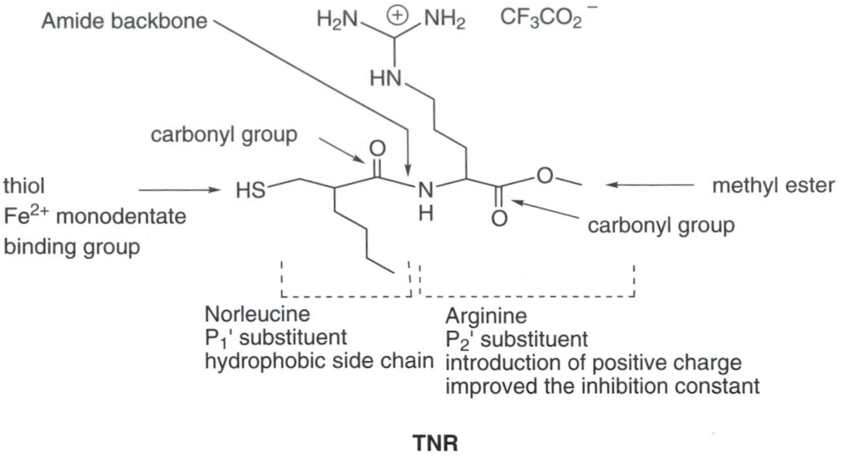


**Figure 27.7** Structure of 2-thiomethyl-Nle-Arg-$OCH_3$ (TNR) as the trifluorocetate salt.

evidence that TNR binds inside the active-site cavity of PDF while keeping intact the three-dimensional fold of the protein. A fingerprint of the interactions of the inhibitor with the residues of the enzyme was obtained.

The identification of a PDF inhibitor from a rationally designed combinatorial library was reported recently (Wei et al., 2000). The most potent inhibitor identified from the combinatorial library was *N*-[(3-mercapto-2-*n*-butyl)propionyl]-L-leucyl-anthramide (compound 1) (Fig. 27.8). To improve the aqueous solubility of compound 1, these researchers replaced the hydrophobic side chain of leucine at the $P_2$' position by the more polar side chains of lysine or arginine to give inhibitors 2 and 3, respectively.

In a subsequent publication, Huntington et al. (2000) reported the synthesis and antibacterial activity of a series of peptide thiols that act as potent, reversible inhibitors of peptide deformylase from *E. coli* and *Bacillus subtilis* (Fig. 27.8, compounds 4 and 5). These inhibitors showed potent antibacterial activity against gram-positive bacteria and moderate activity against gram-negative bacteria.

The rationale behind these thiol derivatives was based on the knowledge that a thiol group is a good ligand for transition metal ions such as $Zn^{2+}$ metallopeptidases.

## Hydroxamate Inhibitors of Peptide Deformylase

Recently, Yuan and coworkers at Versicor, Inc. (Fremont, Calif.) reported an investigation employing a strategy that relies on the fact that PDF is a metallohydrolase. In this investigation, a chemical library collection of more than 25,000 "drug-like" low-molecular-weight compounds, about 20% of which contained a metal ion-chelating group, was screened in the PDF assay. As a result of this focused screening, four compounds, containing different hydroxamate-chelating groups, were identified as hits, with a 50% inhibitory concentration of less than $10^{-5}$ M. One of these compounds was actinonin (the structure is given in Fig. 27.9), a naturally occurring antibacterial agent that possesses a hydroxamate metal-chelating group and a methionine-like structure at the $P_1$' site.

Actinonin is a tight-binding inhibitor of PDF with a $K_i$ value of 0.28 nM. From a constructed model of PDF-

**Figure 27.8** Chemical structures of compounds 1 through 5.

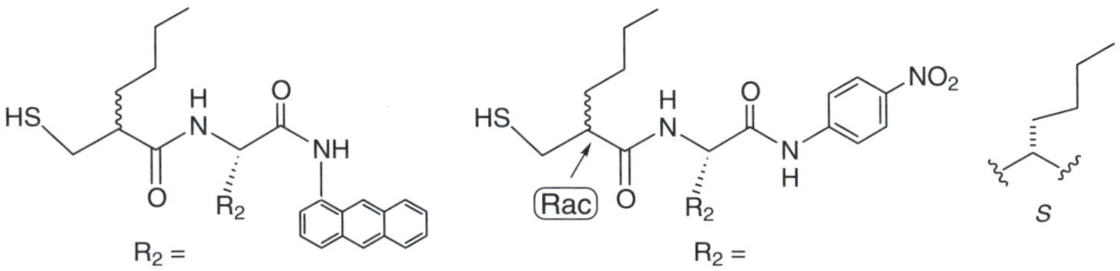


$R_2$ =

**1** $CH_2CH(CH_3)_2$

**2** $(CH_3)_4NH_2$

**3** $(CH_3)_3NHC(=NH)NH_2$

$R_2$ =

**4** $CH_2CH(CH_3)_2$

**5** $(CH_2)_4NH_2$

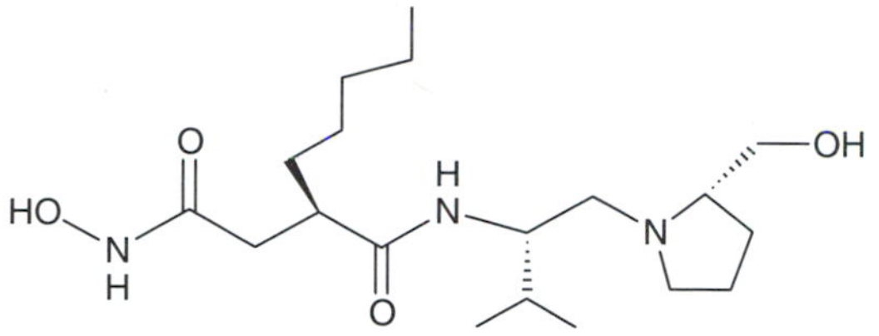

Actinonin

**Figure 27.9** Chemical structure of actinonin.

bound actinonin and by analogy to previous known inhibitors, the *n*-pentyl chain defines the $P_1'$ site of the inhibitor and functions as an analogue of methionine, the preferred amino acid at $P_1'$ in PDF substrates. The hydroxamate group of actinonin evidently acts as a chelating group to bind the metal ion of the enzyme. On the basis of all experimental results and from molecular modeling studies, the researcher at Versicor concluded that actinonin is a reversible, tight-binding, competitive PDF inhibitor acting via effective metal chelation.

Actinonin is active against gram-positive bacteria, including *S. aureus* and *S. pneumoniae*. It is also active against fastidious gram-negative bacteria such as *H. influenzae, Moraxella catrrhalis,* and *Neisseria gonorrheae*. Since actinonin has such a high affinity for deformylase, it is logical to suspect that the antibacterial activity of this compound is mainly due to its ability to inhibit PDF, which would result in the accumulation of newly synthesized formylmethionine-capped inactive peptides and the subsequent cessation of bacterial growth.

In 2000, Hubschwerlen and coworkers at Hoffmann-La Roche (Basel, Switzerland) identified low-molecular-weight β-sulfonyl- and β-sulfinylhydroxamic acid derivatives as inhibitors of *E. coli* PDF. They mentioned that these compounds were designed based on the discovery of actinonin by screening inhibitors of *E. coli* PDF. The highest activity was observed with the compound (*R*)-3-(phenylsulfonyl)heptanoic acid hydroxamide (the structure is shown in Fig. 27.10). An X-ray crystal structure of this compound bound to PDF was determined.

The researchers mentioned that most of the compounds synthesized and tested displayed antibacterial activities that cover several pathogens found in respiratory tract infections, including *Chlamydia pneumoniae, Mycoplasma pneumoniae, H. influenzae,* and *M. catarrhalis.*

**Figure 27.10** Chemical structure of (*R*)-3-(phenylsulfonyl)heptanoic acid hydroxamide.

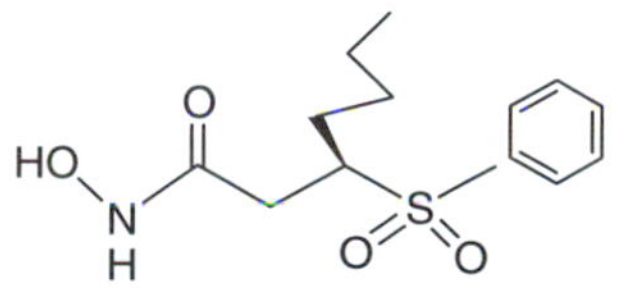

(*R*)-3-(Phenylsulfonyl)heptanoic acid hydroxamide

In 2001, Hubschwerlen and coworkers identified the compound (5-chloro-2-oxo-1,4-dihydro-2*H*-quinazolin-3-yl)acetic acid hydrazide by high-throughput screening in an *E. coli* PDF-based assay. On the basis of their X-ray and modeling studies, they proposed the general structure shown in Fig. 27.11, in which the hydrazide group was replaced by a hydroxamic acid function, resulting in stronger metal coordination. However, these compounds showed only weak antibacterial activity.

In 2001, Thorarensen and coworkers at Pharmacia (Kalamazoo, Mich.) reported that on screening their compound collection using *S. aureus* PDF, they identified a potent inhibitor with a 50% inhibitory concentration in the low nanomolar range but found that, unfortunately, this compound, which contained a hydroxamic acid, did not exhibit antibacterial activity. They then prepared several analogues and identified a compound with an *N*-hydroxyurea functionality (shown on the right in Fig. 27.12), with both PDF-inhibitory and antibacterial activity.

In 2002, Gupta et al. reported quantitative structure-activity relationship studies that were carried out with a series of β-sulfonyl- and β-sulfinyl hydroxamic acid derivatives for their PDF-inhibitory and antibacterial activities against *E. coli* DC2 and *M. catarrhalis* RA21, which demonstrate that the PDF-inhibitory activity in cell-free and whole-cell systems increases with an increase in molar refractivity and hydrophobicity. The comparison of the quantitative structure-activity relationships between the cell-free and whole-cell systems indicates that the active binding sites in PDF isolated from *E. coli* and *M. catarrhalis* RA21 are similar and that the whole-cell antibacterial activity is mainly due to inhibition of PDF.

## *N*-Formylhydroxylamine Inhibitors of Peptide Deformylase

In 2001, Clements and coworkers at British Biotech Pharmaceuticals, Ltd. (Oxford, United Kingdom), the

**Figure 27.11** General structure of (5-chloro-2-oxo-1,4-dihydro-2*H*-quinazolin-3-yl)acetic acid hydrazide derivatives.

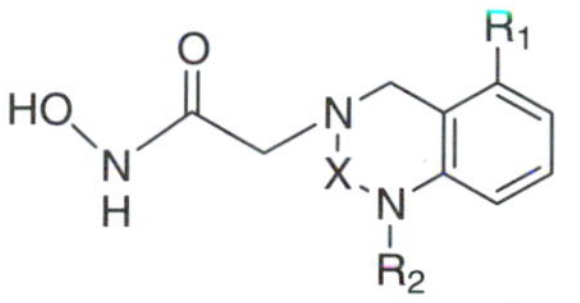

X = CO; $SO_2$ $R_1$ = H, F, Cl, Br, $CF_3$
$R_2$ = H, alkyl, benzyl, etc.

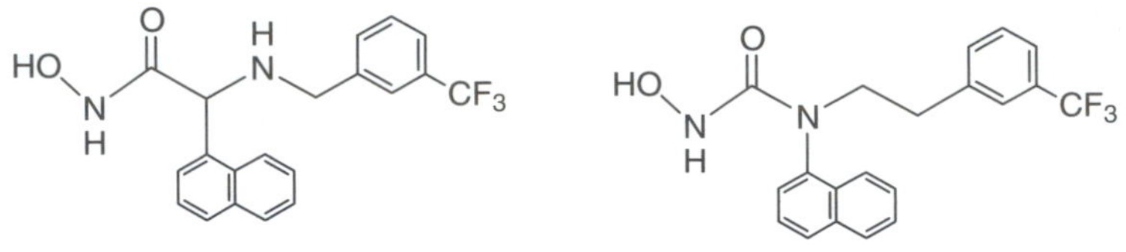

**Figure 27.12** Structures of the compounds developed by Thorarensen and coworkers.

Hydroxamic acid derivative

*N*-Hydroxyurea derivative

Krebs Institute for Biomolecular Research (University of Sheffield, Sheffield, United Kingdom), and Naeja Pharmaceutical, Inc. (Edmonton, Canada) reported that from a screening program of a compound library using PDF, they identified an *N*-formylhydroxylamine derivative (called BB-3497) (Fig. 27.13) as a potent and selective inhibitor of PDF. To elucidate the interactions, these scientists determined the crystal structure of BB-3497 and actinonin bound to *E. coli* PDF-Ni at resolutions of 2.1 and 1.75 Å, respectively. The inhibitors bound in a similar fashion in each complex, lying in a cleft on the enzyme surface and approximately within the active site, confirming the location of the S1', S2', and S3' binding pockets in PDF.

BB-3497 had activity against gram-positive bacteria, including MRSA and vancomycin-resistant *E. faecalis* (VREF), and against some gram-negative bacteria. Time-kill analysis showed that the mode of action of BB-3497 was primarily bacteriostatic.

In 2002, Smith et al. reported the synthesis of a series of analogues of the potent PDF inhibitor BB-3497 containing alternative metal binding groups. Enzyme inhibition and antibacterial activity data for these compounds revealed that the bidentate hydroxamic acid and *N*-formylhydroxylamine structural motifs represent the optimum chelating group on the pseudopeptidic BB-3497 backbone.

### Peptide Aldehyde Inhibitors of Peptide Deformylase

In 1999, Grant and coworkers at Merck Research Laboratories (Rahway, N.J.) reported that peptide aldehydes containing methional or norleucinal inhibited recombinant PDF from gram-negative *E. coli* and gram-positive *B. subtilis*. The most potent inhibitor was calpeptin, *N*-CBZ-Leu-norleucinal (Fig. 27.14) which was a competitive inhibitor of the zinc PDF.

**Figure 27.13** Chemical structure of BB-3497.

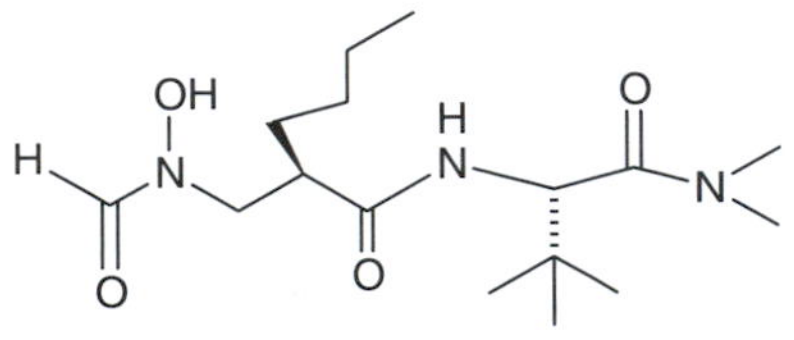

BB-3497

### Biaryl Acid Analog Inhibitors of Peptide Deformylase

In 2000, Green and coworkers at Merck identified a series of biaryl acid analogs as inhibitors of PDF activity by screening a diverse chemical library. Each compound is composed of a head group of fused heterocyclic aromatic rings, a biaryl group (e.g., biphenyl), and an acidic group on the biaryl B ring (e.g., tetrazole). Two examples of these biaryl acid analogs, a biphenyl tetrazole (compound 1) and a biphenyl acyl sulfonamide (compound 4) (Fig. 27.15) were found to be competitive inhibitors of PDF, with $K_i$ values of 1.2 and 6.0 μM, respectively.

## Two-Component Signal Transduction in Bacteria

As society and the environment change, infectious-disease agents also change. Microorganisms adapt to the changing environment and are carried throughout the world by growing numbers of travelers.

Rapid adaptation to environmental change is essential for bacterial survival. To orchestrate the adaptive response to changes in their surroundings, bacteria mainly use **two-component regulatory systems** (TCS). These signal transduction systems are intimately involved

**Figure 27.14** Chemical structure of *N*-CBZ-Leu-norleucinal (calpeptin).

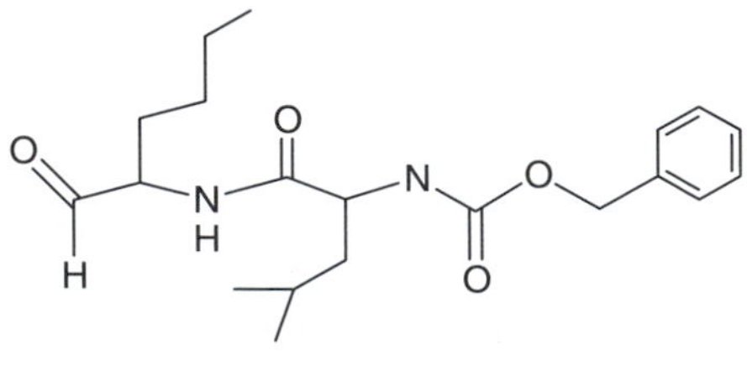

*N*-CBZ-leu-norleucinal (calpeptin)

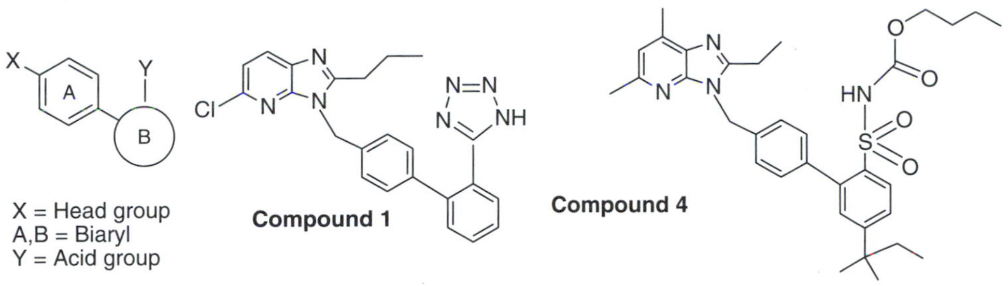

**Figure 27.15** (Left) General structure of biaryl acid analogs. (Middle and right) Chemical structures of compounds 1 and 4.

in the maintenance of bacterial cell homeostasis and the expression of virulence determinants.

TCSs consist of a sensor **histidine kinase** and a **response regulator** (Fig. 27.16). Sensor kinases (or histidine kinases) usually possess two domains: an input domain, which monitors environmental stimuli, and a transmitter domain, which autophosphorylates following stimulus detection. The chemistry involves three phosphotransfer reactions and two phosphoprotein intermediates (equations 1 to 3). These reactions work to transduce information.

ATPases catalyze the hydrolysis of ATP. The $\gamma$-phosphoryl group in ATP is first transferred to a histidine side chain. The ATP-dependent phosphorylation of histidine is generally regulated in response to environmental signals by a family of histidine protein kinases (equation 1):

**Figure 27.16** A typical TCS. Reprinted from J. F. Barrett and J. A. Hoch, *Antimicrob. Agents Chemother.* **42:**1529–1536, 1998, with permission from the American Society for Microbiology.

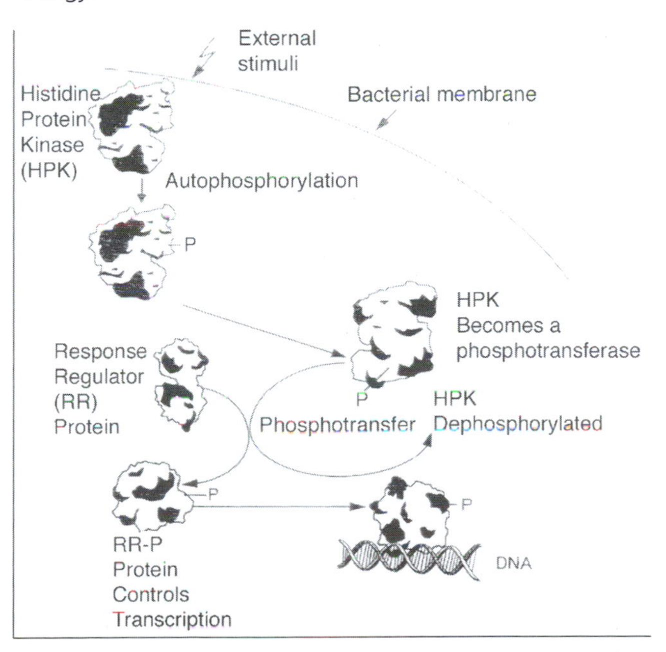


$$\text{ATP} + \text{histidine} \rightarrow \text{ADP} + \text{His—P} \qquad (1)$$

The phosphoryl group is then transferred from the phosphohistidine residue to an aspartate side chain (Fig. 27.16) (equation 2). The phosphorylated aspartate is contained within another type of protein, called a response regulator:

$$\text{His—P} + \text{Asp} \rightarrow \text{His} + \text{Asp—P} \qquad (2)$$

Phosphorylation of the response regulator domain activates the transcription-regulating functions of the DNA-binding domain. Response regulators are the "on-off" switch in this system, depending on their state of phosphorylation.

Finally, in the third reaction, the phosphorylated aspartate transfers the phosphoryl group to water (equation 3):

$$\text{Asp—P} + H_2O \rightarrow \text{Asp} + P_i \qquad (3)$$

## Inhibitors of Two-Component Regulatory Systems in Bacteria

In the search for new antibacterial agents with a novel mechanism of action to combat emerging drug-resistant bacteria, the two-component regulatory systems of bacteria have become a logical target. Bacterial cells must sense and respond to their environment, a process that requires signal transduction across their membranes. This signal transduction pathway is used by bacteria to sense and respond to environmental changes. These signal transduction systems are common to bacteria but absent from mammals. Over 100 TCSs have been identified to date in such important nosocomial pathogens as *S. aureus, E. faecium, E. coli,* and *Pseudomonas aeruginosa.*

The potential of a bacterium to grow and divide in an ecological niche is determined genetically by its repertoire of genes and by its capacity to induce new gene expression to cope with the new environment. Often, a bacterial infection occurs when an organism moves from an environment where its presence is benign (e.g., the gastrointestinal system) to another environment where it poses a serious problem (e.g., the urinary tract).

In a 2002 review, Matsushita and Janda summarized our current knowledge of the structure and function of histidine kinase and the development of antibiotics with a new mode of action targeting histidine kinase-promoted signal transduction and its subsequent regulation of the gene expression system.

### SYNTHETIC INHIBITORS OF TWO-COMPONENT REGULATORY SYSTEMS

#### Substituted Salicylanilides

The discovery of a new class of inhibitors of the TCS of bacteria was reported in 1998 by M. J. Macielag and coworkers of the R. W. Johnson Pharmaceutical Research Institute (Raritan, N.J.) when a screening of their chemical library against KinA/Spo0F (this assay measures the ability of test compounds to inhibit phosphorylation of KinA), a model TCS which regulates sporulation in *B. subtilis,* revealed that closantel and 3,3',4',5-tetrachlorosalicylanilide (Fig. 27.17) inhibited autophosphorylation of the KinA kinase. Furthermore, both compounds exhibited potent antibacterial activity against resistant gram-positive pathogens, including MRSA and vancomycin-resistant *E. faecium* (VREF).

In 1998, in an earlier communication about the results of a screening test, D. J. Hlasta and coworkers from R. W. Johnson had identified the salicylanilide closantel as the pharmacophore for inhibition of the bacterial TCS in a structure-activity relationship (SAR) study.

#### Amidino Benzimidazole Inhibitors

In 2001, M. A. Weidner-Wells and coworkers, of the Johnson Pharmaceutical Research Institute, reported that screening of their chemical library in the model KinA/Spo0F TCS from *B. subtilis* resulted in the identification of bisamidino indole derivative 1 as a lead structure (Fig. 27.18). These scientists also mentioned that since the synthesis of this class of indoles is tedious and low yielding, the indole nucleus was replaced with a benzimidazole. A series of amidino benzimidazoles was synthesized, and an SAR study identified these compounds as potent inhibitors of histidine kinase. Compound 2 is one of these novel amidino benzimidazole derivatives. Several of these compounds exhibited good in vitro antibacterial activity against susceptible as well as resistant strains of several gram-positive organisms.

In 1999, J. J. Hilliard and coworkers, also from the Johnson Pharmaceutical Research Institute, reported the multiple mechanisms of action of inhibitors of histidine protein kinases from bacterial TCSs.

In 1998, Z. Sui and coworkers, of the Johnson Pharmaceutical Research Institute, reported that they had identified a series of diaryltriazoles as inhibitors of bacterial TCSs. Both phenyl rings exhibited well defined SARs in the KinA-Spo0F assay. Several compounds possessed good antibacterial activity against gram-positive organisms, including MRSA and VRE. Compound 3 (Fig. 27.18) is a representative compound of this series of diaryltriazoles.

## Novel Antibacterial Agents Based on the Cyclic DL-α-Peptide Architecture

In 2001, S. Fernandez-Lopez and coworkers from The Scripps Research Institute (La Jolla, Calif.) reported that six- and eight-residue cyclic D,L-peptides act preferentially on gram-positive and/or gram-negative bacterial membranes compared to mammalian cells. These peptides were shown to increase membrane permeability, collapse transmembrane ion potentials, and cause rapid cell death. If turned into drugs, these self-assembling "nanotubes" could form a new class of antibacterial agents as an alternative to those rendered useless by some bacterial resistance mechanisms.

Cyclic peptides with an even number of alternating D- and L-α-amino acids can adopt flat, ring-shaped conformations in which the backbone amide functionalities are oriented perpendicular to the side chains and the plane of the ring structure (Fig. 27.19). Under conditions that favor hydrogen bonding, such as adsorption onto lipid membranes, the cyclic peptides can stack to form hollow, β-sheet-like tubular structures that are open ended, presenting the amino acid side chains on the outside surface of the ensemble (Fig. 27.20). Fernandez-Lopez and

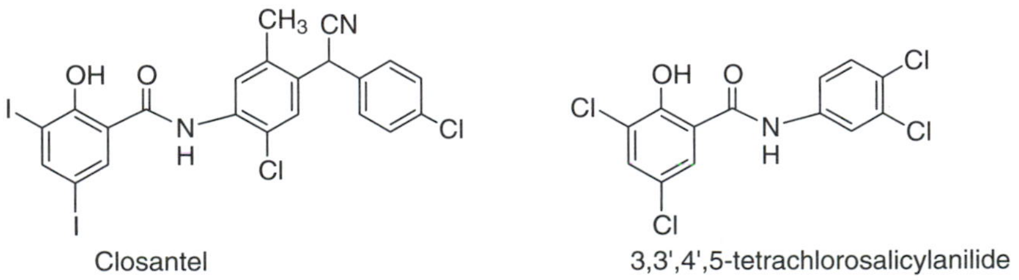

**Figure 27.17** Chemical structures of closantel, and 3,3',4',5-tetrachlorosalicylanilide.

Closantel

3,3',4',5-tetrachlorosalicylanilide

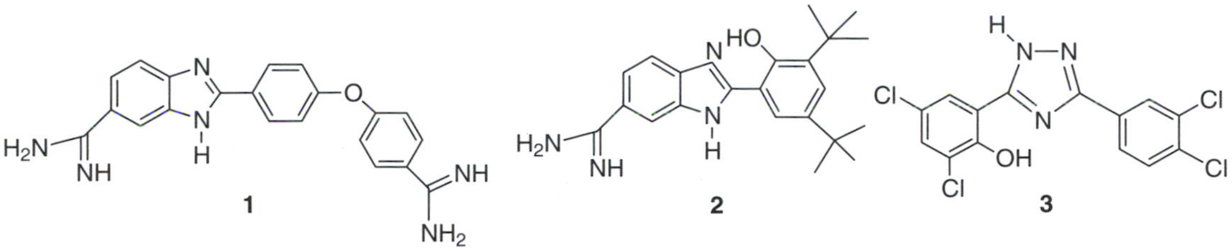

**Figure 27.18** Chemical structures of bisamidino indole derivative 1, amidino benzimidazole derivative 2, and diaryltriazole 3.

coworkers proposed the hypothesis that appropriately designed cyclic DL-α-peptides may be able to selectively target and self-assemble in bacterial membranes and exert antibacterial activity by increasing the membrane permeability. This hypothesis was demonstrated by the effectiveness of this class of materials as selective antibacterial agents, highlighted by the high efficacy observed against lethal MRSA infections in mice. Cyclic DL-α-peptides are proteolytically stable and can be derived from a potentially vast membrane-active sequence space. The cyclic peptide KQRWLWLW (lysine-glutamine-arginine-tryptophan-leucine-tryptophan-leucine-tryptophan), a representative member of this class of cyclic peptides, displayed potent in vitro activity against gram-positive *B. subtilis* and *S. aureus* and against gram-negative *S. pneumoniae* and vancomycin-resistant *E. faecalis*.

With supramolecular assemblies such as the nanotubes designed by M. Reza Ghadiri that involve active molecules within a large sequence space, it is expected that the mode of action of these peptides will be maintained despite some structural changes. This cannot be achieved by the molecular approach to drug design based on single molecules interacting with bacterial cells in very specific ways. Because these interactions typically involve molecular recognition, the structure and function of a drug molecule are interconnected. The specific interaction, which depends on the structure of the molecule, ensures that the drug acts only against its target. Functions can be lost with just a slight change in either

**Figure 27.19** Peptide stack self-assembly of flat, cyclic, eight-residue D,L-α-peptides forms β-sheet-like, tubular, open-ended supramolecular structures. Reprinted from S. Fernandez-Lopez, H. S. Kim, E. C. Choi, M. Delgado, J. R. Granja, A. Khasanov, K. Kraehenbuehl, G. Long, D. A. Weinberger, K. M. Wilcoxen, and M. R. Ghadiri, *Nature* **412:**452–455, 2001, with permission from the publisher.

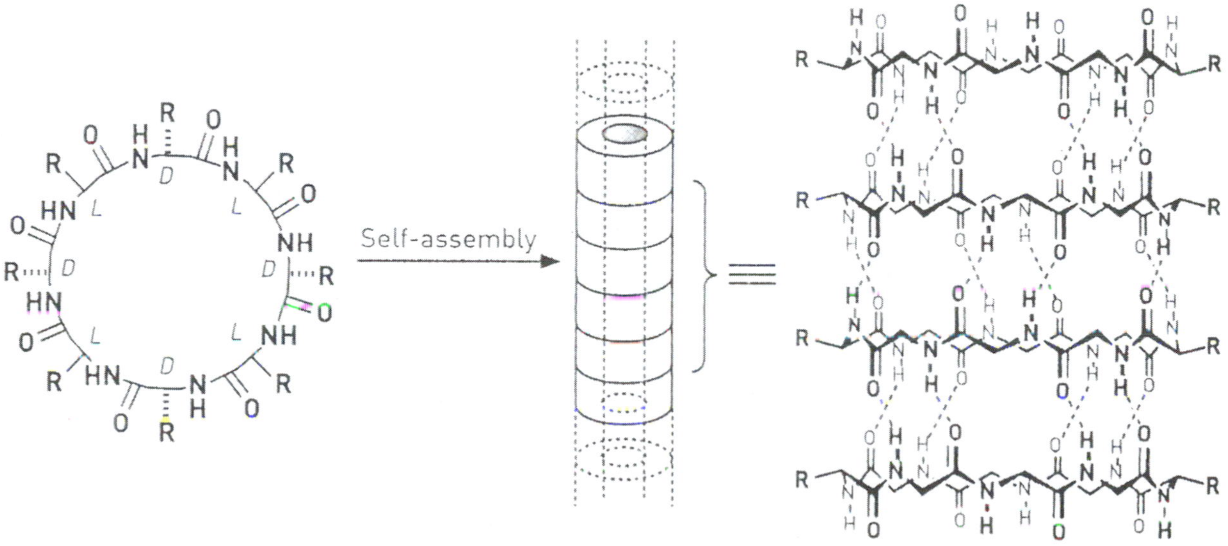

a

Peptide **1**:

$R_1 = R_2 =$

Peptide **2**:

$R_1 =$ , $R_2 =$

Peptide **3**:

$R_1 = R_2 =$

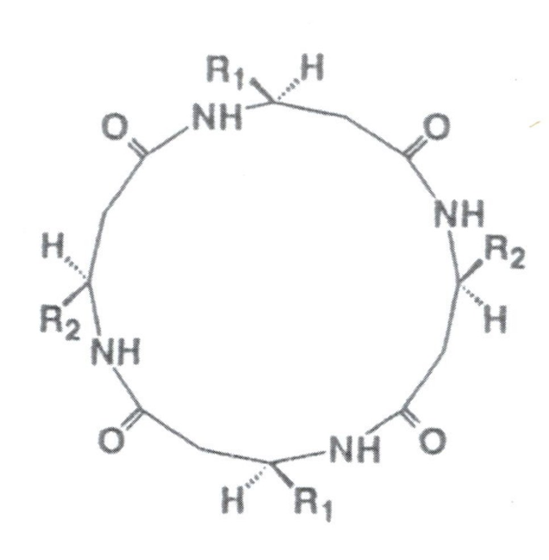


b

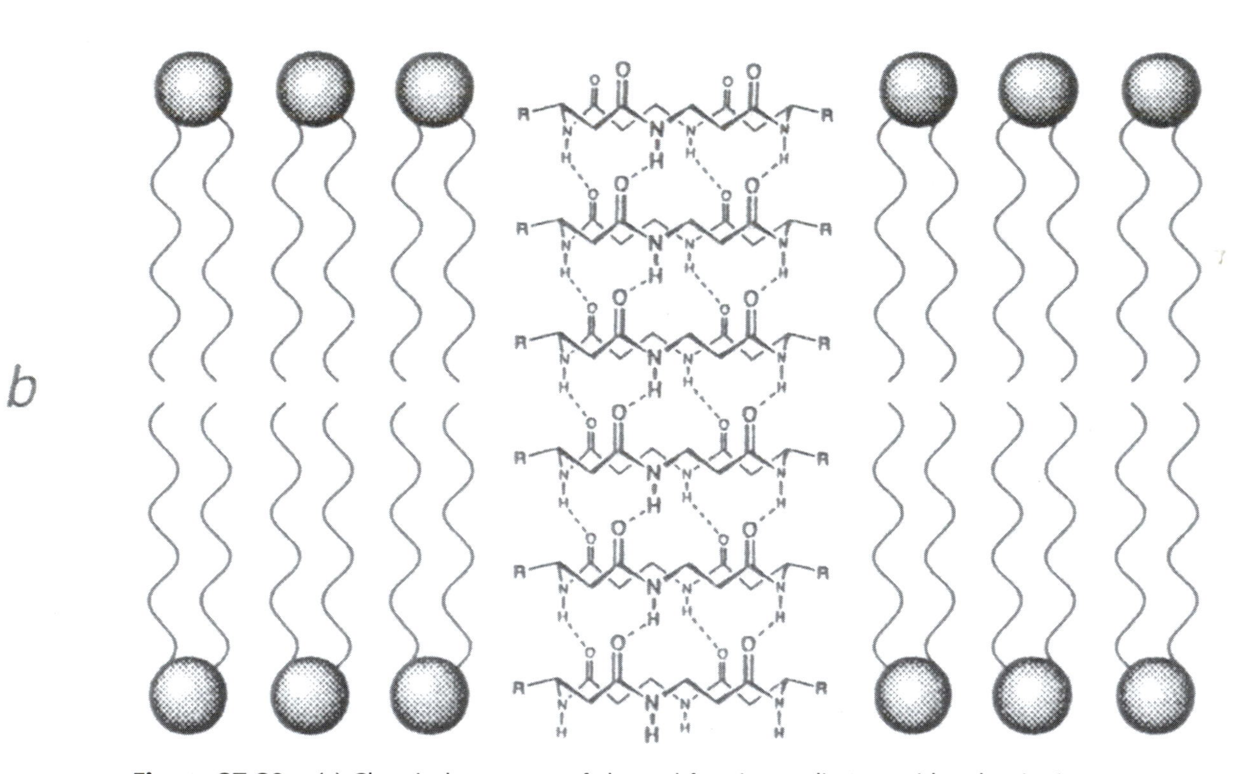

**Figure 27.20** (a) Chemical structures of channel-forming cyclic β-peptide subunits 1 through 3 represented in a flat ring-shaped conformation. (b) Putative structure of self-assembled transmembrane channels formed from cyclic β-peptides 1 through 3. The tubular channel ensemble is represented with the expected parallel ring stacking and extensive intersubunit hydrogen bonding. (For clarity, most side chains are omitted.) Reprinted from T. D. Clark, L. K. Buehler, and M. R. Ghadiri, *J. Am. Chem. Soc.* **120:**651–656, 1998, with permission from the publisher.

of the interacting moieties. As we have already seen, such changes occur easily in bacteria, leading to resistance to many antibacterial drugs. In recent years, bacterial resistance to antibacterial drugs has become a global public health threat and has been increasing due to the use, overuse, and misuse of broad-spectrum antibiotics and the ability of bacteria to exchange resistance genes.

## Antimicrobial Cationic Peptides

A large number of antimicrobial cationic peptides composed of 12 to 50 amino acid residues and folded into a variety of different structures, including α-helices, β-sheets, extended helices, and loops, are found in natural host defense systems directed against various invading pathogenic microorganisms. Their distribution is widespread and includes bacteria, fungi, plants, insects, birds, crustaceans, amphibians, and mammals. Their broad antimicrobial spectra and highly selective toxicity make these cationic peptides promising candidates as novel antibiotics for clinical use.

It is widely thought that the mechanism of action of these cationic peptides involves electrostatic interactions with anionic lipids, which are abundant in bacterial membranes, forming a peptide-lipid supramolecular complex pore.

Despite their structural variation and extensive sequence variability, most antimicrobial cationic peptides share two unique features: (i) they are rich in cationic residues Arg and Lys and are polycationic, with a net positive charge of more than +2, and (ii) they fold into amphipathic structures with both hydrophobic and hydrophilic domains. These characteristics allow them to interact with anionic lipids such as phosphatidylglycerol and lipopolysaccharides (LPSs), which are abundant in bacterial membranes. The outer surfaces of the membranes of mammalian cells are almost exclusively composed of zwitterionic phospholipids, such as phosphatidylcholine and sphingomyelin, except for minor, negatively charged gangliosides.

Cationic peptides exhibit a broad spectrum of activity against various targets, including gram-negative and gram-positive bacteria, fungi, enveloped viruses, and parasites. Certain linear amphipathic α-helical antimicrobial peptides, such as the cecropins, magainins, and melittins, permeabilize model membrane systems, cause leakage of fluorescent dyes, and/or induce ion transport across lipid bilayers. This has led to the general conclusion that lysis or leakage of essential molecules due to the formation of channels in the cytoplasmic membrane is the mechanism of killing.

In 2003, Strøm et al. reported that the cationic antibacterial peptides which have been proclaimed as new drugs against multiresistant bacteria have had limited success so far, partially due to the size of the peptides, which gives rise to unresolved issues regarding administration, bioavailability, metabolic stability, and immunogenicity. These scientists have systematically investigated the minimum antibacterial motif of cationic antibacterial peptides regarding the charge and lipophilicity/bulk and have found that the pharmacophore was surprisingly small, opening the opportunity for the development of short antibacterial peptides for systemic use.

### Antimicrobial Cationic Peptides and Toll-Like Receptors

Gram-negative LPSs are well-known endotoxins with diverse structures; they are displayed on bacterial outer membranes. LPSs are among the most potent bacterial signal molecules that activate host defenses to release proinflammatory mediators, cytokines, chemokines, and lipoproteins. Mechanisms that regulate host responses to LPS are known to involve plasma lipoproteins and the LPS-binding receptor CD14. These include the LPS-binding protein (LBP), bactericidal/permeability-increasing protein, phospholipid transfer protein, serum amyloid P component, and antimicrobial proteins secreted by neutrophils.

LBP is a lipid transfer protein whose facilitation of LPS binding to CD14 on cell membranes or soluble CD14 in plasma activates a wide variety of cells through **Toll-like receptors** (these receptors were originally discovered in *Drosophila*, where they play the dual function of being involved in both embryonic development and the response of the insect to fungi and bacteria). Cationic antimicrobial peptides with diverse structures have been reported to bind LPSs and thus suppress their ability to stimulate the production of proinflammatory cytokines. Recently, Scott et al. (2000) reported a good correlation between the activity of certain antimicrobial peptides against gram-negative bacteria and their ability to block the interaction of LPS with LBP.

## Bacterial Fatty Acid Biosynthesis

Bacterial fatty acid biosynthesis (FAS) is an essential process by which organisms make fatty acids for use in the assembly of important cellular components, including phospholipids, lipoproteins, and LPSs.

Since FAS is organized differently in bacteria and mammals, there is good potential for selective inhibition of the bacterial system. The bacterium most thoroughly understood at the molecular level is *E. coli*, and lipid metabolism has been most extensively studied with this organism. FAS in *E. coli* is catalyzed by the type II fatty acid synthase system (FAS-II) in which each step is carried out by an individual enzyme. Each of the individual enzymes is encoded by a distinct gene, and some chemical reactions are often catalyzed by multiple enzymes.

A key feature of bacterial FAS is that the intermediates in the pathway are covalently attached to acyl carrier proteins (ACP). The precursors of fatty acid biosynthesis are derived from the acetyl coenzyme A (acetyl-CoA) pool. Malonyl-CoA is required for all the elongation steps. Acetyl-CoA carboxylase catalyzes the first step, the formation of malonyl-CoA (Fig. 27.21, step 1). The malonyl group is transferred to ACP by malonyl-CoA–ACP transacylase (FabD) (step 2), which provides the successive two-carbon units to the growing acyl chain. FAS is initiated by the condensation of acetyl-CoA with malonyl-ACP by β-ketoacyl-ACP synthetase III (FabH) (step 3). Subsequent rounds of elongation (steps 4 to 7) begin by the condensation of malonyl-ACP with acyl-ACP, catalyzed by either β-ketoacyl-ACP synthetase I (FabB) or II (FabF) (step 4). The resulting β-ketoacyl-ACP is reduced by the NADPH-dependent reductase, FabG (step 5). The β-hydroxyacyl-ACP is then dehydrated to enoyl-ACP by either of two dehydratases (FabA or FabZ) (step 6). Enoyl ACP reductase (FabI) is the enzyme of FAS-II that catalyzes the final reaction in the enzymatic sequence. In contrast, eukaryotes synthesize fatty acids via a multifunctional enzyme complex (FAS-I) in which all enzymatic activities reside on a single polypeptide. Thus, there exists a potential for selective inhibition of gram-positive and gram-negative bacterial cell growth by the inhibition of the FabI enzyme. In short, the importance of FAS to bacterial growth and the fact that prokaryotes and eukaryotes use different types of FAS make this pathway an attractive target for the discovery of new antibacterial agents.

## Inhibitors of Bacterial Fatty Acid Biosynthesis

Several fatty acid inhibitors, including cerulenin, thiolactomycin, diazaborine, isoniazid, and triclosan, as well as 2,9-disubstituted 1,2,3,4-tetrahydropyrido[3,4-*b*]indoles, 1,4-disubstituted imidazoles, and the aminopyridine derivative (structure 9) (the chemical structures are given in Fig. 27.22), have potential as lead compounds for the development of new antibacterials. Triclosan (2,4,4'-trichloro-2'-hydroxydiphenyl ether) is a commercial broad-spectrum antibacterial agent with widespread application in a multitude of contemporary consumer products, including soaps, detergents, toothpastes, skin care products, cutting boards, and mattress pads. Recent work has revealed that triclosan, as well as other 2-hydroxydiphenyl ethers, inhibits FAS in vivo and FabI in vitro.

Triclosan is an inhibitor of FabI, the enoyl reductase enzyme from FAS-II. In 2003, Sivaraman et al. presented the results of their investigation of the mechanism of triclosan inhibition and selectivity in *E. coli* FabI. They characterized triclosan as a slow reversible tight-binding inhibitor of *E. coli* FabI. Triclosan binds preferentially to the E-NAD$^+$ form of the wild-type enzyme with a $k_i$ value of 23 nM. They also characterized three mutations of FabI, G93V, M159T, and F203L, that have been found to correlate with increases in the triclosan MIC of 95-, 12-, and 6-fold, respectively.

Scientists at GlaxoSmithKline (Collegeville, Pa.) are currently addressing strategies to control the emergence

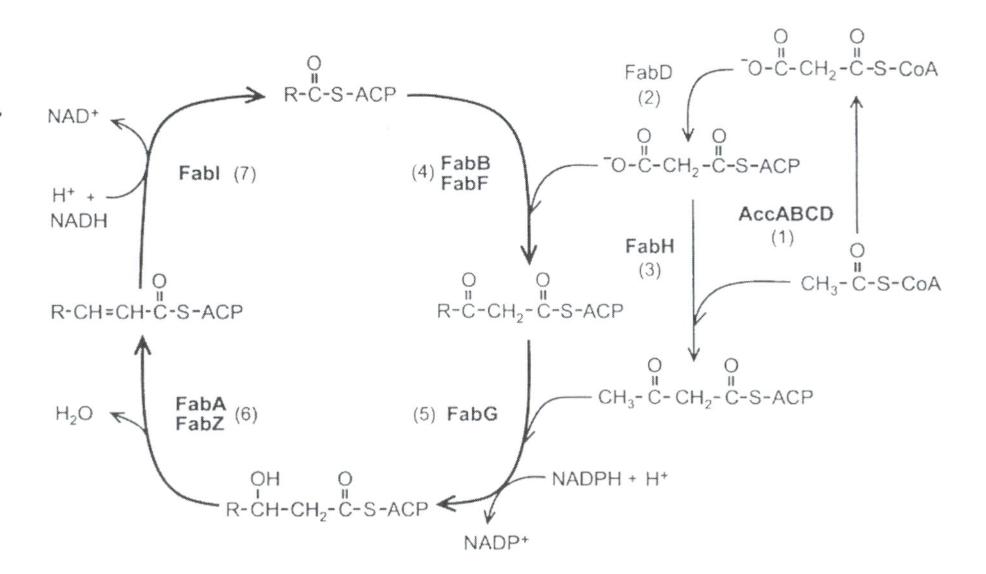


**Figure 27.21** Fatty acid synthesis in *E. coli*.

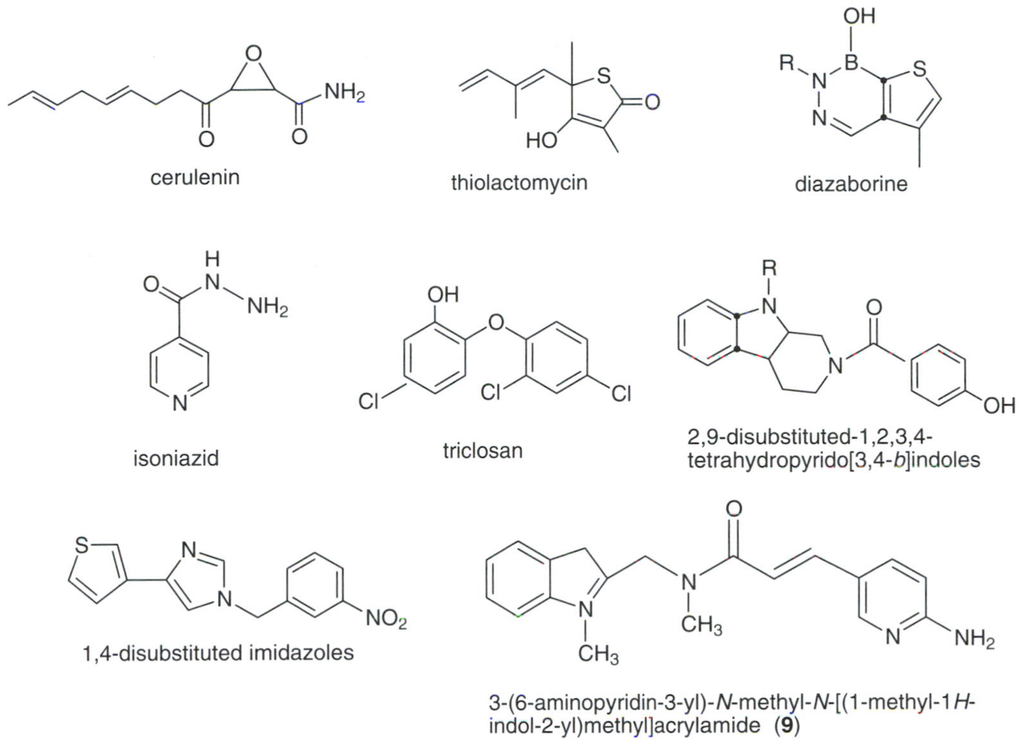

cerulenin

thiolactomycin

diazaborine

isoniazid

triclosan

2,9-disubstituted-1,2,3,4-tetrahydropyrido[3,4-*b*]indoles

1,4-disubstituted imidazoles

3-(6-aminopyridin-3-yl)-*N*-methyl-*N*-[(1-methyl-1*H*-indol-2-yl)methyl]acrylamide (**9**)

**Figure 27.22** Chemical structures of cerulenin, thiolactomycin, diazaborine, isoniazid, triclosan, 2,9-disubstituted 1,2,3,4-tetrahydropyrido[3,4-*b*]indoles, 1,4-disubstituted imidazoles, and the aminopyridine derivative (compound 9).

of bacterial resistance, in particular by developing new antibacterial agents that inhibit bacterial enzymes related to protein synthesis. In 2001 and 2002, they published three reports describing the identification of small-molecule FabI inhibitors with good antibacterial activity against *S. aureus* through a combination of iterative medicinal chemistry and X-ray crystal structure.

In 2003, Seefeld et al., working at GlaxoSmithKline (Collegeville, Pa.), mentioned that in previous communications they had described a series of small molecule-FabI inhibitors that were optimized from screening leads obtained from the GlaxoSmithKline compound collection. Additional research from these labs led to the discovery of the aminopyridine-based Fab inhibitor 4 (Fig. 27.23), which exhibited in vivo efficacy against *S. aureus*. They now report the discovery of a naphthyridinone-based series of FabI inhibitors that demonstrates improvements in potency, spectrum, and in vivo efficacy

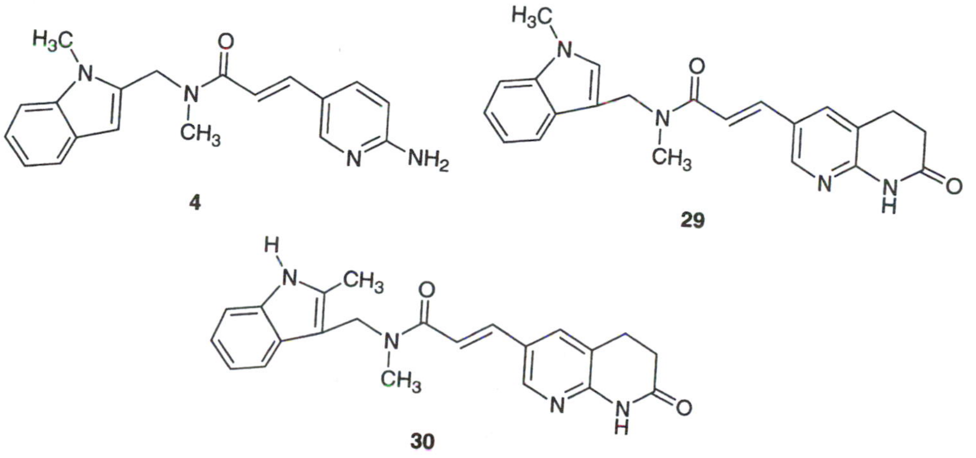

**4**

**29**

**30**

**Figure 27.23** Chemical structures of compounds 4, 29, and 30.

over previously reported selective FAS-II inhibitors. They found that selected compounds from this class of inhibitors display dual FabI-FabK inhibition. Several of these new inhibitors have potent antibacterial activity against multidrug-resistant strains of *S. aureus,* and compound 30 demonstrated exceptional oral in vivo efficacy in an *S. aureus* infection model in rats. While optimizing FabI inhibitory activity, compounds 29 and 30 were identified as having low micromolar FabK inhibitory activity, thereby increasing the antimicrobial spectrum of these compounds to include the FabK-containing pathogens *S. pneumoniae* and *E. faecalis.* These scientists mentioned that the results they had reported supported the hypothesis that bacterial enoyl-ACP reductases are valid targets for antibacterial agents.

Campbell and Cronan (2001) published an excellent comprehensive review covering most of these known inhibitors. The reader should consult this reference for more detailed information on FAS as a target for antibacterial agents.

In 2002, Daines et al. of GlaxoSmithKline reported the determination of the first cocrystal of a bacterial FabH condensing enzyme and small-molecule inhibitor 5. β-Ketoacyl–ACP synthetase III, also known as FabH, plays an essential role in bacterial fatty acid synthesis. FabH is the initiator of the fatty acid chain elongation cycle and plays a key regulatory role in the entire biosynthetic pathway via feedback inhibition by long-chain acyl ACPs.

A high-throughput screening effort directed against *S. pneumoniae* FabH identified indole compound 1 (Fig. 27.24) as a potent inhibitor of the enzyme. This compound, in addition to possessing antibacterial activity against both gram-positive and gram-negative bacteria, demonstrated good selectivity for the bacterial FAS-II over the human FAS-I. Cocrystallization studies yielded an X-ray structure for the enzyme-inhibitor complex between compound 5 (Fig. 27.24) and *E. coli* FabH. The authors of this report concluded by stating that the information obtained from this cocrystal will be used for the rational design of more potent inhibitors of this novel and unexploited antibacterial target.

## Inhibitors of Bacterial Tyrosyl-tRNA Synthetase

Aminoacyl-tRNA synthetases play a crucial role in protein synthesis in all organisms, and selective inhibition of the bacterial enzymes has potential for the discovery of new antibacterial agents.

In a screening program at GlaxoSmithKline, it was found that SB-219383 (structure 1 in Fig. 27.25) is a potent and selective inhibitor of the bacterial tyrosyl-tRNA synthetase from *S. aureus* but not of the mammalian counterpart. Following the discovery of this compound, several structural analogues have been synthesized to elucidate the SAR. At least two synthetic analogues, compounds 2 and 3, were identified as highly potent stereoselective inhibitors of bacterial tyrosyl-tRNA synthetase. However, it was found that the synthetic compounds showed poor antibacterial activity, attributed by the GlaxoSmithKline scientists to the overall polarity of these compounds, which contributed to reduced penetration through the bacterial cell wall. These results are now being used for further development of other synthetic analogues.

In 2002, these researchers reported the structures of nanomolar inhibitors of *S. aureus* methionyl-tRNA synthetase with potent antibacterial activity against gram-positive pathogens. Compound 11 was the more potent and selective inhibitor of *S. aureus* methionyl-tRNA synthetase (Fig. 27.25).

## Inhibitors of the Enzymes of the Diaminopimelate Pathway

A recent review by Cox et al. (2000) discusses the effort being directed toward the identification of an efficient inhibitor of the enzymes of the diaminopimelate pathway. These are key enzyme in the bacterial biosynthetic pathway of L-lysine, an essential cross-linking amino acid for the peptidoglycan macromolecule in gram-positive bacteria, whereas many gram-negative bacteria employ *meso*-diaminopimelic acid (DAP) as a cross-linking amino acid.

Since mammals do not make or use DAP and require L-lysine as a dietary component, inhibitors of the DAP

**Figure 27.24** Chemical structures of compounds 1 and 5.

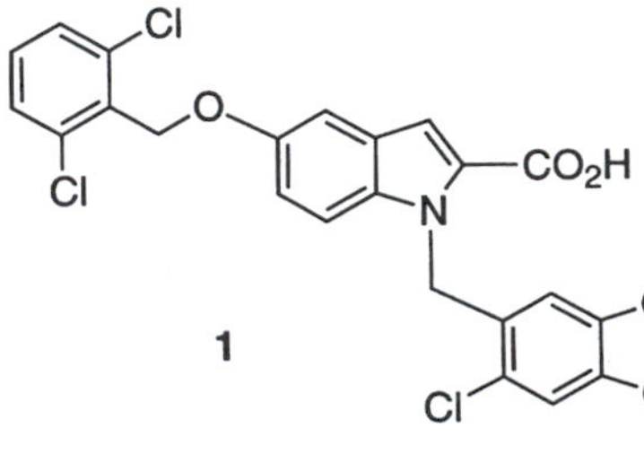

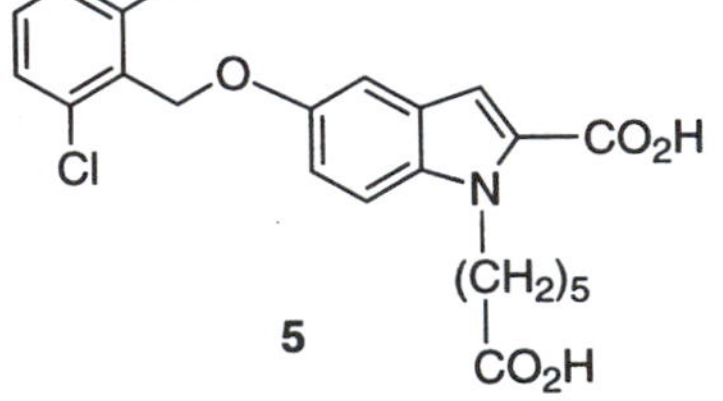

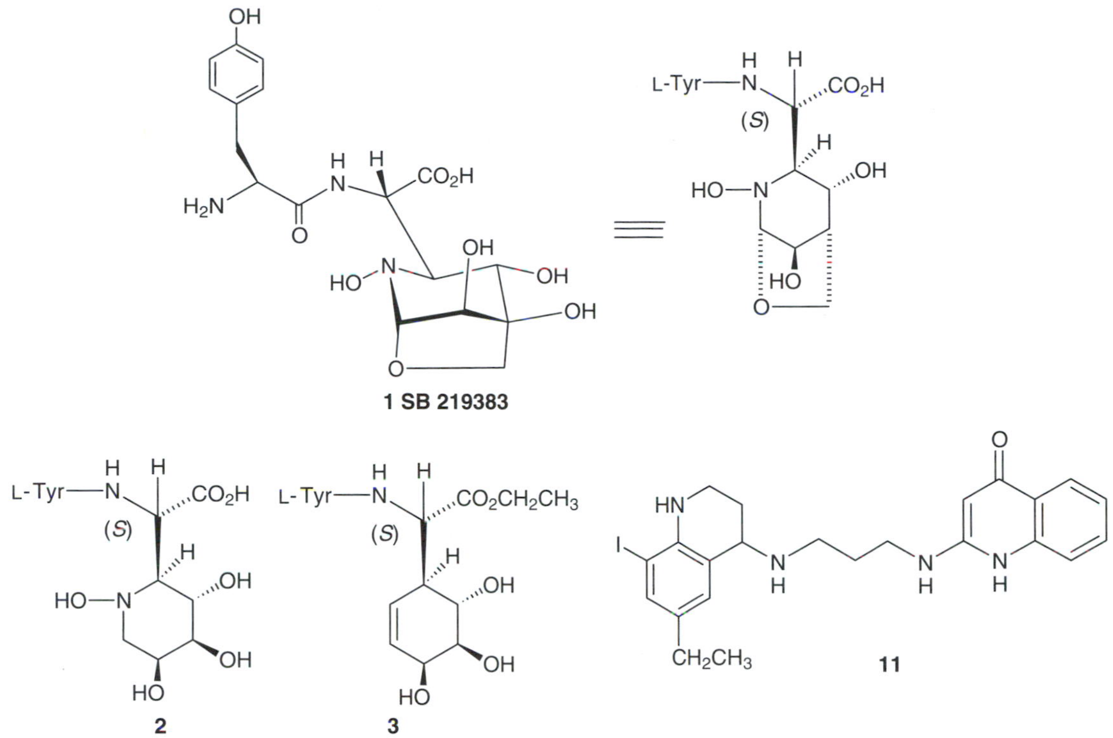

**Figure 27.25** Chemical structures of compounds 1 (SB-219383), 2, 3, and 11.

biosynthetic pathway appear to be novel candidates as antibacterial agents. The authors concluded the review by mentioning that "while no potent broad spectrum antimicrobial compounds have yet emerged, studies toward this goal are underway."

In 2002, Ray et al. reported the cocrystal structures of *Methanococcus jannaschii* diaminopimelate decarboxylase (DAPDC) bound to a substrate analog, azelaic acid, and its L-lysine product determined at 2.6- and 2.0-Å resolution, respectively. These authors discussed the implications for rational design of broad-spectrum antimicrobial agents targeted against DAPDCs of drug-resistant strains of bacterial pathogens, such as *S. aureus*.

## Inhibitors of UDP-*N*-Acetylglucosamine Reductase (MurB)

Scientists at Pharmacia Corp. (Kalamazoo, Mich.) in collaboration with scientists at Human Genome Sciences, Inc. (Rockville, Md.), recently reported that the X-ray crystal structure of the substrate-free form of *S. aureus* UDP-*N*-acetylenoylpyruvylglucosamine reductase (MurB) had been solved to 2.3-Å resolution. They pointed out that this structure demonstrates the importance of conducting structural and biochemical analyses on the target from the bacterial species of interest when the closest possible fit of inhibitor to the enzyme is desired to facilitate meaningful drug discovery. These scientists are currently developing inhibitors for *S. aureus* UDP-*N*-acetylglucosamine reductase.

## Inhibitors of the Bacterial Cell Wall Biosynthesis Enzyme MurC

Targeting the biosynthesis of peptidoglycan during bacterial cell wall synthesis has resulted in very useful antibacterial drugs, as already discussed. One key enzyme, the ligase MurC, which catalyzes the conversion of UDP-MurNAc to UDP-MurNAc-Ala in the assembly of the disaccharide-peptide unit required for peptidoglycan biosynthesis, was selected by scientists at AstraZeneca (Waltham, Mass.) for the design and synthesis of phosphinates that mimic the tetrahedral reaction center of the transition state.

A series of phosphinate transition state analogues of the L-alanine-adding enzyme (MurC) of bacterial peptidoglycan biosynthesis was prepared and tested as inhibitors of the *E. coli* enzyme. Compound 1 (Fig. 27.26) was identified as a potent inhibitor of MurC from *E. coli*, with a 50% inhibitory concentration of 49 nM.

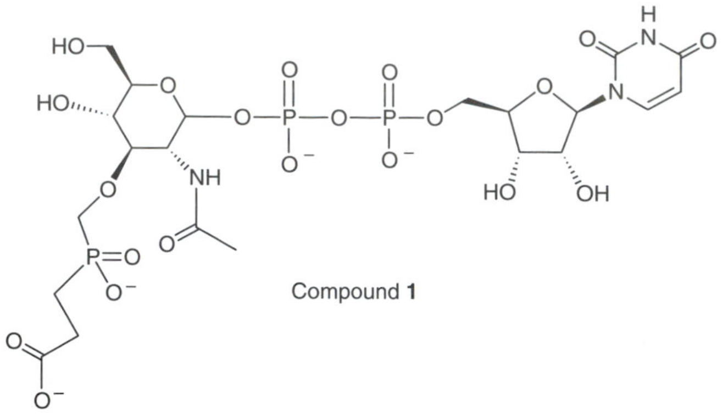

Compound 1

**Figure 27.26** Chemical structure of phosphinate derivative 1 (compound 1).

## Selective Cleavage of D-Alanyl–D-Lactic Acid Termini of Peptidoglycan Precursors

Here I discuss molecules that catalytically and selectively cleave the altered termini of cell wall peptidoglycan precursors and therefore disable the mechanism of antibiotic resistance of gram-positive bacterial pathogens to the glycopeptide antibiotics vancomycin and teicoplanin.

In a recent publication, Chiosis and Boneca (2001) described the synthesis and biological activity of small molecules with well-oriented nucleophile-electrophile assemblies and complementary chirality to the peptidoglycan termini as selective cleaving agents for the peptidoglycan precursor depsipeptide. The authors tested these molecules in combination with vancomycin, and at least three of them were found to resensitize vancomycin-resistant bacteria to this antibiotic.

As discussed in chapter 14, vancomycin and teicoplanin are considered to be the last resort for the treatment of resistant strains of *S. aureus.* Unfortunately, enterococci are becoming increasingly resistant to vancomycin, raising fears that the high-level resistance genes will be transferred to staphylococci. In VRE bearing the *vanA* or *vanB* gene cluster, some of the D-Ala–D-Ala termini of the peptidoglycan precursors are replaced by D-Ala–D-Lac. As a result, the affinity of vancomycin for the peptidoglycan layer is diminished by a factor of over 1,000.

To identify small molecules that cleave the altered cell wall peptidoglycan precursors, these scientists prepared a D-Ala–D-Lac probe labeled with a red dye. Screening of nonbiased combinatorial libraries against the labeled depsipeptide yielded three active compounds with the sequences X–L-Lys–L-Ser dimethylurea (where X is variable), X–D-Lys–D-Ser dimethylurea, and L-Lys–D-Pro–L-Ser dimethylurea. They then designed the SProC5 molecule (Fig. 27.27), which was the most active of the compounds tested. They tested the activity of SProC5 combined with vancomycin against the high-level-resistant *vanA*-positive *E. faecium* strain EF228 and found an MIC of 65.5 μg/ml, an eightfold decrease relative to the MIC of vancomycin alone. Consequently, a simple molecule, SProC5, was found to have structural features that cleave the depsipeptide and increase the sensitivity of VRE to vancomycin. The results obtained suggested to these scientists that SProC5 enhanced vancomycin activity because of its D-Ala–D-Lac hydrolytic activity.

These authors concluded their publication by expressing the hope that restoration of vancomycin activity with small molecules that cleave the altered peptidoglycan component is a promising strategy.

## Inhibitors of Bacterial Adhesins

Adhesion is a key step in the colonization of host tissues by pathogenic bacteria. The topic of bacterial adhesion is briefly covered in chapter 2. Some basic aspects are reviewed here. Bacteria produce different adhesins, which typically are protein molecules that mediate bacterial attachment by interacting with specific receptors. Some bacterial adhesins function as **lectins,** mediating bacterial interactions with carbohydrate moieties on glycoproteins and/or glycolipids. Carbohydrate-binding proteins, other than enzymes and immunoglobulins, are generally called lectins. Adhesins are present on bacterial surfaces, such as pili or fimbriae, that are fibrous organelles.

Uropathogenic strains of *E. coli* are the primary causative agents of urinary tract infections in humans. It has been demonstrated that the pilus in these strains contains a distally located adhesin. P pili are adhesive organelles encoded by genes (*papA* through *papK*) in the *pap* (for "pilus-associated with pyelonephritis") gene cluster found in the chromosome of uropathogenic strains of *E. coli.* The biogenesis of P pili occurs via the highly conserved chaperone-usher pathway. PapD is the prototypical periplasmic chaperone in a family that includes more than 30 members. A drug inhibiting pilus

**Figure 27.27** Chemical structure of the compound SProC5.

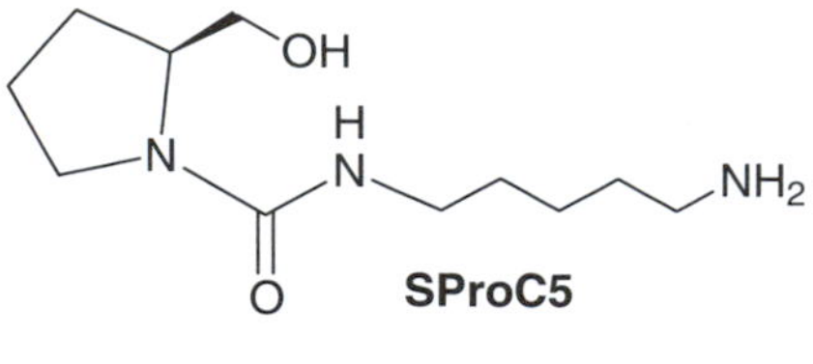

SProC5

formation would therefore have the potential to be an effective antibacterial agent.

Binding of the class II PapG adhesin, found at the tips of filamentous pili on *E. coli*, to the carbohydrate moiety of the globoseries of glycolipids (see Fig. 2.2 and 2.3) in the human kidney is a key step in the development of pyelonephritis, a severe form of urinary tract infection. In 2003, Larsson et al., as part of a program aimed at the development of carbohydrate-based inhibitors of pyelonephritic *E. coli*, described the development of an assay based on surface plasma resonances for direct studies of the binding of the class II PapG adhesin to saccharide ligands. The assay was used to determine the dissociation constants for binding of PapG to the saccharide moieties of the globoseries of glycolipids and fragments thereof, as well as to a panel of substituted derivatives of the disaccharide galactobiose. The adhesin was found to bind to saccharides from the globoseries of glycolipids, which function as ligands for *E. coli* in the upper urinary tract with $K_d$ values ranging from 80 to 540 μM.

In 2002, Mitchell et al. reported that *Pseudomonas aeruginosa* galactose- and fucose-binding ligands (PA-IL and PA-IIL) contribute to the virulence of this pathogenic bacterium, which is a major cause of morbidity and mortality in cystic fibrosis patients. The crystal structure of PA-IIL in complex with fucose reveals a tetrameric structure. Each monomer displays a nine-stranded, antiparallel β-sandwich arrangement and contains two calcium cations, located close to each other, that mediate the binding of fucose in a recognition mode unique among carbohydrate-protein interactions. Experimental binding studies, together with theoretical docking of fucose-containing oligosaccharides, are consistent with the assumption that antigens of the Lewis series may be the preferred ligands of this lectin. The authors of the report conclude that precise knowledge of the lectin-binding site should allow a better design of new antibacterial-adhesion prophylactics.

### β-Lactams as Potential Inhibitors of Pilus Assembly in Pathogenic Bacteria

A recently published report from a collaborative project of scientists of Umeå University (Umeå, Sweden) and Washington University (St. Louis, Mo.) describes the stereoselective synthesis of penicillin aldehydes exemplified by the general formula shown in Fig. 27.28. Based on the examination of the crystal structure of peptide-PapD complexes, these scientists predicted that these compounds would superimpose well on the two C-terminal amino acids. However, the authors found that penicillin aldehydes were unstable and decomposed; this prevented them from testing the antibacterial activity of the synthesized compounds. I had also found, almost 15 years ago, that substituted penam 3-aldehydes are very prone to decomposition.

## Inhibitors of Lipid A Biosynthesis

LPS is the major lipid component of the outer leaflet of the outer membrane of gram-negative bacteria (see chapter 1). The hydrophobic anchor of LPS is lipid A, a unique amphiphilic lipid. Lipid A from *E. coli* is a hexa-acetylated disaccharide of glucosamine sugar, with phosphate residues at positions 1 and 4' (see Fig. 1.19). It is essential for bacterial growth.

The enzymes and genes involved in *E. coli* lipid A synthesis have been extensively studied and characterized. The first step of this pathway involves the *lpxA* gene, which catalyzes the transfer of an *R*-3-hydroxymyristoyl moiety from *R*-3-hydroxymyristoyl–ACP to the third position of the glucosamine ring of UDP-*N*-acetylglucosamine. The second step is catalyzed by UDP-3-*O*-(*R*-3-hydroxymyristoyl)-*N*-acetylglucosamine deacetylase (LpxC), which removes an acetyl moiety from the N-2 position of glucosamine in the lipid A precursor UDP-3-*O*-(*R*-3-hydroxymyristoyl)-*N*-acetylglucosamine (Fig. 27.29). *E. coli* LpxC contains a bound zinc ion that is required for catalytic activity.

**Figure 27.28** Bicyclic β-lactam compounds of the general structure 1 superimpose well with the structure of a peptide whose crystal structure complexed with PapD was determined by X-ray crystallography.

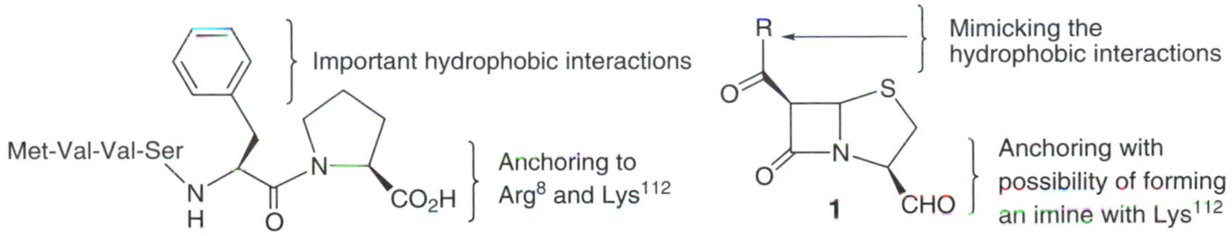

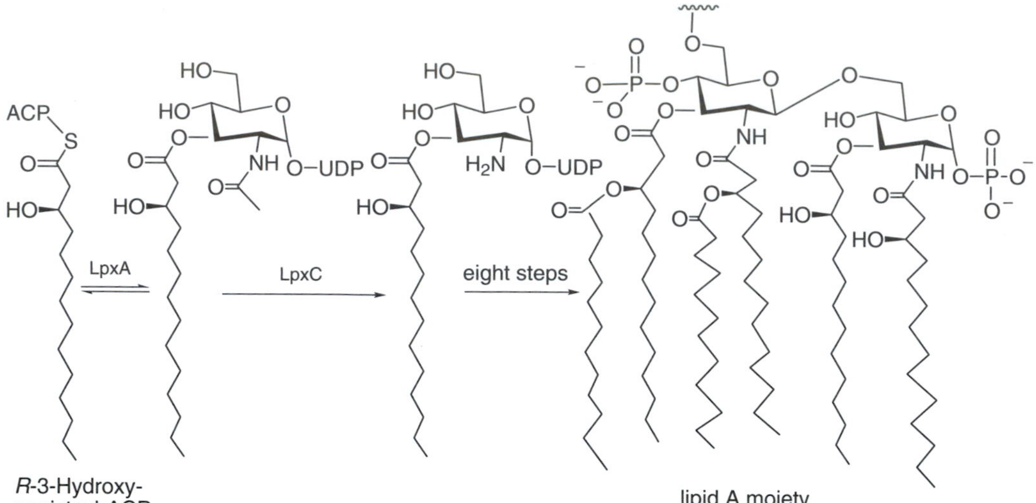

**Figure 27.29** A unique deacetylase catalyzes the second step of lipid A biosynthesis. The LpxA-catalyzed acylation that occurs before deacetylation is reversible and has an unfavorable equilibrium constant.

In 1996, Onishi et al. reported that phenyloxazoline-based hydroxamates have significant antibacterial activities and that they inhibit lipid A biosynthesis. In 1999, Chen et al. reported a series of carhydroxamido-oxazolidine inhibitors of LpxC. In 2000, Jackman et al. described hydroxamate- and phosphinate-based LpxC inhibitors, although they do not possess significant antibacterial activities. In 2002, Clemens et al. reported on the screening of an *lpxC* mutant of *E. coli,* predicted to be hypersusceptible to LpxC inhibitors, with a collection of low-molecular-weight compounds containing metal-chelating groups. Several compounds with MICs of $<1$ mg/liter were identified by this screen, including two sulfonamide derivatives of α-(*R*)-aminohydroxamic acid, BB-78484 and BB-78485 (Fig. 27.30).

**Figure 27.30** Structures of the LpxC inhibitors BB-78484 and BB-78485.

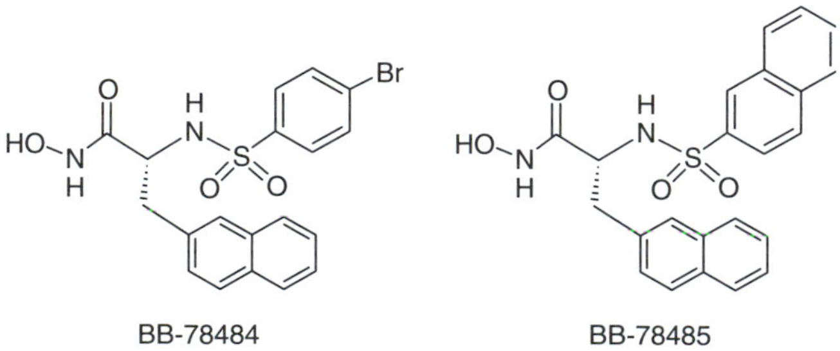

## Bacterial Genome Sequencing and Antibacterial Drug Discovery

Bacterial genomic sequencing is now being used to find previously unidentified genes and their corresponding proteins. Genes and proteins essential for the survival of bacteria can be regarded as potential drug targets. Bacterial DNA sequences from a wide variety of bacteria are now available. Particularly important are those of bacterial pathogens with emerging or reemerging antibacterial resistance. Very recently, the complete genome sequences of a virulent isolate of *S. pneumoniae,* MRSA, *P. aeruginosa, H. influenzae* and *S. enterica* were reported. Currently, 79 completely sequenced bacterial genomes are publicly available, including those of more than 40 human pathogens.

## Novel Antibacterial Compounds Identified by Using Synthetic Combinatorial Library Technology

Combinatorial chemistry has had a significant impact on the discovery of new antibacterial drugs. Most of the successes have come from the use of small libraries to explore a specific pharmacophore. This kind of application has been exemplified in this chapter with the discovery of actinonin, a selective peptide deformylase inhibitor. Combinatorial chemistry has also been used to optimize new antibacterial agents, such as in the case of oxazolidinone lead compounds.

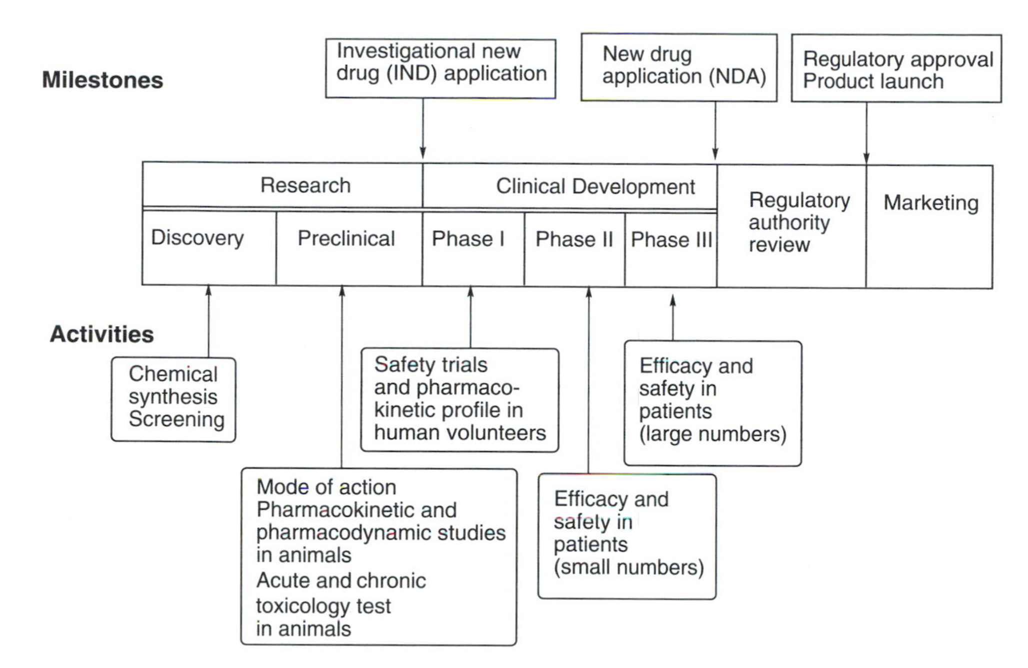


**Figure 27.31** Steps in the research and development process for antibacterial agents. Reprinted from I. Chopra, *Curr. Opin. Microbiol.* **1**:495–501, 1998, with permission from the publisher.

## How the Pharmaceutical Industry Brings an Antibacterial Drug to the Market in the United States

The topic of how the pharmaceutical industry brings antibacterial drugs to the market is quite well covered in a review published in 1994 by S. A. Billstein of Roche Laboratories (Nutley, N.J.). Following the design and synthesis of novel antibacterial agents, the preclinical and clinical development phases (Fig. 27.31) can each take several years to complete. Therefore, the period between discovery of the candidate antibacterial drug in the research laboratory and its final approval by the regulatory authorities, the Food and Drug Administration in the United States, generally takes at least 10 years.

## Concluding Remarks and Overview of This Chapter

The traditional method for obtaining new antibacterial drugs has been to synthesize analogues of existing antibacterial drugs and evaluate them for improved therapeutic activity by using in vitro and in vivo methods that detect antibacterial activity against gram-positive and gram-negative organisms.

Over the last 10 years, several new strategies have been developed, including

- the use of libraries of synthetic and natural product compounds
- combinatorial chemistry
- high-throughput screening
- new biological assays based on new insights into the molecular biology and biochemistry of target organisms
- mechanism-based and target-directed strategies
- genomic research

## REFERENCES AND FURTHER READING


### Standard Nomenclature of the Active Site in Proteases

**Schechter, J., and A. Berger.** 1967. On the site of the active site in proteases. I. Bapain. *Biochem. Biophys. Res. Commun.* **27**:157–162.

### Peptide Deformylase

**Baldwin, E. T., M. S. Harris, A. W. Yem, C. L. Wolfe, A. F. Vosters, K. A. Curry, R. W. Murray, J. H. Bock, V. P. Marshall, J. I. Cialdella, M. H. Merchant, G. Choi, and M. R. Deibel.** 2002. Crystal structure of type II peptide deformylase from *Staphylococcus aureus. J. Biol. Chem.* **277**:31163–31171.

**Becker, A., I. Schlichting, W. Kabsch, S. Schultz, and A. F. V. Wagner.** 1998. Structure of peptide deformylase and identification of the substrate binding site. *J. Biol. Chem.* **273**:11413–11416.

**Becker, A., I. Schlichting, W. Kabsch, D. Groche, S. Schultz, and A. F. V. Wagner.** 1998. Iron center, substrate recognition and mechanism of peptide deformylase. *Nat. Struct. Biol.* **5**:1053–1058.

**Chan, M. K., W. Gong, P. T. Ravi Rajagopalan, B. Hao, C. M. Tsai, and D. Pei.** 1997. Crystal structure of the *Escherichia coli* peptide deformylase. *Biochemistry* **36**:13904–13909.

**Dardel, F., S. Ragusa, C. Lazennec, S. Blanquet, and T. Meinnel.** 1998. Solution structure of nickel-peptide deformylase. *J. Mol. Biol.* **280:**501–513.

**Giglione, C., M. Pierre, and T. Meinnel.** 2000. Peptide deformylase as a target for new generation, broad spectrum antimicrobial agents. *Mol Microbiol.* **36:**1197–1205.

**Groche, D., A. Becker, I. Schlichting, W. Kabsch, S. Schultz, and A. F. V. Wagner.** 1998. Isolation and crystallization of functionally competent *Escherichia coli* peptide deformylase forms containing either iron or nickel in the active site. *Biochem. Biophys. Res. Commun.* **246:**342–346.

**Mazel, D., S. Pochet, and P. Marlière.** 1994. Genetic characterization of polypeptide deformylase, a distinctive enzyme of eubacterial translation. *EMBO J.* **13:**914–923.

**Meinnel, T., S. Blanquet, and F. Dardel.** 1996. A new subclass of the zinc metalloproteases superfamily revealed by the solution structure of peptide deformylase. *J. Mol. Biol.* **262:**375–386.

**Ragusa, S., S. Blanquet, and T. Meinnel.** 1998. Control of peptide deformylase activity by metal cations. *J. Mol. Biol.* **280:**515–523.

**Ravi Rajagopalan, P. T., A. Datta, and D. Pei.** 1997. Purification, characterization, and inhibition of peptide deformylase from *Escherichia coli. Biochemistry* **36:**13910–13918.

**Ravi Rajagopalan, P. T., S. Grimme, and D. Pei.** 2000. Characterization of cobalt(II)-substituted peptide deformylase: function of the metal ion and the catalytic residue Glu-133. *Biochemistry* **39:**779–790.

**Ravi Rajagopalan, P. T., and D. Pei.** 1998. Oxygen-mediated inactivation of peptide deformylase. *J. Biol. Chem.* **273:**22305–22310.

**Ravi Rajagopalan, P. T., X. C. Yu, and D. Pei.** 1997. Peptide deformylase: a new type of mononuclear iron protein. *J. Am. Chem. Soc.* **119:**12418–12419.

### Inhibitors of Peptide Deformylase

#### H-phosphonate derivatives

**Hao, B., W. Gong, P. T. Ravi Rajagopalan, Y. Zhou, D. Pei, and M. K. Chan.** 1999. Structural basis for the design of antibiotics targeting peptide deformylase. *Biochemistry* **38:**4712–4719.

**Hu, Y. J., P. T. Ravi Rajagopalan, and D. Pei.** 1998. H-phosphonate derivatives as novel peptide deformylase inhibitors. *Bioorg. Med. Chem. Lett.* **8:**2479–2482.

#### Thiol derivatives

**Huntington, K. M., T. Yi, Y. Wei, and D. Pei.** 2000. Synthesis and antibacterial activity of peptide deformylase inhibitors. *Biochemistry* **39:**4543–4551.

**Meinnel, T., L. Patiny, S. Ragusa, and S. Blanquet.** 1999. Design and synthesis of substrate analogue inhibitors of peptide deformylase. *Biochemistry* **38:**4287–4295.

**Wei, Y., T. Yi, K. M. Huntington, C. Chaudhury, and D. Pei.** 2000. Identification of a potent peptide deformylase inhibitor from a rationally designed combinatorial library. *J. Comb. Chem.* **2:**650–657.

#### Hydroxamate derivatives

**Apfel, C., D. W. Banner, D. Bur, M. Dietz, T. Hirata, C. Hubschwerlen, H. Locher, M. G. P. Page, W. Pirson, G. Rossé, and J. L. Specklin.** 2000. Hydroxamic acid derivatives as potent peptide deformylase inhibitors and antibacterial agents. *J. Med. Chem.* **43:**2324–2331.

**Apfel, C., D. W. Banner, D. Bur, M. Dietz, C. Hubschwerlen, H. Locher, F. Marlin, R. Masciadri, W. Pirson, and H. Stalder.** 2001. 2-(2-Oxo-1,4-dihydro-2*H*-quinazolin-3-yl)- and 2-(2,2-dioxo-1,4-dihydro-2*H*-2λ6-benzo[1,2,6]thiazidin-3-yl)-*N*-hydroxy-acetamides as potent and selective peptide deformylase inhibitors. *J. Med. Chem.* **44:**1847–1852.

**Chen, D. Z., D. V. Patel, C. J. Hackbarth, W. Wang, G. Dreyer, D. C. Young, P. S. Margolis, C. Wu, Z. J. Ni, J. Trias, R. J. White, and Z. Yuan.** 2000. Actinonin, a naturally occurring antibacterial agent, is a potent deformylase inhibitor. *Biochemistry* **39:**1256–1262.

**Gupta, M. K., P. Mishra, P. Prathipati, and A. K. Saxena.** 2002. 2D-QSAR in hydroxamic acid derivatives as peptide deformylase inhibitors and antibacterial agents. *Bioorg. Med. Chem.* **10:**3713-3716.

**Thorarensen, A., M. R. Douglas, D. C. Rohrer, A. F. Vosters, A. W. Yem, V. D. Marshall, J. C. Lynn, M. J. Bohanon, P. K. Tomich, G. E. Zurenko, M. T. Sweeney, R. M. Jensen, J. W. Nielsen, E. P. Seest, and L. A. Dolak.** 2001. Identification of novel potent hydroxamic acid inhibitors of peptidyl deformylase and the importance of the hydroxamic acid functionality on inhibition. *Bioorg. Med. Chem. Lett.* **11:**1355–1358.

#### *N*-Formylhydroxylamine derivatives

**Clements, J. M., R. P. Beckett, A. Brown, G. Catlin, M. Lobell, S. Palan, W. Thomas, M. Whittaker, S. Wood, S. Salama, P. J. Baker, H. F. Rodgers, V. Barynin, D. W. Rice, and M. G. Hunter.** 2001. Antibiotic activity and characterization of BB-3497, a novel peptide deformylase inhibitor. *Antimicrob. Agents Chemother.* **45:**563–570.

**Smith, H. K., R. P. Beckett, J. M. Clements, S. Doel, S. P. East, S. B. Launchbury, L. M. Pratt, Z. M. Spavold, W. Thomas, R. S. Todd, and M. Whittaker.** 2002. Structure-activity relationships of the peptide deformylase inhibitor BB-3497: modification of the metal binding group. *Bioorg. Med. Chem. Lett.* **12:**3595–3599.

#### Peptide aldehyde derivatives

**Durand, D. J., B. G. Green, J. F. O'Connell, and S. K. Grant.** 1999. Peptide aldehyde inhibitors of bacterial peptide deformylase. *Arch. Biochem. Biophys.* **367:**297–302.

#### Biaryl acid analogs

**Green, B. G., J. H. Toney, J. W. Kozarich, and S. K. Grant.** 2000. Inhibition of bacterial peptide deformylase by biaryl acid analogs. *Arch. Biochem. Biophys.* **375:**355–358.

### Two-Component Signal Transduction

**Barrett, J. F., and J. A. Hoch.** 1998. Two-component signal transduction as a target for microbial anti-infective therapy. *Antimicrob. Agents Chemother.* **42:**1529–1536.

**Hilliard, J. J., R. M. Goldschmidt, L. Licata, E. Z. Baum, and K. Bush.** 1999. Multiple mechanisms of action for inhibitors of histidine protein kinases from bacterial two-component systems. *Antimicrob. Agents Chemother.* **43:**1693–1699.

**Rodrigue, A., Y. Quentin, A. Lazdunski, V. Méjean, and M. Foglino.** 2000. Two-component systems in *Pseudomonas aeruginosa:* why so many? *Trends Microbiol.* **8:**498–504.

**Stock, J. B., M. G. Surette, M. Levit, and P. Park.** 1995. Two-component signal transduction systems: structure-function relationship and mechanisms of catalysis, p. 25–51. *In* J. A. Hoch and T. J. Silhavy (ed.), *Two-Component Signal Transduction.* ASM Press, Washington, D.C.

**Inhibitors of bacterial two-component systems**

**Hlasta, D. J., J. P. Demers, B. D. Foleno, S. A. Fraga-Spano, J. Guan, J. J. Hilliard, M. J. Macielag, K. A. Phemeng, C. M. Sheppard, Z. Sui, G. C. Webb, M. A. Weidner-Wells, H. Werblood, and J. F. Barrett.** 1998. Novel inhibitors of bacterial two-component systems with gram-positive antibacterial activity: pharmacophore identification based on the screening hit closantel. *Bioorg. Med. Chem. Lett.* **8:**1923–1928.

**Macielag, M. J., J. P. Demers, S. A. Fraga-Spano, D. J. Hlasta, S. G. Johnson, R. M. Kanojia, R. K. Russell, Z. Sui, M. A. Weidner-Wells, H. Werblood, B. D. Foleno, R. M. Goldschmidt, M. J. Loeloff, G. C. Webb, and J. F. Barrett.** 1998. Substituted salicylanilides as inhibitors of two-component regulatory systems in bacteria. *J. Med. Chem.* **41:**2939–2945.

**Matsushita, M., and K. D. Janda.** 2002. Histidine kinases as targets for new antimicrobial agents. *Bioorg. Med. Chem. Lett.* **10:**855–867.

**Sui, Z., J. Guan, D. J. Hlasta, M. J. Macielag, B. D. Foleno, R. M. Goldschmidt, M. J. Loeloff, G. C. Webb, and J. F. Barrett.** 1998. SAR studies of diaryltriazoles against bacterial two-component regulatory systems and their antibacterial activities. *Bioorg. Med. Chem. Lett.* **8:**1929–1934.

**Weidner-Wells, M. A., K. A. Ohemeng, V. N. Nguyen, S. Fraga-Spano, M. J. Macielag, H. M. Werblood, B. D. Foleno, G. C. Webb, J. F. Barrett, and D. J. Hlasta.** 2001. Amidino benzimidazole inhibitors of bacterial two-component systems. *Bioorg. Med. Chem. Lett.* **11:**1545–1548.

### Antibacterial Agents Based on Cyclic DL-α-Peptide Architecture

**Bong, D. T., T. D. Clark, J. R. Granja, and M. R. Ghadiri.** 2001. Self-assembling organic nanotubes. *Angew. Chem. Int. Ed.* **40:**988–1011.

**Clark, T. D., L. K. Buehler, and M. R. Ghadiri.** 1998. Self-assembling cyclic β-peptide nanotubes as artificial transmembrane ion channels. *J. Am. Chem. Soc.* **120:**651–656.

**Fernandez-Lopez, S., H. S. Kim, E. C. Choi, M. Delgado, J. R. Granja, A. Khasanov, K. Kraehenbuehl, G. Long, D. A. Weinberger, K. M. Wilcoxen, and M. R. Ghadiri.** 2001. Antibacterial agents based on the cyclic DL-α-peptide architecture. *Nature* **412:**452–455.

**Ghadiri, M. R., J. R. Granja, and L. K. Buehler.** 1994. Artificial transmembrane ion channels from self-assembling peptide nanotubes. *Nature* **369:**301–304.

**Ghadiri, M. R., J. R. Granja, R. A. Milligan, D. E. McRee, and N. Khazanovich.** 1993. Self-assembling organic nanotubes based on a cyclic peptide architecture. *Nature* **366:**324–327.

**Kim, H. S., J. D. Hartgerink, and M. R. Ghadiri.** 1998. Oriented self-assembly of cyclic peptides nanotubes in lipid membranes. *J. Am. Chem. Soc.* **120:**4417–4424.

### Antimicrobial Cationic Peptides

**Dutton, C. J., M. A. Haxell, H. A. I. McArthur, and R. G. Wax.** 2002. *Peptide Antibiotics. Discovery, Modes of Action, and Applications.* Marcel Dekker, Inc., New York, N.Y.

**Friedrich, C. L., D. Moyles, T. J. Beveridge, and R. E. W. Hancock.** 2000. Antibacterial action of structurally diverse cationic peptides on gram-positive bacteria. *Antimicrob. Agents Chemother.* **44:**2086–2092.

**Scott, M. G., C. M. Rosenberg, M. R. Gold, B. B. Finlay, and R. E. Hancock.** 2000. An α-helical cationic antimicrobial peptide selectively modulates macrophage responses to lipopolysaccharide and directly alters macrophage gene expression. *J. Immunol.* **164:**549–553.

**Strøm, M. B., B. E. Haug, M. L. Skar, W. Stensen, T. Stiberg, and J. S. Svendsen.** 2003. The pharmacophore of short cationic antibacterial peptides. *J. Med. Chem.* **46:**1567–1570.

**Zhang, L., and R. E. W. Hancock.** 2001. Peptide antibiotics, p. 209–257. *In* D. Hughes and D. L. Andersson (ed.), *Antibiotic Development and Resistance.* Taylor & Francis, London, United Kingdom.

**Zhang, L., A. Rozek, and R. E. W. Hancock.** 2001. Interaction of cationic antimicrobial peptides with model membranes. *J. Biol. Chem.* **276:**35714–35722.

### Toll and Toll-Like Receptors

**Wilson, M., R. McNab, and B. Henderson.** 2002. *Bacterial Disease Mechanisms. An Introduction to Cellular Microbiology,* p. 307–309. Cambridge University Press, Cambridge, United Kingdom.

### New Targets and Strategies for the Development of Antibacterial Agents

**Desnottes, J. F.** 1996. New targets and strategies for the development of antibacterial agents. *Trends Biochem. Technol.* **14:**134–140.

**Domagala, J. M., and J. P. Sanchez.** 1997. New approaches and agents to overcome bacterial resistance. *Annu. Rep. Med. Chem.* **32:**111–120.

**Setti, E. L., L. Quattrochio, and R. G. Micetich.** 1997. Current approaches to overcome bacterial resistance. *Drugs Future* **22:**271–284.

### Bacterial Fatty Acid Biosynthesis

**Campbell, J. W., and J. E. Cronan.** 2001. Bacterial fatty acid biosynthesis: targets for antibacterial drug discovery. *Annu. Rev. Microbiol.* **55:**305–332.

**Cronan, J. E., and C. O. Rock.** 1996. Biosynthesis of membrane lipids, p. 612–636. *In* F. C. Neidhardt, R. Curtiss III, J. L. Ingraham, E. C. C. Lin, K. B. Low, B. Magasanik, W. S. Reznikoff, M. Riley, M. Schaechter, and H. E. Umbarger (ed.), *Escherichia coli* and *Salmonella: Cellular and Molecular Biology,* 2nd ed. ASM Press, Washington, D.C.

**Daines, R. A., I. Pendrak, K. Sham, G. S. Van Aller, A. K. Konstantinidis, J. T. Lonsdale, C. A. Janson, X. Qiu, M. Brandt, S. S. Khandekar, C. Silverman, and M. S. Head.** 2002. First X-ray cocrystal structure of a bacterial FabH condensing enzyme and a small molecule inhibitor achieved using rational design and homology modeling. *J. Med. Chem.* **46:**5–8.

**Magnuson, K., S. Jackowski, C. O. Rock, and J. E. Cronan.** 1993. Regulation of fatty acid biosynthesis in *Escherichia coli. Microbiol. Rev.* **57:**522–542.

**Rock, C. O.** 2000. Lipid biosynthesis, p. 55–61. *In* J. Lederberg (ed.), *Encyclopedia of Microbiology,* 2nd ed., vol. 3. Academic Press, Inc., San Diego, Calif.

**Seefeld, M. A., W. H. Miller, K. A. Newlander, W. J. Burgess, P. A. Elkins, M. S. Head, D. R. Jakas, C. A. Janson, T. D. Moore, D. J. Payne, S. Pearson, B. J. Polizzi, X. Qiu, S. F. Rittenhouse, I. N. Uzinskas, N. G. Wallis, and W. F. Huffman.** 2003. Indolenaphthyridinones as inhibitors of bacterial enoyl-ACP reductases FabI and FabK. *J. Med. Chem.* **46:**1627–1635.

**Sivaraman, S., J. Zwahlen, A. F. Bell, L. Hedstrom, and P. J. Tonge.** 2003. Structure-activity studies of the inhibition of FabI, the enoyl reductase from *Escherichia coli,* by triclosan: kinetic analysis of mutant FabIs. *Biochemistry* **42:**4406-4413.

### Inhibitors of Bacterial Enoyl Acyl Carrier Protein Reductase

**Bergler, H., S. Fuchsbichler, G. Högenauer, and F. Turnowsky.** 1996. The enoyl-[acyl-carrier-protein] reductase (FabI) of *Escherichia coli,* which catalyzes a key regulatory step in fatty acid biosynthesis, accepts NADH and NADPH as cofactors and is inhibited by palmitoyl-CoA. *Eur. J. Biochem.* **242:** 689–694.

**Heath, R. J., and C. O. Rock.** 1995. Enoyl-acyl carrier protein reductase (*fabI*) plays a determinant role in completing cycles of fatty acid elongation in *Escherichia coli. J. Biol. Chem.* **270:**26538–26542.

**Heath, R. J., and C. O. Rock.** 2000. A triclosan-resistant bacterial enzyme. *Nature* **406:**145–146.

**Heath, R. J., J. R. Rubin, D. R. Holland, E. Zhang, M. E. Snow, and C. O. Rock.** 1999. Mechanism of triclosan inhibition of bacterial fatty acid synthesis. *J. Biol. Chem.* **274:**11110–11114.

**Heerding, D. A., G. Chan, W. E. DeWolf, A. P. Fosberry, C. A. Janson, D. D. Jaworski, E. McManus, W. H. Miller, T. D. Moore, D. J. Payne, X. Qiu, S. F. Rittenhouse, C. Slater-Radosti, W. Smith, D. T. Takata, K. S. Vaidya, C. C. K. Yuan, and W. F. Huffman.** 2001. 1,4-Disubstituted imidazoles are potential antibacterial agents functioning as inhibitors of enoyl acyl carrier protein reductase (FabI). *Bioorg. Med. Chem. Lett.* **11:**2061–2065.

**Levy, C. W., A. Roujeinkova, S. Sedelnikova, P. J. Baker, A. R. Stuitje, A. R. Slabas, D. W. Rice, and J. B. Rafferty.** 1999. Molecular basis of triclosan activity. *Nature* **398:**383–384.

**Miller, W. H., M. A. Seefeld, K. A. Newlander, I. N. Uzinskas, W. J. Burgess, D. A. Heerding, C. C. K. Yuan, M. S. Head, D. J. Payne, S. F. Rittenhouse, T. D. Moore, S. C. Pearson, V. Berry, W. E. DeWolf, P. M. Keller, B. J. Polizzi, X. Qiu, C. A. Janson, and W. F. Huffman.** 2002. Discovery of aminopyrimidine based inhibitors of bacterial enoyl-ACP reductase (FabI). *J. Med. Chem.* **45:**3246–3256.

**Seefeld, M. A., W. H. Miller, K. A. Newlander, W. J. Burgess, D. J. Payne, S. F. Rittenhouse, T. D. Moore, W. E. DeWolf, P. M. Keller, X. Qiu, C. A. Janson, K. Vaidya, A. P. Fosberry, M. G. Smyth, D. D. Jaworski, C. Slater-Radosti, and W. F. Huffman.** 2001. Inhibitors of bacterial enoyl acyl carrier protein reductase (FabI): 2,9-disubstituted 1,2,3,4-tetrahydropyrido[3,4-b]indoles as potential antibacterial agents. *Bioorg. Med. Chem. Lett.* **11:**2241–2244.

### Inhibitors of Bacterial Tyrosyl-tRNA Synthetase

**Brown, P., D. S. Eggleston, R. C. Haltiwanger, R. L. Jarvest, L. Mensah, P. J. O'Hanlon, and A. J. Pope.** 2001. Synthetic analogues of SB-219383, novel *C*-glycosyl peptides as inhibitors of tyrosyl tRNA synthetase. *Bioorg. Med. Chem. Lett.* **11:** 711–714.

**Jarvest, R. L., J. M. Berge, V. Berry, E. F. Boyd, M. J. Brown, J. S. Elder, A. K. Forrest, A. P. Fosberry, D. R. Gentry, M. J. Hibbs, D. D. Jaworski, P. J. O'Hanlon, A. J. Pope, S. Rittenhouse, R. J. Sheppard, C. Slater-Radosti, and A. Worby.** 2002. Nanomolar inhibitors of *Staphylococcus aureus* methionyl tRNA synthetase with potent antibacterial activity against Gram-positive pathogens. *J. Med. Chem.* **45:**1959–1952.

**Jarvest, R. L., J. M. Berge, P. Brown, D. W. Hamprecht, D. J. McNair, L. Mensah, P. J. O'Hanlon, and A. J. Pope.** 2001. Potent synthetic inhibitors of tyrosyl tRNA synthetase derived from *C*-pyranosyl analogues of SB-219383. *Bioorg. Med. Chem. Lett.* **11:**715–718.

**Jarvest, R. L., J. M. Berge, C. S. V. Houge-Frydrych, L. M. Mensah, P. J. O'Hanlon, and A. J. Pope.** 2001. Inhibitors of bacterial tyrosyl tRNA synthetase: synthesis of carbocyclic analogues of the natural product SB-219383. *Bioorg. Med. Chem. Lett.* **11:**2499–2502.

### Inhibitors of Diaminopimelate Aminotransferase

**Cox, R. J., A. Sutherland, and J. C. Vederas.** 2000. Bacterial diaminopimelate metabolism as a target for antibiotic design. *Bioorg. Med. Chem.* **8:**843–871.

**Ray, S. S., J. B. Bonanno, K. R. Rajashankar, M. G. Pinho, G. He, H. De Lencastre, A. Tomasz, and S. K. Burley.** 2002. Cocrystal structures of diaminopimelate decarboxylase: mechanism, evolution, and inhibition of an antibiotic resistance accessory factor. *Structure* **10:**1499–1508.

### Inhibitors of UDP-*N*-Acetylglucosamine Reductase (MurB)

**Benson, T. E., M. S. Harris, G. H. Choi, J. I. Cialdella, J. T. Herberg, J. P. Martin, and E. T. Baldwin.** 2001. A structural variation for MurB: X-ray crystal structure of *Staphylococcus aureus* UDP-*N*-acetyl-enoylpyruvylglucosamine reductase (MurB). *Biochemistry* **40**:2340–2350.

**van Heijenoort, J.** 2001. Recent advances in the formation of the bacterial peptidoglycan monomer unit. *Nat. Prod. Rep.* **18**:503–519.

### Inhibitors of the Bacterial Cell Wall Biosynthesis Enzyme MurC

**Reck, F., S. Marmor, S. Fisher, and M. A. Wuonola.** 2001. Inhibitors of the bacterial cell wall biosynthesis enzyme MurC. *Bioorg. Med. Chem. Lett.* **11**:1451–1454.

### Selective Cleavage of D-Ala–D-Lac Termini of the Peptidoglycan Precursors

**Chiosis, G., and I. G. Boneca.** 2001. Selective cleavage of D-Ala–D-Lac by small molecules: re-sensitizing resistant bacteria to vancomycin. *Science* **293**:1484–1487.

### Inhibitors of Bacterial Adhesins

**Entenäs, H., D. Soto, S. J. Hultgren, G. R. Marshall, and F. Almqvist.** 2000. Stereoselective synthesis of optically active β-lactams, potential inhibitors of pilus assembly in pathogenic bacteria. *Org. Lett.* **2**:2065–2067.

**Larsson, A., J. Ohlsson, K. W. Dodson, S. J. Hultgren, U. Nilsson, and J. Kihlberg.** 2003. Quantitative studies of the binding of the class II PapG adhesin from uropathogenic *Escherichia coli* to oligosaccharides. *Bioorg. Med. Chem.* **10**:2255–2261.

**Mitchell, E., C. Houles, D. Sudakevitz, M. Wimmerova, C. Gautier, S. Pérez, A. M. Wu, N. Gilboa-Garber, and A. Imberty.** 2002. Structural basis for oligosaccharide-mediated adhesion of *Pseudomonas* aeruginosa in the lungs of cystic fibrosis patients. *Nat. Struct. Biol.* **9**:918–921.

### Inhibitors of Lipid A Biosynthesis

**Chen, M. H., M. G. Steiner, S. E. de Laszlo, A. A. Patchett, M. S. Anderson, S. A. Hyland, H. R. Onishi, L. L. Silver, and C. R. Raetz.** 1999. Carbohydroxamido-oxazolidines: antibacterial agents that target lipid A biosynthesis. *Bioorg. Med. Chem. Lett.* **9**:313–318.

**Clements, M., F. Coignard, I. Johnson, S. Chandler, S. Palan, A. Waller, J. Wijkmans, and M. G. Hunter.** 2002. Antibacterial activities and characterization of novel inhibitors of LpxC. *Antimicrob. Agents Chemother.* **46**:1793–1799.

**Jackman, J. E., C. A. Fierke, L. N. Tumey, M. Pirrung, T. Uchiyama, S. H. Tahir, O. Hindsgaul, and C. R. H. Raetz.** 2000. Antibacterial agents that target lipid A biosynthesis in gram-negative bacteria. Inhibition of diverse UDP-3-*O*-(*R*-3-hydroxymyristoyl)-*N*-acetylglucosamine deacetylases by substrate analogs containing zinc binding motifs. *J. Biol. Chem.* **275**:11001–11009.

**Jackman, J. E., C. R. H. Raetz, and C. A. Fierke.** 1999. UDP-3-*O*-(*R*-3-hydroxymyristoyl)-*N*-acetylglucosamine deacetylase of *Escherichia coli* is a zinc metalloenzyme. *Biochemistry* **38**:1902–1911.

**Jackman, J. E., C. R. H. Raetz, and C. A. Fierke.** 2001. Site-directed mutagenesis of the bacterial metalloamidase UDP-(3-*O*-acyl)-*N*-acetylglucosamine deacetylase (LpxC). Identification of the zinc binding site. *Biochemistry* **40**:514–523.

**Onishi, H. R., B. A. Pelak, L. S. Gerckens, L. L. Silver, F. M. Kahan, M. H. Chen, A. A. Patchett, S. M. Galloway, S. A. Hyland, M. S. Anderson, and C. R. H. Raetz.** 1996. Antibacterial agents that inhibit lipid A biosynthesis. *Science* **274**:980–982.

### Bacterial Genome Sequencing and Antibacterial Drug Discovery

General

**Brown, T. A.** 1999. *Genomes.* John Wiley & Sons, Inc., New York, N.Y.

Bacterial genomes

**Atlas, R. M., D. Drell, and C. Fraser.** Bacterial genomes. Accepted for publication in *Encyclopedia of Life Sciences,* Nature Publishing Group, London, United Kingdom. [Online.] http://www.els.net.

**Buysse, J. M.** 2001. The role of genomics in antibacterial target discovery. *Curr. Pharm. Des.* **8**:1713–1726.

**Mills, S. D.** 2003. The role of genomics in antimicrobial discovery. *J. Antimicrob. Chemother.* **51**:749–752.

**Rosamond, J., and A. Allsop.** 2000. Harnessing the power of the genome in the search for new antibiotics. *Science* **287**:1973–1976.

Genome sequence

**Bentley, S. D., K. F. Chater, A. M. Cerdeño-Tearraga, G. L. Challis, N. R. Thomson, K. D. James, D. E. Harris, M. A. Quail, H. Kieser, D. Harper, A. Bateman, S. Brown, G. Chandra, C. W. Chen, M. Collins, A. Cronin, A. Fraser, A. Goble, J. Hidalgo, T. Hornsby, S. Howarth, C. H. Huang, T. Kieser, L. Larke, L. Murphy, K. Oliver, S. O'Neil, E. Rabbinowitsch, M. A. Rajandream, K. Rutherford, S. Rutter, K. Seeger, D. Saunders, S. Sharp, R. Squares, S. Squares, K. Taylor, T. Warren, A. Wietzorrek, J. Woodward, B. G. Barrel, J. Parkhill, and D. A. Hopwood.** 2002. Complete genome sequence of the model actinomycete *Streptomyces coelicolor* A4(2). *Nature* **417**:141–147.

**Fleischmann, R. D., M. D. Adams, O. White, R. A. Clayton, E. F. Kirkness, A. R. Kerlavage, C. J. Bult, J. F. Tomb, B. A. Dougherty, J. M. Merrick, K. McKenney, G. Sutton, W. FitzHugh, C. Fields, J. D. Gocayne, J. Scott, R. Shirley, L. Liu, A. Glodek, J. M. Kelley, J. F. Weidman, C. A. Phillips, T. Spriggs, E. Hedblom, M. D. Cotton, T. R. Utterback, M. C. Hanna, D. T. Nguyen, D. M. Saudek, R. C. Brandon, L. D. Fine, J. L. Fritchman, J. L. Fuhrmann, N. S. M. Geoghagen, C. L. Gnehm, L. A. McDonald, K. V. Small, C. M. Fraser, H. O. Smith, and J. C. Venter.** 1995. Whole-genome random sequencing and assembly of *Haemophilus influenzae* Rd. *Science* **269**:496–512.

**Kuroda, M., T. Ohta, I. Uchiyama, T. Baba, H. Yuzawa, I. Kobayashi, L. Cui, A. Oguchi, K. Aoki, Y. Nagai, J. Lian, T. Ito, M. Kanamori, H. Matsumaru, A. Maruyama, H. Murakami, A. Hosoyama, Y. Mizutani-Ui, N. K. Takahashi, T. Sawano, R. Inoue, C. Kaito, K. Sekimizu, H. Hirakawa, S.

**Kuhara, S. Goto, J. Yabuzaki, M. Kanehisa, A. Yamashita, K. Oshima, K. Furuya, C. Yoshino, T. Shiba, M. Hattori, N. Ogasawara, H. Hayashi, and K. Hiramatsu.** 2001. Whole genome sequencing of methicillin-resistant *Staphylococcus aureus*. *Lancet* **357**:1225–1240.

**McClelland, M., K. E. Sanderson, J. Spieth, S. W. Clifton, P. Latreille, L. Courtney, S. Porwollik, J. Ali, M. Dante, F. Du, S. Hou, D. Layman, S. Leonard, C. Nguyen, K. Scott, A. Holmes, N. Grewal, E. Mulvaney, E. Ryan, H. Sun, L. Florea, W. Miller, T. Stoneking, M. Nhan, R. Waterston, and R. K. Wilson.** 2001. Complete genome sequence of *Salmonella enterica* serovar Typhimurium LT2. *Nature* **413**:852–856.

**Parkhill, J., G. Dougan, K. D. James, N. R. Thomson, D. Pickard, J. Wain, C. Churcher, K. L. Mungall, S. D. Bentley, M. T. G. Holden, M. Sebalhla, S. Baker, D. Basham, K. Brooks, T. Chillingworth, P. Connerton, A. Cronin, P. Davis, R. M. Davis, L. Dowd, N. White, J. Farrar, T. Feltwell, N. Hamlin, A. Haque, T. T. Hien, S. Holroyd, K. Jagels, A. Krogh, T. S. Larsen, S. Leather, S. Moule, P. O'Gaora, C. Parry, M. Quail, K. Rutherford, M. Simmonds, J. Skelton, K. Stevens, S. Whitehead, and B. G. Barrel.** 2001. Complete genome sequence of a multiple drug resistant *Salmonella enterica* serovar Typhi CT18. *Nature* **413**:848–852.

**Stover, C. K., X. Q. Pham, A. L. Erwin, S. D. Mizoguchi, P. Warrener, M. J. Hickey, F. S. L. Brinkman, W. O. Hufnagle, D. J. Kowalik, M. Lagrou, R. L. Garber, L. Goltry, E. Tolentino, S. Westbrock-Wadman, Y. Yuan, L. L. Brody, S. N. Coulter, K. R. Folger, A. Kas, K. Larbig, R. Lim, K. Smith, D. Spencer, G. K. S. Wong, Z. Wu, I. T. Paulsen, J. Reizer, M. H. Saier, R. E. W. Hancock, S. Lory, and M. V. Olson.** 2000. Complete genome sequence of *Pseudomonas aeruginosa* PAO1, an opportunistic pathogen. *Nature* **406**:959–964.

**Tettelin, H., K. E. Nelson, I. T. Paulsen, J. A. Eisen, T. D. Read, S. Peterson, J. Heidelberg, R. T. DeBoy, D. H. Haft, R. J. Dodson, A. S. Durkin, M. Gwinn, J. F. Kolonay, W. C. Nelson, J. D. Peterson, L. A. Umayam, O. White, S. L. Salzberg, M. R. Lewis, D. Radune, E. Holtzapple, H. Khouri, A. M. Wolf, T. R. Utterback, C. L. Hansen, L. A. McDonald, T. V. Feldblyum, S. Anginoli, T. Dickinson, E. K. Hickey, I. E. Holt, B. J. Loftus, F. Yang, H. O. Smith, J. C. Venter, B. A. Dougherty, D. A. Morrison, S. K. Hollingshead, and C. M. Fraser.** 2001. Complete genome sequence of a virulent isolate of *Streptococcus pneumoniae*. *Science* **293**:498–506.

### Identification of Antibacterial Agents through Combinatorial Chemistry

**Blondelle, S. E., and R. A. Houghten.** 1996. Novel antimicrobial compounds identified using synthetic combinatorial library technology. *Trends Biotechnol.* **14**:60–65.

**Trias, J.** 2001. The role of combichem in antibiotic discovery. *Curr. Opin. Microbiol.* **4**:520–525.

### How the Pharmaceutical Industry Brings an Antibacterial Drug to the Market in the United States

**Billstein, S. A.** 1994. How the pharmaceutical industry brings an antibiotic drug to market in the United States. *Antimicrob. Agents Chemother.* **38**:2679–2682.

**Chopra, I.** 1998. Research and development of antibacterial agents. *Curr. Opin. Microbiol.* **1**:495–501.

# Epilogue

## The Debt of Humanity to Oxford's Scientists

Humanity owes a debt to the Oxford scientists who discovered the chemical structure of the penicillin antibiotic and its application to treat human infections caused by bacteria.

Howard Florey, Ernest Chain, and other scientists at the Sir William Dunn School of Pathology, Oxford University, Oxford, United Kingdom, succeeded in partly purifying penicillin for the first time from a culture of Fleming's *Penicillium notatum* and demonstrated, first in mice and then in humans, that it had remarkable therapeutic properties. They published their results in 1940. Evidence of the great clinical usefulness of penicillin in humans followed in 1941. It was this demonstration of the clinical value of penicillin that stimulated the governments of Britain and the United States to support efforts by pharmaceutical companies to develop methods for large-scale production. Also in 1940, Abraham and Chain described the first bacterial enzyme that destroyed penicillin; they termed it penicillinase, since the β-lactam structure of penicillin was not then known.

One of the most important landmarks in the story of penicillin was the X-ray crystallographic determination of the structure of penicillin (currently named penicillin G) by Dorothy Hodgkin and Barbara W. Roger-Low in May 1945 at the Crystallography Laboratory of Oxford University. They established unequivocally for the first time the fused nature of the bicyclic β-lactam–thiazolidine nucleus and the relative stereochemistry of the substituents of the penicillin molecule.

The Albert and Mary Lasker Foundation presented a tribute to many of these scientists in June 1953 with a monument in the Botanic Garden of Oxford University (see figure) with the following inscription.

*This rose garden was given in honor of the research workers in the University who discovered the clinical importance of penicillin for saving life, relief of suffering and inspiration to further research. All mankind is in their debt. Those who did this work were:*

| | |
|---|---|
| *E. P. Abraham* | *E. Chain* |
| *C. M. Fletcher* | *H. W. Florey* |
| *N. G. Heatley* | *A. D. Gardner* |
| *I. Orr-Ewing* | *M. A. Jennings* |
| *M. E. Florey* | *A. G. Sanders* |

Currently the Albert and Mary Lasker Foundation Inc., is at 110 East 42nd St., Room 1300, New York, NY 10017. The web site is http://www.laskerfoundation.org.

# Appendix A

# Amino Acids, the Building Blocks of Proteins

## A Few Words about Nomenclature

The IUPAC-IUB rules for the nomenclature and symbolism for amino acids and peptides can be retrieved free of charge from http://www.chem.qmw.ac.uk/iupac/AminoAcid/ (website prepared by G. P. Moss of the Department of Chemistry, Queen Mary and Westfield College, London, United Kingdom).

The one- and three-letter abbreviations for the 20 amino acid residues are given in Fig. A1. Amino acid residues in polypeptides are named by dropping the suffix "-ine" in the name of the amino acid and replacing it by "-yl." Polypeptide chains are described by starting at the amino terminus (known as the **N terminus**) and sequentially naming each residue until the carboxyl terminus (the **C terminus**) is reached. When using the one- or three-letter abbreviations, the name of the polypeptide chain is always written so that the N terminus is at the left and the C terminus is at the right.

## Configuration at the α-Carbon Atom

In 19 of the 20 standard amino acids, the α-carbon atom is **chiral,** or asymmetric, since it has four different groups bonded to it. (The exception is glycine, in which the R group is simply a hydrogen atom, so that two hydrogen atoms are bonded to the α-carbon). Therefore, amino acids can exist as stereoisomers, compounds with the same molecular formula that differ in their **configuration,** the arrangement of their atoms in space. Two stereoisomers that are nonsuperimposable mirror images of each other can exist for each chiral amino acid. Such stereoisomers are called **enantiomers.** All amino acids with at least one chiral center are also **optically active.** In other words, enantiomers are optically active; i. e., they rotate the plane of polarized light. A minus sign in parentheses (–) is used for levorotatory (rotating light to the left) compounds, and a plus sign in parentheses (+) is used for dextrorotatory (rotating light to the right) compounds.

A molecule may have multiple asymmetric centers. The term "enantiomer" still refers to a molecule that is the nonsuperimposable mirror image of the one under consideration. Since each asymmetric center in a chiral molecule can have two possible configurations, a molecule with $n$ chiral centers has $2^n$ different possible stereoisomers and $2^{n-1}$ enantiomeric pairs. The amino acids threonine and isoleucine each have two chiral centers and hence $2^2$ (i.e., 4) possible stereoisomers. Pairs of stereoisomers that are not mirror images of each other are called **diastereoisomers.** A special kind of diastereoisomer exists when the two asymmetric centers are chemically identical. This is called a *meso* form; it is a stereoisomer which is not chiral because the two stereocenters cancel each other out.

Special nomenclature has been developed to specify the configuration of the four substituents of chiral (or

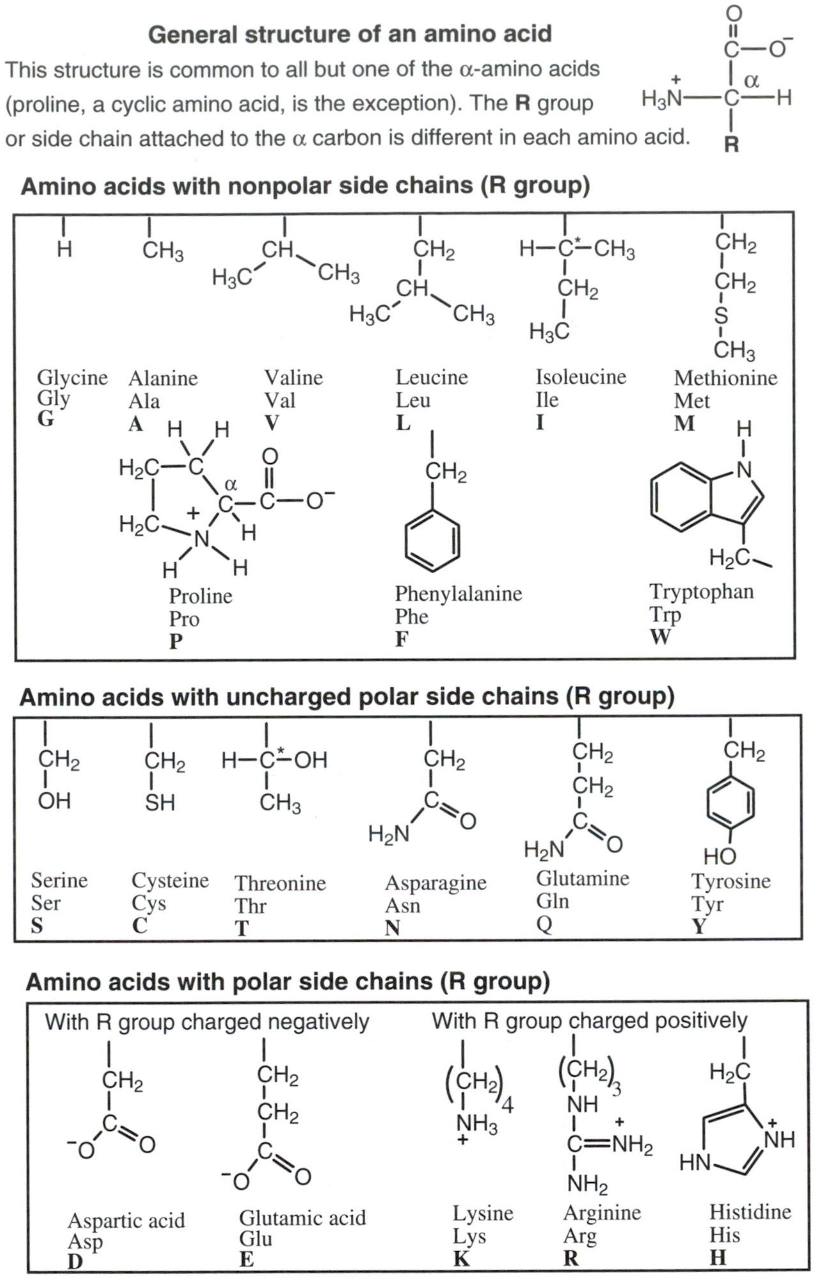

**General structure of an amino acid**

This structure is common to all but one of the α-amino acids (proline, a cyclic amino acid, is the exception). The **R** group or side chain attached to the α carbon is different in each amino acid.

**Amino acids with nonpolar side chains (R group)**

Glycine Gly **G**; Alanine Ala **A**; Valine Val **V**; Leucine Leu **L**; Isoleucine Ile **I**; Methionine Met **M**; Proline Pro **P**; Phenylalanine Phe **F**; Tryptophan Trp **W**

**Amino acids with uncharged polar side chains (R group)**

Serine Ser **S**; Cysteine Cys **C**; Threonine Thr **T**; Asparagine Asn **N**; Glutamine Gln Q; Tyrosine Tyr **Y**

**Amino acids with polar side chains (R group)**

With R group charged negatively: Aspartic acid Asp **D**; Glutamic acid Glu **E**

With R group charged positively: Lysine Lys **K**; Arginine Arg **R**; Histidine His **H**

**Figure A1** Structures and abbreviations of the 20 "standard" amino acids. The ionic forms (zwitterions) shown are those predominant at pH 7.0. The Cα atoms, as well as those atoms marked with an asterisk, are chiral centers (also called stereocenters), but the spatial orientation is not indicated. The one-letter symbol for an undetermined or "nonstandard" amino acid is X.

asymmetric) carbon atoms. The configurations of amino acids are specified by the DL **system** of Fischer (Fig. A2).

## The Fischer Convention

Based on the configuration of the three-carbon compound glyceraldehyde, a convention was proposed by Emil Fischer in 1891. In this system, the configuration of the groups around an asymmetric center is related to glyceraldehyde. Fischer assumed that the configurations of these molecules were as shown in Fig. A3. In the Fischer convention, horizontal bonds extend above the plane of the paper and vertical bonds extend below the plane of the paper.

## The *RS* (Cahn-Ingold-Prelog) Convention for Designating Stereoisomer Configurations

In the *RS* system, the four groups surrounding a chiral carbon center are prioritized according to a specific scheme, as follows. (i) Atoms of higher atomic number bonded to a chiral center are ranked above those of

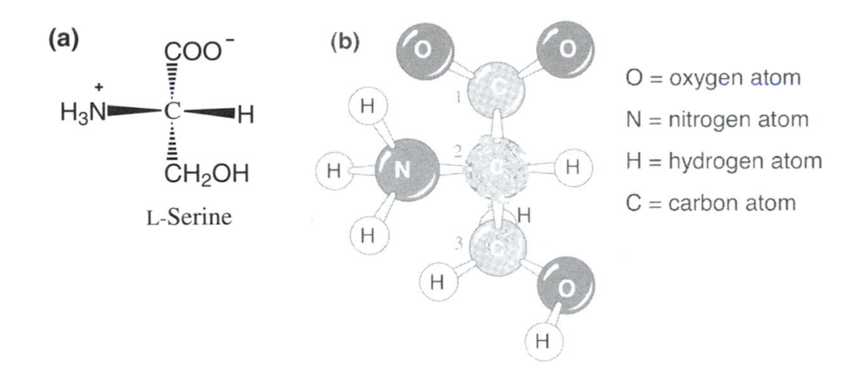


**Figure A2** Two representations of an α-amino acid at neutral pH. (a) General structure. (b) Three-dimensional ball-and-stick drawing of serine. Note the alternative numbering and lettering systems for designating carbon atoms.

lower atomic number. If any of the first substituent atoms are of the same element, the priority between groups is established from the atomic number of the second, third, etc., atoms outward from the asymmetric center. Hence, a -COOH group takes precedence over $-CH_2OH$. (ii) If an atom is bound by a double or triple bond, the atom is counted once for each formal bond. Thus, -CHO, with a double-bonded oxygen atom, has a higher priority than $-CH_2OH$.

To establish the configuration of the chiral center, it is viewed from the asymmetric center toward the lowest-priority group (numbered 1) (Fig. A4). If the movement

**Figure A3** The Fischer convention for naming the enantiomers of glyceraldehyde and serine as represented by Fischer projection formulae. Note that in a Fischer projection, all horizontal bonds point above the page and all vertical bonds point below the page. The mirror planes relating the enantiomers are represented by a vertical broken line. Sometimes Fischer projection formulae omit the central chiral carbon atom.

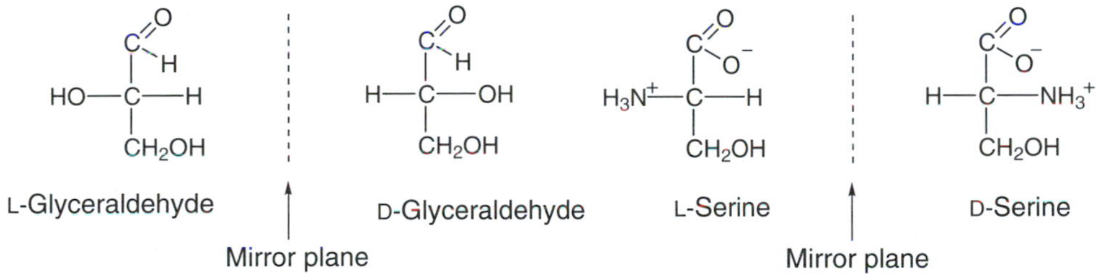

L-Glyceraldehyde | D-Glyceraldehyde | L-Serine | D-Serine

Mirror plane | Mirror plane

**Figure A4** Configuration based on the *RS* system. (a) Using the rules described in the text, each group attached to a chiral carbon is assigned a priority, with 1 being the lowest priority and 4 being the highest. (b) By orienting the molecule with the priority 1 group pointing away and tracing the path from the highest-priority group to the lowest, the configuration can be established. If the sequence 4 → 3 → 2 is clockwise, the configuration is *R*. If the sequence 4 → 3 → 2 is counterclockwise, the configuration is *S*. L-Serine has the *S* configuration.

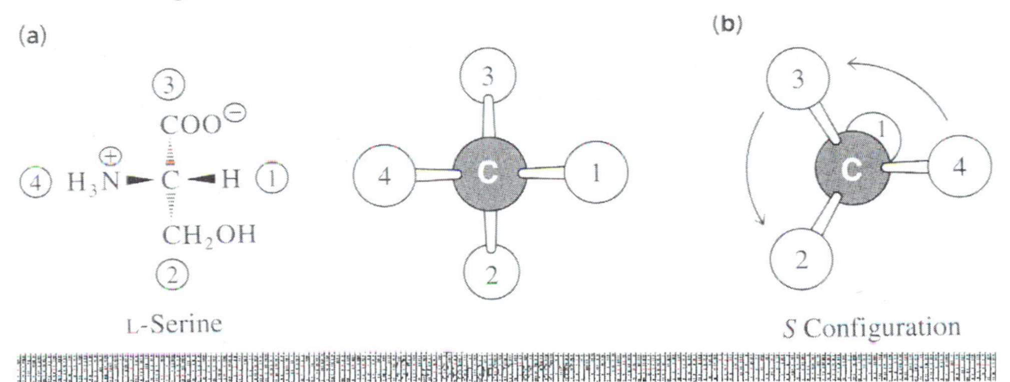

{—Phe—Ala—Phe—Ala—Ser—Thr—Tyr—Lys—{

**Figure A5** The primary structure describes the linear sequence of amino acid residues in a protein.

from highest to lower priority with the three higher-priority groups facing the observer (4 → 3 → 2) is counterclockwise, the molecule is defined as *S* (Latin, *sinister*, left-handed). If the movement is clockwise, the molecule is defined as *R* (Latin, *rectus*, right-handed). Fig. A4 demonstrates the assignment of the configuration to L-serine in the *RS* system.

## The Structural Hierarchy of Proteins

There are four levels of protein structure. **Primary structure** describes the linear sequence of amino acid residues in a protein (Fig. A5). The three-dimensional structure of a protein is described by three additional levels: **secondary structure, tertiary structure,** and **quaternary structure.** The forces responsible for maintaining or stabilizing these three levels are primarily noncovalent. By formation of tertiary and quaternary structures, amino acids far apart in the sequence can be brought close together in three dimensions to form a functional region, an **active site.** The major secondary structures are α-helices and β-strands (including β-sheets). Schematic representations showing the structures of folded proteins usually represent α-helical regions by helices and β-strands by broad arrows pointing in the N-terminal to C-terminal direction (Fig. A6). **Motifs** or supersecondary structures are combinations of the α and/or β structure.

Tertiary structure describes the completely folded polypeptide chain. Several motifs usually combine to form compact globular structures, which are called **domains** (Color Plate A1 [see color insert]).

Protein molecules that have only one chain are called **monomeric** proteins. However, a fairly large number of proteins have a quaternary structure, which consists of at least two polypeptide chains (subunits) that associate into a **multimeric** molecule in a specific way. For example, human immunodeficiency virus type 1 protease has two identical β-subunits that bind symmetrically (Color Plate A2 [see color insert]). Human immunodeficiency virus protease is the target of many new drugs designed to treat AIDS patients.

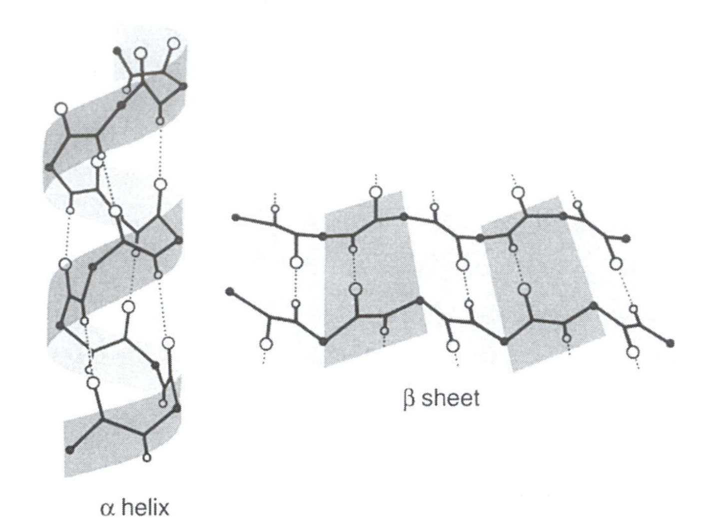


**Figure A6** The secondary structure consists of regions of regularly repeating conformations of the peptide chain, such as α-helices and β-sheets.

### REFERENCES AND FURTHER READING

#### Books

**Branden, C., and J. Tooze.** 1999. *Introduction to Protein Structure*, 2nd ed. Garland Publishing, New York, N.Y.

**Fersht, A.** 1999. *Structure and Mechanism in Protein Science.* W. H. Freeman & Co., New York, N.Y.

**Lesk, A. M.** 2001. *Introduction to Protein Architecture.* Oxford University Press, Oxford, United Kingdom.

#### Human Immunodeficiency Virus Type 1 Protease

**De Clercq, E.** 1995. Toward improved anti-HIV chemotherapy: therapeutic strategies for intervention with HIV infections. *J. Med. Chem.* 38:2491–2517.

#### Crystal Structure of Orotidine 5'-Monophosphate Decarboxylase

**Houk, K. N., J. K. Lee, D. J. Tantillo, S. Bahmanyar, and B. H. Hietbrink.** 2001. Crystal structures of orotidine monophosphate decarboxylase: does the structure reveal the mechanism of nature's most proficient enzyme? *ChemBioChem* **2:** 113–118.

# Appendix B

# Enzyme Catalysis and Enzyme Inhibition

This appendix presents the basic principles of enzyme inhibition in drug design. For details, readers are referred to specialist texts on the subject.

## Enzyme Catalysis

### Enzyme Terminology

A **catalyst** is a substance that accelerates a chemical reaction without itself undergoing any net change. An **enzyme** is a catalyst, and consequently it cannot alter the equilibrium of a chemical reaction. This means that $k_{cat}$, which is a function of the reaction rate ($V_{max}$) and the binding constant ($k_M$), is the same for the forward and the reverse reaction. The reactant(s) whose reaction is catalyzed by an enzyme is known as the **substrate(s).** Enzymes are protein molecules. A number of enzymes contain metal ions, coordinated to the peptide chain; these enzymes are called **metalloenzymes.** Enzymes are broadly classified into six major types (Table B1).

An enzyme-catalyzed reaction proceeds rapidly under mild conditions because the enzyme lowers the activation energy; enzymes are usually highly specific for the reaction they catalyze. Although enzymes may contain several hundred amino acid residues, only a few of them are usually chemically involved in the bond-making and bond-breaking steps of the transformation catalyzed by the enzyme. The specific part of an enzyme that binds a substrate is called the **active site.** Active sites are usually visualized as pockets or clefts. These physical features are determined by the conformation of the protein's peptide chain. The main purpose of the bulk of the amino acid residues is to provide the **three-dimensional framework** required to maximize the binding energy between the substrate and the enzyme.

### Enzymatic Reaction Mechanisms Are Described by the Terminology of Organic Chemistry

The mechanism of both nonenzymatic and enzyme-catalyzed reactions is a description of the atomic or molecular events that occur during the reaction, in as much useful detail as possible. The reactant(s), product(s), and any intermediates must be identified. The same symbolism used in organic chemistry to represent chemical-bond transformations is employed in representing enzymatic reaction mechanisms. In ionic reactions, two species are involved: one is electron rich, or **nucleophilic,** and the other is electron poor, or **electrophilic.** A nucleophile, which has a negative charge or an unshared electron pair, attacks the electrophilic center of the other reactant. For example, in a nucleophilic substitution reaction at an $sp^3$ carbon center, the reaction involves two molecules reacting in the initial, slow step (nucleophilic bimolecular substitution reaction, or $S_N2$ mechanism). The nucleophilic attacking group is added to the face of the carbon atom opposite the leaving group to form a transition state that has five groups attached to

**Table B1** Classification of enzymes

| Code | Classification | Type of reaction catalyzed |
|---|---|---|
| 1 | Oxidoreductases | Oxidations and reductions |
| 2 | Transferases | Transfer of a group from one molecule to another |
| 3 | Hydrolases | Hydrolysis of various functional groups |
| 4 | Lyases | Cleavage of a bond by nonoxidative and nonhydrolytic mechanisms |
| 5 | Isomerases | Interconversion of all types of isomers |
| 6 | Ligases (synthases) | Formation of a bond between molecules |

the central carbon atom. This **transition state,** shown in square brackets in Fig. B1, is an unstable high-energy state. It has a structure between those of the reactant and the product.

**Figure B1** Hypothetical transition state for an enzymatically catalyzed $S_N2$ reaction.

Transition state

## Enzymes Accelerate Reactions by Stabilizing Transition States

In the transition state, chemical bonds are in the process of being made and broken. The transition state occurs at the peak in the reaction diagram (Fig. B2), in which the energy of the species involved is plotted against the course of the reaction. In contrast, **intermediates,** in which the bonds are fully formed, occupy troughs in the reaction diagram. Transition states are not detectable experimentally, but their structures can be predicted. They have a lifetime of about $10^{-13}$ s, the time for one bond vibration. Intermediates are compounds that can be detected or isolated if they are sufficiently stable.

## Formation of an Enzyme-Substrate Complex Is the First Step in Enzymatic Catalysis

An enzyme-catalyzed reaction always is initiated by the formation of an **enzyme-substrate complex** (E-S complex), from which the catalysis proceeds. In 1958, Koshland proposed the **induced-fit hypothesis,** which predicts that when a substrate begins to bind to an enzyme, interactions between various groups in the substrate and particular enzyme functional groups are initiated and that these mutual interactions induce a **conformational change** in the enzyme. This results in a change of the enzyme from a low catalytic form to a high catalytic form by destabilizing the enzyme and/or induc-

**Figure B2** (a) Reaction diagram for a single-step reaction. This curve shows the lowest-energy path between the substrate and the product. The upper double-headed arrow shows the energy of activation. Whether it is spontaneous depends on the activation energy. The reaction is exergonic, i.e., the equilibrium favors the product. (b) Reaction diagram for a reaction with an intermediate. The metastable intermediate occurs in the trough between two transition states. The rate-determining step in the forward reaction is the formation of the first transition state, the step that has the higher activation energy.

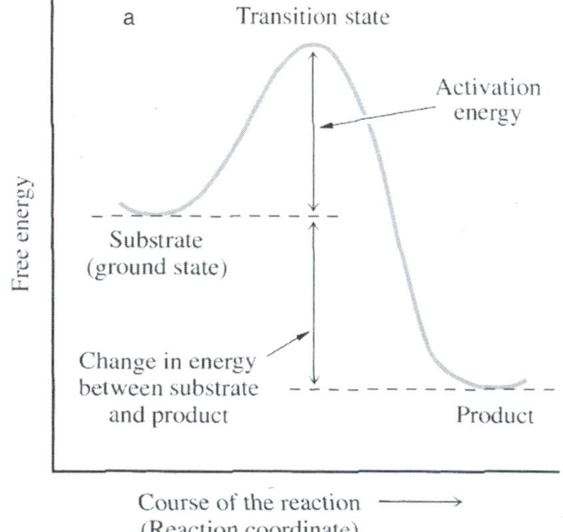


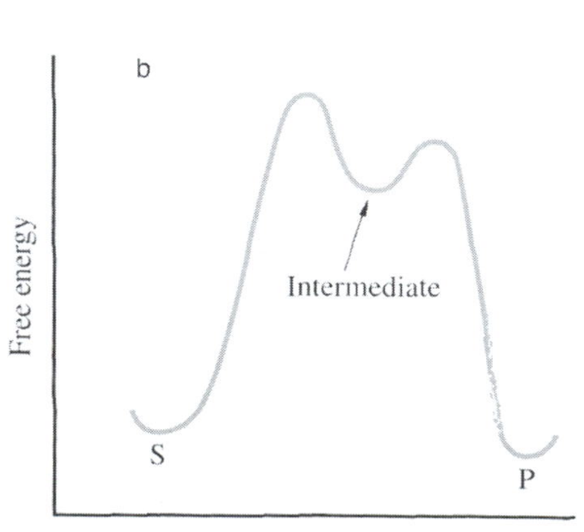

ing proper alignments of the groups involved in catalysis.

**Site-directed mutagenesis** refers to a process by which an amino acid residue in the enzyme is genetically changed to a different amino acid.

## Enzyme Kinetics (Definitions Only)

For derivations of the equations of enzyme kinetics, the reader is referred to a textbook of biochemistry. As mentioned above, enzyme catalysis is initiated by an interaction between the enzyme and the substrate, to form an E-S complex or Michaelis complex (Fig. B3). The driving force for the interaction of the substrate with the enzyme results from covalent and noncovalent interactions (discussed below).

When $k_2 << k_{-1}$, we refer to the term $k_2$ as $k_{cat}$ (the catalytic rate constant); this means that the dissociation of the E-S complex to E and S is more rapid than the formation of enzyme and product. The dissociation constant, $K_s = k_{-1}/k_1$, is called the $K_m$ (the Michaelis-Menten constant). When this condition is met, $K_m$ is a measure of the strength of the E-S complex: a high $K_m$ value indicates weak binding, and a low $K_m$ value indicates strong binding.

**Figure B3** Equation showing the formation of the Michaelis (E-S) complex.

$$E + S \underset{k_{-1}}{\overset{k_1}{\rightleftharpoons}} E\text{—}S \underset{k_{-2}}{\overset{k_2}{\rightleftharpoons}} E + P$$

The **turnover number** of an enzyme is the number of substrate molecules converted into product by an enzyme molecule in a unit of time when the enzyme is fully saturated with substrate.

## Basis for Enantioselectivity of Enzymes

When an enzyme is exposed to a racemic mixture of substrate, the binding energy for E-S complex formation with one enantiomer may be much higher than that with the other enantiomer, either because of differential binding interactions or for steric reasons. For example, the racemic mixture of lactic acid is represented by the three-dimensional shapes shown in Fig. B4 and the sections of the active site that bind to these groups are represented as the correspondingly shaped sockets. In Fig. B4a, the relevant groups fit into these sockets; that is, the structures are complementary. The groups of L-lactic acid are correctly aligned for binding. However, in Fig. B4b, two of the groups are misaligned. The wedge and cylinder cannot fit their sockets, and so the groups of D-lactic acid are not correctly aligned for binding to the active site.

## Specific Forces Involved in Enzyme-Substrate Complex Formation

The forces acting between a substrate and an enzyme in forming the E-S complex are predominantly weak nonco-

**Figure B4** Schematic representation of the stereospecific nature of enzymes (see the text for an explanation). Reprinted from G. Thomas, *Medicinal Chemistry* (John Wiley & Sons, Inc., New York, N.Y., 2000), with permission from the publisher.

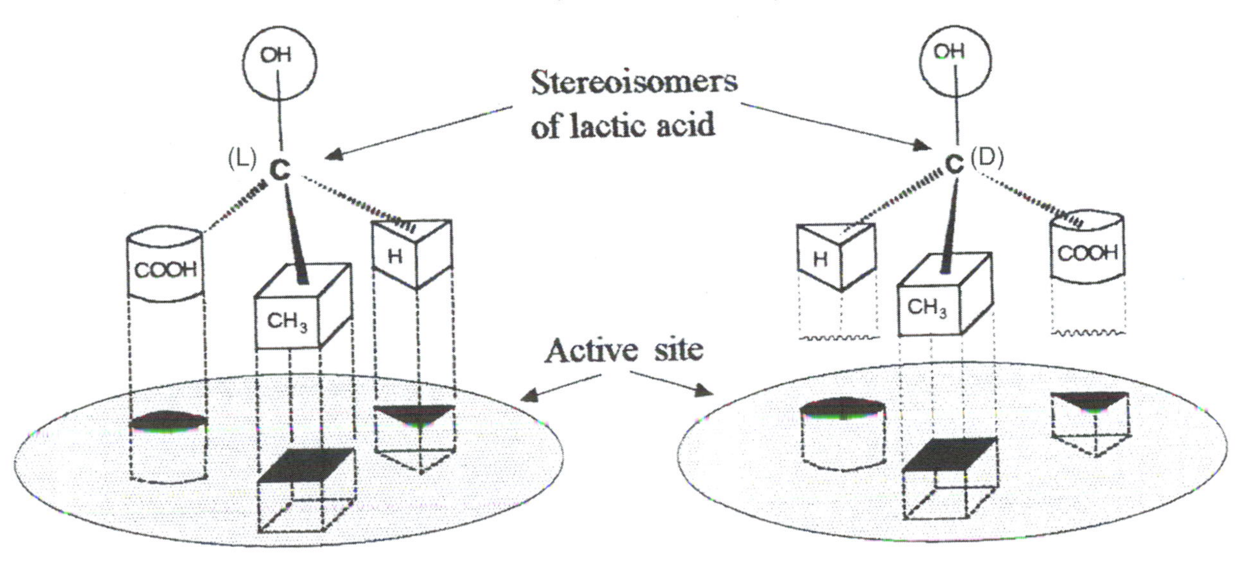

valent interactions; consequently, the effects produced are reversible, which is very important for product release.

### Covalent Bonds

The covalent bond is the strongest bond, generally equivalent to –49 to 110 kcal/mol in stability. All the enzymes that utilize pyridoxal-5'-phosphate as a cofactor form a covalent E-S complex.

### Interactions between Noncovalently Bonded Atoms

The interactions between nonbonded atoms are not discussed, except to mention that they include

- ionic (or electrostatic) interactions
- ion-dipole and dipole-dipole interactions
- hydrogen bonds
- charge transfer complexes
- hydrophobic interactions
- van der Waals forces

Examples of these noncovalent interactions are shown in Fig. B5. Because several different types of interactions are involved, selectivity in enzyme-substrate interactions can result.

## Enzyme Inhibition

Any compound that slows or prevents enzyme catalysis is an enzyme inhibitor. This appendix reviews the basic concepts of enzyme inhibition and ignores aspects dealing with absorption, transport, metabolism, excretion, and toxicological problems associated with the development of an enzyme inhibitor as a successful drug.

A useful classification of enzyme inhibitors is based on three criteria:

1. what they structurally mimic (a single substrate or a transition state formed during catalysis)
2. where they bind (a catalytic [active] site or a noncatalytic site)
3. how they inhibit covalent binding (suicide [mechanism based] or group-directed) and noncovalent binding

In the first category, an inhibitor structurally resembles substrates or a transition state formed during catalysis. In the second category, enzyme inhibitors most

**Figure B5** Noncovalent interactions.

electrostatic (ionic)
ion-dipole
dipole-dipole
H-bonding
charge transfer
hydrophobic

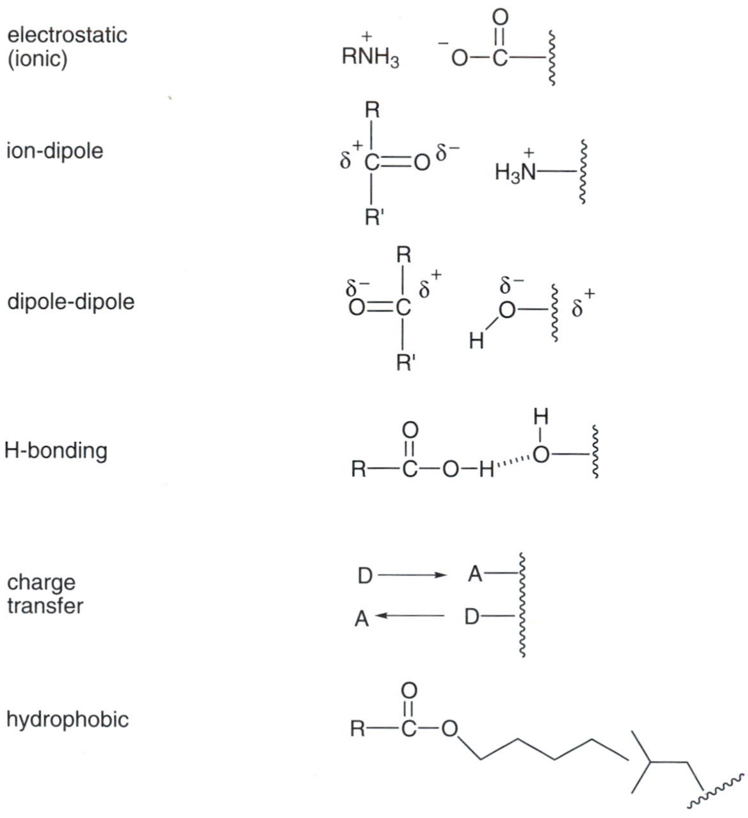

commonly bind to the active site. Among the covalently binding inhibitors, two types can be distinguished: suicide and group-directed inhibitors.

Inhibitors can also be grouped into reversible and irreversible inhibitors. As the names imply, inhibition of enzyme activity by a reversible inhibitor is reversible, suggesting the presence of noncovalent interactions between the enzyme and the inhibitor. An irreversible inhibitor (also called an inactivator) is one that prevents the return of enzyme activity for an extended period, suggesting the presence of covalent bond formation.

## Reversible Inhibitors

The kinetics of the reversible inhibition of enzymes are that the enzyme-inhibitor (E-I) complex regenerates the inhibitor and enzyme as follows:

$$E + I \underset{k_{-1}}{\overset{k_1}{\rightleftharpoons}} E\text{-}I$$

Generally, reversible inhibition is associated with noncovalent bond formation. There are three main types of reversible inhibitors, which can be distinguished experimentally by their effects on the kinetic behavior of enzymes: competitive inhibitors, noncompetitive inhibitors, and uncompetitive inhibitors.

### COMPETITIVE INHIBITORS BIND ONLY TO FREE ENZYME

Competitive inhibitors are the most commonly encountered inhibitors in biochemistry. In competitive inhibition, the inhibitor can bind only to unliganded molecules of enzyme. If the inhibitor binds to the active site of the enzyme and prevents substrate binding and vice versa, the inhibitor and substrate compete for the active site and the inhibitor is said to be a **competitive inhibitor.**

Substrate analogs are compounds that can bind to enzymes because of their similarity to substrates. These compounds are usually competitive inhibitors. An example of a nonmetabolizable substrate analog is benzamidine, a competitive inhibitor of trypsin. Trypsin catalyzes the hydrolysis of the peptide bond on the carboxyl side of arginine and lysine, and benzamidine is an analog of arginine that resembles the alkylguanidyl side chain of the amino acid. Benzamidine is a competitive inhibitor since it competes with its analogous substrate for the binding pocket in the active site. By systematically varying the structure of the substrate analog and measuring the inhibitory potencies, one can sometimes draw conclusions about the structure of the active site of an enzyme.

### NONCOMPETITIVE INHIBITORS BIND TO BOTH ENZYMES AND THE ENZYME-SUBSTRATE COMPLEX

Noncompetitive inhibitors bind to both the enzyme and the E-S complex, so that E-I and E-S-I complexes—both inactive—can be formed. In noncompetitive inhibition, the inhibitor and substrate bind simultaneously to the enzyme instead of competing for the same binding site. Noncompetitive inhibiton is rare, but examples are known among allosteric enzymes.

### UNCOMPETITIVE INHIBITORS BIND ONLY TO THE ENZYME-SUBSTRATE COMPLEX

Uncompetitive inhibitors bind only to E-S complexes, not to the free enzyme. Uncompetitive inhibition is rare in reactions that involve a single substrate but more common in reactions with multiple substrates.

## Irreversible Inhibitors

The various types of inhibition that we have discussed so far are all reversible. There are, however, numerous inhibitors that react essentially irreversibly with enzymes, usually by the formation of a covalent bond to the functional group of an amino acid side chain or to a bound coenzyme.

An unreactive compound with structural similarity to a substrate for an enzyme becomes reactive after it has been modified chemically by the enzyme to a species that usually forms a covalent bond to the enzyme, producing inactivation. Such a reagent is termed a **mechanism-based inhibitor** or **suicide substrate,** to emphasize that the enzyme brings about its own inhibition.

Mechanism-based enzyme inactivators are most useful for the study of enzyme mechanisms, because they are really nothing more than just substrates for the enzymes.

Figures 11.2, 11.4, and 11.6 present examples of β-lactamase inactivation by mechanism-based reagents (clavulanic acid, sulbactam, and tazobactam).

## REFERENCES AND FURTHER READING

**Beynon, R., and J. S. Bond.** 2001. *Proteolytic Enzymes,* 2nd ed. Oxford University Press, Oxford, United Kingdom.

**Bugg, T. D. H.** 1997. *An Introduction to Enzyme and Coenzyme Chemistry.* Blackwell Scientific Publications, Oxford, United Kingdom.

**Bugg, T. D. H.** 2001. The development of mechanistic enzymology in the 20th century. *Nat. Prod. Rep.* **18:**465–493.

**Copeland, R. A.** 2000. *Enzymes. A Practical Introduction to Structure, Mechanism, and Data Analysis.* John Wiley & Sons, Inc., New York, N.Y.

**Faber, K.** 2000. *Biotransformation in Organic Chemistry,* 4th ed. Springer-Verlag KG, Berlin, Germany.

**Fersht, A.** 1999. *Structure and Mechanism in Protein Science.* W. H. Freeman & Co., New York, N.Y.

**Palmer, T.** 1985. *Understanding Enzymes,* 2nd ed. Ellis Horwood, Chichester, United Kingdom.

**Silverman, R. B.** 2000. *The Organic Chemistry of Enzyme Catalyzed Reactions.* Academic Press, Inc., San Diego, Calif.

**Suckling, C. J., C. L. Gibson, and A. R. Pitt.** 1998. *Enzyme Chemistry. Impact and Applications,* 3rd ed. Blackie Academic & Professional/Kluwer Academic Publishing, London, United Kingdom.

**Thomas, G.** 2000. *Medicinal Chemistry.* John Wiley & Sons, Inc., New York, N.Y.

# Appendix C

# Lipids and Membranes

## Lipids

It seems appropriate to consider briefly the lipid biomolecules. Lipids are either hydrophobic (nonpolar) or amphipathic (containing both nonpolar and polar groups). Figure C1 shows the major types of lipids and their relationships to one another. Note that lipids containing phosphate moieties are grouped into the general category of phospholipids. A variety of amphipathic lipids, including glycerophospholipids and sphingolipids, serve as important structural components of all biological membranes.

In some cases, lipids perform their biological functions as individual molecules; in other cases, they interact with other biomolecules to function as part of a complex or aggregate. Lipid complexes include lipoproteins (composed of lipid and protein), and lipid aggregates include biological membranes (thin bilayers containing lipids, proteins, and sometimes carbohydrates).

## Fatty Acids

The term "fatty acids" denotes aliphatic monocarboxylic acids, generally with chains from 14 to 20 carbon atoms long (structures are shown in Table C1). The fatty acids are saturated or unsaturated (i.e., containing double bonds). The term "acyl" is used in chemical nomenclature to denote the group formed by loss of the hydroxyl group of the carboxylic acid function of any aliphatic acid. Fatty acids are named according to the International Union of Pure and Applied Chemistry (IUPAC) system for the nomenclature of organic chemistry. A list of trivial names is given in Table C1. Fatty acids are numbered starting with the carbon atom of the carboxyl group as C-1. By standard chemical convention, the ending "ate" (e.g., in palmitate) denotes any anion or ester. The nomenclature of unsaturated fatty acids involves a system of abbreviations exemplified as follows: *cis*-Δ9-octadecenoate means an 18-carbon chain length, with one double bond. The Greek letter Δ indicates a double bond, here between carbons C-9 and C-10, and *cis* denotes a double bond with a *cis* configuration. The fatty acid composition defines the physical properties of the membrane bilayer. Unsaturated fatty acids increase the fluidity of the membrane.

## Bacterial Fatty Acid Biosynthesis

In most bacteria, the biosynthesis of fatty acids is catalyzed by a multienzyme complex, denoted as a type II fatty acid synthetase. The enzymes of lipid synthesis are compartmentalized between the cytoplasm and the cytoplasmic membrane. Fatty acid synthesis is catalyzed by a multienzyme pathway in *Escherichia coli* and by a multifunctional enzyme in mammals.

Fatty acid synthesis proceeds on the acyl carrier protein (ACP) (Fig. C2). ACP carries the growing acyl chain from one enzyme to another and supplies precursors for

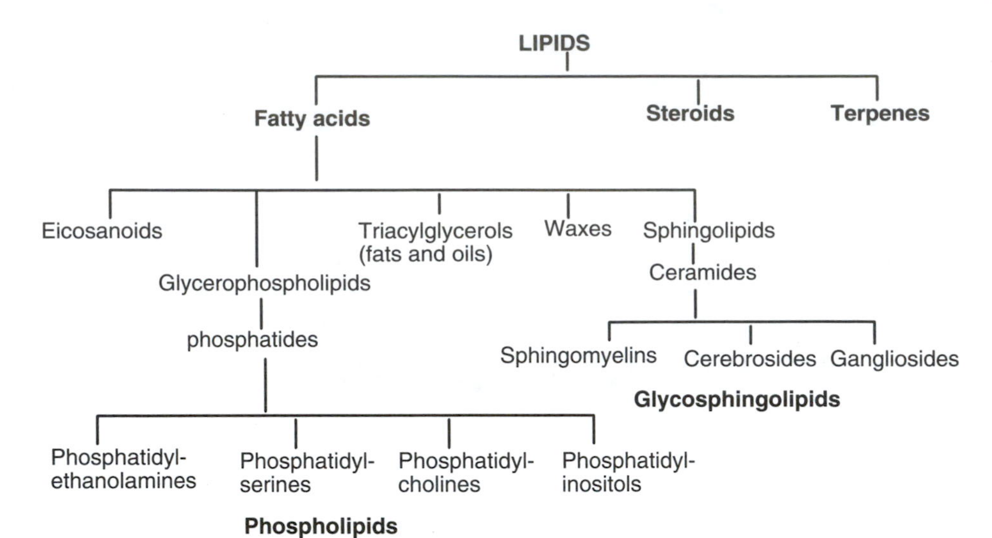


**Figure C1** Organization of the major types of lipids based on structural relationships.

the condensation reaction. The acyl intermediate is bound to the protein as a thioester attached to the prosthetic group phosphopantheine. The cycle of fatty acid synthesis consists of the following steps: transacylation, initiation, and elongation. In the transacylation step, acetyl coenzyme A (acetyl-CoA) and malonyl-CoA are transacylated to ACP by acyltransferases. Elongation involves four steps. These steps begin by the condensation of malonyl-ACP with acyl-ACP catalyzed by a β-ketoacyl-ACP synthetase (FabB or FabG). The resulting β-ketoacyl-ACP is then dehydrated to enoyl-ACP by either of two dehydratases (FabA or FabZ). The final steps in each cycle are catalyzed by the NADH-dependent enoyl-ACP reductase (FabI).

## Glycerophospholipids Are Major Components of Biological Membranes

In glycerophospholipids, the phosphate group is esterified to both glycerol and another compound bearing an OH group. Figure C3 identifies some of the families of glycerophospholipids. Note that glycerophospholipids are **amphipathic** molecules, having a polar head (with an anionic phosphate group and often one or two other charged groups) and a long nonpolar tail.

## Membranes

This appendix briefly outlines some useful information about the composition and structure of bacterial biological membranes. However, further analyses of the structure, permeability and transport properties of membranes are not considered here. In general, biological membranes are composed of proteins associated with a lipid bilayer matrix. The cell, the basic biological unit, is essentially defined by its enveloping cytoplasmic membrane.

## The Bacterial Cytoplasmic Membrane

The bacterial cytoplasmic membrane is composed of a phospholipid bilayer and proteins, and it encloses the content of the bacterial cell. Hydrophobic in nature, it

**Table C1** Some common fatty acids (anionic forms) incorporated in membrane lipids

| No. of carbons | No. of double bonds | Common name | IUPAC name | Structure |
|---|---|---|---|---|
| 12 | 0 | Laurate | Dodecanoate | $CH_3(CH_2)_{10}COO^-$ |
| 14 | 0 | Myristate | Tetradecanoate | $CH_3(CH_2)_{12}COO^-$ |
| 16 | 0 | Palmitate | Hexadecanoate | $CH_3(CH_2)_{14}COO^-$ |
| 18 | 0 | Stearate | Octadecanoate | $CH_3(CH_2)_{16}COO^-$ |
| 20 | 0 | Arachidate | Eicosanoate | $CH_3(CH_2)_{18}COO^-$ |
| 16 | 1 | Palmitoleate | *cis*-$\Delta^9$-hexadecenoate | $CH_3(CH_2)_5CH{=}CH(CH_2)_7COO^-$ |
| 18 | 1 | Oleate | *cis*-$\Delta^9$-octadecenoate | $CH_3(CH_2)_7CH{=}CH(CH_2)_7COO^-$ |
| 18 | 2 | Linoleate | *cis,cis*-$\Delta^{9,12}$-octadecadienoate | $CH_3(CH_2)_4(CH{=}CHCH_2)_2(CH_2)_6COO^-$ |
| 18 | 3 | Linolenate | All-*cis*-$\Delta^{9,12,15}$-octadecatrienoate | $CH_3CH_2(CH{=}CHCH_2)_3(CH_2)_6COO^-$ |
| 20 | 4 | Arachidonate | All-*cis*-$\Delta^{5,8,11,14}$-eicosatetraenoate | $CH_3(CH_2)_4(CH{=}CHCH_2)_4(CH_2)_2COO^-$ |

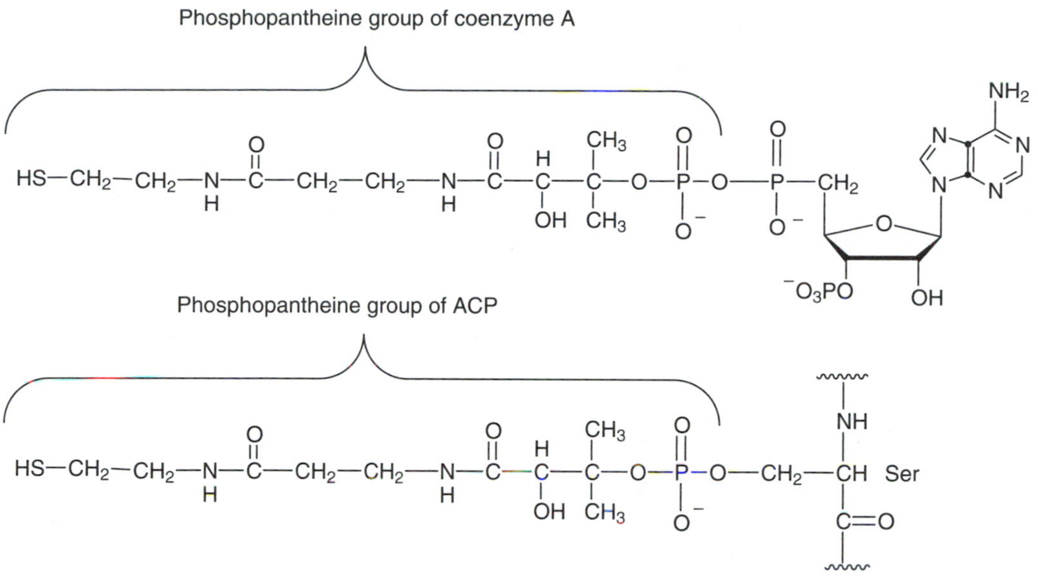

Phosphopantheine group of coenzyme A

Phosphopantheine group of ACP

**Figure C2** Common structure of the phosphopantheine prosthetic group of CoA and ACP.

acts as a barrier, preventing the leakage of the hydrophilic cytoplasmic constituents and protecting the inside of the cell from the environment.

The cytoplasmic membrane, together with the outer membrane in gram-negative bacteria and the peptidoglycan layer, helps to maintain the shape of the gram-negative cell. The cell membrane serves many other important functions, including electron transport and protein translocation from the cytoplasm to the periplasm or outer membrane.

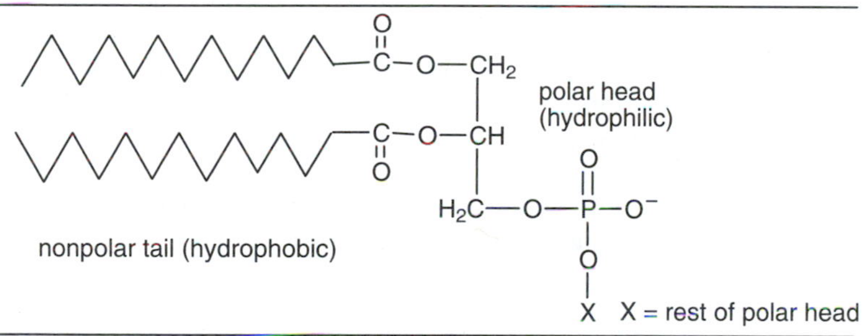


| Precursor of X | Formula of X | Name of resulting glycerophospholipid family |
|---|---|---|
| Water | —H | Phosphatidate |
| Choline | $—CH_2CH_2\overset{+}{N}(CH_3)_3$ | Phosphatidylcholine |
| Ethanolamine | $—CH_2CH_2\overset{+}{N}H_3$ | Phosphatidylethanolamine |
| Serine | $—CH_2-CH(\overset{+}{N}H_3)COO^-$ | Phosphatidylserine |
| Glycerol | $—CH_2CH(OH)—CH_2OH$ | Phosphatidylglycerol |
| *myo*-Inositol | bond site; 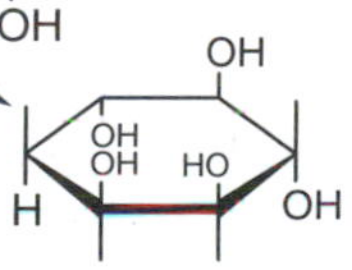 OH, OH, HO, OH, OH, H | Phosphatidylinositol |

**Figure C3** Some common substituents attached to the phosphate group of glycerophospholipids.

The unique properties of the phospholipid constituents of the cytoplasmic membrane allow the formation of the bilayer sheet and modulate its characteristics to ensure the function of the membrane under a variety of environmental conditions. In *E. coli,* the three major phospholipids present are phosphatidylethanolamine (75%), phosphatidylglycerol (20%), and diphosphatidylglycerol (1 to 5%). In addition to these major phospholipids, there are small amounts of phosphatidic acid, phosphatidylserine, lysophospholipids, and diacylglycerols.

In addition to maintaining the overall bilayer architecture, the phospholipids must allow the inclusion of proteins that are vital to cellular metabolic function. The general architecture of the cytoplasmic membrane is shown in Fig. 1.2; see also Fig. 1.23 and 1.24.

The cytoplasmic membrane is said to be differentially permeable. It allows free passage of certain molecules such as water and oxygen across the membrane, but it is impermeable to electrolytes. Membranes are impermeable to protons; this property is exploited by the cell in energizing of the membrane and generation of proton transfer gradients. ATP is synthesized using the proton electrochemical potential ($\Lambda\mu_d$).

Efflux is usually associated with the energy-dependent excretion of antibacterial agents (e.g., tetracyclines and fluoroquinolones). Typical efflux pumps such as AcrE of *E. coli* consist of 12 to 14 transmembrane helices and are driven by the transmembrane potential $\Lambda\mu_d$. Another class of efflux pumps consists of very small proteins with only four transmembrane helices such as Smr or Emr from *Staphylococcus* and *E. coli,* respectively.

## Outer Membrane

In addition to the cytoplasmic membrane, gram-negative bacteria are surrounded by an outer membrane. The outer membrane protects gram-negative bacteria against harsh environments. At the same time, the embedded proteins fulfill a number of tasks that are crucial to the bacterial cell, such as solute and protein translocation and signal transduction.

Lipopolysaccharides are a family of complex macromolecules functioning as integral membrane components in the cell envelopes of all gram-negative bacteria. The outer membrane harbors numerous proteins, including transmembrane pore-forming proteins (**porins**). In recent years, the atomic structures of several outer membrane proteins, belonging to six families, have been determined. They include the OmpA membrane domain, the OmpX protein, phospholipase A, general porins (OmpF and PhoE), substrate-specific porins (LamB and ScrY), and the TonB-dependent porins from siderophore transporters FhuA and FepA. These crystallographic studies have yielded valuable insight into the functions of these proteins.

Molecules cross the outer membrane while entering the cell by several different routes, including specific and nonspecific pores, specific receptor complexes, and, under special conditions, a hydrophobic pathway. The general architecture of the major surface structures of the cells of gram-positive and gram-negative bacteria is shown in Fig. 1.7 and 1.8 (see also Fig. 1.18).

## REFERENCES AND FURTHER READING


### General Texts on Lipid Composition and Structure

**Moran, L. A., K. G. Scrimgeour, H. R. Horton, R. S. Ochs, and J. D. Rawn.** 1994. *Biochemistry,* 2nd ed. Prentice-Hall, Englewood Cliffs, N.J.

**Voet, D., and J. G. Voet.** 1995. *Biochemistry,* 2nd ed. John Wiley & Sons, Inc., New York, N.Y.

### Nomenclature of Lipids

**IUPAC-IUB Commission on Biochemical Nomenclature of Lipids.** 1976. Recommendations. These can be obtained from http://www.chem.qmul.ac.uk./iupac/lipid/. This World Wide Web version was prepared by G. P. Moss, Department of Chemistry, Queen Mary and Westfield College, London, United Kingdom.

### Bacterial Cytoplasmic Membrane

**Weiner, J. H., D. Sambasivaran, and R. A. Rothery.** Bacterial cytoplasmic membrane. Accepted for publication in *Encyclopedia of Life Science.* Nature Publishing Group, London, United Kingdom. [Online.] http://www.els.net.

### Lipopolysaccharides

**Helander, I. M., U. Manat, and E. T. Rietschel.** Lipopolysaccharides. Accepted for publication in *Encyclopedia of Life Science.* Nature Publishing Group, London, United Kingdom. [Online.] http://www.els.net.

### Outer Membrane

**Hancock, R. E. W.** 1984. Alterations in outer membrane permeability. *Annu. Rev. Microbiol.* **38**:237–264.

**Koebnik, R., K. P. Locher, and P. Van Gelder.** 2000. Structure and function of bacterial outer membrane proteins: barrels in a nutshell. *Mol. Microbiol.* **37**:239–253.

**Nikaido, H.** 1979. Permeability of the outer membrane of bacteria. *Angew. Chem. Int. Ed.* **18**:337–350.

**Nikaido, H.** 1989. Outer membrane barrier as a mechanism of antimicrobial resistance. *Antimicrob. Agents Chemother.* **33**:1831–1836.

**Nikaido, H., and M. Vaara.** 1985. Molecular basis of bacterial outer membrane permeability *Microbiol. Rev.* **49**:1–32.

**Preston, A., R. E. Mandrell, B. W. Gibson, and M. A. Apicella.** 1996. The lipooligosaccharides of pathogenic gram-negative bacteria. *Crit. Rev. Microbiol.* **22**:139–180.

# Appendix D

# Antibiotic Influx and Efflux Systems

## Structure of the Cell Envelope of Gram-Negative Bacteria

The aim of this appendix is to provide a general introduction to intracellular transport and active efflux of antibiotics in bacteria. Before these topics are addressed, a brief summary of the structures of bacterial outer and cytoplasmic membranes is provided.

The cell wall of gram-negative bacteria has two membranes, the cytoplasmic and outer membranes, separated by a fluid-filled region, known as the periplasm or periplasmic space, which contains the peptidoglycan (see Fig. 1.8). The peptidoglycan is highly hydrated and is attached to the outer membrane by the so-called Braun lipoprotein.

The cytoplasmic membrane is generally a phospholipid bilayer. Its major function is active transport of nutrients and enzymatic synthesis and translation of cell wall components. Lipid bilayers usually are quite permeable to lipophilic compounds. However, the outer membrane (OM) of gram-negative bacteria forms an effective permeability barrier to various lipophilic substances, including antibiotics.

The OM (see Fig. 1.18) of gram-negative bacteria consists of the inner leaflet, containing mainly phospholipids, and the outer leaflet, with lipopolysaccharide (LPS) as its major lipid molecule. LPS is an amphipathic molecule comprising three regions (see Fig. 1.19):

- a glycolipid (lipid A) that is responsible for most of the biological action of LPS
- a core oligosaccharide containing a characteristic sugar acid, 2-keto-3-deoxyoctulonic acid (KDO), and other sugars including a heptose
- an antigenic polysaccharide composed of a repeating oligosaccharide subunit (O antigen side chain)

LPS binds cations, since it is a polyanion with a number of negative charges in its lipid A and core parts. Adjacent polyanionic LPS molecules are linked electrostatically by divalent cations ($Mg^{2+}$ and $Ca^{2+}$). It is known that the polymyxin group of antibiotics forms complexes with LPS and disorganizes the whole OM. This is not the basis of their lethality, but it is the means of permeating the OM to reach their final target, the cytoplasmic membrane. Also, the cationic peptides are effective against gram-negative bacteria because they penetrate the OM by interacting with LPS.

The OM also contains a large number of outer membrane proteins (OMPs), including porins that form trimers spanning the OM to create a narrow channel with a diameter of approximately 1 nm.

## Antibacterial Agents That Inhibit Lipid A Biosynthesis

Lipid A constitutes the outer membrane of gram-negative bacteria and is essential for bacterial growth.

Synthetic antibacterials were identified that inhibit the second enzyme of lipid A biosynthesis.

In 1996 Raetz and coworkers from Merck Research Laboratories (Rahway, N.J.) employed a screening program that selectively blocks [U-$^{14}$C]galactose incorporation into acid-precipitable material in living cells and found that a compound (the *R* enantiomer of L-573,655) which is a chiral hydroxamic acid attached to 2-phenyloxazoline (Fig. D1) inhibited a deacetylase enzyme of LPS biosynthesis in growing *Escherichia coli* cells. This enzyme acts on UDP-3-O-(*R*-3-hydroxymyristoyl)-*N*-acetylglucosamine, and it was demonstrated that there is a metal atom at the deacetylase active site (Fig. D1). Raetz et al. noted that deacetylase inhibitors could be developed to treat infections caused by gram-negative bacteria and also to retard the development of gram-negative sepsis by reducing the amount of endotoxin released from bacteria, a problem with other antibiotics used as the primary therapy.

## Antibiotic Influx Transport and Efflux in Gram-Negative Bacteria

Depending on the chemical structure of the antibiotic (hydrophilic or hydrophobic), penetration may occur by use of transmembrane porins or by diffusion through the lipid bilayer. The intrinsic permeability of antibiotics in gram-positive and gram-negative bacteria is a balance between the uptake into the bacterial cell (influx) and the drug efflux pathways (active efflux). A related term is **antibiotic accumulation,** which implies that concentrations of the molecule inside the bacterial cell are higher than those outside. One of the most frequent strategies for antibacterial resistance employed by bacterial cells is active efflux (out of the cell), reducing the intracellular concentration to subtoxic levels.

Because of the intimate association between uptake of antibiotics into bacterial cells and the drug efflux pathway, an understanding of the molecular details of bacterial influx and efflux systems is essential to the rational design of antibacterial agents for treating antibacterial-resistant bacteria.

The major modes of uptake of antibiotics by bacterial cells are

- uptake by the hydrophilic pathway
- self-promoted uptake
- energy-requiring uptake
- illicit uptake

### Uptake by the Hydrophilic Pathway

It is now well established that the passage of small, relatively hydrophilic molecules through the porins of the



**Figure D1** A unique deacetylase catalyzes the second step of lipid A biosynthesis. The structure of the *R*-isomer of compound L-573,655, a deacetylase inhibitor, is shown.

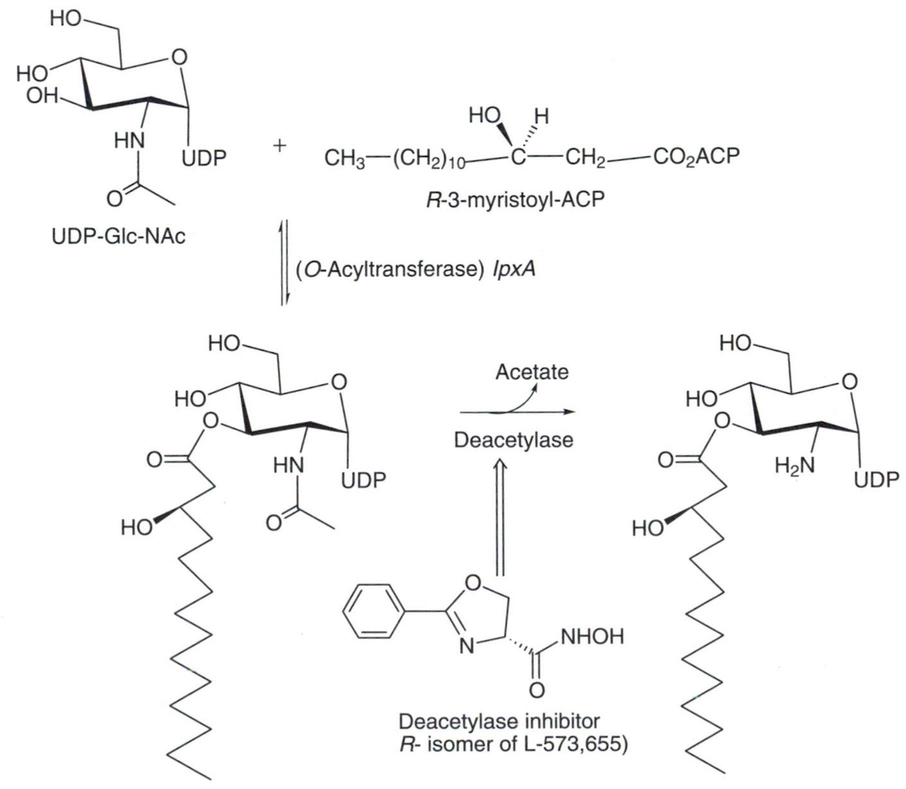

OM is allowed. Exclusion limits are typically around 600 Da in size.

Porins are integral membrane proteins (see Fig. 1.21) capable of forming transmembrane-OM channel "pores" that have an aqueous interior. They are generally divided into two classes: (i) nonspecific porins, e.g., OmpC and OmpF (Omp means "outer membrane protein") found in *E. coli,* which permit the general diffusion of small polar molecules ($\leq$600 Da), and (ii) specific porins (e.g., LamB), which facilitate the diffusion of maltodextrins. In *E. coli* these porins are found together with PhoE. OmpC and OmpF are weakly cation selective, allowing the entry only of molecules with a positive or no charge, whereas PhoE porin is weakly anion selective.

Although a detailed discussion of the function of porins is beyond the scope of this discussion, it is worth noting that protein D2 of the *Pseudomonas aeruginosa* OM facilitates the specific penetration of imipenem and meropenem across this membrane barrier. These molecules (see the structures in Fig. 10.2 and 10.3) carry positively charged substituents on the C-2 position of the carbapenem substituent. As discussed in chapter 7, the main targets of β-lactam antibiotics are the penicillin-binding proteins, which are located in the cytoplasmic membrane. Thus, to reach their target in gram-negative bacteria the β-lactam antibiotics first have to penetrate the OM.

Gram-negative bacteria in general, and *P. aeruginosa* in particular, present a formidable physical barrier to antibiotics as a consequence of the low-permeability characteristics of the OM. Many antibiotics exert their bactericidal or inhibitory action only after reaching the cytoplasm; these include those that affect protein synthesis, such as aminoglycosides, tetracyclines, and macrolides, and those that affect DNA synthesis, such as fluoroquinolones. Thus, the importance of transport to the site of the target receptor is clear. Lipid bilayers usually are quite permeable to lipophilic compounds. However, the OM of gram-negative bacteria forms a rather effective permeability barrier to various lipophilic substances, including antibiotics. The rates of permeation of lipophilic compounds through the OM are at least 10- to 100-fold lower than those through the cytoplasmic membrane because of the highly charged LPS-based OM and the stabilization of this layer by divalent cations.

## The Self-Promoted Pathway

Polycationic antibiotics such as polymyxin B, aminoglycosides, and cationic antimicrobial peptides access their target by using a self-promoted pathway involving the displacement of $Mg^{2+}$ from LPS-$Mg^{2+}$ cross-bridges. The resulting rearrangement of LPS allows the entry of these antibiotics into the periplasm. The general phenomenon of self-induced uptake was reviewed by Hancock (1984) and Vaara (1992) in the context of modification of OM permeability for enhancement of antibiotic efficacy.

Polymyxin B (see chapter 15) is a decapeptide antibiotic characterized by a heptapeptide ring containing four 2,4-diaminobutyric acid residues. An additional peptide chain covalently bound to the γ-amino group carries an aliphatic chain attached to the peptide through an amide bond. The molecule carries five positively charged diaminobutyric acid residues. Because of its molecular mass (about 1,200 Da), charge, and amphiphilicity, polymyxin B should be excluded by the bacterial OM. However, polymyxin B is known to penetrate the OM, using the self-promoted pathway.

## Energy-Requiring Uptake

In accord with the thermodynamics of diffusion processes, accumulation of antibiotics inside the bacterial cell occurs against a concentration gradient. Uptake of some antibiotics depends either on ATP or on the proton motive force as a source of energy. Either mode requires that a mechanism exist in the OM for coupling the utilization of energy to transmembrane movement of the antibiotic.

Understanding the energetics of antibiotic uptake will require a comparable understanding of the energetics of efflux. The dependence of aminoglycoside antibiotic uptake on energy production by bacterial cells is well established. Leviton et al. (1995) showed that uptake of tobramycin in *E. coli* is consistent with diffusion across the cytoplasmic membrane through a voltage-gated channel, which closes following uptake as a result of decreased membrane potentials associated with the effects of aminoglycosides themselves on the cytoplasmic membrane. Other compounds, such as albomycin (Fig. D2), depend on transporters of the ATP-binding cassette (ABC) type.

## Illicit Uptake: the Use of Metabolic Uptake Systems for Antibiotic Entry

If an antibiotic bears a sufficiently close structural relationship to a molecule for which bacteria have a specific uptake system, then the antibiotic may be carried into the cell by that system. One example is the transport of iron-binding molecules (siderophores) across the outer membrane of gram-negative bacteria. $Fe^{2+}$ is insoluble, so iron is transported into bacteria as soluble $Fe^{3+}$ complexes bound to bacterially synthesized low-molecular-weight siderophores. OM proteins with requisite specificity bind the siderophore-iron complexes, transport them across the OM by energy-dependent processes, and deposit them in the periplasm, where the complexes bind to soluble binding proteins, which in

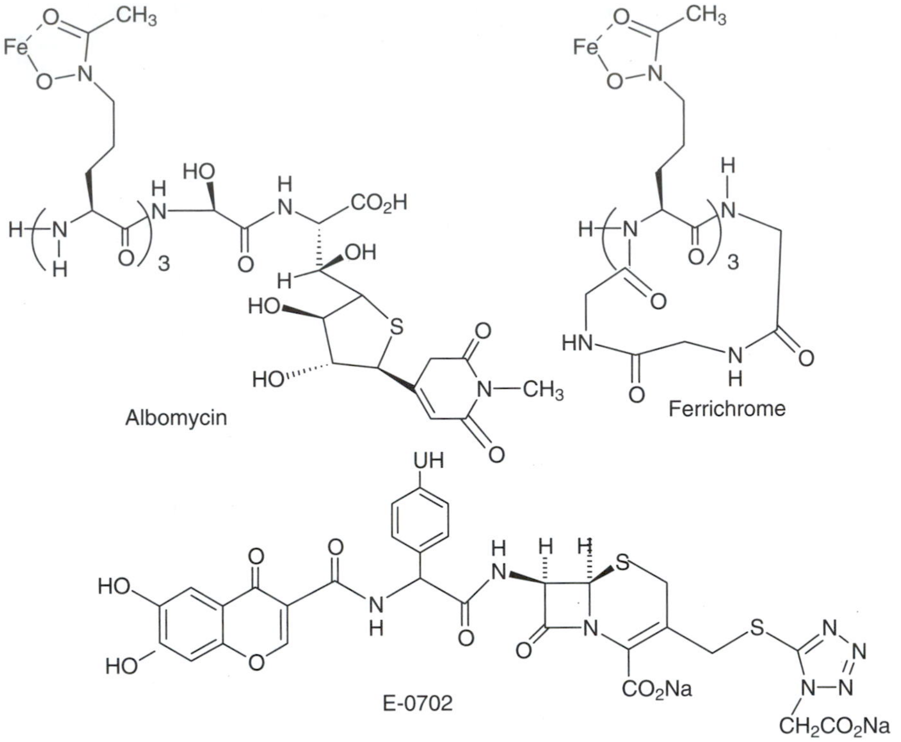

**Figure D2** Structures of albomycin, ferrichrome, and the catechol-containing cephalosporin E-0702.

turn deliver them to ABC transporters in the cytoplasmic membrane.

Many natural siderophores contain two or three catecholate groups as chelating ligands based on di- or triamines. A well-studied example of synthetic derivatives of $Fe^{3+}$-siderophore complexes is albomycin (Fig. D2), which has structural similarities to the natural substrate ferrichrome; albomycin and ferrichrome are both transported by the FhuA protein. FhuA is the OM-binding protein, FhuD is the periplasm-binding protein, and FhuB and FhuC form a cytoplasmic membrane protein complex that, when energized by ATP, catalyzes the entry of ferrichrome-$Fe^{3+}$ or albomycin-$Fe^{3+}$ complexes into the cytoplasmic compartment. Active transport across the outer membrane via FhuA is thought to depend on input of energy from the cytoplasmic membrane via the TonB-ExbB-ExbD complex.

## ACTIVE TRANSPORT OF SIDEROPHORE-MIMICKING ANTIBACTERIALS ACROSS THE OUTER MEMBRANE

The OM permeability barrier of bacteria is an important component of the resistance of gram-negative bacterial pathogens to β-lactam antibiotics. This applies especially to *P. aeruginosa* and *Stenotrophomonas maltophilia*. To overcome this membrane-mediated resistance, siderophore structures covalently bound to the antibiotic can function as a shuttle for active transport into the bacterial cell. A number of $Fe^{3+}$-siderophore carriers of the hydroxamic acid and catechol types linked to cephalosporins and monobactams have been synthesized. In *E. coli* and several *Pseudomonas* species, the uptake of ferrisiderophores requires TonB, an essential protein that couples the cytoplasmic membrane proton motive force to active transport across the OM. As an example, compound E-0702 (Fig. D2) is a semisynthetic iron-chelating antipseudomonal cephalosporin derivative which is incorporated into *E. coli* cells by the TonB-dependent iron transport system.

## CRYSTAL STRUCTURE OF THE OUTER MEMBRANE-ACTIVE TRANSPORTER FepA FROM *E. COLI*

Integral OM receptors for iron chelates carry out specific ligand transport against a concentration gradient. Energy for active transport is obtained from the proton motive force of the inner membrane (IM) through physical interaction with TonB-ExbB-ExbD, an IM complex. In 1999, Buchanan et al. reported the crystal structure of an active-transport OM receptor at 2.4-Å resolution (Color Plate D1 [see color insert]). The ferric enterobactin receptor from *E. coli* (FepA) is a 724-residue integral OMP that transports ferric enterobactin (a cyclic triester of 2,3-dihydroxybenzylserine, with a molecular mass of 719 Da) into the periplasm. It is taken up by a periplasmic binding protein and transported through the IM by an ABC-type transport complex. In the cytoplasm, ferric enterobactin is hydrolyzed by an esterase to provide the iron necessary for cell metabolism.

## Efflux Pump Proteins

In recent years, interest in efflux-mediated resistance in bacteria has been triggered by the growing amount of data implicating efflux pumps in antibiotic resistance of clinical isolates. The antibiotic is removed from intracellular compartments by efflux proteins. An overview of the chemical and biological properties of the major efflux proteins found within *E. coli* and *P. aeruginosa* and their activity in establishing resistance is provided here.

Resistance by efflux is widespread in many bacterial species. Some pumps are specific for a particular antibiotic, such as tetracycline-, chloramphenicol-, and macrolide-specific transporters. Other translocases, called **multidrug resistance pumps** (MDRs), are capable of recognizing multiple antibiotics. One such efflux system, the MexAB-OprM multidrug efflux system in *P. aeruginosa,* encoded by the *mexAB-oprM* operon, effluxes a range of antibiotics, including tetracycline, macrolides, trimethoprim, chloramphenicol, quinolones, and most β-lactam antibiotics, but not imipenem.

Five families of bacterial drug efflux pumps have been identified, based primarily on amino acid sequence homology and on the energy source used to export their substrates. Transport can be driven by ATP hydrolysis (for pumps from the ABC), or pumps can utilize the proton motive force. The four proton motive force-dependent families are

- the small multidrug resistance (SMR) family
- the major facilitator superfamily (MFS)
- the multidrug and toxic compound extrusion (MATE) family
- the resistance-nodulation-division (RND) family

The structural and functional aspects of these five families of drug pumps have been reviewed recently and are not discussed here (Lomovskaya et al., 2001; Lewis and Lomovskaya, 2002; Lomovskaya and Watkins, 2001).

### Multicomponent Efflux Pumps from Gram-Negative Bacteria

In gram-negative bacteria, the majority of multidrug transporters have a common three-component organization:

- a transporter located in the IM
- an OM channel
- a periplasmic linker protein to link the other two components

In this arrangement, efflux complexes traverse both the IM and OM and thus facilitate direct passage of the substrate from the cytoplasm or cytoplasmic membrane into the external medium.

Genetic studies have suggested that TolC, a multifunctional OM channel, may associate with AcrAB to form a functional tripartite complex (Fig. D3). Details of the recognition events between pumps and substrates remain unclear. The model in Fig. D3 is based on results obtained by Zgurskaya and Nikaido (1999) and the theoretical analysis by Johnson and Church (1999).

The recent determination of the crystal structure of TolC (Color Plate D2 [see color insert]) indicates its function and has provided insight into its mechanism for export of a wide range of molecules across the periplasmic space and OM of gram-negative bacteria. The structure was compared to those of other proteins that are embedded in the bacterial OM or that traverse the periplasmic space.

### Efflux Pumps and Drug Resistance

Nowadays, efflux mechanisms are broadly recognized as major components of resistance to many classes of antibiotics. Both antibiotic-specific and MDR pumps are known to confer clinically significant resistance.

#### TETRACYCLINES

Active efflux and ribosomal protection are the two major mechanisms of tetracycline resistance. Efflux pumps involved in tetracycline efflux can be either specific or multidrug resistant. TetA through TetE, TetG, TetH, TetK, TetL, TetA(P), and OtrB are tetracycline-specific efflux pumps found in the majority of pathogenic gram-negative and gram-positive bacteria. In *E. coli* (and in other enteric bacteria), tetracyclines are substrates of multiple endogenous MDR systems that belong to the RND family.

#### MACROLIDES

Two major mechanisms of macrolide resistance are target modification (dimethylation of bacterial 23S rRNA within the macrolide-binding site by the products of the *erm* genes) and efflux. The *mef* genes encode efflux pumps specific for 14- and 15-member macrolides. Mef pumps belong to the MFS family of transporters. A phenotype of resistance to 14- and 15-member macrolides and streptogramin B (the $MS_B$ phenotype) but not 16-member macrolides has been identified in several staphy-

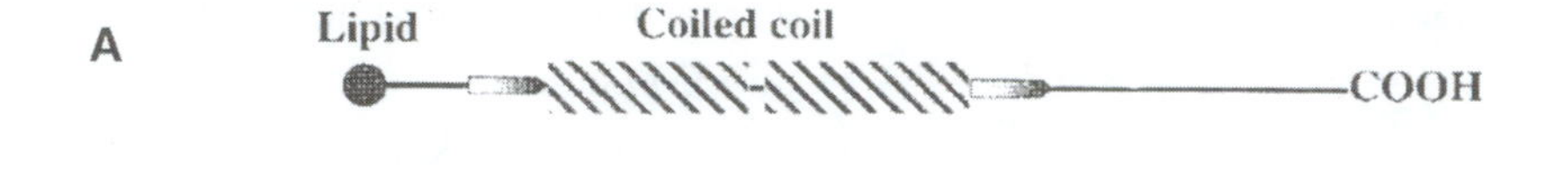


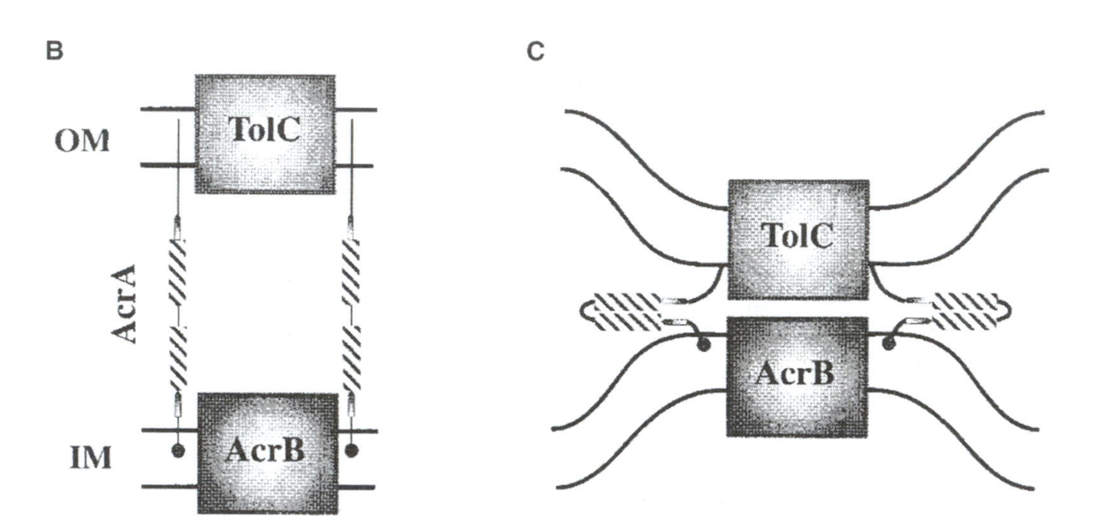


**Figure D3** The hypothetical structure and mechanism of tripartite efflux pumps in gram-negative bacteria, using as an example the AcrAB-TolC efflux pump from *E. coli.* (A) Domains within AcrA, with its N-terminal lipid, an interrupted coiled-coil domain, and a pair of lipoyl arm domains (shown as arrows flanking the coiled-coil domain). (B) In the nonfunctioning state, the inner membrane-associated transporter AcrB simply exists as a complex with the periplasmic component AcrA, which is anchored to the outer surface of the IM through its N-terminal lipid moiety and is presumably loosely associated with OM through its C-terminal domain. The OM channel TolC is also shown as part of the tripartite complex. (C) AcrA might bring TolC and AcrB together by folding back on itself, using the interrupted coiled-coil domain and the two lipoyl arm domains. This will allow drug efflux across the two membranes directly into the medium. Reprinted from H. I. Zgurskaya and H. Nikaido, *Mol. Microbiol.* **37:**219–225, 2000, with permission from the publisher.

lococcal species. Efflux pumps belonging to the ABC transporter family confer this phenotype.

CHLORAMPHENICOL
*cmlA* genes encode efflux pumps that belong to the TMF exporters from the MFS. These genes appear to be widespread among gram-negative bacteria.

β-LACTAMS
In *P. aeruginosa,* most of the β-lactam antibiotics (with the exception of imipenem) are substrates of the MexAB-OprM RND MRD efflux pump.

FLUOROQUINOLONES
Two resistance mechanisms found in gram-negative and gram-positive bacteria are target modification and efflux by MDR pumps. Fluoroquinolone resistance attributable to efflux has been reported for a number of gram-negative bacteria. *Staphylococcus aureus* NorA is a transmembrane multidrug efflux pump that confers low-level resistance to hydrophilic fluoroquinolones.

AMINOGLYCOSIDES
Two main mechanisms of aminoglycoside resistance are specific enzymatic inactivation by modifying enzymes and a mechanism often referred to as permeability resistance. Recently, an MDR pump, MexXY, capable of effluxing aminoglycosides was identified in *P. aeruginosa.* The MexXY pump appears to play a key role in low-level resistance to aminoglycoside antibiotics in *P. aeruginosa.*

## STRATEGIES TO OVERCOME EFFLUX-MEDIATED RESISTANCE BY MODIFICATION OF EXISTING ANTIBIOTICS

### New Tetracyclines

In chapter 17, the development of a new class of semisynthetic tetracyclines, the glycylcyclines, is discussed. These compounds exhibit potent activity against a broad spectrum of gram-negative and gram-positive bacteria, including those that carry TetA through TetD and TetK determinants. Glycylcyclines overcome efflux-mediated resistance because they are not recognized by the transporter proteins.

### New Macrolides and Ketolides

Ketolides are still affected by the MEF pumps in streptococci, although to a much lesser degree than are the macrolides.

### New Fluoroquinolones

For the new quinolones such as levofloxacin, moxifloxacin, and gatifloxacin, it appears that NorA from *S. aureus* and efflux pumps from *Streptococcus pneumoniae* do not confer the same increase in resistance as that for the older fluoroquinolones.

### Novel Pump Inhibitors

The development of novel pump inhibitors to potentiate by coadministration with various clinically used antibiotics while minimizing the efflux mechanims is a particularly attractive feature that is being pursued in some pharmaceutical and technological research centers.

## REFERENCES AND FURTHER READING


#### Antibiotic Permeability

**Taber, H. W.** 2002. Antibiotic permeability, p. 193–208. *In* K. Lewis, A. A. Salyers, H. W. Taber, and R. G. Wax (ed.), *Bacterial Resistance to Antimicrobials.* Marcel Dekker, Inc., New York, N.Y.

#### Outer Membrane

**Hancock, R. E. W.** 1984. Alterations in outer membrane permeability. *Annu. Rev. Microbiol.* **38:**237–264.

**Nikaido, H.** 1996. Outer membrane, p. 29–47. *In* F. C. Neidhardt, R. Curtiss III, J. L. Ingraham, E. C. C. Lin, K. B. Low, B. Magasanik, W. S. Reznikoff, M. Riley, M. Schaechter, and H. E. Umbarger (ed.), Escherichia coli *and* Salmonella*: Cellular and Molecular Biology,* 2nd ed., vol. 1. ASM Press, Washington, D.C.

**Vaara, M.** 1992. Agents that increase the permeability of the outer membrane. *Microbiol. Rev.* **56:**395–411.

#### Bacterial Lipopolysaccharide

**Raetz, C. R. H.** 1996. Bacterial lipopolysaccharide: a remarcable family of bioactive macroamphiphiles, p. 1035–1063. *In* F. C. Neidhardt, R. Curtiss III, J. L. Ingraham, E. C. C. Lin, K. B. Low, B. Magasanik, W. S. Reznikoff, M. Riley, M. Schaechter, and H. E. Umbarger (ed.), Escherichia coli *and* Salmonella*: Cellular and Molecular Biology,* 2nd ed., vol. 1. ASM Press, Washington D.C.

#### Antibacterial Agents That Inhibit Lipid A Biosynthesis

**Onishi, H. R., B. A. Pelak, L. S. Gerckens, L. L. Silver, F. M. Kahan, M. H. Chen, A. A. Patchett, S. M. Galloway, S. A. Hyland, M. S. Anderson, and C. R. H. Raetz.** 1996. Antibacterial agents that inhibit lipid A biosynthesis. *Science* **274:**980–982.

#### Uptake of Aminoglycosides

**Leviton, I. M., H. S. Fraimow, N. Carrasco, T. J. Dougherty, and M. H. Miller.** 1995. Tobramycin uptake in *Escherichia coli* membrane vesicles. *Antimicrob. Agents Chemother.* **39:**467–475.

#### Peptide Antibiotics

**Zhang, L., and R. E. W. Hancock.** 2001. Peptide antibiotics, p. 209–232. *In* D. Hughes and D. Andersson (ed.), *Antibiotic Development and Resistance.* Taylor & Francis, London, United Kingdom.

#### Active Transport of Siderophore-Mimicking Antibacterials

**Roosenberg, J. M., Y. M. Lin, Y. Lu, and M. J. Miller.** 2000. Studies and syntheses of siderophores, microbial iron chelators, and analogs as potential drug delivery agents. *Curr. Med. Chem.* **7:**159–197.

#### Crystal Structure of FepA

**Buchanan, S. K., B. S. Smith, L. Venkatramani, D. Xia, L. Esser, M. Palnitkar, R. Chakraborty, D. van der Helm, and J. Deisenhofer.** 1999. Crystal structure of the outer membrane active transporter FepA from *Escherichia coli. Nat. Struct. Biol.* **6:**56–63.

#### Efflux Mechanisms

**Johnson, J. M., and G. M. Church.** 1999. Alignment and structure prediction of divergent protein families: periplasmic and outer membrane proteins of bacterial efflux pumps. *J. Mol. Biol.* **287:**695–715.

**Lewis, K., and O. Lomovskaya.** 2002. Drug efflux, p. 61–90. *In* K. Lewis, A. A. Salyers, H. W. Taber, and R. G. Wax (ed.), *Bacterial Resistance to Antimicrobials.* Marcel Dekker, Inc., New York, N.Y.

**Lomovskaya, O., M. S. Warren, and V. Lee.** 2001. Efflux mechanisms: molecular and clinical aspects, p. 65–90. *In* D. Hughes and D. Andersson (ed.), *Antibiotic Development and Resistance.* Taylor & Francis, London, United Kingdom.

**Lomovskaya, O., and W. J. Watkins.** 2001. Efflux pumps: their role in antibacterial drug discovery. *Curr. Med. Chem.* **8:**1699–1711.

**McKeegan, K. S., M. I. Borges-Walmsley, and A. R. Walmsley.** 2003. The structure and function of drug pumps: an update. *Trends Microbiol.* **11:**21–29.

**Murakami, S., R. Nakashima, E. Yamashita, and A. Yamaguchi.** 2002. Crystal structure of bacterial multidrug efflux transporter AcrB. *Nature* **419:**587–593.

**Van Bambeke, F., E. Balzi, and P. M. Tulkens.** 2000. Antibiotic efflux pumps. *Biochem. Pharmacol.* **60:**457–470.

**Zgurskaya, H. I., and H. Nikaido.** 1999. AcrA from *Escherichia coli* is a highly asymmetric protein capable of spanning the periplasm. *J. Mol. Biol.* **285:**409–420.

**Zgurskaya, H. I., and H. Nikaido.** 2000. Multidrug reistance mechanisms: drug efflux across two membranes. *Mol. Microbiol.* **37:**219–225.

### Bacterial Membrane Protein TolC

**Koronakis, V., A. Sharff, E. Koronakis, B. Luisi, and C. Hughes.** 2000. Crystal structure of the bacterial membrane protein TolC central to multidrug efflux and protein export. *Nature* **405:**916–919.

**Wiener, M. C.** 2000. Bacterial export takes its Tol. *Structure* **8:**R171–R175.

# Index

## B

## D



## E

## F

## Q



## R

## S

## T

## U



## V